Wireless Networking

Wireless Networking
Understanding Internetworking Challenges

Jack L. Burbank
Julia Andrusenko
Jared S. Everett
William T.M. Kasch

IEEE PRESS

Library of Congress Cataloging-in-Publication Data:
Burbank, Jack L.
  Wireless networking : understanding internetworking challenges / Jack L. Burbank, Julia Andrusenko, Jared S. Everett, William T.M. Kasch.
      pages cm
  ISBN 978-1-118-12238-9 (hardback)
  1. Internetworking (Telecommunication)   2. Computer networks.   3. Wireless communication systems.   I. Andrusenko, Julia.   II. Everett, Jared S.   III. Kasch, T.M. William.   IV. Title.
  TK5105.5.B855 2013
  004.6–dc23
                                    2012043405

Printed in the United States of America.

10 9 8 7 6 5 4 3 2 1

This book is dedicated to Helen Janine Clark,
the kindest and most generous person that I have ever known.
Cancer can take your life but your spirit remains strong in my heart.
Everything I am, and will be, is because of everything that
you have been for me. I have been truly blessed to
have you as a mother.

Love, Jack

Contents

Preface

During the initial periods of the current information era, obtaining and sharing information was a wired experience. The masses gravitated toward home personal computers and began experiencing "The Internet" through a variety of wired technologies. As our hunger for information increased, the sophistication and capability of these wired technologies improved. Over time, however, we began to desire high capability and convenience. With the proliferation of laptop computers and the advent of mainstream wireless networking technologies, we began to enjoy the Internet on our own terms. Initial wireless networking technologies provided only rudimentary capability, but they showed us all what the potential information experience could be. In the subsequent years we have been rapidly untethering our information devices, increasingly accessing information on our terms. The era of "The Wireless Internet" is now upon us all. "Broadband to the Masses" has become a battle cry for our generation.

However, effective wireless internetworking introduces several key technical challenges. Users constantly demand more capacity, more functional capabilities, and fewer constraints and restrictions on usage. These are not easily achievable simultaneously—if possible at all. Yet this is what we all demand from wireless technology. And at the same time, the industry faces the ever-increasing issue of spectrum scarcity, which demands ever-increasingly efficient and inventive ways to utilize this scarce resource. This creates an extremely challenging era in which to work in the networking industry. Some argue that in the rush to satisfy growing user demands for capacity, the market has become saturated with a glut of too many competing wireless technologies.

The authors find this to be a very exciting time. We rejoice in the technological advances that we have been able to witness and sometimes even contribute to over the past 15 years. While sometimes unpleasant to witness, we are thankful for the competitive forces that we believe are healthy for the industry. Indeed, even if a particular technology experiences little or no market success, it may still be highly influential. Our recent history is filled with examples of commercial failures that helped spawn innovation across the entire industry.

It is undeniable that the current wireless networking landscape is highly complex, with seemingly dozens of competing technologies attempting to solve the same problem. Maintaining knowledge and expertise in the industry is certainly a daunting task; it is difficult to not feel overwhelmed by it all. We the authors also face this challenge. If you were to look in our offices, you would find a glut of textbooks on every conceivable topic and technology related to the wireless industry. We have our favorite texts on particular topics, with particular pages bookmarked because those are the "killer tables" or "killer figures" we find ourselves revisiting time and again. However, we have long been frustrated that there is a general lack of comprehensive texts that cover all the topics that we are interested in. Why can't a book cover Wi-Fi and Global System for Mobile Communications (GSM)? Why can't a book consider Bluetooth and Transmission Control Protocol (TCP) performance in wireless environments? After all, in this complex world of wireless internetworking, it is difficult for any of us to limit our scope of work to a particular technology. For as long as we have been working in this industry, we have sought, with futility, a text that discusses everything that we care about. And too many times, when we have thought we found one, we find that it provides only a superficial treatment of what we are interested in, after dedicating the majority of its pages to general engineering topics that have been covered hundreds of times before in other texts. Other texts simply provide a long-winded regurgitation of a technical specification. They give all the facts of the matter, stating *what* a technology is. But they do not provide any insight regarding *why* a technology is what it is or *how* it got that way. Some of us teach college courses on topics related to wireless networking and for years have struggled with poor textbooks. Simply put, we have not found *the* wireless networking book that we have long sought after. That is why we decided to write this book.

The goal of this book is to provide a comprehensive overview of the entire wireless networking landscape. This is an oxymoron because it is not possible to be completely comprehensive in a single text; a comprehensive text would be unwieldy. We have consolidated 10,000+ pages of specification into a 500+ page text. This book accomplishes this by focusing on what we consider to be the "important" and "interesting" aspects of wireless technologies. This book does not spend much time covering "fundamental" topics. For fundamental topics, there already exist many key texts that a reader would be better suited to consider. This book does not regurgitate every detail from a technology specification. Rather, this book focuses on the "interesting" parts of those technology specifications. Those key figures, those key tables, those things that we have bookmarked because we keep returning to them—those are what we put in this book.

Not only do we provide the "whats" of a technology, but also we try to help answer (at least in part) the "whys" and "hows" of a technology. What we hope we have accomplished is to have created a book that is filled with "killer tables" and "killer figures." Our hope is that this book is worthy of dozens and dozens of bookmarks, that this book becomes the first place you go to find an answer. We hope that this book will help you understand not only the current wireless networking landscape, but also how we have gotten to where we are and where we are going. We hope you find as much pleasure in reading this book as we found in writing it.

JACK L. BURBANK
JULIA ANDRUSENKO
JARED S. EVERETT
WILLIAM T.M. KASCH

Acknowledgments

$\mathbf{W}$e would like to acknowledge the numerous individuals who have helped make this book a reality. We would like to thank Dr. Brian Haberman and Dr. Phil Chimento for their tremendous expertise in networking and the years of collaboration with them that have proven invaluable in our own professional growth and in the writing of this book. We would like to thank Dr. Feng Ouyang for his expertise in multiple-input, multiple-output (MIMO) communications techniques and his assistance in the writing of this book.

We would like to thank The Johns Hopkins University Applied Physics Laboratory for their support during the writing of this book.

We would like to give special thanks to Robert Nichols for his longtime support of our activities in this field.

Most of all, we would like to thank all of our friends and family for their patience and support during the writing of this book.

J.L.B.
J.A.
J.S.E.
W.T.M.K.

About the Authors

Jack L. Burbank (jack.burbank@jhuapl.edu) received his B.S. and M.S. degrees in Electrical Engineering from North Carolina State University in 1994 and 1998, respectively. As part of the Communications and Network Technologies Group of The Johns Hopkins University Applied Physics Laboratory, he works with a team of engineers and scientists focused on assessing and improving the performance of wireless networking technologies through test, evaluation, and technical innovation. His primary expertise is in the areas of wireless networking and modeling and simulation, with research interests in cognitive networking, cross-layer design, and ad hoc networking. He has published numerous technical papers and book chapters on topics of wireless networking, and regularly acts as a technical reviewer for journals and magazines. He is an associate technical editor of the *IEEE Communications Magazine* and is the coauthor of a book on modeling and simulation of wireless networks. He teaches courses on the topics of networking and wireless networking in The Johns Hopkins University Part Time Engineering Program and is a member of the Institute of Electrical and Electronics Engineers (IEEE) and the American Society for Engineering Education (ASEE).

Julia Andrusenko (julia.andrusenko@jhuapl.edu) received her B.S. and M.S. degrees in Electrical Engineering in 2002 from Drexel University, Philadelphia. She currently works as a communications engineer at The Johns Hopkins University Applied Physics Laboratory. Her background is in communications theory, wireless networking, computer simulation of communications systems, evolutionary computation, genetic algorithms, and programming. Her recent work has focused on radio frequency (RF) propagation prediction, satellite communications, wireless networking, communications vulnerability, and multiple-input, multiple-output (MIMO) technologies. Ms. Andrusenko is a member of the IEEE Communications Society.

Jared S. Everett (jared.everett@jhuapl.edu) received his B.S. degrees in Electrical Engineering and Computer Engineering as well as a B.A. degree in Humanities in 2007, and his M.S. degree in Electrical Engineering in 2009, all from North Carolina State University. Since 2009, he has worked as a wireless communications research and development (R&D) engineer in the Communications and Networking Technologies Group of The Johns Hopkins

University Applied Physics Laboratory, where he specializes in cellular air interface technologies including Long-Term Evolution (LTE), Wideband Code Division Multiple Access (WCDMA), and Global System for Mobile Communications (GSM). This work is supported by Mr. Everett's experience in GSM mobile phone development at Sony Ericsson Mobile Communications North American R&D Headquarters in Research Triangle Park, NC. His research interests span a diverse range of topics in wireless communications that include next-generation cellular networks, error-control coding, physical layer security, and underwater free-space optical communication. He is an active member of the IEEE Communications Society and Information Theory Society.

William T.M. Kasch (william.kasch@jhuapl.edu) received his bachelor of science in Electrical Engineering from the Florida Institute of Technology in 2000 and his M.S. in Electrical and Computer Engineering at The Johns Hopkins University in 2003. He is currently a member of the Senior Professional Staff and the Assistant Group Supervisor for the Communications and Networking Technology Group of The Johns Hopkins University Applied Physics Laboratory. His current interests include software-defined networking and development of robust mobile routing protocols for heterogeneous multi-exit networks. He has collaborated on multiple publications, including serving as a co-author on another book on network modeling and simulation best practices and as a co-author in the IEEE Communications Magazine.

List of Acronyms

Acronym	Definition	Context
1G	First generation	Cellular
1xRTT	One times radio transmission technology	Cellular/3G/CDMA2000
2G	Second generation	Cellular
3G	Third generation	Cellular
3GPP	Third Generation Partnership Project	Standards Bodies
3GPP2	Third Generation Partnership Project 2	Standards Bodies
4G	Fourth generation	Cellular
A2DP	Advanced Audio Distribution Profile	Networking/WPAN/Bluetooth
AAA	Authentication , authorization, and accounting	Cellular
AAS	Advanced antenna system	Networking/WMAN/IEEE 802.16
AB	Access burst	Cellular/2G/GSM
ABM	Asynchronous Balanced Mode	Cellular/2G/GSM
ABS	Almost-black subframe	Cellular/4G/LTE
AC	Access category	Networking/WLAN/IEEE 802.11
ACK	Acknowledgement	Networking
ACL	Asynchronous Connection Oriented	Networking/WPAN/Bluetooth
AD	Area Director	Standards Bodies/IETF/IESG
ADC	Analog-to-digital converter	Electronics
ADPM	Adaptive differential phase modulation	Communications/Modulation
ADSL	Asymmetric digital subscriber line	Networking/Wired
AES	Advanced encryption standard	Security
AFS	Adaptive full-rate	Cellular/2G/GSM
AGC	Automatic gain control	Communications
AGCH	Access grant channel	Cellular/2G/GSM
AIB	APS information base	Networking/WPAN/ZigBee

AICH	Acquisition indicator channel	Cellular/3G/UMTS
AID	Association identifier	Networking/WLAN/IEEE 802.11
AIFS	Arbitration interframe space	Networking/WLAN/IEEE 802.11
AIMD	Additive increase/multiplicative decrease	Networking/TCP
AIPN	All-IP network	Cellular/4G
ALMP	Air link management protocol	Cellular/3G/CDMA2000
ALOHANET	Refers to the ALOHA Network (Not an acronym)	Networking
AM	Acknowledged mode	Cellular/3GPP
AMC	Adaptive modulation and coding	Communications
AMPS	Advanced Mobile Phone Service	Cellular/1G
AMR	Adaptive multirate	Cellular/2G/GSM
AMR-WB	Wideband adaptive multirate	Cellular/2G/GSM
ANR	Automatic neighbor relations	Cellular/4G/LTE
ANSI	American National Standards Institute	Standards Bodies
AODV	Ad hoc on-demand distance vector	Networking/Routing Protocols
AP	Access point	Networking/WLAN/IEEE 802.11
APAC	Asia Pacific	General
APDU	Application protocol data unit	Networking
APL	Application layer	Networking/WPAN/ZigBee
APN	Access point name	Cellular/2G/GSM
APS	Application support sublayer	Networking/WPAN/ZigBee
APSDE	Application layer service data entity	Networking/WPAN/ZigBee
APSME	Application support sublayer management entity	Networking/WPAN/ZigBee
AQPSK	Adaptive quadrature phase shift keying	Communications/Modulation
ARFCN	Absolute RF channel number	Cellular/2G/GSM
ARIB	Association of Radio Industries and Business	Standards Bodies
ARQ	Automatic repeat request	Networking
AS	Access stratum	Cellular/3GPP
ASB	Active Slave Broadcast	Networking/WPAN/ Bluetooth
ASCI	Advanced speech call items	Cellular/2G/GSM
ASDU	Application service data unit	Networking
ASK	Amplitude shift keying	Communications/Modulation

ASMDE	Application layer service management data entity	Networking/WPAN/ZigBee
ASN	Access service network	Networking/WMAN/IEEE 802.16
ASN-GW	Access service network gateway	Networking/WMAN/IEEE 802.16
AT	Access terminal	Cellular/3G/CDMA2000
ATIS	Alliance for Telecommunications Industry Solutions	Standards Bodies
ATM	Asynchronous transfer mode	Networking/Data Link Layer
ATT	Attribute Protocol	Networking/WPAN/ Bluetooth
AuC	Authentication Center	Cellular
AV	Audio/Video	General
AVRCP	Audio/Video Remote Control Profile	Networking/WPAN/ Bluetooth
AWS	Advanced Wireless Services	Communications/Spectrum
BAAA	Broker AAA (server)	Cellular
BCCH	Broadcast control channel	Cellular/3GPP
BCH	Broadcast channel	Cellular/3GPP
BCS	Block check sequence	Communications/Coding
BDP	Bandwidth-delay product	Networking/TCP
BET	Bluetooth Ecosystem Team	Networking/WPAN/ Bluetooth
BGP	Border gateway protocol	Networking/Routing Protocols
BIP	Basic Imaging Profile	Networking/WPAN/ Bluetooth
BLER	Block error rate	General
BMC	Broadcast/multicast control	Cellular/3G/UMTS
BPM	Burst position modulation	Networking/WPAN/UWB
BPP	Basic Printing Profile	Networking/WPAN/ Bluetooth
BPSK	Binary phase shift keying	Communications/Modulation
BQP	Bluetooth Qualification Program	Networking/WPAN/ Bluetooth
BR	Basic Rate *or* Radiocommunication Bureau	Networking/WPAN/ Bluetooth Standards Bodies/ITU
BS	Base station	Networking/WMAN/IEEE 802.16
BSA	Basic service area	Networking/WLAN/IEEE 802.11
BSC	Base station controller	Cellular/2G/GSM

BSN	Block sequence number	Networking
BSS	Basic service set	Networking/WLAN/IEEE
	or Base station subsystem	802.11
		Cellular/2G/GSM
BSSGP	Base station system GPRS	Cellular/2G/GSM
	protocol	
BT	Bluetooth	Networking/WPAN
BTC	Block turbo code	Communications/Coding
BTS	Base transceiver station	Cellular/2G/GSM
BTTI	Basic transmission time interval	Cellular/2G/GSM
BU	Binding update	Networking/Mobility/MIP
BWA	Broadband wireless access	Networking
CA	Cell allocation	Cellular/2G/GSM
CAC	Channel Access Code	Networking/WPAN/
		Bluetooth
CAMEL	Customized applications for	Cellular/2G/GSM
	mobile networks enhanced logic	
CAZAC	Constant amplitude zero	General
	autocorrelation	
CBC	Cell broadcast center	Cellular/2G/GSM
CBCH	Cell broadcast channel	Cellular/2G/GSM
CBC-MAC	Cipher block chaining with	Networking/IEEE 802
	message authentication codes	
CC	Convolutional code	Communications/Coding
	or Channel combination	Cellular/2G/GSM
CCA	Clear channel assessment	Networking/WLAN/IEEE
		802.11
CCCH	Common control channel	Cellular/3GPP
CCE	Control channel entity	Cellular/4G/LTE
CCF	CDMA Certification Forum	Trade Associations
CCI	Cochannel interference	Communications
CCK	Complimentary Code Keying	Communications/Modulation
CCMP	Counter mode with CBC-MAC	Networking/IEEE 802
	Protocol	
CCO	Coverage and capacity	Cellular/4G/LTE
	optimization	
CCS	Common channel signaling	Networking/ ATM
CCSA	China Communications Standards	Standards Bodies
	Association	
CDD	Cyclic delay diversity	Communications/Antenna
		Systems
CDMA	Code Division Multiple Access	Communications/Multiple
		Access
CDMA2000	Code Division Multiple Access	Cellular/3G/CDMA2000
	2000	

CEPT	Conférence Européenne des administrations des Postes et des Télécommunications	Standards Bodies/Europe
CGI	Cell global identifier	Cellular/2G/GSM
CI	Cell identifier	Cellular/2G/GSM
CID	Connection identifier	Networking/WMAN/IEEE 802.16
CINR	Carrier-to-interference-and-noise ratio	Communications
CM	Connection management	Cellular/2G/GSM
CN	Core network	Cellular
CoA	Care-of address	Networking/Mobility/MIP
CoMP	Coordinated multipoint	Cellular/4G/LTE
CONCERTO	Control Over Network Coding for Enhanced Radio Transport	Networking/Mobility/ MANET
CP	Cyclic prefix	Communications
CPCH	Common packet channel	Cellular/3G/UMTS
	or Common physical channel	Cellular/3G/CDMA2000
CPICH	Common pilot channel	Cellular/3G/UMTS
CPS	Common part sublayer	Networking/IEEE 802
CQI	Channel quality indicator	Cellular/3GPP
CQICH	Channel quality information channel	Networking/WMAN/IEEE 802.16
CR	Cognitive radio	Communications
CRC	Cyclic Redundancy Check	Communications/Coding
C-RNTI	Cell RNTI	Cellular/4G/LTE
CS	Convergence sublayer	Networking/WMAN/IEEE 802.16
	or Circuit-switched	Cellular
CSD	Circuit-switched data	Cellular/2G/GSM
CSG	Closed subscriber group	Cellular
CSI	Channel state information	Communications
CSMA/CA	Carrier Sense Multiple Access with Collision Avoidance	Communications/Multiple Access
CSMA/CD	Carrier Sense Multiple Access with Collision Detection	Communications/Multiple Access
CSN	Connectivity service network	Networking/WMAN/IEEE 802.16
CTC	Convolutional turbo code	Communications/Coding
CTCH	Common traffic channel	Cellular/3G/UMTS
CTIA	Cellular Telecommunications Industry Association (Acronym deprecated in 2004)	Trade Associations
CTS	Clear-to-send	Networking/WLAN/IEEE 802.11

DAB	Digital Audio Broadcast	Broadcast Standards
DAC	Digital-to-analog converter	Electronics
	or Device Access Code	Networking/WPAN/
		Bluetooth
D-AMPS	Digital Advanced Mobile Phone	Cellular/2G
	Service	
DARP	Downlink advanced receiver	Cellular/2G/GSM
	performance	
DB	Dummy burst	Cellular/2G/GSM
D-BPSK	Differential binary phase shift	Communications/Modulation
	keying	
DC	Direct current	General
DCCH	Dedicated control channel	Cellular/3GPP
DCF	Distributed coordination function	Networking/WLAN/IEEE
		802.11
DCH	Dedicated channel	Cellular/3G/UMTS
DCI	Downlink control information	Cellular/4G/LTE
DCS	Digital Cellular Service	Cellular
DECT	Digital Enhanced Cordless	Communications
	Telecommunications	
DeNB	Donor evolved Node B	Cellular/4G/LTE
DES	Data Encryption Standard	Security
DFS	Dynamic Frequency Selection	Networking/WLAN/IEEE
		802.11
DFT	Discrete Fourier transform	DSP
DFTS-	DFT-spread OFDM	Communications/Multiple
OFDM		Access
DHCP	Dynamic host configuration	Networking/Application
	protocol	Layer
DIAC	Dedicated Inquiry Access Code	Networking/WPAN/
		Bluetooth
DID	Device Identification	Networking/WPAN/
		Bluetooth
DIFS	DCF interframe space	Networking/WLAN/IEEE
		802.11
DIUC	Downlink interval usage code	Networking/WMAN/IEEE
		802.16
DL	Downlink	General
DLC	Data Link Control	Networking
DLCI	Data link connection identifier	Cellular/2G/GSM
DLDC	Downlink dual carrier	Cellular/2G/GSM
DL-MAP	Downlink map	Networking/WMAN/IEEE
		802.16
DLP	Direct link protocol	Networking/WLAN/IEEE
		802.11

DLS	Direct Link Setup	Networking/WLAN/IEEE 802.11
DL-SCH	Downlink shared channel	Cellular/4G/LTE
DM-RS	Demodulation reference signal	Cellular/4G/LTE
DoD	Department of Defense	General
DoS	Denial-of-service	Security
DPCCH	Dedicated physical control channel	Cellular/3G/UMTS
DPCH	Dedicate physical control channel	Cellular/3G/UMTS
DPDCH	Dedicated physical data channel	Cellular/3G/UMTS
DPSK	Differential Phase Shift Keying	Communications/Modulation
DQPSK	Differential Quadrature Phase Shift Keying	Communications/Modulation
D-QPSK	Differential quadrature phase shift keying	Communications/Modulation
DRC	Data rate control	Cellular/3G/CDMA2000
DRNC	Drift radio network controller	Cellular/3G/UMTS
DRX	Discontinuous reception	Cellular
DS	Distribution system	Networking/WLAN/IEEE 802.11
DSA	Dynamic spectrum access	Communications/Cognitive Radio
DSCH	Downlink shared channel	Cellular/3G/UMTS
DSL	Digital subscriber line	Networking/Wired
DSMIP	Dual-stack mobile IP	Networking/Mobility
DSP	Digital signal processing	General
DSR	Dynamic source routing	Networking/Routing Protocols
DSRC	Dedicated Short Range Communications	Networking/WLAN/IEEE 802.11
DSSS	Direct Sequence Spread Spectrum	Communications
DTCH	Dedicated traffic channel	Cellular/3GPP
DTM	Dual-transfer mode	Cellular/2G/GSM
DTX	Discontinuous transmission	Cellular/2G/GSM
DUN	Dial-up Networking Profile	Networking/WPAN/Bluetooth
DVB	Digital Video Broadcasting	Broadcast Standards
DVB-H	Digital Video Broadcasting–Handheld	Broadcast Standards
E-AGCH	Enhanced access grant channel	Cellular/3G/UMTS
EAP	Extensible authentication protocol	Security
eBMSC	Evolved broadcast multicast service center	Cellular/4G/LTE
EC	Executive Committee	Standards Bodies/IEEE
ECGI	E-UTRAN cell global identifier	Cellular/4G/LTE

ECM	EPS connection management	Cellular/4G/LTE
ECN	Explicit congestion notification	Networking/Cross-layer design
ECSD	Enhanced circuit-switched data	Cellular/2G/GSM
EDCF	Enhanced distributed coordination function	Networking/WLAN/IEEE 802.11
E-DCH	Enhanced uplink dedicated channel	Cellular/3G/UMTS
EDGE	Enhanced Data rates for Global Evolution	Cellular/2G/GSM
E-DPCCH	Enhanced dedicated physical control channel	Cellular/3G/UMTS
E-DPDCH	Enhanced dedicated physical data channel	Cellular/3G/UMTS
EDR	Enhanced Data Rate	Networking/WPAN/Bluetooth
EFR	Enhanced full-rate	Cellular/2G/GSM
EGPRS	Enhanced General Packet Radio Service	Cellular/2G/GSM
E-HICH	Enhanced hybrid indicator channel	Cellular/3G/UMTS
EIA	Electronic Industries Association	Standards Bodies
eICIC	Enhanced intercell interference coordination	Cellular/4G/LTE
EIFS	Extended interframe space	Networking/WLAN/IEEE 802.11
EIR	Equipment identity register	Cellular
eMBMS	Evolved multimedia broadcast/multicast service	Cellular/4G/LTE
EMM	EPS mobility management	Cellular/4G/LTE
eNB	Evolved Node B	Cellular/4G/LTE
EPC	Evolved Packet Core	Cellular/4G/LTE
ePDG	Enhanced packet data gateway	Cellular/4G/LTE
EPS	Evolved Packet System	Cellular/4G/LTE
EQM	Equal modulation	Networking/WLAN/IEEE 802.11
E-RAB	E-UTRAN radio access bearer	Cellular/4G/LTE
E-RNTI	Enhanced radio network temporary identity	Cellular/3G/UMTS
eSCO	Extended Synchronous Connection Oriented	Networking/WPAN/Bluetooth
ESM	EPS session management	Cellular/4G/LTE
ESS	Extended service set	Networking/WLAN/IEEE 802.11
ETSI	European Telecommunications Standards Institute	Standards Bodies

E-UTRA	Evolved Universal Terrestrial Radio Access	Cellular/4G/LTE
E-UTRAN	Evolved Universal Terrestrial Radio Access Network	Cellular/4G/LTE
EV-DO	Evolution—Data Optimized	Cellular/3G/CDMA2000
EV-DV	Evolution-Data/Voice	Cellular/3G/CDMA2000
EVRC	Enhanced Variable Rate Codec	Cellular/3G/CDMA2000
EVRC-B	Enhanced Variable Rate Codec B	Cellular/3G/CDMA2000
FA	Foreign agent	Networking
FACCH	Fast associated control channel	Cellular/2G/GSM
FACH	Forward access channel	Cellular/3G/UMTS
FANR	Fast ACK/NACK reporting	Cellular/2G/GSM
FB	Frequency correction burst	Cellular/2G/GSM
F-BCCH	Forward broadcast control channel	Cellular/3G/CDMA2000
F-CAPICH	Forward common auxiliary pilot channel	Cellular/3G/CDMA2000
FCC	Federal Communications Commission	Regulatory Bodies/United States
F-CCCH	Forward common control channel	Cellular/3G/CDMA2000
FCCH	Frequency correction channel	Cellular/2G/GSM
F-CCHT	Forward common channel type	Cellular/3G/CDMA2000
FCH	Frame control header	Networking/WMAN/IEEE 802.16
FCS	Frame check sequence	Communications/Coding
F-DAPICH	Forward dedicated auxiliary pilot channel	Cellular/3G/CDMA2000
F-DCCH	Forward dedicated control channel	Cellular/3G/CDMA2000
FDD	Frequency Division Duplex	Communications/Duplex Techniques
FDE	Frequency-domain equalization	Communications
FEC	Forward error correction	Communications/Coding
FET	Frame early termination	Cellular/3G/CDMA2000
F-FCH	Forward fundamental channel	Cellular/3G/CDMA2000
FFD	Full Function Device	Networking/WPAN/ZigBee
FFR	Fractional frequency reuse	Cellular/4G/LTE
FFT	Fast Fourier transform	DSP
FH	Frequency Hopping	Communications
FHS	Frequency Hop Synchronization	Networking/WPAN/Bluetooth
FHSS	Frequency Hopping Spread Spectrum	Communications
FIN	Finish	Networking/TCP
FLO	Flexible layer one	Cellular/2G/GSM
FN	Frame number	Cellular/2G/GSM
F-PCH	Forward paging channel	Cellular/3G/CDMA2000

F-PICH	Forward pilot channel	Cellular/3G/CDMA2000
FR	Full-rate	Cellular/2G/GSM
FRAMES	Future radio wideband multiple access systems	Cellular/3G/UMTS
F-SCHT	Forward supplemental channel type	Cellular/3G/CDMA2000
FSK	Frequency Shift Keying	Communications/Modulation
FSTD	Frequency-switched transmit diversity	Communications/Antenna Systems
FTP	File Transfer Protocol *or* File Transfer Profile	Networking/Application Layer Networking/WPAN/ Bluetooth
FUSC	Full usage of subcarriers	Networking/WMAN/IEEE 802.16
FWT	Fast Walsh transform	DSP
GAN	Generic access network	Cellular/2G/GSM
GAP	Generic Access Profile	Networking/WPAN/ Bluetooth
GBR	Guaranteed bit rate	Cellular/4G/LTE
GEA	GPRS encryption algorithm	Cellular/2G/GSM
GERAN	GSM/EDGE Radio Access Network	Cellular/2G/GSM
GF	Galois field	Communications/Coding
GFSK	Gaussian Frequency Shift Keying	Communications/Modulation
GGSN	Gateway GPRS support node	Cellular/3GPP
GI	Guard interval	General
GIAC	General Inquiry Access Code	Networking/WPAN/ Bluetooth
GMM	GPRS mobility management	Cellular/2G/GSM
GMSC	Gateway MSC	Cellular
GMSK	Gaussian Minimum Shift Keying	Communications/Modulation
GPC	Grant-per-connection	Networking/WMAN/IEEE 802.16
GPCS	Generic packet convergence sublayer	Networking/WMAN/IEEE 802.16
GPRS	General Packet Radio System	Cellular/2G/GSM
GPS	Global Positioning System	Communications
GPSS	Grant-per-subscriber-station	Networking/WMAN/IEEE 802.16
GSA	Global mobile Suppliers Association	Trade Associations
GSM	Global System for Mobile communications	Cellular/2G/GSM
GSMA	GSM Association	Trade Associations

GTP	GPRS Tunneling Protocol	Cellular/3GPP
GTS	Guaranteed Time Slot	Networking/WPAN/ZigBee
GUMMEI	Globally unique MME identifier	Cellular/4G/LTE
GUTI	Globally unique temporary identity	Cellular/4G/LTE
HA	Home agent	Networking
HAAA	Home AAA (server)	Cellular
HARQ	Hybrid automatic repeat request	Communications/Coding
HCCA	HCF controlled channel access	Networking/WLAN/IEEE 802.11
HCF	Hybrid coordination function	Networking/WLAN/IEEE 802.11
HCI	Host Controller Interface	Networking/WPAN/ Bluetooth
HCRP	Hardcopy Cable Replacement Profile	Networking/WPAN/ Bluetooth
HCR-TDD	High chip rate time division duplex	Cellular/3G/UMTS
HCS	Header check sequence	Communications/Coding
HDLC	High-Level Data Link Control	Networking/Data Link Protocols
HDMI	High-definition multimedia interface	General
HDP	Health Device Profile	Networking/WPAN/ Bluetooth
HEC	Header error check	Networking
HeNB	Home evolved Node B	Cellular/4G/LTE
HFP	Hands-Free Profile	Networking/WPAN/ Bluetooth
HID	Human Interface Device	Networking/WPAN/ Bluetooth
HII	High interference indicator	Cellular/4G/LTE
HIPERLAN	High performance radio local area network	Networking/WLAN
HIPERMAN	High Performance Metropolitan Area Network	Networking/WMAN
HL	Hybrid location	Cellular/2G/GSM
HLR	Home location register	Cellular
HNB	Home Node B	Cellular/3G/UMTS
HoA	Home address	Networking/Mobility/MIP
HR	High rate	Networking/WLAN/IEEE 802.11
HRPD	High-rate packet data	Cellular/3G/CDMA2000
HS	High Speed	Networking/WPAN/ Bluetooth

HSCSD	High-speed circuit-switched data	Cellular/2G/GSM
HSDPA	High-Speed Downlink Packet Access	Cellular/3G/UMTS
HS-DPCCH	High-speed dedicated physical control channel	Cellular/3G/UMTS
HS-DSCH	High-speed downlink shared channel	Cellular/3G/UMTS
HSGW	HRPD serving gateway	Cellular/3G/CDMA2000
HSP	Headset Profile	Networking/WPAN/ Bluetooth
HSPA	High-Speed Packet Access	Cellular/3G/UMTS
HSPA+	Evolved High Speed Packet Access	Cellular/3G/UMTS
HS-PDSCH	High-speed physical downlink shared channel	Cellular/3G/UMTS
HSS	Home subscriber server	Cellular/3GPP
HS-SCCH	High-speed shared control channel	Cellular/3G/UMTS
HSUPA	High-Speed Uplink Packet Access	Cellular/3G/UMTS
IAC	Inquiry Access Code	Networking/WPAN/ Bluetooth
IAPP	Inter-Access-Point Protocol	Networking/WLAN/IEEE 802.11
IBSS	Independent basic service set	Networking/WLAN/IEEE 802.11
ICIC	Intercell interference coordination	Cellular/4G/LTE
ID	Identification	General
iDEN	Integrated Digital Enhanced Network	Cellular/2G
IDFT	Inverse discrete Fourier transform	DSP
IEEE	Institute of Electrical and Electronics Engineers	Standards Bodies
IESG	Internet Engineering Steering Group	Standards Bodies/IETF
IETF	Internet Engineering Task Force	Standards Bodies
IFFT	Inverse fast Fourier transform	DSP
IFS	Interframe space	Networking/WPAN/ZigBee
IMEI	International MS equipment identity	Cellular
IMS	IP multimedia subsystem	Cellular/3GPP
IMSI	International mobile subscriber identity	Cellular
IMT-2000	International Mobile Telecommunications 2000	Cellular/3G
IMT-Advanced	International Mobile Telecommunications Advanced	Cellular/4G

IP	Internet Protocol	Networking/Network Layer
IPTV	Internet Protocol Television	General
IPv4	Internet Protocol version 4	Networking/Network Layer
IPv6	Internet Protocol version 6	Networking/Network Layer
IR	Infrared	General
	or Incremental redundancy	Communications/Coding
IRTF	Internet Research Task Force	Standards Bodies/IETF
IS-2000	Interim Standard 2000	Cellular/3G/CDMA2000
IS-95	Interim Standard 95	Cellular/2G
ISDN	Integrated Services Digital Network	Communications
ISI	Inter-Symbol Interference	Communications
ISM	Industrial, Scientific, and Medical	Communications/Spectrum
ISO	International Organization for Standardization	Standards Bodies
ISOC	Internet Society	Non-profit Organizations
ITU	International Telecommunications Union	Regulatory Bodies/United Nations
ITU-D	ITU Telecommunication Development Sector	Standards Bodies/ITU
ITU-R	ITU Radiocommunication Sector	Standards Bodies/ITU
ITU-T	ITU Telecommunication Standardization Sector	Standards Bodies/ITU
IV	Initialization vector	Security
JD	Joint detection	Communications/Multiuser detection
JTACS	Japanese Total Access Communications System	Cellular/1G
L1	Layer one	Networking
L2	Layer two	Networking
L2CAP	Logical Link Control and Adaptation	Networking/WPAN/ Bluetooth
LA	Location area	Cellular/2G/GSM
LAC	Link access control	Cellular/3G/CDMA2000
LAI	Location area identifier	Cellular/2G/GSM
LAN	Local Area Network	Networking
LAPD	Link Access Protocol Channel D	Networking/Data Link Layer
LAR	Location-aware routing	Networking/Routing Protocols
LBR	Life-based routing protocol	Networking/Routing Protocols
LC	Link Control	Networking/WPAN/ Bluetooth
LCH	Long transport channel	Networking/WLAN/ HIPERLAN

LCLS	Local call local switch	Cellular/2G/GSM
LCR-TDD	Low chip rate time division duplex	Cellular/3G/UMTS
LCS	Location services	Cellular/3GPP
LDPC	Low-density parity-check	Communications/Coding
LE	Low Energy	Networking/WPAN/ Bluetooth
LIFS	Long interframe space	Networking/WPAN/ZigBee
LLC	Logical Link Control	Networking
LLME	Logical link management entity	Cellular/2G/GSM
LMP	Link Manager Protocol	Networking/WPAN/ Bluetooth
LMSC	LAN/MAN Standards Committee	Standards Bodies/IEEE
LNA	Low noise amplifier	Electronics
LOS	Line-of-sight	Communications
LQI	Link quality indicator	Networking/WPAN/ZigBee
LSB	Least Significant Bit	General
LTE	Long Term Evolution	Cellular/4G/LTE
LUP	Location update protocol	Cellular/3G/CDMA2000
M&S	Modeling and simulation	General
M2M	Machine-to-machine	Cellular
MAC	Medium Access Control	Networking
MAN	Metropolitan Area Network	Networking
MANET	Mobile ad hoc network	Networking/Mobility
MAP	Message Access Profile	Networking/WPAN/ Bluetooth
MBMS	Multimedia broadcast and multicast service	Cellular/3GPP
MBS	Multicast and broadcast service	Networking/WMAN/IEEE 802.16
MBSFN	Multimedia broadcast over single frequency network	Cellular/4G/LTE
MBWA	Mobile Broadband Wireless Access	Networking
MC	Multicarrier *or* Message center	Communications Cellular/3G/CDMA2000
MCC	Mobile country code	Cellular
MCCH	Multicast control channel	Cellular/3GPP
MCH	Multicast channel	Cellular/4G/LTE
MCM	Multicarrier modulation	Communications
MCPS	MAC common part sublayer	Networking/WPAN/ZigBee
MCS	Modulation and coding scheme	Communications
ME	Mobile equipment	Cellular
MEMS	Microelectromechanical system	Electronics

MIB	Management information base or Master information block	Networking/IEEE 802 Cellular/4G/LTE
MIC	Message Integrity Check	Security
MID	Multiple interface declaration	Networking/Routing/OLSR
MIMO	Multiple Input, Multiple Output	Communications/Antenna Systems
MIP	Mobile IP	Networking/Mobility
MISO	multiple input, single output	Communications/Antenna Systems
MLB	Mobility load balancing	Cellular/4G/LTE
MLME	MAC layer management entity	Networking/WPAN/ZigBee
MM	Mobility Management	Cellular
MMDS	Multichannel multipoint distribution services	Communications/Spectrum
MME	Mobility management entity	Cellular/4G/LTE
MMPDU	MAC management protocol data unit	Networking/WLAN/IEEE 802.11
MMS	Multimedia messaging service	Cellular
MN	Mobile node	Networking/Mobility/MIP
MNC	Mobile network code	Cellular
MOCN	Multioperator core network	Cellular/2G/GSM
MORE	MAC-independent Opportunistic Routing and Encoding	Networking/Network Coding
MoU	Memorandum of Understanding	Cellular/2G/GSM
MPDU	MAC Protocol Data Unit	Networking
MPR	Multipoint relay	Networking/Routing/OLSR
MR	Mobile router	Networking/Mobility/NEMO
MRD	Mobile receive diversity	Cellular/3G/CDMA2000
MRO	Mobility robustness optimization	Cellular/4G/LTE
MS	Mobile station	Networking/WMAN/IEEE 802.16
MSB	Most significant bit	General
MSC	Mobile services switching center	Cellular
MSCH	MBMS point-to-multipoint scheduling channel	Cellular/3G/UMTS
MSDU	MAC service data unit	Networking
MSIN	Mobile subscriber identification number	Cellular
MSISDN	Mobile subscriber ISDN number	Cellular
MSR	Multistandard radio	Cellular/3GPP
MSRD	Mobile station receiver diversity	Cellular/2G/GSM
MSRN	Mobile station roaming number	Cellular
MT	Mobile termination	Cellular/3G/UMTS
MTCH	Multicast traffic channel	Cellular/3GPP

MTD	Mobile transmit diversity	Cellular/3G/CDMA2000
M-TMSI	MME Temporary Mobile Subscriber Identity	Cellular/4G/LTE
MU-MIMO	Multiuser MIMO	Communications/Antenna Systems
MUROS	Multi-user reusing one slot	Cellular/2G/GSM
NACC	Network-assisted cell change	Cellular/2G/GSM
NACK	Negative Acknowledgement	Networking
NAP	Network access provider	Networking/WMAN/IEEE 802.16
NAS	Non-access stratum	Cellular/3GPP
NAV	Network allocation vector	Networking/WLAN/IEEE 802.11
NB	Node B (base station) or Normal burst	Cellular/3G/UMTS Cellular/2G/GSM
NBPS	Network-based positioning support	Cellular/4G/LTE
NCH	Notification channel	Cellular/2G/GSM
NCMS	Network control and management system	Networking/WMAN/IEEE 802.16
NDP	Neighbor Discovery Protocol	Networking/WLAN/IEEE 802.11
NEMO	Network Mobility	Networking/Mobility
NET	Network Layer	Networking
NFC	Near-field communication	Communications
NGN	Next Generation Networks	Networking
NHLE	Next higher layer entity	Networking
NIB	NWK information base	Networking/WPAN/ZigBee
NIC	Network interface card	Electronics
NLDE	Network layer data entity	Networking/WPAN/ZigBee
NLME	Network layer management entity	Networking/WPAN/ZigBee
NLOS	Non-line-of-sight	Communications
NMSI	National mobile subscriber identity	Cellular
NMT	Nordic Mobile Telephony	Cellular/1G
NOC	Network operations center	Networking/Infrastructure
NPDU	Network-level packet data unit	Networking
NRM	Network reference model	Networking/WMAN/IEEE 802.16
NSAPI	Network layer service access point identifier	Cellular/2G/GSM
NSDU	Network service data unit	Networking
NSP	Network service provider	Networking/WMAN/IEEE 802.16
NSS	Network switching subsystem	Cellular/2G/GSM

NTIA	National Telecommunications and Information Administration	Regulatory Bodies/United States
NWK	Network layer	Networking/WPAN/ZigBee
O&M	Operations and Maintenance	Cellular
OFDM	Orthogonal Frequency Division Multiplexing	Communications
OFDMA	Orthogonal Frequency Division Multiple Access	Communications/Multiple Access
OI	Overload indicator	Cellular/4G/LTE
OLSR	Optimized link-state routing	Networking/Routing Protocols
OMS	Operations and maintenance subsystem	Cellular
OPP	Object Push Profile	Networking/WPAN/ Bluetooth
O-QPSK	Offset quadrature phase-shift keying	Communications/Modulation
OSC	Orthogonal subchannel	Cellular/2G/GSM
OSI	Open Standards Interconnect	Networking
OSPF	Open shortest path first	Networking/Routing Protocols
OSS	Operation subsystem	Cellular
OVSF	Orthogonal variable speeding factor	Cellular/3G/UMTS
PA	Power amplifier	Electronics
PACCH	Packet associated control channel	Cellular/2G/GSM
PAGCH	Packet access grant channel	Cellular/2G/GSM
PAPR	Peak-to-average power ratio	Communications
PAR	Power-aware routing	Networking/Routing Protocols
PBAP	Phonebook Access Profile	Networking/WPAN/ Bluetooth
PBCCH	Packet broadcast control channel	Cellular/2G/GSM
PBCH	Physical broadcast channel	Cellular/4G/LTE
PCC	Policy and charging control	Cellular/3GPP
	or Primary component carrier	Cellular/4G/LTE
PCCCH	Packet common control channel	Cellular/2G/GSM
PCCH	Paging control channel	Cellular/3GPP
PCCPCH	Primary common control physical channel	Cellular/3G/UMTS
PCEF	Policy and charging enforcement function	Cellular/3GPP
PCF	Point coordination function	Networking/WLAN/IEEE 802.11
	or Packet control function	Cellular/3G/CDMA2000

PCFICH	Physical control format indicator channel	Cellular/4G/LTE
PCH	Paging channel	Cellular/3GPP
PCI	Protocol Control Information	Networking
PCM	Pulse code modulation	Communications/Modulation
PCRF	Policy and charging rules function	Cellular/3GPP
PCS	Personal Cellular Service	Cellular
PCU	Packet control unit	Cellular/2G/GSM
PDA	Personal data assistant	General
PDC	Personal Digital Cellular	Cellular/2G
PDCCH	Packet dedicated control channel *or* Physical downlink control channel	Cellular/2G/GSM Cellular/4G/LTE
PDCP	Packet data convergence protocol	Cellular/3GPP
PDN	Packet data network	General
PDP	Packet data protocol	Cellular/2G/GSM
PDSCH	Physical downlink shared channel	Cellular/4G/LTE
PDSN	Packet data serving node	Cellular/3G/CDMA2000
PDTCH	Packet data traffic channel	Cellular/2G/GSM
PDU	Protocol Data Unit	Networking
PEP	Performance enhancing proxy	Networking
P-GW	PDN gateway	Cellular/4G/LTE
PHICH	Physical HARQ indicator channel	Cellular/4G/LTE
PHS	Packet header suppression *or* Personal Handy-phone System	Networking/ATM Cellular/2G
PHY	Physical Layer	Networking
PIB	PAN information base	Networking/WPAN/ZigBee
PICH	Paging indicator channel	Cellular/3G/UMTS
PIFS	PCF interframe space	Networking/WLAN/IEEE 802.11
PKM	Privacy key management	Security
PLCP	Physical layer convergence protocol	Networking/WLAN/IEEE 802.11
PLME	Physical layer management entity	Networking/WPAN/ZigBee
PLMN	Public land mobile network	Cellular/2G/GSM
PMCH	Physical multicast channel	Cellular/4G/LTE
PMI	Precoding matrix indicator	Cellular/4G/LTE
PMIP	Proxy Mobile IP	Networking/Mobility
PMP	Point-to-multipoint	Networking/WMAN/IEEE 802.16
PN	Pseudorandom noise	Cellular/3G/UMTS
PNCH	Packet notification channel	Cellular/2G/GSM
PoC	Push-to-talk over cellular	Cellular/3G/UMTS
PPCH	Packet paging channel	Cellular/2G/GSM

PPDU	PHY protocol data unit	Networking/WPAN/ZigBee
	or PLCP protocol data unit	Networking/WLAN/IEEE 802.11
PPM	Pulse position modulation	Communications/Modulation
PPP	Point-to-point protocol	Networking/Data Link Layer
PRACH	Physical random access channel	Cellular/3G/UMTS, or 4G/LTE
	or Packet random access channel	Cellular/2G/GSM
PRB	Physical resource block	Cellular/4G/LTE
PRBS	Pseudorandom binary sequence	General
PROPHET	Probabilistic routing protocol using history of encounters and transitivity	Networking/Routing Protocols
PS	Packet-switched	Cellular
PSB	Parked Slave Broadcast	Networking/WPAN/ Bluetooth
PSC	Primary scrambling code	Cellular/3G/UMTS
PSD	Power spectral density	General
PSDU	PHY service data unit	Networking/WPAN/ZigBee
	or PLCP service data unit	Networking/WLAN/IEEE 802.11
PSK	Phase Shift Keying	Communications/Modulation
PSS	Primary synchronization signal	Cellular/4G/LTE
PSSS	Parallel sequence spread spectrum	Communications
PSTN	Public switched telephone network	General
PTCCH	Packet timing advance control channel	Cellular/2G/GSM
PTCP	Point-to-consecutive point	Networking/WMAN/IEEE 802.16
PTM	Push-to-media	Cellular
P-TMSI	Packet temporary mobile subscriber identity	Cellular
PTP	Point-to-point	General
PTS	Profile Tuning Suite	Networking/WPAN/ Bluetooth
PTT	Push-to-talk	Cellular
PUCCH	Physical uplink control channel	Cellular/4G/LTE
PUSC	Partial usage of subcarriers	Networking/WMAN/IEEE 802.16
PUSCH	Physical uplink shared channel	Cellular/4G/LTE
QAM	Quadrature Amplitude Modulation	Communications/Modulation
QCI	QoS class identifiers	Cellular/3GPP
QLIC	Quasi-linear interference cancellation	Cellular/3G/CDMA2000

QOF	Quasi-orthogonal function	Cellular/3G/CDMA2000
QoS	Quality of Service	Networking
QPSK	Quadrature Phase Shift Keying	Communications/Modulation
R&D	Research and Development	General
RA	Routing area	Cellular/3GPP
RAB	Radio access bearer	Cellular/4G/LTE
RABR	Route-lifetime assessment-based routing	Networking/Routing Protocols
RACH	Random access channel	Cellular
R-ACH	Reverse access channel	Cellular/3G/CDMA2000
RADIUS	Remote authentication dial-in user service	Networking/Application Layer
RAI	Routing area identifier	Cellular/2G/GSM
RAN	Radio Access Network	Cellular
RAT	Radio access technology	Cellular
RB	Resource block	Cellular/4G/LTE
RBG	Resource block group	Cellular/4G/LTE
R-CCCH	Reverse common control channel	Cellular/3G/CDMA2000
RCM	Ranging and communications module	Networking/WPAN/UWB
R-DDCH	Reverse dedicated control channel	Cellular/3G/CDMA2000
RE	Resource element	Cellular/4G/LTE
R-EACH	Reverse enhanced access channel	Cellular/3G/CDMA2000
REG	Resource element group	Cellular/4G/LTE
RERR	Route Error	Networking/Routing/AODV
RF	Radio Frequency	General
RF4CE	Radio frequency for consumer electronics	Networking/WPAN/ZigBee
RFC	Request For Comments	Standards Bodies/IETF
R-FCH	Reverse fundamental channel	Cellular/3G/CDMA2000
RFD	Reduced Function Device	Networking/WPAN/ZigBee
RI	Rank indicator	Cellular/4G/LTE
RIFS	Reduced interframe space	Networking/WLAN/IEEE 802.11
RLC	Radio link control	Cellular/2G/GSM
RLIC	Reverse link interference cancellation	Cellular/3G/CDMA2000
RLP	Radio link protocol	Cellular/3G/CDMA2000
RN	Relay node	Cellular/4G/LTE
RNC	Radio network controller	Cellular/3G/UMTS
RNS	Radio network subsystem	Cellular/3G/UMTS
RNTI	Radio network temporary identifier	Cellular/4G/LTE
RNTP	Relative narrowband transmit power	Cellular/4G/LTE

ROHC	Robust header compression	Networking
R-PICH	Reverse pilot channel	Cellular/3G/CDMA2000
RR	Radio Resource	Cellular
RRC	Radio resource control	Cellular/3GPP
RREP	Route reply	Networking/Routing/AODV
RREQ	Route request	Networking/Routing/AODV
RRH	Remote radio head	Cellular
RRM	Radio resource management	Cellular
RS	Reed-Solomon	Communications/Coding
	or Reference signal	Cellular/4G/LTE
RSA	Ravist Shamir Adleman (public key encryption algorithm)	Security
R-SCHT	Reverse supplemental channel type	Cellular/3G/CDMA2000
RSS	Radio subsystem	Cellular/3G/UMTS
RTG	receive/transmit transition gap	Networking/WMAN/IEEE 802.16
RTS	Request-to-send	Networking/WLAN/IEEE 802.11
RTT	Round trip time	Networking/TCP
RTTI	Reduced transmission time interval	Cellular/2G/GSM
RUIM	Removable user identity module	Cellular/3G/CDMA2000
SACCH	Slow associated control channel	Cellular/2G/GSM
SACK	Selective acknowledgement	Networking
SAE	System architecture evolution	Cellular/4G/LTE
SAIC	Single antenna interference cancellation	Cellular/2G/GSM
SAP	Service Access Point	Networking
	or SIM Access Profile	Networking/WPAN/ Bluetooth
SAPI	Service access point identifier	Cellular/2G/GSM
SAR	Segmentation and reassembly	Cellular/3G/CDMA2000
SATCOM	Satellite communication	Communications
SB	Synchronization burst	Cellular/2G/GSM
SC	Single-carrier	Communications
SCADA	Supervisory control and data acquisition	General
SCC	Secondary component carrier	Cellular/4G/LTE
SCCPCH	Secondary common control physical channel	Cellular/3G/UMTS
SC-FDMA	Single-carrier frequency division multiple access	Communications/Multiple Access
SCH	Synchronization channel	Cellular/3GPP

	or Short transport channel	Networking/WLAN/HIPERLAN
SCO	Synchronous Connection Oriented	Networking/WPAN/Bluetooth
SCPIR	Subchannel power imbalance ratio	Cellular/2G/GSM
SDCCH	Stand-alone dedicated control channel	Cellular/2G/GSM
SDMA	Space division multiple access	Communications/Multiple Access
SDP	Service Discovery Protocol	Networking/WPAN/Bluetooth
SDU	Service Data Unit	Networking
SF	Subchannelization factor	Networking/WMAN/IEEE 802.16
	or Spreading factor	Cellular/3G/UMTS
SFBC	Space-frequency block code	Communications/Coding
SFID	Service flow ID	Networking
SFN	Single frequency network	Broadcast Standards/DVB-H
SFR	Soft frequency reuse	Cellular/4G/LTE
SGSN	Serving GPRS support node	Cellular/3GPP
S-GW	Serving gateway	Cellular/4G/LTE
SHA	Secure Hash Algorithm	Security
SHCCH	Shared channel control channel	Cellular/3G/UMTS
SI	System information	Cellular
SIB	System information block	Cellular/4G/LTE
SIC	Successive interference cancellation	Communications/Multiuser detection
SIFS	Short interframe space	Networking/IEEE 802
SIG	Special Interest Group	Networking/WPAN/Bluetooth
SIGTRAN	Signaling transport	Cellular/2G/GSM
SIM	Subscriber identity module	Cellular/2G/GSM
SIMO	single input, multiple output	Communications/Antenna Systems
SIR	Signal-to-interference ratio	General
SI-RNTI	System information RNTI	Cellular/4G/LTE
SISO	Single input, single output	Communications/Antenna Systems
SLP	Signaling link protocol	Cellular/3G/CDMA2000
SM	Session management	Cellular
	or Spatial multiplexing	Communications/Antenna Systems
SME	Short message entity	Cellular/3G/CDMA2000
SMG	Special Mobile Group	Standards Bodies/ETSI
SMS	Short Message Service	Cellular

SMSS	Sender's maximum segment size	Networking/TCP
SNDCP	Subnetwork-dependent convergence protocol	Cellular/2G/GSM
SNP	Signaling network protocol	Cellular/3G/CDMA2000
SNR	Signal-to-noise ratio	General
S-OFDMA	Scalable OFDMA	Networking/WMAN/IEEE 802.16
SON	Self-organizing network	Cellular/4G/LTE
SPDU	SSCS protocol data units	Networking/WPAN/ZigBee
SPP	Serial Port Profile	Networking/WPAN/ Bluetooth
SPS-RNTI	Semipersistent scheduling RNTI	Cellular/4G/LTE
SRB	Signaling radio bearer	Cellular/4G/LTE
SRNC	Serving radio network controller	Cellular/3G/UMTS
SRS	Sounding reference signal	Cellular/4G/LTE
SRVCC	Single-radio voice call continuity	Cellular/3GPP
SS	Subscriber station *or* Supplementary Services	Networking/WMAN/IEEE 802.16 Cellular
SS#7	Signaling System No. 7	Cellular
SSC	Secondary scrambling code	Cellular/3G/UMTS
SSCS	Service-specific convergence sublayer	Networking/WPAN/ZigBee
SSID	Service set identity	Networking/WLAN/IEEE 802.11
SSS	Secondary synchronization signal	Cellular/4G/LTE
STBC	Space-time block code	Communications/Coding
STC	space-time coding	Communications/Coding
STF	Short training field	Networking/WLAN/IEEE 802.11
STIRC	Space-time interference rejection combining	Communications/Multiuser detection
S-TMSI	SAE temporary mobile subscriber identity	Cellular/4G/LTE
SU-MIMO	Single-user MIMO	Communications/Antenna Systems
SVDO	Simultaneous voice and data	Cellular/3G/CDMA2000
SYN	Synchronize	Networking/TCP
SYNC	Synchronization Profile	Networking/WPAN/ Bluetooth
TA	Tracking area	Cellular/4G/LTE
TACS	Total Access Communications System	Cellular/1G
TAG	Technical Advisory Group	Standards Bodies/IEEE
TAI	Tracking area identity	Cellular/4G/LTE

TAPS	TETRA Advanced Packet Services	Cellular/2G/GSM
TB	Transport block	Cellular/4G/LTE
TBF	Temporary block flow	Cellular/2G/GSM
TBRPF	Topology-based reverse path forwarding	Networking/Routing Protocols
TC	Turbo code	Communications/Coding
	or Topology control	Networking/Routing/OLSR
	or Traffic category	Networking/WLAN/IEEE 802.11
TCH	Traffic channel	Cellular/2G/GSM
TCP	Transmission Control Protocol	Networking/Transport Layer
TDD	Time Division Duplex	Communications/Duplex Techniques
TD-LTE	Time division Long Term Evolution	Cellular/4G/LTE
TDM	Time division multiplexing	Communications
TDMA	Time Division Multiple Access	Communications/Multiple Access
TD-SCDMA	Time Division Synchronous Code Division Multiple Access	Cellular/3G/UMTS
TE	Terminal equipment	Cellular/3G/UMTS
TEID	Tunnel end point identifier	Cellular/3GPP
TEK	Traffic encryption key	Security
TETRA	Terrestrial trunked radio	Cellular/2G/GSM
TFI	Temporary flow identity	Cellular/2G/GSM
TFT	Traffic flow template	Cellular/4G/LTE
TIA	Telecommunications Industry Association	Standards Bodies
TKIP	Temporal key integrity protocol	Networking/WLAN/IEEE 802.11
TLLI	Temporary logical link identifier	Cellular/2G/GSM
TM	Transparent mode	Cellular/3GPP
TMSI	Temporary mobile subscriber identity	Cellular
TN	Time slot number	Cellular/2G/GSM
TNC	Terminal Node Controller	Networking/ALOHANET
TND	TBRPF neighbor discovery	Networking/Routing/TBRPF
TPC	Transmit Power Control	Communications
TR	Technical report	Standards Bodies/3GPP
TRAU	Transcoding rate adaptation unit	Cellular/2G/GSM
TRM	TBRPF route management	Networking/Routing/TBRPF
TRX	Transceiver	Cellular/2G/GSM
TS	Technical specification	Standards Bodies/3GPP
TSC	Technical steering committee	Trade Associations/WiMAX Forum
	or Training sequence code	Cellular/2G/GSM

TSG	Technical Specification Group	Standards Bodies/3GPP, 3GPP2
TTA	Telecommunications Technology Association	Standards Bodies
TTC	Telecommunication Technology Committee	Standards Bodies
TTG	Transmit/receive transition gap	Networking/WMAN/IEEE 802.16
TTI	Transmission time interval	Cellular
TV	Television	General
TVBD	Television band device	Communications/Cognitive Radio
TVWS	Television white space	Communications/Cognitive Radio
TXOP	Transmission opportunity	Networking/WLAN/IEEE 802.11
UARFCN	UTRA absolute RF channel number	Cellular/3G/UMTS
UCI	Uplink control information	Cellular/4G/LTE
UDP	User Datagram Protocol	Networking/Transport Layer
UE	User equipment	Cellular/3GPP
UEQM	Unequal modulation	Networking/WLAN/IEEE 802.11
UGS	Unsolicited grant service	Networking/WMAN/IEEE 802.16
UHF	Ultra high frequency	Communications/Spectrum
UICC	Universal Integrated Circuit Card	Electronics
UIM	User identity module	Cellular/3G/CDMA2000
UIUC	Uplink interval usage code	Networking/WMAN/IEEE 802.16
UL	Uplink	General
UL-MAP	Uplink map	Networking/WMAN/IEEE 802.16
UL-SCH	Uplink shared channel	Cellular/4G/LTE
UM	Unacknowledged mode	Cellular/3GPP
UMA	Unlicensed Mobile Access	Cellular
UMB	Ultra Mobile Broadband	Cellular/4G
UMTS	Universal Mobile Telecommunications System	Cellular/3G/UMTS
UN	United Nations	General
UNII	Unlicensed National Information Infrastructure	Communications/Spectrum
UP	User priorities	Networking/WLAN/IEEE 802.11
US	United States	General

USB	Universal Serial Bus	General
USCH	Uplink shared channel	Cellular/3G/UMTS
USF	Uplink state flag	Cellular/2G/GSM
USIM	Universal subscriber identity module	Cellular/3GPP
U-TDOA	Uplink time difference of arrival	Cellular
UTRA	Universal Terrestrial Radio Access	Cellular/3G/UMTS
UTRAN	Universal Terrestrial Radio Access Network	Cellular/3G/UMTS
UUID	Universally Unique Identifier	Networking/WPAN/Bluetooth
UWB	Ultra wideband	Networking/WPAN
VAAA	Visited AAA (server)	Cellular
VAMOS	Voice services over adaptive multiuser channels on one slot	Cellular/2G/GSM
VDP	Video Distribution Profile	Networking/WPAN/Bluetooth
VHF	Very high frequency	Communications/Spectrum
VLR	Visitor location register	Cellular
VoIP	Voice over IP	Networking
VRB	Virtual resource block	Cellular/4G/LTE
VT	Video telephony	Cellular
WAVE	Wireless Access in Vehicular Environments	Networking/WLAN/IEEE 802.11
WCDMA	Wideband Code Division Multiple Access	Cellular/3G/UMTS
WEP	Wired equivalent privacy	Networking/WLAN/IEEE 802.11
WFA	Wi-Fi Alliance	Trade Associations
WG	Working Group	General
WiBRO	Wireless Broadband	Networking/WMAN
WiMAX	Worldwide Interoperability for Microwave Access	Networking/WMAN
WLAN	Wireless Local Area Network	Networking
WLL	Wireless local loop	Networking
WMAN	Wireless Metropolitan Area Network	Networking
WMM	Wi-Fi Multimedia	Networking/WLAN/IEEE 802.11
WPA	Wi-Fi Protected Access	Networking/WLAN/IEEE 802.11
WPA2	Wi-Fi Protected Access 2	Networking/WLAN/IEEE 802.11
WPAN	Wireless Personal Area Networking	Networking

WRAN	Wireless Regional Area Network	Networking
WRC	World Radiocommunications Conference	Standards Bodies/ITU
WTSA	World Telecommunications Standardization Assembly	Standards Bodies/ITU
XOR	exclusive OR	General
ZC	Zadoff-Chu	Cellular/4G/LTE
ZDO	ZigBee device object	Networking/WPAN/ZigBee
ZDP	ZigBee device profile	Networking/WPAN/ZigBee

Chapter 1

Introduction

If you were in a state of cryogenic suspension for the past two decades and suddenly awoke, you would find yourself inundated with information coming from the use of cellular phones, personal computers connected to the Internet, tablets, digital music players, e-book readers, high-definition televisions, digital cameras, and a variety of other digital gadgets that have entered common households while you were asleep. Dear reader (drumroll, please), welcome to the information age, an era that can be characterized by one's ability to disseminate and access knowledge freely and instantly, something that would have been extremely difficult to attain in the past. When the traditional industry in the 1800s shifted into the age of the Industrial Revolution, the effect on global socioeconomic and cultural conditions was phenomenally profound. The information era we find ourselves in now leads us into another time of changing socioeconomic and cultural conditions. "Information is king" is the new mantra of our generation: how we acquire it, how we share it, and how we protect it.

As we continue to experience the information technology revolution, the requirements for broadband access will become even more challenging with the growing popularity of various multimedia applications. To satisfy this expanding market need, current broadband networks need to extend their reach while new broadband infrastructures need to be deployed in developing countries without existing wired broadband infrastructures.

When one considers how essential the wireless Internet is to our everyday life today, it is hard to fathom that wireless Internet access was virtually nonexistent a mere two decades ago. In that time, wireless Internet technology has grown at a tremendous rate. In this book, we attempt to provide a comprehensive overview of today's diverse wireless networking landscape. This text attempts to provide a good

Wireless Networking: Understanding Internetworking Challenges, First Edition.
Jack L. Burbank, Julia Andrusenko, Jared S. Everett, and William T.M. Kasch.
© 2013 the Institute of Electrical and Electronics Engineers, Inc. Published 2013 by John Wiley & Sons, Inc.

starting point for those unfamiliar with the numerous wireless networking standards and also provides a concise refresher or gap-filler for the seasoned professional.

1.1 DATA NETWORKS VERSUS CELLULAR NETWORKS

The predominant wireless networking standards in today's commercial wireless networking landscape can be divided into two general categories: wireless computer networking standards and cellular networking standards. Wireless computer networking standards are those standards that emerged from the discipline of computer networking. These are networks that, from their inception, served the primary purpose of supporting computer communications and Internet access. Cellular networking standards are those standards that emerged from the discipline of cellular telephony, a traditionally voice-oriented field. Today, the line between these two categories has become increasingly blurred with the convergence of computers and mobile communications devices. Many of today's mobile cellular handsets contain one or more radios to communicate with wireless computer networks, such as IEEE 802.11 (Wi-Fi) or Bluetooth. Similarly, cellular data modems for laptop computers have been available for over a decade. This distinction was blurred even further in 2007, when the International Telecommunications Union (ITU) recognized mobile WiMAX, an IEEE-based wireless metropolitan area network standard with mobility enhancements, as an official third-generation (3G) cellular standard. Moving forward, the distinction between wireless computer networking standards and cellular networking standards is more a historical distinction than a functional one. As we will see later, however, these historical differences play a significant role in how these standards came to be the way they are today.

A key historical difference between wireless computer networks and cellular networks is the degree of mobility support they are designed to provide. Early wireless computer networks were designed for fixed or nomadic use. The term "nomadic" refers to those terminals that may move between sessions but typically are stationary during communication with the network. This is in contrast to a mobile terminal, which is one that may be in motion as communication takes place. Cellular networks, on the other hand, were designed from the beginning to support a high degree of mobility (e.g., in a moving vehicle). Modern wireless communications devices have shifted toward increased mobility across the board. However, today's systems build upon their predecessors and often inherit design decisions from previous generations. In this way, a modern technology's heritage (i.e., wireless computer network vs. cellular network) impacts the manner in which mobility is implemented and how well it is supported.

There is, additionally, a trend toward seamless mobility—this is also breaking down the boundaries between "wireless computer networking" and "cellular networking" technologies as internetworking (and handoff) is implemented between all these standards. The user may jump from Bluetooth to Wi-Fi at home, then make a transition to a combination of 2G/3G/4G cellular outside the home. All the while, the user never loses his or her connection and, depending on implementation, may be unaware that any handover occurred.

1.2   THE HISTORY OF THE WIRELESS INTERNET

Wireless networking finds its origins in the early 1970s. In 1971, the University of Hawaii introduced the first wireless network of record. The ALOHANET research project interconnected computers on seven University of Hawaii campuses spread across four islands via wireless connectivity. However, wireless networking through much of the 1970s, 1980s, and 1990s remained alive only due to the efforts of amateur radio hobbyists in the United States and Canada, who developed terminal node controllers (TNCs) that could interconnect various sites around the world. These TNCs were analogous to computer modems. A historic date came in 1985, when the U.S. Federal Communications Commission (FCC) authorized the public use of the industrial, scientific, and medical (ISM) frequency bands. This led the way to an increased commercial interest in wireless networking, and in the late 1980s the IEEE 802 Local Area Network (LAN) and Metropolitan Area Network (MAN) Working Group authorized a project for the development of a wireless local area network (WLAN) standard. The resulting 802.11 Working Group published the original 802.11 WLAN standard on November 18, 1997 (see Reference [1]). That original WLAN standard provided raw link data rates of 1 and 2 Mbps. Since that original specification, the IEEE 802.11 Working Group has published thousands of additional specification pages spread over dozens of modifications to their WLAN technology. IEEE 802.11-based WLANs now consistently provide data rates on the order of 300 Mbps with a vastly increased feature set.

Other communities began development activities in parallel. In 1994, Ericsson initiated a project to study the feasibility of a low-power, low-cost radio interface to eliminate cable from mobile phones and their accessories. However, the wider applicability of this technology was quickly realized, and in 1998 the Bluetooth Special Interest Group (SIG) was formed; the founding members were Ericsson, Intel, Nokia, and Toshiba. In July 1999, Version 1.0 of the Bluetooth specification was published [2]. Over the next dozen years, the Bluetooth specification experienced numerous revisions (Bluetooth 4.0 was published in 2010 [3]), with thousands of additional pages of technology specification that has produced orders of magnitude improvement in Bluetooth capability over that timeframe.

The IEEE 802.16 Wireless Metropolitan Area Network (WMAN) Working Group met for the first time in 1999, with the goal of developing a standardized approach to point-to-point wireless backhaul communications. In 2001, the original IEEE 802.16 specification was published [4]. In the decade since that original specification, thousands of additional pages of technology specification have been published over numerous technology revisions. The result has been a complete transformation of the technology from a "niche" technology (wireless backhaul) to a legitimate choice as a "4G" technology.

At the same time, cellular telephony has experienced tremendous growth, in terms of its deployment and usage and also in terms of its capabilities. One needs to go back only to 1992 to find a world in which GSM had not yet been deployed outside Europe. Today, GSM spans the globe with over 5 billion subscribers. One needs to travel back in time only a little over a decade to find a world in which second-generation cellular networks are at the cutting edge, with data capabilities comparable to dial-up modems. The past decade has seen multiple generations of technology development, deployment, and adoption that have provided several orders of magnitude of improvement in data rates and general capabilities. Technologies such as Universal Mobile Telecommunications System (UMTS), Wideband Code Division Multiple Access (WCDMA), High-Speed Packet Access (HSPA), CDMA2000, CDMA2000 Evolution–Data Optimized (EV-DO), and Long-Term Evolution (LTE) have brought cellular networks into the forefront of the data revolution and the rise of the wireless Internet. To put this all into perspective, think about it this way: over the course of a decade, cellular data networks have transformed from a dial-up modem to a Wi-Fi network interface card.

Before we move on, let us first take a step back and marvel at the timeline over which this has all taken place. Heinrich Hertz first produced radio waves in 1888. Nikola Tesla developed the first radiofrequency (RF) transmitter in 1893. Guglielmo Marchese Marconi conducted the first transatlantic RF transmission in 1901. During World War II, the United States Army conducted the first known data transmissions using radio signals. Approximately 60 years elapsed from the time when radio waves were first produced to the time when they were used to convey data. Then it took another 20 years before anything more significant took place. Then the ALOHANET project was conducted in 1971, and Martin Cooper also created the first portable cell phone in 1973. Then, once again, it took another 20+ years before anything more significant took place. Then, only a little over a decade ago, many wireless technologies came into being. And in the decade since, they have grown at astronomical rates: grown in terms of technical capabilities, in terms of the ecosystems they have created, and in terms of the impact they have on all our lives. We hope the readers of this book have a strong appreciation of how much has happened so quickly. We now live in a world with a glut of wireless networking technologies that perpetually change and evolve. It is indeed a daunting task to even attempt to be aware of everything in this landscape, much less have familiarity or expertise across it. That is the motivation for this book, to provide the reader with a single source from which he or she can become familiar with a large portion of the entire wireless networking landscape.

The authors also place a premium on the history of wireless networking. Too often, texts focus solely on the what, here, and now, and less on the journey of how we got to where we are. We contend that it is important not only to understand the current state of the art, but to also have insight into how we have gotten to where we are at, to understand the long, complex journey. For this reason, you will find that throughout this book, we emphasize the historical perspective of technology development efforts. For this reason, you will find references throughout the book to sources that discuss the history of a particular technology development effort, when available. For a good overall treatment of the history of wireless communications, we recommend Tapan Sarker's book *History of Wireless* (2006), which provides a very good timeline of developments in this field [5].

1.3   THE DIFFERENCE BETWEEN WIRELESS AND WIRED

So why are there so many different wireless networking technologies? And why do they seem to change and evolve so rapidly? The answer is that the overall problem space of wireless networking is large and diverse. There does not yet exist a "silver bullet," the end-all technology that solves every part of the problem space. This is due in part to the fact that rarely do large monolithic solutions work. This is particularly true when the problem space faces many fundamental technical challenges, which is definitely the case in the problem space of wireless networking.

Wireless networks are typically disadvantaged in various aspects and present many challenging cases that simply are not concerns in the wired domain. Some of the negative characteristics that must be overcome include the following:

Low Bandwidth That Limits Raw Link Capacity: This is perhaps the greatest difference between the wired and wireless networking domains. In the wired domain, the solution almost always comes down to provisioning. When a network becomes congested, as the user base grows, more bandwidth is applied to the problem. This is not a viable solution in the wireless domain. RF spectrum is a precious, finite resource. RF spectrum cannot be invented. RF spectrum cannot be created. All wireless network developers can do is hope to use existing RF spectrum more efficiently. While no technology developer, either for wired or wireless networks, wishes to develop an inefficient solution, inefficiency can be tolerated in the wired domain.

Poor Channel Quality That Results in Data Loss: In the wired domain, loss in the channel (in this case, on the wire) is rare. Error rates across a wired infrastructure are typically very low. When losses do occur, they are usually because of congestion, too much data overwhelming the infrastructure to the point that not all can be processed in a timely manner. In this case, more bandwidth is applied to the problem and the congestion problem typically goes away. In the wireless domain, errors can and do happen on a regular basis. A wireless channel is inherently unreliable. Technology developers must develop robust solutions to these channel errors. This challenge is exacerbated by the simultaneous desire for bandwidth efficiency. These two characteristics are very difficult to simultaneously achieve.

Intermittent Connectivity Induced by Terrain, Environment, and Mobility (Potentially Extreme) That Produces a (Potentially Rapidly) Fluctuating Network Topology: The wired networking domain must be robust to device failure. That is one of the greatest qualities of the Internet Protocol (IP), its ability to quickly reroute traffic around a network failure point. However, the wireless domain has the potential to introduce much greater volatility in network topology compared with wired networks. If a network or network user is highly mobile, the changes in topology can be extreme. Consequently, most of the approaches employed in the wired domain do not necessarily work well in the wireless domain. Furthermore, if the wireless network is attached to a wired infrastructure, as is typically the case, mobility and topology change in the wireless domain can actually cause severe degradation in the wired domain as well.

Platform Constraints That Place Limitations on Size, Weight, Power, and Complexity: No technology developer wishes to develop a product that is large, heavy, and power-inefficient. However, the wired domain is much more forgiving of these traits. In the wired domain, devices typically have access to power infrastructure. They are often fixed devices, so while a large size is undesirable, it is tolerable. In the wireless domain, devices need to be small and light or they will not succeed in the consumer marketplace. They are also expected to operate for relatively long time periods on battery power, or else they will not succeed in the consumer marketplace. This limits the types of solutions that can be implemented and makes the wireless domain much more difficult than the wired domain.

It is factors such as these that make providing useful wireless network capability technically challenging, especially when considering how to integrate these wireless networks into a larger network context.

1.4   THE WIRELESS INTERNET: DIFFERENT MODELS

Today's wireless networking standards can be broadly divided into three main categories based on range:

- Wireless personal area networks (WPANs)
- Wireless local area networks (WLANs)
- Wireless metropolitan area networks (WMANs)

The conventional differentiator in technology classification is geographic range of operation. WPANs typically spans up to tens of meters. WLANs typically can span hundreds of meters up to a few kilometers. WMANs typically span tens to hundreds of kilometers. The hierarchical relationship of these three categories is shown in Figure 1.1, along with the most significant standards in each category in

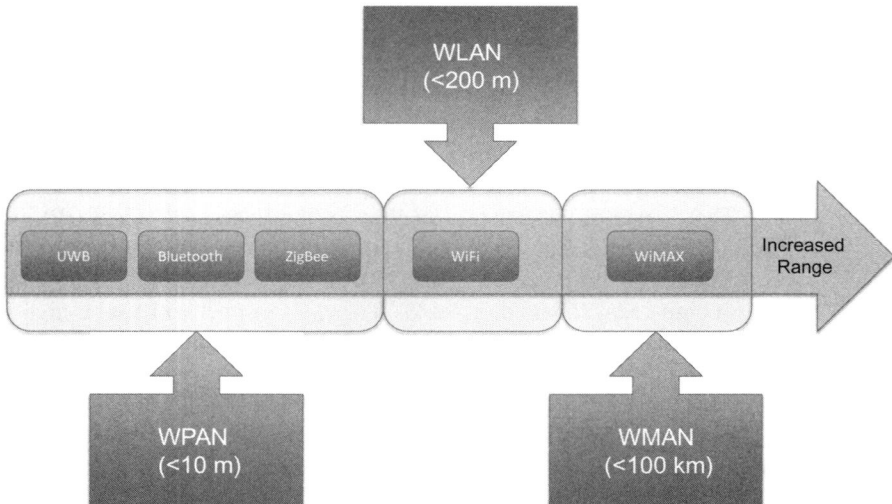

Figure 1.1   Hierarchy of wireless network technologies.

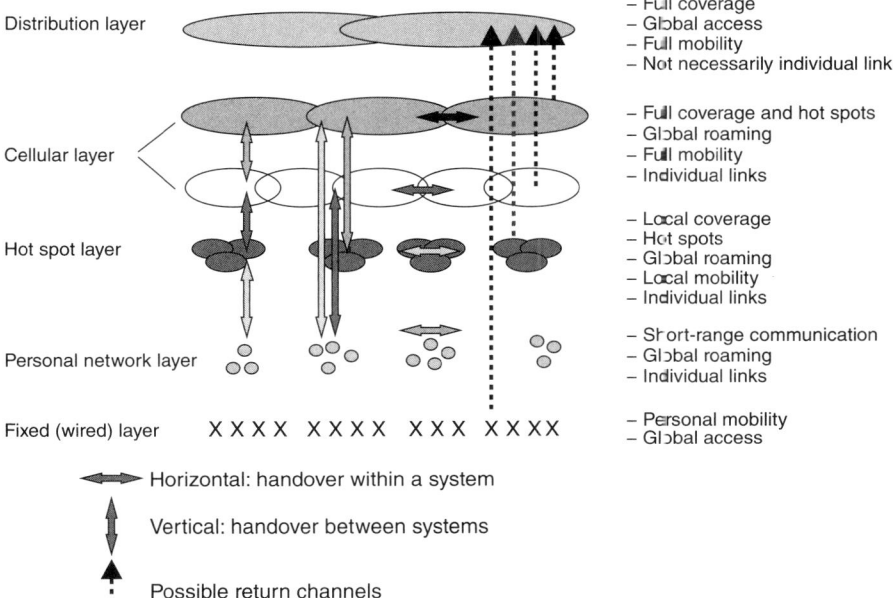

Distribution layer — — Full coverage
— Global access
— Full mobility
— Not necessarily individual links

Cellular layer — — Full coverage and hot spots
— Global roaming
— Full mobility
— Individual links

Hot spot layer — — Local coverage
— Hot spots
— Global roaming
— Local mobility
— Individual links

Personal network layer — — Short-range communication
— Global roaming
— Individual links

Fixed (wired) layer   X X X X   X X X X   X X X   X X X X — — Personal mobility
— Global access

◄───► Horizontal: handover within a system

↕ Vertical: handover between systems

▲ Possible return channels

Figure 1.2   ITU model of complimentary access systems. Reprinted from Reference [6].

terms of mass-market deployments. The ranges listed are approximate and will vary between standards and environments in which the technology is deployed.

The ITU defines a different taxonomy in which wireless access technologies can be binned. The ITU model of complementary access systems is illustrated in Figure 1.2 [6].

These layers of wireless access technologies are defined in Reference [6] as:

- **Distribution Layer.** This layer comprises digital broadcast type systems to distribute the same information to many users simultaneously through unidirectional links. Other access systems can be used as a return channel.

- **Cellular Layer.** The cellular layer may comprise several cell layers with different cell size and/or different access technologies.

- **Hot Spot Layer.** This layer may be used for very high data rate applications, very high traffic density, and individual links, such as in very dense urban areas, campus areas, conference centers, and airports.

- **Personal Network Layer.** Personal area networks will support short-range direct communication between devices.

- **Fixed (Wired) Layer.** This layer includes any fixed wireline access system.

The ITU defines in Reference [6] various attributes wireless technologies should possess at each of these layers to meet expected future user demands. Through this

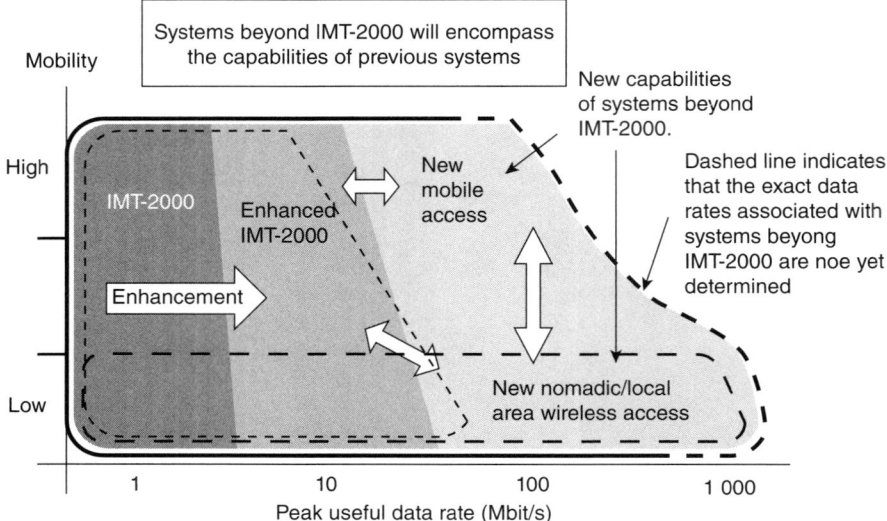

Figure 1.3   Capabilities of future wireless access systems as envisioned by the ITU. Dark shading indicates existing capabilities, medium shading indicates enhancements to the IMT-2000, and the lighter shading indicates new capabilities of systems beyond IMT-2000. The degree of mobility as used in this figure is described as follows: low mobility covers pedestrian speed and high mobility covers high speed on highways or fast trains (from 60 to ~250 km/h or more). Reprinted from Reference [6].

process, the authors believe the ITU has had a profound impact on shaping technologies and their technical capabilities. ITU provides a high-level view of the needs of future wireless access technologies in Reference [7], illustrating the trade-off future technologies must face in terms of mobility support and data rates. This relationship is shown in Figure 1.3 [6].

As you progress through this book, you may wonder why some of the design choices were made for a particular standard, and why they differ from other standards. For instance, Bluetooth WPAN devices are usually quite small (mobile phones, hands-free devices), while WiMAX-compliant 802.16 WMAN devices can be a bit larger (1 rack-unit base stations). The features built into each are designed specifically for a set of applications or use cases conceived during the standardization process. In the Bluetooth example, the maximum range is expected to operate around 10 meters, and so the transmit power required to achieve this range is much smaller than a WiMAX base station with a maximum operating range of 2 km.

Furthermore, guard intervals designed to improve transmission reception and minimize interference between transmitted frames are often set based on achievable tolerances in RF front-end systems that handle the modulation of the bit stream and the expected maximum propagation delay as a function of range—remember that the speed of light is constant, and so setting guard intervals greater than the maximum expected range may be unnecessary and reduce throughput. However, if one desires an inexpensive receiver or transmitter front end, one may want to increase those intervals to minimize implementation cost.

Finally, consider in depth the need for a wireless networking standard to achieve goals relative to the applications for which it was designed. In the case of 802.16, this standard aims to provide a wireless metropolitan area network capability that enables users at many different ranges to operate with maximum data rate and efficiency. To achieve this, 802.16 devices often perform ranging, which is a method to synchronize all the subscribers in a coverage area so that the efficiency of the data rates achievable in the shared medium is maximized. Without ranging, guard intervals and contention periods would need to be increased to enable users to have a chance to access the network, and the achievable throughput per user would go down. Ranging is a key part of the 802.16 architecture. But in Bluetooth, this is not necessarily the case. While Bluetooth basic rate does perform some frequency hop synchronization methods to align master and slave devices to hop at the same rate and on the same channels, it needs to achieve synchronization only on a much smaller timescale because of the shortened expected operating range. Thus, the algorithms designed in a Bluetooth device to achieve this functionality are less complex, which is one of the reasons why a Bluetooth device is much smaller than an 802.16 base station.

As you read this book, keep in mind this type of information as you consider why a particular standard chooses one design or method over another.

1.5 A REVIEW OF LAYERED COMMUNICATIONS MODELS

Generally, any communications system may be described in the context of the Open Standards Interconnect (OSI) seven-layer communications model. This model, first published as the ISO 7498 standard in 1984 and depicted in Figure 1.4, is a framework for communications between two end points. The reader is referred to Reference [307] for additional information.

The model employs a hierarchical structure of seven layers, where each layer provides service for the next highest layer and requests service from the next lower layer. The layers are summarized as follows:

- *Application Layer.* This layer consists of the applications that communicate between end points (e.g., instant messaging, Web browsing)
- *Presentation Layer.* This layer is responsible for formatting and presenting data. Functions performed include syntax negotiation and data compression.

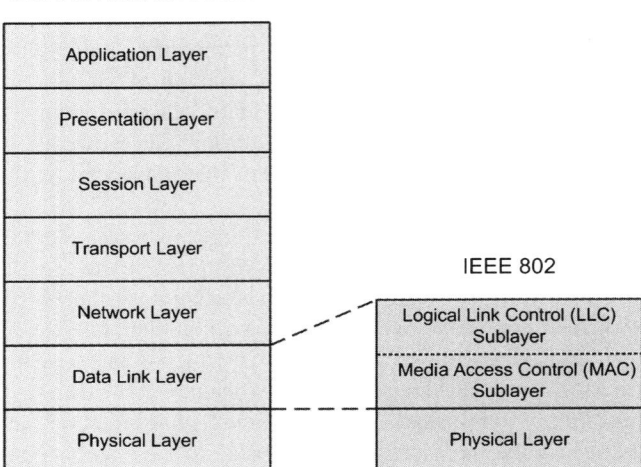

ISO/OSI Reference Model

Figure 1.4 Mapping of IEEE 802 layers to ISO/OSI reference model.

- *Session Layer.* This layer is responsible for establishing and maintaining stateful sessions between entities.
- *Transport Layer.* This layer is responsible for handling transport of data between entities in a transparent way. Multiple services can be provided, including reliable or cost-effective service.
- *Network (NET) Layer.* This layer is responsible for maintenance of connectionless or connection-oriented transmission between entities. Services provided include routing and addressing to ensure transmission across a network of nodes.
- *Data Link Layer.* This layer is responsible for handling error correction and multiplexing/multiple access of data across one or more physical links, among other functions.
- *Physical (PHY) Layer.* This layer consists of the physical devices and transmission medium used for communication.

Furthermore, the data link layer is typically further subdivided into two distinct sublayers: data link control (DLC) and media access control (MAC), as depicted in Figure 1.5.

These sublayers perform distinct functions that will be described later. Layers 4–7 represent the end-to-end communications functions between two communications entities; however, layers 1 and 2 represent the point-to-point (PTP) communications functions between any two end points on the composite path between the two communicating entities. The network layer provides an end-to-end path between communicating end hosts (in packet-switched networks in which end-to-end path is

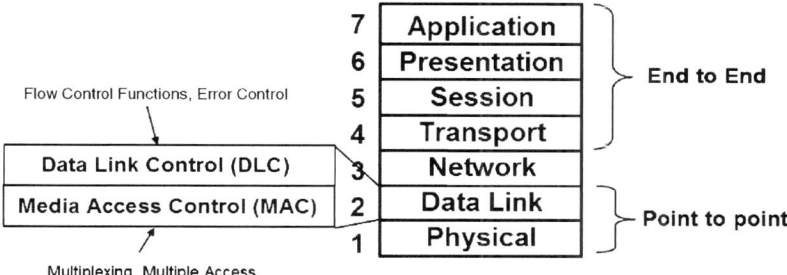

Figure 1.5   OSI communications reference model.

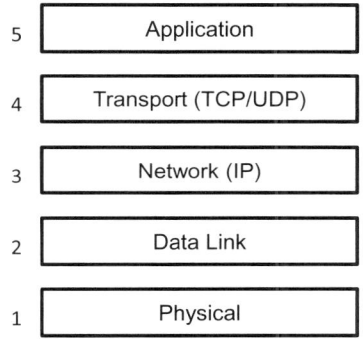

Figure 1.6   TCP/IP five-layer model.

established on a packet-by-packet basis) and is the "glue" that binds the set of end-to-end and PTP functions together.

It should be noted that the seven-layer model has not become the norm, as the session and presentation layers are not typically implemented as separate layers in most modern architectures but rather are distributed between the application and transport layers or omitted entirely. The most common model used in today's Internet is known as the TCP/IP five-layer model. This model collapses the seven-layer model down to five layers. Figure 1.6 shows an illustration of the TCP/IP five-layer model.

Here, the layers are named similarly to the seven-layer model. However, the presentation and session layers have disappeared. The transport layer actually provides multiplexing for differentiation of data streams to different applications running on a single end point. In this sense, the transport layer, via the TCP and User Datagram Protocol (UDP), provides the ability to maintain session-like flows. TCP provides a connection-oriented, reliable, acknowledgment-based method to transmit and receive data so it is provided in order, error-free to the application layer. UDP provides a connectionless, unreliable method to transmit and receive data in a

Table 1.1   Wireless Networking Problem Space within OSI Layer Model Context

(Sub)layer	Technical issues and challenges
Application	Rate control, application-level error behavior, delay-tolerant behavior
Transport	Error and delay tolerance, congestion control, error control, multihoming
Network	Network configuration management, routing, authentication, multicast support, multihoming, encryption
DLC	Error correction and detection in presence of lossy and/or intermittent network links
MAC	Authentication, network admission, on-demand resource provisioning, routing, multicast and broadcast support, topology management, encryption
Physical	Timing and synchronization in presence of Doppler and multipath effects caused by mobility, capacity provision.

best-effort fashion. In this sense, UDP does not guarantee that packets will arrive in order or error-free. The common network layer protocol, IP, is ubiquitous across the entire Internet, and it is the reason that the end points on the Internet can exchange data. In IP, data streams are broken up into IP datagrams or packets.

Each layer of the protocol stack plays an intricate part in the overall communications process. Subsequently, each layer has unique technical issues and challenges that are either uniquely associated with or exacerbated by the wireless networking problem space. Many of these technical issues are summarized in Table 1.1.

All layers across the protocol stack are equally important, and all pose a significant technical challenge. An effective wireless network solution must address issues at all layers of the protocol stack; single mechanisms at particular layers can mitigate particular technical issues but not the general wireless network problem space.

Different technical communities focus on different portions of the protocol stack. The scope of IEEE 802 encompasses the bottom two layers of the ISO/OSI reference model. Although the IEEE standard designation (e.g., IEEE 802.11) and the corresponding trade name (e.g., Wi-Fi) are often used synonymously, it should be noted that the IEEE standard encompasses only the PHY and MAC, whereas the trade name refers to the entire protocol stack, including the higher layers. Other communities, such as cellular standards bodies, one of which is the 3GPP, typically define all aspects of the communications process. The standardization process employed by the IEEE 802 LAN/MAN Standards Committee (LMSC), other standards bodies, as well as the roll of industry consortia that define trade names, will be addressed in greater detail in Chapter 2.

1.5.1   Layer Definitions

A layer standard consists of two elements: a service definition and a protocol specification. The service definition describes the functions that the layer performs and what services it provides to the higher layers of the protocol stack. The protocol specification describes the procedures used within the layer and between peer entities to execute the functions defined by the service definition.

A protocol is an agreed-upon set of rules through which information is conveyed between two entities. One example of a protocol is human language. In data communications, this generally defines the format of data, how those data are processed, and how those data are interested (i.e., the meaning of the data). There are three components of a protocol:

(a) The service data unit (SDU) consists of user data and control information created in the upper layers of the protocol stack and is the information that is to be conveyed by that particular layer of the protocol stack.

(b) The protocol control information (PCI) is information exchanged between peer entities to perform certain tasks or settle on a format. This is information that must be appended by the layer so that the layer can provide its prescribed service and successfully complete its required functions.

(c) The protocol data unit (PDU) is the combination of the SDU and the PCI. Also referred to as the encapsulated SDU, it is the information that is passed to the next lower layer.

Note that these definitions are layer-dependent. For example, the PDU of one layer is the SDU of the next lower layer. Furthermore, each unit is generally limited in size—that is, the information contained within a single SDU or PDU is not unlimited. Because data are transmitted using binary (in bits), each of these units is generally some limited number of bits. Multiple units are sent in order to represent a single full transmission (e.g., a file transfer that may consist of many of these units).

Layers interact with one another by passing these PDUs and SDUs, as depicted in Figure 1.2. Layer N + 1 generates PDUs and passes them down to layer N for servicing. Layer N adds the appropriate PCI, generating a new PDU and passing it down to layer N − 1. Layer N − 1 accepts the SDU and appends the PCI onto it, thus creating a different PDU. This process repeats until the bottom of the protocol stack is reached and the data are transmitted across the transmission medium. Layers communicate through service access points (SAPs), which are standardized interfaces between the layers. These layer interactions are depicted in Figure 1.7 [4].

1.6 WIRELESS DATA NETWORKING TECHNOLOGIES AT A GLANCE

The remainder of this book provides a rather thorough view of the wireless networking landscape. However, before we begin, let us first provide a quick summary of key wireless networking technologies. Here, an emphasis is placed upon key standardized technologies. There is a plethora of proprietary wireless technologies—so many that an exhaustive treatment would not be practical in a single text, even in a volume of texts. Any attempt to do so would be further complicated by the fact that technical details of proprietary technologies are often not freely available, making

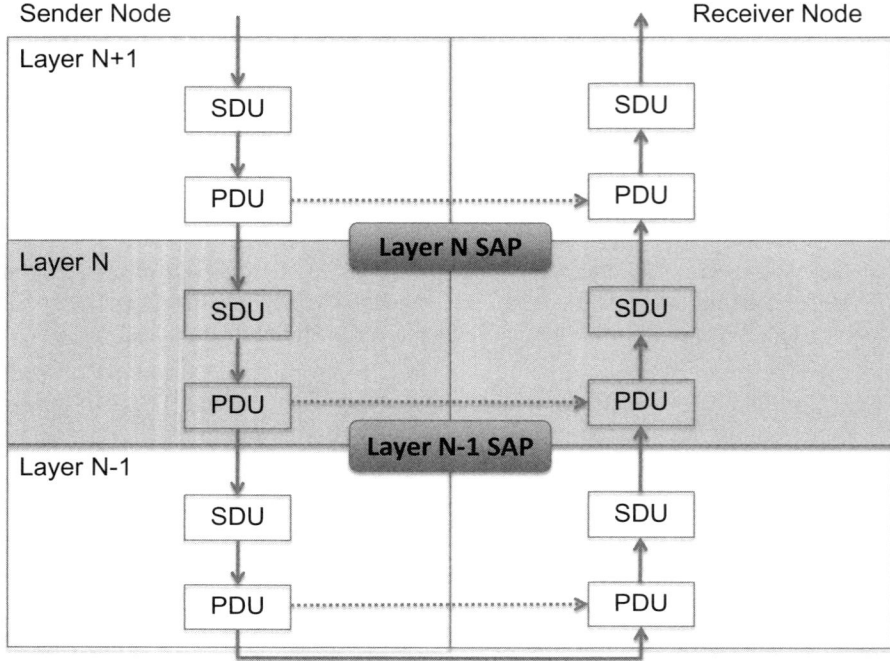

Figure 1.7 Layer Interactions within OSI communications reference model. Reprinted from Reference [4].

it difficult to provide a useful overview. Therefore, we focus our examination on those technologies that are based on open standards and are deemed the "most important" within the context of the global wireless networking landscape.

The evolution of wireless networking standards can be characterized by a few attributes:

1. Fast-moving evolution with incremental updates developed in more rapid time succession.

2. Increasing complexity of technology with each incremental update.

3. Increasing capability of technology with each incremental update.

These attributes are largely true across every major wireless technology sector. Figure 1.8 shows a qualitative summary of many of the key wireless networking technologies, attempting to depict the portion of the overall problem space that technology is attempting to solve.

1.6.1 Wireless Personal Area Networks

This section provides a brief overview of key WPAN technologies. A more thorough examination of these technologies is provided in Chapter 3.

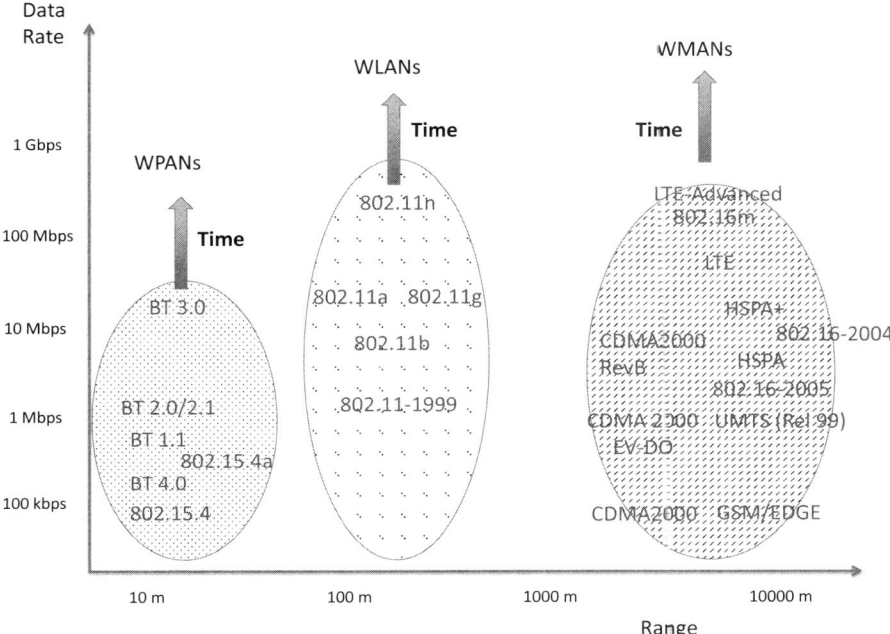

Figure 1.8 Qualitative summary of the many existing wireless networking technologies. BT, Bluetooth.

1.6.1.1 *Bluetooth*

The original standard specified a Basic Rate (BR) system, while the current version 4.0 of the standard (as of 2011) indicates support for a Low Energy (LE) system. These systems perform typical functions for personal area networking technologies, including device discovery and connection establishment and maintenance. The BR system now includes some enhancements beyond the initial scope of the first standard in Bluetooth, known as Enhanced Data Rate (EDR) and associated MAC and PHY extensions to support improved data rates. The Bluetooth 4.0 system was developed to support low energy consumption requirements. Table 1.2 summarizes the Bluetooth standards to date.

1.6.1.2 *IEEE 802.15.4*

The IEEE 802.15 Working Group develops technology standards for WPAN applications. While there are many activities within IEEE 802.15, IEEE 802.15.4 is the technology upon which ZigBee is based, a branded technology family that is focused on applications such as automation and sensor networking. Table 1.3 summarizes some of the major IEEE 802.15.4 standards releases.

Table 1.2 Bluetooth Standards Summary

Year	Reference	Specification	Description	Data rate (Mbps)	Maximum application throughput (Mbps)
1999	[2]	Bluetooth 1.0	Original Bluetooth specification	1	0.7
2002	[8]	Bluetooth 1.1	Corrected many errors found in original specification	1	0.7
2004	[9]	Bluetooth Core Version 2.0+EDR	Enhanced data rates, up to 3 Mbps	3	2.1
2007	[10]	Bluetooth Core Version 2.1+EDR	Improves pairing process	3	2.1
2009	[11]	Bluetooth Core Version 3.0+HS	Uses 802.11 for high-speed transport	24	24
2010	[3]	Bluetooth Core Version 4.0	New protocol stack for low energy applications	1	0.26

1.6.2 Wireless Local Area Networks

While there have been many updates to IEEE 802.11 over the past 12 years, we consider there to be four major updates to the technology:

- *IEEE 802.11b:* High-rate extension of IEEE 802.11, extending data rates to 11 Mbps in the 2.4 GHz frequency band.
- *IEEE 802.11a:* High-rate extension of IEEE 802.11, extending data rates to 54 Mbps in the 5 GHz frequency band.
- *IEEE 802.11g:* High-rate extension of IEEE 802.11, extending data rates to 54 Mbps in the 2.4 GHz frequency band.
- *IEEE 802.11n:* High-rate extension of IEEE 802.11, extending data rates up to 600 Mbps in the 2.4 and 5 GHz frequency bands.

Table 1.4 summarizes these four major updates.

These "major releases" represent only a small portion of the overall work conducted by the IEEE 802.11 Working Group. There have been dozens of amendments to the original specification over the past dozen years. These various amendments

Table 1.3   Summary of IEEE 802.15.4 Standards

Year	Reference	Specification	Type of modification	Description
2003	[12]	802.15.4-2003	PHY and MAC	Low-rate WPAN standard providing low data rate and long battery life with low complexity
2006	[13]	802.15.4-2006	PHY and MAC	Corrected numerous errors in 802.15.4-2003, improved security, frequency considerations, and many other changes
2007	[14]	802.15.4a-2007	PHY	New PHYs to provide higher precision ranging and location capability, higher aggregate throughput, improved scalability, and lower power consumption
2009	[15]	802.15.4c-2009	PHY	PHY modification to address Chinese regulation
2009	[16]	802.15.4d-2009	PHY and MAC	Modification to support operations in Japan

are summarized in Table 1.5. Table 1.6 summarizes the various ongoing activities within the IEEE 802.11 Working Group.

These amendments are periodically "rolled up" into a new major release of the base IEEE 802.11 specification. As an example, IEEE 802.11a, IEEE 802.11b, IEEE 802.11d, IEEE 802.11e, IEEE 802.11g, IEEE 802.11h, IEEE 802.11i, and IEEE 802.11j were all combined with the previous base IEEE 802.11-1999 specification to form the IEEE 802.11-2007 specification [36]. Likewise, the new IEEE 802.11-2012 specification [37] was recently published, which brings together many of the amendments completed since the publication of IEEE 802.11-2007. The result is a base specification that has much more capability and a much larger feature set than previous versions of IEEE 802.11. However, this comes with the cost of increased complexity. As an example, the original IEEE 802.11 specification was slightly over 500 pages in length. The new 2012 specification is over 2700 pages in length. Figure 1.9 shows an illustration of the (past and projected future) growth of the IEEE 802.11 specification (data obtained from Reference [38]).

Many of the key IEEE 802.11 standards are considered in detail in Chapter 4.

1.6.3   Wireless Metropolitan Area Networks

Perhaps no technology has undergone as radical a transformation since its inception as has IEEE 802.16, the technology standard that is the basis for the branded

Table 1.4 A High-Level Comparison between 802.11 Standards

	802.11b	802.11a	802.11g	802.11n
Standard approval	September 1999	February 2001	June 2003	October 2009
Available bandwidth (MHz)	83.5	580	83.5	83.5/580
Frequency band of operation (GHz)	2.4	5	2.4	2.4/5
Number of nonoverlapping channels (United States)	3	24	3	3/24
Data rate per channel (Mbps)	1–11	6–54	1–54	1–600
Modulation type	Direct sequence spread spectrum (DSSS), complimentary code keying (CCK)	Orthogonal frequency-division multiplexing (OFDM)	DSSS, CCK, OFDM	DSSS, CCK, OFDM, MIMO-OFDM
Typical ranges (indoor/outdoor) (m)	~35/~110	~30/~100	~35/~110	~70/~160

technology WiMAX. IEEE 802.16 originated as a standardized way to provide high data rate point-to-point backhaul communications. Since then, it has morphed into a more cellular-like technology that now stands as a viable 4G candidate technology. Figure 1.10 provides a qualitative representation of this transformation, showing that IEEE 802.16 has, over time, sought to improve scalability and mobility support at the expense of range and, until the newest incantation, 802.16m, data rate.

Table 1.7 provides a summary of the history of modifications to the IEEE 802.16 standard.

WMAN technology is the subject of Chapter 5.

1.7 CELLULAR NETWORKING TECHNOLOGIES AT A GLANCE

While technologies emerging from the 802 LMSC community have rapidly evolved over the past 10–15 years, cellular technologies have evolved at an equally rapid rate. One need travel only a dozen years or so to find a world where 2G technologies such as GSM and IS-95 (CDMA) were the state of the art in cellular communications, where data support was limited to the equivalent of dial-up Internet service.

Table 1.5 History of IEEE 802.11 Standardization Activates

Task group	Reference	Type of specification	Description	Status
Original	[1]	PHY and MAC	Original standard (1997): 1 and 2 Mbps, 2.4 GHz, frequency hopping (FH), direct sequence (DS), and infrared (IR was never implemented)	Published 1997
a	[17]	PHY	Up to 54 Mbps in the 5+ GHz band	Published 2001
b	[18]	PHY	Up to 11 Mbps in the 2.4 GHz band	Published 1999
c	[19]	MAC	MAC layer bridging	Published 2001
d	[20]	PHY	Additional regulatory domains for .11b	Published 2001
e	[21]	MAC	Quality-of-service (QoS) improvements	Published 2005
f	N/A	MAC	Inter-Access Point Protocol (IAPP) (*Recommended Practice*)	Published 2003, withdrawn 2006
g	[22]	PHY	Up to 54 Mbps in the 2.4 GHz band	Published 2003
h	[23]	MAC	Dynamic frequency selection/transmit power control (DFS/TPC) for .11a	Published 2004
i	[24]	MAC	Security enhancements	Published 2004
j	[25]	PHY	4.9–5 GHz regulatory issues for Japan	Published 2004

(*Continued*)

Table 1.5 (*Continued*)

Task group	Reference	Type of specification	Description	Status
k	[26]	MAC	Radio resource measurement	Published 2008
m		PHY and MAC	Maintenance and technical corrections to the original standard	Published 2009
n	[27]	PHY and MAC	Performance ≥ 100 Mbps	Published 2009
p	[28]	PHY and MAC	Vehicular speeds (up to 200 km/h, to 1000 m, 5.85–5.925 GHz). Also known as dedicated short-range communications (DSRC)/wireless access in vehicular environment (WAVE)	Published 2010
r	[29]	MAC	"Fast roaming"—reduction of latency during handoffs	Published 2008
s	[30]	MAC	Meshes (infrastructure and Client)	Published 2011
t	N/A	MAC	Performance prediction—benchmarking	Cancelled
u	[31]	MAC	Interworking with external networks (handoff)	Published 2011
v	[32]	MAC	Network management	Published 2011
w	[33]	MAC	Management frame (and other control functions) security	Published 2009
y	[34]	PHY	Operation in the 3650–3700 MHz band (contention-based protocols)	Published 2011
z	[35]	MAC	Direct link setup (DLS)	Published 2010

Table 1.6   Current IEEE 802.11 Standardization Activities

Group	Type	Description	Status
aa	MAC	Audio/video transport streams	Working
ac	PHY and MAC	Very high throughput below 6 GHz	Working
ad	PHY and MAC	Very high throughput—60 GHz	Working
ae	MAC	QoS management	Working
af	PHY and MAC	TV whitespace operation	Working
ah	PHY and MAC	Below 1 GHz operations	Working
ai	MAC	Fast initial link setup	Working

(a)

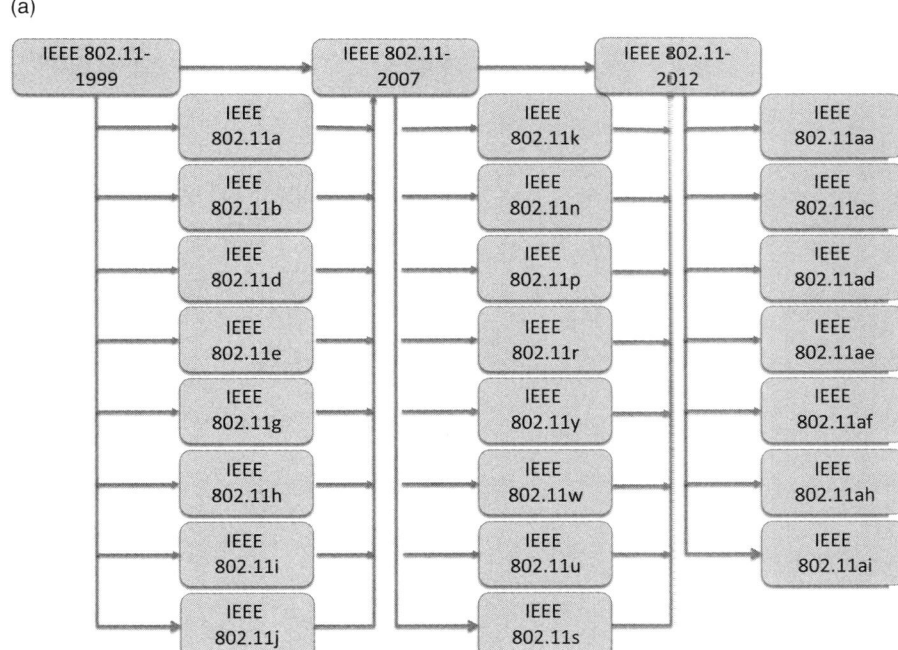

Figure 1.9   Growth of IEEE 802.11 specification. (a) Evolution of IEEE 802.11 specifications; (b) growing complexity of IEEE 802.11 specifications.

(b)

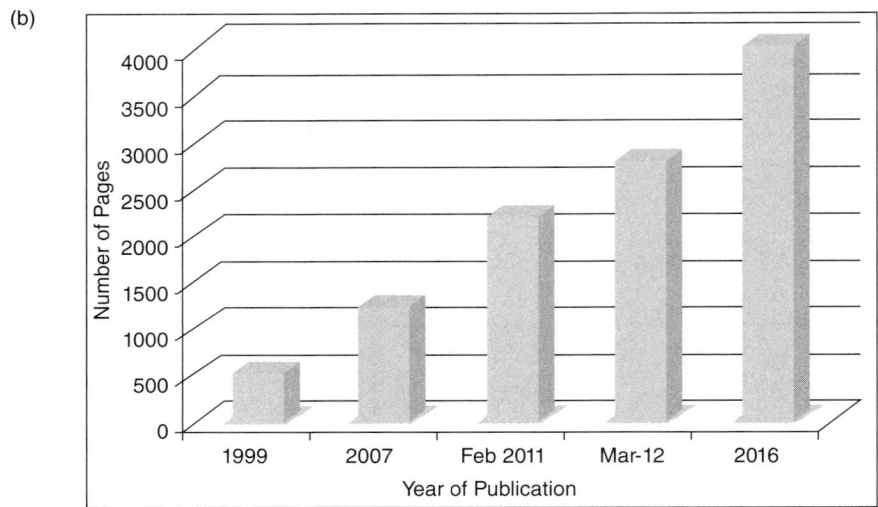

Figure 1.9 (*Continued*)

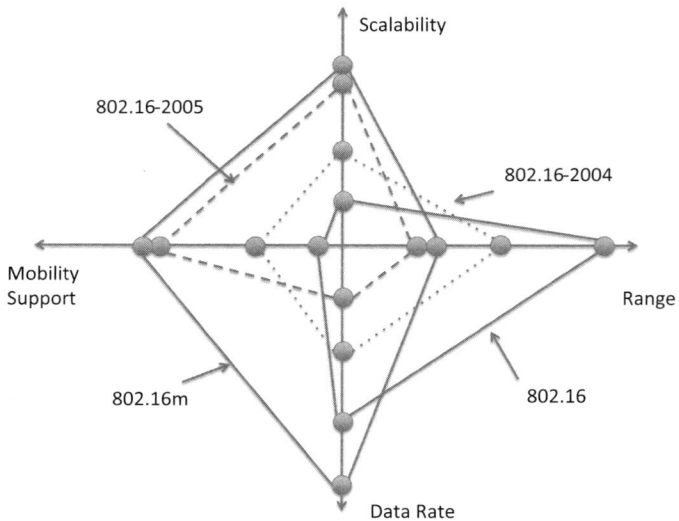

Figure 1.10 Qualitative summary of IEEE 802.16 over the years.

Table 1.7   IEEE 802.16 Technology Specifications and Other Related Standards

Specification	Description	Reference	Release date
IEEE 802.16	MAC and PHY definition for fixed broadband wireless access in the 10- to 66-GHz frequency bands	[4]	2001
IEEE 802.16a	Amendment to the original specification, containing new PHY definitions for the 2- to 11-GHz frequency bands. Also includes mesh network modes of operation	[39]	2003
IEEE 802.16c	System profiles for 10- to 66-GHz operations	[40]	2003
IEEE 802.16d (Fixed WiMAX)	Also known as IEEE 802.16-2004, this specification is considered to be the base IEEE 802.16 standard for fixed broadband wireless communications. Contains IEEE 802.16, IEEE 802.16a, and various MAC enhancements	[41]	2004
IEEE 802.16e-2005 (Mobile WiMAX Release 1)	Amendment to the IEEE 802.16d specification to provide explicit support for mobility	[42]	2006
IEEE 802.16f	Amendment to IEEE 802.16, Air Interface for Fixed Broadband Wireless Access Systems—Management Information Base	[43]	2005
IEEE 802.16g	Amendment to IEEE 802.16, Air Interface for Fixed Broadband Wireless Access Systems—Management Plane Procedures and Services	[44]	2007
IEEE 802.16-2009	IEEE Standard for Local and Metropolitan Area Networks—Part 16: Air Interface for Broadband Wireless Access Systems Revision of IEEE 802.16-2004. It consolidates and obsoletes IEEE standards 802.16-2004, 802.16e-2005, 802.16-2004/Cor1-2005, 802.16f-2005, and 802.16g-2007.	[45]	2009
IEEE 802.16h	Amendment of IEEE 802.16 standard— Improved Coexistence Mechanisms for License-Exempt Operation	[46]	2010
IEEE 802.16i	Amendment to IEEE 802.16, Air Interface for Fixed Broadband Wireless Access Systems—Management Plane Procedures and Services	N/A	2008 (withdrawn)

(Continued)

Table 1.7 (*Continued*)

Specification	Description	Reference	Release date
IEEE 802.16j	Amendment to IEEE 802.16 standard—Multihop Relay Specification	[47]	2009
IEEE 802.16k	Amendment of IEEE 802.1D, Standard for Local and Metropolitan Area Networks: Media Access Control (MAC) Bridges—Bridging of 802.16	[48]	2007
IEEE 802.16m (Mobile WiMAX Release 2, also known as 4G WiMAX or WirelessMAN-Advanced	Amendment to IEEE 802.16-2009, Air Interface for Fixed and Mobile Broadband Wireless Access Systems—Advanced Air Interface	[49]	2011
IEEE 802.20 Mobile Broadband Wireless Access	Mobile broadband wireless access standards group. Initially formed as a standards group within the IEEE 802.16 Working Group, it consisted of a group of individuals who wished to develop a new technology focused solely on mobility. The 802.20 Working Group specifies PHY and MAC layers of an air interface for interoperable mobile wireless access systems operating in licensed bands below 3.5 GHz and optimized for IP data transport with user data rates above 1 Mbps. The standard supports vehicular velocities up to 250 km/h in a MAN environment.	[50]	2008 (base standard publication)
Wireless Broadband (WiBRO)	Korean wireless broadband standard that was incorporated into the IEEE 802.16e standard	N/A	N/A

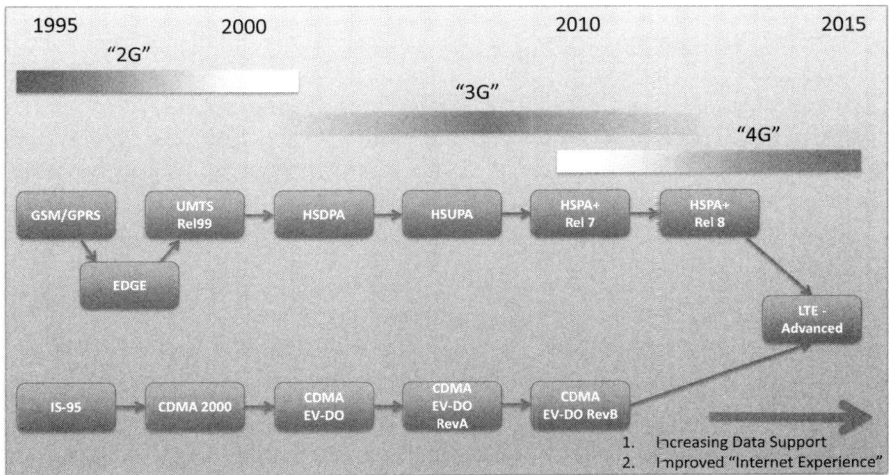

Figure 1.11 Evolution of cellular communications: moving to the wireless Internet.

Figure 1.11 depicts the evolution that cellular communications has experienced over the past 15 years.

Table 1.8 provides a summary of all major cellular technologies, comparing and contrasting many of their attributes. These technologies are considered in great detail in Chapter 6 (2G), Chapter 7 (3G), and Chapter 8 (4G). Unless noted, all data rates are theoretical maximum values; actual data rates will vary as they are a function of numerous factors, such as the number of antennas in MIMO systems, the number of network users, the geometric distribution of users, and wireless channel conditions.

Table 1.8 Summary of Cellular Network Technologies

Standard	Generation	Standard family	Standards body	Initial deployment	Multiple access	Bandwidth	Modulation	DL data rate	UL data rate	Notes
GSM	2G	GSM	CEPT/ ETSI/3GPP	1991 [51]	TDMA	200 kHz	GMSK	9.6 kbps [52]	9.6 kbps	Data rates are for Phase 2 release (1995)
GPRS (R97)	2G+ (2.5G)	GSM	ETSI/3GPP	2000 [53]	TDMA	200 kHz	GMSK	115 kbps [52]	Symmetry	Evolution of GSM
EDGE (R98)	2G+ (2.75G)/3G	GSM	ETSI/3GPP	2003 [53]	TDMA	200 kHz	GMSK, 8PSK	384 kbps [52]	Symmetry	Evolution of GSM
EDGE Evolution (Rel-7)	3G	GSM	3GPP	a	TDMA	200 kHz	GMSK, QPSK, 8PSK, 16-QAM, 32-QAM	1.9 Mbps [GSA]	947 kbps [GSA]	Evolution of GSM (Evolved GERAN)
IS-95A	2G	CDMA	TIA	1995	CDMA	1.25 MHz	QPSK, OQPSK	14.4 kbps [52]	Symmetry	
IS-95B	2G+ (2.5G)	CDMA	TIA	1998	CDMA	1.25 MHz	QPSK, OQPSK	64 kbps [52]	Symmetry	Not widely deployed due to release of CDMA2000 1xRTT
WCDMA(R99)	3G	GSM/ UMTS	3GPP	2001 [53]	CDMA	5 MHz	QPSK, DQPSK	384 kbps [54]	Symmetry	Primary air interface of UMTS family
TD-SCDMA	3G	GSM/ UMTS	3GPP	2009	CDMA	1.6 MHz	QPSK, 8-PSK, 16-QAM	2.048 Mbps	2.048 Mbps	TDD air interface of UMTS family, developed and primarily deployed in China

Name	Generation	Family	Standards Body	Year	Access	Bandwidth	Modulation	Peak Downlink	Peak Uplink	Notes
HSPA	3G+ (3.5G)	GSM/UMTS	3GPP	HSDPA: 2005 [53], HSUPA: 2007 [53]	CDMA/FDD	5 MHz	QPSK, 16-QAM, 64-QAM	14.4 Mbps [55]	5.7 Mbps [55]	Evolution of WCDMA
HSPA+	3G+ (3.9G)	GSM/UMTS	3GPP	2009	CDMA/FDD	5, 10 MHz	Up to 64 QAM	84 Mbps [54]	22 Mbps	Evolution of WCDMA
CDMA2000 1xRTT	3G	CDMA	3GPP2	2000	CDMA	1.25 MHz	QPSK	144 kbps [52]	Symmetry	Backward compatible with IS-95A/B
CDMA2000 EV-DO Rel. 0	3G+ (3.5G)	CDMA	3GPP2	2002	CDMA/FDD	1.25 MHz	QPSK, 8-PSK, 16-QAM	2.45 Mbps	150 kbps	Not backward compatible with 1xRTT
CDMA2000 EV-DO Rev. A	3G+ (3.5G)	CDMA	3GPP2	2006	CDMA/FDD	1.25 MHz	BPSK, QPSK, 8-PSK, 16-QAM	3.1 Mbps	1.8 Mbps	Evolution of EV-DO
CDMA2000 EV-DO Rev. B	3G+ (3.9G)	CDMA	3GPP2	2010	CDMA/FDD	1.25–5 MHz	BPSK, QPSK, 8-PSK, 16-QAM	9.3 Mbps	5.4 Mbps	Evolution of EV-DO
Mobile WiMAX	3G+ (3.9G)/ pre-4G	WiMAX	IEEE LMSC/ WiMAX Forum	2005	OFDMA	Scalable: from 1.25 to 40 MHz	BPSK, QPSK, 16-QAM, 64-QAM	141 Mbps	138 Mbps	Recognized by ITU as an official 3G cellular standard. Also sometimes termed "pre-4G."

(*Continued*)

Table 1.8 (*Continued*)

Standard	Generation	Standard family	Standards body	Initial deployment	Multiple access	Bandwidth	Modulation	DL data rate	UL data rate	Notes
LTE (Rel-8)	Pre-4G	GSM/ UMTS	3GPP	2009	OFDMA/ SC-FDMA	Scalable: from 1.4 to 20 MHz [55]	QPSK, 16-QAM, 64-QAM	100 Mbps [55] or 150+ Mbps [54]	50 Mbps [55]	Sometimes termed pre-4G, Rel-8 LTE does not meet IMT-Advanced requirements
LTE-Advanced (Rel-10)	4G	GSM/ UMTS	3GPP	2013	OFDMA/ SC-FDMA	Scalable: from 1.4 to 100 MHz	QPSK, 16-QAM, 64-QAM, 128-QAM	3 Gbps	1.5 Gbps	Accepted by the ITU into the IMT-Advanced family of standards (i.e., "4G")
802.16m (WMAN-Advanced)	4G	WiMAX	IEEE LMSC/ WiMAX Forum	2013 (anticipated)	OFDMA	Scalable: from 5 to 40 MHz	BPSK, QPSK, 16-QAM, 64-QAM	365 Mbps	376 Mbps	Accepted by the ITU into the IMT-Advanced family of standards (i.e., "4G")

[a]First successful live network trial of EDGE Evolution conducted by Nokia Siemens Networks in China in 2009. The GSA has stated that this technology is "commercially available," but no commercial network launches have been announced as of 2012.

28

Chapter 2

The Wireless Ecosystem

Wireless networks and the devices people use to access them have been a part of modern life for many years. Because of the increasing number of users and diverse set of applications that wireless networks provide, bandwidth demands continue to grow every year. Evolving from the early voice-only cellular networks of the 1980s, today's diverse high-bandwidth wireless data networks provide many compelling reasons for consumers to use them: applications like social networking, video chat, multiplayer online gaming, and streaming television and their associated content streams continue to expand. Furthermore, consumers have an ever-growing choice of technologies, handsets, and applications. This chapter will discuss the wireless technology ecosystem that has been leveraged to build the foundations of today's networks. We will discuss the standards organizations that develop the technology specifications, including their motivations, structures, interrelationships, and processes for developing standardized technologies. Finally, we will discuss the role of industry players in implementing these standards, highlighting some of the key alliances between companies to facilitate interoperability among different vendors that build wireless networking equipment based on the same technology standard.

2.1 WIRELESS STANDARDIZATION PROCESS

A key driver for the success of wireless networking at all levels—from Bluetooth headsets connecting to a user's cellular phone device all the way to nationwide wireless data networks based on WiMAX-compliant equipment—is the wireless standardization process. Standardization is the process by which technology specifications are defined. Many different organizations and their members participate in the standardization process, with key players including the Institute of Electrical and Electronics Engineers (IEEE), the Internet Engineering Task Force (IETF), the 3rd Generation Partnership Project (3GPP), the 3rd Generation Partnership Project 2 (3GPP2), the International Organization for Standardization (ISO), and the International Telecommunications Union (ITU). This chapter will discuss these organizations' roles in

Wireless Networking: Understanding Internetworking Challenges, First Edition.
Jack L. Burbank, Julia Andrusenko, Jared S. Everett, and William T.M. Kasch.
© 2013 the Institute of Electrical and Electronics Engineers, Inc. Published 2013 by John Wiley & Sons, Inc.

developing wireless networking standards, highlighting each organization's goals, scope, and major contributions to the wireless ecosystem today.

2.2 IEEE

The IEEE has long been a leader in developing wireless networking standards. The IEEE 802 Local Area Networking/Metropolitan Area Networking (LAN/MAN) Standards Committee (LMSC) has developed the vast majority of the wireless networking standards work accomplished through IEEE. It is arguable which standard may be more famous: IEEE 802.3 (wired Ethernet) or IEEE 802.11 (wireless Ethernet)—generally, if you were born after 1990, you're probably more familiar with IEEE 802.11 than 802.3, although both of these standards continue to serve major roles in networking applications today.

The LMSC was formed in February 1980, and while some claim the "802" designation is a reference to this date (80 for the year and 2 for the month), the IEEE LMSC overview [56] actually states this number was next in the sequence ready to be assigned to a committee group.

The IEEE 802 LMSC develops networking standards that are focused on defining specifications for both the physical layer and data link layer in the OSI seven-layer model. Participation by individuals and organizations is completely voluntary, and international involvement is extensive. The committee meets regularly throughout the year to advance a wide variety of networking standards, ranging from the smaller scale personal area networking technologies to the larger scale of metropolitan area networking. Many large corporate organizations, including Intel, Samsung, and Motorola, have participated in the past or actively participate today. Such companies generally have large stakes—in particular, technology threads—within their own organizations. As a result, these organizations often participate in the IEEE LMSC with the intention of ratifying a particular standard that aligns well with their own technology investments, expertise, and market strategies. Because such organizations are very large and have committed substantial funding to research and development for particular technologies, they exert a strong influence on the standards developed.

The LMSC consists of a number of working groups and technical advisory groups (TAGs). Working groups are chartered with developing particular networking technologies and maintaining/enhancing those technologies as appropriate. For instance, the widely known 802.11 Working Group was responsible for developing the wireless local area networking standard. This standard has been implemented widely across many consumer devices, including laptops and mobile phones. Furthermore, it has been enhanced over the years to provide increased data rates, quality-of-service support, and improved security.

Technical advisory groups exist within the LMSC to assess particular technical challenges or to act as liaisons between the LMSC and other organizations. For instance, the 802.18 Radio Regulatory TAG offers its services to the LMSC and all working groups by monitoring the progress of each working group and national and

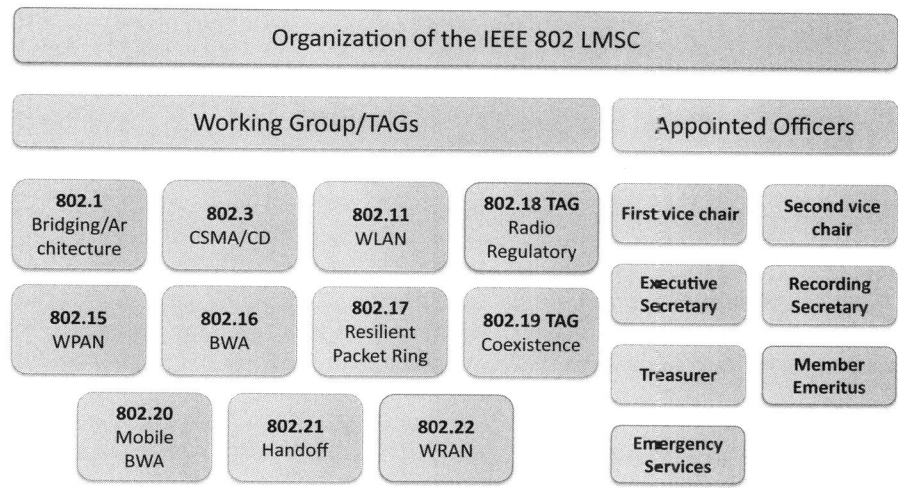

Figure 2.1 Organizational structure of IEEE 802 LMSC.

international radio regulatory organizations such as the Federal Communications Commission (FCC) and ITU, and recommending policies and providing comments to regulators that are consistent with the desired goals of the LMSC. The 802.19 Coexistence TAG focuses on wireless coexistence challenges for the various wireless networking technologies developed in the LMSC (e.g., 302.11, 802.16).

Figure 2.1 shows an illustration of the organizational structure of the IEEE 802 LMSC as of June 2011. The executive committee (EC) consists of a chair and a set of appointed officers. Working group/TAG chairs are named for various technology threads, and these chairs maintain voting rights as EC members as well. There are currently 11 Working Groups/TAGS in the LMSC:

- 802.1, Bridging and Architecture
- 802.3, Carrier Sensing Multiple Access/Collision Detection (CSMA/CD)
- 802.11, Wireless Local Area Networks (WLANs)
- 802.18 TAG, Radio Regulatory
- 802.15, Wireless Personal Area Networks (WPANs)
- 802.16, Broadband Wireless Access (BWA)
- 802.17, Resilient Packet Ring
- 802.19 TAG, Coexistence
- 802.20, Mobile Broadband Wireless Access (MBWA)
- 802.21, Handoff
- 802.22, Wireless Regional Area Networks (WRANs)

Older working groups in the past, such as 802.4 (Token Bus) and 802.5 (Token Ring) have been disbanded. Furthermore, each working group today is broken down into task groups, some of which have been disbanded as the working group's technology evolves.

Generally, working groups originate from a group of people who have an interest in developing a particular standard. In order to authorize a project to standardize a technology, a Project Authorization Request (PAR) is submitted to the EC for approval. Once the EC votes to approve, it sends the PAR to the IEEE Standards Board New Standards Committee for approval as an official IEEE Standards project. Once the standard is approved, the working group evaluates and votes on proposals to be incorporated into a draft standard. Voting members within the working group decide whether to continue a particular proposal, to downselect to a "best of breed," or to debate details. When there is sufficient consensus in the working group on a draft standard document, a working group letter ballot is conducted to release the document from the working group. During this process, the EC approves the draft standard and then advances the document to the sponsor letter ballot process. Once this process has passed and any "No" votes are resolved sufficiently, the document is received by the IEEE Standards Board Standards Review Committee. Once this committee approves the document, it is published as an IEEE standard. This process can take years for new standards, while updates to existing standards generally take less time.

To become a voting member of the LMSC, one must attend at least two of the last four working group plenary sessions. Interim sessions are also scheduled between plenaries, and these sessions may sometimes count toward earning voting member status. Most voting members are allowed to vote only in one particular working group (e.g., 802.11), as 75% attendance is required in each meeting. Because most of the existing working groups meet concurrently during a scheduled meeting, it is impossible to attend each working group's meetings 75% of the entire time.

2.3   IETF

The IETF is a large open community of people who develop technical standards for the Internet and related networks. The mission of the IETF is as follows [57]:

- Identify and propose solutions to pressing operational and technical problems in the Internet
- Specify the development or usage of protocols and the near-term architecture to solve such problems for the Internet
- Make recommendations to the Internet Engineering Steering Group (IESG) regarding the standardization of protocols and protocol usage in the Internet
- Facilitate technology transfer from the Internet Research Task Force (IRTF) to the wider Internet community
- Provide a forum for the exchange of information within the Internet community between vendors, users, researchers, agency contractors, and network managers

IETF meets three times a year; however, many of its participants do not attend meetings. It is similar to the IEEE 802 LMSC in that there are working groups, but instead of relying on formal voting processes and procedures, rough consensus among working group participants is considered acceptable for advancing standards. Rough consensus is defined subjectively—WG chairs have substantial leeway in determining what constitutes rough consensus. Furthermore, there is no explicit membership in IETF for participants. The organizations that are involved in the standards process for the IETF are discussed in Reference [58]. Besides the IETF, these include:

- The IETF Secretariat, which serves as an administrative body to support IETF activities. This organization consists of the IETF Executive Director and associated staff. The Executive Director is the formal point of contact for all aspects of the Internet standards process.

- The Internet Society (ISOC), an international organization focused on evolution and growth of the Internet, along with sociological and technical issues that arise from its use. Standardization is an organized activity of the ISOC.

- The IESG, part of the ISOC responsible for managing IETF technical activities. This group is responsible for associated actions for progressing "standards track" technical specifications, including initial approval of new working groups and final approval of Internet standards.

- The Internet Architecture Board (IAB), chartered by the ISOC trustees to oversee the architecture of the Internet and associated protocols. This board appoints the IETF chair and approves other IESG candidates proposed by the IETF nominating committee.

The IESG consists of a number of Area Directors (ADs), who are selected by the IETF Nominations Committee (NomCom) for appointments lasting 2 years. An AD is responsible for the particular area and working groups that fall under that area. The current areas are as follows:

- *Applications (APP).* Protocols seen by user programs (chat, Web)
- *General (GEN).* WGs that don't fit into any other category
- *Internet (INT).* Addresses IP packet movement and DNS information
- *Operations and Management (OPS).* Operational aspects of Internet systems, including network monitoring and configuration
- *Real-Time Applications and Infrastructure (RAI).* Focuses on delay-sensitive interpersonal communications (e.g., voice, video chat)
- *Routing (RTG).* Focuses on moving packets to their intended destinations
- *Security (SEC).* Authentication, privacy, authorization
- *Transport (TSV).* Special services for special packets

Working groups are where most of the work gets done in the IETF. In particular, WGs have charter documents that specify the scope of the WG and associated goals. WG chairs are volunteers who determine consensus goals and milestones, and take responsibility for keeping the charter up to date and ensuring technical and nontechnical quality of WG output. During WG meetings, chairs often set agendas and manage otherwise unwieldy debates during periods of contention. When a WG fulfills its charter, it is expected to stop operating.

In order for a person or group of people to develop a protocol or standardized approach that attempts to solve a problem related to the Internet, they can write an Internet draft document that details their approach and petition to present it within an existing working group, or to use it as the basis for a "Birds-of-a-Feather" (BOF) session at the next scheduled IETF meeting. Because of limited space during a particular meeting, BOF sessions require approval. However, meetings are not the only method of proposing approaches at the IETF. Internet drafts can be submitted on working group mailing lists for discussion as well.

Once an Internet draft is accepted by a working group as a working item or a BOF garners sufficient interest to propose chartering a new working group, the process for turning an Internet draft into a standard Request for Comments (RFC) document starts, if the WG or AD deems it to be relevant and useful. When an RFC is published, it is assigned a number (e.g., RFC 2028). However, the RFC name is misleading—generally, by the time a document has been assigned an RFC number by the RFC editor, it has gone through many rounds of comments and debate. The IETF has kept the RFC system in name although the specific words do not necessarily apply as in the past. There are six types of RFCs:

- Proposed standards
- Draft standards
- Internet standards
- Informational documents
- Experimental protocols
- Historic documents

The first three are considered official standards within the IETF. Generally, standards-track documents are fully changeable by the IETF community (with consensus) regardless of their origin. However, some authors find it difficult to relinquish control of their protocol designs, and as such are reluctant to pursue standards-track status.

After an Internet draft has been discussed and is considered to be useful as a potential standard, it is submitted to the IESG. If the draft is sponsored by a WG (i.e., it is an official WG draft), the WG chair will send it to the AD responsible after it has been submitted for "working group last call." This last call process gives the participants in the WG a final chance to comment and debate the details of the Internet draft. If the draft is an individual submission not sponsored by a particular WG, the author will submit the document to the appropriate AD for consideration. Once submitted to the IESG, there is an IETF last call issued for the draft. After this process, if the IESG approves the draft to become an Internet standard, they will request the RFC editor to publish it as a Proposed standard. After 6 months, the

Individual/group submission via AD or WG official document

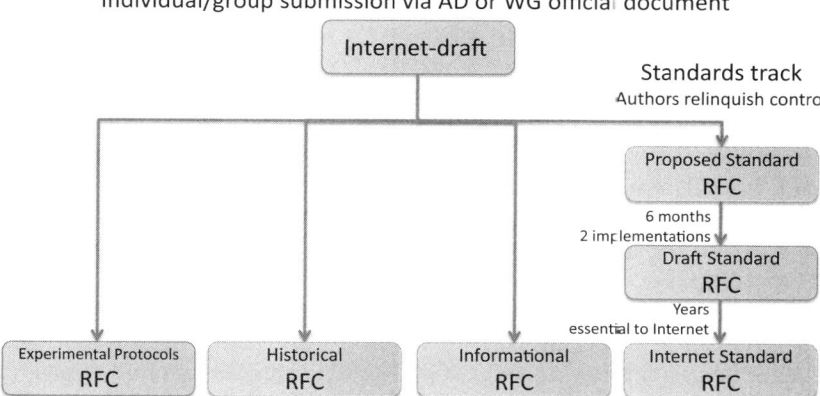

Figure 2.2 IETF Internet draft to RFC process.

author or WG chair can request the document be moved to Draft standard status, with the additional requirement that there must be at least two independent, interoperable implementations of the standard that must be verified by the AD. Requiring tested implementations generally improves the quality of the document during this process as separate implementers may interpret text in the standard differently. After a few years, a Draft standard can become an Internet standard, although this rarely happens—as Internet standards are considered essential for the Internet to function. An example of such a standard is RFC 791, the original Internet Protocol. Figure 2.2 shows an illustration of the potential RFC flows from Internet draft to RFC form.

2.4 3GPP

The 3rd Generation Partnership Project (3GPP) was established in December 1998 as a collaboration between multiple regional telecommunications standards bodies: the Association of Radio Industries and Business (ARIB) in Japan, the Alliance for Telecommunications Industry Solutions (ATIS) in the United States, the China Communications Standards Association (CCSA), the European Telecommunications Standards Institute (ETSI), the Telecommunications Technology Association (TTA) in Korea, and the Telecommunication Technology Committee(TTC) in Japan. Together, these standards bodies comprise the Organizational Partners for 3GPP. The 3GPP Project Agreement originally signed by all the Organizational Partners stated that they would cooperate in producing "globally applicable" technical specifications and reports for a 3G mobile system, based primarily on Global System for Mobile Communications (GSM) core networks and the radio access technologies they support, such as Enhanced Data rates for GSM Evolution (EDGE), High-Speed Data Packet Access (HSDPA), and Universal Terrestrial Radio Access (UTRA). The

3GPP was established primarily for preparation, approval, and maintenance of technical specifications and reports for 3G networks based on the GSM core structure. Furthermore, 3GPP is not considered a legal entity. However, the 3GPP organization has since changed the language in their scope to include "3G and beyond." Long-Term Evolution (LTE) is a fourth-generation (4G) cellular access technology that has been standardized by 3GPP, and this technology is considered a key component of many 4G cellular networks moving forward.

The 3GPP task group structure is illustrated in Figure 2.3. The Project Coordination Group (PCG) oversees time frame and management of technical work to ensure 3GPP specifications are produced "in a timely manner as required by the market place according to the principles and rules contained in the Project reference documentation." Technical specification development work within 3GPP is done largely by Technical Specification Groups (TSGs). Each TSG prepares, approves, and maintains specifications within its defined scope, and may organize working groups to address topics deemed relevant to the TSG. TSGs report to the PCG periodically. The PCG has the following responsibilities:

- Determination of overall time frame and management of overall work progress.
- Final adoption of work items within the agreed 3GPP scope.
- Allocation of budgeted human and financial resources to each TSG as provided by Organizational Partners.
- Allocation of additional voluntary human and/or financial resources to each TSG as provided by Individual Members.
- Appointment of TSG Chairmen.
- Appointment of PCG Chairman.

There are four active TSGs: TSG GSM/EDGE Radio Access Network (GERAN), TSG Radio Access Networks (RAN), TSG Services and System Aspects (SA), and TSG Core Networks and Terminals (CT). These groups are summarized as follows:

- *TSG GERAN.* This group is responsible for specification of the radio access part of the GSM/EDGE network, including the radiofrequency (RF) layer, layers 1, 2, and 3 internal and external interfaces, conformance test specifications, and GERAN-specific operations and maintenance (O&M) specifications.
- *TSG RAN.* This group is responsible for definition of requirements, functions, and interfaces of the UTRA/evolved UTRA (E-UTRA) network in both frequency division duplex (FDD) and time division duplex (TDD) modes, including radio performance, layer 1, 2, 3 specifications, and UTRAN/E-UTRAN conformance and O&M requirements.
- *TSG SA.* This group focuses on service capabilities and overall architecture of systems based on 3GPP specifications; coordinates across multiple TSGs.

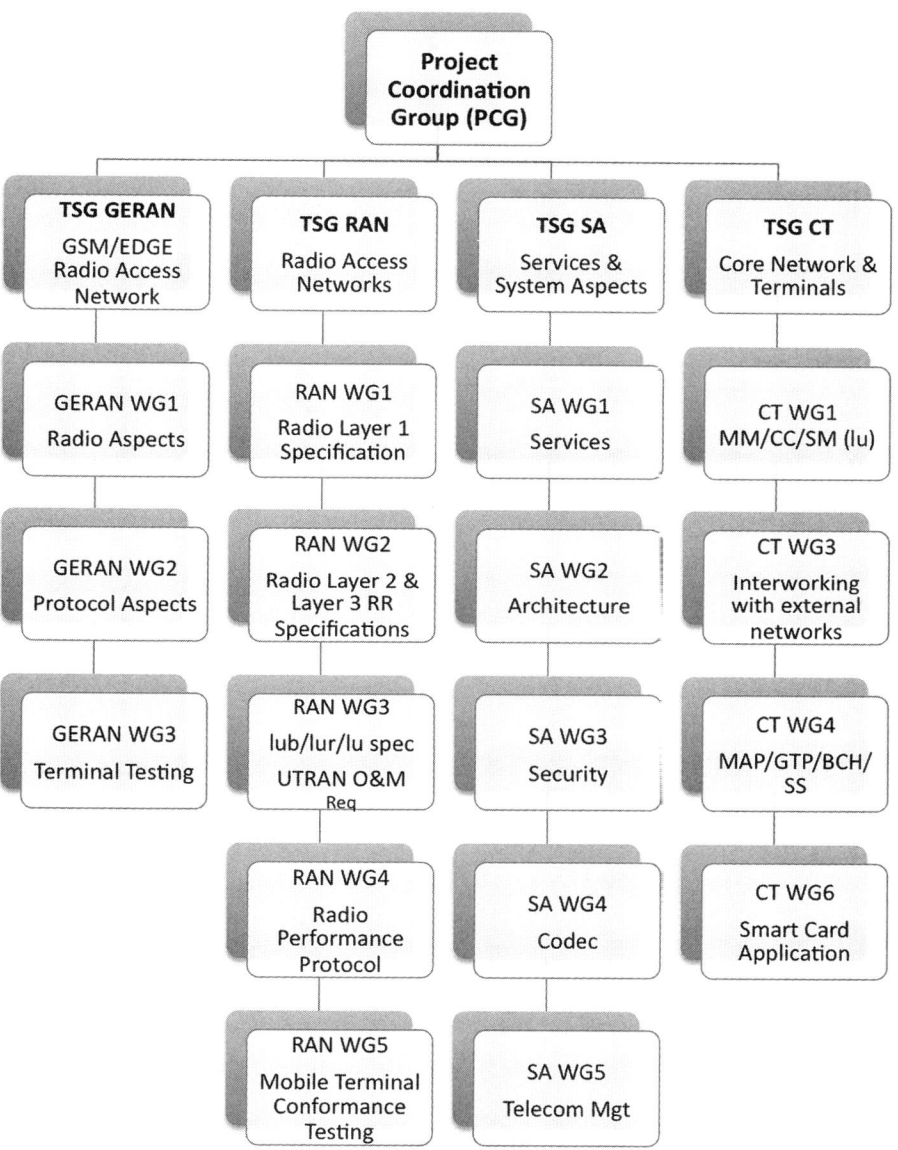

Figure 2.3   3GPP organization.

- *TSG CT.* This group specifies logical and physical terminal interfaces, capabilities, and core network part of 3GPP-based systems. This group also focuses on layer 3 radio protocols, signaling, interconnection with external networks, O&M requirements, and QoS, among others.

3GPP has been a leader in defining the evolving capabilities of cellular technology standards. In particular, 3GPP was key in standardizing the original GSM standard and has played a key role in standardizing its evolution from 2G (GSM/GPRS) to 4G (LTE).

There are multiple "releases" for specifications defined by 3GPP. Each of these releases generally maps to a particular technology—for example, Release 99 contained the original UMTS standard, while Release 5 contained enhancements to the UMTS air interface to include High Speed Packet Access–Downlink (HSPA-DL). Table 2.1 summarizes some of the major 3GPP releases and associated technologies since 2000. All release standards documents are freely available online at http://www.3gpp.org.

It is expected that 3GPP will continue to develop enhancements to existing standards such as LTE-Advanced while developing new approaches to increase capability and data rates in the future.

Table 2.1   Major 3GPP Releases

Year	Release	Technology	Description
2000	R99	Universal Mobile Telephone System (UMTS)	Code Division Multiple Access-based air layer (data rates up to 384 Kbps)
2002	R5	High-Speed Packet Access–Downlink (HSPA-DL)	Increased downlink data rates (up to 14 Mbps) compared to UMTS
2004	R6	High-Speed Packet Access–Uplink (HSPA-UL)	Increased uplink data rates (up to 5.8 Mbps) compared to UMTS
2006	R7	HSPA+	Multiple-input, multiple-output (MIMO) technology, voiceover HSPA. Uplink data rates up to 22 Mbps, downlink data rates up to 84 Mbps
2008	R8	LTE	All-IP core network, downlink data rates up to 292 Mbps, uplink data rates up to 71 Mbps, improved latency performance, packet-switched radio interface
2011	R10	LTE-Advanced	Fully compliant with IMT-Advanced requirements, multicell HSDPA, downlink data rates up to 1 Gbps, uplink data rates up to 500 Mbps.

2.5 3GPP2

The 3rd Generation Partnership Project 2 (3GPP2) was established in December 1998 as a collaboration between multiple regional telecommunications standards bodies: the ARIB in Japan, CCSA in China, Telecommunications Industry Association (TIA) in North America, TTA in Korea, and TTC in Japan. Together, these standards bodies comprise the Organizational Partners for 3GPP2. Moreover, Market Representation Partners include the CDMA Development Group, the IPv6 Forum, and the International 450 Association. These Market Representation Partners offer market advice and a consensus view on market requirements. 3GPP2 was established primarily for preparation, approval, and maintenance of technical specifications and reports for 3G networks based on the cdma2000 network structure. Like 3GPP, 3GPP2 is not considered a legal entity.

3GPP2 standard networks employ the cdma2000® cellular network technology and its associated migration technologies. 3GPP2 aims to employ a full IP core, with full IP functionality on every BS and mobile user node. There are four TSGs in 3GPP2: TSG-A (Access Network Interfaces), TSG-C (cdma2000), TSG-S (Services and Systems Aspects), and TSG-X (Core Networks). Note that these TSGs are similar in scope to the 3GPP structure. Furthermore, releases have been defined in 3GPP2 in a similar fashion to 3GPP.

3GPP2 has defined an evolutionary path from the original CDMA standard (cdmaOne, also known as IS-95) to cdma2000 and beyond. Table 2.2 summarizes some of the major capabilities developed within 3GPP2.

Table 2.2 Major 3GPP2 Releases

Year	Technology	Description
1993	IS-95A (cdmaOne)	Original CDMA standard that provided voice and circuit-switched data (14.4 kbps) capabilities
1995	IS-95B revision	Similar to IS-95A with improved circuit-switched data (64 kbps) capability
2000	IS-2000 (CDMA2000 1x)	First 3G CDMA standard, supports voice and packet-switched data, with rates up to 153.6 kbps
2001	CDMA2000 1xEV-DO Release 0	Supporting the IMT-2000 vision, offers data rates up to 2.4 Mbps downlink and 153 kbps uplink in a single 1.25 MHz channel
2006	CDMA2000 1xEV-DO Revision A	Peak data rates of 3.1 Mbps downlink and 1.8 Mbps uplink in a single 1.25 MHz channel
2007	CDMA2000 1xEV-DO Revision B	Peak data rates of 9.3 Mbps downlink and 5.4 Mbps uplink in a 5 MHz, 3-carrier aggregated channel

Indeed, 3GPP2 was also on the path to developing a 4G standard technology to compete with LTE, known as Ultra Mobile Broadband (UMB). Qualcomm was the lead sponsor of this technology, but abandoned further development in November 2008 in favor of LTE. Today, IEEE 802.16 is the major competitor to LTE, namely IEEE Wireless Metropolitan Area Network (WMAN) standards 802.16e and 802.16m.

2.6 INTERNATIONAL TELECOMMUNICATIONS UNION

What is now known as the International Telecommunications Union (ITU) originated back in May 17, 1865, the date of the first International Telegraph Convention. This convention was established to develop a framework agreement among 20 European countries due to the rapid expansion of telegraph networks. The group made decisions on common rules and standardization of equipment to improve international connectivity, set up international charging rates and accounting rules, and adopted uniform instructions for operation. The ITU was born after this initial arrangement to develop amendments to the original agreement. Today, even in the face of a vastly different networking environment, the ITU's goals largely remain the same.

International consensus and cooperation are paramount to ITU's success as an organization. Countries around the world recognize the ITU's importance in developing and maintaining international standards and recommendations for communications networks. These same countries enact national laws and international treaties that enforce ITU recommendations to manage scarce RF spectrum at the world level. Because of the large number of nations represented, standards work in the ITU often proceeds at a slower pace compared with the IETF, IEEE, or other standards organizations. Proposed documents that are slated to become ITU recommendations must go through an extensive review process, often limited to periodic meeting schedules. Because of the high impact ITU recommendations can have on national communications systems and associated policy, extensive, multinational review and long time frames (years!) before ratification are common.

The first International Telegraph Union began in 1885. With the advent of wireless telegraphy and the implications worldwide, the first International Radiotelegraph Convention was held in 1906 in Berlin. During this convention, an agreement was signed that contained the first governing regulations on wireless telegraphy. These regulations are known as Radio Regulations.

Fast forward to today, and you'll find the ITU has evolved substantially. In 1992, a reorganization of the ITU led to today's three-sector structure. The three sectors that comprise the ITU today include:

- *ITU-T*. Telecommunication Standardization
- *ITU-R*. Radiocommunication
- *ITU-D*. Telecommunication Development

ITU-T is focused specifically on standardization methods for telecommunications networks. The recommendations developed in ITU-T address issues including

IP television (IPTV), core network functionality, and broadband networks. The framework of ITU-T includes a World Telecommunications Standardization Assembly (WTSA), which sets structure and direction for the ITU-T. The WTSA meets every 4 years. During these meetings, the WTSA sets general policy for ITU-T, establishes study groups, approves these groups' work programs for the next 4-year period, and appoints chairmen and vice chairmen. There are currently 10 active study groups in ITU-T, each focused on a particular set of telecommunications network aspects:

- *Study Group 2.* Operational aspects of service provision and telecommunications management
- *Study Group 3.* Tariff and accounting principles including related telecommunication economic and policy issues
- *Study Group 5.* Environment and climate change
- *Study Group 9.* Television and sound transmission integrated broadband cable networks
- *Study Group 11.* Signaling requirements, protocols, and test specifications
- *Study Group 12.* Performance, quality of service (QoS), and quality of experience (QoE)
- *Study Group 13.* Future networks including mobile and next-generation networks (NGNs)
- *Study Group 15.* Optical transport networks and access network infrastructures
- *Study Group 16.* Multimedia coding, systems and applications
- *Study Group 17.* Security

The ITU-R sector plays a key role in global management of RF spectrum and satellite communications. As the Federal Communications Commission (FCC) administers similar issues for the United States, ITU-R is the international body that administers the same issues for the world. ITU-R products include Radio Regulations and Regional Agreements, both of which help to ensure effectively interference-free wireless operations. Figure 2.4 is an illustration of the organization of ITU-R.[1]

ITU-R schedules periodic meetings in a similar fashion to ITU-T—the World Radiocommunications Conference (WRC) is the primary ITU-R meeting that is held every 3–4 years. Scope and agenda for WRC meetings are generally scheduled 4–6 years in advance, with final agendas agreed upon 2 years in advance by the ITU Council. Activities at a WRC include:

- Revising Radio Regulations and associated frequency assignment and allotment plans

---

[1] Reproduced with the kind permission of ITU.

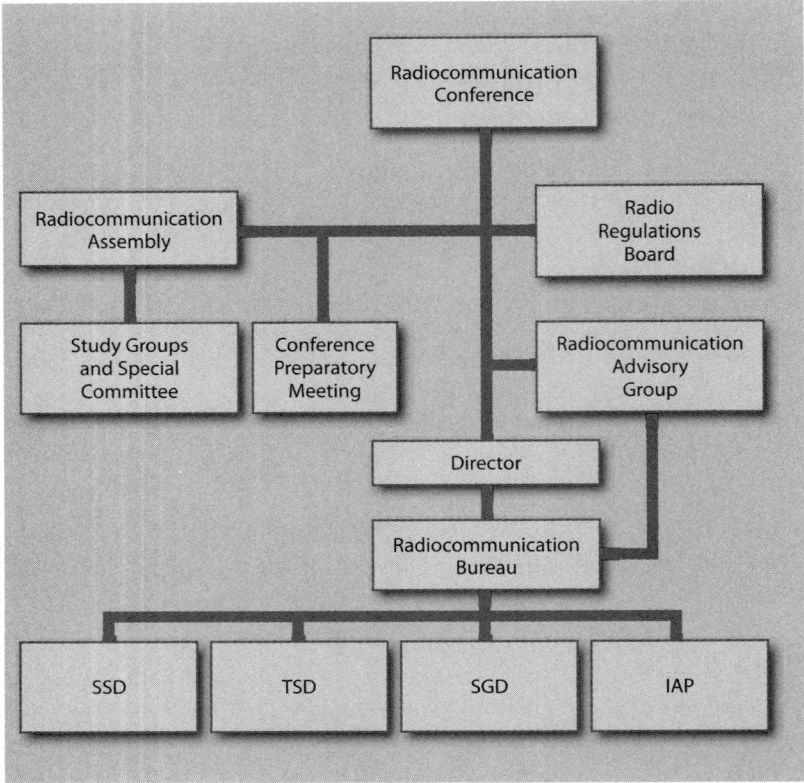

Figure 2.4   Illustration of ITU-R organization. SSD, Space Services Department; TSD, Terrestrial Services Department; SGD, Study Group Department; IAP, Informatics, Administration, and Publications Department.

- Provide instruction to the Radio Regulations Board (RRB) and Radiocommunications Bureau, reviewing activities of both
- Determine what questions should be considered for study by the Radiocommunication Assembly and its associated Study Groups, to prepare for future Radiocommunication Conferences

In addition to the WRC, there are Regional Radiocommunications Conferences that address similar issues for particular regions in the world.

The RRB consists of 12 members who are elected. This board conducts the following activities:

- Approval of procedural rules
- Addressing matters unresolved through the application of existing Radio Regulations and associated rules of procedure

- Consideration of unresolved interference investigation reports at the request of one or more administrations, and associated formulation of recommendations
- Advise Radiocommunication Conferences and Radiocommunication Assemblies
- Handles appeals process for adverse decisions regarding frequency assignments
- Any additional duties required by a conference or the ITU Council

The Radiocommunications Advisory Group (RAG) is tasked to conduct the following activities:

- Review priorities and strategies adopted within ITU-R
- Monitor Study Group work progress and provide guidance
- Provide recommendations for coordination across other ITU sectors and external organizations

The Radiocommunication Bureau (BR) handles executive actions for ITU-R. It is headed by a Director who is responsible for coordinating work across the sector. The Director has a staff of engineers, computer scientists, managers, and administrative staff. The BR is responsible for:

- Providing technical and administrative support to conferences, assemblies, and study groups
- Applying the provisions of Radio Regulations and Regional Agreements
- Registering frequency assignments and space plans for satellites worldwide
- Advises member states on equal, effective, economic use of RF spectrum and satellite orbits, including investigations and recommendations for mitigating interference
- Coordinates document distribution in ITU-R
- Assists developing countries by providing technical information, seminars, and working with the ITU Telecommunication Development Bureau

There are four departments within the BR:

- The Space Services Department (SSD, which is focused on space-related issues, including satellite orbits and interference
- The Terrestrial Services Department (TSD), which is focused on issues pertaining to communications that are not space based
- The Study Groups Department (SGD), which is responsible for administration of the ITU-R Study Groups
- The Informatics, Administration, and Publications (IAP) Department, which is responsible for internal administration of the BR.

Table 2.3   Active ITU-R Study Groups as of 2011

Study group #	Description
1	*Spectrum management*: Handles principles and techniques related to spectrum management, including utilization, monitoring, economics, and assisting developing countries
3	*Radio wave propagation*: Focuses on radio wave propagation effects in ionized and nonionized media, as well as radio noise characteristics, to improve performance of RF communications systems
4	*Satellite services*: Focuses on systems and associated networks for fixed, mobile, broadcasting, and radio-determination satellite services
5	*Terrestrial services*: Networks and systems for mobile, fixed, amateur, radio-determination, and amateur-satellite services
6	*Broadcasting service*: Focused on public delivery broadcasting using RF, including audio, video, and multimedia and data services.
7	*Science services*: Systems for space research, remote sensing, and radio astronomy and other applications, such as meteorology and Earth exploration

The Radiocommunication Assemblies (RA) maintain responsibility for program, structure, and approval of radiocommunication studies. These RAs are convened every 3–4 years and often coincide with WRCs. Assemblies specifically handle the following:

- Assignment of conference preparatory work and other Study Group questions
- Responding to other requests from other ITU conferences
- Suggesting appropriate topic content for future WRC agendas
- Approving and issuing ITU-R Recommendations and Questions
- Creating, dismantling, and setting the program for Study Groups

As of 2011, there are six active Study Groups. These groups are summarized in Table 2.3 [59].

2.6.1 International Mobile Telecommunications-2000 and International Mobile Telecommunications-Advanced

The International Mobile Telecommunications (IMT)-2000 vision was put forth by the ITU to guide standards organizations, companies, and wireless technologies to achieve specific performance thresholds for future mobile communications networks. This concept started in the mid-1980s with ITU working groups focused on developing requirements for 3G mobile communications. With the first-generation and 2G cellular networks well defined, ITU finally ratified the technical specifications for 3G networks in 2000. This vision was a result of many entities within and external to the ITU, including ITU-R, ITU-T, 3GPP, and 3GPP2, among others.

The IMT-2000 vision includes capabilities of providing value-added services and applications via a single standard that converges voice, video, and data on a single platform. Another focus for IMT-2000 is to enable ubiquitous global roaming, so mobile users may be able to move from country to country and maintain seamless services. IMT-2000 was not limited to cellular bearers—in fact, it specified multiple media types for achieving data rates and availability, including satellite. One of the major features of IMT-2000 is the vision for increased transmission rates over second-generation cellular: a minimum speed of 2 Mbps for slow or stationary users, and 348 kbps in a vehicle. Indeed, these increased data rates are now common across 3G networks in many countries and have helped enable a smartphone revolution in the cellular industry.

The IMT-Advanced vision was put forth by the ITU as the next generation after IMT-2000 for wireless network performance threshold requirements and major features. Other standards organizations such as 3GPP and IEEE 802 are working on or have released standards that aim to achieve the requirements for IMT-Advanced. Like IMT-2000, IMT-Advanced provides a framework for delivering services based on stationary, low-, and high-mobility scenarios. It also specifies capabilities for multimedia across a diverse set of platforms and services, with the intention of improving performance and QoS. The key features of IMT-Advanced include:

- Common but flexible functionality worldwide
- Service compatibility within fixed networks and IMT
- Internetworking capabilities with other radio networks
- Improved quality of mobile services compared with the past
- World-capable user equipment and roaming
- Improved user friendliness for services, applications, and equipment
- Enhanced peak data rates compared with IMT-2000: 1 Gbps for low-mobility scenarios, 100 Mbps for high mobility

Standards organizations expressing interest in developing wireless networking technologies that meet the IMT-Advanced requirements can submit proposals to the ITU for consideration. IEEE 802.16m is one example of an enhancement to the IEEE 802.16 series of WMAN standards that aims to provide IMT-Advanced capabilities. LTE-Advanced has been proposed by 3GPP toward the same end.

2.7 WI-FI ALLIANCE

The term "wireless fidelity," or Wi-Fi, is a marketing term that has become synonymous with 802.11 WLANs and WLANs in general. The Wi-Fi Alliance (WFA) was formed in 1999 as a nonprofit international organization to certify interoperability of 802.11 WLAN products [60]. The WFA is also heavily involved as an advocate for 802.11 technologies. The WFA includes hundreds of member companies and has certified thousands of WLAN products. Its trademarked Wi-Fi Certified logo, which is placed on the package of certified products, is a key criterion for market viability.

There are three primary components of certification for a Wi-Fi-certified product:

- *Conformance*, to ensure that the product conforms to specific behaviors dictated by the series of IEEE 802.11 standards
- *Performance*, to ensure that the product meets performance thresholds for sufficient user experience
- *Compatibility*, to ensure that the product will interoperate and communicate effectively with other vendors' products that conform to the same standard

Together, these components ensure that Wi-Fi-certified products meet acceptable standards for performance and usage by end users. Tests for certification are conducted at independent locations around the world, known as Wi-Fi Alliance Authorized Test Laboratories (ATLs). Certification profiles have been defined that test different IEEE 802.11 standards functionality, such as IEEE 802.11b, 802.11g, and 802.11n, security features such as Wireless Protected Access (WPA), and mobile convergence, among other profiles that have been defined. Vendors voluntarily submit their products for certification by the Wi-Fi Alliance, and consider the process critical to their own market success, given many enterprises and consumers desire Wi-Fi-certified products.

Certification programs have been defined by the Wi-Fi Alliance to provide vendors a path to interoperability with 802.11-based products. Since 2000, the Wi-Fi Alliance has certified over 11,000 products. Currently, the Wi-Fi Alliance defines both Core and Optional program categories for certification:

- Core programs
  - Products based on IEEE 802.11 radio standards a, b, and g, in single, dual mode, or multiband.
  - **Wi-Fi Protected Access 2 (WPA2™)**: This is for government-grade wireless network security for personal and enterprise purposes.
  - **Extensible Authentication Protocol (EAP)**: This is for authentication used to validate network device identities.
  - **Wi-Fi Certified n**: This is focused on products that conform to the 802.11n standard, and also includes Wi-Fi multimedia (WMM) testing profiles.
- Optional programs
  - **Wi-Fi Direct™**: This is focused on certifying products that connect via 802.11 without infrastructure such as access points for the purposes of printing, content sharing, and wireless display applications. Any certified product in this category will conform to the Wi-Fi Alliance Peer-to-Peer Technical Specification.
  - **Wi-Fi Protected Setup™**: This is focused on providing ease of configuration for security features using a personal identification number or other

defined method. Products certified to this standard will implement technology defined in the Wi-Fi Simple Configuration Technical Specification.

○ *Wi-Fi Multimedia*™: This is focused on supporting multimedia content over Wi-Fi networks, with traffic prioritization to support improved performance. Products certified to this standard will implement technologies defined in the WMM Technical Specification.

○ *WMM-Power Save*: This is focused on providing power savings over Wi-Fi networks for multimedia content.

○ *Voice-Personal*: This is focused on providing conformance to voice application performance over Wi-Fi.

○ *CWG-RF*: This is focused on convergence of devices with both cellular and Wi-Fi technologies. This profile is mandatory for Wi-Fi-enabled handsets that require Cellular Telecommunications Industry Association (CTIA) certification. Features tested include indication for Wi-Fi service availability, selection of cellular or Wi-Fi interface during times when both are available, successful operation of voice and data when Wi-Fi is available, and operation without interference from the cellular network. Handover is not tested between Wi-Fi and cellular.

Currently, certification testing occurs in 15 independent laboratories across 8 countries. Locations and the associated company names that host the laboratories follow:

• *Taiwan.* Allion Test Labs, Inc.; Bureau Veritas ADT; SGS Group

• *Spain.* AT4 Wireless

• *Germany.* CETECOM ICT Services GmbH

• *United States.* CETECOM, Inc.; TUV Rheinland Group

• *Japan.* SGS Group; TUV Rheinland Group

• *Korea.* SGS Group; TTA; TUV Rheinland Group

• *China.* TA Technology; Telecommunication Metrology Station (TMS)

• *India.* Wipro Technologies

In addition to the certification labs, a precertification test lab has been established at the University of New Hampshire.

The Wi-Fi Alliance maintains a Wi-Fi Certified™ database containing a list of product models, associated vendors, and interoperability certificates that summarize the standards the products has been tested and certified against.

2.8 WIMAX FORUM

The term "WiMAX" is a marketing term that has become synonymous with 802.16-based BWA networks in much the same manner as Wi-Fi has become synonymous with IEEE 802.11-based WLANs. The WiMAX Forum was formed in April 2001 as a nonprofit international organization to certify conformance and interoperability of products based on the IEEE 802.16 and ETSI HIPERMAN standards. The WiMAX Forum is also heavily involved as an advocate for 802.16 technologies. The WiMAX Forum is comprised of hundreds of member companies. Its WiMAX Certified logo is placed on the package of certified products, and is a key criterion for market viability in the same way that the Wi-Fi Certified logo of the WFA is a key criterion for market viability of 802.11 WLAN products.

There is typically much confusion regarding "the WiMAX standard." WiMAX is not a standard. WiMAX is a marketing term trademarked by the WiMAX Forum to describe 802.16-based technology. IEEE 802.16 is a wireless networking technology standard. However, it is not uncommon for 802.16 and WiMAX to be referred to as separate standards. The WiMAX standard refers to the set of capabilities within 802.16 that the WiMAX Forum will test against when performing conformance and interoperability testing in its equipment certification process. In this sense, the WiMAX Forum will indeed have significant impact into what functionality within the 802.16 standard will be brought to market by vendors, but it in of itself is not a technical standard. Rather, the WiMAX forum specifies conformance and interoperability profiles that refer to the subset of 802.16 capabilities that are likely to experience widespread implementation.

2.9   BLUETOOTH SPECIAL INTEREST GROUP

The Bluetooth Special Interest Group (SIG) was formed in 1998 as a privately held, nonprofit trade association. It is comprised of thousands of companies that are leaders in the fields of mobile communications, consumer electronics, computing, and automotive industries, among others. These member companies drive the development of the family of Bluetooth standards while implementing these standards in their own products. The SIG is responsible for publishing the Bluetooth specification standards documents, administering the qualification program for Bluetooth certification, protecting Bluetooth trademarks, and promoting Bluetooth technology [61].

Within the SIG, there are multiple subgroups tasked to carry on the business of the organization. These subgroups include:

- Working Groups, responsible for developing and maintaining Bluetooth technical specifications and standards documents.
- Policy Committees, responsible for handling external organization relationships and overarching policy for the Bluetooth SIG
- Bluetooth Ecosystem Teams (BETs), responsible for promoting and proliferating Bluetooth-enabled devices in specific markets (e.g., automotive, mobile phone, smart energy)
- Study Groups, formed to represent companies that desire to extend Bluetooth technologies to support new applications
- Country user groups, responsible for assisting member companies' influence of Bluetooth technology across multiple languages

Table 2.4   Bluetooth SIG Standardization

Year	Specification	Description
1999	Bluetooth 1.0	Original Bluetooth specification
2004	Bluetooth Core Version 2.0+EDR	Enhanced data rates, up to 3 Mbps
2007	Bluetooth Core Version 2.1+EDR	Improves pairing process
2009	Bluetooth Core Version 3.0+HS	Uses 802.11 for high-speed transport up to 24 Mbps
2010	Bluetooth Core Version 4.0	New protocol stack for low energy applications

During the SIG's existence, multiple standards and enhancements for Bluetooth technologies have been developed and released. Table 2.4 summarizes some of the key technologies and provides a history of Bluetooth standards.

In order to use the Bluetooth name, products must go through a qualification process. The Bluetooth Qualification Program (BQP) is used by the SIG as the framework for this process. Once a product has successfully completed the program, it is listed on the End Product Listing (EPL). It is important to note that membership in the Bluetooth SIG is required, and is limited to member companies only. Individuals cannot become members, and hence cannot submit designs without a company.

The purpose of the BQP is to provide interoperability, compatibility, and conformance to the Bluetooth specifications. In addition, there are other requirements for licensees with regards to compliance and intellectual property. In order for a product to become qualified, the following steps are taken according to the Bluetooth SIG website at http://www.bluetooth.org:

1. Upon conceptual development, start the project online at http://www.bluetooth.org and obtain a Qualification Design ID (QD ID) for a new project.

2. Create a Bluetooth test plan using the Test Plan Generator (TPG).

3. Use the SIG-developed Profile Tuning Suite (PTS), Device Library, and UnPlugFest (UPF) events to assist in testing and preparing the product for qualification.

4. When the product is listed via the Qualified Listing Interface (QLI), it can be listed on the EPL. EPL displays the products to the general public via the Bluetooth Gadget Guide and Product Directory on http://www.bluetooth.com.

2.10   SUMMARY OF THE WIRELESS ECOSYSTEM

Wireless networks have transformed the way the world communicates. A large, ever-growing ecosystem has spawned from this transformation. We've covered an overview of some of the key organizations whose members define the technical standards that drive the wireless marketplace and impact the consumer's experience. We've seen that these organizations' processes for defining standards are as diverse as the organizations themselves, yet each organization's impact in the worldwide wireless market is undeniable. We've also seen that many wireless networking standards rely on adoption by industry consortiums to test interoperability and compatibility while advocating for the market proliferation of a particular technology. Furthermore, new standards are released every year, pushing the market to increase data rates and further enhance users' experiences. It is challenging to keep track of the ecosystem's evolution as market trends and expectations change so quickly, but the organizations listed here are likely to have a major impact in the wireless world for the foreseeable future.

Chapter 3

Wireless Personal Area Networks

This chapter provides an overview of wireless personal area networks (WPANs), including an overview of some of the key technologies, usage cases, and evolution. WPAN technologies enable communications between devices on the order of tens of feet apart. Furthermore, WPANs generally do not rely on complex infrastructure, reducing power consumption and cost compared with other wireless networking technologies. Historically, the ad hoc networking concepts driving WPANs started in the late 1990s. Today, there are multiple WPAN standards that have been implemented for a wide range of applications, including wireless headsets, wireless gaming, wireless display mirroring, and home automation.

We will discuss three WPAN technologies in this chapter: ultra wideband (UWB), IEEE 802.15.4/ZigBee, and Bluetooth. Each section will provide a technical overview and some usage cases.

3.1 BLUETOOTH

The Bluetooth Special Interest Group (SIG) was established in 1998 as a privately held nonprofit trade association, specifically focused on developing personal area networking technology. The Bluetooth name originated from a byname for the tenth-century king, Harald I of Denmark. Bluetooth is a short-range, personal area networking wireless standard that is intended to replace cables connecting portable or fixed devices. Some examples of Bluetooth-enabled devices include mobile phones, headsets, and laptop computers. Bluetooth operates in the 2.4 GHz ISM band, and features include low power consumption and low cost.

There are two types of Bluetooth systems. The original standard specified a Basic Rate (BR) system, while the current version 4.0 of the standard (as of 2011) indicates support for a Low Energy (LE) system. These systems perform typical functions for personal area networking technologies, including device discovery and

Wireless Networking: Understanding Internetworking Challenges, First Edition.
Jack L. Burbank, Julia Andrusenko, Jared S. Everett, and William T.M. Kasch.
© 2013 the Institute of Electrical and Electronics Engineers, Inc. Published 2013 by John Wiley & Sons, Inc.

Table 3.1   Bluetooth Standards Summary

Year	Reference	Specification	Description	Data rate (Mbps)	Maximum application throughput (Mbps)
1999	[2]	Bluetooth 1.0	Original Bluetooth specification	1	0.7
2002	[8]	Bluetooth 1.1	Corrected many errors found in original specification	1	0.7
2004	[9]	Bluetooth Core Version 2.0+EDR	Enhanced data rates, up to 3 Mbps	3	2.1
2007	[10]	Bluetooth Core Version 2.1+EDR	Improves pairing process	3	2.1
2009	[11]	Bluetooth Core Version 3.0+HS	Uses 802.11 for high-speed transport	24	24
2010	[3]	Bluetooth Core Version 4.0	New protocol stack for low energy applications	1	0.26

connection establishment and maintenance. The BR system now includes some enhancements beyond the initial scope of the first standard in Bluetooth, known as Enhanced Data Rate (EDR) and associated MAC and PHY extensions to support improved data rates. The LE system was developed to support low-energy consumption requirements. Table 3.1 summarizes the Bluetooth standards to date.

The Bluetooth protocol stack is illustrated in Figure 3.1 [3]. The Radio layer is defined and represents the PHY. The baseband layer is similar to the OSI Link Layer or MAC layer. The link manager layer is used for physical link setup and control. The L2CAP layer is used for logical link control (multiple logical links can exist within a single physical link) and provides both connection-oriented and connectionless data services to upper layer protocols.

A core Bluetooth system will consist of a host and one or more controllers. The host is logically defined as all layers below noncore profiles but above the Host Controller Interface (HCI). The controller is the opposite—all layers below the HCI.

The BR system in Bluetooth supports data rates up to 721.2 kbps, with EDR support up to 2.1 Mbps. High-speed operation can be supported up to 24 Mbps using an 802.11 waveform. This system uses a frequency-hop spread-spectrum method to enhance robustness against interference. BR operation uses a binary frequency modulation method to minimize receiver complexity, with a symbol rate of 1 Ms/s. EDR uses PSK modulation methods.

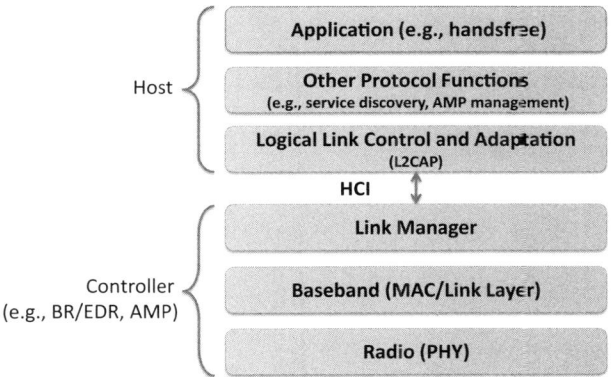

Figure 3.1   Bluetooth protocol stack. Reproduced from Reference [3].

In the BR system, the primary modulation mode is Gaussian frequency shift keying (GFSK) with a bandwidth-bit period product of 0.5. The modulation index is specified to be between 0.28 and 0.35. Symbol timing is required to be within ±20 ppm. The mathematical expression for frequency shift keying (FSK) is shown in Equation (3.1):

$$s(t) = \sqrt{\frac{2E_s}{T}} \cos\left(2\pi f_c t + \frac{a_n h(t - nT)}{T} + \theta \right), nT \le t \le (n+1)T, \quad (3.1)$$

where t is the independent time variable, E_s is the energy per symbol for the modulation, T is the bit period, f_c is the carrier frequency, n is the integer index consistent with digital sampling at equally spaced distances T apart, a_n is the nth data bit (0 or 1), h is the modulation index, and θ is a constant phase shift. An FSK waveform can be converted into a GFSK waveform by passing the signal expression in Equation (3.1) through a Gaussian filter, whose form follows in Equation (3.2):

$$g(t) = \frac{1}{\sqrt{2\pi}\sigma T} \exp\left(\frac{-t^2}{2\sigma^2 T^2} \right); \sigma = \frac{\sqrt{\ln(2)}}{2\pi BT}, \quad (3.2)$$

where t is the independent time variable, T is the bit period, and BT is the bandwidth-bit period product. The GFSK waveform is derived by convolving $s(t)$ with $g(t)$:

$$S_{\text{GFSK}}(t) = s(t) * g(t) = \int_{-\infty}^{\infty} s(\tau) g(t - \tau) d\tau. \quad (3.3)$$

GFSK was chosen as the desired modulation because it is more spectrally efficient than FSK alone. Employing a matched Gaussian filter extends slope transitional times while minimizing intersymbol interference (ISI).

Table 3.2 Modulation Mapping in Bluetooth EDR

Modulation mode	Bit sequence	Phase mapping
π/4-DQPSK	0,0	π/4
π/4-DQPSK	0,1	3π/4
π/4-DQPSK	1,1	−3π/4
π/4-DQPSK	1,0	−π/4
8DPSK	0,0,0	0
8DPSK	0,0,1	π/4
8DPSK	0,1,1	π/2
8DPSK	0,1,0	3π/4
8DPSK	1,1,0	π
8DPSK	1,1,1	−3π/4
8DPSK	1,0,1	−π/2
8DPSK	1,0,0	−π/4

Reproduced from Reference [3].

In the EDR system, the modulation mode is changed within each Bluetooth packet sent. The access code and packet header are transmitted with BR 1 Mbps GFSK, while the additional fields are transmitted with EDR PSK. This improves synchronization performance by increasing the probability that the packet header will be received successfully.

For the 2 Mbps EDR waveform, the modulation format is defined as π/4-rotated differentially encoded quadrature phase-shift keying (π/4-DQPSK). For the 3 Mbps EDR waveform, the modulation format is defined as 8-ary differentially encoded phase-shift keying (8DPSK). Raised cosine pulse shaping is used. To differentially encode the bits, a binary data stream is mapped into symbols that represent the phase of the signals. Table 3.2 [3] summarizes the mapping for the two types of modulation.

Within the 2.4 GHz band, the Bluetooth specification has defined specific channels. This channelization is used in the frequency hopping scheme. Channels are defined by Equation (3.4):

$$F = 2402 + k \text{ MHz}, k = 0, 1, 2, \ldots, 78. \tag{3.4}$$

In Bluetooth, connections can be defined as point to point or point to multipoint. For point-to-point connections, the physical channel is shared between two devices. For point-to-multipoint connections, the physical channel is shared between more than two devices. These devices form a piconet when they share the same channel. Within a piconet, there is a single master node that coordinates all communications. The other nodes are known as slaves. Slaves can join multiple piconets by multiplexing their time across multiple masters, with each master corresponding to a separate piconet. Furthermore, a master for one piconet can be a slave in another. Because

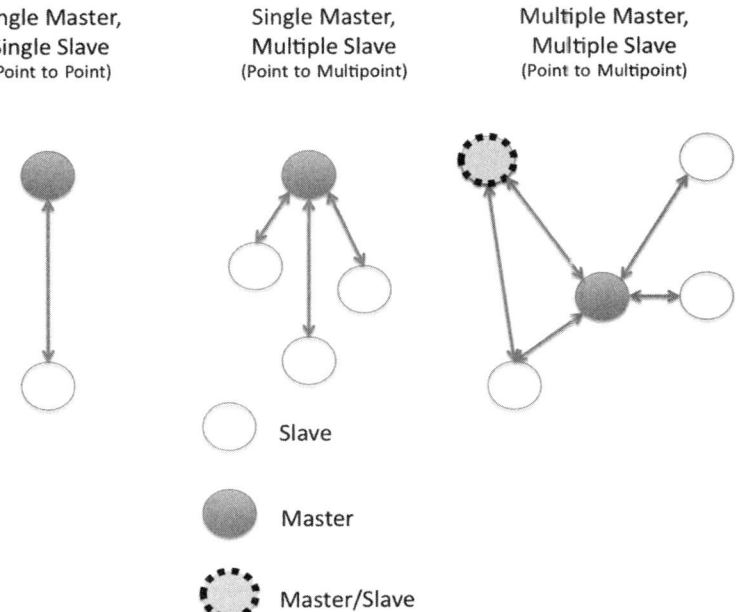

Single Master,
Single Slave
(Point to Point)

Single Master,
Multiple Slave
(Point to Multipoint)

Multiple Master,
Multiple Slave
(Point to Multipoint)

Slave

Master

Master/Slave

Figure 3.2   Piconet concept. Reproduced from Reference [3].

Company Assigned	Company ID
24 bits	24 bits

Figure 3.3   Bluetooth addressing. Reproduced from Reference [3].

of Bluetooth's hopping capability, multiple piconets within the 2.4 GHz band will operate with unsynchronized hopping patterns, which mitigate interference between the piconets. Figure 3.2 [3] illustrates a conceptual view of some piconets.

There are three different configurations for the piconets shown in the figure. Note that the multiple master/multiple slave configuration illustrates a node that serves as a master in one piconet while also serving as a slave in another.

Each Bluetooth device on a network is allocated a 48-bit device address. This includes a 3-octet Company ID, also referred to as the upper address part (UAP), to identify the vendor of the Bluetooth device, and a company-assigned portion of 3 octets, known as the lower address part (LAP), that is used to distinguish the device from other devices made by the same vendor. Figure 3.3 [3] illustrates the structure of a Bluetooth address.

Within the company-assigned portion of the address, certain values are reserved for inquiry operations.

Access Code 68/72 bits	Header 54 bits	Payload 0–2745 bits

Figure 3.4   Bluetooth Basic Rate packet structure. Reproduced from Reference [3].

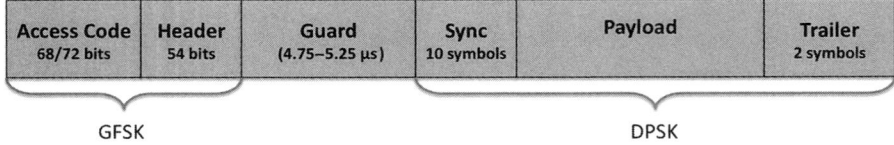

Access Code 68/72 bits	Header 54 bits	Guard (4.75–5.25 µs)	Sync 10 symbols	Payload	Trailer 2 symbols

GFSK   DPSK

Figure 3.5   Bluetooth Enhanced Data Rate packet structure. Reprinted from Reference [3].

Bluetooth devices generate and receive Bluetooth packets. The packet format for basic rate is shown in Figure 3.4 [3].

As shown in the figure, the access code can be 68 or 72 bits long, the header is 54 bits, and the payload can vary between 0 and 2745 bits. If the header field follows, the access code length is 72 bits. Otherwise, it is 68 bits, indicating a shortened access code used in and of itself for synchronization or other functions. A single access code is used to identify all packets on the same physical channel. This enables receivers to use sliding correlators to rapidly identify packets on the channel. There are four types of access codes used in Bluetooth: the channel access code (CAC), the device access code (DAC), the general inquiry access code (GIAC), and the dedicated inquiry access code (DIAC). The CAC is used during periods of user data transmission, while the DAC is used to communicate directly to a device. The general and dedicated inquiry access codes are used for inquiry functions.

The packet format for enhanced data rate is shown in Figure 3.5. Note this format contains more fields than basic rate.

As shown in the figure, the first two fields are identical to the basic rate fields and modulation method. However, to support the higher rate DPSK modulation, a guard interval is inserted between the header and the start of the synchronization symbols before the data payload. There is also a trailer of two DPSK symbols appended to the payload.

Within the header for both BR and EDR packets, there are six fields:

- LT_ADDR (3 bits) logical transport address indicating the logical transport type
- TYPE (4 bits) type code indicating the type of packet
- FLOW (1 bit) flow control field
- ARQN (1 bit) acknowledgment indicator
- SEQN (1 bit) sequence number indication
- HEC (8 bits) header error check

A Bluetooth physical channel is defined by four elements: a pseudorandom hopping sequence that changes the center frequency of the modulated baseband signal over time, a specific slot transmission time, the access code, and the packet header. In order for devices to communicate on a physical channel, they must maintain tight time synchronization to ensure they hop at the same rate and maintain the same center frequency simultaneously. A 28-bit clock is used on the devices to provide this synchronization, with each clock "tick" corresponding to 312.5 μs of time. A Bluetooth master will choose the hopping sequence used and command the slaves in the piconet to synchronize to this sequence. The maximum hopping rate is 1600 hops per second during connection states, while it can reach 3200 hops per second in the inquiry and page states.

There are four physical channels defined in Bluetooth:

- *Basic Piconet Physical Channel:* for user data transmission
- *Adapted Piconet Physical Channel:* for user data transmission
- *Inquiry Scan Physical Channel:* to discover devices
- *Page Scan Physical Channel:* to establish connections between devices

The basic and adapted piconet physical channels are used for user data transmission. The adapted piconet physical channel differs from the basic piconet physical channel in its hopping sequence. In the adapted channel, the hopping sequence may use fewer than the 79 hopping frequencies defined, while the basic sequence uses the full 79 frequencies.

Time slots are used in Bluetooth to share the medium between the master and the slaves. All devices are assigned particular time slots by the master. Figure 3.6 [3] illustrates a notional example of time slot assignment.

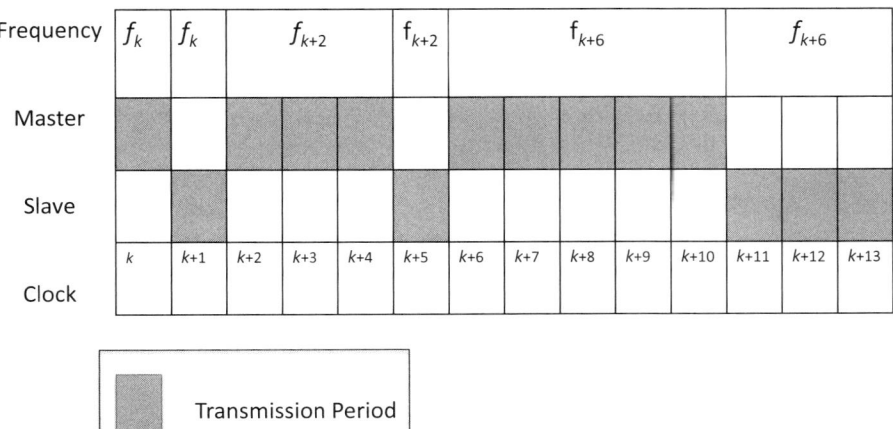

Figure 3.6 Time slot assignment in Bluetooth. Reproduced from Reference [3].

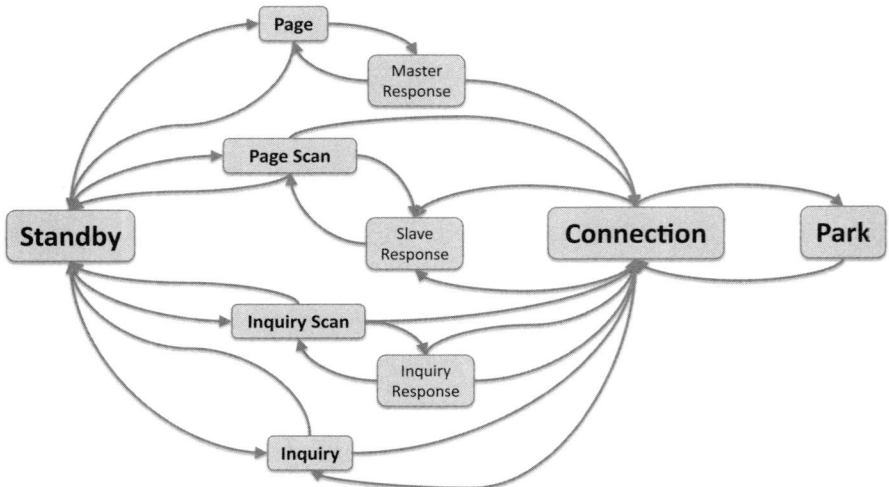

Figure 3.7 State diagram for Bluetooth links. Reproduced from Reference [3].

Note that the master chooses the frequency based on the current clock value (represented by an index, k). Once the master has completed a transmission, the slave then uses the next adjacent time slot to transmit on the same frequency. The time slots may be allocated based on the size of the packets being transmitted, and the frequency remains the same for an entire packet length.

For each Bluetooth device, there are three primary states: *connection*, where the device is passing user data and is active; *park*, where the link may be idle and transmissions are silent; and *standby*, the default mode, where the device may be in a low-power configuration with only its clock running. Figure 3.7 [3] illustrates the possible states.

Establishing new connections requires paging procedures. Any device that initiates a connection becomes the master of the piconet. The page scan substate is used by the device to scan specific frequencies for the existence of any other devices. The page substate is used for the master device to connect to any slave device that is in a page scan substate. Table 3.3 [3] summarizes the steps required for a connection to be established.

In the table, the packet type ID refers to an identity packet, which consists of a DAC or IAC and has a fixed length of 68 bits. The packet type FHS refers to a frequency hop synchronization packet, which contains the Bluetooth device address and clock information from the master. FHS packets are critical for ensuring that slaves in a piconet follow the master's clock and maintain hopping synchronization. The POLL type does not have a payload, but requires confirmation from the slave. This is so the master can confirm that the slave is indeed matching its frequency-hopping pattern. Otherwise, both devices revert back to page/page scan substates.

For device discovery, the inquiry substates are used. For a device to discover other devices, it must enter the inquiry substate. While in this state, a device will

Table 3.3 Initial Messaging for Master–Slave Connection

Step	Message	Packet type	Direction	Hopping sequence	Access code and clock
1	Page	ID	Master to slave	Page	Slave
2	First slave page response	ID	Slave to master	Page response	Slave
3	Master page response	FHS	Master to slave	Page	Slave
4	Second slave page response	ID	Slave to master	Page response	Slave
5	First packet master	POLL	Master to slave	Channel	Master
6	First packet slave	Any type	Slave to master	Channel	Master

Reproduced from Reference [3].

transmit an inquiry message (ID packet) at different frequencies, driven by a hopping pattern. This pattern is derived from the lower address portion of the GIAC (a single value defined for all devices). Any device that is "discoverable" will periodically enter the inquiry scan substate, looking for this message and using it as the basis for negotiating a connection if desired.

In the connection substate, devices may be active or parked. There can be up to seven active devices in the connected mode at any given time. These active devices can participate on the channel freely. In the park state, up to 255 parked devices can reside. During this state, devices stay synchronized with the master but do not participate in regular transmissions/receptions on the channel. Beacons are transmitted from the master at regular intervals for parked devices to maintain synchronization.

Bluetooth defines a series of logical transports that are used to carry different types of traffic in different modes of operation. These types include:

- Synchronous connection oriented (SCO)
- Extended SCO (eSCO)
- Asynchronous connected oriented (ACL)
- Active slave broadcast (ASB)
- Parked slave broadcast (PSB)

Both SCO and eSCO are used for point-to-point transports between a master and a slave. They include traffic such as voice or synchronization information, and are generally reserved at regular intervals by the master. The ACL is used for transports between master and slave, but time slots are not reserved. ASB is used by a master to communicate with active slaves, while the PSB is used by the master to communicate with parked slaves.

There are five logical links defined in Bluetooth:

- Link control (LC)
- ACL control (ACL-C)
- User asynchronous/isochronous (ACL-U)
- User synchronous (SCO-S)
- User extended synchronous (eSCO-S)

Both LC and ACL-C are control links, used to control and manage the links, respectively. The LC link is mapped onto the packet header and is carried in every packet except ID. ACL-U is used as the carrying link for ACL traffic, while SCO-S and eSCO-S are used to carry both SCO and eSCO traffic, respectively. Table 3.4 [3] summarizes the transports and associated links supported.

Table 3.5 [3] summarizes most of the packet types in Bluetooth. The one packet type that is not shown is the ID packet because it does not have a header (hence no type code).

Table 3.4   Transports, Links, and Description

Logical transport	Links supported	Overview
Asynchronous connection oriented (ACL)	Control (LMP) ACL-C or (PAL) AMP-C	BR/EDR, AMP
Synchronous connection oriented (SCO)	Stream (unframed) SCO-S	Bidirectional, symmetric, point-to-point, AV channels. Used for 64 Kb/s constant rate data.
Extended synchronous connection oriented (eSCO)	Stream (unframed) eSCO-S	Bidirectional, symmetric or asymmetric, point to point, general regular data, limited retransmission. Used for constant rate data synchronized to the master Bluetooth clock.
Active slave broadcast (ASB)	User (L2CAP) ASB-U	Unreliable, unidirectional broadcast to any devices synchronized with the physical channel. Used for broadcast L2CAP groups.
Parked slave broadcast (PSB)	Control (LMP) PSB-C, User (L2CAP) PSB-U	Unreliable, unidirectional broadcast to all piconet devices. Used for LMP and L2CAP traffic to parked devices, and for access requests from parked devices.

Reproduced from Reference [3].

Table 3.5 Packet Types in Bluetooth

Type code $(b_3b_2b_1b_0)$	Slot occupancy	SCO logical transport (1 Mbps)	eSCO logical transport (1 Mbps)	eSCO logical transport (2–3 Mbps)	ACL logical transport (1 Mbps)	ACL logical transport (2–3 Mbps)
0000	1	NULL	NULL	NULL	NULL	NULL
0001	1	POLL	POLL	POLL	POLL	POLL
0010	1	FHS	Reserved	Reserved	FHS	FHS
0011	1	DM1	Reserved	Reserved	DM1	DM1
0100	1	Undefined	Undefined	Undefined	DH1	2-DH1
0101	1	HV1	Undefined	Undefined	Undefined	Undefined
0110	1	HV2	Undefined	2-EV3	Undefined	Undefined
0111	1	HV3	EV3	3-EV3	Undefined	Undefined
1000	1	DV	Undefined	Undefined	Undefined	3-DH1
1001	1	Undefined	Undefined	Undefined	AUX1	AUX1
1010	3	Undefined	Undefined	Undefined	DM3	2-DH3
1011	3	Undefined	Undefined	Undefined	DH3	3-DH3
1100	3	Undefined	EV4	2-EV5	Undefined	Undefined
1101	3	Undefined	EV5	3-EV5	Undefined	Undefined
1110	5	Undefined	Undefined	Undefined	DM5	2-DH5
1111	5	Undefined	Undefined	Undefined	DH5	3-DH5

Reproduced from Reference [3].

The packet types from Table 3.5 are defined as follows:

- **NULL.** This has no payload, only contains the CAC and packet header. This can be used to send acknowledgments or status of receive buffers.
- **POLL.** This is similar to NULL but requires acknowledgment.
- **FHS.** This contains the Bluetooth device address and the clock of the sender, and assists in synchronizing frequency hop sequences.
- **DM1.** This supports control messages in any logical transport that allows DM1 packet type, but can also support user data.
- **HV1.** This contains 10 information bytes, used by the SCO logical transport, typically used for user voice or data.
- **HV2.** This contains 20 information bytes, used by the SCO logical transport, typically used for user voice or data.
- **HV3.** This contains 30 information bytes, used by the SCO logical transport, typically used for user voice or data.
- **DV.** This is a combined voice–data packet, with a voice field of 80 bits and a data field of up to 150 bits.

- **EV3.** This contains 1–30 information bytes, used by eSCO logical transport, typically used for user voice or data.

- **EV4.** This contains 1–120 information bytes, used by eSCO logical transport, typically used for user voice or data.

- **EV5.** This contains 1–180 information bytes, used by eSCO logical transport, typically used for user voice or data.

- **2-EV3.** This is similar to EV3 except that payload is modulated using $\pi/4$-DQPSK, and contains 1–60 information bytes, typically used for user voice or data.

- **2-EV5.** This is similar to EV5 except that payload is modulated using $\pi/4$-DQPSK, and contains 1–360 information bytes, typically used for user voice or data.

- **3-EV3.** This is similar to EV3 except that payload is modulated using 8DPSK, and contains 1–90 information bytes, typically used for user voice or data.

- **3-EV5.** This is similar to EV5 except that payload is modulated using 8DPSK, and contains 1–540 information bytes, typically used for user voice or data.

- **DM1.** This is used by the ACL logical transport, contains 1–18 information bytes, and carries data information only.

- **DH1.** This is used by ACL logical transport, similar to DM1 except that information in payload is not forward error correction coded.

- **DM3.** This is used by ACL logical transport, can use up to three time slots, and contains 2–123 bytes of information in payload.

- **DH3.** This is similar to DM3 except not forward error correction coded.

- **DM5.** This can use up to five time slots, 2–226 information bytes in payload.

- **DH5.** This is similar to DM5 except not forward error correction coded.

- **AUX1.** This resembles DH1, but does not have a cyclic redundancy check (CRC). Contains 1–30 information bytes and uses 1 time slot.

- **2-DH1.** This is similar to DH1 except modulated with $\pi/4$-DQPSK

- **2-DH3.** This is similar to DH3 except modulated with $\pi/4$-DQPSK

- **2-DH5.** This is similar to DH5 except modulated with $\pi/4$-DQPSK

- **3-DH1.** This is similar to DH1 except modulated with 8DPSK

- **3-DH3.** This is similar to DH3 except modulated with 8DPSK

- **3-DH5.** This is similar to DH5 except modulated with 8DPSK

The Link Manager Protocol (LMP) is used as the mechanism to manage links in Bluetooth. It controls all the aspects of a Bluetooth connection. The messages that define the LMP are transferred on the ACL-C logical link using the default ACL transport. Figure 3.8 illustrates the LMP procedures involved in establishing a connection. It is seen in the figure that the baseband page procedures are followed initially, and then negotiations take place for clock offset corrections, what version of LMP is supported, and what features are supported. A host connection request is

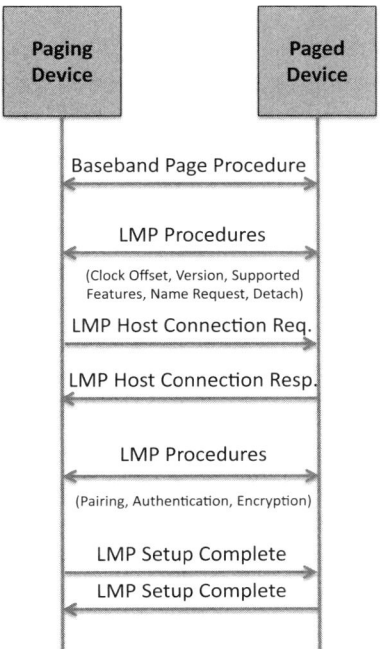

Figure 3.8   LMP connection procedure. Reprinted from Reference [3].

then generated from the paging device to the paged device, which responds with an accept or deny message. Assuming the connection request is accepted, features such as pairing, authentication, and encryption are then negotiated. Once this is complete, the LMP setup complete message is generated on both sides to confirm the connection setup.

To ensure interoperability of applications, Bluetooth profiles are specified to define the necessary functions for each layer in the Bluetooth stack. The profiles are defined in Reference [3] as a definition for the vertical interactions between each layer. All devices must support the Generic Access Profile (GAP), which defines the basic requirements for a Bluetooth device. Within the GAP, there is an attribute protocol (ATT) that enables a device to share a set of attributes for read/write purposes. These attributes can be used to support functions defined by GAP.

In order for a device to support a particular application, the Service Discovery Protocol (SDP) is used to enable applications to discover services and associated characteristics that are available [62]. For instance, in the case of a Bluetooth headset connecting to a mobile phone, SDP would be used by the Bluetooth core systems on both these devices to enable the headset (voice) application to operate. Specific services are assigned particular Universally Unique Identifiers (UUIDs) from the Bluetooth assigned numbers authority. These IDs are used to indicate particular Service Classes and/or profiles. Table 3.6 summarizes the service classes/profiles.

Table 3.6 Service Class

Service class name	UUID	Specification	Allowed usage
ServiceDiscoveryServerServiceClassID	0x1000	*Bluetooth* Core Specification	Service Class
BrowseGroupDescriptorServiceClassID	0x1001	*Bluetooth* Core Specification	Service Class
SerialPort	0x1101	Serial Port Profile (SPP)	Service Class/Profile
DialupNetworking	0x1103	Dial-Up Networking Profile (DUN)	Service Class/Profile
IrMCSync	0x1104	Synchronization Profile (SYNC)	Service Class/Profile
OBEXObjectPush	0x1105	Object Push Profile (OPP)	Service Class/Profile
OBEXFileTransfer	0x1106	File Transfer Profile (FTP)	Service Class/Profile
IrMCSyncCommand	0x1107	Synchronization Profile (SYNC)	
Headset	0x1108	Headset Profile (HSP)	Service Class/Profile
AudioSource	0x110A	Advanced Audio Distribution Profile (A2DP)	service Class
AudioSink	0x110B	Advanced Audio Distribution Profile (A2DP)	Service Class
A/V_RemoteControlTarget	0x110C	Audio/Video Remote Control Profile (AVRCP)	Service Class
AdvancedAudioDistribution	0x110D	Advanced Audio Distribution Profile (A2DP)	Profile
A/V_RemoteControl	0x110E	Audio/Video Remote Control Profile (AVRCP)	Service Class/Profile
A/V_RemoteControlController	0x110F	Audio/Video Remote Control Profile (AVRCP)	Service Class
Headset—Audio Gateway (AG)	0x1112	Headset Profile (HSP)	Service Class
PAN User (PANU)	0x1115	PAN Profile	Service Class/Profile
Network Access Point (NAP)	0x1116	PAN Profile	Service class/profile
Group ad-hoc network (GN)	0x1117	PAN Profile	Service Class/Profile
DirectPrinting	0x1118	Basic Printing Profile (BPP)	Service Class
ReferencePrinting	0x1119	See Basic Printing Profile (BPP)	Service Class
Basic Imaging Profile	0x111A	Basic Imaging Profile (BIP)	Profile
ImagingResponder	0x111B	Basic Imaging Profile (BIP)	Service Class
ImagingAutomaticArchive	0x111C	Basic Imaging Profile (BIP)	Service Class
ImagingReferencedObjects	0x111D	Basic Imaging Profile (BIP)	Service Class
Handsfree	0x111E	Hands-Free Profile (HFP)	Service Class/Profile

HandsfreeAudioGateway	0x111F	Hands-free Profile (HFP)	Service Class
DirectPrintingReferenceObjectsService	0x1120	Basic Printing Profile (BPP)	Service class
ReflectedUI	0x1121	Basic Printing Profile (BPP)	Service Class
BasicPrinting	0x1122	Basic Printing Profile (BPP)	Profile
PrintingStatus	0x1123	Basic Printing Profile (BPP)	Service Class
HumanInterfaceDeviceService	0x1124	Human Interface Device (HID)	Service Class/Profile
HardcopyCableReplacement	0x1125	Hard Copy Cable Replacement Profile (HCRP)	Profile
HCR_Print	0x1126	Hard Copy Cable Replacement Profile (HCRP)	Service Class
HCR_Scan	0x1127	Hard Copy Cable Replacement Profile (HCRP)	Service Class
SIM_Access	0x112D	SIM Access Profile	Service Class/Profile
Phonebook Access—Phone Book Client Equipment (PCE)	0x112E	Phone Book Access Profile (PBAP)	Service Class
Phonebook Access—Phone Book Server Equipment (PSE)	0x112F	Phone Book Access Profile (PBAP)	Service Class
Phonebook Access	0x1130	Phone Book Access Profile (PBAP)	Profile
Headset—HS	0x1131	Headset Profile (HSP)	Service Class
Message Access Server	0x1132	Message Access Profile (MAP)	Service Class
Message Notification Server	0x1133	Message Access Profile (MAP)	Service Class
Message Access Profile	0x1134	Message Access Profile (MAP)	Profile
PnPInformation	0x1200	Device Identification (DID)	Service Class/Profile
GenericNetworking	0x1201	N/A	Service Class
GenericFileTransfer	0x1202	N/A	Service Class
GenericAudio	0x1203	N/A	Service Class
GenericTelephony	0x1204	N/A	Service Class
VideoSource	0x1303	Video Distribution Profile (VDP)	Service Class
VideoSink	0x1304	Video Distribution Profile (VDP)	Service Class
VideoDistribution	0x1305	Video Distribution Profile (VDP)	Service Class
HDP	0x1400	Health Device Profile (HDP)	Profile
HDP Source	0x1401	Health Device Profile (HDP)	Service Class
HDP Sink	0x1402	Health Device Profile (HDP)	Service Class

User parameters are made available for Bluetooth device control. For instance, a user may define a device name for a particular device (such as a smartphone). Furthermore, when pairing, a user may define a six-digit code known as the passkey that provides a basic authentication mechanism between two devices as they proceed with the pairing process.

3.1.1 Bluetooth Low Energy

The Bluetooth Low Energy (LE) specification was defined in the Core 4.0 Version of the Bluetooth standard [3]. This specification defines physical and link layer characteristics for improved energy efficiency in Bluetooth devices. Unlike the BR/EDR specification, LE is more lightweight in nature—reducing power consumption substantially over a BR/EDR implementation. Furthermore, LE is not generally suited for what users may consider a typical Bluetooth application (voice, video, data). Rather, LE is suited for low-bandwidth, low-latency applications, such as periodic or triggered sensor transmissions.

Bluetooth LE is defined in the 2.4 GHz ISM band. Modulation in Bluetooth LE is GMSK, with a bandwidth-bit period of 0.5 and modulation index between 0.45 and 0.55, improving the range of operation over BR/EDR for equal transmit power. Output power ranges from 0.01 to 10 mW.

Channelization in Bluetooth LE is substantially different from BR/EDR. In the 2.4 GHz band, there are a total of 40 channels with 2 MHz spacing in between. Three channels are assigned as advertising channels, and there are 37 data channels. Table 3.7 summarizes the channelization of Bluetooth LE. Note that channel numbers

Table 3.7 Channelization in Bluetooth LE

Channel number	Frequency (MHz)	Channel number	Frequency (MHz)	Channel number	Frequency (MHz)
37	**2402**	12	2430	26	2458
0	2404	13	2432	27	2460
1	2406	14	2434	28	2462
2	2408	15	2436	29	2464
3	2410	16	2438	30	2466
4	2412	17	2440	31	2468
5	2414	18	2442	32	2470
6	2416	19	2444	33	2472
7	2418	20	2446	34	2474
8	2420	21	2448	35	2476
9	2422	22	2450	36	2478
10	2424	23	2452	**39**	**2480**
38	**2426**	24	2454		
11	2428	25	2456		

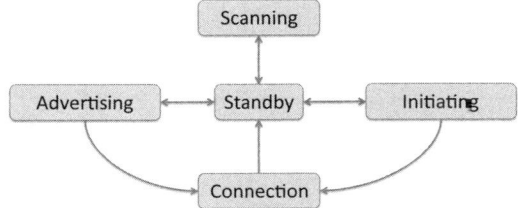

Figure 3.9 States for Bluetooth LE operation. Reproduced from Reference [3].

Preamble 8 bits	Access Address 32 bits	PDU 8–312 bits	CRC 24 bits

Figure 3.10 LE link layer packet format. Reproduced from Reference [3].

37, 38, and 39 correspond to the advertising channels (indicated in **bold**) while the numbers 0–36 correspond to the data channels (indicated in *italics*).

There are five states that characterize LE operation. These are shown in Figure 3.9 [3].

In Bluetooth LE, a device can only be in one state at a time. In the standby state, a device does not send or receive any packets. In the advertising state, a device transmits advertising channel packets and may listen for responses. In the scanning state, a device will listen for advertising channel packets. In the initiating state, the device will listen for advertising channel packets from a particular device and may respond to that device for the purpose of initiating a connection. In the connection state, a device actually maintains an active connection with another device. Like BR/EDR, LE allows for masters and slaves in the connection state.

Advertising channels are used in LE to discover devices, initiate connections, and broadcast data. Data channels are used for communications between connected devices. Unlike BR/EDR, LE does not frequency hop—only one physical channel is used at a given time. However, Access Addresses are used as identifiers at the beginning of every packet to mitigate interference between multiple devices using the same RF channel.

There is a single link layer packet format for LE, shown in Figure 3.10 [3]. The preamble bits for advertising channel packets is indicated in the figure as 0xAA, while the data channel can be either 0xAA if the least significant bit (LSB) of the access address is 0, or 0x55 if the LSB is 1. For all advertising packets, the access address is 0x8E89BED6. For data packets, the address is different for each connection between two nodes. The PDU contains the user data and the CRC is the cyclic redundancy check for data integrity.

Within the PDU field, there are separate definitions for advertising and data channel PDUs. Figure 3.11 [3] illustrates the advertising channel PDU format.

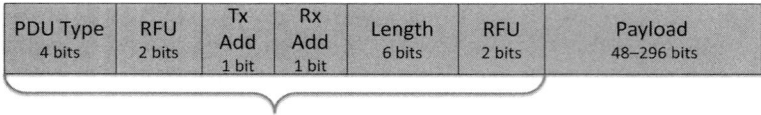

Header

Figure 3.11   Advertising PDU format. Reproduced from Reference [3].

Table 3.8   PDU Type Field for Advertising Channel Packet

PDU type value (binary)	Packet name and description
0000	ADV_IND. Connectable undirected advertising event.
0001	ADV_DIRECT_IND. Connectable directed advertising event.
0010	ADV_NONCONN_IND. Nonconnectable undirected advertising event.
0011	SCAN_REQ. Scanning request.
0100	SCAN_RSP. Scanning response.
0101	CONNECT_REQ. Connection request.
0110	ADV_SCAN_IND. Scannable undirected advertising event.
0111-1111	Reserved

Reproduced from Reference [3].

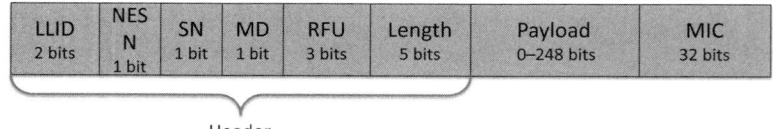

Header

Figure 3.12   Data PDU format. Reproduced from Reference [3].

As shown in the figure, the header consists of six fields. The PDU Type field indicates the type of advertising channel packet. Table 3.8 [3] summarizes these types.

The RFU fields are "Reserved for Future Use," indicating they have not been defined as of the Core 4.0 standard. The TxAdd and RxAdd fields are unique to each of the advertising channel packet types. The length field indicates the length of the payload in octets (6–37 octets or 48–296 bits).

The data channel PDU format is used for data packets. This format is shown in Figure 3.12 [3].

Table 3.9 [3] summarizes the fields in the Data PDU.

Figure 3.13 illustrates a connection initiation sequence from one device to another. As shown in the figure, Device 1 issues a command to initiate a connection to Device 2. The link layer for Device 1 receives an advertisement from Device 2,

Table 3.9   Data PDU Fields

Field name	Description
LLID	Indication of control or data packet type. Values (binary) shown are as follows: 01 = data PDU, L2CAP fragment 10 = data PDU, start of L2CAP message or complete message 11 = control PDU
NESN	Next expected sequence number
SN	Sequence number
MD	More data
Length	Indicates size of the payload + MIC, in octets
MIC	Message integrity check, used for encryption functions

Reproduced from Reference [3].

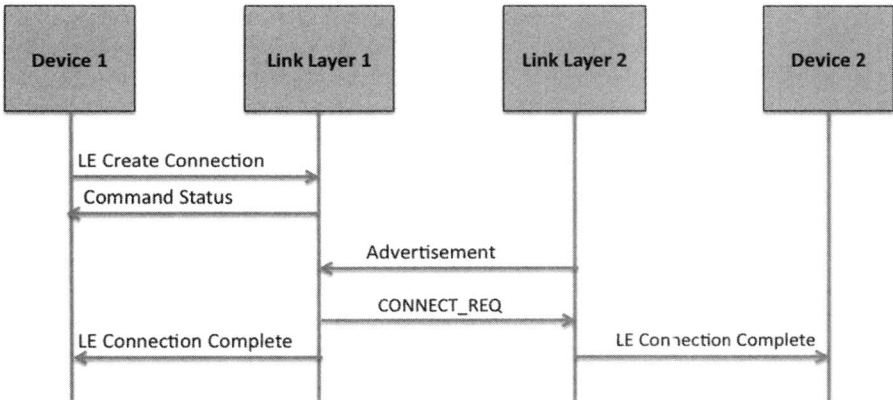

Figure 3.13   Connection initiation sequence. Reproduced from Reference [3].

and then responds with a CONNECT_REQ message. An LE Connection Complete response is issued to each device after this message is sent.

3.1.2   Examples of Bluetooth Technology

Perhaps the most common example of Bluetooth in operation today is the ubiquitous hands-free mode of operation employed by most mobile phones. In this scenario, a mobile phone acts as a Bluetooth core system, along with a hands-free device, such as a car stereo system with a Bluetooth core or a standalone headset. In this example, the device pair formed by the mobile phone and the hands-free device enables the user to perform actions on the mobile phone with their voice instead of using their hands to dial numbers or hold the phone. The hands-free device now handles the microphone and speaker functionality normally handled exclusively by the phone, with the Bluetooth link passing the data between the two devices for a seamless experience.

An additional but less common example for Bluetooth technology is wireless computer networking. In this example, consider two users in the same small room with two separate laptops. If both laptops are Bluetooth enabled, they may be able to pair and exchange files with each other using the Bluetooth waveform.

Bluetooth technology is increasingly used in gaming applications. For instance, the Nintendo Wii and Sony PlayStation 3 both use Bluetooth technology to connect the gaming console to the wireless controllers.

Bluetooth can also be employed as a local wireless bridge to tether a laptop to a mobile phone Internet connection. Assuming the mobile phone has a Bluetooth radio and tethering capabilities, a laptop may pair with the mobile phone to route its own IP traffic over the Bluetooth link, into the phone, and over the cellular network. In this case, the phone acts as a gateway between the laptop and the cellular network, using Bluetooth as the link to pass IP-based traffic between the two systems.

For Bluetooth LE, streaming data such as voice or video is not generally suitable. However, periodic sensor data applications may be more appropriate. Take, for example, a pedometer installed inside a shoe that is Bluetooth LE enabled. The pedometer sensor triggers an event when the pressure inside the shoe passes a certain threshold. In this sense, the pedometer can be used as a counter to determine the number of steps taken in a given timeframe. A runner who buys these shoes may be able to pair the sensor with a Bluetooth LE-enabled smartphone and count the number of steps taken during a run, for instance. The technology could be employed in a connected mode, with the pedometer sensor sending event updates via the data channel, or in a connectionless mode, using the advertising channel.

3.1.3 Bluetooth Performance

Back in 2006, the Bluetooth BR capability was assessed in Reference [63]. In particular, the authors tested several scenarios, each involving a file transfer of a fixed size ranging from 200 to 5000 kB. In each test, the authors separated two Bluetooth devices by a fixed range (2, 4, or 6 m), and also tested a case where the two devices were separated by two walls. The results of these tests are presented in Figure 3.14 [63]. It is relatively clear to see that the file sizes do not seem to impact the performance substantially from the 2 to the 6 m separation distance, showing achievable throughput on the order of 280–400 kbps. However, at low file sizes, the results suggest increased variance. For the two walls case, the throughput is noticeably reduced, ranging from 220 to 280 kbps. The results presented show that the original BR physical layer performs well for personal-area communications ranges, given applications that can reasonably utilize data rates ranging from 200 to 400 kbps.

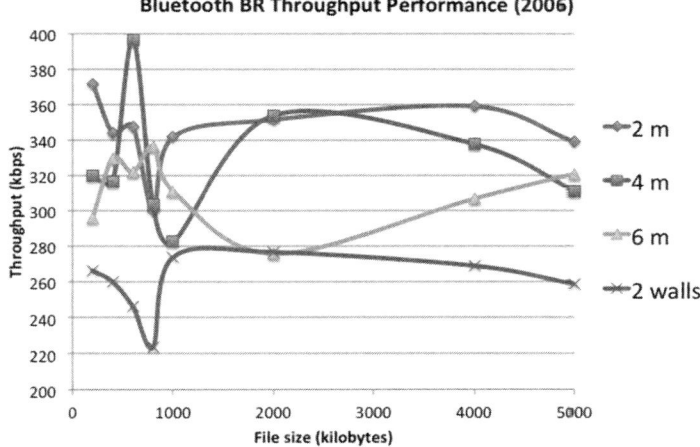

Figure 3.14   Bluetooth performance. Reproduced from Reference [63].

These rates are suitable for hands-free voice and low-rate Internet tethering (Web browsing/e-mail). While the maximum rate advertised in BR is 1 Mbps, it is clear that achievable user throughput rates are quite smaller, on the order of 20–40% of the maximum. However, it should be noted that the results presented were conducted with specific Bluetooth transceiver devices, and as such could differ with other device manufacturers and improved transceiver front ends developed since 2006.

3.2   ZIGBEE

3.2.1   IEEE 802.15.4

The IEEE 802.15.4-2006 standard [13] defines a MAC and PHY for Low-Rate Wireless Personal Area Networks (LR-WPANs). Such technologies are suited for data sensors, wireless automation, or other applications that do not require high data rates, such as high-quality video teleconferencing. Compare this to the IEEE 802.11 family of technologies that operate at much higher data rates and can readily support bandwidth-intensive applications. The standard is a revision to the IEEE 802.15.4-2003 standard, which is the primary technology employed in ZigBee products. The main objectives of IEEE 802.15.4 include:

- Easy user installation
- Reliable transfer of data
- Low cost
- Short range
- Low power consumption
- Simple, flexible protocol design and implementation

These objectives are essential to the commercial viability of IEEE 802.15.4-compliant products. WPAN technologies often exist on small devices that exhibit or desire these characteristics for an acceptable user experience.

IEEE 802.15.4 has defined several key technical characteristics:

- Wireless data rates ranging from 20 to 250 kbps
- Peer-to-peer or star topology operation
- Flexible 16- or 64-bit addressing schemes
- Optional support for allocating guaranteed time slots (GTS)
- Carrier sensing multiple access with collision avoidance (CSMA/CA) media access method
- Reliable protocol features, including full acknowledgments
- Low power consumption
- Energy level detection
- Link quality estimation and indication
- 16 channels defined in the 2.4 GHz band, 30 channels defined in the 915 MHz band, 3 channels defined in the 868 MHz band

Within this section, much of the content presented is based on or derived from the IEEE 802.15.4-2006 standard [13] or the ZigBee standard [64].

In IEEE 802.15.4, there are two types of devices: a full-function device (FFD) and a reduced-function device (RFD). In a WPAN, there must be at least one FFD. FFDs can operate in three modes, while RFDs can only operate in one mode. The FFD modes are as follows:

- *PAN coordinator*, which is responsible for initial PAN setup and principal coordinator in the network.
- *Coordinator*, a device capable of relaying messages.
- *Device*, which does not relay messages.

All FFDs can communicate with RFDs or other FFDs, but RFDs can only communicate with FFDs. Because of this definition, RFDs are considered simple devices. Examples of RFDs include passive sensors that do not send large volumes of data or coordinate network functions among multiple devices. Figure 3.15 [13] illustrates the types of devices and topologies considered.

In the star topology, all devices must communicate via the PAN coordinator FFD, and all star topologies operate independently from each other. However, in the peer-to-peer topology, FFDs can communicate directly to each other, while the RFD shown must exchange communications with the PAN coordinator only. While not shown, peer-to-peer PAN networks can scale to clusters; FFD gateways can be identified and assigned roles within the cluster of networks to pass data from one cluster (identified by a single FFD PAN coordinator and at least one other device) to another.

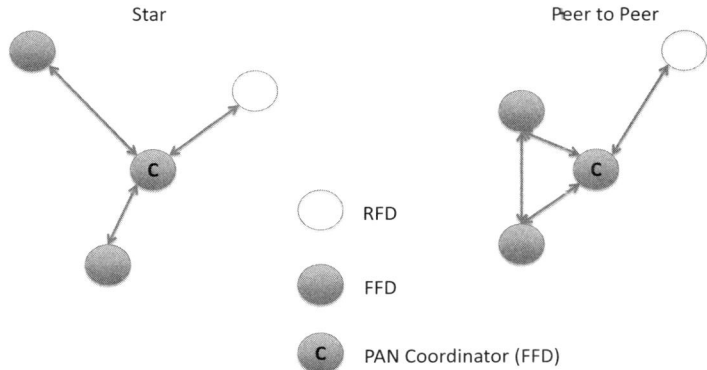

Figure 3.15 IEEE 802.15.4 topologies and device types. Reproduced from Reference [13].

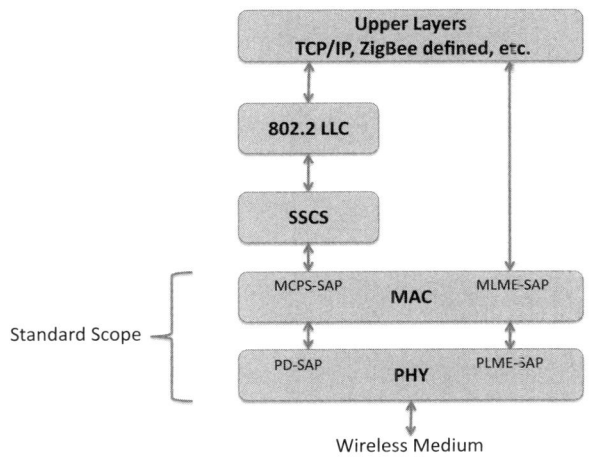

Figure 3.16 IEEE 802.15.4-2006 layered architecture. Reproduced from Reference [13].

The IEEE 802.15.4 standard specifies both the PHY and MAC layers of the LR-WPAN technology. Figure 3.16 [13] illustrates the layered architecture for this standard. It can be seen that the physical layer has two service access points (SAPs) that are used to interact with the local MAC layer and remote PHY/MAC layers: the PHY data service access point (PD-SAP) and the physical layer management entity SAP (PLME-SAP). The PD-SAP supports transport of data, via MAC protocol data units (MPDUs), while the PLME-SAP allows for management and coordination with the MAC layer. The service-specific convergence sublayer (SSCS) is defined above the MAC layer to handle specific services. For instance, the ZigBee Alliance defines particular SSCSs and network layers to support their desired applications.

IEEE 802.2 Logical Link Control (LLC) can be employed above the SSCS to provide a common interface to any higher layer protocols that may be implemented on a device.

The PHY definition includes the following functions: activation/deactivation of the RF transceiver, energy detection within a defined channel, link quality estimation for received packets, clear channel assessment for CSMA/CA, channel frequency selection, and data transmission/reception. Furthermore, IEEE 802.15.4-2006 defines four PHY schemes:

- Direct sequence spread spectrum (DSSS) with binary phase-shift keying (BPSK) modulation in the 868 MHz (20 kbps data rate) and 915 MHz (40 kbps data rate) frequency bands

- DSSS offset quadrature phase-shift keying (O-QPSK) modulation in the 868 MHz (100 kbps) and 915 MHz (250 kbps) frequency bands

- Parallel sequence spread spectrum (PSSS) BPSK and amplitude shift keying (ASK) modulation in the 868 and 915 MHz bands (250 kbps for both bands)

- DSSS O-QPSK modulation in the 2.4 GHz (250 kbps) band

The original IEEE 802.15.4-2003 specification defined a BPSK PHY in 868 and 915 MHz bands operating at 20 and 40 kbps, respectively, and an O-QPSK PHY in the 2450 MHz frequency band. These specifications form the basis for ZigBee-enabled devices today. Table 3.10 [13] summarizes the frequency bands and data rates for IEEE 802.15.4-2006.

Channel assignments in IEEE 802.15.4 are handled by the assignment of channel "pages," with each page having a particular definition for channel assignments. Channel pages are used because of the expected growth of frequency bands and allocations for these devices. Channel page 0 currently contains the highest number of channel assignments and is the most common. Channel page 0 defines $k_{max} = 27$ total channels in both bands, with 16 channels in the 2450 MHz band, 10 in 915 MHz, and 1 in 868 MHz. Center frequencies for channel page 0 are defined as follows:

Table 3.10   Frequency Bands and Data Rates

PHY (MHz)	Frequency band (MHz)	Chip rate (kchips/s)	Modulation	Bit rate (kbps)	Symbol rate (ksymbols/s)	Symbols
868/915	868–868.6	300	BPSK	20	20	Binary
	902–928	600	BPSK	40	40	Binary
868/915	868–868.6	400	ASK	250	12.5	20-bit PSSS
(optional)	902–928	1600	ASK	250	50	5-bit PSSS
868/915	868–868.6	400	O-QPSK	100	25	16-ary
(optional)	902–928	1000	O-QPSK	250	62.5	16-ary
2450	2400–2483.5	2000	O-QPSK	250	62.5	16-ary

Reproduced from Reference [13].

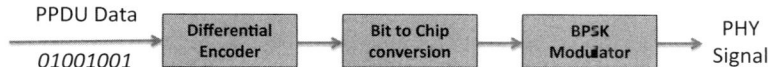

PPDU Data

01001001

PHY
Signal

Figure 3.17 A 868/915 MHz BPSK PHY block diagram. Reproduced from Reference [13].

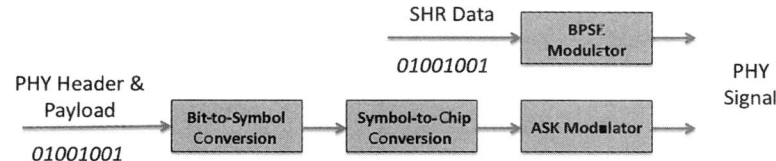

SHR Data

01001001

PHY Header &
Payload

01001001

PHY
Signal

Figure 3.18 A 868/915 MHz ASK PHY block diagram. Reproduced from Reference [13].

$$F_c = 868.3 \text{ MHz } (k = 0),$$
$$F_c = 906 + 2(k - 1) \text{ MHz } (k = 1 \text{ through } k = 10),$$
$$F_c = 2405 + 5(k - 11) \text{ MHz } (k = 11 \text{ through } k = 26).$$

Additional channel pages are defined in Reference [13], with an expectation that more will be defined as frequency assignments change.

The 868/915 MHz BPSK PHY uses a chip modulation sequence and differential bit encoding. This is accomplished by a three-stage operation shown in Figure 3.17 [13].

In the first stage, the differential encoder is an exclusive-OR operation of the current input bit with the immediately prior input bit. The bit-to-chip conversion provides a conversion of the bit to a DSSS chip at a much faster rate than the user data rate. In this case, the chip rate is 300 kchips/s for the 868 MHz band and 600 kchips/s for the 915 MHz band. The symbol rate is 20 and 40 ksymbols/s for the 868 and 915 MHz bands, respectively. The chip representation for input bit "0" is {111101011001000}, while the chip representation for input bit "1" is {000010100110111}, the 2's complement of the chip representing bit "0." A raised-cosine filter is applied at the BPSK modulator to shape the signal.

The 868/915 MHz ASK PHY uses a multicode modulation method known as parallel sequence spread spectrum (PSSS). During the transmission period for each data symbol, there are 20 or 5 information bits for the 868 or 915 MHz bands, respectively, each of which are modulated into 20 or 5 orthogonal pseudorandom (PN) sequences. Each of these PN sequences is linearly added: each constitutes a 32-chip symbol. The 32-chip symbol is equal to a multilevel 64-chip half-symbol for 868 MHz, and a full 32-chip symbol for 915 MHz. Figure 3.18 [13] illustrates the block diagram for this configuration.

As shown in the figure, the PHY header and payload bits are sent through a bit-to-symbol conversion, then a symbol-to-chip conversion, and finally to an ASK modulator. However, the synchronization header is sent through a BPSK modulator

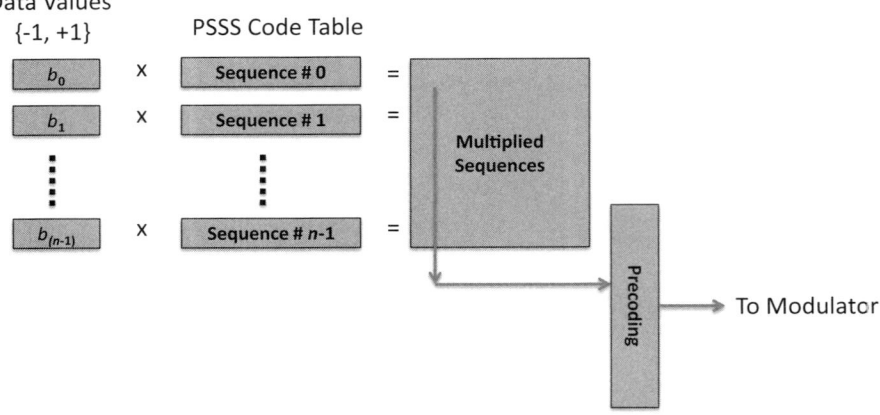

Figure 3.19 Symbol-to-chip mapping in 868/915 MHz ASK PHY. Reproduced from Reference [13].

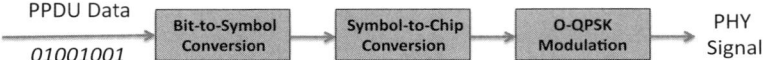

Figure 3.20 A 868/915 and 2450 MHz O-QPSK PHY block diagram. Reproduced from Reference [13].

separately. The symbol–to-chip mapping is a bit more complex in this configuration compared with the others. Figure 3.19 [13] illustrates the symbol-to-chip mapping.

In the figure, each bit (0/1) is converted into a data "value," with $\{-1\}$ corresponding to a 0 bit and $\{+1\}$ corresponding to a 1 bit. The corresponding data values are multiplied by set sequences. These multiplied sequences are then linearly added and a precoding operation is applied. The resulting chip sequences are modulated using ASK, with a square root-raised cosine pulse-shaping method. The ASK symbol rate is 12.5 ksymbol/s for 868 MHz and 50 ksymbol/s for 915 MHz. More detail on this method, including the actual sequence values and raised cosine filter equation, may be found in Reference [13].

The 868/915 MHz O-QPSK PHY configuration operates at a data rate of 100 kbps and uses a 16-ary modulation method. For each data symbol period, 4 information bits are utilized to choose 1 of 16 PN sequences to be transmitted. These PN sequences for additional successive symbols are concatenated together, and the aggregate resulting sequence is modulated using the O-QPSK method. Figure 3.20 [13] illustrates the block diagram for this PHY. It should be noted that the 2450 MHz O-QPSK PHY uses the same block diagram shown.

Table 3.11 provides a mapping between a data symbol of 4 bits and the associated chip values. These sequences are modulated on the O-QPSK carrier with a half-sine pulse-shaping method. Chips that have an even index value are modulated

Table 3.11 Symbol-to-Chip Mapping for 868/915 MHz O-QPSK PHY

Data symbol (binary) $(b_0\ b_1\ b_2\ b_3)$	Chip values $(c_0\ c_1, \ldots, c_{15})$
0000	0011111000100101
1000	0100111110001001
0100	0101001111100010
1100	1001010011111000
0010	0010010100111110
1010	1000100101001111
0110	1110001001010011
1110	1111100010010100
0001	0110101101110000
1001	0001101011011100
0101	0000011010110111
1101	1100000110101101
0011	0111000001101011
1011	1101110000011010
0111	1011011100000110
1111	1010110111000001

Reproduced from Reference [13].

on the in-phase (I) carrier, while those with odd index values are modulated on the quadrature (Q) carrier. The chip rate is 400 kchips/s for 868 MHz and 1 Mchips/s for 915 MHz. Q-phase chips are delayed by a single chip period compared to the I-phase chips.

The 2450 MHz O-QPSK PHY uses a similar 16-ary modulation method as compared with the 868/915 MHz O-QPSK PHY, except that the chips are 32 values long instead of 16. Table 3.12 [13] summarizes the symbol-to-chip mapping for the 2450 MHz O-QPSK PHY.

In the PHY, the primary unit that holds data is known as the PHY protocol data unit (PPDU). This packet consists of three basic components

- Synchronization header (SHR), which is used for receiver synchronization and lock
- PHY header (PHR), which contains frame length information
- Variable length payload, which contains other layer headers and user data

Figure 3.21 [13] illustrates the structure of the PPDU. As shown in the figure, the Preamble field is used for chip and symbol synchronization, and has a variable length of 3.75–5 octets depending on the frequency band and modulation used. The start-of-frame delimiter (SFD) field indicates the end of the SHR and the start of the packet data. This field also has a variable length between 0.625 and 2.5 octets. The payload is the PHY service data unit (PSDU), which contains the user data and

Table 3.12   Symbol-to-Chip Mapping for 2450 MHz O-QPSK PHY

Data symbol (binary) (b_0 b_1 b_2 b_3)	Chip values (c_0 c_1, . . . , c_{32})
0000	11011001110000110101001000101110
1000	11101101100111000011010100100010
0100	00101110110110011100001101010010
1100	00100010111011011001110000110101
0010	01010010001011011011001110000011
1010	00110101001000101110110110011100
0110	11000011010100100010111011011001
1110	10011100001101010010001011101101
0001	10001100100101100000011101111011
1001	10111000110010010110000001110111
0101	01111011100011001001011000000111
1101	01110111101110001100100101100000
0011	00000111011110111000110010010110
1011	01100000011101111011100011001001
0111	10010110000001110111110111000110
1111	11001001011000000111011110111000

Reproduced from Reference [13].

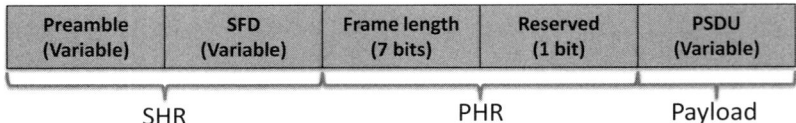

Figure **3.21**   PPDU structure. Reproduced from Reference [13].

the associated layer headers for the communications stack implemented on the device.

Table 3.13 [13] summarizes the preamble and SFD field lengths in octets and symbols as defined in the standard.

In the PHY layer, each SAP is specified via a set of primitives—these primitives are particular functions that are supported by the SAP. For instance, there are primitives defined for the PD-SAP to request transfer of MPDUs from the MAC to the PHY, confirm the end of a transfer of an MPDU from a local node (local PHY entity) to a remote node (peer PHY entity), or confirmation of a primitive request sent from the MAC to the PHY via the SAP. These primitives supply the basic functions for transferring user data between the MAC and PHY in a controlled way. The set of primitives in IEEE 802.15.4 are very small compared with other technologies (i.e., IEEE 802.16), which keeps the protocol complexity and implementation costs low. Table 3.14 [13] summarizes the PD-SAP primitives and their functions. Within each

Table 3.13   Preamble and SFD Field Lengths

PHY	Preamble length		SFD length	
868–868.6 MHz BPSK	4 octets	32 symbols	1 octet	8 symbols
902–928 MHz BPSK	4 octets	32 symbols	1 octet	8 symbols
868–868.6 MHz ASK	5 octets	2 symbols	2.5 octets	1 symbol
902–928 MHz ASK	3.75 octets	6 symbols	0.625 octets	1 symbol
868–868.6 MHz O-QPSK	4 octets	8 symbols	1 octet	2 symbols
902–928 MHz O-QPSK	4 octets	8 symbols	1 octet	2 symbols
2400–2483.5 MHz O-QPSK	4 octets	8 symbols	1 octet	2 symbols

Reproduced from Reference [13].

Table 3.14   PD-SAP Primitives and Functions

Primitive	Types/description
PD-DATA	*Used for passing MPDUs (data packets) between the local MAC layer entity and a remote PHY entity.*
	Request: MAC layer requests transmission of MPDU
	Confirm: Local PHY entity confirms end of transmission of MPDU to a peer PHY entity
	Indication: Indicates transfer of MPDU from the PHY to the local MAC sublayer

Reproduced from Reference [13].

entry for a primitive, a general description is provided in *italics*, and the types are indicated in ***bold italics***, along with a brief description of the functions of each type.

Table 3.15 [13] summarizes the PLME-SAP primitives and associated functions. Within each entry for a primitive, a general description is provided in *italics*, the types are indicated in ***bold italics***, along with a brief description of the functions of each type.

The MAC layer specified in IEEE 802.15.4-2006 is divided into two parts: the MAC common part sublayer (MCPS) and the MAC layer management entity (MLME). The following functions are part of the MAC layer:

- Coordinator generation of network beacons
- Synchronization to network beacons
- Support of PAN association and disassociation
- Security
- CSMA/CA channel access mechanism
- Handling/maintaining guaranteed time slot (GTS) availability to devices
- Maintaining a reliable link between peer devices (at the MAC layer)

Table 3.15   PLME-SAP Primitives and Functions

Primitive	Types/description
PLME-CCA	*Used to perform a clear channel assessment, which is a necessary function of CSMA/CA before transmission of a MPDU.* **Request**: Requests that the PLME conduct a CCA **Confirm**: Reports the results of the CCA
PLME-ED	*Utilized to perform an energy detection measurement. This measurement can be used to determine the level of energy in the channel at the moment of measurement, and is useful for determining whether or not to move to a different channel. For instance, if the energy detection measurement comes in above a certain threshold, the device may decide to choose another channel whose energy detection measurement is lower.* **Request**: Requests the PLME conduct an energy detection measurement **Confirm**: Reports the results of the energy detection measurement
PLME-GET	*Used to gather information from the PHY PAN Information Base (PIB) (Table 3.16). The PIB contains various metrics that can be used to determine the state of the network, similar to the management information base (MIB) concept used in routers and other network devices. This primitive is used by passing a particular PIB attribute as an input.* **Request**: Requests information about a particular PHY PIB attribute **Confirm**: Reports the results of the information request for a particular PHY PIB attribute
PLME-SET- TRX- STATE	*Handles transmitter and receiver states. Can turn on the transmitter function alone, receiver function alone, or disable both.* **Request**: Generated by the MLME and sent to the PLME when the transceiver operation state needs to be changed **Confirm**: Confirms the results of the request function
PLME-SET	*Used to set particular PHY PIB attributes. Inputs include the PHY PIB attribute and value to be set.* **Request**: Generated by MLME, sent to PLME to write a particular PHY PIB attribute value **Confirm**: Confirms that the request to write an attribute was successful or not

Adapted from Reference [13].

Figure 3.22 [13] illustrates the high level architecture for the MAC layer. Here, the MLME contains the MAC PIB, which is a set of attributes that can be accessed or set via the MLME through its two SAPs. There are two primary services provided by the MAC layer:

- Data service, which is accessed via the CPS and associated SAPs (MCPS, PD)
- Management, accessed via the MLME-SAP.

Table 3.16 PHY PIB Attributes

Attribute	Identifier	Type	Range	Description
phyCurrentChannel	0x00	Integer	0–26	The RF channel to use for all following transmissions and receptions.
phyChannelsSupported	0x01	Array	An $R \times$ 32-bit array $(1 \leq R \leq 32)$	The array is composed of R rows, each of which is a bit string with the following properties: The 5 most significant bits (MSBs) ($b27, \ldots, b31$) indicate the channel page, and the 27 least significant bits (LSBs) ($b0, b1, \ldots, b26$) indicate the status (1 = available, 0 = unavailable) for each of the up to 27 valid channels (bk shall indicate the status of channel k as in 6.1.2) supported by that channel page. The device only needs to add the rows (channel pages) for the PHY(s) it supports.
phyTransmitPower	0x02	Bitmap	0x00–0xbf	The 2 MSBs represent the tolerance on the transmit power: 00 = ±1 dB, 01 = ±3 dB, 10 = ±6 dB, and shall be read-only. The 6 LSBs, which may be written to represent a signed integer in twos-complement format, correspond to the nominal transmit power of the device in decibels relative to 1 mW. The lowest value of phyTransmitPower is interpreted as less than or equal to –32 dBm
phyCCAMode	0x03	Integer	1–3	The CCA mode.
phyCurrentPage	0x04	Integer	0 31	This is the current PHY channel page. This is used in conjunction with phyCurrentChannel to uniquely identify the channel currently being used.
phyMaxFrameDuration	0x05	Integer	55, 212, 266, 1064	The maximum number of symbols in a frame: = phySHRDuration + ceiling([aMaxPHYPacketSize + 1] × phySymbolsPerOctet)
phySHRDuration	0x06	Integer	3, 7, 10, 40	The duration of the synchronization header (SHR) in symbols for the current PHY.
phySymbolsPerOctet	0x07	Float	0.4, 1.6, 2, 8	The number of symbols per octet for the current PHY.

Reprinted from Reference [13].

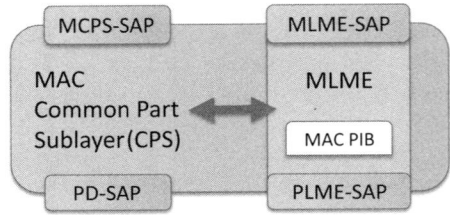

Figure 3.22   IEEE 802.15.4-2006 MAC layer architecture. Reproduced from Reference [13].

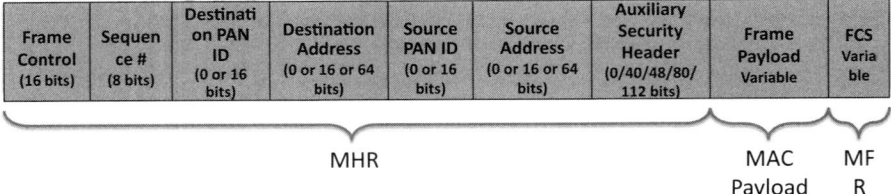

Figure 3.23   MAC frame format for IEEE 802.15.4-2006. Reproduced from Reference [13].

The PD-SAP and PLME-SAP interface to the PHY, while the MCPS- and MLME-SAP interface to a service-specific convergence sublayer (SSCS) such as that defined by the ZigBee specification [64].

At the SSCS, data units are passed in the form of SSCS protocol data units (SPDUs) between peer SSCS entities. It should be noted that these SPDUs are encapsulated in MAC service data units (MSDUs) that are handled by the MAC layer.

Figure 3.23 [13] illustrates the generalized MAC frame format. The MAC header (MHR) consists of the frame control field, sequence number, destination PAN ID, destination address, source PAN ID, source address, and auxiliary security header. The MAC payload contains the frame payload, which contains information from higher layers including the user data, and the MAC footer (MFR), which contains the frame check sequence.

- Frame control indicates the type of frame, addressing fields, and other control flags. There are four primary types of frames: beacon, data, acknowledgment, and MAC command. Not all frames will contain all the fields shown in Figure 3.22. More detail about the frame types can be found in Reference [13].

- The sequence number field indicates a sequence ID for each frame. For beacon frames, it specifies a beacon sequence number (BSN), while for data, acknowledgment, or MAC command frames, it specifies a data sequence number (DSN).

- When present, the destination PAN ID field indicates the unique ID of the PAN identifier for the intended receiver.

- When present, the destination address is the unique address ID for the receiver device.
- When present, the source PAN ID field specifies the unique ID of the PAN identifier for the transmitter.
- When present, the source address is the unique address ID for the transmitter device.
- When present, the Auxiliary Security Header field provides information to be processed by security mechanisms, including frame protection (security-level) information and which keying is used from the MAC security PIB.
- The Frame Payload field contains the information specific to the four types of frames. This could include user data, beacon identifier information, or MAC commands.
- The frame check sequence (FCS) field is 16 bits long and uses an ITU-T cyclic redundancy check (CRC) that is calculated over the MHR and MAC payload frame portions. It is used for error detection.

MAC Command frames provide the primary method for MAC control in IEEE 802.15.4. Specifically, there are nine MAC command frames:

- *Association Request.* This is generated to request association to a PAN via a PAN coordinator or a coordinator
- *Association Response.* This allows the PAN coordinator or a coordinator to respond to an association request
- *Disassociation Notification.* The PAN coordinator, or coordinator, or associated device can send disassociation notification to indicate device disassociation from a PAN
- *Data Request.* This is sent to the device to request data from a PAN coordinator or coordinator
- *PAN ID Conflict.* This is sent by the device to PAN coordinator when PAN ID is already used or conflicted
- *Orphan Notification.* This is used by the associated device that has lost synchronization with coordinator
- *Beacon Request.* This is used by the device to locate all coordinators within range during scanning process
- *Coordinator Realignment.* This is sent by the PAN coordinator or coordinator following receipt of orphan notification or when any PAN configuration attributes change due to MLME-START/Request primitive
- *Guaranteed Time-Slot (GTS) Request.* This is used by the associated device to request allocation of a new GTS or deallocation of existing GTS, sent to the PAN coordinator

For the data service, there are two primitives that can be used. These are summarized in Table 3.17 [13].

Table 3.17   MCPS Data Service Primitives [13]

Primitive	Types/description
MCPS-DATA	*Used for passing SPDUs (user data packets) between the local SSCS layer entity and a remote SSCS entity.* ***Request***: SSCS layer requests transmission of SPDU ***Confirm:*** Local MAC entity confirms results of the request to transfer an SPDU to a peer SSCS entity ***Indication***: Indicates transfer of MPDU from the MAC to the local SSCS sublayer
MCPS-PURGE	*Enables the next higher layer to purge a MAC Service Data Unit (MSDU) from transaction queue—optional for RFDs.* ***Request***: SSCS layer requests MSDU to be purged from queue ***Confirm***: MAC layer notifies the SSCS of request success

Adapted from Reference [13].

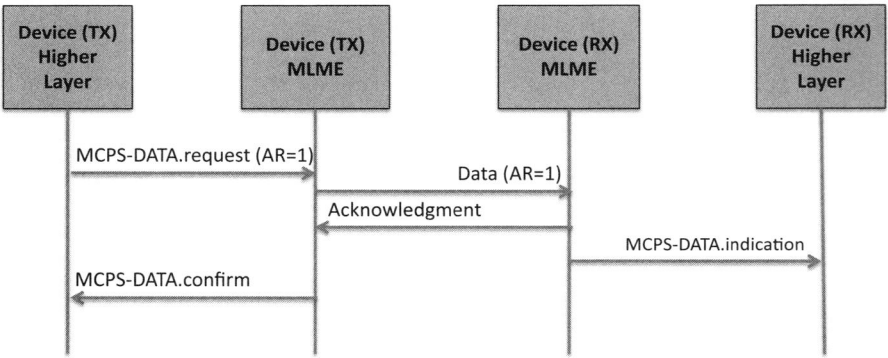

Figure 3.24   MAC data transfer sequence. Reproduced from Reference [13].

Within the MCPS, the major function is data transfer. Figure 3.24 [13] illustrates the data transfer process from a sender to a receiver. Here, the sending device forms an MCPS-DATA.request from its higher layer to the MAC, which then creates a data frame and passes this frame via its PHY to the recipient MAC. The receiving device MAC will send back an acknowledgment if requested, and pass on an MCPS-DATA.indication message to indicate that the MAC has passed an SPDU to the higher layer.

For the management service, there are 15 primitives. These are summarized in Table 3.18 [13]. Table 3.19 [13] summarizes the MAC PIB attributes that can be accessed to perform MAC functions.

Table 3.18 MLME Management Service Primitives

Primitive	Types/description
MLME-ASSOCIATE	*Used to facilitate device association (joining) to a particular PAN* ***Request***: Allows device to request association with coordinator ***Indication***: Indicates reception of association command ***Response***: Initiates response to MLME-ASSOCIATE.indication primitive ***Confirm:*** Informs higher layer of initiating device whether or not association request succeeded
MLME-DISASSOCIATE	*Used to facilitate device disassociation (leaving) of a PAN* ***Request***: Used by a device to notify coordinator of intent to leave PAN, or used by a coordinator to tell a device to leave the PAN ***Indication***: Indicates reception of disassociation command ***Confirm***: Reports result of disassociation request
MLME-BEACON-NOTIFY	***Indication***: Sends parameters contained in beacon frame received by MAC sublayer to the next higher layer, and includes a link quality indicator (LQI) and the beacon receive time
MLME-GET	*Used to request information on a particular MAC PIB attribute* ***Request***: Sends a request to obtain the set value of a particular PIB attribute ***Confirm***: Confirms the results of an information request from the PIB
MLME-GTS	*Handles GTS requests and maintenance.* ***Request***: Device can send this primitive to request the PAN coordinator allocate a new GTS, or deallocate an existing one. Also used by the PAN coordinator to initiate a GTS deallocation ***Confirm***: Reports results of request primitive ***Indication***: Indicates whether or not a GTS has been allocated or a prior GTS has been deallocated
MLME-ORPHAN	*Defines how a coordinator can send a notification indicating an orphaned device.* ***Indication***: Coordinator MLME uses to notify next higher layer of existence of an orphaned device ***Response***: Enables next higher layer of coordinator to respond to indication message
MLME-RESET	*Resets the MAC sublayer to its default values.* ***Request***: Enables next higher layer to request MLME reset the MAC layer ***Confirm***: Confirms the request
MLME-RX-ENABLE	*Handles device receiver control (enabled/disabled).* ***Request***: Enables next higher layer to request that the receiver is enabled for a fixed time frame or disabled ***Confirm***: Reports results of request

(Continued)

Table 3.18   (*Continued*)

Primitive	Types/description
MLME-SCAN	*Indicates how device can determine energy levels and PANs in particular channels.* **Request**: Initiates channel scan over a list of channels. These scans can be used to measure energy on the channel, search for associated coordinators, or look for coordinators sending beacon frames within range **Confirm**: Confirms the request
MLME-COMM-STATUS	*Indicates the communications status of the MLME.* **Indication**: Provides indication of the communication status of the MLME. Parameters that are reported include the PAN ID, addressing mode, and communication status
MLME-SET	*Handles writing particular values to PIB attributes.* **Request**: Attempt to write a value to a MAC PIB attribute **Confirm**: Reports result of request message
MLME-START	*Defines how an FFD requests a new superframe configuration to initiate a PAN, begin transmission of beacons on an existing PAN, or stop transmission of beacons.* **Request**: PAN coordinator uses this message to initiate a new PAN or use a new superframe configuration. Can also be generated by an associated device on the PAN to request a new superframe configuration. **Confirm**: Reports results of the request
MLME-SYNC	*Handles device synchronization with a coordinator.* **Request**: Requests device synchronization with a coordinator by acquiring and tracking beacons if specified.
MLME-SYNC-LOSS	*Handles loss of synchronization and notification to higher layer.* **Indication**: Indicates the loss of synchronization with coordinator
MLME-POLL	*Handles device request of data from a coordinator.* **Request**: Used by the device to request data from a coordinator. **Confirm**: Confirms the request

Adapted from Reference [13].

Figure 3.25 [13] illustrates the detailed association process for a new device to associate to a particular PAN in IEEE 802.15.4. Here, the device higher layer will send an association request to the device MLME, which then formulates an association request to send to the coordinator's MLME. Once the coordinator receives this request, it will send back an acknowledgment, and then an MLME.ASSOCIATE. indication message to its higher layer. The higher layer will respond with an MLME. ASSOCIATE.response, indicating it confirms the receipt of the indication message. After a set wait time, a data request is sent by the device MLME to the coordinator,

Table 3.19 MAC PIB Attributes

Attribute	Description
macAckWaitDuration	The maximum number of symbols to wait for an acknowledgment frame to arrive following a transmitted data frame.
macAssociatedPAN-Coord	Indication of whether the device is associated to the PAN through the PAN coordinator. A value of TRUE indicates the device has associated through the PAN coordinator. Otherwise, the value is set to FALSE.
macAssociation-Permit	Indication of whether a coordinator is currently allowing association. A value of TRUE indicates that association is permitted.
macAutoRequest	Indication of whether a device automatically sends a data request command if its address is listed in the beacon frame. A value of TRUE indicates that the data request command is automatically sent.
macBattLifeExt	Indication of whether BLE, through the reduction of coordinator receiver operation time during the CAP, is enabled. A value of TRUE indicates that it is enabled.
macBattLifeExtPeriods	In BLE mode, the number of backoff periods during which the receiver is enabled after the IFS following a beacon.
macBeaconPayload	The contents of the beacon payload.
macBeaconPayload-Length	The length, in octets, of the beacon payload.
macBeaconOrder	Specification of how often the coordinator transmits its beacon. If BO = 15, the coordinator will not transmit a periodic beacon.
macBeaconTxTime	The time that the device transmitted its last beacon frame, in symbol periods. The measurement shall be taken at the same symbol boundary within every transmitted beacon frame, the location of which is implementation specific.
macBSN	The sequence number added to the transmitted beacon frame.
macCoordExtended-Address	The 64-bit address of the coordinator through which the device is associated.
macCoordShort-Address	The 16-bit short address assigned to the coordinator through which the device is associated. A value of 0xfffe indicates that the coordinator is only using its 64-bit extended address. A value of 0xffff indicates that this value is unknown.
macDSN	The sequence number added to the transmitted data or MAC command frame.
macGTSPermit	TRUE if the PAN coordinator is to accept GTS requests. FALSE otherwise.
macMaxBE	The maximum value of the backoff exponent, BE, in the CSMA/CA algorithm.
macMaxCSMABackoffs	The maximum number of backoffs the CSMA/CA algorithm will attempt before declaring a channel access failure.

(Continued)

Table 3.19 *(Continued)*

Attribute	Description
macMaxFrameTotal-WaitTime	The maximum number of CAP symbols in a beacon-enabled PAN, or symbols in a non-beacon-enabled PAN, to wait either for a frame intended as a response to a data request frame or for a broadcast frame following a beacon with the Frame Pending subfield set to one.
macMaxFrameRetries	The maximum number of retries allowed after a transmission failure.
macMinBE	The minimum value of the backoff exponent (BE) in the CSMA-CA algorithm.
macMinLIFSPeriod	The minimum number of symbols forming a LIFS period.
macMinSIFSPeriod	The minimum number of symbols forming a SIFS period.
macPANId	The 16-bit identifier of the PAN on which the device is operating. If this value is 0xffff, the device is not associated.
macPromiscuous-Mode	Indication of whether the MAC sublayer is in a promiscuous (receive all) mode. A value of TRUE indicates that the MAC sublayer accepts all frames received from the PHY.
macResponseWaitTime	The maximum time, in multiples of aBaseSuperframeDuration, a device shall wait for a response command frame to be available following a request command frame.
macRxOnWhenIdle	Indication of whether the MAC sublayer is to enable its receiver during idle periods. For a beacon-enabled PAN, this attribute is relevant only during the CAP of the incoming superframe. For a non-beacon-enabled PAN, this attribute is relevant at all times.
macSecurityEnabled	Indication of whether the MAC sublayer has security enabled. A value of TRUE indicates that security is enabled, while a value of FALSE indicates that security is disabled.
macShortAddress	The 16-bit address that the device uses to communicate in the PAN. If the device is the PAN coordinator, this value shall be chosen before a PAN is started. Otherwise, the address is allocated by a coordinator during association.
	A value of 0xfffe indicates that the device has associated but has not been allocated an address. A value of 0xffff indicates that the device does not have a short address.
macSuperframe-Order	The length of the active portion of the outgoing superframe, including the beacon frame. If superframe order, SO, = 15, the superframe will not be active following the beacon.
macSyncSymbolOffset	The offset, measured in symbols, between the symbol boundary at which the MLME captures the timestamp of each transmitted or received frame, and the onset of the first symbol past the SFD, namely, the first symbol of the Length field.
macTimestamp-Supported	Indication of whether the MAC sublayer supports the optional timestamping feature for incoming and outgoing data frames.
macTransaction-PersistenceTime	The maximum time (in unit periods) that a transaction is stored by a coordinator and indicated in its beacon.

Reproduced from Reference [13].

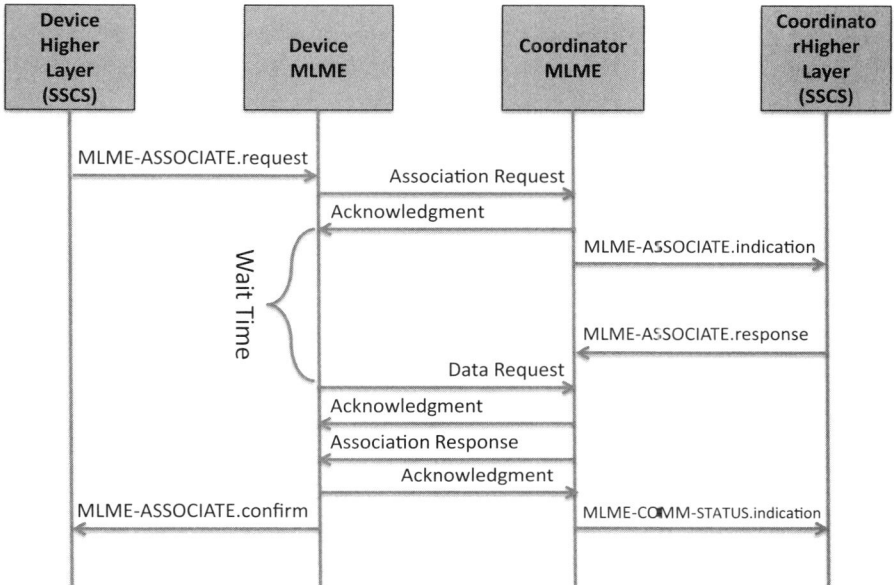

Figure 3.25 Association process. Reproduced from Reference [13].

and the coordinator MLME sends back an acknowledgment. After this, the coordinator MLME confirms association with an association response sent to the device MLME. The now-associated device MLME sends an acknowledgment of this response, while also passing an MLME-ASSOCIATE.confirm message to the higher layer of the device. The coordinator MLME finally sends an MLME-COMM-STATUS.indication message to the higher layer, informing the higher layer of the successful association for the new device. Figure 3.26 [13] illustrates the device-originated disassociation process.

In order for a PAN to form, an FFD must establish itself as the PAN coordinator and create a PAN. This process is summarized in Figure 3.27 [13]. Here, the PHY, MAC, and higher layer entities are shown to exchange messages and drive events to establish the PAN. An MLME-RESET.request message is generated and sent from the higher layer to the FFD MAC, which then sends a PLME-SET-TRX-STATE. request message to turn its receiver off. Once this is confirmed, the RESET message is confirmed to the higher layer, and a MLME-SET.request message with a PAN ID is sent from the higher layer to the MAC to establish the initial PAN ID. Once this message is confirmed, an MLME-SCAN.request is sent to scan potential channels and detect existing PANs. Once this scan is confirmed, another scan to perform energy detection is executed. It is at this point the higher layer will issue another MLME-SET.request with an addressing mechanism (short addressing is shown in the figure), and a logical channel is also decided based on the results of the scans. Once this set command is confirmed, then the MLME-START.request is issued to

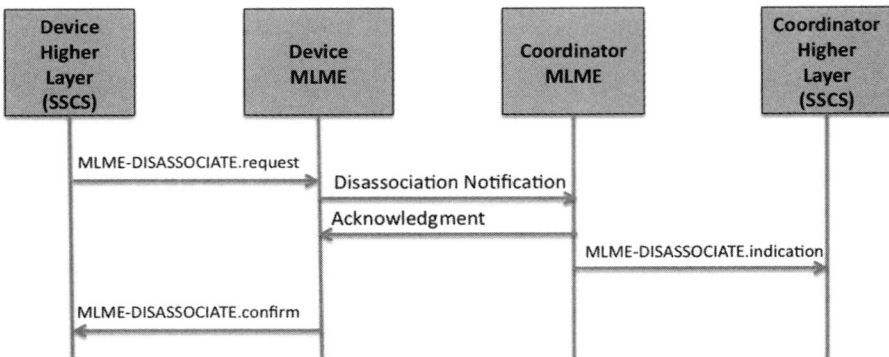

Figure 3.26   Device-initiated disassociation process. Reproduced from Reference [13].

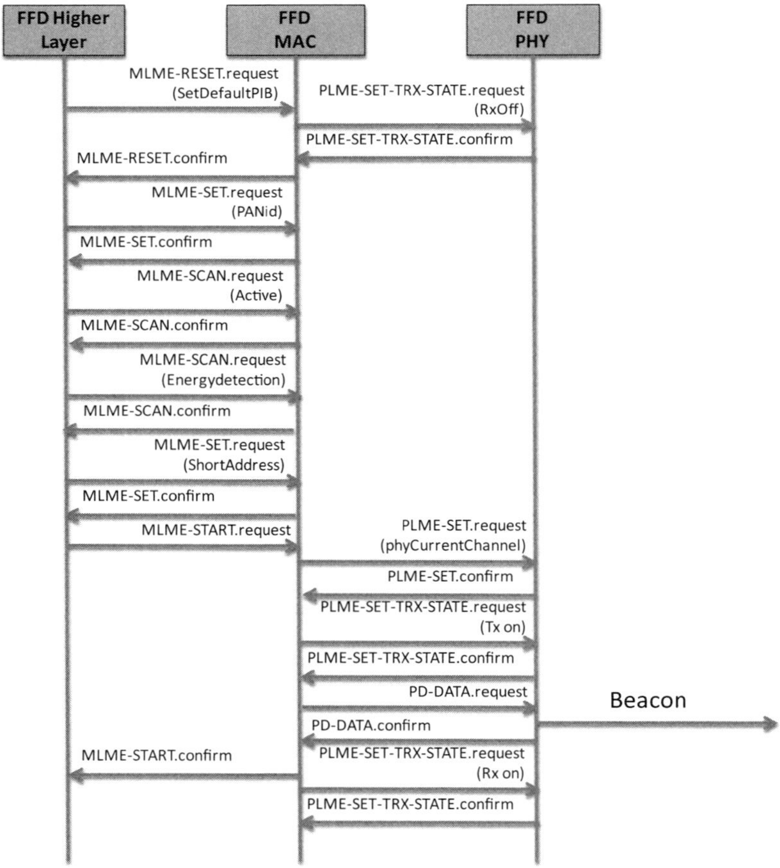

Figure 3.27   PAN formation message exchange for FFD. Reproduced from Reference [13].

start the PAN. The MAC sends a PLME-SET.request to the PHY to request a particular physical channel, at which point the PHY will respond with a PLME-SET. confirm message indicating success. Then, the MAC will send a PLME-SET-TRX-STATE.request message to turn the transmitter on, and will receive a confirmation. Then, PD-DATA.request and confirm messages are sent so that beacons can be transmitted. Additionally, an MLME-START.confirm message is passed from the MAC to the higher layer, and the receiver is turned on by the PLME-SET-TRX-STATE.request message and subsequently confirmed. It is at this point that the PAN is fully established and nodes can join and send/receive messages.

3.2.2   ZigBee Alliance and Specification

The ZigBee Alliance® was established in 2002 as a nonprofit association of business, academia, and government agencies with the aim of developing standards to "deliver greater freedom and flexibility for a smarter, more sustainable world" [16]. Specifically, the Alliance is focused on development of green, low-power, and open global low-power consumption wireless networking standards for sensor, control, and monitoring applications, while maintaining a high degree of simplicity and ease of use. There are multiple specifications and stacks defined by the ZigBee Alliance. ZigBee PRO uses the IEEE 802.15.4-2003 standard as the basis for the PHY and MAC layers, while defining a specification for layers above the MAC. ZigBee RF4CE uses the IEEE 802.15.4-2006 standard. In the remainder of this section, we'll discuss the ZigBee PRO specification [64] in greater detail to complete the discussion of IEEE 802.15.4. As of 2011, over 55% of IEEE 802.15.4-compliant devices in the market are ZigBee devices and growing (see the ZigBee Web site, http://www.ZigBee.org).

Figure 3.28 [64] illustrates the ZigBee layered architecture. Note that the PHY and MAC are defined by IEEE 802.15.4-2003, while the network layer and above are defined by the ZigBee specification. We'll discuss the Network (NWK) and Application (APL) layers here, and then provide a few examples that show how this technology is used today.

The NWK layer in ZigBee PRO is used to connect the IEEE 802.15.4 MAC to an associated APL layer. Just as with the IEEE 802.15.4 MAC and PHY, there are two primary service entities that exist at the NWK layer: one for passing data and the other for management. These are known as the network layer data entity (NLDE) and the network layer management entity (NLME). SAPs are defined to connect these entities to the MAC layer. The NLDE provides the following services:

- Generation of network-level PDUs (NPDUs)
- Routing based on network topology

The NLME provides the following services:

- New device configuration
- Network startup

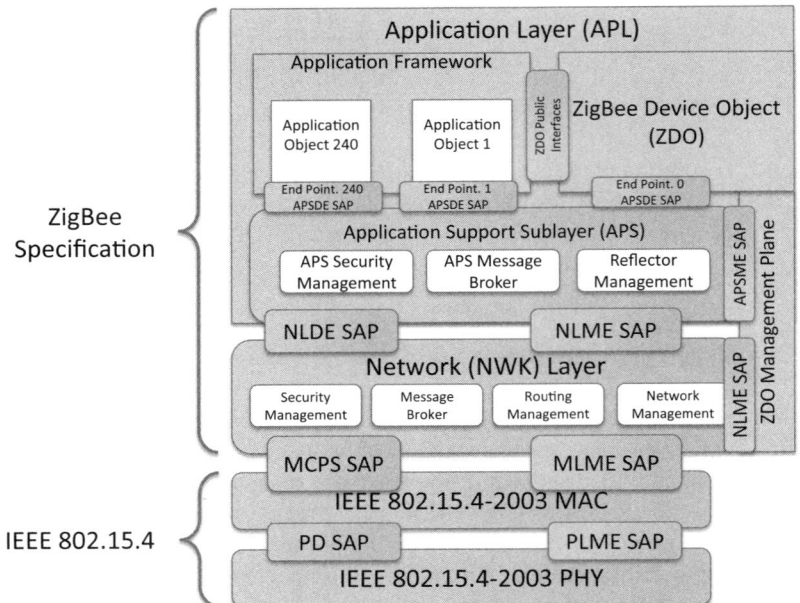

Figure 3.28   ZigBee layered architecture. Reproduced from Reference [64].

- Joining, leaving, and rejoining a network
- Addressing
- Neighbor discovery
- Route discovery
- Reception control
- Routing

The ZigBee specification discusses in detail the service primitives that are used for the NWK layer; these are similar in scope to the primitives for IEEE 802.15.4, with the added functions of network-level routing and characteristics. More detail can be found in Reference [64].

Table 3.20 [64] summarizes the NLDE primitives used for the NWK layer, while Table 3.21 [64] summarizes the NLME primitives. Table 3.22 [64] summarizes and provides a description of the NIB attributes.

Figure 3.29 [64] illustrates the NPDU frame format. In the figure, the Frame Control field indicates the type of frame. Frame types include Data and NWK Command, and each can be identified as a unicast, broadcast, multicast, or source-routed frame. Destination and source address fields contain the 16-bit network address (unicast or multicast). The radius field specifies the limit of range of transmission. The sequence number identifies the NPDU and increases by 1 for every additional

Table 3.20   NLDE Primitives for NWK Layer

Primitive	Types/description
NLDE-DATA	*Used for passing application protocol data units (APDUs) between peer application layer entities.* **Request**: Requests transfer of a data PDU from the application layer **Confirm**: Confirms the request **Indication**: Indicates transfer of network service data unit (NSDU) from the NWK to the local APS sublayer

NPDU generated. The destination and source IEEE address fields contain the 64-bit IEEE 802.15.4 MAC addresses, corresponding to a 16-bit network layer address. If the 16-bit destination network address is broadcast or multicast, then the destination IEEE address is not present. The multicast control field is 8 bits long when present, and handles multicast-related parameters including range, radius, and particular multicast modes. The source route subframe is variable in length and contains information related to relaying messages, including relay count and relay lists. Finally, the frame payload contains the user data or NWK command information.

The message sequence for creating a network is shown in Figure 3.30 [64]. Here, we see the APL makes a network formation request that starts the process off at the NWK and MAC layers to create the network and execute the required functions to do so. Note the MAC layer primitives that are invoked.

The message sequence for joining a network using the association process is shown in Figure 3.31 [64]. Here, the NLME-NETWORK-DISCOVERY primitive is used to start the joining process. The NWK layer then passes an MLME-SCAN request to instruct the MAC layer to search for available networks within range. As beacons are received that advertise the networks available, the results of the request are sent back to the APL. The APL then sends a NLME-JOIN request to the NWK layer, which then formulates an MLME-ASSOCIATE request to the MAC layer to request association with a particular network. Once this NLME-JOIN request is confirmed, if the device is a router or coordinator, it can request to start the routing functions.

A major function provided by the ZigBee NWK layer is routing. Per the specification, the routing algorithm is based on a link cost metric. This cost function is defined by Equation (3.5) from Reference [64]:

$$C = \left\{ \begin{matrix} 7, \\ \min\left[7, \text{round}\left(\dfrac{1}{p^4} \right) \right] \end{matrix} \right\} , \tag{3.5}$$

where C is the cost of a single link. This can be defined as a fixed cost of 7, or the minimum value of 7 or the rounded value of the fourth power of the inverse

Table 3.21   NMDE Primitives for NWK Layer

Primitive	Types/description
NLME-NETWORK-DISCOVERY	*Enables network layer to find other networks in operation within range.* *Request*: Requests network layer to discover networks within range *Confirm:* Reports the results of the discovery operation
NLME-NETWORK-FORMATION	*Handles network creation functions.* *Request*: Enables next higher layer to request device start new ZigBee network with itself as coordinator *Confirm*: Reports results of request to form network
NLME-PERMIT-JOINING	*Configures coordinator device to accept other devices joining the network.* *Request*: Enables next higher layer to set MAC layer association permit flag for a fixed period, when devices may be accepted into the network *Confirm*: Informs next higher layer of results to request device permissions to join network
NLME-START-ROUTER	*Handles starting of routing functions on the network.* *Request*: Enables next higher layer to request starting of routing functions, including data frame routing, route discovery, and accepting join requests from other devices *Confirm*: Reports the result of this request
NLME-ED-SCAN	*Handles energy scan requests to evaluate channel conditions* *Request*: Enables next higher layer to request energy scan of channels *Confirm*: Provides the results of the energy scan
NLME-JOIN	*Handles network join functions and channel operations.* *Request*: Enables next higher layer to request to join/rejoin network, or change the channel of operation. *Indication*: Enables next higher layer to be notified when a new device joins or rejoins the network *Confirm*: Enables next higher layer to be notified of results for join request.
NLME-DIRECT-JOIN	*Optional; enables the coordinator or router device to force another device to join its network.* *Request*: Enables the next higher layer to request another device join its network *Confirm*: Enables next higher layer to be notified of success of request

Table 3.21   (*Continued*)

Primitive	Types/description
NLME-LEAVE	*Handles device functions associated with leaving a network.*
	Request: Enables next higher layer to request a device (local or remote) leave the network
	Indication: Enables next higher layer notification of local or remote device leaving the network
	Confirm: Enables next higher layer to be notified of the results of the request to leave the network
NLME-RESET	*Handles network layer reset functions.*
	Request: Enables next higher layer to request NWK layer reset itself
	Confirm: Confirms the results of the request
NLME-SYNC	*Handles synchronization or extraction of data from coordinator/ router.*
	Request: Enables next higher layer to request data or synchronization from a coordinator/router
	Confirm: Confirms the results of the request
NLME-SYNC-LOSS	*Manages synchronization loss events.*
	Indication: Enables next higher layer to be notified of loss of synchronization at the MAC layer
NLME-GET	*Handles obtaining network information base (NIB) attributes.*
	Request: Enables next higher layer the ability to read a value from a particular NIB attribute
	Confirm: Reports the results of the request
NLME-SET	*Handles setting NIB attributes.*
	Request: Enables next higher layer the ability to set a particular value for a NIB attribute
	Confirm: Reports the results of the request
NLME-NWK-STATUS	*Handles notification of network failures.*
	Indication: Enables the next higher layer to be notified of a network failure.
NLME-ROUTE-DISCOVERY	*Handles route discovery functions.*
	Request: Enables the next higher layer to start route discovery functions.
	Confirm: Reports the results of the request

probability of packet delivery, p, on the link. It is up to the implementer to decide which of these methods to use—and if using the latter choice, they must estimate and supply the p value for the link. This can be estimated using the link quality indicator (LQI) functions. Routing across the network is achieved by determining the lowest cost path.

Table 3.22 NIB Attributes

Attribute	Description
nwkSequenceNumber	A sequence number used to identify outgoing frames.
nwkPassiveAckTimeout	The maximum time duration in milliseconds allowed for the parent and all child devices to retransmit a broadcast message (passive acknowledgment timeout).
nwkMaxBroadcastRetries	The maximum number of retries allowed after a broadcast transmission failure.
nwkMaxChildren	The number of children a device is allowed to have on its current network. Note that when nwkAddrAlloc has a value of 0x02, indicating stochastic addressing, the value of this attribute is implementation dependent.
nwkMaxDepth	The depth a device can have.
nwkMaxRouters	The number of routers any one device is allowed to have as children. This value is determined by the ZigBee coordinator for all devices in the network. If nwkAddrAlloc is 0x02, this value not used.
nwkNeighborTable	The current set of neighbor table entries in the device.
nwkNetworkBroadcastDeliveryTime	Time duration in seconds that a broadcast message needs to encompass the entire network. This is a calculated quantity based on other NIB attributes.
nwkReportConstantCost	If this is set to 0, the NWK layer shall calculate link cost from all neighbor nodes using the LQI values reported by the MAC layer; otherwise, it shall report a constant value.
nwkRouteDiscoveryRetriesPermitted	The number of retries allowed after an unsuccessful route request.
nwkRouteTable	The current set of routing table entries in the device.
nwkSymLink	The current route symmetry setting: TRUE means that routes are considered to be comprised of symmetric links. Backward and forward routes are created during one-route discovery and they are identical. FALSE indicates that routes are not considered to be comprised of symmetric links. Only the forward route is stored during route discovery.
nwkCapabilityInformation	This field shall contain the device capability information established at network joining time.

nwkAddrAlloc

A value that determines the method used to assign addresses:

0x00 = use distributed address allocation

0x01 = reserved

0x02 = use stochastic address allocation

nwkUseTreeRouting

A flag that determines whether the NWK layer should assume the ability to use hierarchical routing:

TRUE = assume the ability to use hierarchical routing.

FALSE = never use hierarchical routing.

nwkManagerAddr

The address of the designated network channel manager function.

nwkMaxSourceRoute

The maximum number of hops in a source route.

nwkUpdateId

The value identifying a snapshot of the network settings with which this node is operating with.

nwkTransactionPersistenceTime

The maximum time (in superframe periods) that a transaction is stored by a coordinator and indicated in its beacon. This attribute reflects the value of the MAC PIB attribute macTransactionPersistenceTime and any changes made by the higher layer will be reflected in the MAC PIB attribute value as well.

nwkNetworkAddress

The 16-bit address that the device uses to communicate with the PAN. This attribute reflects the value of the MAC PIB attribute macShortAddress and any changes made by the higher layer will be reflected in the MAC PIB attribute value as well.

nwkStackProfile

The identifier of the ZigBee stack profile in use for this device.

nwkBroadcastTransactionTable

The current set of broadcast transaction table entries in the device.

nwkGroupIDTable

The set of group identifiers, in the range 0x0000–0xffff, for groups of which this device is a member.

nwkExtendedPANID

The Extended PAN Identifier for the PAN of which the device is a member. The value 0x0000000000000000 means the Extended PAN Identifier is unknown.

nwkUseMulticast

A flag determining the layer where multicast messaging occurs.

TRUE = multicast occurs at the network layer.

FALSE = multicast occurs at the APS layer and using the APS header.

nwkRouteRecordTable

The route record table.

nwkIsConcentrator

A flag determining if this device is a concentrator.

TRUE = Device is a concentrator.

FALSE = Device is not a concentrator.

(Continued)

Table 3.22 (*Continued*)

Attribute	Description
nwkConcentratorRadius	The hop count radius for concentrator route discoveries.
nwkConcentratorDiscoveryTime	The time in seconds between concentrator route discoveries. If set to 0x0000, the discoveries are done at startup and by the next higher layer only.
nwkSecurityLevel	Indicates security level.
nwkSecurityMaterialSet	Contains network keying material.
nwkActiveKeySeqNumber	Frame counter identifier associated with a particular security key.
nwkAllFresh	Indicates whether or not NWK frames should be checked to make sure they are indeed new frames.
nwkSecureAllFrames	Indicates whether or not frames should be secured.
nwkLinkStatusPeriod	The time in seconds between link status command frames.
nwkRouterAgeLimit	The number of missed link status command frames before resetting the link costs to zero.
nwkUniqueAddr	A flag that determines whether the NWK layer should detect and correct conflicting addresses: TRUE = assume addresses are unique. FALSE = addresses may not be unique.
nwkAddressMap	The current set of 64-bit IEEE to 16-bit network address map.
nwkTimeStamp	A flag that determines if a time stamp indication is provided on incoming and outgoing packets. TRUE = time indication provided. FALSE = no time indication provided.
nwkPANId	This NIB attribute should, at all times, have the same value as macPANId.
nwkTxTotal	A count of unicast transmissions made by the NWK layer on this device. Each time the NWK layer transmits a unicast frame, by invoking the MCPS-DATA.request primitive of the MAC sublayer, it shall increment this counter. When either the NHL performs an NLME- SET.request on this attribute or if the value of nwkTxTotal rolls over past 0xffff the NWK layer shall reset to 0x00 each Transmit Failure field contained in the neighbor table.

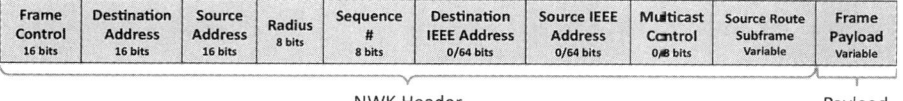

| Frame Control 16 bits | Destination Address 16 bits | Source Address 16 bits | Radius 8 bits | Sequence # 8 bits | Destination IEEE Address 0/64 bits | Source IEEE Address 0/64 bits | Multicast Control 0/8 bits | Source Route Subframe Variable | Frame Payload Variable |

NWK Header   Payload

Figure 3.29   NPDU frame format. Reproduced from Reference [64].

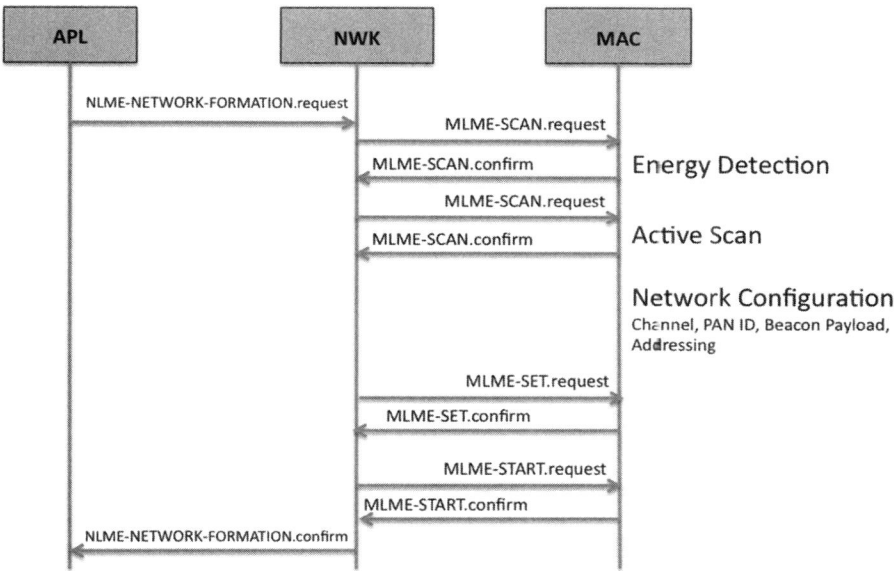

Figure 3.30   Creation of ZigBee network. Reproduced from Reference [64].

The APL layer defined in the ZigBee specification consists of the application support sublayer (APS), and the ZigBee device object (ZDO). Manufacturer-defined application objects are also present at this layer to provide application-specific functionality using ZigBee. There are two services supported at this layer: data and management. Like the NWK layer defined in ZigBee and PHY/MAC defined in IEEE 802.15.4, there are two service entities: the application layer service data entity (APSDE) and the application layer service management data entity (ASMDE). The APS information base (AIB) is the set of layer attributes that can be accessed/controlled to support various functions at this layer. At this layer, application protocol data units (APDUs) are passed via the APSDE to the NWK layer and vice versa. The APSDE is responsible for generating APDUs, enabling data transfer between bound devices, filtering group addresses based on end point membership, handling reliable transport functions, rejecting duplicate frames, and fragmentation of frames longer than a single NWK layer frame. The APSME allows an application to interact

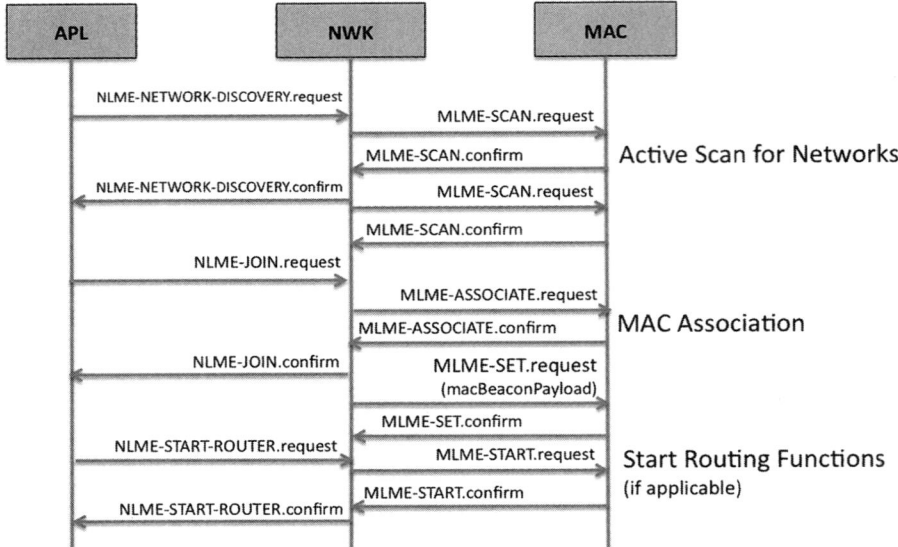

Figure 3.31 Joining a ZigBee network. Reproduced from Reference [64].

with the protocol stack. Functions that the APSME provide include binding management, which is the ability to match devices together based on services and needs. Also, the APSME provides AIB management, security, and group management functions.

Table 3.23 and Table 3.24 provide a list of the primitives associated with the APSDE and APSME. Table 3.25 [64] summarizes the AIB attributes that are available.

Figure 3.32 [64] illustrates the APDU frame format. Here, the frame control field defines the type of APDU—data, command, or acknowledgment. The destination end point field specifies the end point of the final destination for the frame, but it is not used for indirect or group-addressed frames. The group address field is only used for group-addressed frames, and contains the unique identifier for a particular group of devices. The cluster ID field specifies the identifier of the application profile-specific cluster for the frame, but is not present for command frames. The profile ID is present only for data or acknowledgment frames and is unique to the ZigBee profile and is used for filtering messages at each device that receives the frame. The source end point field specifies the initial originator of the frame. The APS counter is used to prevent receiving duplicate frames. The extended header can be used to handle fragmentation or extended frame control functions. Finally, the frame payload contains user data or other information specific to the three frame types.

The ZigBee specification provides for application profiles—these profiles are agreements for message formats, processing actions, and message types that help

Table 3.23 APSDE Primitives

Primitive	Types/description
APSDE-DATA	*Used for passing next higher layer entity (NHLE) PDU from the local NHLE to one or more peer NHLEs.* **Request**: Requests transfer of a NHLE from the next higher layer **Confirm**: Confirms the request **Indication**: Indicates transfer of data PDU (ASDU) from the APS to the local NHLE

Reproduced from Reference [64].

Table 3.24 APSME Primitives

Primitive	Types/description
APSME-BIND	*Handles binding functions for applications and peer devices.* **Request**: Enables next higher layer to request to bind two devices together, or optionally to bind a device to a group by creating an entry in its local binding table **Confirm**: Enables next higher layer to be notified of results of request
APSME-UNBIND	*Handles unbinding functions for applications and peer devices.* **Request**: Enables next higher layer to request to unbind two devices or remove a device from a particular group. **Confirm**: Enables next higher layer to be notified of results of the request
APSME-GET	*Handles reading AIB attributes.* **Request**: Enables next higher layer to read the value of an AIB attribute **Confirm**: Reports the results of the request
APSME-SET	*Handles writing AIB attributes.* **Request**: Enables next higher layer to write the value of an AIB attribute **Confirm**: Reports the results of the request
APSME-ADD-GROUP	*Handles group membership functions.* **Request**: Enables next higher layer to request an end point group membership assignment **Confirm**: Reports the results of the request
APSME-REMOVE-GROUP	*Handles group membership functions.* **Request**: Enables next higher layer to request an end point group membership removal **Confirm**: Reports the results of the request
APSME-REMOVE-ALL-GROUPS	*Handles group membership functions.* **Request**: Enables next higher layer to request all group membership assignments for a particular end point be removed **Confirm**: Reports the results of the request

Reproduced from Reference [64].

Table 3.25   AIB Attributes

Attribute	Description
apsBindingTable	The current set of binding table entries in the device.
apsDesignatedCoordinator	TRUE if the device should become the ZigBee Coordinator on startup, FALSE if otherwise.
apsChannelMask	The mask of allowable channels for this device to use for network operations.
apsUseExtendedPANID	The 64-bit address of a network to form or to join.
apsGroupTable	The current set of group table entries.
apsNonmemberRadius	The value to be used for the NonmemberRadius parameter when using NWK layer multicast.
apsPermissionsConfiguration	The current set of permission configuration items.
apsUseInsecureJoin	A flag controlling the use of insecure join at startup.
apsInterframeDelay	Fragmentation parameter—the standard delay, in milliseconds, between sending two blocks of a fragmented transmission.
apsLastChannelEnergy	The energy measurement for the channel energy scan performed on the previous channel just before a channel change.
apsLastChannelFailureRate	The latest percentage of transmission network transmission failures for the previous channel just before a channel change (in percentage of failed transmissions to the total number of transmissions attempted).
apsChannelTimer	A countdown timer (in hours) indicating the time to the next permitted frequency agility channel change. A value of NULL indicates the channel has not been changed previously.

Reproduced from Reference [64].

developers with creating interoperable, distributed applications. In this sense, the different vendors that develop devices that conform to specific application profiles can maintain some sense of interoperability, thus improving and diversifying the marketplace for ZigBee products. Some examples of profiles are as follows: home automation, device discovery, and voice calling. ZigBee defines profiles in two classes: manufacturer-specific and public. Device descriptions and cluster identifiers are both defined within a profile. The ZigBee Alliance is responsible for assigning particular profile IDs.

Similar to the port mechanism in TCP/UDP/IP, the application layer in ZigBee must differentiate between different application types. Particular application "flows" are identified in ZigBee as "clusters." These cluster identifiers are unique to a particular application profile.

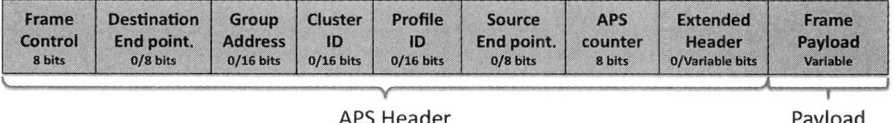

Frame Control 8 bits	Destination End point. 0/8 bits	Group Address 0/16 bits	Cluster ID 0/16 bits	Profile ID 0/16 bits	Source End point. 0/8 bits	APS counter 8 bits	Extended Header 0/Variable bits	Frame Payload Variable

APS Header   Payload

Figure 3.32   APDU frame format. Reproduced from Reference [64].

A single device may support many profiles, providing for subsets of cluster IDs defined within each profile, and can therefore support multiple device descriptions. There is a hierarchy of addressing within the device that enables this differentiation, based on both the unique IEEE MAC address and the NWK address, as well as an end point identifier. The end point value, as illustrated in Figure 3.32, describes different applications supported by a single radio. The end point value 0x00 is used to address the device profile, 0xff is used to address all active end points, and those in between are reserved. A single radio can support up to 240 applications on end points 0x01 through 0xf0. In the case of service discovery, profile ID, input cluster ID lists, and output cluster ID lists are used as the basis for service availability on a particular device.

Discovering services can be achieved via ZigBee descriptors. The node descriptor contains information about the ZigBee device capabilities, and it is mandatory for each node. The node power descriptor provides a dynamic indication of power status for each node. The simple descriptor contains information specific to each end point in a node, including application profile IDs. The complex descriptor contains extended information for each device description contained within a node, including manufacturer information. Additionally, user descriptors may be defined to provide other information. These descriptors are transmitted during service discovery periods.

The ZigBee device profile (ZDP) provides three interdevice communications functions:

- *Device and Service Discovery:* This pertains to the device determination of identities of other devices and associated services on the network.

- *Binding and Associated Functions:* This refers to creating, appending, and removing binding table entries that map control messages to intended destinations

- *Network Management:* This refers to the ability to obtain management information from devices including network discovery results, link quality to neighbors, routing tables, binding tables, discovery cache, and energy detection scan information, as well as providing ability of devices to leave or join networks, permit joining onto their own network if they are the coordinator, and provide fault notification.

The device profile uses a client/server methodology. Any device that performs discovery functions by sending device profile messages, for instance, is behaving as a client. Those devices that respond to these messages are behaving as servers.

Table 3.26   ZigBee Device Objects

Object Name	Description
:Device_and_Service_Discovery	Device and service discovery functions.
:Network_Manager	Network activities including network discovery, leaving and joining networks, resetting network connections and creation of a network.
:Binding_Manager	End device binding and associated activities.
:Security_Manager	Security services.
:Node_Manager	Management functions.
:Group_Manager	Group management functions.

Reproduced from Reference [64].

Device profile messages can include obtaining a network address, or service description, among other parameters.

The ZDP is used to implement the lower layer primitives to provide a particular application function for a ZigBee device. Table 3.26 [64] summarizes the ZDPs. ZDPs are responsible for the following:

- Initialization of the APS, NWK, and security service provider (SSP).

- Using configuration information from end applications to implement various application layer functions, including instantiation of ZigBee routers, coordinators, and devices.

The ZigBee Alliance Web site (http://www.ZigBee.org) contains a list of various ZigBee Certified products for different applications. Here, the following application standards are defined: health care, building/home automation, lighting, remote controls, and retail services.

For health care, the ZigBee Alliance has defined a ZigBee Health Care™ standard. According to Reference [16], this standard "offers a global standard for interoperable products enabling secure and reliable monitoring and management of non-critical, low-acuity healthcare services targeted at chronic disease, aging independence and general health, wellness and fitness."

For home automation, the ZigBee Alliance has defined a ZigBee Home Automation™ standard. Devices that support this profile include the following:

- "Smart" electrical outlets that can monitor energy loads and turn on/off power wirelessly

- Home automation controllers that can be used for energy management, security, and media management

- Small-size switches that can be installed behind a regular wall switch and wirelessly controlled to turn on/off the circuit driving the wall switch

- Door sensors that can wirelessly indicate to a central controller whether a door is open or closed

- "Smart" door locks that can be wirelessly accessed via a home controller connected to the Internet for remote locking/unlocking of doors
- Light on/off and dimmer switches that can be controlled locally or via wireless home automation controllers
- Home automation controllers with built-in wireless radios that can be interfaced to standard personal computers via Universal Serial Bus (USB)
- Connected thermostats controllable locally or wirelessly via home automation controllers

The ZigBee Remote Control™ standard is another profile that is supported by the ZigBee RF4CE standard. ZigBee RF4CE is simpler and lightweight compared to ZigBee PRO. In this application, a remote can be programmed to issue commands to a device such as a television or cable box wirelessly, using ZigBee Remote Control™. By using RF instead of infrared, the performance and feature sets of such remote control systems can be greatly enriched.

The ZigBee Alliance has also defined the ZigBee Telecom Services™ standard that aims to enable rich user services such as pay by mobile, creation of gaming networks, and gathering information about public spaces without GPS, ideal for places like subway tunnels or underground areas. Furthermore, the voice-over ZigBee cluster definition has enabled ZigBee devices to serve as headsets, speakers, and microphones.

3.2.3 802.15.4 Performance

In Reference [65], the author analyzed effective data rate performance parametrically, varying the number of devices and the data payload size. Figure 3.33 [65]

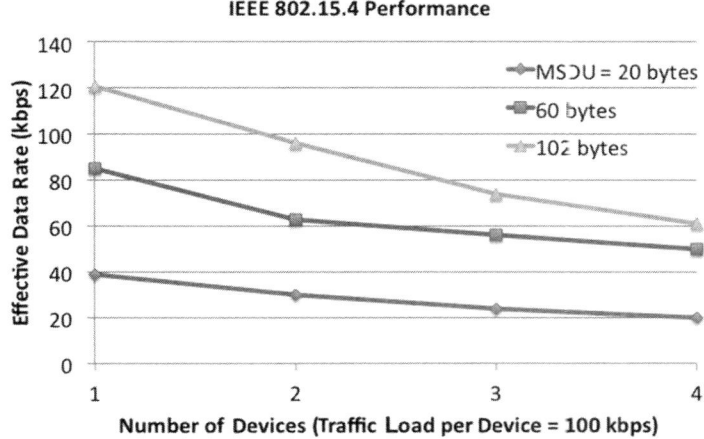

Figure 3.33   802.15.4 Performance per number of devices and MSDU size. Reproduced from Reference [65].

illustrates one of the findings from Reference [65]. Note here that as the number of devices increases, the effective data rate decreases as expected because of increased contention for the shared medium. Also, a larger MSDU size performs better than a smaller one, due to the reduced overhead (both headers and a smaller number of guard intervals per unit time). Also, the data rates achievable are suitable for low-rate applications such as home security sensors; high rate video would not perform well in this particular instance. This is reasonable in that 802.15.4 is advertised as a low-rate PHY/MAC—higher rates can be achieved through other protocol definitions (such as the WiMedia solution presented in Section 3.3).

3.3   ULTRA WIDEBAND

Ultra wideband (UWB) is a wireless technology that employs very short duration pulses of energy in time, resulting in very large bandwidth consumption in frequency. We'll discuss the concept and application of UWB in the context of personal area networks;, however, the technology has been employed in other applications, including radar and ranging. UWB signals occupy large segments of bandwidth, unlike traditional communications signals. For example, the IEEE 802.11g OFDM waveform occupies a 20 MHz channel, while a pulsed UWB signal may occupy a 4 GHz channel, a 200-fold increase in bandwidth. In general, UWB signals are considered to occupy a bandwidth of 500 MHz or greater.

In order to modulate UWB signals to carry information, the phase or the amplitude of the pulse can be set. Figure 3.34 illustrates a notional UWB pulse signal. Note in this example that the pulse time lasts on the order of single-digit nanoseconds. If we assume the case of 1 ns pulse length, in the frequency domain this would represent a bandwidth of about 1 GHz. Note that the pulse can be modulated in amplitude and in phase.

One characteristic of these short-duration pulses is the inherent resistance to multipath fading. In a continuous narrowband signal, constructive and destructive interference takes place at the receiver due to multipath. However, the duration of

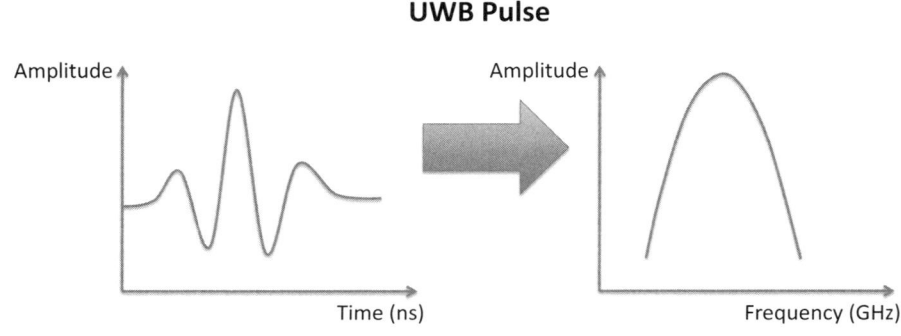

Figure 3.34   UWB signal.

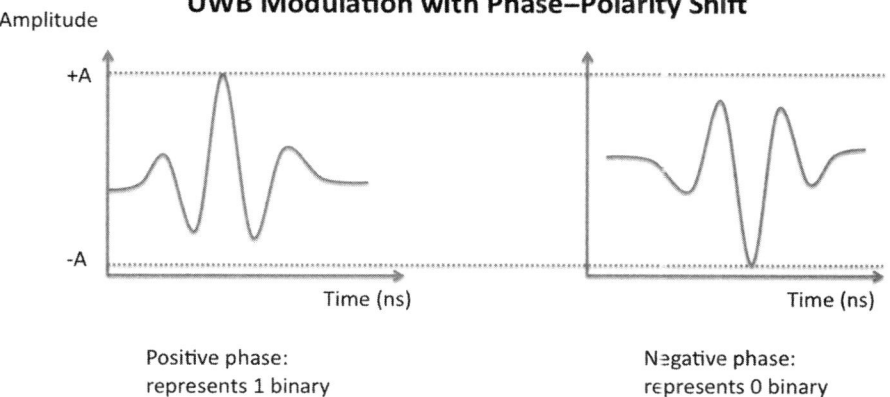

Positive phase:
represents 1 binary

Negative phase:
represents 0 binary

Figure 3.35   UWB modulation with phase/polarity shift.

UWB signals is so short in comparison that constructive and destructive interference is less of an issue. This is because any scattered and delayed energy due to multipath from the UWB signal has very little time to interfere with itself at the receiver because of the short duration.

One method for modulating UWB symbols is by shifting the polarity of the pulse. Figure 3.35 illustrates this concept.

As shown in the figure, a pulse that peaks at an amplitude of +A represents a "1" in binary, while a pulse whose polarity is reversed peaks at an amplitude of –A, representing a "0" in binary. Detection systems can be developed to determine the initial slope of the received pulse for deciding which pulse was transmitted. In this case, the positive phase symbol represents a positive slope, while the negative phase symbol represents a negative slope.

The Federal Communications Commission (FCC) has defined a UWB signal per FCC-02-48 [66] as one whose fractional bandwidth is greater than 0.25 or occupies at least 1.5 GHz of spectrum. The fractional bandwidth equation is as follows:

$$B_{fr} = \frac{2(f_h - f_l)}{f_h + f_l}, \tag{3.6}$$

where B_{fr} is the fractional bandwidth, f_h is the upper frequency limit of the signal at the –10 dB point relative to the maximum level, and f_l is the lower frequency limit of the signal at the –10 dB point relative to the maximum level. Figure 3.36 illustrates this more clearly.

For example, a UWB signal of 1 GHz, centered at 2 GHz with –10 dB points located at f_l and f_h, corresponds to $f_l = 1.5$ GHz and $f_h = 2.5$ GHz. Therefore, fractional bandwidth for this case is 1.0, corresponding to 100%. However, an 802.11g signal centered at 2.45 GHz and 20 MHz wide corresponds to $f_l = 2.44$ GHz and $f_h = 2.46$ GHz, resulting in a fractional bandwidth of 0.008, or 0.8%.

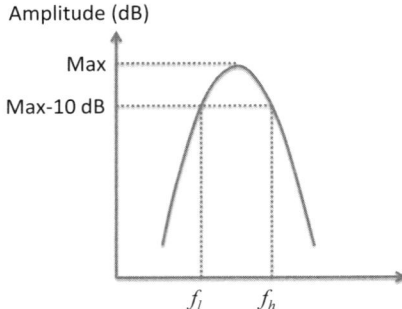

Figure 3.36   Fractional bandwidth definition.

To minimize interference, UWB signals are not permitted to exceed noise-like power densities in the bands they operate. This reduction in power density for UWB signals substantially limits their range on the order of 10 meters or so. Because of this limitation, UWB has generally been considered an ideal candidate for PAN applications.

3.3.1   UWB for Wireless Personal Area Networks

The original intention for UWB in PANs focused on short-range, very high data rate communications. Applications in this context include wireless video and peripheral connections to PCs. The WiMedia Alliance developed a standard as a potential solution for the IEEE 802.15.3a group, known as multiband OFDM [67] that spanned a large bandwidth. While it did not employ the pulse method presented in Figure 3.34, it used a large number of OFDM carriers spread across a wide bandwidth to achieve high throughput. Wireless USB technology is based on the WiMedia standards.

The history of UWB technology in the IEEE 802 forums is an interesting one. In the IEEE 802.15 PAN Working Group, UWB technology was pursued as an alternative high data rate physical layer in Task Group 3a, which was withdrawn in 2006. Two competing groups with different technical proposals were not able to come to agreement on a single proposal to move forward. However, in Task Group 4a, a continuous-phase UWB standard was developed as an alternative physical layer option for PANs.

3.3.1.1   UWB PHY in 802.15.4a

The UWB PHY defined in Reference [13] is based on burst position modulation (BPM) and BPSK. There are three frequency bands defined for operation:

- 249.6–749.6 MHz (single channel)
- 3.1–4.8 GHz (four channels)
- 6.0–10.6 GHz (11 channels)

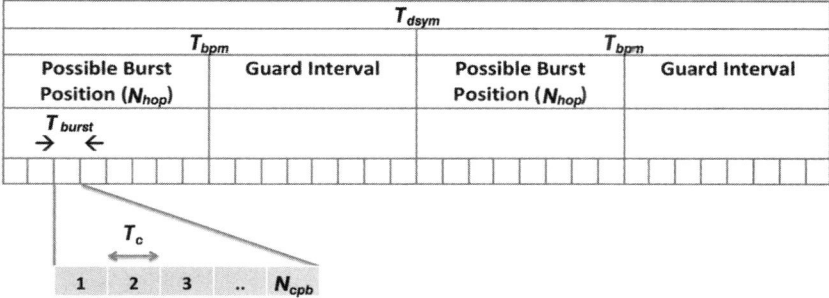

Figure 3.37   UWB symbol structure for IEEE 802.15.4a. Reproduced from Reference [14].

BPM is a modulation technique used in UWB to modulate the pulse to represent binary data. Bits are encoded based on where a burst of pulses is in time along a symbol period.

For the UWB definition in IEEE 802.15.4a, a symbol carries two bits of information: one is used to determine the position of a series of pulse bursts while another modulates the polarity of the burst. Figure 3.37 [14] illustrates the structure of a UWB symbol for IEEE 802.15.4a.

As shown in the figure, T_{bpm} is the length of time for a BPM portion of the symbol. The phase of the burst and the position of the burst, whether in the first or second T_{bpm} interval of the total symbol, are used to represent information bits. T_{dsym} is the total symbol duration, $N_{hop} = T_{dsym}/[4T_{burst}]$, used for hopping between positions for interference mitigation, and T_c is the length in time of each BPM chip. N_{cpb} is the number of chips within the burst.

3.3.1.2   UWB PHY in WiMedia PHY 1.5

The WiMedia PHY 1.5 specification [67] includes a description of the UWB waveform used for the Wireless USB standard. However, this UWB signal is not based on short duration pulses, but rather on OFDM. In this sense, a large number of OFDM carriers spread across a wide bandwidth can represent a signal that has the characteristics of UWB by the fractional bandwidth definition. The following equation (Eq. 3.7) summarizes the signal representation as per the standard:

$$S(t) = \text{Re}\left\{ \sum_{n=0}^{N_{\text{packet}}-1} s_n\left(t - nT_{\text{sym}}\right)e^{j2\pi f_c(q(n)*t)} \right\}, \tag{3.7}$$

where Re{ } is the function that represents the real part of the equation, N_{packet} is the number of symbols within a packet, $f_c(m)$ is the center frequency of the mth frequency band, $q(n)$ is the function that maps the nth symbol to a particular frequency band, and $s_n(t)$ is the complex baseband representation of the nth symbol.

Table 3.27   PHY Parameters for WiMedia PHY

Parameter	Value
N_{ST}: Number of total subcarriers (including 6 NULL subcarriers)	128
N_{SD}: Number of data subcarriers	100
N_{SDP}: Number of pilot subcarriers	12
N_{SG}: Number of guard subcarriers	10
Δ_F: Subcarrier frequency spacing (=528 MHz/128)	4.125 MHz
T_{FFT}: IFFT/FFT period (=$1/\Delta_F$)	242.42 ns
T_{ZP}: Zero suffix duration (=32/528 MHz)	60.61 ns
T_{GI}: Guard Interval duration (=5/528 MHz)	9.47 ns
T_{SYM}: Symbol period (=$T_{FFT}+T_{ZP}+T_{GI}$)	312.5 ns

Reproduced from Reference [67].

The WiMedia UWB OFDM PHY version 1.5 operates in the 3.1–10.6 GHz band. Center frequencies are defined for particular band numbers, as in Equation (3.8) per the standard:

$$f_c(n_b) = 2904 + 528 * n_b(\text{MHz}), n_b = 1, 2, 3, \ldots, 14. \qquad (3.8)$$

Each band is 528 MHz wide. Table 3.27 [67] summarizes the parameters of the WiMedia PHY.

3.3.1.3 Examples

The Time Domain Corporation has long worked in the UWB domain and has developed multiple products based on pulsed UWB technology. One of their products advertised today is the PulsON 400 Ranging and Communications Module (RCM). This module is a circuit board designed to integrate into a system and contains the necessary functions and interfaces for UWB ranging and communications applications.

Wireless USB based on the WiMedia standard has been used in multiple applications, including:

- Laptop-to-projector multimedia interfaces to project laptop screen content on larger screen without need for video cable

- Laptop-to-wireless docking station interface to use external keyboard, mouse, and monitor, among other peripherals, without having to connect cables to the laptop for each of the functions

- Audio streaming for devices to connect to wireless speakers or systems with speakers attached for the purposes of streaming audio without the need for cables

Another application for UWB includes High-Definition Multimedia Interface (HDMI) wireless connectivity for full HD (1080p) resolution. The Gefen Corporation has developed a Gefen Wireless for HDMI Extender device that enables up to three separate HDMI sources (e.g., DVD player, laptop, gaming console) to connect to a single HDMI input (e.g., large flat screen television). Given the limitations of the UWB power density levels, ranges can be achieved up to 30 feet with line of sight. This may be more than sufficient for home or conference room applications, however.

Chapter 4

Wireless Local Area Networks

WLAN technologies have truly changed the way in which we all live our lives. Prior to IEEE 802.11, we all lived in a wired world. IEEE 802.11 put wireless networks in virtually every home and office, and untethered the Internet experience. It was simple to install and use. It was affordable. It was arguably the first wireless technology truly developed for the masses. One could go even further and make a compelling argument that IEEE 802.11, more than any other technology, is what started the entire trend toward mobile data networking and ushered in the era of the wireless Internet. It was the technology that allowed us all to unhook our laptops and experience the Internet at our convenience. And we liked it! Once our laptops were unhooked, we then wanted the same experience with our handheld devices. WLAN technologies like IEEE 802.11 have also proven highly influential to other technologies. Over the years, IEEE 802.11 has often found itself as an early adopter of new technology, serving as a pioneer of innovation in the consumer wireless marketplace. The importance of IEEE 802.11, in terms of both its success and its influence, is undeniable.

It sometimes seems as if the world has become wireless very quickly, like we all went to sleep one night to awake to a brave new wireless world. However, the journey to today's wireless society has been a long, sometimes slow-moving process that dates back 40 years. Wireless data networking finds its origins in the early 1970s. In 1971, the University of Hawaii introduced the first wireless network of record. The ALOHANET research project interconnected computers on seven University of Hawaii campuses spread across four islands via wireless connectivity. However, wireless networking through much of the remainder of the 1970s, 1980s, and 1990s remained alive only because of the efforts of amateur radio hobbyists in the United States and Canada, who developed terminal node controllers (TNCs) that could interconnect various sites around the world.

A historic date was 1985, when the U.S. Federal Communications Commission (FCC) authorized the public use of the industrial, scientific, and medical (ISM) frequency bands. This led the way to an increased commercial interest in wireless

Wireless Networking: Understanding Internetworking Challenges, First Edition.
Jack L. Burbank, Julia Andrusenko, Jared S. Everett, and William T.M. Kasch.
© 2013 the Institute of Electrical and Electronics Engineers, Inc. Published 2013 by John Wiley & Sons, Inc.

networking, and in the late 1980s, the IEEE 802 Local Area Network (LAN) and Metropolitan Area Network (MAN) Working Group authorized a project for the development of a WLAN standard. The resulting 802.11 Working Group published the original IEEE 802.11 WLAN standard on November 18, 1997 and revised it in 1999 [1].

There have been numerous subsequent amendments to the original IEEE 802.11 standard, most notably the IEEE 802.11a [17], IEEE 802.11b [22], IEEE 802.11g [22], IEEE 802.11i [24], and IEEE 802.11n [27] technology standards, with many other already completed and ongoing 802.11 technologies expected (802.11e, 802.11s, 802.11r, and many more). A thorough listing of complete and ongoing IEEE 802.11 standardization efforts can be found in Table 1.5 and Table 1.6, respectively.

Concurrent to initial IEEE 802.11 standardization efforts, ETSI published several WLAN technologies such as HIPERLAN [68] and HIPERLAN/2 [69]. Moreover, there have been separate WLAN technology development efforts within China, resulting in the WAPI WLAN technology standard. However, IEEE 802.11, or Wi-Fi, as it has grown to be known, has become synonymous with wireless local area networking. It has certainly been the most successful of all WLAN technologies, with it hitting the milestone of 1 billion units shipped in 2012, making it only the third technology with that distinction, the other two being GSM and Bluetooth.

The remainder of this section provides an overview of the various IEEE 802.11 technologies that have been developed in largely chronological order of publication, with particular emphasis on those that have experienced significant deployment.

4.1 THE ORIGINAL 802.11 SPECIFICATION

The original IEEE 802.11 specification was published in 1997, but achieved only very modest commercial success. However, it did lay the groundwork for an astonishing next decade of growth in this area.

4.1.1 An 802.11 Wireless Network

An 802.11 network consists of four primary components: a distribution system, access points (APs), a wireless medium, and mobile stations (MSs). The distribution system is the component of 802.11 used to forward frames to their destination when several APs are connected together to form a large coverage area. The 802.11 standard does not specify the distribution system method, but such functionality is usually realized via a wired backbone network that connects the multiple APs. APs are the devices that bridge the wireless 802.11 network to the wired network, typically 802.3 (Ethernet). However, MSs are the users of the network. They generate and/or receive the data that are being transported across the wireless medium.

The basic service set (BSS) is the foundation of an 802.11 network. The BSS is a group of stations that communicate with one another. These communications take place in the basic service area (BSA). A station that is within the BSA can communicate with other members of the BSS. There are two types of BSSs:

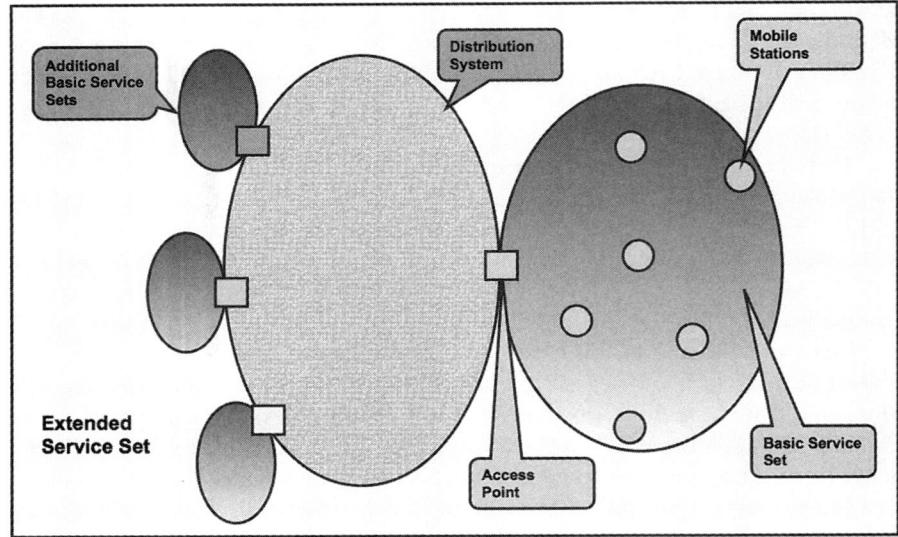

Figure 4.1   An IEEE 802.11 network (infrastructural view).

infrastructural and ad hoc. An ad hoc BSS, also known as an independent BSS (IBSS), is one in which stations communicate directly to one another. IBSSs are typically short lived in nature, and are thus referred to as "ad hoc." These are the least common type of 802.11 networks. An infrastructural BSS is one in which all communications take place through the AP within that BSS. This is the most common type of 802.11 network. Multiple BSSs can be interconnected into an extended service set (ESS). An ESS is formed by chaining BSSs together with a backbone network. The 802.11 does not specify the backbone network, but rather that this backbone network provide a certain set of services. It should be noted that from the perspective of the LLC sublayer, an ESS appears identical to a larger BSS (i.e., the concept of BSS vs. ESS is transparent to the higher LLC sublayer). The basic 802.11 network architecture is depicted in Figure 4.1, from an infrastructural mode perspective. Figure 4.2 depicts the basic 802.11 network when operating in an ad hoc manner.

An ESS or BSS is identified by its service set identity (SSID). The SSID is a 0-32 byte identifier that is typically assigned a human readable ASCII character string. As a result, it is alternatively known as the 802.11 network name.

The first thing an MS wishing to join an 802.11 network must do is to detect the presence of the network. There are two methods by which this can be accomplished: passive and active. In the passive case, the MS scans all frequency channels listening for the presence of network beacons, which are periodically transmitted by the stations of the network to announce their presence. These beacons contain required information about that network, such as its SSID. The station can then begin the authentication and association procedures required to join the network (see later

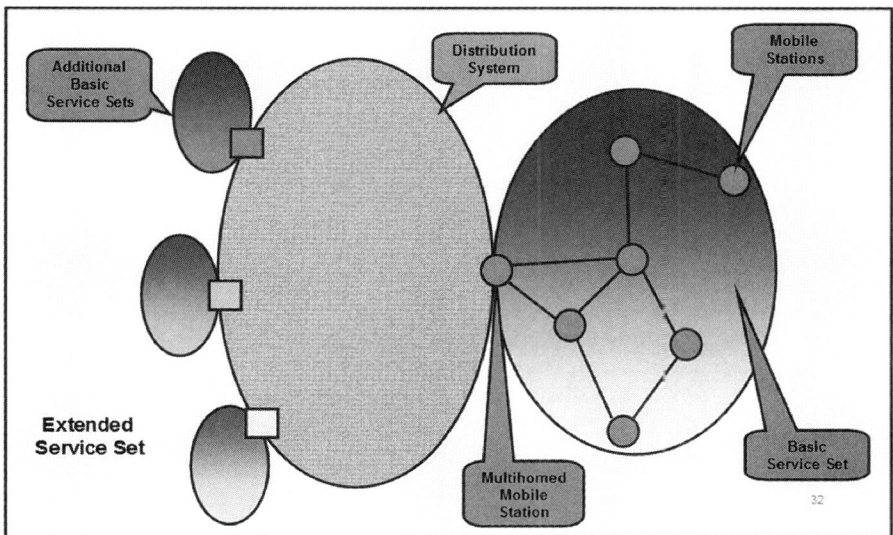

Figure 4.2   An IEEE 802.11 network (ad hoc view).

discussion on authentication and association). In the active case, the MS begins transmitting probes with the SSID of the network it wishes to join and then waits for a response to the probes. Upon receipt of a probe response, the MS can then begin joining the network. In fact, this active method is the method that is required if SSID broadcast is suppressed (see IEEE 802.11 security discussions).

4.1.2 The Original IEEE 802.11 Media Access Control

A common misperception is that the IEEE 802.11 MAC design is "Ethernet-over-the-air." While certain aspects of the IEEE 802.11 MAC are similar to the IEEE 802.3 LAN standard, the 802.11 MAC differs significantly from MAC designs of wired networks due to the inherently unique problems that must be overcome due to the wireless interface. There are four primary elements of the IEEE 802.11 MAC:

- Authentication and association procedures
- Channel access procedures
- Data link control (DLC)
- Framing mechanisms

4.1.2.1 Authentication and Association Procedures

Before an MS is allowed to utilize data services (i.e., join the network), the MS must perform authentication and association. This is illustrated in Figure 4.3. The state of

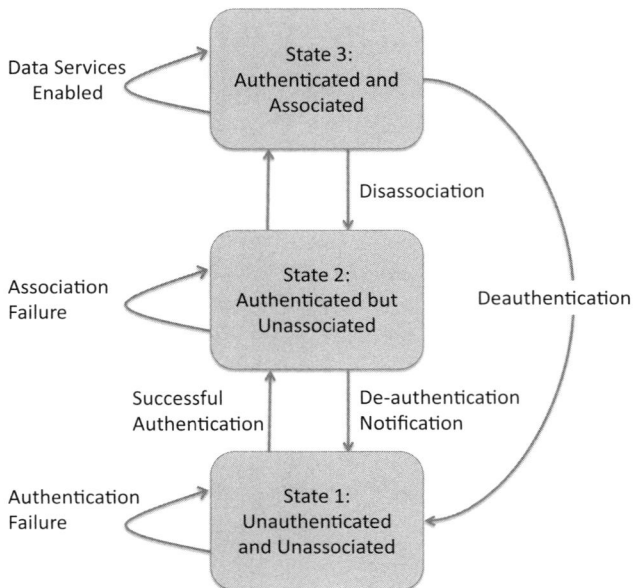

Figure 4.3   IEEE 802.11 high-level state machine.

the MS dictates the type of network access and frame transmission privileges. State 1 is the initial state of a MS. Class 1 frames are allowed while in State 1. Class 1 and 2 frames are allowed while in State 2. Class 1, 2, and 3 frames are allowed while in State 3. Once an MS has reached State 3, data distribution system services are available for its use and the MS can now reach destinations beyond its access point.

There are two basic methods by which an MS can perform authentication in the original IEEE 802.11 security model: open-system authentication and shared-key authentication. In both methods, the MS initiates authentication. Open-system authentication is the only method required by the 802.11 standard. This method is the weakest form of authentication, as the access point trusts the MS when asked its identity (no credentials are required). In this method, the exchange of two management frames is required. This is illustrated in the authentication model illustrated in Figure 4.3. The MS initiates authentication with an authentication request frame. The AP processes the authentication request and returns a response. Upon receipt of this response, authentication is complete and association can begin. The shared-key authentication approach utilizes a cryptographic key. This method requires that a shared key be distributed to a station prior to attempting authentication. This method involves an exchange of four management frames. Upon successful authentication, the MS can now associate with an access point to gain full access to the network and data services. The basic IEEE 802.11 authentication process is depicted in Figure 4.4.

Upon successful authentication, the MS can now associate with an access point to gain full access to the network and data services. The requesting MS

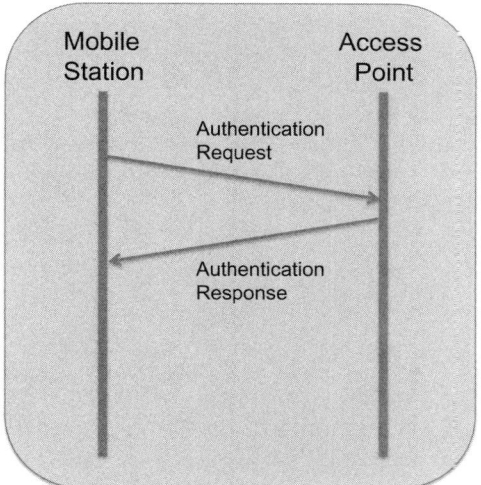

Figure 4.4   IEEE 802.11 authentication.

initiates the process with an association request, which includes the MAC address of the requesting node, the MAC address of the AP, and the SSID value. The AP then responds with an association response message, which includes the result of the association attempt and an association identifier (AID). This association process must be successfully completed before generated MAC layer frames can be passed up to higher networking layers. The process of association enables network book-keeping; the network will have knowledge of the MS position in order to properly forward frames destined to the MS to the correct AP. The basic association procedure consists of the exchange of two frames. This is depicted in the figure. An MS that attempts association without having successfully authenticated will receive a deauthentication frame from the AP in response. Reassociation occurs when an MS moves to a new AP within the 802.11 network or if that MS temporarily loses coverage to its AP. The reassociation process is almost identical to the association process. Reassociation consists of the exchange of two messages, as is the case for association. In the case of reassociation, however, those messages are a reassociation request and a reassociation response. The association process is depicted in Figure 4.5.

This is a highly simplified view of the association and authentication processes utilized in IEEE 802.11 WLANs, with actual procedures varying across 802.11 specifications as security models and mechanisms have evolved. Consequently, this area will be revisited numerous times in subsequent sections of the text.

4.1.2.2   Channel Access Procedures

There are two methods by which an MS can access the wireless 802.11 channel: contention-based access or contention-free access. The corresponding coordination functions, which control access to the wireless medium, are the distributed

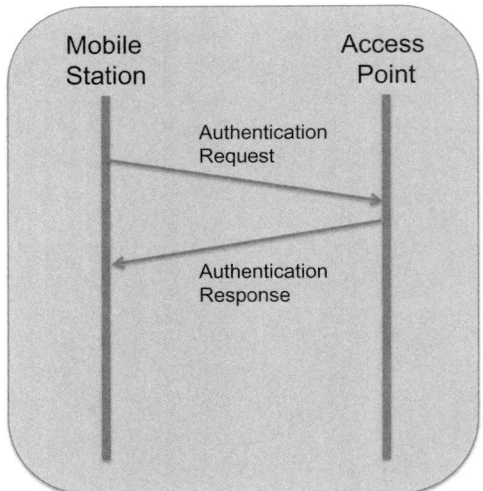

Figure 4.5 IEEE 802.11 association.

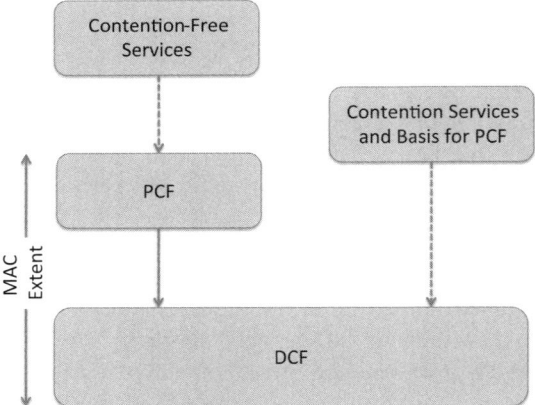

Figure 4.6 IEEE 802.11 channel access.

coordination function (DCF) and the point coordination function (PCF), respectively. PCF is built on top of the DCF and is only used by infrastructural networks. This is illustrated in Figure 4.6.

The IEEE 802.11 Distributed Coordination Function Carrier sensing multiple access/collision avoidance (CSMA/CA) is provided by the DCF. The DCF allows multiple independent MSs to interact without central management. Like Ethernet, the DCF first checks to ensure that the wireless medium is available before transmitting a frame. To avoid collisions, MSs use a random backoff after each frame, with the first transmitter seizing the channel. It should be noted that 802.11 does not employ CSMA/collision detection (CSMA/CD), but rather employs CSMA/CA. CSMA/CD, as employed by the IEEE 802.3 LAN standard, is not used because collisions are considered too costly due to the limited amount of bandwidth present in 802.11 networks. Two types of carrier sensing functions are employed by 802.11 to determine if the medium is available: physical carrier sensing and virtual carrier sensing. If either of these functions determines that the medium is not available, the MAC reports this unavailability to the higher layers. The "hidden node problem" (discussed later in this chapter) can introduce problems when performing physical carrier sensing. Virtual carrier sensing is performed through the use of the Network Allocation Vector (NAV). Most 802.11 frame types contain a duration field, which can be used to reserve the wireless medium for a period of time. The NAV is a timer that indicates the amount of time the medium will be reserved. MSs set the NAV to the time they anticipate channel usage, including any frames required to complete the current operation. All other MSs perform a countdown from the NAV value to 0. When the NAV is nonzero, the virtual carrier sensing function reports the medium as in use, and other MSs will not attempt to contend for the medium. When the NAV reaches zero, the virtual carrier sensing function reports the medium is idle, and other MSs can then contend for usage of the medium.

An important mechanism in coordinating access to the medium is interframe spacing. IEEE 802.11 employs four different interframe spaces: short interframe space (SIFS), PCF interframe space (PIFS), DCF interframe space (DIFS), and extended interframe space (EIFS). These varying interframe spacings provide the ability for 802.11 to support different priority levels for different types of traffic. This is illustrated in Figure 4.7. As can be seen from Figure 4.7, the interframe spacings are used in conjunction with the CSMA contention protocol. Upon observation of an idle medium, the MS then waits some prescribed period of time before contending for channel access. That period of time is defined by the interframe spacing. SIFS is used for highest priority transmissions because its length in time is the shortest of all interframe spacings, allowing a MS with a high priority frame to "seize" the channel before any other MSs have the opportunity, giving them priority over frames that can be transmitted after the longer PIFS, DIFS, or EIFS interval. Frames transmitted using SIFS includes request-to-send/clear-to-send (RTS/CTS)

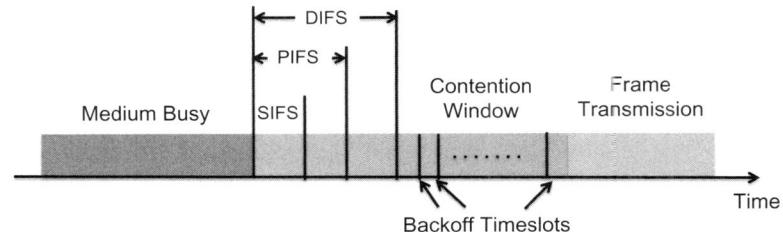

Figure 4.7   IEEE 802.11 channel access.

frames (described later) and positive acknowledgements (discussed later). PIFS is used by the PCF during contention-free operations. MSs that have data to transmit in the contention-free period can transmit after the PIFS has elapsed and effectively preempt contention-based traffic: the PIFS is 30 μs for the 802.11b PHY. DIFS is the minimum medium idle time for contention-based services; the DIFS is 50 μs for 802.11b. MSs operating in the contention-based mode may have access to the medium if it has been free for a period longer than the DIFS. The EIFS (not shown in Figure 4.7) is not a fixed interval. The EIFS is only used when there is an error during frame transmission (an error is detected in a received frame).

After the interframe spacing has been observed, there is a period called the "contention window" or "backoff window." This window is divided into time slots. Slot length is PHY dependent, and is smaller for higher data rate PHYs. MSs pick a random slot and wait for that slot before attempting to access the medium. All slots are equally likely. As in Ethernet, the backoff time (i.e., number of slots) is selected from a larger range each time a transmission error occurs. Contention windows are always sized $2n - 1$, $n \geq 1$. For each transmission error, n is incremented by 1. Each PHY has a maximum backoff time: the direct-sequence spread spectrum (DSSS) PHY has a maximum backoff time of 1023 time slots. For the 802.11b PHY, the minimum contention window size is 31 and the maximum contention window size is 1203. The slot time for the 802.11b PHY is 20 μs. Once the MS has waited the selected backoff period of time, the MS will determine if the medium is in use. If the medium is not idle, the MS defers access. The MS will then wait for the medium to become idle and then employ an exponential backoff procedure. If the medium is still idle after the interframe spacing, transmission can begin immediately. If the previous frame was received without error, the medium must be free of transmissions for at least the interframe spacing. If the previous transmission contained an error, the medium must be free for the amount of the EIFS.

The operation of the DCF access function is shown in Figure 4.8 [70].

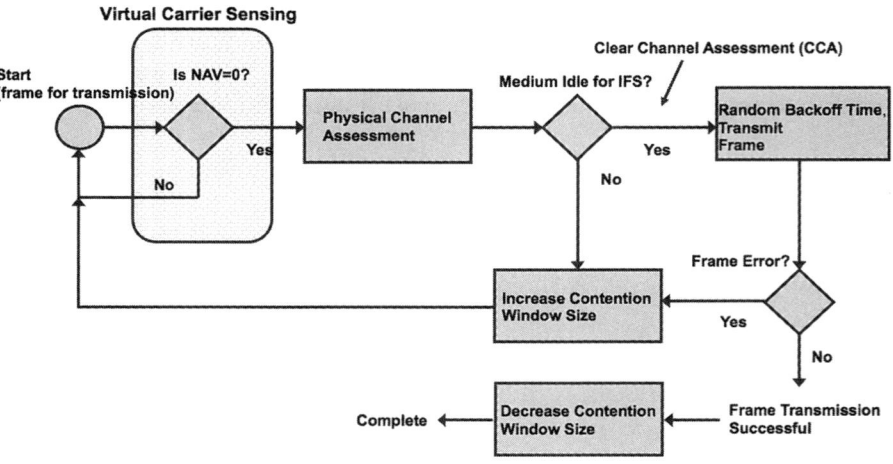

Figure 4.8   IEEE 802.11 DCF operation. Modified from Reference [70].

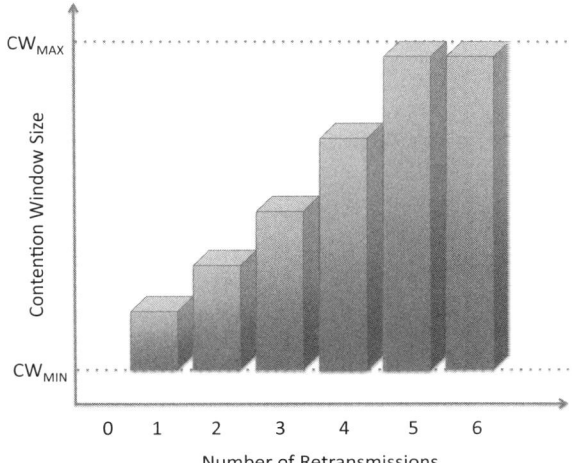

Figure 4.9   Illustration of IEEE 802.11 contention window dynamic resizing. Reproduced from Reference [1].

The CW size increases exponentially within the constraints set by the maximum and minimum window size, as depicted in Figure 4.9 [1]. Correspondingly, this also increases the amount of time that is required to gain access to the channel to transmit data, increasing the latency of the system as seen by that data.

IEEE 802.3 networks rely on the reception of transmissions to perform the carrier sensing functions within the CSMA/CD protocol. The wireless network, however, is more complicated than its wired counterpart in which messages are distributed to all nodes along the wire. In the wireless network, not every node can communicate with all other nodes. This can be due to physical obstructions or exceedingly great distances between the MSs. This causes the "hidden node problem" that is depicted in Figure 4.10.

To combat the hidden node problem, 802.11 allows stations to use request to send (RTS) and clear to send (CTS) to effectively "clear" out an area. The transmitting MS will initially transmit a RTS frame. The intended recipient will then respond by sending a CTS frame. Both the RTS and CTS are used to set the NAV of MSs that are within reception of those messages. Using this approach, stations can essentially reserve resources even within the contention-based access period. When using the RTS/CTS procedure, any frames must be positively acknowledged. The RTS/CTS message exchange is shown in Figure 4.11. This approach has been shown to significantly alleviate the hidden node problem [71]. Literature shows that when RTS/CTS procedures are employed, the performance of DCF is only marginally dependent on the minimum contention window size and the number of active MSs [72]. Not surprisingly, this RTS/CTS approach is relatively inefficient from a capacity usage perspective because of the additional latency introduced before the transmission can begin. As a result, it is employed in certain cases. Otherwise, it is not

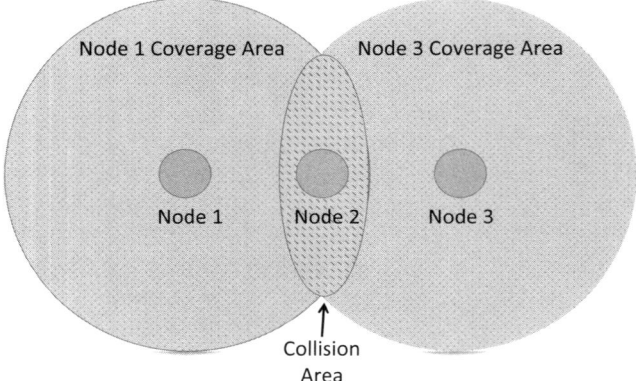

Figure 4.10   The hidden node problem.

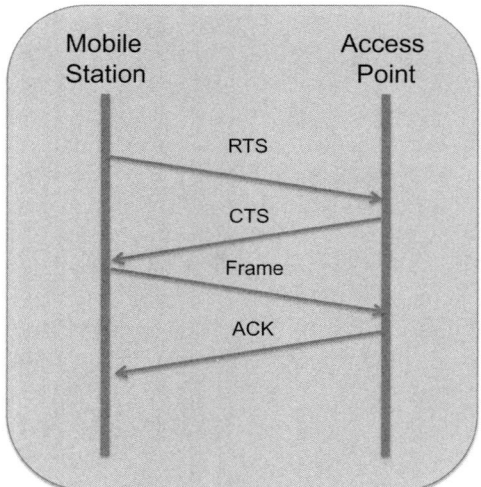

Figure 4.11   Request to send/clear to send.

employed. Exactly when the RTS/CTS procedure is employed is controlled by a parameter known as the RTS Threshold. This parameter is often configurable through the 802.11 device driver (if configuration is allowed by the 802.11 device manufacturer). This threshold is in bytes and is compared against the size of a frame to be transmitted. If the size of the frame is larger than the RTS Threshold, then the RTS/CTS procedure is employed. Otherwise, standard DCF access procedures are employed.

The IEEE 802.11 Point Coordination Function   The 802.11 MAC also provides the PCF, a second coordination function to provide a low-latency method to access the medium. The PCF is an enforced access scheme by which the AP arbitrates medium access among MSs. The PCF is an optional component of the 802.11 standard and is not widely implemented. The PCF within 802.11 actually resembles the token concept found in IEEE 802.4 and IEEE 802.5 token LAN specifications (token bus and token ring, respectively). The AP transmits a beacon at the beginning of the contention-free period. The duration of the contention-free period is one component of that beacon. All MSs set the NAV to the duration of the contention-free period to lock out DCF-based access to the medium. Additionally, all contention-free transmissions are separated by the shorter PIFS interframe space to ensure that contention-free transmissions will begin before DCF-based transmissions, which will employ the DIFS so that no MS can gain access to the wireless media using DCF. After the AP has gained control of the medium, it polls any associated MSs that it has on its polling list to query for data transmissions. MSs may only transmit during the contention-free period if the AP solicits the transmission with a polling frame. Stations get on the polling list when they associate with the AP. Part of the association request message that is exchanged during association is a field that indicates whether the MS is capable of responding to polls during the contention-free period. Each poll from the AP is a token to transmit exactly one frame. When the contention-free period ends, the AP transmits a frame signaling the end of contention-free interval in order to release MSs from the PCF access rules and return to the DCF.

It should be noted that the PCF is an optional component of the IEEE 802.11 standard and is not widely implemented. Other than outlier cases, IEEE 802.11 is synonymous with the DCF and CSMA/CA.

4.1.2.3 Data Flow Control

IEEE 802.11 provides point-to-point reliability through data retransmission mechanisms. IEEE 802.11 is somewhat unique in that only positive acknowledgements are employed. The detection and correction of errors within 802.11 is solely the responsibility of the sender. The sender must track the receipt of positive acknowledgements to determine if transmission was successful. If transmission was not successful, the sender must take appropriate steps (i.e., retransmission until success).

Not all 802.11 frames, however, are subject to retransmission. Any transmissions that are part of an "atomic process" must be acknowledged. An atomic process is a process that is a multiframe exchange, in which success is judged on the set of frame exchanges, not just a single frame. Association is an example of an atomic process. Authentication is another example. The RTS/CTS procedure is another example of an atomic process. The success of atomic operations is judged by the completion of the entire operation, not just a single packet. If any single frame within the process is not successfully received and positively acknowledged, the entire process is repeated. By using the NAV, MSs can ensure that "atomic operations" are not interrupted, essentially "seizing" the medium for as long as the atomic process transpires.

Additionally, all unicast data must be acknowledged. If the sender does not receive a positive acknowledgement within a certain timeout period, the sender will assume an error has occurred and retransmit the frame. The sender does, however, maintain a retry counter and will only retry frame transmissions up to some configurable maximum number of retries. The maximum number of frame transmission retries is usually a configurable item within an 802.11 device's driver software.

Another important aspect of 802.11 data transmission is fragmentation. 802.11's MAC performs fragmentation on higher-layer packets and some larger management frames in order to optimize frame transmissions across the wireless channel. Fragmentation takes place when a network layer packet is larger than the fragmentation threshold configured by the network administrator. Fragments all have the same frame sequence number but will have ascending fragment numbers to aid in reassembly at the receiver. All fragments that belong to a single frame are normally transmitted in a fragmentation burst. It is common for the RTS/CTS access procedure to be employed to support the transmission of fragmentation bursts, and it is quite common for the fragmentation threshold to be configured to the same value as the RTS Threshold.

4.1.2.4 Framing Mechanisms

IEEE 802.11 MAC frames have significant differences from Ethernet MAC frames. The generic 802.11 frame format is depicted in Figure 4.12 [1].

802.11 MAC frames do not include any type of length field. 802.11 MAC frames also do not carry any type of preamble, which is part of the physical layer. 802.11 MAC frames include four separate address fields: destination address, source address, receiver address, and transmitter address. The destination address is analogous to the destination address in Ethernet, while the destination address is a 48-bit MAC identifier that corresponds to the final recipient (i.e., the station that processes the frames and passes data up to the network layer). The source address is a 48-bit MAC identifier corresponding to the source of the transmission. The receiver address is a 48-bit MAC identifier that indicates which wireless station should process the frame. For frames that are destined to a wired network connected to an AP, the receiver is the wireless interface in the AP. The transmitter address is a 48-bit MAC address to identify the wireless interface that transmitted the frame onto the wireless medium. This address is optional and is only used in wireless bridging applications. The duration/ID field, or NAV, identifies the duration of the next frame transmission. Stations in the network use this field to perform virtual carrier sensing. 802.11 MAC addresses can be either individual or group addresses, where group addresses can

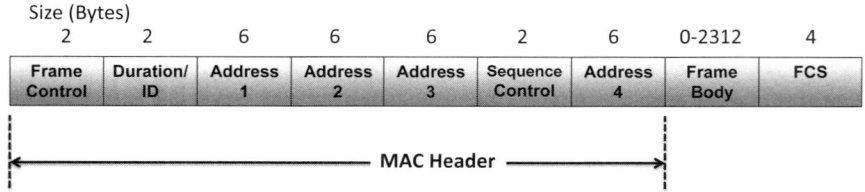

Figure 4.12   IEEE 802.11 generic frame format. Reproduced from Reference [1].

Size (Bits)

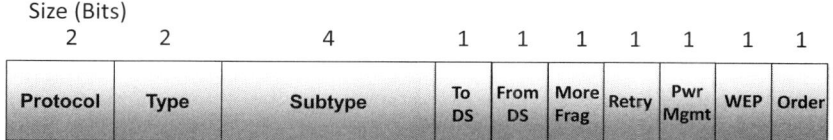

Figure 4.13   IEEE 802.11 frame control field. Reproduced from Reference [1].

be either multicast or broadcast group addresses. The sequence control field is used to indicate the fragment number and MSDU identification. The FCS field contains a CRC-32 error detection field that is used by the recipient to determine correct frame reception.

A detailed breakdown of the 802.11 Frame Control Field is shown in Figure 4.13 [1].

A brief description of the various data field within the IEEE 802.11 Frame Control Field is provided:

- *Protocol Version:* Currently set to 0 (only one MAC)
- *Type:* Indicates whether the frame is management, control, or data
- *Subtype:* Indicates the particular message
  - ○ For example, subtype = 0001 corresponds to an Association Response frame
- *To DS:* Indicates if frame is destined for DS
- *From DS:* Indicates if frame originates from the DS
- *More Frag:* Indicates if additional fragments of an MSDU follows in subsequent frames
- *Retry:* Indicates whether this is a retransmission
- *Power Management:* Indicates the power management mode the sending station will assume after the current exchange
- *More Data:* Indicates if the sending station has more data to send to the receiving station
- *WEP:* Indicates whether the frame body has been WEP encrypted
- *Order:* Indicates that frames must be processed in order

The 802.11 MAC provides three types of frames: management frames, control frames, and data frames. The frame type and subtype field that is contained within the frame control field, which is common to all frame types, delineates the type of 802.11 MAC frame.

- *Management Frames:* These include association-related frames, reassociation-related frames, probe-related frames, beacons, disassociation frames, and authentication-related frames.
- *Control Frames:* These include CTS frames, RTS frames, and acknowledgement frames.

802.11 MAC frames are also divided into different classes: class 1, class 2, and class 3.

- *Class 3 Frames:* These can only be transmitted once the MS has successfully completed authentication and association procedures. Class 3 frames include select control frames, select management frames, and most data frames.

- *Class 2 Frames:* These can be transmitted once the MS has successfully completed the authentication procedure. Class 2 frames include no control frames, select management frames, and no data frames. Class 2 management frames consist of association requests and responses, reassociation requests and responses, and disassociation frames.

- *Class 1 Frames:* These can be transmitted at any state of the authentication and association process. Class 1 frames include most types of control frames, many management frames, and select data frame types. Class 1 control frames include RTS, CTS, and ACK. Class 1 management frames include probe requests and responses, beacons, and authentication-related messages.

4.1.3 The Original IEEE 802.11 PHY

The IEEE 802.11 WLAN standard uses the 2400–2483.4 MHz frequencies within the United States. This bandwidth within the ISM frequency band is subdivided into 11 (United States) 22 MHz frequency channels with a 5-MHz frequency separation between channels. The 802.11 frequency plan for use within the United States is shown in Table 4.1. In addition to this overlapping frequency plan, there is also a frequency plan for allowing nonoverlapping channels. These channels are 22 MHz wide with 5 MHz separations, and are centered at 2412, 2437, and 2463, respectively.

Table 4.1 IEEE 802.11 Frequency Plan (United States)

Channel ID	Frequency (MHz)
1	2412
2	2417
3	2422
4	2427
5	2432
6	2437
7	2442
8	2447
9	2452
10	2457
11	2462

The original IEEE 802.11 WLAN standard [1] specified three PHY designs:

- Frequency-hopped spread spectrum (FHSS)
- Direct sequence spread spectrum (DSSS)
- Infrared

All PHYs provide link data rates of 1 and 2 Mbps.

4.1.3.1   The Original IEEE 802.11 FHSS PHY

This physical layer is defined for transmission rates of 1 and 2 Mbps, and is included in the original IEEE 802.11 standard. In this physical layer, data bits are modulated using M-Gaussian frequency shift keying (M-GFSK) modulation, where $M = 2$ or 4, for 1 or 2 Mbps transmissions, respectively. The resulting symbol stream is then upconverted and transmitted within its frequency channel. Spreading occurs by changing the carrier frequency of the channel as a function of time. The IEEE 802.11 standard defines 79 hop positions (i.e., center frequencies) in the 2400 MHz frequency band, with each hop staying at a carrier frequency for 20 ms. There are 78 different hopping patterns defined by the standard, with each pattern constituting a channel, as these patterns are nonoverlapping.

4.1.3.2   The Original IEEE 802.11 Infrared PHY

The infrared physical layer is defined for transmission rates of 1 and 2 Mbps, and is included in the original IEEE 802.11 standard. In this physical layer, data bits are modulated using M-pulse position modulation (PPM), where $M = 8$ or 16, for 1 or 2 Mbps transmissions, respectively. The resulting symbol stream is then upconverted and transmitted via infrared techniques in the 850–950 nm wavelength range. One benefit to this physical layer implementation is its inherent security because of the very narrow transmission beamwidth. Any type of malicious activity would require a direct line-of-sight position.

4.1.3.3   The Original IEEE 802.11 DSSS PHY

The DSSS waveforms are created by first applying an 11-bit Barker sequence to each data bit to create 11 chips from every 1 data bit. The 11-bit Barker sequence, $Bi, i = \{1,2,3, \ldots, 11\}$, is:

$$[+1, -1, +1, +1, -1, +1, +1, +1, -1, -1, -1].$$

The total length in time of this 11-bit sequence is equal to a data bit period, $Tb = 1$ μs. The 11-bit Barker sequence is then exclusive ORed with the data bit:

$$c(t) = d(t) \oplus b_i(t). \tag{4.1}$$

This converts the 1 and 2 Mbps data bit stream, $d(t)$ into 11 and 22 Mcps chip streams, $c(t)$, respectively.

Table 4.2   IEEE 802.11 1 Mbps Waveform D-BPSK Encoding Scheme

Bit input	Phase change $(+j\omega)$
0	0
1	π

Table 4.3   IEEE 802.11 2 Mbps Waveform D-QPSK Encoding Scheme

Bit input	Phase change $(+j\omega)$
00	0
01	$\pi/2$
11	π
10	$3\pi/2$

This chip stream is then modulated using differential binary phase-shift keying (D-BPSK) modulation for the 1 Mbps case and differential quaternary phase-shift keying (D-QPSK) modulation for the 2 Mbps case. The D-BPSK encoding scheme is shown in Table 4.2, while the D-QPSK encoding scheme is shown in Table 4.3.

Both of the 1 and 2 Mbps results in a symbol rate is 11 Msps. Thus, the symbol time, $Ts = 9.090909 \cdot 10^{-8}$. The resulting PSK symbol stream is then upconverted to the proper frequency, 2412 MHz, for transmission over channel 1, for example. The spectrum occupied by the resulting binary phase-shift keying (BPSK) sequence is given by the power spectral density (PSD) equation for BPSK:

$$c(f) = \frac{\sin(\pi f Ts)}{\pi f Ts}. \tag{4.2}$$

The null-to-null bandwidth of an 11 Msps BPSK or QPSK symbol stream is approximately 22 MHz. The received signal, $r(t)$, is then downconverted and fed into a DPSK or D-QPSK hard-decision demodulator, using the same encoding table as given in the previous tables. Each hard-decision output of the demodulator is then exclusive ORed with the same 11-bit Barker sequence as used previously. This will effectively "despread" the received signal. In the event of channel errors, it is possible that there will be discontinuities within a bit. Because of this, a simple 0.5 threshold rule is applied. If a particular despread bit holds a 0 value a higher percentage of its period, then it is declared a 0. Likewise, if a particular despread bit holds a 1 value a higher percentage of its period, then it is declared a 1.

4.1.4   Deployment of Original IEEE 802.11

Products based upon the original IEEE 802.11 standard widely implemented both FHSS and DSSS waveforms. In fact, IEEE 802.11 products can still sometimes be found in the marketplace that supports both DSSS and FHSS waveforms. However, later products largely implemented only DSSS waveforms. Consequently, the original IEEE 802.11 (and IEEE 802.11b) has become synonymous with DSSS. The infrared PHY never experienced significant deployment.

Overall, the original IEEE 802.11 standard was not widely deployed due to relatively small data rates (up to 2 Mbps). IEEE 802.11 deployment did not become significant until the emergence of the IEEE 802.11b standard.

4.2   IEEE 802.11B

The IEEE 802.11b standard was introduced in 1999 [18] as the high-rate (HR) extension to the original 802.11 WLAN standard. The 802.11b standard operates within the same 2.4 GHz ISM frequency band, following the same frequency plan, as the original 802.11 standard. In fact, the 802.11b standard is completely backward compatible with the original standard, specifying the same waveforms. The IEEE 802.11b PHY modification utilizes the same frequency structure as the original IEEE 802.11 specification (Table 4.1).

The 802.11b standard also introduces additional HR rate waveforms that provide data rates of 5.5 and 11 Mbps through the introduction of a new high-rate DSSS (HR-DSSS) PHY. The 802.11b amendment to original IEEE 802.11 specification introduced no changes to the MAC layer.

These new HR-DSSS waveforms employ complementary code keying (CCK). In this encoding scheme, CCK code words form the "spreading" sequence. The 5.5 Mbps waveform employs D-QPSK modulation. The 11 Mbps waveform employs QPSK modulation.

The resulting QPSK symbol stream is then spread in frequency through the use of complementary code keying (CCK), which are polyphase complementary codes. Golay first introduced these codes in 1951 [73]. These codes consist of a set of equal finite length sequences. These codes have the property that their periodic autocorrelative vector sum is zero everywhere except at the zero shift. The IEEE 802.11b standard uses 8-bit CCK code words to spread the QPSK symbol stream. These 8-bit code words are derived from Equation 4.3:

$$c = \begin{bmatrix} e^{j(\phi_1+\phi_2+\phi_3+\phi_4)}, e^{j(\phi_1+\phi_3+\phi_4)}, e^{j(\phi_1+\phi_2+\phi_4)}, -e^{j(\phi_1+\phi_4)}, \\ e^{j(\phi_1+\phi_2+\phi_3)}, e^{j(\phi_1+\phi_3)}, -e^{j(\phi_1+\phi_2)}, e^{j(\phi_1)} \end{bmatrix}. \tag{4.3}$$

The code word shown in the equation is least significant bit (LSB) first, most significant bit (MSB) last. The four CCK phase terms, φ_1, φ_2, φ_3, and φ_4 determine the phase values of the complex code set. The first two bits $\{d0, d1\}$ determine the φ_1 phase value, which is common to all code word elements. Thus, it represents the basic QPSK symbol. The remaining two bits $\{d2, d3\}$ are used to determine the remaining three phase elements, φ_2, φ_3, and φ_4, as described in the previous table. It should be noted that the table is obtained by setting $\varphi_2 = (d2*\pi) + \pi/2$, $\varphi_3 = 0$, and $\varphi_4 = d3*\pi$. This constraint creates a subset of the 64 code words in the CCK

Table 4.4   IEEE 802.11b D-QPSK Encoding Table

Bit input	Phase change $(+j\omega)$	Phase change $(+j\omega)$
$\{d0, d1\}$	Even symbols	Odd symbols
00	0	π
01	$\pi/2$	$3\pi/2$
11	π	0
10	$3\pi/2$	$\pi/2$

Table 4.5   IEEE 802.11b CCK Code Word Generation

$\{d2, d3\}$	$c1$	$c2$	$c3$	$c4$	$c5$	$c6$	$c7$	$c8$
00	1j	1	1j	−1	1j	1	−1j	1
01	−1j	−1	−1j	1	1j	1	−1j	1
10	−1j	1	−1j	−1	−1j	1	1j	1
11	1j	−1	1j	1	−1j	1	1j	1

code set, as only two bits are available to select a code word. Once the basic PSK symbol phase and the CCK code word have been derived, eight identical copies of the PSK symbol are produced, with each symbol time equal to one-eighth of the original PSK symbol time. The phase of each PSK symbol is then modified according to the corresponding code word.

4.2.1   The IEEE 802.11b 5.5 Mbps HR-DSSS PHY

This waveform is generated by first grouping data bits by groups of four, $\{d0, d1, d2, d3\}$, $d0$ first in time. The first two bits, $\{d0, d1\}$, are then used to generate a D-QPSK symbol, where the D-QPSK encoding table is given in Table 4.4.

It should be noted that in this modulation scheme, every odd symbol receives an additional π phase shift. In this modulation scheme, the phase change is relative to the previous QPSK symbol defined by the $\{d0, d1\}$ bits of the previous group of data bits.

The remaining two bits $\{d2, d3\}$ are used to generate a CCW according to Table 4.5.

The complex code set is then used to spread the D-QPSK symbol into eight symbols phase-adjusted per the complex CCK code word.

For example, let us assume the previous QPSK symbol phase was 0 degrees, and the next four data bits are $\{0, 0, 0, 0\}$. From the previous table, that would mean that the phase of the current QPSK symbol would be either 0 or π, depending on whether it's an even or odd symbol. For this example, let us assume its value is 0. Values of $\{0, 0\}$ for $\{d2, d3\}$ would lead to the CCK code word $\{1j, 1, 1j, −1, 1j, 1, −1j, 1\}$. Thus, the resulting PSK symbols would be transmitted with phases $\{90,$

0, 90, 0, 90, 0, −90, 0}. Eight PSK channel symbols are transmitted for every four information bits. Information bits are streaming at a rate of 5.5 Mbps, thus the PSK channel symbol rate is 11 Msps. Thus, the symbol time, $Ts = 9.090909 \cdot 10^{-8}$. The resulting PSK symbol stream is then upconverted to the proper frequency (2412 MHz) for transmission over channel 1, for example. The spectrum occupied by the resulting QPSK sequence is equivalent to the expression for BPSK. The null-to-null bandwidth of an 11 Msps QPSK symbol stream is approximately 22 MHz.

4.2.2   IEEE 802.11b 5.5 Mbps HR-DSSS Receiver

The received signal, $r(t)$, is then downconverted and fed into a matched filter which generates transmit phase estimates. These estimates are then fed into a fast Walsh transform (FWT) circuit, which employs a bank of 64 correlators, which correlate the set of 8 phases to the known 64 code words that are defined by a set of 8 phases. A largest correlation value decision rule is applied to decide upon the transmitted CCK code word. From this, the $\{d2, d3\}$ data bits are determined. The CCK code word is then removed, producing the original QPSK symbol stream, which is then fed into a QPSK demodulator that produces hard decision outputs, providing the $\{d0, d1\}$ bits.

4.2.3   The IEEE 802.11b 11 Mbps HR-DSSS PHY

This waveform is created by first grouping data bits by groups of eight, $\{d0, d1, d2, d3, d4, d5, d6, d7\}$, $d0$ first in time. The first two bits, $\{d0, d1\}$, are then used to generate a D-QPSK symbol, where the D-QPSK encoding table is shown in Table 4.6.

In this modulation scheme, the phase change is relative to the previous QPSK symbol defined by the $\{d0, d1\}$ bits of the previous group of data bits. The resulting QPSK symbol stream is then "spread" in frequency through the use of CCK, as in the 5.5 Mbps case. The four CCK phase terms, φ_1, φ_2, φ_3, and φ_4, determine the phase values of the complex code set. The first two bits $\{d0, d1\}$ determine the φ_1 phase value, which is common to all code word elements. Thus, it represents the basic PSK symbol. The remaining six bits $\{d2, d3, d4, d5, d6, d7\}$ are used to

Table 4.6   IEEE 802.11b 11 Mbps D-QPSK Encoding Table

Bit input $\{di + 1, di\}$	Phase
00	0
01	$\pi/2$
10	π
11	$3\pi/2$

Table 4.7 IEEE 802.11b 11 Mbps CCK Phase Element Mapping

Bit positions	Phase parameter
$\{d1, d0\}$	Φ_1
$\{d3, d2\}$	Φ_2
$\{d5, d4\}$	Φ_3
$\{d7, d6\}$	Φ_4

determine the remaining three phase elements, φ_2, φ_3, and φ_4, as described in Table 4.7.

Once the basic PSK symbol phase and the CCK code word have been derived, eight identical copies of the PSK symbol are produced, with each symbol time equal to one-eighth of the original PSK symbol time. The phase of each PSK symbol is then modified according to the corresponding code word:

$$c = \begin{bmatrix} e^{j(\phi_1+\phi_2+\phi_3+\phi_4)}, e^{j(\phi_1+\phi_3+\phi_4)}, e^{j(\phi_1+\phi_2+\phi_4)}, -e^{j(\phi_1+\phi_4)}, \\ e^{j(\phi_1+\phi_2+\phi_3)}, e^{j(\phi_1+\phi_3)}, -e^{j(\phi_1+\phi_2)}, e^{j(\phi_1)} \end{bmatrix}. \tag{4.4}$$

Eight PSK channel symbols are transmitted for every eight information bits (i.e., no spreading). Information bits are streaming at a rate of 11 Mbps, thus the PSK channel symbol rate is 11 Msps. Thus, the symbol time, $Ts = 9.090909*10 - 8$. The resulting PSK symbol stream is then upconverted to the proper frequency. The spectrum occupied by the resulting QPSK sequence is equivalent to the expression for BPSK. The null-to-null bandwidth of an 11 Msps QPSK symbol stream is approximately 22 MHz. From the FWT, the $\{d2, \ldots, d7\}$ data bits are determined. The CCK code word is then removed, producing the original QPSK symbol stream, which is then fed into a QPSK demodulator that produces hard decision outputs, providing the $\{d0, d1\}$ bits.

4.2.4 IEEE 802.11b DSSS Frame Structure

The IEEE 802.11b WLAN standard transfers data between points in discrete frames called physical layer convergence protocol (PLCP) protocol data units (PPDUs). The PPDU format is shown in Figure 4.14.

There are two types of PPDUs: (1) long preamble PPDU (128-bit SYNC word) and (2) short preamble PPDU (56-bit SYNC word). These different PPDU formats are used to support different data rates within the 802.11 network and provide different robustness for varying levels of protection of PPDU control fields. The long preamble PPDU is also intended to provide better acquisition and tracking performance in noisy environments, as it provides a longer synchronization sequence. Each PPDU has a header error check (HEC) field that serves as an indicator of frame

(a)

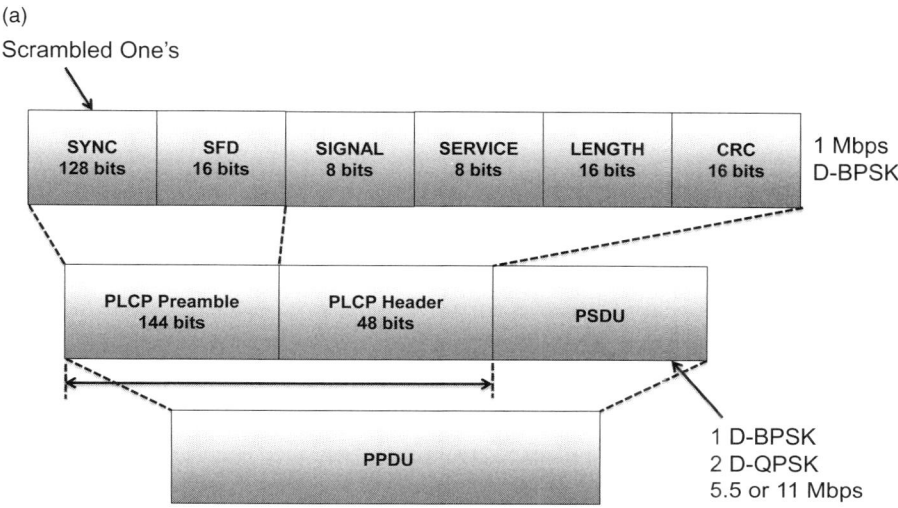

(b)

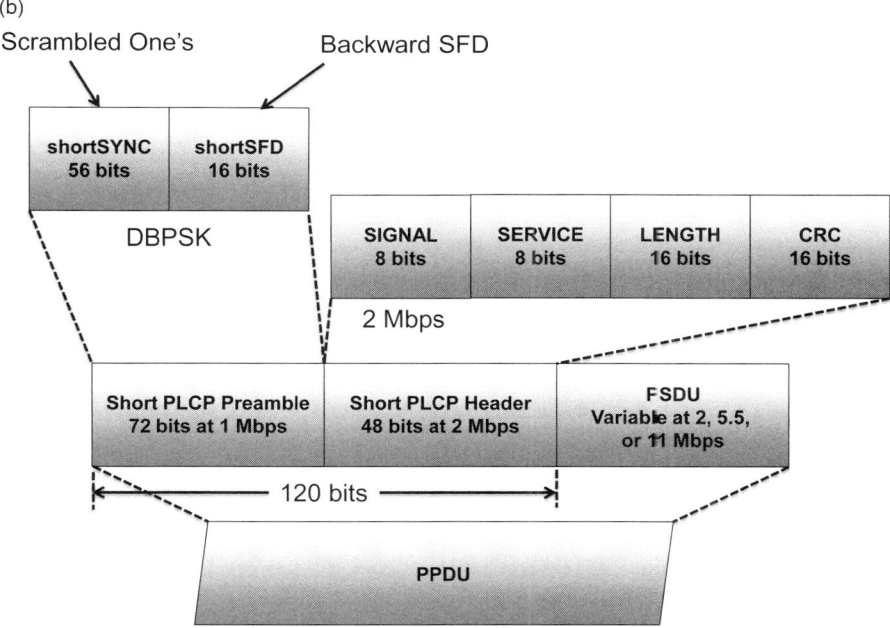

Figure 4.14   IEEE 802.11b DSSS frame structure. (a) Long PLCP preamble frame structure;
(b) short PLCP preamble frame structure. Reproduced from Reference [18]

quality. The IEEE 802.11 standard specifies the use of a CRC as the HEC for PPDUs. In particular, the CCITT CRC-16 is employed. The CCITT CRC-16 is a 16-bit sequence that is generated by the generator polynomial. This HEC protects the SIGNAL, SERVICE, and LENGTH fields of the PPDU. This HEC and PPDU frame format is common across all physical layers defined within the 802.11 standard and 802.11b HR extension to the original standard. Also common to all physical layers defined by the IEEE 802.11b standard, all transmitted bits, data, and overhead are scrambled prior to transmission according to a specified data scrambler polynomial, $G(z)$.

4.3   IEEE 802.11A

As part of the Telecommunications Act in 1996, the FCC allocated bands in the 5-GHz frequency range for unlicensed use [74]. Then, in 1997, the FCC also made available the 5.47- to 5.725-GHz frequency band [75]. This larger amount of frequency spectrum is significantly increased from the 83.5-MHz-wide ISM frequency band.

The IEEE 802.11a WLAN specification was published in 1999 [17] as an HR alternative PHY to operations in the ISM frequency band. The 802.11a specifies the PHY and MAC layer for operation in the U-NII frequency band at 5.2 and 5.7 GHz and provides link data rates of 6–54 Mbps through the employment of an OFDM waveform. However, IEEE 802.11a was a primarily PHY-specific amendment to the IEEE 802.11 specification. Multiple modulation modes and coding rates are defined to support the various data rates of the 802.11a specification.

The 802.11a PHY frequency plan consists of four 20-MHz channels in the U-NII lower band (5.15–5.25 GHz), four 20-MHz channels in the U-NII midband (5.25–5.35 GHz), and four 20-MHz frequency channels in the U-NII upper band (5.725–5.825 GHz). Each frequency channel is separated by 20 MHz. The 802.11a frequency plan is illustrated in Figure 4.15 [17] and summarized in Table 4.8.

Note that the channel IDs in the 802.11a frequency plan come from the fact that the 5-GHz frequency band in the United States is numbered every 5 MHz. Each 20-MHz 802.11a channel occupies four of these 5-MHz channels in the U-NII band. Thus, the resulting channel IDs increment by four.

The spectrum of each 802.11a channel is further subdivided. Each frequency channel is divided into 52 independent subchannels, referred to as subcarriers. Four of these subcarriers are pilot carriers and convey pilot tones that are used at the receiver for synchronization. Forty-eight of these subcarriers are data carriers. Data to be transmitted are then transmitted in parallel bit streams modulated onto a separate data subcarrier. The actual OFDM is constituted by the Fourier coefficients of the data, which are transmitted and demodulated rather than the standard time-domain amplitudes. Operation on all subcarriers is simultaneous. A depiction of the spectrum and placement on OFDM subcarriers is shown in Figure 4.16.

There are 14 steps that must be followed to fully encode a PPDU for transmission [3]:

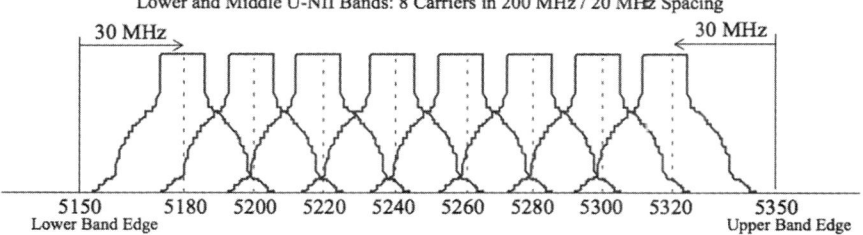

Lower and Middle U-NII Bands: 8 Carriers in 200 MHz / 20 MHz Spacing

30 MHz   30 MHz

5150   5180   5200   5220   5240   5260   5280   5300   5320   5350
Lower Band Edge   Upper Band Edge

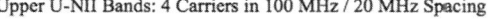

Upper U-NII Bands: 4 Carriers in 100 MHz / 20 MHz Spacing

20 MHz   20 MHz

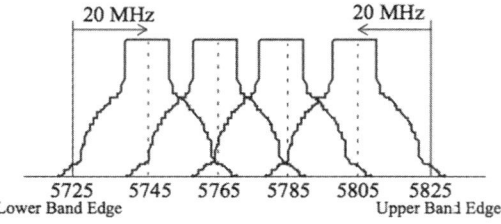

5725   5745   5765   5785   5805   5825
Lower Band Edge   Upper Band Edge

Figure 4.15   IEEE 802.11a frequency structure (United States). Reproduced from Reference [17].

Table 4.8   802.11a Frequency Plan (United States)

Channel ID	Frequency (GHz)
36	5.180
40	5.200
44	5.220
48	5.240
52	5.260
56	5.280
60	5.300
64	5.320
149	5.745
153	5.765
157	5.785
161	5.805

(a) Produce the PLCP preamble field, composed of 10 repetitions of a "short training sequence" (used for automatic gain control [AGC] convergence, diversity selection, timing acquisition, and coarse frequency acquisition in the receiver) and two repetitions of a "long training sequence" (used for channel estimation and fine frequency acquisition in the receiver) preceded by a guard interval (GI).

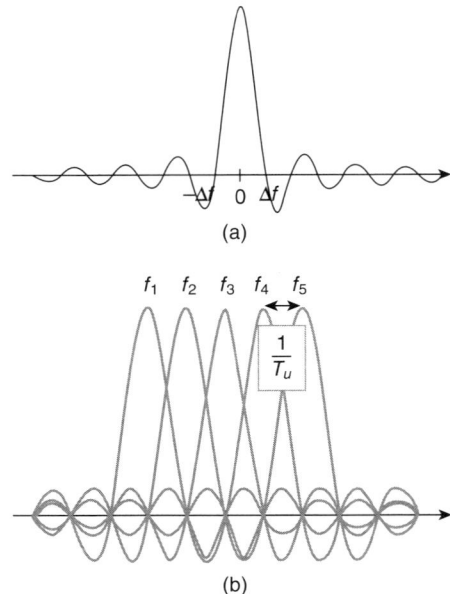

Figure 4.16 OFDM spectra. (a) Spectrum of single subcarrier; (b) composite OFDM spectra.

(b) Produce the PLCP header field from the rate, length, and service fields of the TXVECTOR by filling the appropriate bit fields. The rate and length fields of the PLCP header are encoded by a convolutional code at a rate of $R = 1/2$ and are subsequently mapped onto a single BPSK-encoded OFDM symbol, denoted as the signal symbol. To facilitate a reliable and timely detection of the rate and length fields, six "zero" tail bits are inserted into the PLCP header. The encoding of the signal field into an OFDM symbol follows the same steps for convolutional encoding, interleaving, BPSK modulation, pilot insertion, Fourier transform, and prepending a GI as described subsequently for data transmission at 6 Mbps. The contents of the signal field are not scrambled.

(c) Calculate from rate field of the TXVECTOR the number of data bits per OFDM symbol (N DBPS), the coding rate (R), the number of bits in each OFDM subcarrier (N BPSC), and the number of coded bits per OFDM symbol (N CBPS).

(d) Append the PSDU to the service field of the TXVECTOR. Extend the resulting bit string with zero bits (at least 6 bits) so that the resulting length will be a multiple of N DBPS. The resulting bit string constitutes the data part of the packet.

(e) Initiate the scrambler with a pseudorandom nonzero seed, generate a scrambling sequence, and exclusive OR (XOR) it with the extended string of data bits.

(f) Replace the six scrambled zero bits following the data with six nonscrambled zero bits. (Those bits return the convolutional encoder to the "zero state" and are denoted as "tail bits.")

(g) Encode the extended scrambled data string with a convolutional encoder ($R = 1/2$). Omit (puncture) some of the encoder output string (chosen according to "puncturing pattern") to reach the desired "coding rate."

(h) Divide the encoded bit string into groups of N CBPS bits. Within each group, perform an "interleaving" (reordering) of the bits according to a rule corresponding to the desired rate.

(i) Divide the resulting coded and interleaved data string into groups of N CBPS bits. For each of the bit groups, convert the bit group into a complex number according to the modulation encoding tables.

(j) Divide the complex number string into groups of 48 complex numbers. Each such group will be associated with one OFDM symbol. In each group, the complex numbers will be numbered 0 to 47 and mapped hereafter into OFDM subcarriers numbered −26 to −22, −20 to −8, −6 to −1, 1 to 6, 8 to 20, and 22 to 26. The subcarriers −21, −7, 7, and 21 are skipped and subsequently used for inserting pilot subcarriers. The "0" subcarrier, associated with center frequency, is omitted and filled with zero value.

(k) Four subcarriers are inserted as pilots into positions −21, −7, 7, and 21. The total number of subcarriers is 52 (48 + 4).

(l) For each group of subcarriers −26 to 26, convert the subcarriers to time domain using inverse Fourier transform. Prepend to the Fourier-transformed waveform a circular extension of itself, thus forming a GI, and truncate the resulting periodic waveform to a single OFDM symbol length by applying time-domain windowing.

(m) Append the OFDM symbols one after another, starting after the signal symbol describing the rate and length.

(n) Upconvert the resulting "complex baseband" waveform to a RF according to the center frequency of the desired channel and transmit.

The modulation and forward error correction (FEC) that are employed are a function of the data rate. Table 4.9 [17] summarizes this dependency.

4.3.1 OFDM PLCP Sublayer

The PPDU frame format of 802.11a is shown in Figure 4.17 [17].

The RATE indicates the data rate that is being used for the frame. This field is necessary for the Rx to know which modulation and coding parameters to apply to

Table 4.9   Rate-Dependent Parameters

Data rate (Mbps)	Modulation	Coding rate (R)	Coded bits per subcarrier (N_{BPSC})	Coded bits per OFDM symbol (N_{CBPS})	Data bits per OFDM symbol (N_{DBPS})
6	BPSK	1/2	1	48	24
9	BPSK	3/4	1	48	36
12	QPSK	1/2	2	96	48
18	QPSK	3/4	2	96	72
24	16-QAM	1/2	4	192	96
36	16-QAM	3/4	4	192	144
48	64-QAM	2/3	6	288	192
54	64-QAM	3/4	6	288	216

Reproduced from Reference [17].

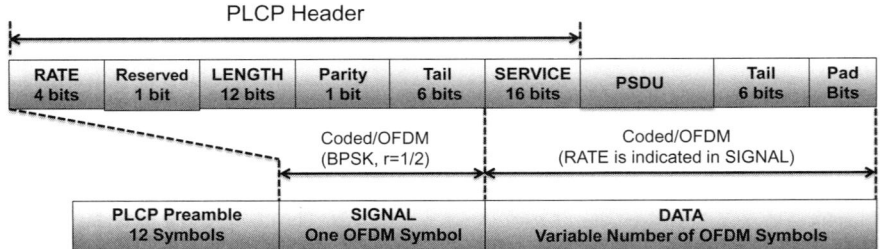

Figure 4.17   IEEE 802.11a OFDM PLCP frame format. Reproduced from Reference [17].

the incoming frame. The LENGTH conveys the length of the frame (0–4095). There is a parity bit check and some tail bits for filling. The SERVICE field contains the scrambler initialization. The PLCP preamble is shown as part of a complete PLCP frame, with the SIGNAL field (one OFDM symbol) essentially including the PLCP header except for the SERVICE field, which is instead included in the DATA field of the complete PLCP frame.

4.4   IEEE 802.11G

The IEEE 802.11g PHY layer was introduced in 2002 as an HR extension to the 802.11b WLAN standard. The 802.11g standard provides data rates of up to 54 Mbps through the addition of new OFDM waveforms in the same 2.4-GHz ISM frequency band as where 802.11b operates. The 802.11g physical layer includes all waveforms of the 802.11b specification and allows for full backward compatibility with existing 802.11b hardware.

While IEEE 802.11g has been one of the most successful variants of IEEE 802.11, we also find it warrants the least discussion. The 802.11g PHY layer is nearly identical to the 802.11a PHY layer. They both share common data rates, modulations, and forward error correction coding schemes. There are several minor differences. For example, the SIFS in 802.11a is 16 µs. In 802.11g, the SIFS is 10 µs, which is extended by 6 µs of silence to provide what is referred to as a "virtual extension of SIFS." However, most major aspects of the technology is similar or identical and the waveform performance is largely equivalent. The primary difference is the frequency of operation (2.4 GHz vs. 5.2, 5.7 GHz). At the risk of oversimplification, there are no other notable differences that warrant discussion.

4.5 IEEE 802.11E

Over the past 10 years, with the growing importance of supporting voice and video communications across data networks, there has been a growing desire for communications networks to provide quality-of-service (QoS) QoS is particularly important to support performance-sensitive multimedia applications, such as interactive voice and video. The Internet, as deployed today, is a best-effort network, not capable of providing effective QoS mechanisms to packets. There has been a significant amount of research on the topic of QoS and many technologies have been developed to facilitate QoS-enabled Internet Protocol (IP) networks. Some notable QoS techniques in the Internet community include integrated services (IntServ), resource ReSerVation Protocol (RSVP), differentiated services (DiffServ), and multiprotocol label switching (MPLS). These technologies have been designed primarily within the context of wired Internet backbone networks. Multimedia applications, however, require end-to-end QoS. WLAN deployment continues to increase with more end hosts attaching to their respective network through WLAN connectivity. Thus, it is important that WLANs provide adequate QoS mechanisms in order to enable end-to-end QoS. There have been numerous research efforts over the years to introduce QoS mechanisms within the 802.11 technology family. These efforts have focused primarily on modifications to the 802.11 MAC design in order to provide differentiation and prioritized treatment of traffic to support multimedia services. The IEEE 802.11e standard, ratified in 2005 [21], modified the legacy 802.11 MAC in order to support applications with QoS requirements, along with enhancing several deficiencies present within the legacy 802.11 MAC.

4.5.1 The IEEE 802.11e MAC

The 802.11e MAC provides a variety of miscellaneous enhancements intended to improve throughput performance relative to the legacy 802.11 MAC. In particular, some of the enhancements that have been made include:

- The transmission opportunity (TXOP) and multiple frame transmissions
- The Direct link protocol (DLP)
- Block acknowledgements.

The 802.11e introduces the TXOP, which is an interval of time during which an MS transmitting is authorized to transmit multiple frames as opposed to a single frame per successful channel contention. Additionally, 802.11e allows the receiving MP or AP to perform positive frame acknowledgements on a block-by-block basis. All frames transmitted within a TXOP are acknowledged by a bit pattern in the block acknowledgement frame, reducing the amount of overhead required. The result is an improvement in overall throughput and network resource usage efficiency as the receiving entity is not required to provide positive acknowledgement for every frame (analogous to a stop-and-go automatic retransmission request [ARQ] approach). In a legacy 802.11 network, MPs can only communicate directly with one another in the ad hoc mode; all transmissions in the infrastructural mode must be sent through the AP of the BSS. The 802.11e introduces the DLP, which allows an MS to transmit frames directly to another MS without involving the AP. In this case, the MS uses the DLP to establish a direct link with another MS before initiating direct frame transmissions. This leads to more efficient use of network resources, as it reduces the number of multihop transmissions.

4.5.2   The Hybrid Coordination Function

The 802.11e MAC introduces a new medium access mechanism, the hybrid coordination function (HCF). HCF combines functions from the DCF and PCF of the legacy 802.11 MAC into a single coordination function that supports both contention-free and contention-based channel access. The HCF concurrently exists with both DCF and PCF to enable backward compatibility. There are two methods by which an MS can access the wireless 802.11 channel within the 802.11e HCF: contention-based access or contention-free access. The corresponding coordination function, which controls access to the wireless medium, is the enhanced DCF (EDCF) and the HCF controlled channel access (HCCA), respectively. The 802.11e MAC coordination functions are illustrated in the Figure 4.18.

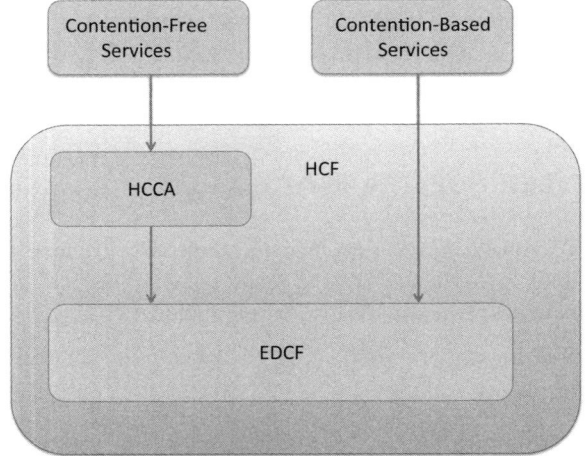

Figure 4.18   The IEEE 802.11e MAC coordination function.

4.5.3 QoS Support within the EDCF

QoS support is realized within the EDCF through the introduction of traffic categories (TCs). There are a total of eight priorities of application traffic for which the MS can provide differentiated channel access. It should be noted that in the EDCF, this is a distributed differentiation mechanism. The EDCF introduces the concept of an access category (AC). The AC mechanism is what provides priority support at the MS. Each MS may have up to four ACs to support eight user priorities (UPs). Table 4.10 [21] provides a summary of how UPs are mapped to ACs and the designation of each UP value.

One or more UPs may be assigned to each AC within the MS. The MS accesses the wireless medium based upon the AC of the frame to be transmitted. The concept of ACs is illustrated in Figure 4.19.

Table 4.10   IEEE 802.11e Access Categories

Access category index	Access category	Designation	CWMIN	CWMAX	AIFSN
00	AC_BE	Best effort	aCWmin	aCWmax	7
01	AC_BK	Background	aCWmin	aCWmax	3
10	AC_VI	Video	(aCWmin+1)/2-1	aCWmin	2
11	AC-VO	Voice	(aCWmin+1)/4-1	(aCWmin+1)/2-1	2

Reproduced from Reference [21].

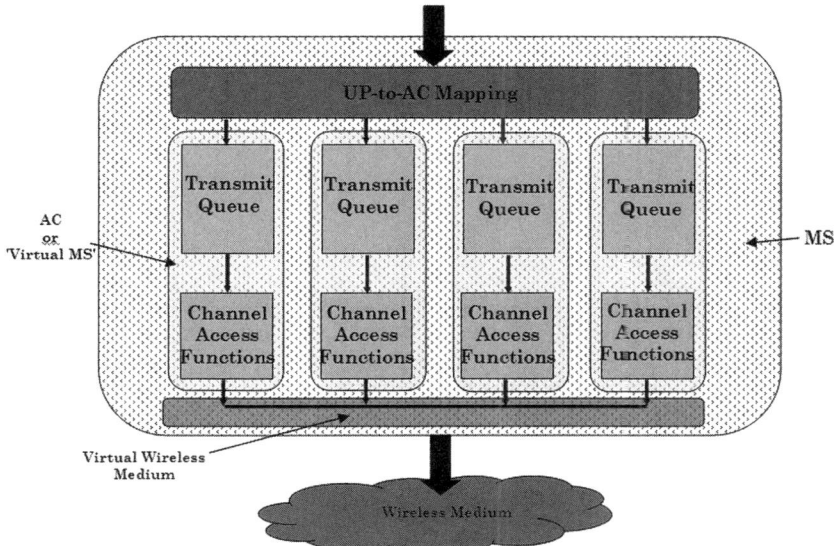

Figure 4.19   The IEEE 802.11e reference model.

Each AC is a "virtual MS" that contends for medium access independently of the other ACs within the MS. The stream of traffic that is offered to the MS is parsed into four separate streams, based upon the UP of the traffic and routed to the AC that corresponds to the UP (according to the mappings shown in Table 4.10). Each AC is an enhanced variant of the DCF from the legacy 802.11 MAC. Each AC must perform channel sensing, which is done much the same way as in the legacy 802.11 MAC using the NAV. Each AC must adhere to a particular interframe spacing before it can enter into the backoff process. Each AC contends for TXOPs using its individual set of EDCF channel access parameters. Each AC is assigned a minimum and maximum contention window size. ACs are differentiated from one another by setting these parameters such that higher priority ACs receive preferential treatment in terms of channel access; higher priority ACs are assigned a shorter contention window. For further differentiation, different IFS values are assigned according to the AC priority levels.

4.5.4 Channel Access in the EDCF

In EDCF, the arbitration IFS (AIFS) is employed. The AIFS is at least the same size as DIFS and can be enlarged for each AC. Similarly to the DCF, if the channel has been idle for at least the IFS time, transmission can begin immediately. Otherwise, the AC must defer access and enter into backoff. The backoff interval is still randomly selected, but is now based upon a contention window size that is unique for an AC. Moreover, a small random time is added at the end of the IFS period to avoid collisions among frames of the same priority class. ACs actually contend for access to the "virtual wireless medium." In the event that there is a collision between ACs within a single MS, the AC with the higher priority is given access. The lower priority AC behaves as if this collision occurred external to the MS on the wireless medium, as the AC isn't aware of the difference between internal and external collisions. The EDCF, with varying AIFS and contention window sizes, is illustrated in Figure 4.20 [21].

4.6 IEEE 802.11N

IEEE 802.11n [27], the most recent major member of the IEEE 802.11 technology family, represents a fundamental leap in WLAN capability, with performance of hundreds of Megabits-per-second throughput (as seen at the network layer) at ranges of several hundred meters. A key feature of 802.11n is that it employs multiple-input, multiple-output (MIMO) techniques, where multiple transmitter and receiver antennas are employed to improve link performance by taking advantage of the resulting spatial diversity. Even before its development was complete, IEEE 802.11n began experiencing significant deployment success from products brought to market based on proprietary predraft versions of the standard. Unfortunately, these early products exhibited very poor interoperability.

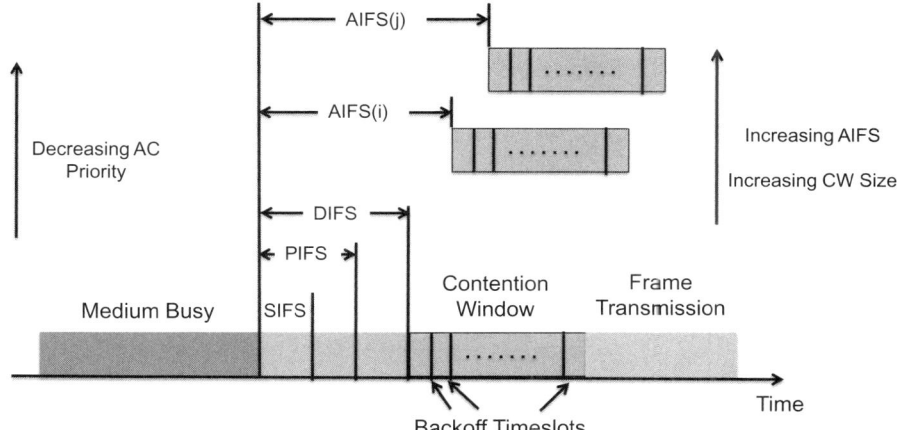

Figure 4.20   Channel Access in the IEEE 802.11e EDCF. Reproduced from Reference [21].

The 802.11n is a MAC and PHY amendment to the IEEE 802.11-2007 baseline specification. The Draft 2.0 802.11n technical specification was approved in March 2007, and the Wi-Fi Alliance began certification testing of products based on the Draft 2.0 IEEE 802.11n specification in June 2007. Consequently, products rapidly began appearing on the market. Final ratification of the IEEE 802.11n specification occurred in 2009 [27].

Many of the commercially available pre-n Draft 1 products provide typical throughput performance of 100–130 Mbps, which is much better than that of 802.11g. Furthermore, Draft 2 products are generally capable of delivering data rates of up to 300 Mbps at moderate ranges, and lower data rates at significantly greater ranges than legacy 802.11g devices. The 802.11n is expected to work best in the 5 GHz band due to the spectral congestion in the 2.4 GHz band. Many key IEEE 802.11n characteristics are summarized in Table 4.11.

4.6.1   IEEE 802.11n PHY Enhancements

Similar to IEEE 802.11g (and IEEE 802.11a), IEEE 802.11n is based on OFDM and introduces new PHY and MAC layer enhancements. The 802.11n high-throughput (HT) PHY is based on the OFDM PHY as defined for IEEE 802.11a extended with up to four spatial streams (i.e., a transmitter/receiver [Tx/Rx] antenna pair at each end of the transmission path) operating with 20-MHz bandwidth and further extended with 40-MHz operation using one to four spatial streams. These features are capable of supporting data rates up to 600 Mb/s (4 spatial streams, 40 MHz bandwidth).

The HT OFDM PHY data subcarriers are modulated using binary phase-shift keying (BPSK), quadrature phase-shift keying (QPSK), 16-quadrature amplitude modulation (16-QAM), or 64-QAM. forward error correction (FEC) coding (convolutional coding) is used with a coding rate of 1/2, 2/3, 3/4, or 5/6.

Table 4.11   Some of the Main Characteristics of 802.11n

Feature	802.11n
Backward compatibility	802.11a/b/g using protection mechanisms
Available bandwidth	83.5/580 MHz
Frequency band of operation	2.4 and 5 GHz
Number of nonoverlapping channels (United States)	3/24
Channel bandwidth	20 or 40 MHz (optional)
Theoretical throughput per channel	From 1 Mbps (b/g) to 600 Mbps (4 × 4 MIMO configuration, 40-MHz channel, 5/6 coding rate, 108 data subcarriers, 2160 data bits per symbol)
Modulation type	DSSS, CCK, OFDM, MIMO-OFDM, and so on. The draft standard lists 77 different modulation and coding schemes (MCSs) indices.
Guard intervals (GIs)	400 and 800 ns
MIMO power save	Limits power consumption penalty of MIMO by utilizing multiple antennas only on as-needed basis (required)
Aggregation	Improves efficiency by allowing transmission bursts of multiple data packets between overhead communication (required)
Reduced interframe spacing (RIFS)	One of several draft-n features designed to improve efficiency. Provides a shorter delay between OFDM transmissions than in 802.11a or g (required).
Greenfield mode	Two new formats are defined for the PLCP (physical layer convergence protocol): the Mixed Mode and the greenfield mode. These two formats are called HT (high-throughput) formats. In addition to the HT formats, there is a legacy duplicate format that duplicates the 20 MHz legacy packet in two 20 MHz halves of a 40 MHz channel. So, the 802.11n physical layer operates in one of 3 modes in the time domain: legacy mode, mixed mode, or greenfield mode. Greenfield mode improves efficiency by eliminating support for 802.11a/b/g devices in an all draft-n network (currently optional).

Two new formats are defined for the physical layer convergence protocol (PLCP): the mixed mode and the greenfield mode. Other features of the PHY are the use of low-density parity-check (LDPC) codes, parallel use of encoders, flexible guard interval (400 or 800 ns), and MIMO support (spatial multiplexing, space–time block coding, and transmit beamforming).

The HT OFDM PHY is mandatory for all equal-modulation rates specified for one and two spatial streams (MCSs 0 through 15) at an access point (AP) and for one spatial stream (MCSs 0 through 7) at a station (STA) using 20-MHz channel width. Support for all other MCSs in two to four spatial streams in 20 MHz, and for all MCSs in one to four spatial streams using 40-MHz channel width is optional.

The maximum HT PSDU (PLCP service data unit) length is 65,535 octets.

Lastly, the HT PHY supports non-HT operation in the 2.4 GHz band and non-HT operation in the 5 GHz band.

4.6.2   MIMO

The 802.11n builds upon previous 802.11 standards by introducing MIMO techniques. MIMO uses multiple Tx and Rx antennas to allow for increased data throughput via SM and increased range by exploiting the spatial diversity (e.g., through the use of coding schemes such as Alamouti coding) and/or beamforming techniques. MIMO antenna configurations are often described with the shorthand "$Y \times Z$" (phonetically "Y by Z"), where Y and Z are integers used to refer to the number of transmitter antennas and the number of receiver antennas. The number of antennas corresponds to the number of simultaneous streams or radiofrequency (RF) chains. The 802.11n specification requires a 2×2 configuration with two streams. The standard also optionally allow for the potential of a 4×4 configuration with four spatial streams.

The basis for MIMO operation is as follows. Each antenna in the system has an RF chain or signal associated with it. Each RF chain is capable of simultaneous reception or transmission, which can drastically improve the throughput. In addition, MIMO has a capability of resolving multipath interference and may improve the quality of the received signal far beyond a simple diversity (when multiple antennas can be used but only a single "best" antenna is selected). Each RF chain and its corresponding antenna are responsible for transmitting a spatial stream. A single frame can be broken up and multiplexed across multiple special streams, which are recombined at the receiver.

4.6.3   Three Types of MIMO: Spatial Diversity, Spatial Multiplexing, and Beamforming

There are three distinct types of MIMO techniques:

- Spatial diversity
- Spatial multiplexing
- Beamforming

The 802.11n supports all three types and allows for various combinations or implementations of these techniques. Spatial diversity and beamforming techniques include but are not limited to the following benefits:

- Increase in communication range
- Boost in signal reception

- Reduction in transmit interference and improvement in receive interference tolerance
- Diversity gain that mitigates multipath effects

The main benefit of spatial multiplexing techniques is the drastic increase in the data throughput. Various MIMO techniques and applications are briefly discussed in the subsequent subsections.

4.6.3.1   Spatial Diversity

Rx Diversity (One Tx Antenna, Multiple Rx Antennas)   Through the use of multiple antennas, this type of MIMO technology demonstrates the ability to coherently resolve information from multiple signal paths using spatially separated *receive* antennas. Multipath signals are the reflected signals arriving at the receiver some time after the line-of-sight (LOS) signal has been received. Typically, multipath is perceived to be the interference that negatively affects the receiver's ability to recover the intelligent information. MIMO, however, uses this opportunity to spatially resolve multipath signals while providing the diversity gain that contributes to the receiver's ability to recover intelligent information.

Tx Diversity (Multiple Tx Antennas, One Rx Antenna)   The use of multiple Tx antennas and a single Rx antenna is another form of spatial diversity. The basic idea behind Tx diversity is illustrated via an example borrowed from.

Consider the case where the transmitter is not "aware" of the channel state. As a result, it must select a data rate (and the appropriate modulation) that is lower than channel capacity in order to be able to achieve error-free transmission. However, the channel capacity changes with channel gain and can get very low during deep fades. Therefore, in order to sustain uninterrupted communication, the data rate must accommodate the worst-case channel gain. With that said, the data rate has a potential of declining virtually to zero.

Now, consider the case where there are two transmit antennas sending the same data stream and one receive antenna. With this arrangement, there are two channels to utilize for transmission. Let us assume that the two transmit antennas are reasonably separated in space, that is, the channels are independently fading. Since the two channels are not likely to go into deep fading at the same time, one might be able to use the "good channel" and thus avoid very low channel capacity. This is the basic idea of Tx diversity.

The Tx diversity can be achieved through the use of the space–time block codes (STBCs). The first and the simplest of all the STBCs is the Alamouti's code, which is still widely used. For more information on the Alamouti's code, please refer to.

4.6.3.2   Spatial Multiplexing

Spatial multiplexing (SM) multiplexes multiple independent data streams that are transferred simultaneously within one spectral channel of bandwidth. SM can

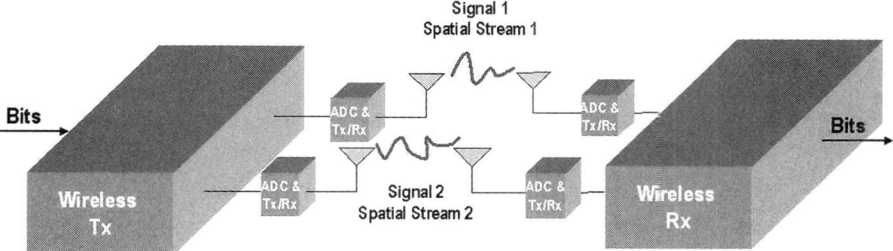

Figure 4.21   Basic two-antenna MIMO system using two-stream SM example.

significantly increase data throughput as the number of resolved data streams is increased. Each spatial stream requires its own Tx/Rx antenna pair at each end of the transmission (Figure 4.21). One must understand that MIMO technology utilizes a separate RF chain and analog-to-digital converter (ADC) for *each* MIMO antenna. Hence, the trade-off for a significantly higher throughput is an increased system complexity, which ultimately translates to higher implementation costs.

The spatial multiplexing in 802.11n MIMO is achieved through the use of the OFDM technique. OFDM is a spread-spectrum technique that distributes the data over a large number of carriers that are spaced apart at precise frequencies. This spacing provides the "orthogonality" that prevents the demodulators from seeing frequencies other than their own. The benefits of OFDM include high spectral efficiency, resiliency to RF interference via minimization of intersymbol interference (ISI), and multipath mitigation. OFDM is sometimes called multicarrier or discrete multitone modulation.

4.6.3.3   Beamforming

There are two types of beamforming: transmit beamforming and receive beamforming/combining. Phased-array transmit beamforming is used for focusing the energy to each Rx, whereas receive combining is used for boosting the reception of standard 802.11 signals. Transmit beamforming offers a power gain due to the simultaneous use of the power amplifiers, up to the regulatory limits, of course. In general, beamforming increases the transmission range through the multipath mitigation and reduction in cochannel interference.

4.6.4   Structure of a Channel

In the previous section, we discussed various benefits of MIMO. However, there is another important tool that can increase the PHY data rate. It is the wider spectral channel bandwidth (B). Increasing channel bandwidth is not a new concept. From Shannon's capacity equation $[C = B*\log_2(1 + SNR)]$ it can be easily seen that the theoretical capacity limit C increases with the channel bandwidth and the signal-to-noise ratio (SNR). From Shannon's equation, it is apparent that one can double the data rate simply by doubling the channel bandwidth.

As previously mentioned, the 802.11n physical layer operates in one of three modes in the time domain: legacy mode, mixed mode, or greenfield mode. Table 4.12 presents timing-related constants for the different 802.11n modes. Figure 4.22 shows the 802.11n transmit procedure for different modes of operation. In the case of a non-HT (legacy) mode and HT mode transmission over a 20-MHz channel, the channel is divided into 64 0.3125-MHz subcarriers. Four pilot signals are inserted in subcarriers -21, -7, 7, and 21. In the non-HT mode, the signal is transmitted on subcarriers -26 to -1 and 1 to 26, with 0 being the center (DC) carrier. In the HT mode, the signal is transmitted on subcarriers -28 to -1 and 1 to 28.

In the case of a 40-MHz HT transmission, two adjacent 20-MHz channels are used. The channel is divided into 128 0.3125 MHz subcarriers. Six pilot signals are inserted in subcarriers -53, -25, -11, 11, 25, and 53. The signal is transmitted on subcarriers -58 to -2 and 2 to 58.

In the case of the HT duplicate and non-HT duplicate mode over 40 MHz, the same data are transmitted over two adjacent 20-MHz channels. Note that the use of the HT duplicate format provides the lowest transmission rate in 40 MHz. It shall only be used for one spatial stream and only with BPSK modulation and rate-1/2 coding. In the HT duplicate and non-HT duplicate formats, the 40-MHz channel is divided into 128 subcarriers and the data are transmitted on subcarriers -58 to -6 and 6 to 58. For example, request-to-send (RTS) messages, clear-to-send (CTS) response, CTS to self, and certain acknowledgements may be transmitted in the non-HT duplicate mode to ensure that devices operating at 20 MHz bandwidth can also receive them.

The transmitted radiofrequency (RF) signal is described in complex baseband signal notation. The actual transmitted signal is related to the complex baseband signal by Equation 4.5 [27]:

$$r_{RF}(t) = \text{Re}\left\{ r(t) \cdot e^{j2\pi f_c t} \right\}, \qquad\qquad (4.5)$$

where Re{ } represents the real part of a complex variable, and f_c is the center frequency of the carrier. The transmitted RF signal is derived by modulating the complex baseband signal, which consists of several fields. The timing boundaries for the various fields of the OFDM PLCP frame or PPDU[1] are shown in Figure 4.23 for the legacy (non-HT), mixed, and greenfield modes [27].

The elements of the PLCP packet are summarized in Table 4.13 [27].

The HT-SIG, HT-STF, and HT-LTFs exist only in HT packets. In non-HT and non-HT duplicate formats, only the L-STF, L-LTF, L-SIG, and Data fields exist.

In both HT mixed format and HT greenfield format frames, there are two types of HT-LTFs: data HT-LTFs (DLTFs) and extension HT-LTFs (ELTFs). DLTFs are always included in HT frames to provide the necessary reference for the receiver to

---

[1] The complete PLCP frame, including PLCP headers, MAC headers, the MAC data field, and the MAC and PLCP trailers.

Table 4.12 Timing-Related Constants

Parameter	Value in non-HT 20-MHz channel	Value in HT 20-MHz channel	Value in 40-MHz channel	
			HT format	MCS32 and non-HT duplicate
N_{SD}: Number of data subcarriers	48	52	108	48
N_{SP}: Number of pilot subcarriers	4	4	6	4
N_{ST}: Total number of subcarriers	52	56	114	104[a]
N_{SR}: The highest data subcarrier index	26	28	58	58
Δ_F: Subcarrier frequency spacing	312.5 kHz (20 MHz/64)	312.5 kHz	312.5 kHz (40 MHz/128)	312.5 kHz (40 MHz/128)
T_{DFT}: inverse discrete Fourier transform (IDFT)/DFT period	3.2 μs	3.2 μs	3.2 μs	3.2 μs
T_{GI}: GI duration	0.8 μs = $T_{DFT}/4$	0.8 μs	0.8 μs	0.8 μs
T_{GI2}: Double GI	1.6 μs	1.6 μs	1.6 μs	1.6 μs
T_{GIS}: Short GI duration	N/A	0.4 μs = $T_{DFT}/8$	0.4 μs	0.4 μs(N/A for non-HT duplicate)
T_{L-STF}: Non-HT Short training field duration	8 μs = $10*T_{DFT}/4$	8 μs	8 μs	8 μs

(*Continued*)

149

Table 4.12 (*Continued*)

Parameter	Value in non-HT 20-MHz channel	Value in HT 20-MHz channel	Value in 40-MHz channel	
			HT format	MCS32 and non-HT duplicate
$T_{HT\text{-}GF\text{-}STF}$: HT greenfield short training field duration	N/A	$8\ \mu s = 10*T_{DFT}/4$	$8\ \mu s$	$8\ \mu s$
$T_{L\text{-}LTF}$: Non-HT long training field duration	$8\ \mu s = 2*T_{DFT} + T_{GI2}$	$8\ \mu s$	$8\ \mu s$	$8\ \mu s$
T_{SYM}: Symbol interval	$4\ \mu s = T_{DFT} + T_{GI}$	$4\ \mu s$	$4\ \mu s$	$4\ \mu s$
T_{SYMS}: Short GI symbol interval	N/A	$3.6\ \mu s = T_{DFT} + T_{GIS}$	$3.6\ \mu s$	$3.6\ \mu s$(N/A for Non-HT Duplicate)
$T_{L\text{-}SIG}$: Legacy (non-HT) signal field	$4\ \mu s = T_{SYM}$	$4\ \mu s$	$4\ \mu s$	$4\ \mu s$
$T_{HT\text{-}SIG}$: HT signal field	N/A	$8\ \mu s = 2*T_{SYM}$	$8\ \mu s$	$8\ \mu s$
$T_{HT\text{-}STF}$: HT Short training field duration	N/A	$4\ \mu s$	$4\ \mu s$	$4\ \mu s$
$T_{HT\text{-}LTF1}$: HT First long training field duration	N/A	4 μs in HT mixed format, 8 μs in HT greenfield format	4 μs in HT mixed format, 8 μs in HT greenfield format	4 μs in HT mixed format, 8 μs in HT greenfield format
$T_{HT\text{-}LTFs}$: HT second and subsequent long training field duration	N/A	$4\ \mu s$	$4\ \mu s$	$4\ \mu s$

[a]Data and pilot tones are replicated in upper and lower 20-MHz portions of 40-MHz signal to make a total of 104 tones used in HT duplicate and non-HT duplicate formats.

N/A, Not applicable.

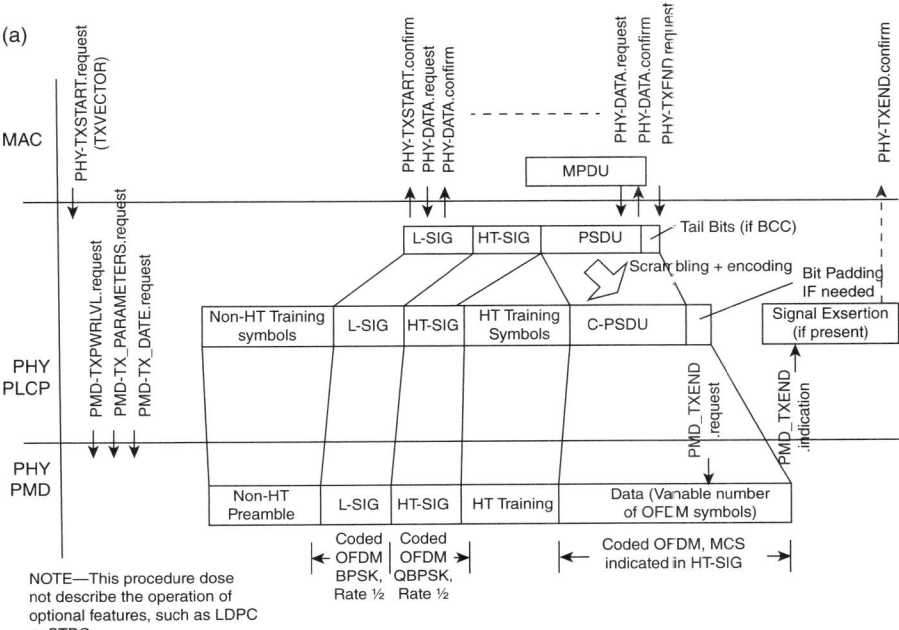

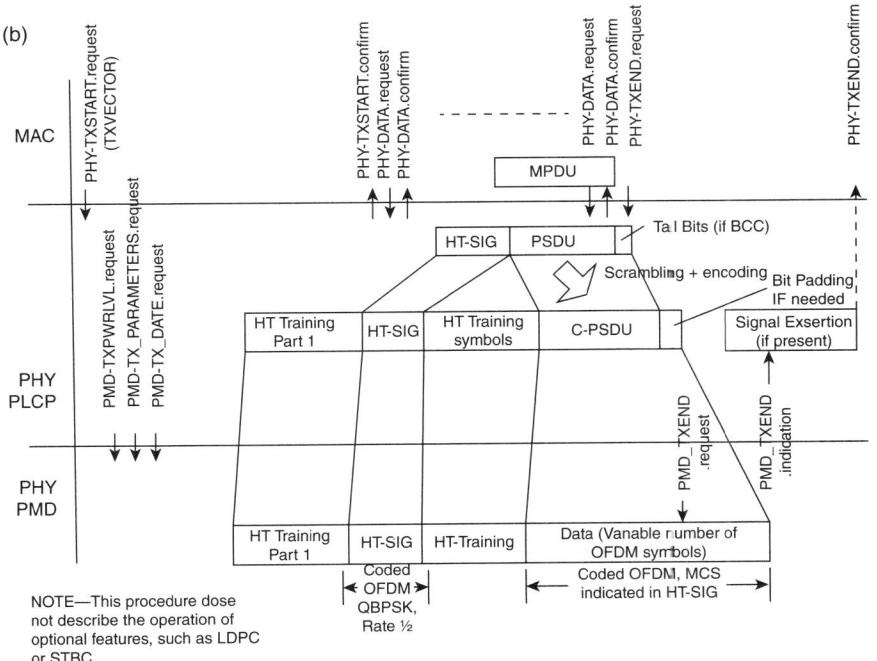

Figure 4.22   802.11n transmit procedure. (a) PLCP transmit procedure (HT-mixed mode); (b) PLCP transmit procedure (HT-greenfield format).

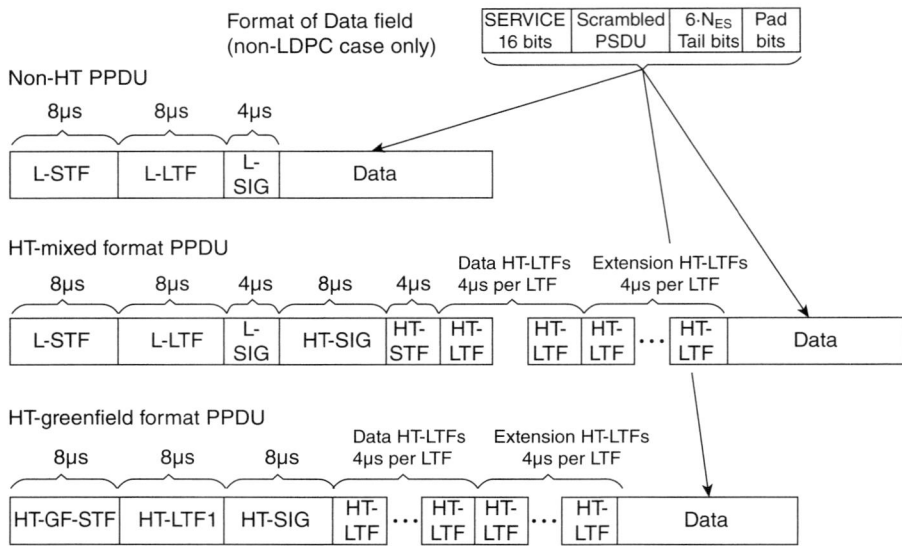

Figure 4.23 PPDU format for different modes. Reproduced from Reference [27].

Table 4.13 Elements of the PLCP Packet

Element	Description
L-STF	Non-HT short training field
L-LTF	Non-HT long training field
L-SIG	Non-HT SIGNAL field
HT-SIG	HT SIGNAL field
HT-STF	HT short training field
HT-GF-STF	HT greenfield short training field
HT-LTF1	First HT long training field (data HT-LTF)
HT-LTFs	Additional HT long training fields (data HT-LTFs and extension HT-LTFs)
Data	The data field includes the PSDU (PHY service data unit)

Reproduced from Reference [27].

form a channel estimate that allows it to demodulate the data portion of the frame. The number of DLTFs may be either 1, 2, or 4, and is determined by the number of space time streams being transmitted in the frame. ELTFs provide additional reference in sounding PPDUs so that the receiver can form an estimate of additional dimensions of the channel beyond those that are used by the data portion of the frame. The number of ELTFs may be either 0, 1, 2, or 4. PLCP preambles in which DLTFs are followed by ELTFs are referred to as staggered preambles. The HT mixed format and HT greenfield format frames both contain staggered preambles for illustrative purposes [27].

4.6.4.1 Legacy Mode

In the legacy mode, frames are transmitted in the legacy 802.11a/g OFDM format.

L-STF   The L-STF is BPSK modulated at 6 Mbps. It contains no channel coding and is not scrambled. The L-STF has a period of 0.8 μs. The entire short training field includes 10 such periods, with a total duration of 8 μs.

- In the 20-MHz transmission mode: the legacy short training OFDM symbols are assigned to subcarriers −24, −20, −16, −12, −8, −4 4, 8, 12, 16, 20, and 24.
- In the 40-MHz transmission mode: the legacy short training OFDM symbols are assigned to subcarriers −58 to −2 and 2 to 58. The tones in the upper subchannel (subcarriers 6–58) are phase rotated by +90°. The 90° rotation helps keep the peak-to-average power ratio (PAPR) of the short training field (STF) in 40 MHz comparable to that in 20 MHz.

L-LTF   The L-LTF is BPSK modulated at 6 Mbps. It contains no channel coding, and is not scrambled.

- In the 20-MHz transmission mode: the legacy long training OFDM symbols are assigned to subcarriers −26 to −1 and 1 to 26.
- In the 40-MHz transmission mode: the legacy long training OFDM symbols are assigned to subcarriers −58 to −2 and 2 to 58. The tones in the upper subchannel (subcarriers 6–58) are phase rotated by +90°. The 90° rotation helps keep the PAPR of the STF in 40 MHz comparable to that in 20 MHz. The subcarriers at ±32 in 40 MHz, which are the DC subcarriers for the legacy 20 MHz transmission, are both nulled in the L-LTF. Such an arrangement allows proper synchronization of the 20 MHz legacy device.

L-SIG   The L-SIG is used for transferring the data rate and length information. The L-SIG consists of one OFDM symbol assigned to all 52 subcarriers. This symbol is BPSK modulated at 6 Mbps and is encoded at a 1/2 rate. L-SIG is interleaved and mapped, and has pilots inserted in subcarriers −21, −7, 7, and 21. The L-SIG is not scrambled. It has different meaning when used in legacy transmission than when used in a nonlegacy transmission.

- In the legacy 20-MHz transmission mode, it is transmitted using the same method and meaning as specified in the IEEE 802.11a standard, shown in Figure 4.24 [27].
- In the legacy 40-MHz transmission mode and greenfield 20-MHz and 40-MHz transmission modes, the bits in the rate field are [1,1,0,1], corresponding to a rate of 6 Mbps in legacy representation. The value in the length field is given through the TX vector. This value is used to spoof legacy devices to defer transmission for a period corresponding to the length of the rest of the packet.

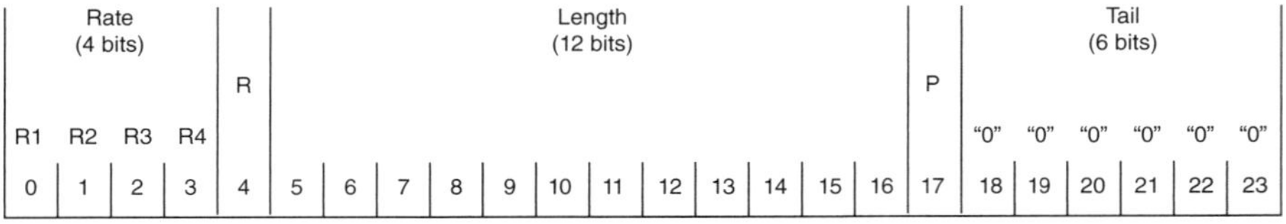

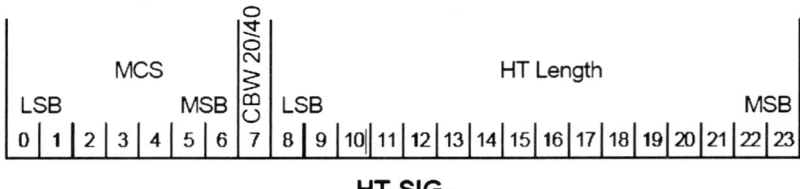

Rate (4 bits)				R	Length (12 bits)												P	Tail (6 bits)					
R1	R2	R3	R4															"0"	"0"	"0"	"0"	"0"	"0"
0	1	2	3	4	5	6	7	8	9	10	11	12	13	14	15	16	17	18	19	20	21	22	23

Figure 4.24 Legacy (non-HT) signal field. Reproduced from Reference [27].

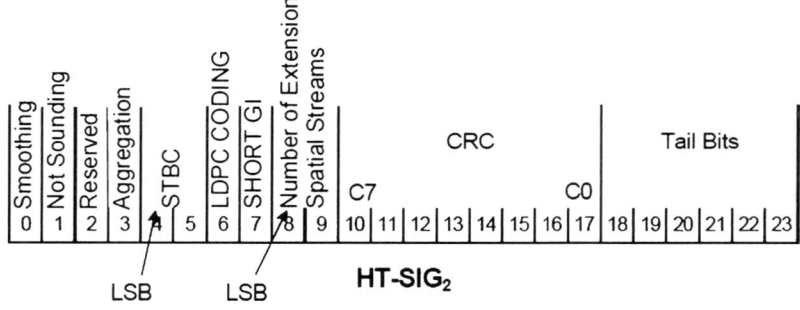

HT-SIG₁

Figure 4.25 Format of HT-SIG1. Reproduced from Reference [27].

HT-SIG₂

Figure 4.26 Format of HT-SIG2. Reproduced from Reference [27].

Data The data field includes the service field, the PSDU, the pad bits, and the tail bits. The data field is scrambled. The data rate, encoding rate, and modulation vary. The supported MSCs are discussed in the subsequent sections.

4.6.4.2 Mixed Mode

In the mixed mode, packets are transmitted with a preamble compatible with the legacy 802.11a/g standards. The L-STF, L-LTF, and L-SIG are transmitted so they can be decoded by legacy 802.11a/g devices (see Figure 4.24). The rest of the packet has a new MIMO training field format.

HT-SIG The HT-SIG is used to carry information required to interpret the HT packet formats. The HT-SIG is composed of two parts: HT-SIG₁ (Figure 4.25) and HT-SIG₂ (Figure 4.26), each containing 24 bits [27]. All the fields in the HT-SIG are transmitted least significant bit (LSB) first, most significant bit (MSB) last. Note that channel bandwidth (CBW) is explicitly signaled within the HT-SIG.

The HT-SIG consists of two OFDM symbols. It is interleaved and mapped. The HT-SIG is not scrambled.

- In the 20-MHz transmission mode: the two OFDM symbols are assigned to subcarriers −28 to 1 and 1 to 28, and have pilot inserted in subcarriers −21, −7, 7 and 21.

- In the 40-MHz transmission mode: the two OFDM symbol are assigned to subcarriers −58 to −2 and 2 to 58, and have pilot inserted in subcarriers −53, −25, −11, 11, 25, 53.

HT-SIGs are BPSK modulated, where the constellation of the data tones is rotated by 90° relative to the legacy signal field. This is shown in Figure 4.27 [27].

HT-STF   The purpose of the HT-STF is to improve AGC training in a multitransmit and multireceive system. The duration of the HT-STF is 4 µs.

- In the 20-MHz transmission mode: the HT-STF is assigned to subcarriers −28 to −1 and 1 to 28; the frequency sequence used to construct the HT-STF is identical to L-STF.

- In the 40-MHz transmission mode: the HT-STF is assigned to subcarriers −58 to −2 and 2 to 58, and constructed from the 20 MHz version by frequency shifting and duplicating, and rotating the upper subcarriers by 90°.

HT-LTF   The HT-LTF assists the receiver in estimating the channel between each spatial mapping input (or spatial stream transmitter if no STBC is applied) and receive chain. The number of training symbols is equal or greater than the number of space–time streams (with an exception in the case of three space–time streams). The HT-LTF portion has one or two parts. The first part consists of one to four HT long training fields that are necessary for demodulation of the HT-Data portion of the PPDU. These HT-LTFs are referred to as data HT-LTFs. The optional second part consists of zero to four HT-LTFs that may be used to probe extra spatial dimensions of the MIMO channel that are not used by the HT-Data portion of the PPDU. These HT-LTFs are referred to as extension HT-LTFs. If a receiver has not advertised

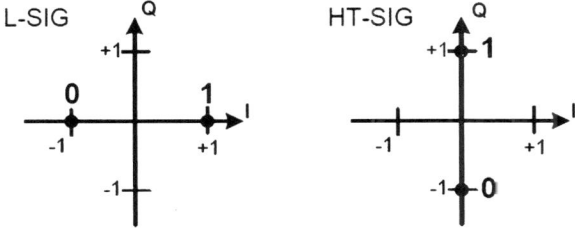

Figure 4.27   Data tone constellations in an HT mixed-format PPDU. Reproduced from Reference [27].

its ability to receive ELTFs, it may discard a frame, including extension HT-LTFs, as an unknown frame type.

- In the 20-MHz transmission mode: the HT-LTF is assigned to subcarriers −28 to −1 and 1 to 28; the training sequence is an extension of the L-LTF where the four extra subcarriers are filled with +1 for negative frequencies and −1 for positive frequencies.

- In the 40-MHz transmission mode: the HT-LTF is assigned to subcarriers −58 to −2 and 2 to 58; the training sequence is also constructed by extending the L-LTF as follows: first, the Legacy LTF is shifted and duplicated for the duplicate legacy mode, then the missing subcarriers are filled: subcarriers [−32 −5 −4 −3 −2 2 3 4 5 32] are filled with the values [1 −1 −1 −1 1 −1 1 1 −1 1], respectively.

Data   The data field of the Mixed Mode includes the service field, the PSDU, the pad bits, and the tail bits. The data field is scrambled. The data rate, encoding rate, and modulation vary. The supported MSCs are discussed in subsequent sections.

4.6.4.3   Greenfield Mode

In the greenfield mode, HT packets are transmitted without a legacy-compatible part. Figure 4.24 shows the greenfield format.

HT Greenfield Format Preamble   For HT greenfield operation, compatibility with legacy devices is not required. Therefore, the portions of the preamble that are compatible with legacy devices are not included. The result is a shorter and more efficient PLCP frame format that includes a short training field, long training fields, and an HT SIGNAL field.

Cyclic Shift Definition for HT Greenfield Format Preamble   Throughout the HT greenfield format preamble, cyclic shift is applied to prevent beamforming when similar signals are transmitted on different spatial streams. The same cyclic shift is applied to these streams during the transmission of the data portion of the frame. The values of the cyclic shift to be used during the HT greenfield format preamble, as well as the data portion of the HT greenfield format frame, are provided in Table 4.14 [27].

4.6.4.4   HT Greenfield Short Training Field

The HT-GF-STF is BPSK modulated at 6 Mbps. It contains no channel coding and is not scrambled. The greenfield waveform has a period of 0.8 µs and the HT-GF-STF includes 10 such periods, with a total duration of 8 µs.

- In the 20-MHz transmission mode: the GF short training OFDM symbols are assigned to subcarriers −24, −20, −16, −12, −8, −4, 4, 8, 12, 16, 20, and 24.

Table 4.14 Cyclic Shift Values of the HT Portion of the Packet

Number of space time streams	Cyclic shift for space-time stream 1 (ns)	Cyclic shift for space-time stream 2 (ns)	Cyclic shift for space-time stream 3 (ns)	Cyclic shift for space-time stream 4 (ns)
1	0	–	–	–
2	0	−400	–	–
3	0	−400	−200	–
4	0	−400	−200	−600

Reproduced from Reference [27].

- In the 40-MHz transmission mode: the GF short training OFDM symbols are assigned to subcarriers −58 to −2 and 2 to 58. The tones in the upper subchannel (subcarriers 6–58) are phase rotated by +90°. The 90° rotation helps keep the peak-to-average-power-ratio (PAPR) of the STF in 40 MHz comparable to that in 20 MHz.

4.6.4.5 First HT Long Training Field

This is the first HT long training field in the greenfield mode. It is 8 µs long consisting of two instances of the sequence plus a double GI insertion.

- In the 20-MHz transmission mode: the HT-LTF is assigned to subcarriers −28 to −1 and 1 to 28. The training field is an extension of the L-LTF where the four extra subcarriers are filled with +1 for negative frequencies and −1 for positive frequencies
- In the 40-MHz transmission mode: the HT-LTF is assigned to subcarriers −58 to −2 and 2 to 58. The training field is also constructed by extending the L-LTF as follows: first, the legacy LTF is shifted and duplicated for the duplicate legacy mode, and then the missing subcarriers are filled: subcarriers [−32 −5 −4 −3 −2 2 3 4 5 32] are filled with the values [1 −1 −1 −1 1 −1 1 1 −1 1], respectively.

4.6.4.6 HT-SIG

The content and format of the HT-SIG field of an HT greenfield format frame is identical to the HT-SIG field in an HT mixed format frame. The placement of the HT-SIG field in an HT greenfield format frame is shown in Figure 4.24 [27].

4.6.4.7 Additional HT-LTFs

The format of the HT-LTF portion of the preamble in an HT greenfield format frame is identical to that of the HT-LTF in an HT mixed format frame with the exception of the first HT-LTF (HT-LTF1), which is twice as long (8 µs) as the other HT-LTFs. The first HT-LTF (HT-LTF1) consists of two periods of the long training symbol, preceded by a double-length (1.6 μs) cyclic prefix. The placement of the first and subsequent HT-LTFs in the greenfield mode is shown in Figure 4.24 [27].

4.6.4.8 Data Field

The Data field consists of the 16-bit service field, the PSDU, either 6 or 12 tail bits, depending on whether there are 1 or 2 encoding streams, and pad bits. The data field is scrambled. The data rate, encoding rate, and modulation vary. The supported MSCs are discussed in subsequent sections.

4.6.5 802.11n Modulation and Coding Schemes

The MCS is a value that determines the modulation, coding, and number of spatial channels. It is a compact representation that is carried in the HT-SIG field. Rate-dependent parameters for the full set of MCSs are described in Table 4.15 [27].

Table 4.16 through Table 4.30 [27] list rate-dependent parameters for MCSs with indices 0 through 76. MCS indices 77–127 are reserved. Table 4.16, Table 4.17, Table 4.18, and Table 4.19 show rate-dependent parameters for equal-modulation (EQM) MCSs for one, two, three, and four streams for 20 MHz operations. Table 4.20, Table 4.21, Table 4.22, and Table 4.23 list rate-dependent parameters for equal-modulation MCSs in one, two, three, and four streams for 40 MHz operations. The same equal-modulation MCSs are used for 20 and 40 MHz operations. Table 4.24 shows rate-dependent parameters for the 40 MHz, 6 Mbps HT duplicate format. The remaining tables—Table 4.25, Table 4.26, Table 4.27, Table 4.28, Table 4.29, and Table 4.30—provide rate-dependent parameters for the MCSs with unequal modulation (UEQM) on the spatial streams for use with the following:

- Transmit beamforming
- STBC modes for which two spatial streams ($N_{SS} = 2$) are encoded into three space time streams ($N_{STS} = 3$), and three spatial streams ($N_{SS} = 3$) that are encoded into four space time streams ($N_{STS} = 4$).

Table 4.15   MCS Parameter Tables

Symbol	Explanation
N_{SS}	Number of spatial streams
R	Coding rate
N_{BPSC}	Number of coded bits per single carrier (total across spatial streams)
$N_{BPSC(iSS)}$	Number of coded bits per single carrier for each spatial stream $i_{SS} = 1, \ldots, N_{SS}$
N_{SD}	Number of data subcarriers
N_{SP}	Number of pilot subcarriers
N_{CBPS}	Number of coded bits per OFDM symbol
N_{DBPS}	Number of data bits per OFDM symbol
N_{ES}	Number of FEC encoders
N_{TBPS}	Total bits per subcarrier

Table 4.16 MCS Parameters for Mandatory 20 MHz, $N_{SS} = 1$, $N_{ES} = 1$, EQM

MCS index	Modulation	R	$N_{BPSCS}(i_{ss})$	N_{SD}	N_{SP}	N_{CBPS}	N_{DBPS}	Data rate (Mb/s) 800 ns GI	400 ns GI[a]
0	BPSK	1/2	1	52	4	52	26	6.5	7.2
1	QPSK	1/2	2	52	4	104	52	13.0	14.4
2	QPSK	3/4	2	52	4	104	78	19.5	21.7
3	16-QAM	1/2	4	52	4	208	104	26.0	28.9
4	16-QAM	3/4	4	52	4	208	156	39.0	43.3
5	64-QAM	2/3	6	52	4	312	208	52.0	57.8
6	64-QAM	3/4	6	52	4	312	234	58.5	65.0
7	64-QAM	5/6	6	52	4	312	260	65.0	72.0

Reproduced from Reference [27].

[a]Support of 400 ns guard interval is optional on transmit and receive.

Table 4.17 MCS Parameters for Optional 20 MHz, $N_{SS} = 2$, $N_{ES} = 1$, EQM

MCS index	Modulation	R	$N_{BPSCS}(i_{ss})$	N_{SD}	N_{SP}	N_{CBPS}	N_{DBPS}	Data rate (Mb/s) 800 ns GI	400 ns GI[a]
8	BPSK	1/2	1	52	4	104	52	13.0	14.4
9	QPSK	1/2	2	52	4	208	104	26.0	28.9
10	QPSK	3/4	2	52	4	208	156	39.0	43.3
11	16-QAM	1/2	4	52	4	416	208	52.0	57.8
12	16-QAM	3/4	4	52	4	416	312	78.0	86.7
13	64-QAM	2/3	6	52	4	624	416	104.0	115.6
14	64-QAM	3/4	6	52	4	624	468	117.0	130.0
15	64-QAM	5/6	6	52	4	624	520	130.0	144.4

Reproduced from Reference [27].

[a]The 400 ns GI rate values are rounded to one decimal place.

Table 4.18 MCS Parameters for Optional 20 MHz, $N_{SS} = 3$, $N_{ES} = 1$, EQM

MCS index	Modulation	R	$N_{BPSCS}(i_{ss})$	N_{SD}	N_{SP}	N_{CBPS}	N_{DBPS}	Data rate (Mb/s) 800 ns GI	400 ns GI
16	BPSK	1/2	1	52	4	156	78	19.5	21.7
17	QPSK	1/2	2	52	4	312	156	39.0	43.3
18	QPSK	3/4	2	52	4	312	234	58.5	65.0
19	16-QAM	1/2	4	52	4	624	312	78.0	86.7
20	16-QAM	3/4	4	52	4	624	468	117.0	130.0
21	64-QAM	2/3	6	52	4	936	624	156.0	173.3
22	64-QAM	3/4	6	52	4	936	702	175.5	195.0
23	64-QAM	5/6	6	52	4	936	780	195.0	216.7

Reproduced from Reference [27].

Table 4.19 MCS Parameters for Optional 20 MHz, $N_{SS} = 4$, $N_{ES} = 1$, EQM

MCS index	Modulation	R	$N_{BPSCS}(i_{ss})$	N_{SD}	N_{SP}	N_{CBPS}	N_{DBPS}	Data rate (Mb/s) 800 ns GI	400 ns GI
24	BPSK	1/2	1	52	4	208	104	26.0	28.9
25	QPSK	1/2	2	52	4	416	208	52.0	57.8
26	QPSK	3/4	2	52	4	416	312	78.0	86.7
27	16-QAM	1/2	4	52	4	832	416	104.0	115.6
28	16-QAM	3/4	4	52	4	832	624	156.0	173.3
29	64-QAM	2/3	6	52	4	1248	832	208.0	231.1
30	64-QAM	3/4	6	52	4	1248	936	234.0	260.0
31	64-QAM	5/6	6	52	4	1248	1040	260.0	288.9

Reproduced from Reference [27].

Table 4.20 MCS Parameters for Optional 40 MHz, $N_{SS} = 1$, $N_{ES} = 1$, EQM

MCS index	Modulation	R	$N_{BPSCS}(i_{ss})$	N_{SD}	N_{SP}	N_{CBPS}	N_{DBPS}	Data rate (Mb/s) 800 ns GI	400 ns GI
0	BPSK	1/2	1	108	6	108	54	13.5	15.0
1	QPSK	1/2	2	108	6	216	108	27.0	30.0
2	QPSK	3/4	2	108	6	216	162	40.5	45.0
3	16-QAM	1/2	4	108	6	432	216	54.0	60.0
4	16-QAM	3/4	4	108	6	432	324	81.0	90.0
5	64-QAM	2/3	6	108	6	648	432	108.0	120.0
6	64-QAM	3/4	6	108	6	648	486	121.5	135.0
7	64-QAM	5/6	6	108	6	648	540	135.0	150.0

Reproduced from Reference [27].

Table 4.21 MCS Parameters for Optional 40 MHz, $N_{SS} = 2$, $N_{ES} = 1$, EQM

MCS index	Modulation	R	$N_{BPSCS}(i_{ss})$	N_{SD}	N_{SP}	N_{CBPS}	N_{DBPS}	Data rate (Mb/s) 800 ns GI	400 ns GI
8	BPSK	1/2	1	108	6	216	108	27.0	30.0
9	QPSK	1/2	2	108	6	432	216	54.0	60.0
10	QPSK	3/4	2	108	6	432	324	81.0	90.0
11	16-QAM	1/2	4	108	6	864	432	108.0	120.0
12	16-QAM	3/4	4	108	6	864	648	162.0	180.0
13	64-QAM	2/3	6	108	6	1296	864	216.0	240.0
14	64-QAM	3/4	6	108	6	1296	972	243.0	270.0
15	64-QAM	5/6	6	108	6	1296	1080	270.0	300.0

Reproduced from Reference [27].

Table 4.22 MCS Parameters for Optional 40 MHz, $N_{SS} = 3$, $N_{ES} = 1$, EQM

MCS index	Modulation	R	$N_{BPSCS}(i_{ss})$	N_{SD}	N_{SP}	N_{CBPS}	N_{DBPS}	N_{ES}	Data rate (Mb/s) 800 ns GI	Data rate (Mb/s) 400 ns GI
16	BPSK	1/2	1	108	6	324	162	1	40.5	45.0
17	QPSK	1/2	2	108	6	648	324	1	81.0	90.0
18	QPSK	3/4	2	108	6	648	486	1	121.5	135.0
19	16-QAM	1/2	4	108	6	1296	648	1	162.0	180.0
20	16-QAM	3/4	4	108	6	1296	972	1	243.0	270.0
21	64-QAM	2/3	6	108	6	1944	1296	2	324.0	360.0
22	64-QAM	3/4	6	108	6	1944	1458	2	364.5	405.0
23	64-QAM	5/6	6	108	6	1944	1620	2	405.0	450.0

Reproduced from Reference [27].

Table 4.23 MCS Parameters for Optional 40 MHz, $N_{SS} = 4$, $N_{ES} = 1$, EQM

MCS index	Modulation	R	$N_{BPSCS}(i_{ss})$	N_{SD}	N_{SP}	N_{CBPS}	N_{DBPS}	N_{ES}	Data rate (Mb/s) 800 ns GI	Data rate (Mb/s) 400 ns GI
24	BPSK	1/2	1	108	6	432	216	1	54.0	60.0
25	QPSK	1/2	2	108	6	864	432	1	108.0	120.0
26	QPSK	3/4	2	108	6	864	648	1	162.0	180.0
27	16-QAM	1/2	4	108	6	1728	864	1	216.0	240.0
28	16-QAM	3/4	4	108	6	1728	1296	2	324.0	360.0
29	64-QAM	2/3	6	108	6	2592	1728	2	432.0	480.0
30	64-QAM	3/4	6	108	6	2592	1944	2	486.0	540.0
31	64-QAM	5/6	6	108	6	2592	2160	2	540.0	600.0

Reproduced from Reference [27].

Table 4.24 MCS Parameters for Optional 40 MHz MCS 32 Format, $N_{SS} = 1$, $N_{ES} = 1$

MCS index	Modulation	R	$N_{BPSCS}(i_{ss})$	N_{SD}	N_{SP}	N_{CBPS}	N_{DBPS}	Data rate (Mb/s) 800 ns GI	Data rate (Mb/s) 400 ns GI
32	BPSK	1/2	1	96	8	48	24	6.0	6.7

Reproduced from Reference [27].

Table 4.25   MCS Parameters for Optional 20 MHz, $N_{SS} = 2$, $N_{ES} = 1$, UEQM

| MCS index | Modulation | | R | N_{BPSC} | N_{SD} | N_{SP} | N_{CBPS} | N_{DBPS} | Data rate (Mb/s) | |
	Stream 1	Stream 2							800 ns GI	400 ns GI
33	16-QAM	QPSK	1/2	6	52	4	312	156	39	43.3
34	64-QAM	QPSK	1/2	8	52	4	416	208	52	57.8
35	64-QAM	16-QAM	1/2	10	52	4	520	260	65	72.2
36	16-QAM	QPSK	3/4	6	52	4	312	234	58.5	65.0
37	64-QAM	QPSK	3/4	8	52	4	416	312	78	86.7
38	64-QAM	64-QAM	3/4	10	52	4	520	390	97.5	108.3

Reproduced from Reference [27].

Unequal modulation MCSs are detailed in the following tables:

- Table 4.25, Table 4.26, and Table 4.27 for 20 MHz operation [27].
- Table 4.28, Table 4.29, and Table 4.30 for 40 MHz operation [27].

MCSs 0 through 15 are mandatory in 20 MHz with 800 ns guard interval at an access point (AP). MCSs 0 through 7 are mandatory in 20 MHz with 800 ns guard interval at all STAs. All other MCSs and modes are optional, specifically including transmit and receive support of 400 ns guard interval, operation in 40 MHz, and support of MCSs with indices 16 through 76.

4.6.6   IEEE 802.11n MAC Enhancements

The MAC enhancements of the IEEE 802.11n mainly aim at overhead reduction and include frame aggregation mechanisms, block acknowledgements, MIMO power-save support, and a reduced interframe space (RIFS). QoS support in IEEE 802.11n is based on IEEE 802.11e (described in Section 4.5).

4.6.6.1   Frame Aggregation

The main purpose behind frame aggregation is to aggregate multiple frames into a single transmission opportunity. Frame aggregation leads to an overhead reduction, which means less headers, channel contention, and acknowledgements. The 802.11n standard also provides additional protection mechanisms to safeguard the aggregated transmission. The proposed aggregation mechanism requires frame sizes to be less than the fragmentation threshold in order to prevent jumbo packet generation.

4.6.6.2   Block Acknowledgements

The Block ACK mechanism is responsible for the channel efficiency improvement by aggregating several ACKs in a single frame. The use of Block ACK is negotiated

Table 4.26 MCS Parameters for Optional 20 MHz, $N_{SS} = 3$, $N_{ES} = 1$, UEQM

MCS index	Modulation			R	N_{BPSC}	N_{SD}	N_{SP}	N_{CBPS}	N_{DBPS}	Data rate (Mb/s)	
	Stream 1	Stream 2	Stream 3							800 ns GI	400 ns GI
39	16-QAM	QPSK	QPSK	1/2	8	52	4	416	208	52	57.8
40	16-QAM	16-QAM	QPSK	1/2	10	52	4	520	260	65	72.2
41	64-QAM	QPSK	QPSK	1/2	10	52	4	520	260	65	72.2
42	64-QAM	16-QAM	QPSK	1/2	12	52	4	624	312	78	86.7
43	64-QAM	16-QAM	16-QAM	1/2	14	52	4	728	364	91	101.1
44	64-QAM	64-QAM	QPSK	1/2	14	52	4	728	364	91	101.1
45	64-QAM	64-QAM	16-QAM	1/2	16	52	4	832	416	104	115.6
46	16-QAM	QPSK	QPSK	3/4	8	52	4	416	312	78	86.7
47	16-QAM	16-QAM	QPSK	3/4	10	52	4	520	390	97.5	108.3
48	64-QAM	QPSK	QPSK	3/4	10	52	4	520	390	97.5	108.3
49	64-QAM	16-QAM	QPSK	3/4	12	52	4	624	468	117	130.0
50	64-QAM	16-QAM	16-QAM	3/4	14	52	4	728	546	136.5	151.7
51	64-QAM	64-QAM	QPSK	3/4	14	52	4	728	546	136.5	151.7
52	64-QAM	64-QAM	16-QAM	3/4	16	52	4	832	624	156	173.3

Reproduced from Reference [27].

Table 4.27 MCS Parameters for Optional 20 MHz, $N_{SS} = 4$, $N_{ES} = 1$, UEQM

| MCS index | Modulation | | | | R | N_{BPSC} | N_{SD} | N_{SP} | N_{CBPS} | N_{DBPS} | Data rate (Mb/s) | |
	Stream 1	Stream 2	Stream 3	Stream 4							800 ns GI	400 ns GI
53	16-QAM	QPSK	QPSK	QPSK	1/2	10	52	4	520	260	65	72.2
54	16-QAM	16-QAM	QPSK	QPSK	1/2	12	52	4	624	312	78	86.7
55	16-QAM	16-QAM	16-QAM	QPSK	1/2	14	52	4	728	364	91	101.1
56	64-QAM	QPSK	QPSK	QPSK	1/2	12	52	4	624	312	78	86.7
57	64-QAM	16-QAM	QPSK	QPSK	1/2	14	52	4	728	364	91	101.1
58	64-QAM	16-QAM	16-QAM	QPSK	1/2	16	52	4	832	416	104	115.6
59	64-QAM	16-QAM	16-QAM	16-QAM	1/2	18	52	4	936	468	117	130.0
60	64-QAM	64-QAM	QPSK	QPSK	1/2	16	52	4	832	416	104	115.6
61	64-QAM	64-QAM	16-QAM	QPSK	1/2	18	52	4	936	468	117	130.0
62	64-QAM	64-QAM	16-QAM	16-QAM	1/2	20	52	4	1040	520	130	144.4
63	64-QAM	64-QAM	64-QAM	QPSK	1/2	20	52	4	1040	520	130	144.4
64	64-QAM	64-QAM	64-QAM	16-QAM	1/2	22	52	4	1144	572	143	158.9
65	16-QAM	QPSK	QPSK	QPSK	3/4	10	52	4	520	390	97.5	108.3
66	16-QAM	16-QAM	QPSK	QPSK	3/4	12	52	4	624	468	117	130.0
67	16-QAM	16-QAM	16-QAM	QPSK	3/4	14	52	4	728	546	136.5	151.7
68	64-QAM	QPSK	QPSK	QPSK	3/4	12	52	4	624	468	117	130.0
69	64-QAM	16-QAM	QPSK	QPSK	3/4	14	52	4	728	546	136.5	151.7
70	64-QAM	16-QAM	16-QAM	QPSK	3/4	16	52	4	832	624	156	173.3
71	64-QAM	16-QAM	16-QAM	16-QAM	3/4	18	52	4	936	702	175.5	195.0
72	64-QAM	64-QAM	QPSK	QPSK	3/4	16	52	4	832	624	156	173.3
73	64-QAM	64-QAM	16-QAM	QPSK	3/4	18	52	4	936	702	175.5	195.0
74	64-QAM	64-QAM	16-QAM	16-QAM	3/4	20	52	4	1040	780	195	216.7
75	64-QAM	64-QAM	64-QAM	QPSK	3/4	20	52	4	1040	780	195	216.7
76	64-QAM	64-QAM	64-QAM	64-QAM	3/4	22	52	4	1144	858	214.5	238.3

Reproduced from Reference [27].

Table 4.28   MCS Parameters for Optional 40 MHz, $N_{SS} = 2$, $N_{ES} = 1$, UEQM

| MCS index | Modulation | | R | N_{BPSC} | N_{SD} | N_{SP} | N_{CBPS} | N_{DBPS} | Data rate (Mb/s) | |
	Stream 1	Stream 2							800 ns GI	400 ns GI
33	16-QAM	QPSK	1/2	6	108	6	648	324	81	90
34	64-QAM	QPSK	1/2	8	108	6	864	432	108	120
35	64-QAM	16-QAM	1/2	10	108	6	1080	540	135	150
36	16-QAM	QPSK	3/4	6	108	6	648	486	121.5	135
37	64-QAM	QPSK	3/4	8	108	6	864	648	162	180
38	64-QAM	16-QAM	3/4	10	108	6	1080	810	202.5	225

Reproduced from Reference [27].

using traffic specification (TSPEC) signaling. The mechanism was introduced in the IEEE 802.11e specification and is further used in IEEE 802.11n.

4.6.6.3 MIMO Power Save

A MIMO device consumes power in all the active Rx (receive) chains. The MIMO power-save feature allows the STA to have only one Rx chain active for most of the time. Multiple Rx chains are enabled only after the STA receives a sequence directed to it. This means that such a sequence should always start with a single-spatial stream frame (e.g., an RTS frame). Then, the station switches back to the mode with one active Rx chain immediately after the TXOP (transmit opportunity) ends. This mechanism ensures that all Rx chains are active only during the active reception.

4.6.6.4 Reduced Interframe Space

RIFS may be used to replace the short interframe space (SIFS) when separating multiple transmissions from a single one. This is mainly a means to reduce the overhead and increase the network efficiency.

Other MAC extensions include: transmit beamforming, antenna selection, link adaptation, and new signaling for bandwidth switching.

4.6.6.5 Security Services

Security services in IEEE 802.11 are provided by the authentication service and the Temporal Key Integrity Protocol (TKIP) and Counter Mode with Cipher Block Chaining Message Authentication Code Protocol (CCMP) mechanisms. The scope of the security services provided is limited to station-to-station data exchange. The data confidentiality service offered by an IEEE 802.11 TKIP implementation is the protection of the MAC service data unit (MSDU) and that by an IEEE 802.11 CCMP implementation is the protection of the MSDU or aggregate MSDU (A-MSDU). In 802.11n, TKIP and CCMP are viewed as logical services located within the MAC

Table 4.29 MCS Parameters for Optional 40 MHz, $N_{SS} = 3$, $N_{ES} = 1$, UEQM

| MCS index | Modulation | | | R | N_{BPSC} | N_{SD} | N_{SP} | N_{CBPS} | N_{DBPS} | N_{ES} | Data rate (Mb/s) | |
	Stream 1	Stream 2	Stream 3								800 ns GI	400 ns GI
39	16-QAM	QPSK	QPSK	1/2	8	108	6	864	432	1	108	120
40	16-QAM	16-QAM	QPSK	1/2	10	108	6	1080	540	1	135	150
41	64-QAM	QPSK	QPSK	1/2	10	108	6	1080	540	1	135	150
42	64-QAM	16-QAM	QPSK	1/2	12	108	6	1296	648	1	162	180
43	64-QAM	16-QAM	16-QAM	1/2	14	108	6	1512	756	1	189	210
44	64-QAM	64-QAM	QPSK	1/2	14	108	6	1512	756	1	189	210
45	64-QAM	64-QAM	16-QAM	1/2	16	108	6	1728	864	1	216	240
46	16-QAM	QPSK	QPSK	3/4	8	108	6	864	648	1	162	180
47	16-QAM	16-QAM	QPSK	3/4	10	108	6	1080	810	1	202.5	225
48	64-QAM	QPSK	QPSK	3/4	10	108	6	1080	810	1	202.5	225
49	64-QAM	16-QAM	QPSK	3/4	12	108	6	1296	972	1	243	270
50	64-QAM	16-QAM	16-QAM	3/4	14	108	6	1512	1134	1	283.5	315
51	64-QAM	64-QAM	QPSK	3/4	14	108	6	1512	1134	1	283.5	315
52	64-QAM	64-QAM	16-QAM	3/4	16	108	6	1728	1296	2	324	360

Reproduced from Reference [27].

Table 4.30 MCS Parameters for Optional 40 MHz, $N_{SS} = 4$, $N_{ES} = 1$, UEQM

MCS index	Modulation				R	N_{BPSC}	N_{SD}	N_{SP}	N_{CBPS}	N_{DBPS}	N_{ES}	Data rate (Mb/s)	
	Stream 1	Stream 2	Stream 3	Stream 4								800 ns GI	400 ns GI
53	16-QAM	QPSK	QPSK	QPSK	1/2	10	108	6	1080	540	1	135	150
54	16-QAM	16-QAM	QPSK	QPSK	1/2	12	108	6	1296	648	1	162	180
55	16-QAM	16-QAM	16-QAM	QPSK	1/2	14	108	6	1512	756	1	189	210
56	64-QAM	16-QAM	QPSK	QPSK	1/2	12	108	6	1296	648	1	162	180
57	64-QAM	16-QAM	QPSK	QPSK	1/2	14	108	6	1512	756	1	189	210
58	64-QAM	16-QAM	16-QAM	QPSK	1/2	16	108	6	1728	864	1	216	240
59	64-QAM	16-QAM	16-QAM	16-QAM	1/2	18	108	6	1944	972	1	243	270
60	64-QAM	64-QAM	QPSK	QPSK	1/2	16	108	6	1728	864	1	216	240
61	64-QAM	64-QAM	16-QAM	QPSK	1/2	18	108	6	1944	972	1	243	270
62	64-QAM	64-QAM	16-QAM	16-QAM	1/2	20	108	6	2160	1080	1	270	300
63	64-QAM	64-QAM	64-QAM	QPSK	1/2	20	108	6	2160	1080	1	270	300
64	64-QAM	64-QAM	64-QAM	16-QAM	1/2	22	108	6	2376	1188	1	297	330
65	16-QAM	QPSK	QPSK	QPSK	3/4	10	108	6	1080	810	1	202.5	225
66	16-QAM	16-QAM	QPSK	QPSK	3/4	12	108	6	1296	972	1	243	270
67	16-QAM	16-QAM	16-QAM	QPSK	3/4	14	108	6	1512	1134	1	283.5	315
68	64-QAM	16-QAM	QPSK	QPSK	3/4	12	108	6	1296	972	1	243	270
69	64-QAM	16-QAM	QPSK	QPSK	3/4	14	108	6	1512	1134	1	283.5	315
70	64-QAM	16-QAM	16-QAM	QPSK	3/4	16	108	6	1728	1296	2	324	360
71	64-QAM	16-QAM	16-QAM	16-QAM	3/4	18	108	6	1944	1458	2	364.5	405
72	64-QAM	64-QAM	QPSK	QPSK	3/4	16	108	6	1728	1296	2	324	360
73	64-QAM	64-QAM	16-QAM	QPSK	3/4	18	108	6	1944	1458	2	364.5	405
74	64-QAM	64-QAM	16-QAM	16-QAM	3/4	20	108	6	2160	1620	2	405	450
75	64-QAM	64-QAM	64-QAM	QPSK	3/4	20	108	6	2160	1620	2	405	450
76	64-QAM	64-QAM	64-QAM	16-QAM	3/4	22	108	6	2376	1782	2	445.5	495

Reproduced from Reference [27].

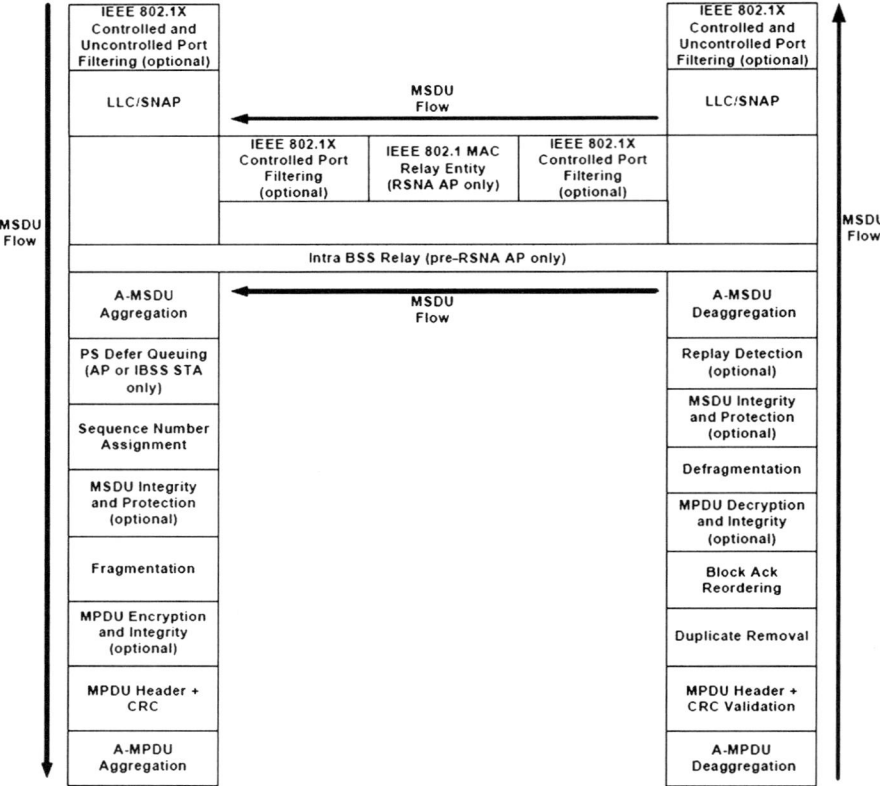

Figure 4.28 MAC data plane architecture. Reproduced from Reference [27].

sublayer. Actual implementations of the TKIP and CCMP services are transparent to the LLC (logical link control) and other layers above the MAC sublayer.

4.6.6.6 MAC Data Service Architecture

The MAC data plane architecture (i.e., processes that involve transport of all or part of an MSDU) is illustrated in Figure 4.28 [27]. During transmission, an MSDU goes through some or all of the following processes: A-MSDU aggregation, frame delivery deferral during power-save mode, sequence number assignment, fragmentation, encryption, integrity protection, frame formatting and A-MPDU (aggregate MPDU) aggregation. IEEE 802.1X may block the MSDU at the Controlled Port. At some point, the data frames that contain all or part of the MSDU are queued per AC/TS (access category/traffic stream).

During reception, a received data frame goes through processes of possible A-MPDU deaggregation, MPDU header and cyclic redundancy code (CRC) validation, duplicate removal, possible reordering if the Block ACK mechanism is used, decryption, defragmentation, integrity checking, and replay detection. After replay

Size (Bytes)

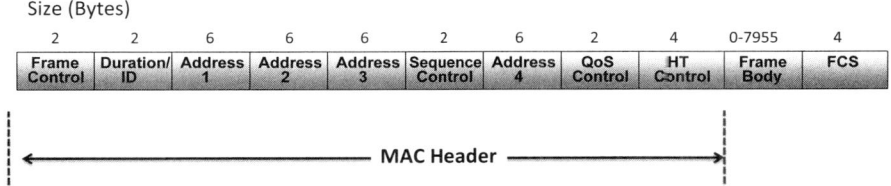

Figure 4.29   General MAC frame format. Reproduced from Reference [27].

detection (or defragmentation if security is used) and possible A-MSDU deaggregation, one or more MSDUs are delivered to the MAC service access point (MAC_SAP) or to the distribution system (DS). The IEEE 802.1X Controlled/Uncontrolled Ports discard any received MSDU if the Controlled Port is not enabled and if the MSDU does not represent an IEEE 802.1X frame. TKIP and CCMP MAC protocol data unit (MPDU) frame order enforcement occurs after decryption, but prior to MSDU defragmentation. Therefore, defragmentation will fail if MPDUs arrive out of order.

4.6.6.7 MAC Frame Format

The MAC frame format comprises a set of fields that occur in a fixed order in all frames. The general MAC frame format is illustrated in Figure 4.29 [27]. The first three fields (Frame Control, Duration/ID, and Address 1) and the last field (FCS [Frame Check Sequence]) in Figure 4.29 constitute the minimal frame format and are present in all frames (for the exception of NDP [Neighbor Discovery Protocol]), including reserved types and subtypes. The fields Address 2, Address 3, Sequence Control, Address 4, QoS Control, HT Control, and Frame Body are present only in certain frame types and subtypes. Fields are transmitted from left to right. A brief description of each field is provided in the subsequent subsections.

A MAC header comprises a Frame Control, Duration/ID, Addresses, optional Sequence Control information, optional QoS Control information (QoS Data frames only), and optional HT Control fields (see Figure 4.2).

Frame Control   Each frame starts with a 2-byte or 2-octet (16 bits) Frame Control field. The components of the Frame Control field are:

- **Protocol version**
  - First 2 bits are allocated for the protocol number. The 802.11n standard does not provide any information about the protocol version, hence it is assumed that it is set to 0.
- **Type and subtype fields**
- **More Fragments field**
- **More Data field**
- **Order field**

Table 4.31   Valid Type and Subtype Combinations

Type value $b3$ $b2$	Type description	Subtype value $b7$ $b6$ $b5$ $b4$	Subtype description
00	Management	1010	Disassociation
00	Management	1011	Authentication
00	Management	1100	Deauthentication
00	Management	1101	Action
00	Management	1110	Action No Ack
00	Management	1111	Reserved
01	Control	0000-0110	Reserved
01	Control	0111	Control Wrapper
01	Control	1000	Block Ack Request (BlockAckReq)
01	Control	1001	Block Ack (BlockAck)
01	Control	1010	PS-Poll
01	Control	1011	RTS
01	Control	1100	CTS
01	Control	1101	ACK
01	Control	1110	CF-End
01	Control	1111	CF-End+CF-Ack

Reproduced from Reference [27].

Ack and ACK, acknowledgment; PS-Poll, power-save poll; RTS, request to send; CTS, clear to send; CF, contention free.

Type and Subtype Fields   Table 4.31 [27] lists various types of management and control frame types. Note that in Table 4.31 bit (b) strings are written with most significant bit first.

The Control Wrapper frame mentioned in Table 4.32 is used to carry any other control frame (excluding the Control Wrapper frame) together with the HT Control field.

More fragments field   The More Fragments field is 1 bit in length and is set to 1 in all data or management type frames that have another fragment of the current MSDU, A-MSDU, or current MMPDU (MAC management protocol data unit) to follow. It is set to 0 in all other frames.

More data field   The More Data field is 1 bit in length and is used for notifying an STA in PS mode that more MSDUs, A-MSDUs, or MMPDUs are buffered for that STA at the AP. The More Data field is valid in directed data or management type frames transmitted by an AP to a STA in PS mode. A value of 1 signifies that at least one additional buffered MSDU, A-MSDU, or MMPDU is present for the same STA.

Table 4.32 QoS Control Field

Applicable frame (sub-) types	Bits 0–3	Bit 4	Bits 5 and 6	Bit 7	Bits 8–15
QoS CF-Poll and QoS CF-Ack+CF-Poll frames sent by HC	TID	EOSP (end of service period)	Ack Policy	Reserved	TXOP Limit
QoS Data+CF-Poll and QoS Data+CF-Ack+CF-Poll frames sent by HC	TID	EOSP	Ack Policy	A-MSDU Present	TXOP Limit
QoS Data and QoS Data+CF-Ack frames sent by HC	TID	EOSP	Ack Policy	A-MSDU Present	AP PS Buffer State
QoS Null frames sent by HC	TID	EOSP	Ack Policy	Reserved	AP PS Buffer State
QoS Data and QoS Data+CF-Ack frames sent by non-AP STAs	TID	0	Ack Policy	A-MSDU Present	TXOP duration requested
	TID	1	Ack Policy	A-MSDU Present	Queue size
QoS Null frames sent by non-AP STAs	TID	0	Ack Policy	Reserved	TXOP duration requested
	TID	1	Ack Policy	Reserved	Queue size

Order field   The Order field is 1-bit long and is set to 1 in any non-QoS Data frame that contains an MSDU, or fragment thereof, which is being transferred using the StrictlyOrdered service class. The Order field is set to 1 in Data or Management frames, except non-QoS Data frames, that are transmitted with a value of HT_GF or HT_MF for the FORMAT parameter of the TXVECTOR, to indicate the presence of the HT Control field. The Order field is set to 0 in Control Wrapper frames. The Order field is set to 0 in all other frames. All QoS STAs that are not HT STAs set the Order field to 0.

Duration/ID   The Duration/ID fields in the MAC headers of MPDUs in an A-MPDU all carry the same value. The reference point for the Duration/ID field is the end of the PPDU carrying the MPDU. If the Duration/ID field is set to the same value in the case of A-MPDU aggregation, each MPDU will consistently specify the same NAV (Network Allocation Vector) setting.

Address Fields   An 802.11n MAC frame may contain up to four address fields. The address fields are numbered because different fields are used for different purposes depending on the frame type. The general rule of thumb is that the Address 1 is used for the receiver, Address 2 for the transmitter, and Address 3 for filtering by the receiver.

Destination Address (DA) Field   The DA field contains an IEEE MAC individual or group address that identifies the MAC entity or entities intended as the final recipient(s) of the MSDU or A-MSDU (or fragment thereof) contained in the frame body field.

Source Address (SA) Field   The SA field contains an IEEE MAC individual address that identifies the MAC entity from which the transfer of the MSDU or A-MSDU (or fragment thereof) contained in the frame body field was initiated. The individual or group bit is always transmitted as a zero in the SA.

Sequence Control Field   The Sequence Control field is a 16-bit field used for both defragmentation and discarding of duplicate frames. It comprises a 4-bit Fragment Number field and a 12-bit Sequence Number field.

Fragment Number Field   The Fragment Number field specifies the number of each fragment of an MSDU, A-MSDU, or MMPDU. The fragment number is set to 0 in the first or only fragment of an MSDU, A-MSDU, or MMPDU, and is incremented by 1 for each successive fragment of that MSDU, A-MSDU, or MMPDU. The fragment number remains constant in all retransmissions of the fragment.

Sequence Control Field   The Sequence Number field specifies the sequence number of an MSDU, A-MSDU, or MMPDU. Each MSDU, A-MSDU, or MMPDU transmitted by a STA is assigned a sequence number. Sequence numbers are not used in control frames because the Sequence Control field is not present. Non-QoS STAs, as well as QoS STAs operating as non-QoS STAs because they are in a non-QoS BSS or non-QoS IBSS, assign sequence numbers to management frames and data frames (the QoS subfield of the Subtype field is set to 0) from a single modulo-4096 counter, starting at 0 and incrementing by 1 for each MSDU, A-MSDU, or MMPDU. QoS STAs associated in a QoS BSS maintain one modulo-4096 counter, per TID (traffic identifier), per unique receiver (specified by the Address 1 field of the MAC header):

> *Sequence numbers for QoS data frames are assigned using the counter identified by the TID subfield of the QoS Control field of the frame, and that counter is incremented by 1 for each MSDU or A-MSDU belonging to that TID. Sequence numbers for management frames, QoS data frames with a broadcast/multicast address in the Address 1 field, and all non-QoS data frames sent by QoS STAs are assigned using an additional single modulo-4096 counter, starting at 0 and incrementing by 1 for each MSDU, A-MSDU or MMPDU. Sequence numbers for QoS (+)Null frames may be set to any value. Each fragment of an MSDU, A-MSDU or MMPDU contains a copy of the sequence number assigned to that MSDU, A-MSDU or MMPDU. The sequence number remains constant in all retransmissions of an MSDU, A-MSDU, MMPDU, or fragment thereof [27].*

QoS Control Field   The QoS Control field is a 16-bit field that identifies the TC (traffic category) or TS (traffic stream) to which the frame belongs and various other

Link Adaptation Control	Calibration Position	Calibration Sequence	Reserved	CSI/ Steering	NDP Announce ment	Reserved	AC Constraint	RDG/ More PPDU
2	2	6	6	6	2	6	2	4

Size (Bits)

Figure 4.30   HT control field. Reproduced from Reference [27].

QoS-related information about the frame that varies by frame type and subtype. The QoS Control field is present in all data frames in which the QoS subfield of the Subtype field is set to 1 (see Table 4.31). Each QoS Control field is composed of five subfields, defined for the particular sender (HC [hybrid coordinator] or non-AP STA) and frame type and subtype (see Table 4.32 [27]). For more information on the usage of these subfields and the various possible layouts of the QoS Control field, please refer to Reference [27].

HT Control Field   The HT Control field is always present in a Control Wrapper frame, and is present in other frames as determined by the Order bit of the Frame Control Field. The format of the 4-octet HT Control Field is illustrated in Figure 4.30 [27]. For more information about individual subfields of the HT Control field, please see Reference [27].

Frame Body Field   The Frame Body is a variable length field that contains information specific to individual frame types and subtypes. The minimum frame body length is 0 octets. The maximum frame body length is 7955 octets. The Frame Body moves the higher-layer payload from one STA to another.

Frame Check Sequence   The FCS is often referred to as the CRC due to the underlying mathematical operations. The FCS allows STAs to check the integrity of received frames. All fields in the MAC header and the body of the frame are included in the FCS.

On the transmitter side of things, the FCS is calculated before frames are sent out over the wireless link. Receivers then can calculate the FCS from the received frame and compare it to the received FCS. If the two match, it is a strong indication that the frame was not corrupted in transit.

4.6.6.8   Transmit Power Limits

The transmit power limits for various regulatory classes listed in 802.11n are provided in the following list:

- 5 GHz: 40, 200, 800, and 1000 mW
- 2.4 GHz: 100 and 1000 mW

4.6.6.9   Receiver Sensitivity Requirements

The receiver minimum input levels must be measured at the antenna connectors and referenced as the average power per receive antenna. The number of spatial streams under test shall be equal to the number of utilized transmitting STA antenna (output)

Table 4.33 Receiver Minimum Input Level Sensitivity

				Minimum sensitivity (dBm)	
Modulation	Rate (R)	Adjacent channel rejection (dB)	Nonadjacent channel rejection (dB)	20 MHz channel spacing	40 MHz channel spacing
BPSK	1/2	16	32	−82	−79
QPSK	1/2	13	29	−79	−76
QPSK	3/4	11	27	−77	−74
16-QAM	1/2	8	24	−74	−71
16-QAM	3/4	4	20	−70	−67
64-QAM	2/3	0	16	−66	−63
64-QAM	3/4	−1	15	−65	−62
64-QAM	5/6	−2	14	−64	−61

Reproduced from Reference [27].

ports and also equal to the number of utilized Device Under Test input ports. Each output port of the transmitting STA shall be connected through a cable to one input port of the Device Under Test [6]. Table 4.33 [27] lists the required receiver minimum input level sensitivity for various MCS.

4.7 IEEE 802.11 SECURITY MODELS

This section provides an overview of the various security architectures and mechanisms that have been standardized and deployed for 802.11 WLANs.

4.7.1 Wired Equivalent Privacy

WEP, introduced in the IEEE 802.11b WLAN technology, aimed to provide data integrity, access control, and confidentiality. WEP security is available only in infrastructure mode 802.11 networks, not in ad hoc mode networks. WEP attempts to provide confidentiality by decrypting user data across the network to prevent eavesdropping. Access control through WEP is achieved in theory because 802.11 stations have the option to discard data that are not properly WEP encrypted, preventing unwanted users from transmitting packets over the wireless network. Data integrity is provided through an integrity value generated using a CRC. If the data have been manipulated in some manner, those data will subsequently be ignored.

The WEP encryption algorithm is based on the well-known RC4 algorithm, which is highly respected in the computer industry [76]. In this algorithm, a long sequence of pseudorandom bytes is generated, the initialization vector (IV). The IV is then processed through a function with a shared secret key. The resulting key stream is used to encrypt user data by performing an XOR operation between the key stream and the unencrypted plaintext. The decryption process performs these operations in reverse using the same IV and shared secret key. The original version of WEP within the IEEE 802.11 standard employed 40-bit shared secret keys.

It is well known that the number of bits constituting such a key is too small, resulting in compromise through brute-force techniques in a matter of minutes utilizing modern computational resources. Even shortly after its release, it was possible to compromise in under 5 hours using modest parallel techniques [76]. Subsequently, WEP2, the next version of WEP, extends the size of the shared secret key to 104 bits. This key size makes brute-force techniques much more difficult. Unfortunately, there are other issues within the 802.11 security architecture that lend itself to compromise even with a 104-bit key. In fact, studies have shown that WEP encapsulation remains insecure with a key length of up to 1000 bits [76]. These problems include the heavy reuse of keys within WEP combined with the ease of data access in a wireless network and the lack of any key management within the protocol. If an adversary possesses two encrypted messages (i.e., ciphertext) that are encrypted with the same key, applying the XOR operation to the two cipher texts will yield a result of the XOR of the two plaintexts. Furthermore, it is possible to ascertain the individual plaintext messages if the XORs of those two plaintexts are known. WEP uses the same shared key for encrypting/decrypting data frames and authentication of MSs. This constitutes a major risk because the encryption and authentication keys are identical. The lack of a key management infrastructure is another problem within WEP. WEP provides no mechanisms for automated exchange or updating of keys. Thus, network administrators are required to manually deliver shared secret keys to all stations. In practice, this has led to WLAN deployments that rarely experience updates to these shared secret keys, making them vulnerable to cryptanalysis. For example, if an AP servicing a single BSS is employing the 11-Mbps 802.11b data rate (assuming a typical packet distribution), the derived key space can be exhausted within an hour.

Wireless networks are quite vulnerable to passive listening or sniffing. Furthermore, very little knowledge is required by an individual to perform sniffing. There is a variety of freely available software products that perform sniffing (i.e., detection and interception) and cryptanalysis that can actually determine the WEP keys in a matter of minutes of an 802.11b network in a fully automated fashion. One of the most infamous examples is one of the very first tools, *AirSnort*, a freeware tool released shortly after WEP vulnerabilities were initially documented. Once this WEP key is determined, that individual has full access to the wireless network just as any other node with knowledge of the shared key. Even without the use of an automated tool, cryptanalysis is easily performed within the 802.11b security paradigm. Upon successful interception of a sufficient number of 802.11b data frames, the adversary possesses the plaintext, the ciphertext, and the IV used to turn the plaintext into ciphertext. This is sufficient information to derive the RC4 key stream.

Another issue associated with WEP is the lack of a sufficiently secure authentication mechanism. The 802.11 stations can perform either open or shared-key authentication. In open authentication, authentication is guaranteed to any station transmitting an authentication request message. Another option is to use the shared secret key of WEP to perform shared-key authentication. This authentication, however, is performed only in one direction. WEP shared-key authentication is a standard challenge–response system. The initiator begins the sequence by transmitting an authentication request. The responder replies with a challenge that is subsequently encrypted by the initiator and returned to the responder. If the decrypted challenge then matches the original sent by the responder, the user is authenticated. The relatively weak encryption of WEP combined with one-way authentication introduces a significant vulnerability to the WLAN.

Furthermore, both mechanisms allow for a variety of attack techniques, such as shared-key attacks or man-in-the-middle attacks, both of which can result in unauthorized users gaining access to the wireless network. The man-in-the-middle attack strategy is one in which the adversary places a node between a valid MS and the AP. The adversary then intercepts the challenge from the responder and sends it on to the initiator. Once the initiator responds with the encrypted response, the adversary once again intercepts that message and then forwards it onto the responder. At this point, the adversary has effectively been authenticated on the network and is a valid network user until the next reauthentication takes place, pending network implementation. This attack strategy is effective primarily because of the poor performance of WEP encryption. Also, this attack strategy would be more difficult if the AP verified the identity of the node that transmitted the authentication request and the node was required to verify the access node to which it is authenticating (i.e., two-way authentication).

The 802.11 WLANs are known to be vulnerable to denial-of-service (DoS) attacks. Some of these vulnerabilities are due to the contention-based MAC, leading to a well-known vulnerability to jamming. Other DoS approaches are made possible through WEP. The association and deassociation messages within 802.11 are not authenticated. As a result, an adversary could continually send streams of association and deassociation frames to the AP. The AP must process all of these, thus potentially affecting the service of legitimate network users. However, if these messages were authenticated, the invalid messages could be ignored. This would significantly reduce the effects of this type of attack.

There are other built-in security mechanisms that can be utilized to provide improved security within the currently deployed WEP security solution: SSID suppression and MAC address filtering. Each 802.11 BSS is identified by an SSID. This SSID is a bit sequence that is typically set to correspond to a readable ASCII text string and is often referred to as the 802.11 network name. To access the network, an MS must know the correct SSID. This SSID, if kept secret, could be used as a simple type of password that is required to gain access to the network. However, this SSID is usually advertised to all potential users by the AP of that BSS through broadcast messages. This is a configurable item that is usually enabled by the network administrator for ease of use but can be disabled. This provides a simple first-line-of-defense type of security, similar to the password mechanism typically employed within computer operating systems (OSs). However, this SSID is still passed in the open through other system management messages, thus easily defeated by passive eavesdropping.

Another method to improve the security of an 802.11 network is to employ what is known as MAC address filtering. An AP uses the unique MAC address of the 802.11 network interface card (NIC) within a station as an identifier. Within an 802.11 network, the AP can be configured to allow only communications with a predefined set of MAC addresses. If a message is received from a MAC address that is not on this list, that client is not allowed to associate with the AP. This is a valid solution for small networks. However, it quickly becomes cumbersome for network administrators managing large networks. Furthermore, an adversary through eavesdropping can easily overcome this approach: an adversary can determine valid MAC addresses through passive eavesdropping and can then set that same MAC address within his network card, thus defeating this security mechanism and gaining access to the WLAN.

4.7.2   802.11i

The IEEE 802.11i WLAN standard [24] is intended to provide significantly enhanced security to 802.11 networks. The 802.11i introduces next-generation authentication, authorization, and encryption capabilities that are generally believed to be much stronger than 802.11b WEP. There are three primary components to the 802.11i security architecture:

- Temporal Key Integrity Protocol (TKIP)
- Counter mode cipher block chaining with message authentication codes (CBC-MAC)
- 802.1x port authentication

TKIP is intended to be an immediate replacement for WEP. It corrects well-known problems associated with small IVs and encryption keys. TKIP uses RC4 encryption, providing a smoother transition path from existing WEP equipment. TKIP employs a 48-bit IV, which is considered much stronger than the current 24-bit IV used in WEP; cryptanalysis is much more difficult with the larger IV [20]. TKIP also employs a larger 128-bit encryption key, which makes brute-force techniques more difficult. TKIP employs per-packet encryption. A shared base key, the MS's MAC address, and a packet's sequence number form a unique key for each packet. This is considered more secure because it mitigates the cryptanalysis threat by making it more impractical by eliminating the threat based upon the capture of large amounts of data encrypted by the same key. Additionally, TKIP periodically rotates the broadcast key to avoid data harvesting attack strategies. TKIP, although an improvement over WEP, is only a temporary solution to provide a quick fix for known WEP vulnerabilities.

There is another protection mechanism: Counter Mode with CBC-MAC Protocol (CCMP). CCMP employs the next-generation 128-bit Advanced Encryption Standard (AES) as a replacement for RC4. CCMP is a required component of any 802.11i-compliant device.

The 802.1x is a port-based authentication mechanism for Ethernet networks. However, 802.1x has been extended to the WLAN environment. In this paradigm, each user must authenticate before receiving full network access. This is also true in the WEP paradigm, but in 802.11i, the AP allows only the client to send authentication information by performing two-way authentication. The introduction of two-way authentication mitigates the man-in-the-middle attacks that have been common in 802.11b networks [18]. The 802.11i also introduces the requirement for an authentication server within the network. The 802.1x employs Extensible Authentication Protocol (EAP), which was originally designed for Point-to-Point Protocol (PPP), to handle authentication requests.

4.7.3   Wi-Fi Protected Access

The first-generation Wi-Fi Protected Access (WPA) standard is a subset of the proposed 802.11i standard and was rushed to market to provide a quick band-aid solution to WEP security. CCMP is also a part of the second-generation WPA encryption standard published by the Wi-Fi Alliance (WFA), WPA2.

4.8   OTHER WLAN TECHNOLOGIES

There have been WLAN technology specifications other than IEEE 802.11 that are worthy of brief discussion, most notably HIPERLAN and WAPI.

4.8.1   HIPERLAN

The HIPERLAN WLAN technology standard was established by the ETSI as a way to enable wireless network connectivity on a variety of platforms: third-generation cellular, home wireless LAN, and corporate wireless LAN, for example. The ETSI Broadband Radio Access Networks (BRAN) group has developed the second-generation HIPERLAN/2 as the follow-on standard to HIPERLAN/1, similar to the IEEE 802.11 evolution of standards (from the 11-Mbps "b" standard to the 54-Mbps "g" standard). HIPERLAN/2 operates in the 5-GHz U-NII band. It supports data rates ranging from 6 to 54 Mbps via an OFDM format. HIPERLAN/2 uses a time division multiple access (TDMA) scheme to share the medium among multiple users.

HIPERLAN topologies are similar to cellular infrastructure topologies, in that there are base stations and wireless users. Because it complies with the BRAN PHY and DLC standards, it is interoperable with a variety of other European core network standards, such as GSM. There are two modes of operation within HIPERLAN: centralized and direct. Centralized mode is analogous to the infrastructural mode within the IEEE 802.11 standard, where cellular-like infrastructure is required to relay packets from users through base stations to other users. The direct mode is analogous to the ad hoc mode in IEEE 802.11, where users can send and receive

Table 4.34   HIPERLAN Specification Summary

Technology	Frequency	Data rates (Mbps)	Nominal ranges (m)	Transmitter power (mW)
ETSI HIPERLAN/2	5.8-GHz U-NII	6, 9, 12, 18, 27, 36, and 54	150–300	1–100

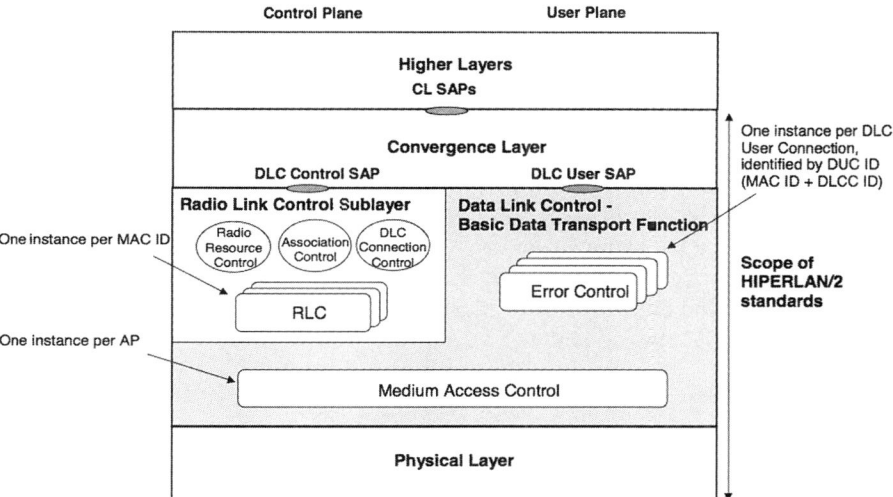

Figure 4.31   HIPERLAN/2 protocol architecture. Reproduced from Reference [77]. Reprinted by permission of the European Telecommunications Standard Institute, Copyright 2000. Further use, modification, copy and/or distribution are strictly prohibited. ETSI standards are available from http://pda.etsi.org/pda/.

packets to and from each other without traversing an infrastructure node. Specifications of HIPERLAN are summarized in Table 4.34.

There are four OFDM subcarrier modulation modes in HIPERLAN/2: BPSK, QPSK, 16-QAM, and 64-QAM. Mandatory error correction code specifications call for rate 1/2, constraint length $k = 7$ convolutional code, with optional rate 9/16 and 3/4 codes for the higher data rates (27–54 Mbps).

One distinguishing feature of HIPERLAN compared to IEEE 802.11 is that it supports multiple-beam antennas (sectoring) for improved link budget performance and reduction in interference. This feature was included primarily for ease of integration into existing cellular infrastructure. Similar to IEEE 802.11, however, HIPERLAN increases or decreases data rate by changing modulation and coding based on PHY and MAC layer metrics (such as signal strength and packet loss ratio).

Figure 4.31 [77] illustrates the packet-based convergence layer. Here, the HIPERLAN/2 PHY and DLC layers are used to connect a user to the HIPERLAN network. The common part of the packet-based convergence layer frames the user data coming from the service-specific convergence sublayer. Service-specific convergence sublayer specifications exist for other various network standards, such as IEEE 802.3 and IEEE 1394.

There are several basic logical channel types in HIPERLAN. The broadcast channel (BCH) is used in the downlink direction and contains information regarding control channel information for a particular coverage area or cell. It is mandatory for APs to support. The frame channel (FCH) conveys information in the downlink direction about the structure of the MAC frame visible at the air interface. The access feedback channel (ACH) is used for sending information to users about success of previous MAC frames sent to an AP. The long transport channel (LCH) and short transport channel (SCH) are used to transmit user data. The random channel (RCH) gives mobile users the opportunity to send control information to the AP when it has not been granted an SCH for user data transport. Figure 4.32 [77] illustrates the MAC frame format for a single-sector implementation of HIPERLAN.

It is shown in the figure that the DL phase contains the user data bound from the AP to the mobile user, where the UL phase contains user data bound from the user to the AP. The DL phase is employed in the direct mode operation, where MAC frames do not have to traverse APs to reach mobile user destination addresses.

Figure 4.33 and Figure 4.34 [77] illustrate the frame formats for user data within the SCH and LCH, respectively.

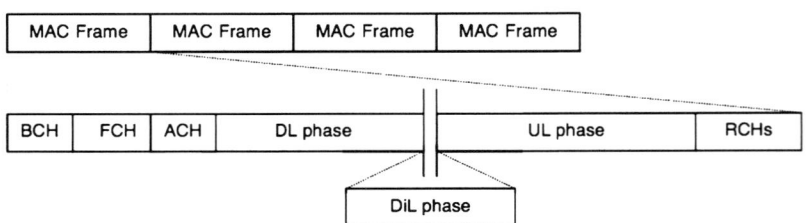

Figure 4.32   MAC frame format for single-sector HIPERLAN. Reproduced from Reference [77]. Reprinted by permission of the European Telecommunications Standard Institute, Copyright 2000. Further use, modification, copy and/or distribution are strictly prohibited. ETSI standards are available from http://pda.etsi.org/pda/.

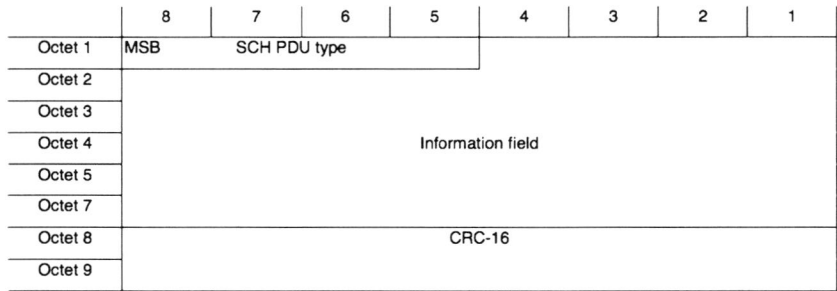

Figure 4.33   Frame format for user data in SCH. Reproduced from Reference [77]. Reprinted by permission of the European Telecommunications Standard Institute, Copyright 2000. Further use, modification, copy and/or distribution are strictly prohibited. ETSI standards are available from http://pda.etsi.org/pda/.

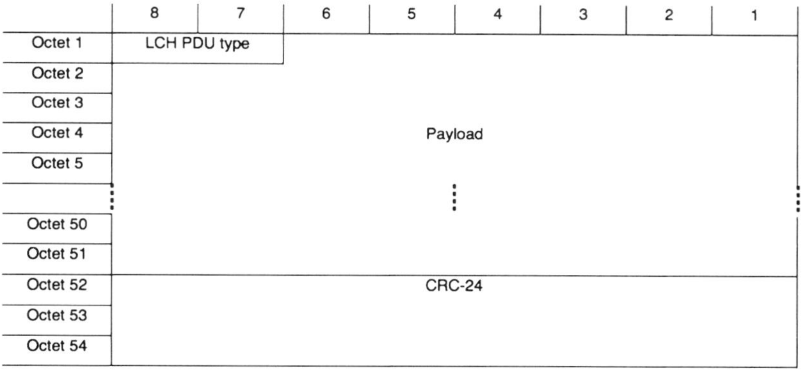

	8	7	6	5	4	3	2	1
Octet 1	LCH PDU type							
Octet 2								
Octet 3								
Octet 4				Payload				
Octet 5								
⋮				⋮				⋮
Octet 50								
Octet 51								
Octet 52				CRC-24				
Octet 53								
Octet 54								

Figure 4.34   Frame format for user data in LCH. Reproduced from Reference [77]. Reprinted by permission of the European Telecommunications Standard Institute, Copyright 2000. Further use, modification, copy and/or distribution are strictly prohibited. ETSI standards are available from http://pda.etsi.org/pda/.

As seen in the figures, the LCH contains a more robust CRC-24 for error detection, as compared with the SCH format, which contains a CRC-16. Up to 6.5 octets of user data can be encapsulated in a SCH frame, whereas the LCH supports up to 51.5 octets of user data. In this sense, SCH would be suited for applications such as VoIP, whereas LCH would operate more efficiently in large file transfers.

HIPERLAN has been shown to have equivalent physical layer performance to the IEEE 802.11a standard. Furthermore, HIPERLAN/2 has been shown to achieve greater throughput performance than 802.11a [78, 79]. However, because HIPERLAN/2 is based on fixed time slots, efficient external scheduling mechanisms are required to enable this improved throughput performance. HIPERLAN is a promising technology for many applications, and there are indeed many lessons that can likely be drawn from HIPERLAN when designing a WLAN. However, HIPERLAN has not achieved significant market presence even in Europe, where it was developed. However, several aspects of the technology have influenced other, more widely deployed, technologies. For example, several components of the HIPERLAN/2 specification, including Dynamic Frequency Selection (DFS) and Transmit Power Control, have been reused in IEEE 802.11 WLANs. Other components of HIPERLAN/2, such as the concept of Dynamic TDMA, can be found in other technologies such as the IEEE 802.16 WMAN specification.

4.8.2 WAPI

Another important, and controversial, development in the WLAN landscape has been the Chinese WLAN standard GB15629.11-2003, which was approved by the SAC in May 2003 and went into effect in December 2003. It is very similar to the IEEE 802.11 standards but employs a different security model, referred to as WAPI. This was motivated due to concerns by the Chinese government of security flaws in IEEE 802.11 technologies. It was decided that widespread deployment and reliance on IEEE 802.11 WLAN technologies could potentially compromise its national information infrastructure. Thus, it was mandated that no WLAN products could be sold in China that did not conform to the GB15629.11-2003 standard. In fact, the WAPI security model is much stronger than original 802.11 security models and is very similar to the 802.11i security model, offering much stronger encryption than WEP and two-way authentication mechanisms. There are no significant differences at the MAC or PHY layers. This technology has been controversial because many have labeled it as a barrier to trade in the Chinese market. However, despite the preference of the Chinese government for WAPI, many reports indicate that IEEE 802.11i-based products have actually dominated the Chinese WLAN market.

4.9 PERFORMANCE OF IEEE 802.11 WLAN TECHNOLOGIES

As evident at this point, IEEE 802.11 WLANs have evolved rapidly since their inception. Combined with their enormous commercial success, this means that there exist a wide range of Wi-Fi networks in existence. Simply stating that a network is "a Wi-Fi network" is not sufficient to understand the network. It could be an older IEEE 802.11b WLAN, which experienced enormous success. It could be a newer IEEE 802.11b/g WLAN. It could be a modern IEEE 802.11n WLAN. It could employ portions of IEEE 802.11e. It could employ a wide range of security features, from WEP to WPA2. This is an important distinction because these various variants of IEEE 802.11 provide a wide range of performance and networking opportunities/ challenges. To illustrate these differences, this section briefly discusses the performance of various types of IEEE 802.11-based WLANs.

4.9.1 Analytical Performance of Original IEEE 802.11 MAC

Consider the throughput performance of the 802.11 MAC. It is generally well known that the original 802.11 MAC exhibited inefficiencies that can significantly degrade 802.11 throughput performance. Throughput performance of the 802.11 MAC is a function of a variety of factors: the distance between communicating nodes, the number of nodes contending for the wireless medium, the conditions of the wireless medium, the offered traffic statistics, and the MAC parameter configuration. Additionally, the overhead associated with the legacy 802.11 MAC inherently introduces inefficiencies. Figure 4.35 [80] illustrates the effect of transmitter–receiver (T-R) separation on throughput for a typical indoor office environment. The result in Figure 4.35 is for a single T-R pair using the 54-Mbps 802.11a PHY. Here, throughput is dependent only on distance.

As can be seen in Figure 4.36, achievable throughput approaches the raw link data rate of the PHY for very small distances excluding MAC overhead. Throughput then decreases as T-R separation increases. When MAC overhead is considered, throughput is limited to approximately 30 Mbps for short T-R separations. Similar results have been frequently observed for all the 802.11 PHY designs. Theoretical

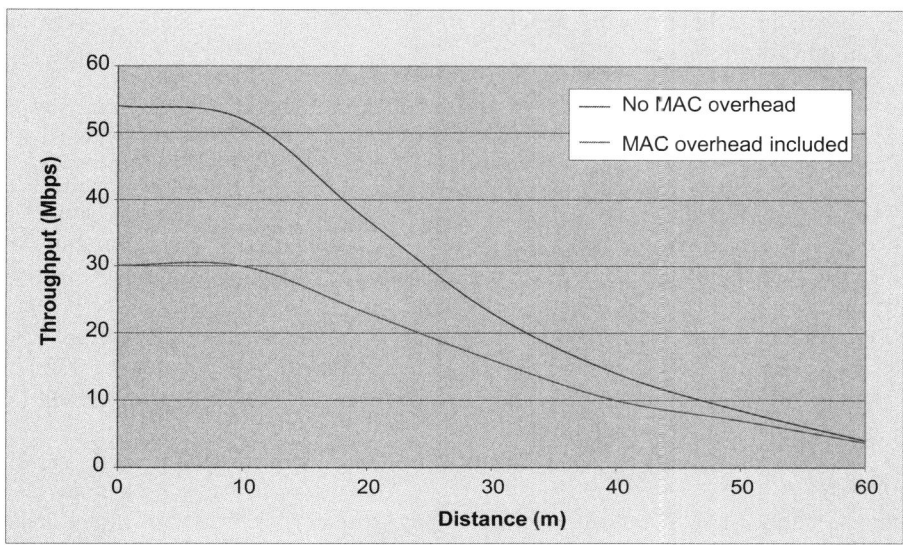

Figure 4.35   Achievable 802.11a throughput versus distance. Reproduced from Reference [80].

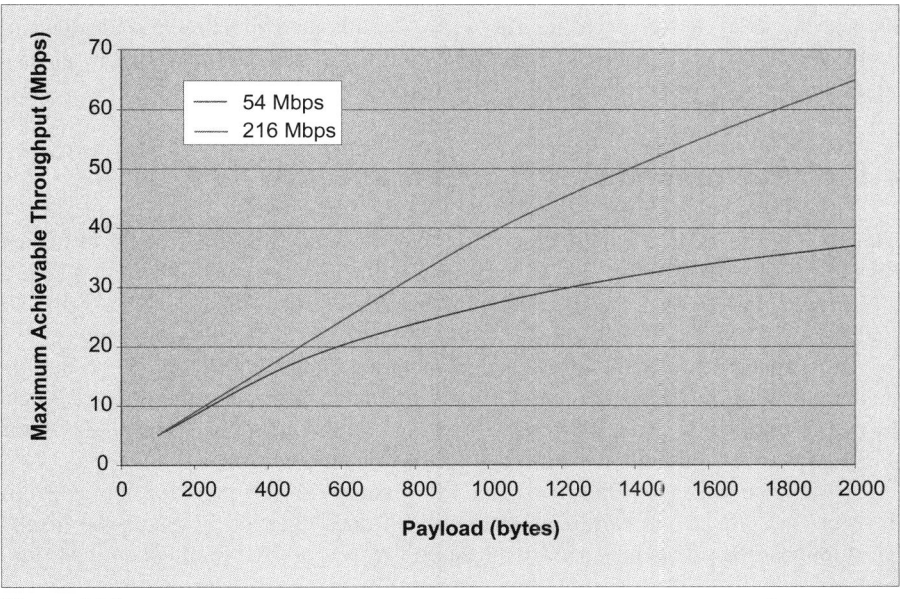

Figure 4.36   Maximum achievable throughput performance versus packet size for different PHY data rates of legacy 802.11 MAC.

maximum throughpts for the legacy 802.11 MAC were derived from the following equation from Reference [80]:

$$Throughput_{MAX} = \frac{8 \cdot L_{DATA}}{\dfrac{CW_{MIN} \cdot T_{SLOT}}{2} + T_{DIFS} + T_{D\_DATA} + \tau + T_{SIFS} + T_{D\_ACK} + \tau}, \quad (3.1)$$

where T_{D_DATA} and T_{D_ACK} refer to the durations required for the transmission of a packet and an acknowledgment, respectively. T_{DIFS} and T_{SIFS} refer to the durations of the DIFS and the SIFS employed during transmission and acknowledgment, respectively. T_{SLOT} is the duration of a time slot. CW_{MIN} is the minimum size of the contention window. τ is the link propagation delay. L refers to the frame size. Figure 4.36 (from Reference [80]) shows the maximum achievable throughputs for the 802.11a 54-Mbps PHY design and a hypothetical 216-Mbps PHY design, as described by equation 3.1 from Reference [80].

For a hypothetical enhanced PHY of 216 Mbps, the expression from Reference [80] shows the throughput remains below 80 Mbps even for large packet sizes. In this scenario, MAC overhead amounts to approximately 75% of transmitted data. These behaviors are more pronounced for small frame sizes because it is well known that the efficiency of the legacy 802.11 MAC design is particularly poor for small frame sizes.

Legacy 802.11 MAC throughput performance further degrades when other factors are considered, such as contention with other MSs. The hidden node problem further degrades MAC performance with or without the RTS/CTS mechanism employed. There are numerous discussions on this topic in the literature that clearly illustrate the poor performance of the original 802.11 MAC design.

4.9.2   Performance of the IEEE 802.11b PHY

The BER-versus-signal energy-to-noise power density ratio (Es/No) performance for the various IEEE 802.11 physical layer waveforms is given in Figure 4.37 [81].

Figure 4.38 [81] provides corresponding PER performance as a function of Es/No.

Note that neither of these results considers the composite (PHY, MAC) performance of 802.11b. Rather, Figure 4.37 and Figure 4.38 provide the error rate performance for the 802.11b PHY design. Thus, although they are still meaningful, their usefulness is limited until combined with MAC layer performance.

Reference [82] provides benign BER-versus-Es/No curves for the 802.11b 11-Mbps waveform. As can be seen, this result does not closely match the results of Reference [81]. This is generally the case of results present in the open literature. This is likely due to the assumptions that must be made regarding receiver aspects that are not strictly specified in the technology standards. In general, the results presented in open literature are based on modeling and simulation (M&S). As a result, the fidelity of presented results must always be scrutinized due to the principle that a result is only as good as the assumptions that have been made in deriving said

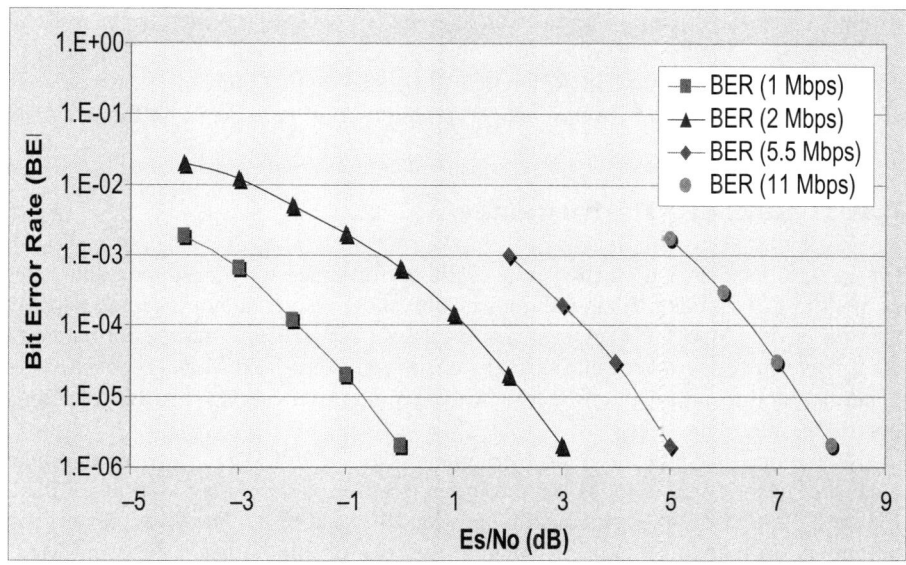

Figure 4.37   BER performance of 802.11b waveforms. Reproduced from Reference [81].

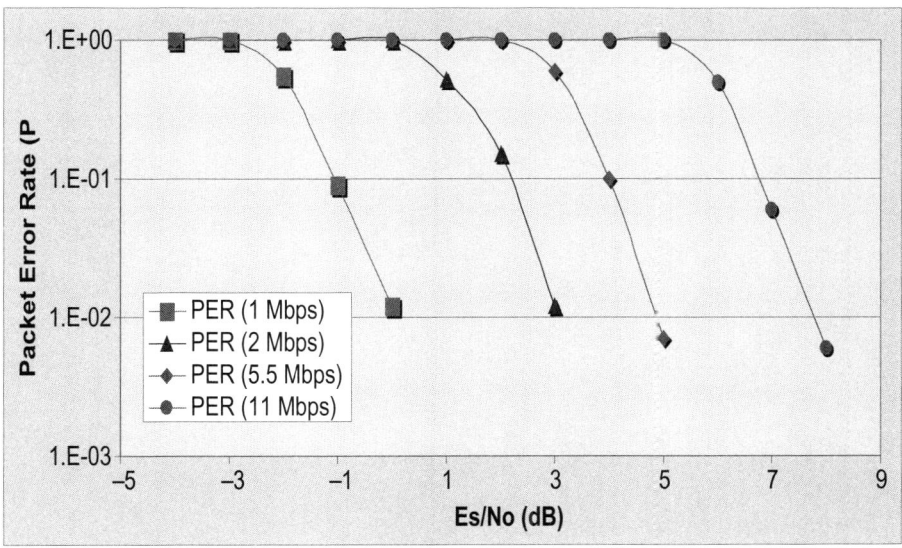

Figure 4.38   PER performance of 802.11b waveforms (packet size of 1000 bytes). Reproduced from Reference [81].

result. In general, error rate versus Eb/No or Es/No results presented in open literature vary by as much as 10 dB.

The results presented in Reference [83], shown in Figure 4.39, agree very
closely with the results of Reference [81], which build confidence in the accuracy
of those results.

4.9.3   802.11g Performance

Figure 4.40 [83] presents PER-versus-SNR performance for the various data rates
of the 802.11g standard. As in previous results, these results do not take into account
the composite (MAC, PHY) performance. These curves show that performance of
each waveform degrades quite quickly once some threshold value is reached. In
general, that threshold value is at SNR values of approximately 3–18 dB, depending
upon data rate.

Reference [84] presents a more complete result of 802.11g performance. Figure
4.41 provides insight into MAC throughput performance as a function of PER.
Figure 4.40 provides the relationship between PER and SNR. These two results can
be applied in tandem to gain insight into the relationship between PER and MAC
throughput, which is the ultimate measure of the utility the WLAN technology is
providing to the network layer above it. These results are provided for a packet size
of 1500 bytes.

It is interesting to note in Figure 4.41 that the MAC throughput degradation is
more marked for the higher data rates as packet error rates increase.

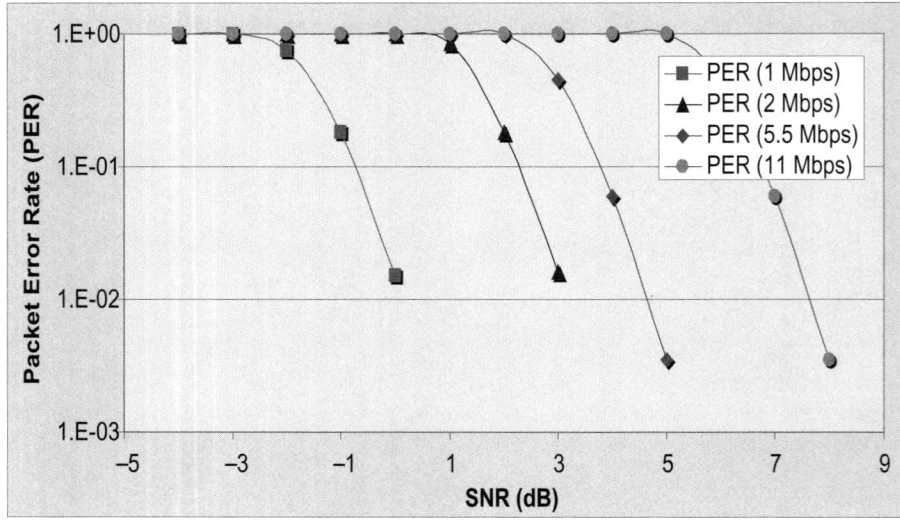

Figure 4.39   PER performance of 802.11b waveforms (packet size of 1000 bytes).

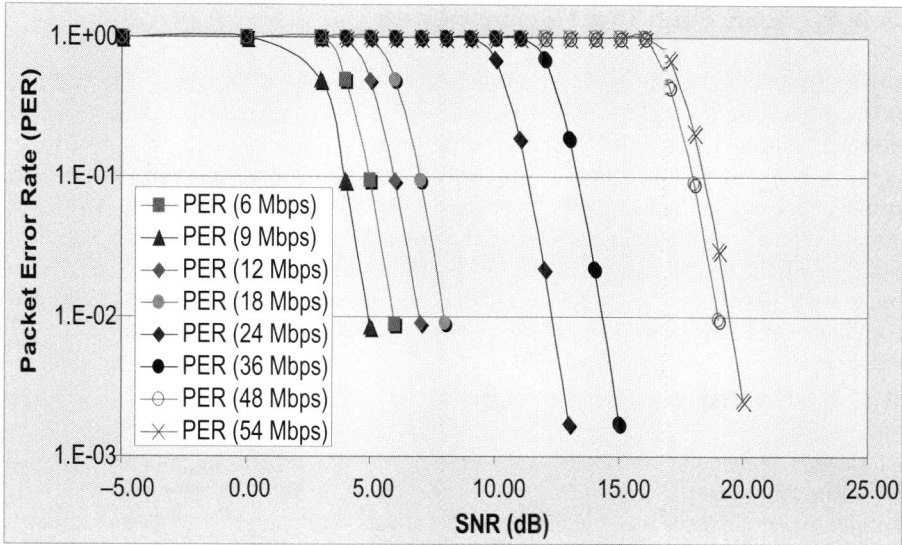

Figure 4.40   PER performance of 802.11g waveforms (packet size of 1000 bytes). Reproduced from Reference [83].

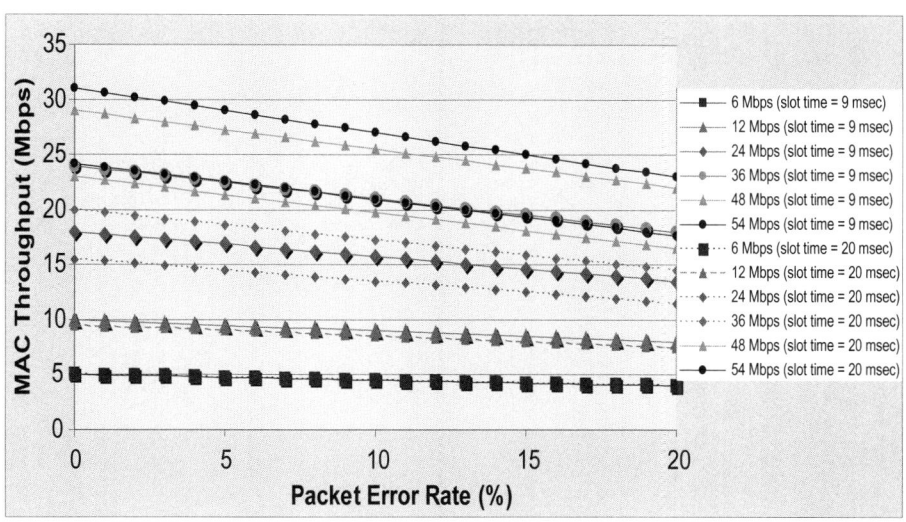

Figure 4.41   MAC throughput performance of 802.11g waveforms (packet size of 1500 bytes).

4.9.4   IEEE 802.11a Performance

IEEE 802.11a performance, in the sense of error rate versus Eb/No, is equivalent to 802.11g because it employs the virtually identical PHY designs and identical MAC designs. The 802.11g will generally outperform 802.11a from the perspective of error rate versus distance because of the higher propagation losses in the 5-GHz frequency band compared with those in the 2.4-GHz frequency band. Thus, the results of Figure 4.39, Figure 4.40, and Figure 4.41 are considered also valid for 802.11a. IEEE 802.11a performance was evaluated in Reference [78]. SNR versus data rate is shown in Figure 4.42 [78]. PER versus carrier-to-noise ratio (C/N) performance for the various 802.11a data rates is given in Figure 4.43.

4.9.5   Performance Variability of IEEE 802.11

The performance results presented thus far are very simplistic in nature. They consider only the error rate performance of the physical layer or they consider only simple two-node networks that are static in nature. These simple two-node scenarios provide the best-case performance of the network. In more realistic scenarios, the WLAN will have multiple nodes contending for the medium, resulting in lost packets due to collisions, degrading the performance of the network. In addition, nodes will often not be fixed but will instead be mobile in nature. This will cause yet further network performance degradation. Interference between WLAN deployments and other devices operating in the same frequency bands, such as microwave ovens and cordless telephones, will further worsen network performance, making the network more susceptible to service disruption. Also, there are complex issues associated

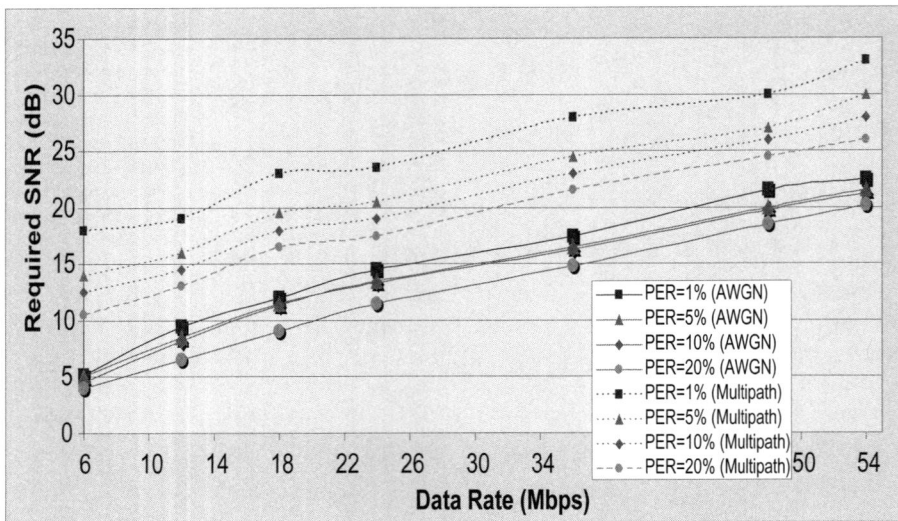

Figure 4.42   Required SNR as a function of data rate and PER performance of 802.11g waveforms (packet size of 1500 bytes).

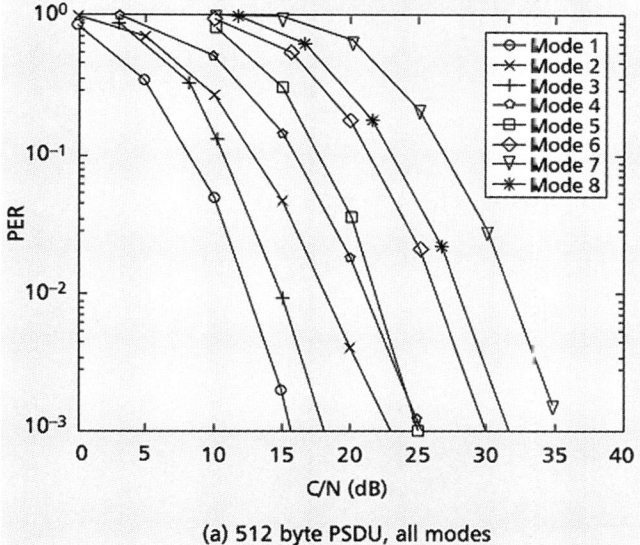

(a) 512 byte PSDU, all modes

Figure 4.43   PER versus C/N performance of 802.11a data rates.

with mixed network populations (e.g., a WLAN with 802.11b and 802.11g network clients) that will further degrade performance. Thus, it is of interest to understand these degradation mechanisms and the resulting degradation in network performance to gauge the variability of performance.

4.9.5.1 Effects of Mobility on IEEE 802.11 Performance

Mobility can have a significant impact on IEEE 802.11 performance. This is illustrated in Figure 4.44 [85].

In Figure 4.44, there are several throughput-versus-mobility rate performance curves provided for various blocking probabilities. Blocking probability refers to the probability that a node is blocked from any other node. Thus, various blocking probabilities can be thought of as varying terrains and its effects on propagation. Interesting to note from Figure 4.45 is the rapid rate at which performance degrades once a particular degree of mobility is achieved. Also interesting is the wide range in performance with different blocking probabilities. This strongly suggests that the effects of mobility are highly terrain dependent, which is intuitively expected. Although the exact results of Figure 4.45 are not of interest, this figure clearly illustrates the degrading effects mobility can have on WLAN throughput performance. The results of Reference [85] are also consistent with the results presented in Reference [86], which are presented in Figure 4.46.

Figure 4.45 [86] shows the same trends as those of Reference [85]. The results in Reference [86] also show that the degradation due to mobility is most severe for the highest data rates; the lower data rates of 802.11a do not exhibit as much degradation as higher data rates. This is the expected result.

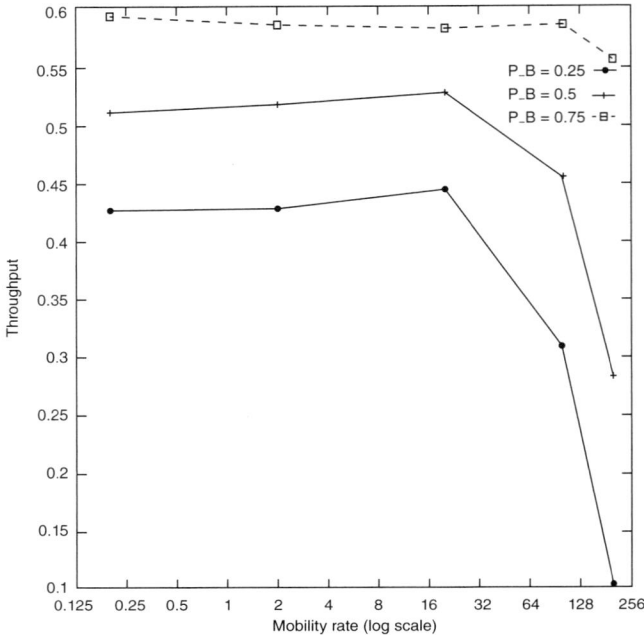

Figure 4.44   Effects of mobility on 802.11 MAC throughput. Reproduced from Reference [85].

The results of Figure 4.44 are also consistent with the performance predictions presented in Reference [87]. In Reference [87], 802.11b throughput performance is assessed for a variety of vehicular scenarios, including suburban, urban, and freeway scenarios. The resulting throughput performance predictions exhibit the trends of Figure 4.44, but performance variation is much more dramatic, varying from tens of kilobits per second to ones of megabits per second at identical velocities. This is due, in part, because different traffic statistics are applied for each scenario. This demonstrates not only the variation in performance due to mobility but also the significant effects of the statistics of the traffic offered to the WLAN. These various scenarios also have a differing number of network nodes. This fluctuation is indicative of the variations of network deployment and scenario that are realistic.

As another example of the degradation caused by mobility, reconsider Figure 4.41 from Reference [84]. Figure 4.41 provides several performance curves in a multipath environment, which is indicative of movement because fading is typically much more spatial than temporal in nature, particularly at these frequency bands. Figure 4.41 suggests that the degradation in network performance varies from 5 to 10 dB, depending upon the desired network error rate performance. Figure 4.42 also suggests that this degradation is greatest for lowest error rate values. For a PER of 1%, the difference between the AWGN channel and the multipath channel is 10 dB. For a PER of 20%, that degradation is 5 dB. This corresponds to 5–10 dB less jammer power to induce that effect on the network. Also, this effect is solely due to

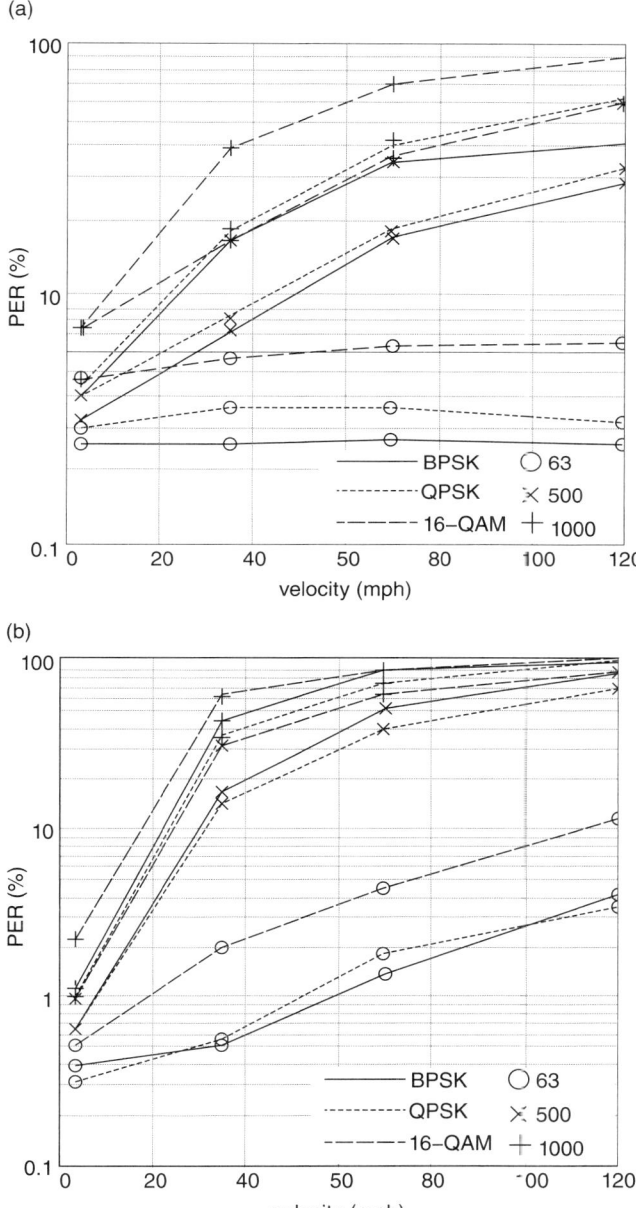

Figure 4.45   Performance degradation of 802.11a with mobility. (a) Performance in Rayleigh channel; (b) performance in Ricean channel.

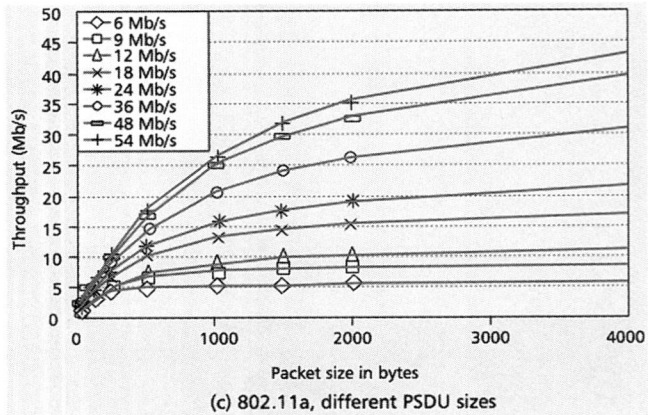

(c) 802.11a, different PSDU sizes

Figure 4.46   802.11a throughput as a function of packet size.

mobility. This still does not consider realistic network sizes, in which contention for the wireless medium will cause additional errors. It should also be noted that the degradation caused by multipath propagation effects appears relatively insensitive to link data rate, at least for the case considered in Figure 4.41.

4.9.5.2   *Variation with Network Traffic Statistics*

Reference [78] presents network throughput performance predictions as a function of packet size. Those results are shown in Figure 4.46 [78].

Figure 4.46 demonstrates that network performance can be very sensitive to the characteristics of the offered traffic. As packet sizes become small, network throughput degrades significantly. This is the expected result.

4.9.5.3   *Coexistence*

As discussed previously, 802.11g WLANs are backward compatible with 802.11b devices. However, there is a performance penalty that is paid for this coexistence. The slot time of an 802.11g-only network is 9 μs. However, when 802.11b devices must be supported, a slot time of 20 μs must be used. Figure 4.42 [84] shows that performance degradation for using this longer slot time can be significant. Figure 4.42 also shows that the performance degradation is most marked for higher data rates.

4.10   THE FUTURE DIRECTION OF IEEE 802.11

IEEE 802.11 development is a never-ending process, with a long line of technology enhancements perpetually in development. Table 4.35 summarizes the current (at the time of this writing) IEEE 802.11 standardization activities.

Table 4.35   Current IEEE 802.11 Standardization Activities

Group	Type	Description
aa	MAC	Audio/video transport streams
ac	PHY and MAC	Very high throughput below 6 GHz
ad	PHY and MAC	Very high throughput—60 GHz
ae	MAC	QoS management
af	PHY and MAC	TV whitespace operation
ah	PHY and MAC	Below 1 GHz operations
ai	MAC	Fast initial link setup

However, not all standardization efforts are equal in terms of excitement and anticipation. This is not a new phenomena; it has always been true that certain standardization efforts can somewhat overshadow other efforts. Anyone familiar with the industry can recall the excitement surrounding IEEE 802.11n in the mid-to-late 2000s as its standardization was ongoing. While arguable, the 802.11 standardization effort that is currently receiving the most attention is IEEE 802.11ac.

IEEE 802.11ac aims to provide extremely high data rates in the 5 GHz band, while providing backward compatibility for 802.11b and 802.11g in the 2.4 GHz band, 802.11a in the 5 GHz band, and 802.11n in the 2.4/5 GHz mode. The official target of the IEEE 802.11ac task group is to provide single link throughput performance of 500 Mbps and multistation throughput of at least 1 Gbps. It is envisioned by the WLAN development community that as we continue moving into the wireless Internet era, applications that have traditionally been wired in nature will continue to be increasingly wireless, including very high bandwidth applications like streaming video.

IEEE 802.11ac is a MAC and PHY amendment to the IEEE 802.11-2007 base specification. MAC layer modifications are expected to largely mirror the various MAC layer modifications introduced by IEEE 802.11n, which has already become part of the IEEE 802.11-2012 specification. Thus, when comparing IEEE 802.11n to IEEE 802.11ac, we can focus on differences in their PHY layers. IEEE 802.11ac intends to primarily achieve increased throughput through a few key mechanisms:

- Increased channel bandwidth
- Expanded MIMO support
- Expanded modulation and coding parameters

IEEE 802.11n supports 20 and 40 MHz channel bandwidths. In contrast, IEEE 802.11ac plans to make channel bandwidths of 20, 40, and 80 MHz as mandatory components of the specification, with an optional 160 MHz channel bandwidth. The benefit here is obvious; more bandwidth equals more achievable data rate.

IEEE 802.11n supports up to four spatial streams, with two spatial streams mandatory for APs. IEEE 802.11ac is expected to support up to eight spatial streams. Again, this will be clearly beneficial for higher throughput.

IEEE 802.11n supports four modulation modes: BPSK, QPSK, 16-QAM, and 64-QAM. IEEE 802.11ac supports these same modulation modes but also adds an optional 256 QAM modulation mode. This clearly can lead to higher throughput as each channel symbol carries additional information bits.

Unfortunately, we are not comfortable going into much more detail on IEEE 802.11ac at this time. As of the time of the writing of this book, IEEE 802.11ac is still undergoing standardization. Any further discussion of this technology would be based upon a draft specification that is still likely facing a year of standardization before it is finalized, and probably not to be ratified until late 2013. Any detailed discussion of IEEE 802.11ac would rely upon a draft specification not publicly available and highly likely to change. For that reason, we believe it premature to go into a detailed technology description. However, we have no doubt in our minds that IEEE 802.11ac is going to experience strong market success and that it will eventually become the de facto technology underlying the WLAN market. In fact, we predict that future revisions of this text will likely focus heavily on IEEE 802.11ac.

ADDITIONAL READING AND ONLINE RESOURCES

While we have attempted to provide you with a solid understanding of the major IEEE 802.11 standards, we are obviously just scratching the surface. However, we can direct the reader toward several high-quality texts that are focused solely on IEEE 802.11 for additional reading. In particular, we recommend the following texts:

T. COOKLEV, *Wireless Communications Standards: A Study of IEEE 802.11, 802.15, and 802.16*, IEEE Press, New York, 2004 [88].

M.S. GAST, *802.11 Wireless Networks: The Definite Guide*, 2nd Edition, O'Reilly, Sebastopol, CA, 2005 [89].

M.S. GAST, *802.11n: A Survival Guide*, O'Reilly, Sebastopol, CA, 2012 [90].

There are also several Web sites that may be of interest to the reader:

802.11 Industry News: http://www.wi-fiplanet.com/

802.11 WG Home Page: http://www.ieee802.org/11/

Wi-Fi Alliance: http://www.wi-fi.org/

Wi-Fi Security: http://www.wardrive.net/

Wireless Networking Statistics and Trends: http://www.itfacts.biz/

Chapter 5

Wireless Metropolitan Area Networks

The introduction of WLAN and WPAN standards led to a widespread adoption of wireless networks for homes and business. Broadband wireless access (BWA) was going to be the next logical step in extending fiber optic backbone networks into areas where no other methods were available. The beauty of BWA lies not only in the fact that it can deliver capacity but also that, as a network infrastructure, it can be deployed significantly faster than its wired counterparts.

The IEEE 802.16 WMAN standards family has been developed by the IEEE 802.16 Working Group to address the market demand for BWA. This working group has been active since 1999, publishing numerous versions of the standard to support evolving user needs and business models.

The nature of IEEE 802.16 is significantly different from other IEEE standards, since the technology is geared more toward cellular service providers. In addition, IEEE 802.16 leaves many of the implementation decisions to the implementer, particularly at the base station side. This flexibility allows the system developer to set up a network that could be cost-effective yet simple or significantly more sophisticated to provide higher-end services to the end user in order to fetch a higher price.

The objective of this standard is a wide deployment of various IEEE 802.16 systems that are interoperable yet differentiated in such way that the market is kept dynamic and novel with a myriad of attractive options to the carriers.

What is broadband wireless? The idea behind broadband wireless is to bring broadband experience to the end user, well, wirelessly, which would provide that user with the convenience and flexibility bundled with other unique benefits that are not available with wired broadband infrastructures.

Two fundamentally different types of broadband wireless services are currently in existence. The first type attempts to provide a set of services comparable to that of the traditional wired broadband (e.g., DSL or cable modem) but doing so wirelessly. The second type of broadband wireless, called *mobile broadband*, offers additional functionality of portability, nomadicity, and mobility. The authors in

Wireless Networking: Understanding Internetworking Challenges, First Edition.
Jack L. Burbank, Julia Andrusenko, Jared S. Everett, and William T.M. Kasch.
© 2013 the Institute of Electrical and Electronics Engineers, Inc. Published 2013 by John Wiley & Sons, Inc.

Reference [91] define *nomadicity* as "the ability to connect to the network from different locations via different base stations" and *mobility* as "the ability to keep ongoing connections active while moving at vehicular speeds."

A broad industry consortium called the Worldwide Interoperability for Microwave Access (WiMAX) Forum was established in June 2001 for the purpose of certifying BWA products for interoperability and compliance with the IEEE WMAN Standard 802.16. The term "WiMAX" is a marketing term that had become synonymous with IEEE 802.16-based BWA networks similarly to how Wireless Fidelity (Wi-Fi) had become synonymous with IEEE 802.11-based wireless local area networks (WLANs). WiMAX standards (used here interchangeably with the 802.16 WMAN standards) accommodate both fixed and mobile broadband applications. The subject of Mobile WiMAX is discussed in Chapter 7 along with other cellular communications standards.

The WiMAX Forum has now grown to include 216 member companies as of August 21, 2011 (see http://www.wimaxforum.org). Its "WiMAX Forum Certified" logo is placed on the package of certified products and is becoming a key criterion for market viability in the same way that the "Wi-Fi Certified" logo of the Wi-Fi Alliance (WFA) is a key criterion for market viability of IEEE 802.11-based WLAN products. Figure 5.1 [92] illustrates the current organization of the WiMAX Forum

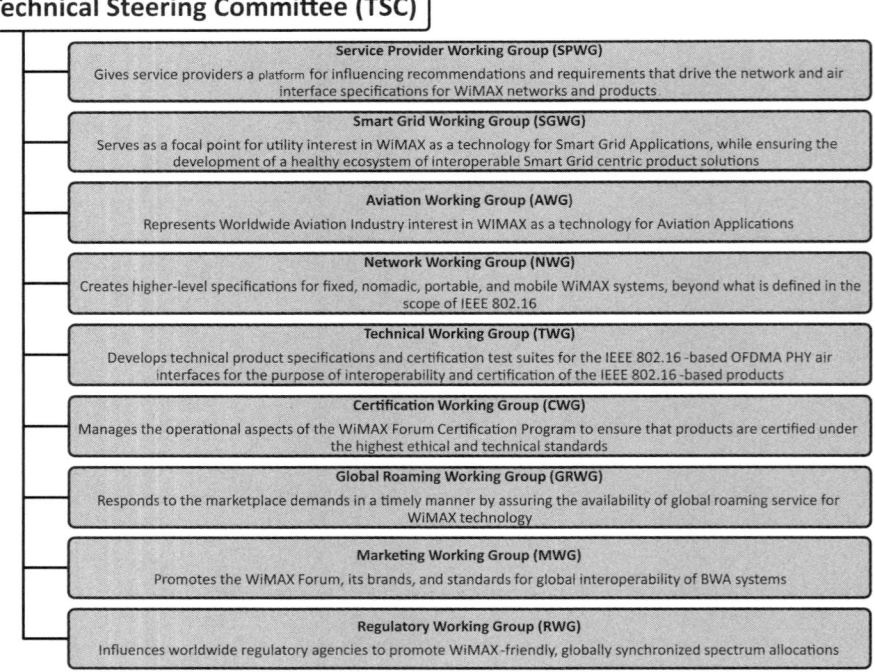

Figure 5.1 WiMAX Forum organization: technical activities and working groups. Reprinted from Reference [92].

and offers a brief description of each Working Group's charter. All WiMAX Forum Working Groups operate under the oversight from a Technical Steering Committee (TSC). The TSC's main functions include management of technical development of the WiMAX Forum standards and specifications, promoting timely and integrated technical programs within the WiMAX Forum, developing/approving roadmaps and various work items, as well as reviewing and acting upon reports delivered by Working Groups [92].

The IEEE 802.16 standard has been viewed by some as a serious threat to the long-standing survivability of several existing wireless technologies, including IEEE 802.11-based WLAN technologies; broadband residential Internet technologies, such as digital subscriber line (DSL) and cable, and even more as a considerable threat to third-generation (3G) cellular technologies. Others take an opposing view and foresee WiMAX becoming a powerful complementary technology to these various tools. Nevertheless, many in the commercial wireless industry view WiMAX as a key enabling technology in the large-scale realization of the wireless Internet, a tool that may potentially allow service providers to deliver high data rates (i.e., tens of megabits per second) to a variety of devices, including handheld devices, and enabling an entire new generation of applications (e.g., handheld, high-resolution videophones). WiMAX already experienced a tremendous deployment success worldwide with 583 networks that are either in active service or active deployment/planned mode across 150 countries [93].

So what really caused the inception of WiMAX? Again, it was a desire to find a cost- and service-feasible alternative to the traditional wired broadband technologies. Many proprietary BWA systems came and went, and, until now, only a few had successful deployments that have been limited to niche markets only. It is probably safe to say that BWA systems suffered a difficult past, partly due to the fragmentation within industry and lack of common standard. Some of the important events related to the BWA development are listed in Table 5.1 [91, 94].

WiMAX technology has evolved through four stages, though they are not clearly distinct from each other or even sequential, for that matter—a seemingly widespread opinion among experts (e.g., see Reference [91]) Nonetheless, they are as follows:

1. Narrowband wireless local-loop systems
2. First-generation line-of-sight (LOS) broadband systems
3. Second-generation non-line-of-sight (NLOS) broadband systems
4. Standards-based broadband wireless systems

The original air interface standard was completed in 2001 [95] but has significantly evolved since then and will continue to do so to address growing needs for BWA.

The main objective of the IEEE 802.16 is to provide protocols and specific technologies to define the air interface of the BWA system. IEEE 802.16 technology aims to provide wireless broadband access (with connectivity to core networks) at

Table 5.1   Notable Events in BWA Systems Development

Date	Event
February 1997	AT&T announces development of fixed broadband wireless technology called "Project Angel"
February 1997	Federal Communications Commission (FCC) sponsors forum on wireless communications service (WCS) in Washington, DC to auction off 30 MHz of spectrum in 2.3 GHz band.
September 1997	American Telecasting, Inc. (later acquired by Sprint) announces wireless Internet access services in the multichannel multipoint distribution services (MMDS) band with telephone dial-up modem upstream and up to 750 kbps on the downstream
September 1998	FCC allows two-way communications on the MMDS band
April 1999	Sprint and MCI acquire several wireless cable operators in order to gain access to MMDS bands
July 1999	IEEE 802.16 Working Group holds its first meeting
March 2000	AT&T finally offers first high-speed fixed wireless broadband service
May 2000	Sprint deploys first MMDS system in Arizona, using first-generation line-of-sight (LOS) technology.
June 2001	WiMAX Forum is established
October 2001	Sprint stops MMDS deployments
December 2001	AT&T stops providing fixed wireless broadband services
December 2001	The IEEE Standard 802.16 is completed for frequencies above 11 GHz
February 2002	Korea allocates spectrum in 2.3-GHz band for wireless broadband (WiBro)
January 2003	The IEEE Standard 802.16a for transmitting data through non-line-of-sight (NLOS) radio channels to and from omnidirectional antennas is completed
June 2004	The IEEE Standard 802.16-2004, which combined the updates from the IEEE 802.16a, 802.16b, and 802.16c regulations, is released
September 2004	Intel initiates the shipment of Rosedale, the first WiMAX chipset
February 2006	The IEEE Standard 802.16e-2005 for the first Mobile WiMAX system is released
January 2006	First WiMAX Forum-certified product is announced for fixed applications
June 2006	Korea launches WiBro
August 2006	Sprint Nextel announces plans of deploying Mobile WiMAX in the United States
May 2009	IEEE Standard 802.16-2009 is completed. It consolidates and obsoletes IEEE Standards 802.16-2004, 802.16e-2005, 802.16-2004/Cor1-2005, 802.16f-2005, and 802.16g-2007.
May 2011	The IEEE Standard 802.16m, also known as Mobile WiMAX Release 2, WirelessMAN-Advanced, and 4G WiMAX, is completed.
August 2011	Sprint starts selling 4G WiMAX to wholesale customers

broadband rates (at least 1.544 Mbps, according to the ITU). As we already mentioned, the IEEE 802.16 is an evolutionary standard. Originally designed to support stationary, enterprise-class deployments, it has eventually evolved to also support residential-class applications and, as of December 2005, mobile applications [95]. The 802.16e amendment brings support for both mobile and fixed terminals. Much of the IEEE 802.16 evolution has occurred within the physical layer (PHY) of the standard, although it is heavily based on the heart of WiMAX—its media access control layer (MAC) applications. MAC, supporting all PHY functionality, was originally designed for enterprise-based telecommunications services. Because of this, the IEEE 802.16 is capable of supporting the most demanding service requirements due to MAC's ability to deliver differentiated quality of service (QoS) among users.

As already noted, the IEEE 802.16 is called the Wireless MAN standard for wireless metropolitan area networks. "Metropolitan" indicates the target network scale (the size of city) rather than the target geography. It is not by any means limited to applications in urban areas. In fact, some of the most likely WiMAX applications are in rural areas in which wired broadband access may not be readily available.

The IEEE 802.16 family of standards defines air interface and is not a manual for service deployment. The fixed and mobile versions of WiMAX have slightly different implementations of their air interfaces. Fixed WiMAX uses a 256 fast Fourier transform (FFT)-based OFDM physical layer, whereas Mobile WiMAX employs a scalable OFDM access (OFDMA)-based PHY with FFT sizes varying between 128 and 2048 bits.

In the subsequent sections, we discuss key technological features of the Fixed WiMAX accompanied by brief descriptions of MAC and its various service aspects such as QoS and security, followed by a brief overview of OFDM PHY. Mobile WiMAX is discussed in Chapter 7 within the context of 3G cellular communications.

5.1 FIXED WIMAX TECHNOLOGY OVERVIEW

The focus of discussion in this section is fixed broadband wireless access systems as described in the harmonized IEEE 802.16-2009 [45].

The IEEE 802.16 technology family is actually composed of several distinct technology specifications, which are summarized in Table 5.2.

Based on the original IEEE 802.16 standard, WiMAX initially emerged as a high data rate, point-to-point backhaul technology between fixed locations. As time went on, the technology has significantly evolved with perceived business opportunities by WiMAX proponents. This evolution first led to Fixed WiMAX, which can be described as a wireless local loop (WLL) technology, providing broadband services to fixed locations (similar to DSL and cable services). This evolution then resulted in Mobile WiMAX, an access technology for mobile users, which is similar to services offered by cellular service providers. This technological evolution of WiMAX has traded capacity and range for mobility support and scalability, attempting to assert itself as a competitor to current cellular technologies. So the question

Table 5.2 IEEE 802.16 Technology Specifications and Other Related Standards

Specification	Description	Reference	Release date
IEEE 802.16	MAC and PHY definition for fixed broadband wireless access in the 10- to 66-GHz frequency bands	[7]	2001
IEEE 802.16a	Amendment to the original specification, containing new PHY definitions for the 2- to 11-GHz frequency bands. Also includes mesh network modes of operation	[39]	2003
IEEE 802.16c	System profiles for 10- to 66-GHz operations	[40]	2003
IEEE 802.16d (fixed WiMAX)	Also known as IEEE 802.16-2004, this specification is considered to be the base IEEE 802.16 standard for fixed broadband wireless communications. Contains IEEE 802.16, IEEE 802.16a, and various MAC enhancements	[41]	2004
IEEE 802.16e-2005 (Mobile WiMAX Release 1)	Amendment to the IEEE 802.16d specification to provide explicit support for mobility.	[42]	2006
IEEE 802.16f	Amendment to IEEE 802.16, Air Interface for Fixed Broadband Wireless Access Systems—Management Information Base	[43]	2005
IEEE 802.16g	Amendment to IEEE 802.16, Air Interface for Fixed Broadband Wireless Access Systems—Management Plane Procedures and Services	[44]	2007
IEEE 802.16-2009	IEEE Standard for Local and Metropolitan Area Networks—Part 16: Air Interface for Broadband Wireless Access Systems Revision of IEEE 802.16-2004. It consolidates and obsolete IEEE Standards 802.16-2004, 802.16e-2005, 802.16-2004/Cor1-2005, 802.16f-2005, and 802.16g-2007.	[45]	2009

IEEE 802.16h	Amendment of IEEE 802.16 standard—Improved Coexistence Mechanisms for License-Exempt Operation	[46]	2010
IEEE 802.16i	Amendment to IEEE 802.16, Air Interface for Fixed Broadband Wireless Access Systems—Management Plane Procedures and Services	[36]	2008 (Withdrawn)
IEEE 802.16j	Amendment to IEEE 802.16 standard—Multihop Relay Specification	[47]	2009
IEEE 802.16k	Amendment of IEEE 802.1D, Standard for Local and Metropolitan Area Networks: Media Access Control (MAC) Bridges—Bridging of 802.16	[48]	2007
IEEE 802.16m (Mobile WiMAX Release 2, also known as 4G WiMAX or WirelessMAN-Advanced	Amendment to IEEE 802.16-2009, Air Interface for Fixed and Mobile Broadband Wireless Access Systems—Advanced Air Interface	[49]	2011
IEEE 802.20 Mobile Broadband Wireless Access	Mobile broadband wireless access standards group. Initially formed as a standards group within the IEEE 802.16 Working Group, it consisted of a group of individuals who wished to develop a new technology focused solely on mobility. The 802.20 Working Group specifies PHY and MAC layers of an air interface for interoperable mobile wireless access systems operating in licensed bands below 3.5 GHz and optimized for IP, data transport with user data rates above 1 Mbps. The standard supports vehicular velocities up to 250 km/h in a MAN environment.	[50]	2008 (base standard publication)
Wireless Broadband (WiBRO)	Korean wireless broadband standard that was incorporated into the IEEE 802.16e standard	N/A	N/A

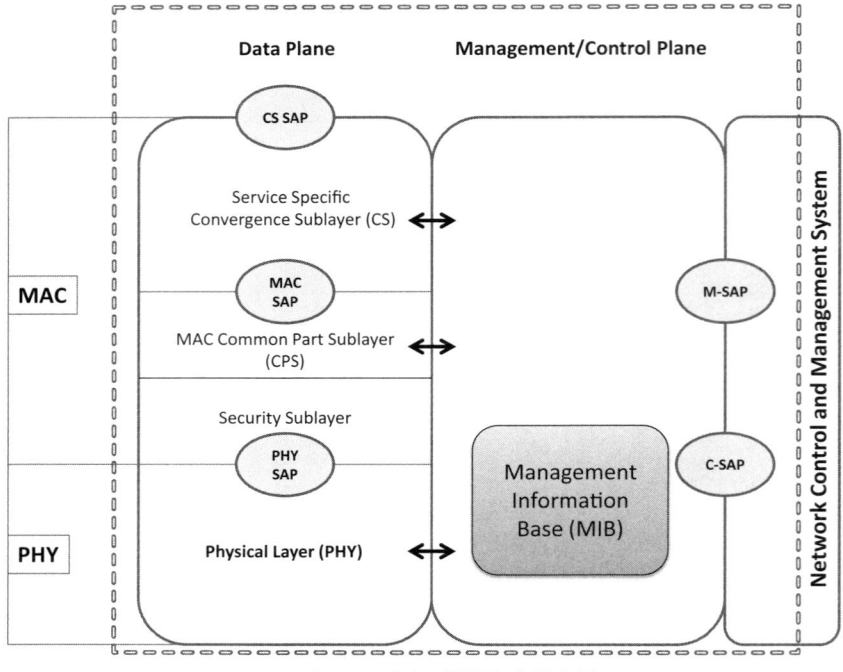

Figure 5.2   IEEE 802.16 logical architecture. Reprinted from Reference [45]. SAP, service access point.

is: Can WiMAX succeed in this market? With the overall success of WiBro (a variant form of WiMAX) in South Korea, many tend to think so.

The logical architecture of IEEE 802.16 is illustrated in Figure 5.2 [45]. The Data Plane is responsible for sending and receiving data, whereas the Management/Control Plane is responsible for managing and controlling all connections. The IEEE 802.16 standard further subdivides the MAC sublayer into three sublayers: the convergence sublayer (CS), the common part sublayer (CPS), and the security sublayer. The CS's main goal is to enable IEEE 802.16 to better accommodate the higher-layer protocols placed above the MAC. Currently, three CS specifications are provided within the standard: the asynchronous transfer mode (ATM) CS, the packet CS, and the generic packet CS (GPCS). The standard might allow for other CS specifications in the future. The service-specific CS resides atop of the MAC CPS and employs the services provided by the MAC CPS through the MAC SAP. The CS is responsible for the following:

- Acceptance of higher-layer protocol data units (PDUs) from the higher layer
- Classification of higher-layer PDUs
- Processing (if required) of the higher-layer PDUs based upon the classification

- Delivery of CS PDUs to the appropriate MAC SAP
- Reception of CS PDUs from the peer entity [45]

As we previously noted, three CS specifications are provided within the standard for interfacing with various protocols. The internal format of the CS payload is unique to the CS, and the MAC CPS is not required to recognize the format of or parse any information from the CS payload.

The MAC CPS resides above the MAC security sublayer, as shown in Figure 5.2 (adapted from Reference [45]), and is responsible for the core MAC functionality of system access, bandwidth allocation, connection establishment, and connection maintenance. It receives data from the various CSs, via the MAC SAP, classified to particular MAC connections. QoS is applied to the transmission and scheduling of data over the PHY.

The MAC security sublayer's main functions include authentication, secure key exchange, and encryption.

The implementation-specific PHY SAP offers means of transferring data, PHY control, and statistics between the MAC CPS and the PHY.

The PHY definition includes multiple specifications for various frequencies and applications. We provide more detail on the various PHY specifications in the subsequent sections.

IEEE 802.16 devices can be classified as subscriber stations (SSs) (for fixed applications), mobile stations (MSs), or base stations (BSs). In anticipation of IEEE 802.16 devices becoming a part of a large network and subsequently needing to interface with various entities for management and control purposes, the standard introduced a Network Control and Management System (NCMS) concept as a "black box" containing such entities. The idea behind the NCMS concept was to allow IEEE 802.16 PHY/MAC layers to be independent of the network architecture, transport network, and protocols used at the back end, therefore allowing a greater flexibility with the network design. NCMS is specified to logically exist at the BS side and the SS/MS side of the radio interface, termed NCMS(BS) and NCMS(SS/MS), respectively. The NCMS(BS) is responsible for performing any necessary inter-BS coordination. The NCMS interfaces with the IEEE 802.16 entity through the control SAP (C-SAP) and management SAP (M-SAP). C-SAP and M-SAP expose control and management plane functions to upper layers. In order to perform the correct MAC operation, the NCMS needs to be present within each SS/MS. The NCMS is a layer-independent entity that can be treated as either a management or control entity. The standard allows general system management entities to perform their functions through NCMS, while letting standard management protocols to be directly implemented within the NCMS.

5.1.1 IEEE 802.16 Networking Overview

The IEEE 802.16 network architecture connotes the presence of fixed infrastructural sites. Indeed, the architectural model of IEEE 802.16 is quite similar to the model employed within cellular telephone networks. Each IEEE 802.16 coverage area

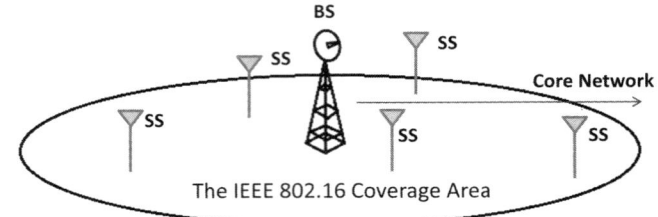

Figure 5.3   IEEE 802.16 coverage area—fixed WiMAX.

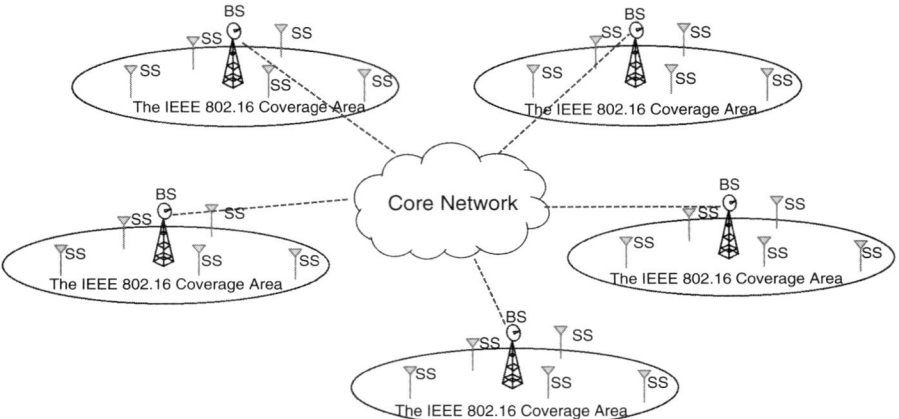

Figure 5.4   A notional IEEE 802.16 network—fixed WiMAX.

comprises one BS and one or more SSs. BSs provide connectivity to core networks, while the SS, the suite of equipment at the customer location or customer-premises equipment, provides access for the end user into the broadband wireless network. A single IEEE 802.16 coverage area is depicted in Figure 5.3. The architecture depicted in Figure 5.3 represents a single cell of network coverage. These IEEE 802.16 cells can then be grouped together to form a larger 802.16 network, where the BS sites are interconnected via a core network, as depicted in Figure 5.4. In the IEEE 802.16 model, channel access is highly centralized, that is, the BS is in complete control over how and when SSs access the wireless medium.

The nature of transmissions may be point to point (PTP), point to multipoint (PMP), or point to consecutive point (PTCP), where PTCP involves the creation of a closed loop through multiple PTP connections (i.e., multihop connectivity with nodes acting as relays for other subscribers to achieve range extension). In addition, IEEE 802.16a offers a mesh networking capability where SSs can act as routers, relaying data to nodes that may not have line of sight (LOS) to the BS. BSs typically utilize one or more wide-beam antennas that may be partitioned into several smaller sectors, with all sectors summing up to complete 360-degree coverage. This is similar to BSs within the cellular model. SSs usually employ highly directional antennas pointed in the direction of the BS. This is a significant departure from the model employed within cellular communications or the IEEE 802.11 WLAN communities, where low-gain, omnidirectional antennas are utilized. This is one of the key reasons why IEEE 802.16 is capable of achieving significantly higher data rates when compared to those technologies.

The BS-to-SS link is referred to as the downlink. The SS-to-BS link is referred to as the uplink. The proper routing of traffic to a BS is a function of the core network, which is not explicitly defined within the IEEE 802.16 standard. As a matter of fact, the IEEE 802.16 standard is capable of accommodating a myriad of existing or future core network technologies. This type of core network can be compared to the distribution system (DS) of the IEEE 802.11 networks.

5.1.2 IEEE 802.16 MAC

As previously discussed, the IEEE 802.16 standard further subdivides the MAC sublayer into the service-specific CS, CPS, and security sublayer. This subsection provides a more detailed description of each of these MAC sublayers.

5.1.2.1 Service-Specific Convergence Sublayer

The service-specific CS allows IEEE 802.16 technology to better accommodate the higher-layer protocols placed above the MAC. These protocols can be Internet Protocol (IP), Ethernet, Time Division Multiplexing (TDM) Voice, ATM, or any future protocols of unknown format. Logically, the CS resides atop of the MAC CPS that delivers its services to the CS via the MAC SAP. The CS is responsible for providing a mapping to and from the higher layers, as well as supporting any necessary MAC service data unit (SDU) header suppression for the higher-layer overhead reduction on each packet. The IEEE 802.16-2009 standard specifies three types of CSs: the asynchronous transfer mode (ATM), the packet CS, and the generic packet CS. Other types of CSs may be specified in the future.

Fundamental to IEEE 802.16 is the notion of a service flow or connection. The IEEE 802.16 standard is connection oriented though not in the traditional meaning of the term.

Traffic flow across the IEEE 802.16 network is associated with connections. An SS can potentially have multiple connections established with the BS, with each connection being assigned a unique connection identifier (CID). The nature of these connections is comparatively static, meaning that each connection possesses a distinctive set of transmission and medium access characteristics. Association of traffic with a CID is the function of the classifier, which is a rule set within the CS that can be based on numerous factors, such as priority, source MAC address, and destination MAC address. Even connectionless traffic, such as IP, is routed through a connection, thus having a unique CID associated with it. The CID is 16 bits in length, allowing for up to 64,000 connections within each uplink and downlink channel.

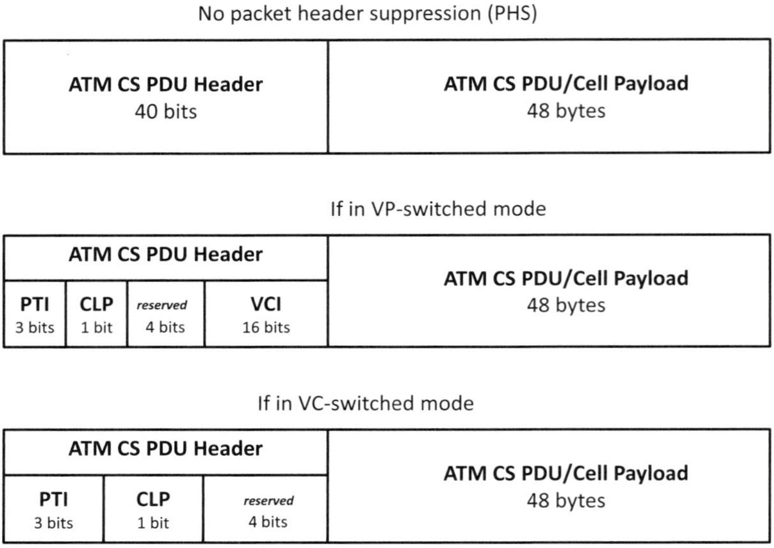

No packet header suppression (PHS)

ATM CS PDU Header 40 bits	ATM CS PDU/Cell Payload 48 bytes

If in VP-switched mode

ATM CS PDU Header				ATM CS PDU/Cell Payload 48 bytes
PTI 3 bits	CLP 1 bit	*reserved* 4 bits	VCI 16 bits	

If in VC-switched mode

ATM CS PDU Header			ATM CS PDU/Cell Payload 48 bytes
PTI 3 bits	CLP 1 bit	*reserved* 4 bits	

Figure 5.5   Three types of ATM CS PDU frames. Reprinted from Reference [45]. VP, virtual path; VC, virtual channel; PTI, payload type indicator; CLP, cell loss priority; VCI, virtual channel indicator.

Upon classification and processing, the data frame is then passed onto the MAC SAP for processing by the MAC CPS. The type of processing performed upon data frames depends on the type of particular CS being employed (ATM, packet, or generic packet).

ATM Convergence Sublayer   The ATM CS supports ATM services flowing across the IEEE 802.16 network, via the MAC CPS SAP. The ATM CS accepts ATM cells, classifies those cells, may perform processing on those cells (such as packet header suppression [PHS]), and, finally, maps the received cells into MAC frames. Two types of ATM connections are supported: virtual path (VP) or virtual channel (VC). Each connection is assigned a unique CID. The ATM CS also supports common channel signaling (CCS), in which signaling messages are carried over a channel independent of user connections. A CCS signaling channel is not, however, limited to a single user connection. Every IEEE 802.16 SS must have a CID that corresponds to CCS messages.

Figure 5.5 [45] illustrates three types of ATM CS packet data unit (PDU) frame formats in the data/control plane, as specified in the IEEE 802.16 standard. Each ATM CS PDU frame consists of an ATM CS PDU header and the ATM CS PDU payload of 48 bytes. The ATM CS PDU payload is equal to the ATM cell payload.

Packet Convergence Sublayer   Logically, the packet CS is located atop the 802.16 MAC CPS. The packet CS is responsible for the transport of all packet-based protocols, such as IP, Point-to-Point Protocol (PPP), and IEEE 802.3 (Ethernet). The packet CS performs functions similar to those of the ATM CS, including classification of the higher-layer protocol PDU into the appropriate transport connection and PHS.

While in the sending mode, the packet CS is responsible for delivering the MAC service data unit (MAC SDU) to the MAC SAP. It is the MAC's responsibility to deliver the MAC SDU to peer MAC SAP, utilizing appropriate functions that are unique to a particular connection's service flow characteristics During the receiving mode, the packet CS accepts the MAC SDU from the peer MAC SAP and delivers it to higher layers.

Generic Packet Convergence Sublayer (GPCS) The generic packet CS (GPCS) is an upper layer, protocol-independent, generic packet convergence sublayer that supports multiple protocols (e.g., IPv4, IPv6, or IEEE 802.3 Ethernet) over the IEEE 802.16 air interface. Some of its main functions are described as follows:

- GPCS employs the MAC SAP and exposes a SAP to GPCS applications.
- GPCS does not redefine or replace other convergence sublayers. Instead, it offers a SAP that is not protocol specific.
- Packet parsing occurs above GPCS, resulting in parameters that are passed onto GPCS SAP for classification. The upper layer packet parsing, however, is performed by the GPCS application.
- In GPCS, multiplexing of multiple layer protocol types (e.g., IPv4, IPv6, Ethernet) is allowed over the same IEEE 802.16 connection.
- The IEEE 802.1D bridging is supported transparently by the IEEE 802.16 air interface, since the GPCS requires the upper layer to provide the MS MAC Address and service flow ID (SFID) with every packet. The MS MAC address and SFID can be representative of a port, either unicast or broadcast.
- PHS is performed in a manner similar to that of the packet CS.

GPCS protocol layering model is illustrated in Figure 5.6 [45].

5.1.2.2 MAC Common Part Sublayer

The CPS is the central piece of the IEEE 802.16 MAC, defining the medium access method. The CPS provides functions related to MAC PDU framing and transmission, duplexing, channelization and channel access, network entry and initialization, and QoS.

MAC Protocol Data Unit Framing and Transmission When the MAC CPS receives service data units (SDUs) from the higher layer, it assembles them to form the MAC PDU, which is the basic payload unit handled by both the MAC and PHY layers [91]. Depending on the payload size, multiple SDUs can be transported on a single MAC PDU, or a single SDU can be fragmented for transport over multiple MAC PDUs, When an SDU is fragmented, each fragment is tagged by a sequence number which enables the MAC layer at the receiver to arrange all of the SDU fragments in the correct order [91].

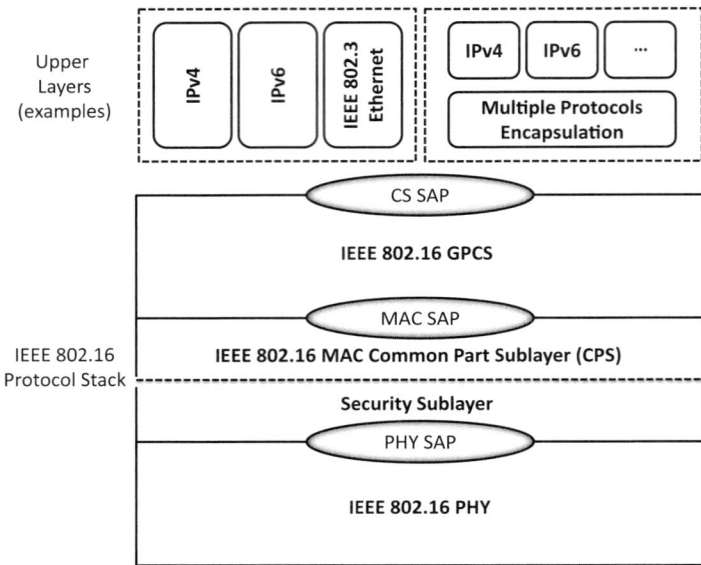

Figure 5.6 GPCS protocol layering model. Adapted from Reference [45].

Each MAC PDU consists of a fixed-length MAC header, which may be followed by the payload and a cyclic redundancy check (CRC). Payload and CRC are optional and are used only if SS asks for them in its QoS parameters. MAC PDUs are transmitted in on-air PHY slots [96]. When present, the payload consists of zero or more subheaders and zero or more MAC SDUs and/or its (MAC SDU's) fragments. The payload information may vary in length to represent a variable number of bytes. Variable-length payload allows the MAC layer to transport various higher-layer traffic types without the need for knowledge of the formats or bit patterns of those messages. A general MAC PDU form is illustrated in Figure 5.7 [45].

Implementation of CRC is mandatory only for the orthogonal frequency division multiplexing (OFDM) and OFDM access (OFDMA) PHYs. The CRC is based on the IEEE 802.3 standard and is calculated via the standard generator polynomial of degree 32 on the entire MAC PDU, which includes the MAC header and the payload. The CRC calculation is performed after encryption to protect the header and the ciphered payload.

WiMAX employs two different types of MAC PDUs, each with a different header structure. To be specific, the IEEE 802.16-2009 standard defines two MAC PDU header formats: generic MAC header and MAC header without payload that further subdivides into two additional subtypes.

Generic MAC PDU Header The *generic* MAC PDU is responsible for carrying data and MAC signaling messages and is used on both the downlink and uplink. It starts with a Generic MAC header, whose structure is illustrated in Figure 5.8 [45],

Figure 5.7   A general MAC PDU form. Adapted from Reference [45]. MSB, most significant bit; LSB, least significant bit; CRC, cyclic redundancy check.

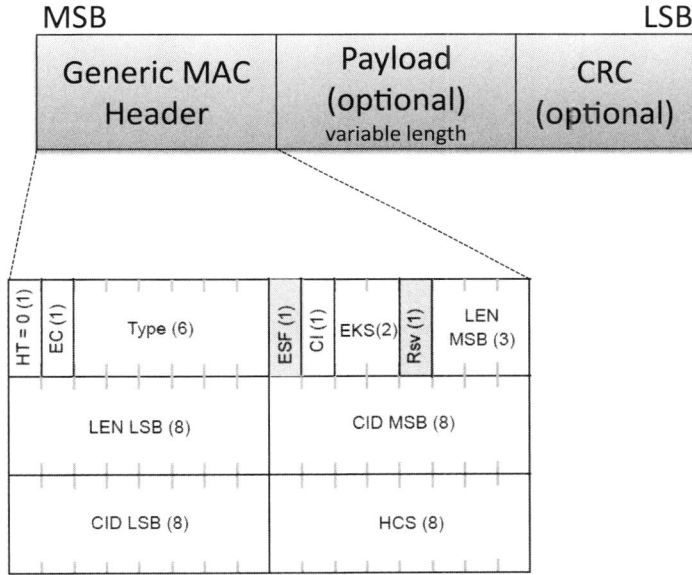

Figure 5.8   Generic MAC header format. Adapted from Reference [45].

followed by a payload and CRC. Table 5.3 [45] describes various fields of the Generic MAC header. The single-bit header type (HT) field is set to 0 for the generic MAC header and to 1 for MAC header without payload. The extended subheader field (ESF) in the Generic MAC header, when set to 1, indicates the presence of an additional subheader that follows immediately after the Generic MAC header and is never encrypted. The Type field of the Generic MAC header is 6 bits long, with each bit having a special meaning as described in Table 5.4 [45].

MAC PDU Header without Payload   The IEEE 802.16-2009 standard defines two types of MAC header without payload: MAC signaling header type I and MAC signaling header type II. These frame header formats are used alone, without a payload and CRC, and are reserved for the uplink use only. Both MAC signaling header types I and II are of the same size as that of the Generic MAC header.

Table 5.3   Generic MAC Header Field Descriptions

Field	Length (bits)	Description
HT	1	Header type (needs to be set to 0 for generic MAC header)
EC	1	Encryption control (0—payload not encrypted; 1—payload encrypted)
Type	6	This field indicates the subheaders and special payload types that are present in the message payload (see Table 5.4)
ESF	1	Extended subheader field (1—ES present; 0—ES not present)
CI	1	CRC indicator (1—CRC included; 0—CRC not included)
EKS	2	Encryption key sequence. The index of the traffic encryption key (TEK) and initialization vector (IV) used to encrypt the payload. This field is only meaningful if the EC field is set to 1.
Rsv	1	Reserved
LEN	11	Length. The length in bytes of the MAC PDU, including the MAC header and the CRC if present.
CID	16	Connection identifier on which the payload is to be sent. It's a unidirectional MAC address that identifies a connection to equivalent peers [96]
HCS	8	Header check sequence is an 8-bit field that is used for detecting errors in the header with the generator polynomial $D^8 + D^2 + D + 1$

Reprinted from Reference [45].

MAC Signaling Header Type I   The format of the MAC signaling header type I is illustrated in Figure 5.9 [45]. Table 5.5 [45] details all the possible encodings of the 3-bit Type field following the encryption control (EC) field. The same table also provides brief descriptions of the seven header formats defined in the IEEE 802.16-2009 standard to specifically conform to the overall MAC signaling header type I structure.

MAC Signaling Header Type II   This type of MAC header is meant for the uplink traffic only. There is no payload or CRC following the MAC header. The structure of the MAC signaling header type II is illustrated in Figure 5.10 [45]. Table 5.5 [45] describes the encodings of the 1-bit Type field following the EC field. In the MAC signaling header type II, the header is changed with regard to the MAC signaling header type I. MAC signaling header type II is used for obtaining channel quality feedbacks specific to OFDMA and multiple input, multiple output (MIMO) [96], and, as a result, shall be regarded as a feature of the Mobile WiMAX.

MAC Subheaders and Special Payloads   Apart from the Generic MAC header and MAC header without payload, the IEEE 802.16 standard also defines

Table 5.4 "Type" Field Encodings

Type bit	Value
#5 most significant bit (MSB)	*Reserved*
#4	Automatic Repeat reQuest[a] (ARQ) feedback payload (1—present, 0—not present)
#3	Extended type
	Indicates whether the present packing subheader (PSH) or fragmentation subheader (FSH) is extended for non-ARQ-enabled connections
	1—extended
	0—not extended
	For ARQ-enabled connections, this bit is set to 1.
#2	Fragmentation subheader (FSH) (1—present, 0—not present)
#1	Packing subheader (PSH) (1—present, 0—not present)
#0 least significant bit (LSB)	Downlink: fast-feedback allocation subheader (FFSH)
	Uplink: grant management subheader (GMSH)
	1—present, 0—not present

Reprinted from Reference [45].

[a]The ARQ (automatic repeat request) is a control mechanism of data link layer where the receiver requests the transmitter to resend a block of data (e.g., MAC SDU fragment) when errors are detected. The ARQ mechanism is based on acknowledgement (ACK) or nonacknowledgment (NACK) messages transmitted by the receiver to the transmitter to indicate a good (ACK) or a bad (NACK) reception of the previous frames.

five subheaders that can be used in a generic MAC PDU. Table 5.6 [45] provides brief descriptions of these subheaders.

After the MAC PDU is constructed, it gets passed onto the scheduler that schedules the MAC PDU over the available PHY resources. The scheduler determines what the MAC PDU's QoS requirements are by the checking the service flow ID and the CID of the MAC PDU. The scheduler then decides, on a frame-to-frame basis, how to optimally allocate PHY resources based on the QoS requirements of the MAC PDU. Scheduling algorithms are left to the implementers to develop as they can have a significant impact on the overall system capacity and performance [91].

Duplexing The MAC protocol supports several duplexing techniques. Duplexing refers to how the uplink and downlink are separated. When compared with other wireless standards, the IEEE 802.16 specification follows a much more centralized architecture, which is expected of a technology that does not support ad hoc networking [88]. The IEEE 802.16 standard offers two methods by which duplexing can be

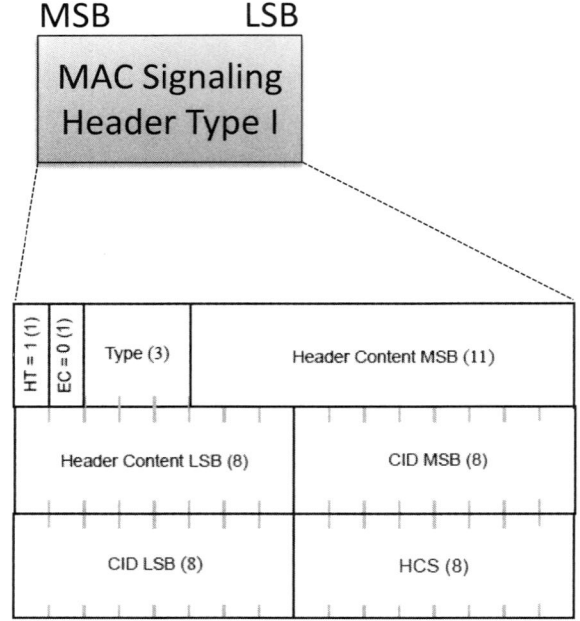

Figure 5.9   MAC signaling header type I. Adapted from Reference [45].

achieved: time division duplexing (TDD) and frequency division duplexing (FDD). In the TDD method, the wireless medium is divided in the time domain, where periods of time are allocated for uplink communications, while other periods of time are allocated for downlink communications. In this method, only a single frequency channel is occupied. The IEEE 802.16-2009 standard defines a frame to be "a structured data sequence of fixed duration used by some physical layer (PHY) specifications. A frame may contain both an uplink (UL) subframe and a downlink (DL) subframe." This definition is somewhat confusing, since the term "frame" is used to describe multiple entities within the system. Hence, in this book, we adopt the terminology "channel frame" and "data frame." Here, channel frame shall refer to the logical grouping of TDMA channel time slots, whereas data frame shall refer to a PDU.

In the case of TDD, the data frame, which is fixed in duration, contains one uplink and one downlink subframe. The frame is divided into an integer number of time slots, and the bandwidth is then allocated between the uplink and downlink in a dynamic fashion, as traffic needs dictate. The fractional bandwidth provided to the uplink and downlink is controlled by the system parameters that are controlled at higher layers within the system. When the TDD mechanism is employed, WiMAX is a half-duplex technology. The TDD mechanism is depicted in Figure 5.11 [45]. Here, the total duration of the frame is configurable, ranging from 2.5 to 20 ms [45]. In addition, guard times (not shown in Figure 5.11) are placed at frame boundaries

Type field encoding	MAC header type	Description
000	Bandwidth request (BR) header—BR incremental	Bandwidth request (in terms of the number of bytes of the uplink bandwidth) by the SS. The BR is necessary for the CID so that the CID can indicate the connection for which the uplink bandwidth is requested. This request is independent of the PHY coding and modulation. The allowed types of BRs are BR incremental (000) and BR aggregate (001).
001	BR header—BR aggregate	
010	PHY channel report header	Report of the uplink transmitted power level for the burst that carries this header.
011	BR and uplink transmit power report header	Incremental bandwidth request along with the report of the uplink transmit power level that carries this header.
100	BR and carrier-to-interference-and-noise ratio (CINR) report header	Incremental bandwidth request along with the CINR report. The CINR is measured by the MS from the BS and is interpreted as a single value from −16.0 to 47.5 dB in 0.5-dB steps.
101	BR and uplink sleep control header	The header is sent by the MS to request activation/deactivation of a certain power saving class. The header also indicates incremental transmission demand.
110	Sequence number (SN) report header	This header is sent by the MS to report the LSB of the next ARQ block sequence number (BSN) or the virtual MAC SDU sequence number for the active connections that are SN Feedback enabled.
111	Channel quality information channel (CQICH) allocation request header	This header is sent by the MS to request the allocation of a CQICH.
0	Feedback header, with additional 4-bit Type field	An MS sends the feedback header either as a response to a Feedback Polling Information Element (IE) or as an unsolicited feedback. When sent as a response to a Feedback Polling IE, the MS sends feedback header utilizing the assigned resource indicated in the Feedback Polling IE. When sent as unsolicited feedback, the MS either sends the feedback header onto currently allocated uplink resource or requests additional uplink resource by sending an Indication flag on the fast-feedback channel or the enhanced fast-feedback channel or by sending a bandwidth request ranging code. The feedback PDU consists of the feedback header only and does not contain a payload or CRC. Two feedback headers are specified in the IEEE 802.16-2009 standard, one with the CID field and another one without. The feedback header with the CID field is used only when the uplink resource used to send the feedback header is requested through BR ranging. Otherwise, the feedback header without the CID field is employed.
1	Reserved	N/A

Reprinted from Reference [45].

213

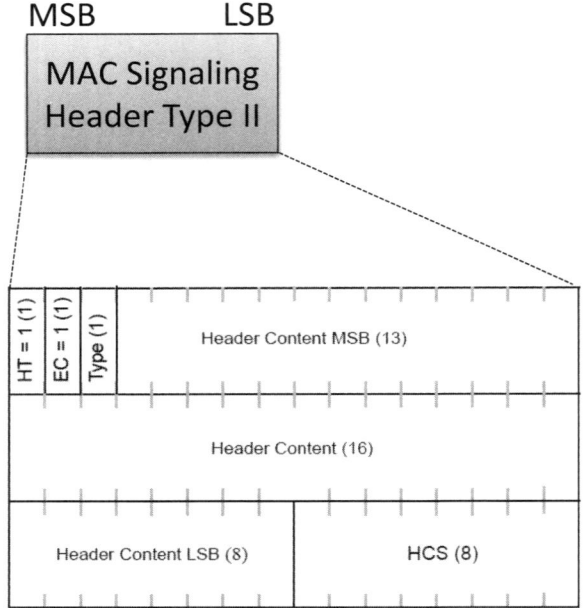

Figure 5.10   MAC signaling header type II. Adapted from Reference [45].

for intersymbol interference prevention. The duration of these guard times is configurable and is typically on the order of 4 or 16 symbol times.

The FDD method provides uplink and downlink transmissions separated in frequency. In this case, uplink and downlink transmissions can overlap in time, making IEEE 802.16 a full-duplex technology. Similar to TDD, the frame duration is fixed. The FDD method has the disadvantage of having static assignments of uplink and downlink, thus making the system somewhat less flexible than TDD. However, the advantage of this method is that it facilitates the use of different modulation types on the uplink and downlink. This method also has the advantage of being able to support full-duplex and half-duplex SSs. The FDD method is illustrated in Figure 5.12 [45]. The fact that the uplink and downlink channels use a fixed duration frame significantly reduces the complexity of the bandwidth allocation algorithms. A full-duplex SS can continuously listen to the downlink channel, whereas a half-duplex SS can listen to the downlink channel only when it is not transmitting in the uplink channel.

FDD frequency separation is not defined by the IEEE 802.16 standard, but rather specified by the BS upon network entry or statically configured by the network administrator (depending on a particular device's capability). This aspect is regulated on a country-by-country basis, with 50 and 100 MHz being the most commonly regulated values for frequency separation.

Table 5.6 Generic MAC PDU Subheaders

Subheader type	Description
Fragmentation subheader (FSH)	This subheader follows the Generic MAC header and indicates that the SDU is fragmented over multiple MAC PDUs
Grant management subheader (GMSH)	This subheader is 2 bytes long and is used by the SS to convey bandwidth management needs to the BS. Using this subheader is more efficient, since it is more compact than the BR PDUs and does not require a transmission of a new PDU. BR PDUs are typically used for the initial bandwidth request.
Packing subheader (PSH)	The presence of this subheader indicates that multiple SDUs are packed into a single MAC PDU.
Fast-feedback allocation subheader (FFSH)	The use of this subheader indicates that the PDU contains feedback from the MS about the downlink channel state information. The channel state information feedback can be used for MIMO and non-MIMO implementations.
Extended subheader group	The extended subheader group always appears immediately after the generic MAC header and before all subheaders. Extended subheaders are never encrypted. In addition, the group further subdivides into the uplink and downlink extended subheader types, with each type serving a different purpose (downlink: feedback request, DL sleep control, SN request, etc., or uplink: MIMO mode feedback, UL transmit power report, error report, etc.)

Reprinted from Reference [45].

Channelization and Channel Access Channelization is achieved through TDM. Frames are formed, representing collections of time slots. In the case of TDD, a frame consists of two distinct portions: an uplink subframe and a downlink subframe. In the case of FDD, uplink and downlink subframes are transmitted concurrently on different frequency channels.

The structure of the downlink subframe using TDD is illustrated in Figure 5.13 [45]. Each downlink subframe begins with Frame Start Preamble used by the PHY for synchronization and equalization, followed by a control message, which informs SSs of the modulation and coding of the various time slots in the downlink frame. This field, referred to as the downlink map (DL-MAP), is required because the IEEE 802.16 technology incorporates adaptive control of modulation and coding to accommodate variations in channel conditions. The subsequent TDM portion carries the data, organized into bursts with different burst profiles. A "burst" is a unit of data transfer for which the transmission characteristics remain constant. "downlink bursts" refer to transfers from BS to SS. "uplink bursts" refer to transfers from SS to BS. The DL-MAP contains "burst profiles," which are a set of parameters describing transmission properties associated with an interval usage code (IUC). A profile

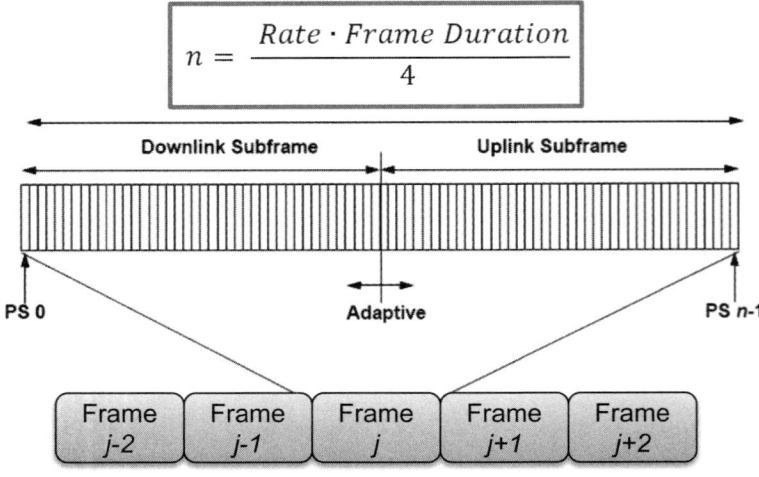

PS – Packet Switched

Figure 5.11   TDD frame structure. Adapted from Reference [45]. PS, packet switched.

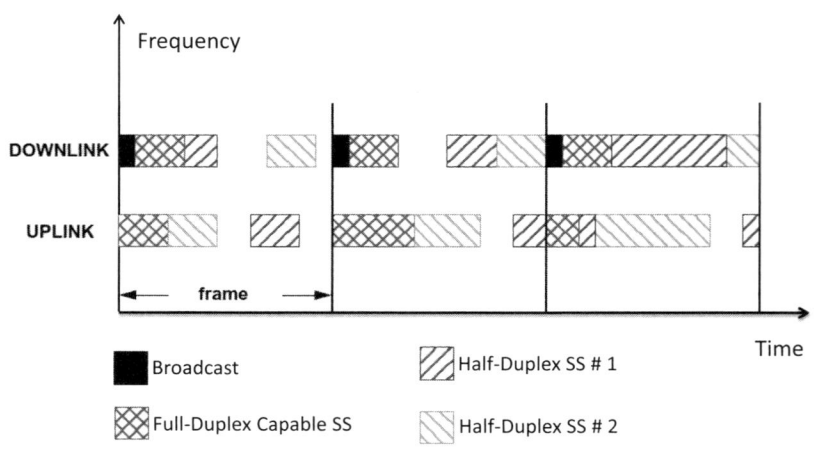

Figure 5.12   FDD within IEEE 802.16. Adapted from Reference [45].

contains parameters, such as modulation type, forward error correction (FEC) type, preamble length, and guard times. The bursts are transmitted in order of decreasing robustness. For instance, in the case of a single FEC type with fixed parameters, data would begin with QPSK modulation, followed by 16-QAM, and then by 64-QAM [IEE09]. These burst profiles are required to be updated for every frame, since profiles can be modified on a burst-by-burst basis. The downlink burst profiles are controlled solely by the BS and are indicated to the SS by downlink interval

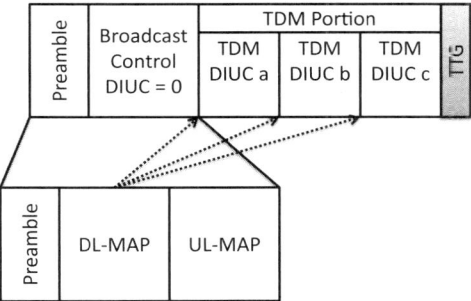

Figure 5.13   TDD downlink subframe structure. Adapted from Reference [45]. DIUC, downlink interval usage code; TDM, time division multiplexing; DL, downlink; UL, uplink; DL-MAP, downlink map; UL-MAP, uplink map; TTG, transmit/receive transition gap.

usage codes (DIUCs), which are the fields within information elements (IEs) of the DL-MAP. The SS is then responsible for determining the proper communications configuration by mapping the DIUC to the locally stored downlink burst profiles. The BS merely informs the SS which burst profile to use for proper reception of data destined for that station. Burst profiles stored within the SSs can be modified, as required, through the periodic transmission of "channel descriptor" messages, which essentially configures an SS with a set of burst profiles. In the case of TDD, transmit/receive transition gap (TTG) is used to separate the downlink subframe from the uplink one.

For the case of FDD, the structure of the downlink subframe is shown in Figure 5.14 [45]. Similar to the TDD case, the downlink subframe begins with a Frame Start Preamble followed by a frame control section and a TDM portion organized into bursts. These bursts are transmitted in the decreasing order of burst profile robustness. The TDM portion of the FDD downlink subframe contains data transmitted to one or more of the following:

- Full-duplex SSs
- Half-duplex SSs scheduled to transmit later in the frame than they receive
- Half-duplex SSs not scheduled to transmit in this frame

The TDM portion of the FDD downlink subframe is then followed by the TDMA portion that is used for transmitting data to any half-duplex SSs scheduled to transmit earlier in the frame than they receive. The presence of the TDMA portion allows an individual SS to decode a specific portion of the downlink without the need to decode the entire downlink subframe. In the TDMA portion, each burst begins with the DL TDMA Burst Preamble for phase resynchronization. Bursts in the TDMA portion are not required to be organized by the burst profile robustness. The FDD frame control section includes a map of both the TDM and TDMA bursts. Lastly, the TDD downlink subframe, which inherently contains data transmitted to

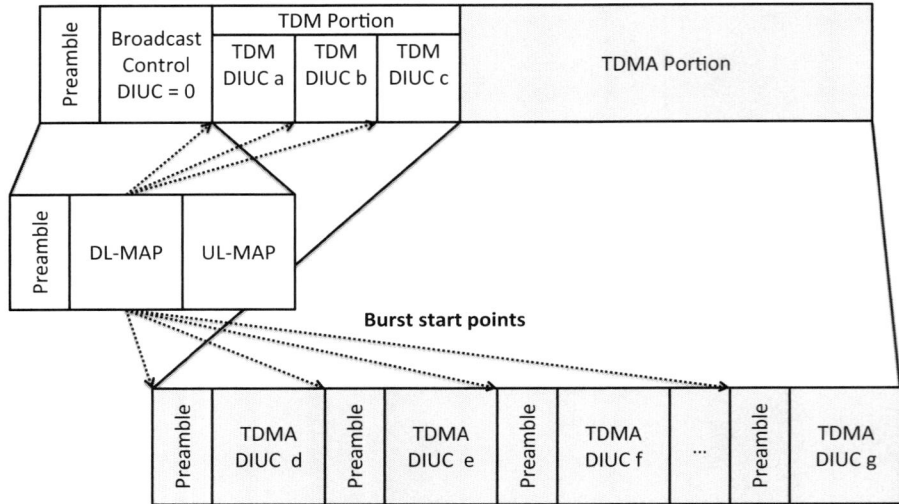

Figure 5.14 FDD downlink subframe structure. Adapted from Reference [45].

SSs that transmit later in the frame than they receive, has the same structure as the FDD downlink subframe for a subframe in which no half-duplex SSs are scheduled to transmit before they receive.

The structure of the uplink subframe that the SS uses to transmit to the BS is illustrated in Figure 5.15 [45]. The IEEE 802.16-2009 standard defines three classes of bursts that can be transmitted by the SS in the uplink subframe:

1. Bursts that are transmitted in contention opportunities reserved for initial ranging.

2. Bursts that are transmitted in contention opportunities defined by Request Intervals reserved for response to multicast and broadcast polls.

3. Bursts that are transmitted in intervals defined by Data Grant IEs specifically allocated to individual SSs.

Each uplink frame begins with a UL-MAP control message that contains uplink burst profile information. However, the UL-MAP is fundamentally different in that it is the mechanism through which bandwidth is allocated to the SS. The UL-MAP informs the SS which time slots it is allowed to use for that uplink frame. Moreover, in the UL-MAP message, there are fields referred to as uplink interval usage codes, which indicate to the SS which uplink burst profiles to use. The BS is in complete control of all bandwidth allocation and communications configuration. The BS may negotiate downlink burst profiles based on received signal quality, but it is still ultimately the decision of the BS which transmission characteristics the SS employs. Figure 5.16 [45] illustrates the process of an SS negotiating a change to a more robust downlink burst profile.

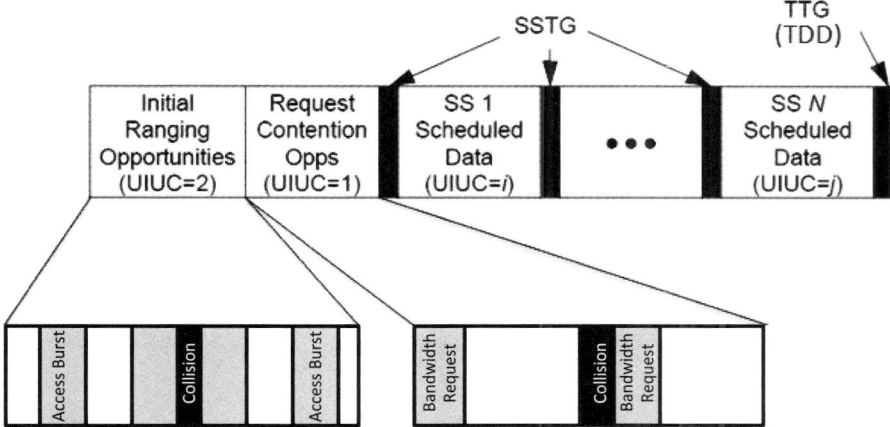

Figure 5.15 Uplink subframe structure. Adapted from Reference [45]. UIUC, uplink interval usage code; SSTG, subscriber station transition gap; TTG, transmit/receive transition gap.

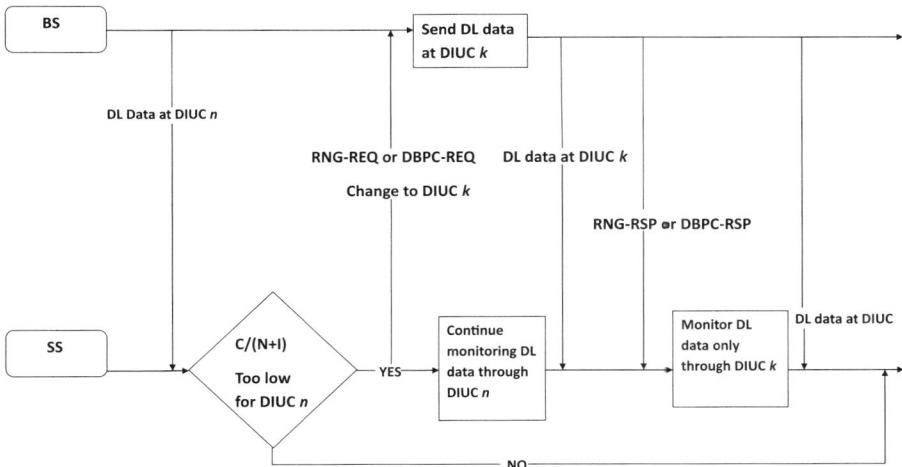

Figure 5.16 Example of a transition to a more robust burst profile. Adapted from Reference [45]. DBPC, downlink burst profile change; REQ, requirement; RNG, range; RSP, response; DL, downlink; DIUC, downlink interval usage code; RNG-REQ, ranging request; DBPC-REQ, DL burst profile change request; RNG-RSP, ranging response; DBPC-RSP, DL burst profile change response.

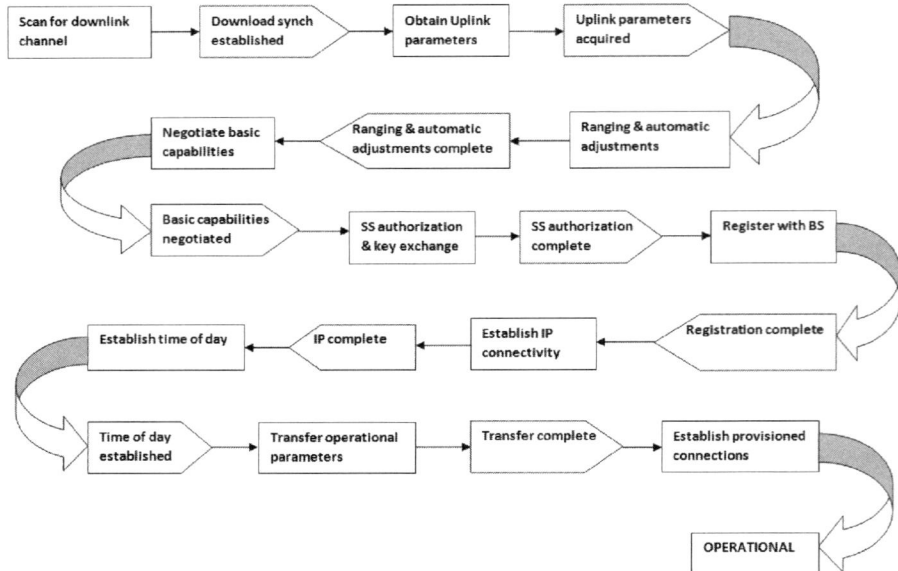

Figure 5.17 IEEE 802.16 initialization and network entry process. Adapted from Reference [45].

Initial Network Access There are many steps that must be taken before an SS can join an IEEE 802.16 network. The network entry process is shown in Figure 5.17 [45].

The first step an SS must take when attempting to join an IEEE 802.16 network is to acquire the downlink signal. Upon entry, the SS will use the last known valid operational parameters to search for the downlink signal. If the downlink signal is not found, the SS will search across all downlink channels. The SS will then obtain the uplink parameters from the UL-MAP messages. After that, the SS will undergo the "ranging" process, which allows stations to calibrate the performance of their PHY layers based upon current channel conditions.

The IEEE 802.16 specification defines "ranging" as the process of acquiring the correct timing offset and power adjustments in such way that the SS's transmissions are aligned with the BS receive frame for OFDM PHY (Fixed WiMAX) or OFDMA PHY (Mobile WiMAX), and are received within the appropriate reception thresholds. The WiMAX technology employs two types of ranging: initial ranging and periodic ranging. During initial ranging, the SS sends a brief ranging request message that allows the BS to send back a ranging response message with the amount of timing offset that the SS must use once it starts transmitting. Initial ranging is the process of time aligning with a communication system *before* establishing a communication session. The initial ranging process involves synchronization with an incoming transmission channel, sending a request at a particular time interval relative to the received channel and obtaining a response that allows for the time synchronization with the device or system.

Periodic ranging, in contrast, is the process of continuous time alignment with a communication system during a communication session. The periodic ranging process involves maintaining synchronization with an incoming transmission channel, periodically transmitting messages (e.g., transmitting user data) and receiving time alignment information from the other device or system.

Once ranging is complete, the SS signals the BS by requesting entry into the network, which initiates the process of authentication and cryptographic key exchange. After the SS gets authenticated as a user of the network, the BS provides IP configuration to the SS via Dynamic Host Configuration Protocol (DHCP). A subchannelized network entry procedure may also be used in which the SS chooses a particular subchannel and sends specialized messages to the BS. This procedure is in place to mitigate the large contention that can take place for initial ranging and logon.

The entire network entry handshaking process typically takes on the order of 20–30 seconds to complete. Once this process is complete, connections can be established rapidly with only a simple message exchange (autonomous if the base station has been provisioned correctly).

Bandwidth Negotiation and Allocation   The WiMAX system supports two types of SSs: those that can accept only bandwidth allocations on a grant-per-connection (GPC) basis and those that can accept bandwidth allocations on a grant-per-subscriber-station (GPSS) basis. In the latter case, the SS is responsible for managing bandwidth allocations to its various data flows. This results in increased flexibility by allowing the SS to "steal" bandwidth from connections to augment other connections, as necessary, and also ties this bandwidth allocation to a priority queuing scheme. GPSS-enabled SSs can request aggregate bandwidth, as required from the BS, and then manage the particular data flows independent of the BS. In the GPC case, the performance that a data flow experiences is largely independent of the SS. Not surprisingly, GPSS generally outperforms GPC. However, GPC SSs are less complex.

SSs typically request bandwidth incrementally, as capacity requirements vary over time. Conversely, SSs can request aggregate bandwidths from the BS, thus providing the BS with a better idea of the SS's actual bandwidth needs. An SS can request bandwidth either during bandwidth request periods dedicated solely to an SS or during contention periods (i.e., periods when all SSs can contend for channel access to make bandwidth requests from the BS). The BS determines how the request is made because the period in which the BS sends the polling message, indicating the upcoming bandwidth request period, dictates whether the SS makes such a request in a dedicated or contention period. In WiMAX, *polling* refers to the process where dedicated or shared uplink resources are provided to the SS for making bandwidth requests. In the Fixed WiMAX case (i.e., OFDM PHY), such bandwidth allocations can be made for either an individual SS or a group of SSs. The polling is always done on an SS basis, meaning that the bandwidth requests are always made using either the *basic* CID of an individual SS or multicast/broadcast CID of a group of SSs. When an SS is polled individually, the polling is called *unicast*, and the dedicated resources in the uplink are allocated for the SS to send a bandwidth request PDU. The BS informs the SS of the uplink allocations for unicast polling opportunities via UL-MAP message in the downlink subframe. SSs that have an active unsolicited grant service (UGS) connection are not polled due to the fact that the bandwidth request can be sent on the UGS allocation either via bandwidth request PDU or by piggybacking on generic MAC PDUs. An example of unicast polling is illustrated in Figure 5.18 [45]. If bandwidth resources are insufficient to poll each

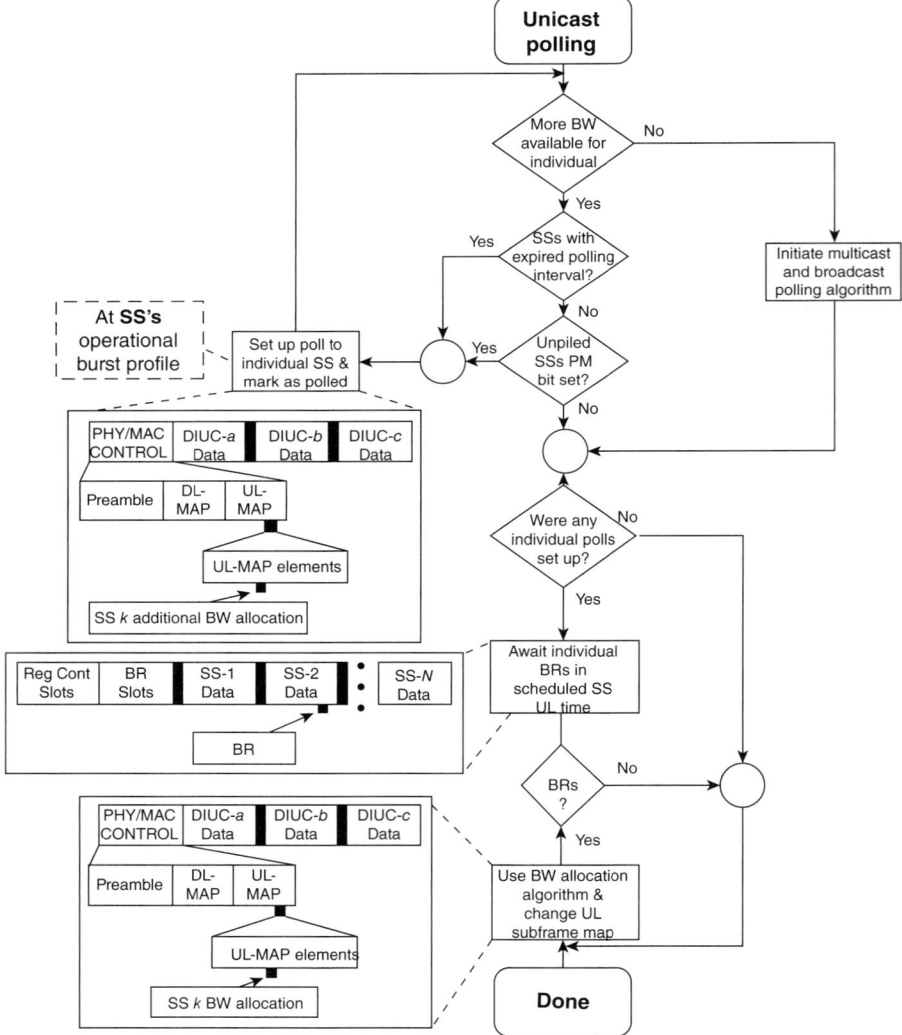

Figure 5.18   An example of unicast polling. Adapted from Reference [45].

SS individually, *multicast* or *broadcast* polling is used to poll a group of SSs or *all* SSs at a time [45, 91]. The WiMAX specification reserves certain CIDs for multicast groups and broadcast messages. As in the case of individual polling, the poll is not an explicit message, but rather bandwidth allocated in the UL-MAP. The difference between unicast and multicast/broadcast polling lies in the fact that instead of associating allocated bandwidth with an SS's basic CID (as in unicast polling case), the allocation is performed to a multicast or broadcast CID. The process of multicast and broadcast polling is shown in Figure 5.19 [45].

Quality of Service (QoS)   In WiMAX, QoS support is provided through the concept of "service classes." Service flow properties are grouped into named service classes (globally well known) so that upper-layer entities can request service flows with desired QoS parameters. This meshes well with upper-layer QoS models such as differentiated services (DiffServ). In the DiffServ case, the MAC service-specific CS sublayer can perform classification of data and association with CIDs while providing a clean mapping from DiffServ code points (DSCPs) to IEEE 802.16 service classes. Nonetheless, it is unclear if the IEEE 802.16 standardization effort will ever produce a DiffServ-aware CS, which means that the implementers may be required to develop such a CS on their own. QoS parameters considered within the IEEE 802.16 standard include priority, minimum traffic rate, maximum sustained traffic rate, maximum traffic burst, maximum latency, and maximum tolerable jitter. When higher layers request a particular type of service flow, the request communicates a variety of performance requirements to the system. These QoS parameters are, in turn, used to influence a variety of system attributes based on the desired QoS, including MAC scheduling functions, bandwidth requests and allocations, and burst profiles. This influence is on the uplink and downlink. Special signaling mechanisms are provided in the IEEE 802.16 specification so that QoS-enabled service flows can be dynamically established, as required. These signaling mechanisms are also used to inform the IEEE 802.16 peer of the level of service required, which will influence the treatment of that service flow.

5.1.2.3   IEEE 802.16 MAC Security Sublayer

The security sublayer, also referred to as the privacy sublayer, provides users with privacy through encryption of the link between the BS and SS. It accomplishes this through application of cryptographic transforms to MAC PDUs transported across connections between SS and BS. The security sublayer also protects operators from theft of service. Service flows are encrypted within the entire network, thus protecting the WiMAX system from unauthorized access.

The security sublayer uses an authenticated client/server key management protocol in which the BS (the server) distributes keying material to the client SS. In addition, the basic security mechanisms are strengthened by adding digital certificate-based SS device authentication to the key management protocol.

The MAC security sublayer consists of two component protocols as defined in the IEEE 802.16-2009 specification:

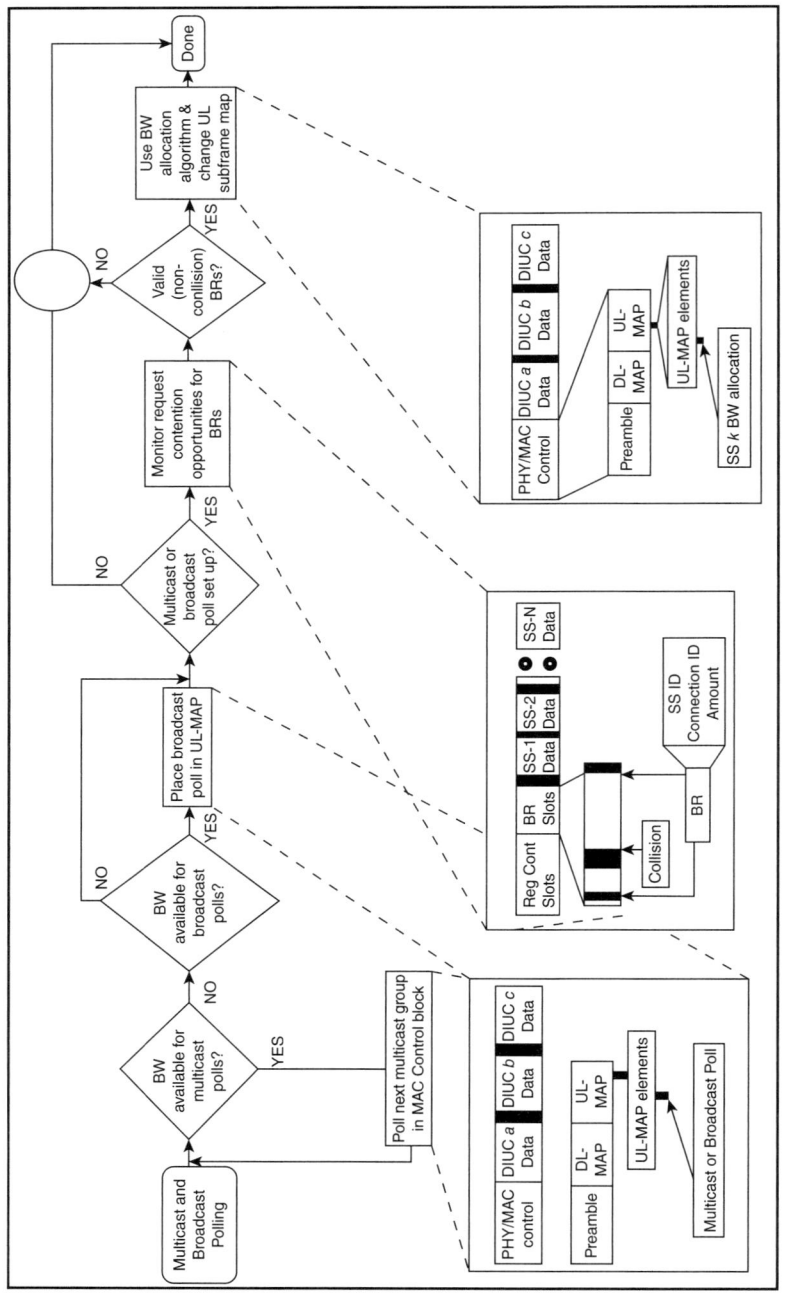

Figure 5.19 An example of multicast and broadcast polling. Adapted from Reference [45].

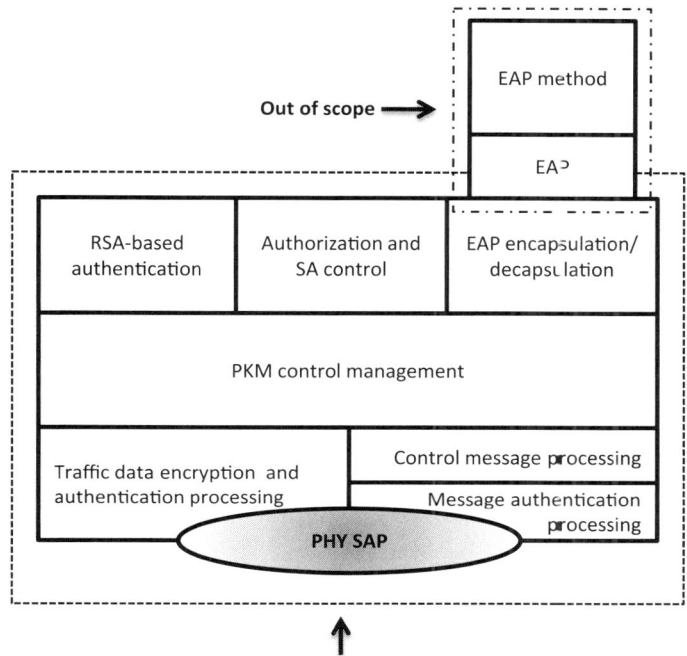

Figure 5.20   The IEEE 802.16 security sublayer protocol stack. Adapted from Reference [45]. EAP, extensible authentication protocol; RSA, Rivest, Shamir, and Adleman publ c key encryption; SA, security association; PKM, privacy key management.

1. ***Encapsulation Protocol.*** This protocol secures packet data across the fixed BWA network and delineates a set of supported cryptographic suites (i.e., pairings of data encryption and authentication algorithms, and the rules for applying those algorithms to a MAC PDU payload).

2. ***Key Management Protocol (PKM).*** This protocol is responsible for the secure distribution of keying data from the BS to the SS. The SS and BS synchronize keying data via PKM. The BS also uses this protocol to enforce conditional access to network services.

The MAC security sublayer protocol architecture is illustrated in Figure 5.20 [45]. Table 5.7 [45] provides descriptions of all security components depicted in Figure 5.20.

The IEEE 802.16 key management protocol is based on the PKM protocol of the Data over Cable Service Internet Specification (DOCSIS) Baseline Privacy Interface (BPI+) specification but has been enhanced to support the Advanced Encryption Standard (AES). PKM is based on the concept of security associations (SAs). Every SS establishes at least one SA during initialization. Every connection is mapped to an SA either at connection setup time or dynamically during operation.

Table 5.7 Security Sublayer Components

Component	Description
PKM Control Management	This stack controls all security components and is responsible for deriving and generating various keys.
Traffic Data Encryption/ Authentication Processing	This stack is responsible for encrypting/decrypting the traffic data as well as executing the authentication function for the traffic data.
Control Message Processing	This stack performs processing of various PKM-related MAC messages.
Message Authentication Processing	This stack executes message authentication function.
RSA-based Authentication	This stack executes the RSA-based authentication function using the X.509 digital certificates from both the SS and BS, whenever the RSA-based authorization is chosen as an authorization policy between the SS and BS.
EAP Encapsulation/ Decapsulation	This stack serves as the interface with the EAP layer, when the EAP-based authorization or the authenticated EAP-based authorization is selected as an authorization policy between an SS and a BS.
Authorization/SA Control	This stack is in charge of the authorization state machine and the traffic encryption key state machine.
EAP and EAP Method Protocol	These stacks are outside of the scope of the IEEE 802.16 standard.

Reprinted from Reference [45].

The PKM protocol uses X.509 digital certificates with Rivest, Shamir, and Adleman (RSA) public key encryption for SS authentication and authorization key exchange. Each SS contains manufacturer-issued and factory-installed X.509 digital certificates. These certificates, which link the 48-bit MAC address of the SS and its public RSA key, are sent to the BS by the SS in authorization request and authentication information messages. The BS verifies the identity of the SS by checking the certificate via database lookup, using this credential to also look up the authorization level of the SS. If the SS is authorized to join the network, the BS transmits an authorization key encrypted with the SSs public key. Authentication in IEEE 802.16-2009 can be either two way (i.e., the SS also authenticates the BS) to mitigate the threat of a rogue BS or one way (e.g., the BS authenticates SS, but not vice versa).

For traffic encryption, the Data Encryption Standard (DES) in the Cipher Block Chaining mode with 56-bit keys is mandated. The initialization vector (IV) is dependent on the frame counter, thus differing from frame to frame. Traffic encryption keys are exchanged using Triple DES (3DES) with a key exchange key derived from the authorization key. The PKM messages are authenticated using the Hashed Message Authentication Code protocol [97] with Secure Hash Algorithm (SHA)-1

[98]. Message authentication for functions considered vital (e.g., connection setup) is provided by the PKM protocol.

5.1.3 IEEE 802.16 Physical Layer

In the 10- to 66-GHz physical layer, line-of-sight (LOS) propagation is assumed required for communications. Subsequently, single-carrier modulation is employed, designated Wireless Mesh and Ad Hoc Network–Single Carrier (WirelessMAN-SC). The WirelessMAN-SC PHY layer can employ quadrature phase-shift keying (QPSK), 16-quadrature amplitude modulation (QAM), or 64-QAM adaptively changing based on channel conditions perceived by a received signal strength indicator (RSSI). The 2- to 11-GHz band, licensed and license exempt, which was originally addressed in the IEEE 802.16a, assumes non-LOS communications. The IEEE 802.16-2009 standard defines two air interfaces for this frequency band:

WirelessMAN-OFDM PHY: Orthogonal frequency division multiplexing (OFDM) modulation with a 256-point fast Fourier transform (FFT) with TDMA channel access

WirelessMAN-OFDMA PHY: OFDM access (OFDMA) is employed with a 2048-point FFT. Multiple access is provided by addressing a subset of carriers to individual receivers

In the 10- to 66-GHz PHY layer specification, channel bandwidths of 20–25 MHz are employed within the United States; and 28-MHz channel bandwidths are employed in Europe. IEEE 802.16 typically employs Reed-Solomon GF 256 FEC coding. However, more robust turbo block codes can be employed to increase link robustness. The 2- to 11-GHz PHY layer specifications also allow for the use of automatic retransmission request as an optional capability. The 2- to 11-GHz PHY layer specifications allow for a wide variety of channel bandwidths to accommodate worldwide regulatory policies. Multiples of 1.25, 1.5, 1.75, and 2.0 MHz are allowed. OFDM subcarrier spacing varies according to channel bandwidth. Within each channel, time is divided into frames of configurable duration (2.5–20 ms).

WiMAX aims to define a global technology that can accommodate spectrum regulatory issues across all key markets. As such, it is anticipated that WiMAX will eventually operate in many frequency bands. However, to date, the majority of certified products operate in the 3.5-GHz frequency band. There are also several certified products that operate in the unlicensed 5.8-GHz Unlicensed National Information Infrastructure (U-NII) frequency band (not certified for operation in this band, which greatly limits deployment but still highly capable). It is currently envisioned that 3.5 GHz will remain the primary frequency band of operation for Fixed WiMAX. For Mobile WiMAX, it is envisioned that 2.5 GHz will be the dominant frequency due to more favorable propagation characteristics at this band as compared with higher bands. Several key proponents hold licenses in the 2.5-GHz frequency band (e.g., Sprint). One notable exception is the WiBRO network deployed by Korea

Table 5.8 Current WiMAX Frequency Bands of Operation versus the Number of Active Deployments

Frequency band	Number of worldwide deployments
2.3 GHz	48
2.5 GHz	112
3.3 GHz	10
3.5 GHz	308
5+ GHz	21

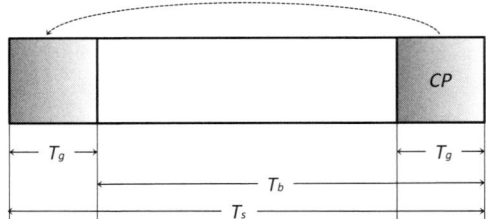

Figure 5.21 OFDM symbol time structure. Adapted from Reference [45]. T_g, guard time; T_b, useful symbol period; T_s, OFDM symbol period; CP, cyclic prefix.

Telecom (KT) in South Korea that operates in the 2.3-GHz frequency band. Current WiMAX frequency bands of operation are listed in Table 5.8 (derived from data from Reference [93]).

Since the topic of this chapter revolves around the Fixed WiMAX technology, in the subsequent sections we will only address one of the three PHY specifications defined in the IEEE 802.16-2009 specification—the WirelessMAN-OFDM PHY, which is also known as OFDM PHY. The topic of WirelessMAN-OFDMA PHY or OFDMA PHY is discussed in the "Mobile WiMAX" subsection of Chapter 7.

5.1.3.1 Wireless MAN-OFDM PHY (OFDM PHY)

The OFDM PHY was designed for NLOS operations in the 2–11 GHz frequency range and is based on OFDM modulation. This PHY specification is applicable to both the licensed and unlicensed bands.

Any OFDM system can be defined by its major design parameter, an OFDM symbol duration. The symbol duration consists of the FFT interval and the cyclic prefix (CP).

Figure 5.21 [45] illustrates the IEEE 802.16 standard OFDM symbol time structure. A copy of the last guard period T_g is a CP, which is responsible for collecting multipath and maintaining the tones' orthogonality. The transmitter energy increases with the length of T_g but the receiver energy remains constant, since the CP is discarded. As a result, the Eb/No suffers the following dB loss:

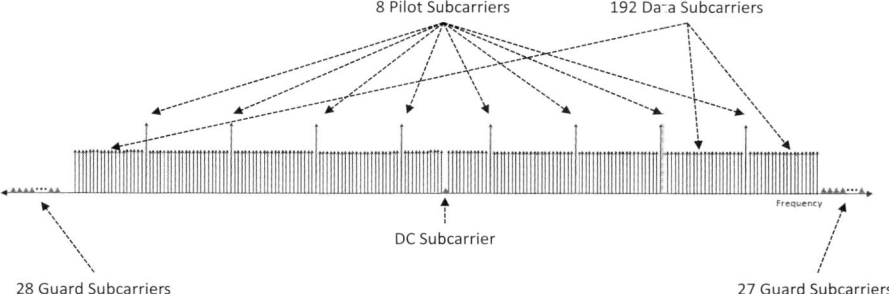

Figure 5.22   OFDM symbol representation in frequency domain. Adapted from Reference [45].

$$10\log_{10}\left(1 - \frac{T_g}{T_b + T_g}\right).$$

As previously mentioned, the presence of CP results in the loss of signal-to-noise ratio (SNR), since the CP at the receiver is always discarded, which wastes part of the transmitter energy. However, the benefits of having a cyclic prefix are numerous and far outweigh this transmitter energy loss. With a cyclic prefix, the samples necessary for performing the FFT at the receiver side can be collected anywhere over the length of the extended symbol, thus providing multipath immunity and a tolerance for symbol time synchronization errors. Note that in the downlink, the BS is not permitted to vary the length of CP, as the SS will lose synchronization [88].

The basic structure of an OFDM symbol in frequency domain is depicted in Figure 5.22 [45]. Each OFDM symbol is composed of data subcarriers, pilot subcarriers, and null subcarriers. Data subcarriers are used for data transmission. Pilot subcarriers serve various channel estimation purposes. Null subcarriers are reserved for DC subcarrier, guard bands, and cases when there is no transmission (all nonactive subcarriers). The number of FFT points in OFDM PHY is fixed at 256. Of these points, 192 are used as data carriers, 8 are used as pilot subcarriers for channel estimation and synchronization purposes, and 55 are used as guard band subcarriers (28 on one side and 27 on the other). The remaining point is reserved for DC subcarrier. The cyclic prefix's length can be 64, 32, 16, or 8 carriers [88].

The OFDM Waveform   The OFDM waveform during any OFDM is specified by Equation 5.1 (from Reference [45]), as a function of time:

$$s(t) = \mathrm{Re}\left\{ e^{j2\pi f_c t} \sum_{\substack{k=-N_{used}/2 \\ k \neq 0}}^{N_{used}/2} c_k \cdot e^{j2\pi k\Delta f(t - T_g)} \right\}, \tag{5.1}$$

where

t: time elapsed since the beginning of the given OFDM symbol, with $0 < t < T_s$.

c_k: data to be transmitted on the subcarrier whose frequency offset index is k, during the given OFDM symbol. In a QAM constellation, it specifies a single data point. For subchannelized transmissions, c_k is zero for all unallocated (null) subcarriers. c_k is a complex number.

f_c: carrier frequency.

BW: transmission bandwidth.

n: oversampling factor.

Δf: subcarrier spacing: $\Delta f = F_s/N_{FFT}$, where sampling frequency $F_s =$ floor($n*BW/8000$) × 8000 and N_{FFT} is the smallest power of two greater than N_{used}.

T_g: Cyclic prefix time: $T_g = G*T_b$, where useful symbol time $T_b = 1/\Delta f$ and G is the ratio of CP time to "useful" time. The allowed values of G are 1/4, 1/8, 1/16, and 1/32.

N_{used}:   number of allocated (nonsuppressed) carriers.

The oversampling factor n creates a sampling bandwidth that is slightly larger than the actual bandwidth BW. Oversampling stretches the spectral shape of the signal by maximally filling the space with the spectral mask and thus reducing the symbol duration. The shortening of the symbol duration leads to slightly higher throughput [95]. The IEEE 802.16 standard specifies the following allowed values of n:

- $n = 8/7$ for BW multiples of 1.75 MHz
- $n = 86/75$ for BW multiples of 1.5 MHz
- $n = 144/125$ for BW multiples of 1.25 MHz
- $n = 316/275$ for BW multiples of 2.75 MHz
- $n = 57/50$ for BW multiples of 2.0 MHz
- $n = 8/7$ for any other BW [45].

Subchannelization   Time division multiple access (TDMA) is the access method for the Fixed WiMAX, which is based on OFDM PHY. However, the IEEE 802.16 specification also provides an optional, integrated OFDMA component. In order to avoid confusion between this optional component of OFDM PHY and the Mobile WiMAX's OFDMA PHY, the OFDMA component in the standard is referred to as *subchannelization*. Subchannelization is a known method for extending the link budget and improving the efficiency of the employed coding and modulation scheme. During subchannelization, the available data subcarriers are binned into several groups of subcarriers or *subchannels*. The IEEE 802.16 specification defines 16 subchannels, each consisting of 12 data subcarriers. In the uplink the subchannels are assigned to the SS in sets of 1, 2, 4, 8, or 16 subchannels. Note that only eight pilot subcarriers are available for all the subchannels. For this very reason, when single subchannel allocations (1/16th of all available subchannels) are employed, data-driven approaches must take full responsibility for phase estimation [91, 95].

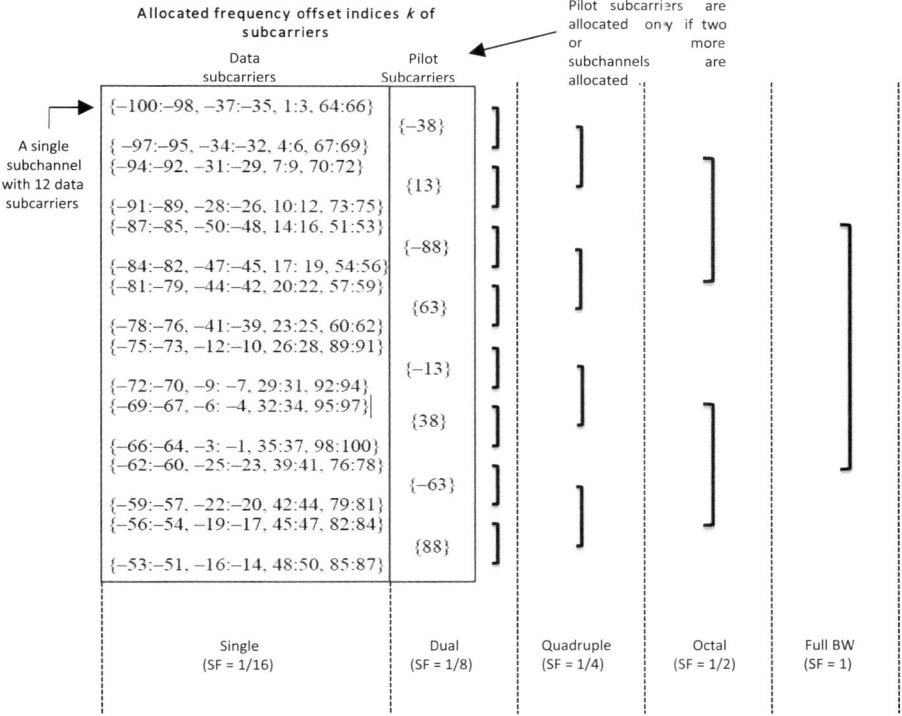

Subchannel Allocations

Figure 5.23   OFDM symbol subchannel structure. Adapted from Reference [45]. SF, subchannelization factor.

The subchannel structure, as defined in the IEEE 802.16 standard, is depicted in Figure 5.23.

Uplink subchannelization in Fixed WiMAX permits SSs to use only a fraction of the total bandwidth allocated to their BS. This fraction can be as low as 1/16th of the total BW. The use of subchannelization improves the link budget that, in turn, can translate into a better range of performance and/or longer battery life of the SS. For example, when an SS employs a single subchannel allocation (subchannelization factor of 1/16), the link budget of that SS improves by 12 dB (though this gain will be slightly reduced by the implementation losses) [91, 95].

The downlink subchannelization in OFDM PHY is restricted to quadruple allocations only (see Figure 5.23 [45]) and is primarily intended for the use in mobile devices.

OFDM PHY Frame Structure   For Fixed WiMAX, the duplexing method in licensed bands can be either TDD or FDD (half-duplex SSs are supported). In license-exempt bands, however, only TDD can be used as a duplexing method to ensure better coexistence with other IEEE 802 standards [95]. Figure 5.24 and Figure

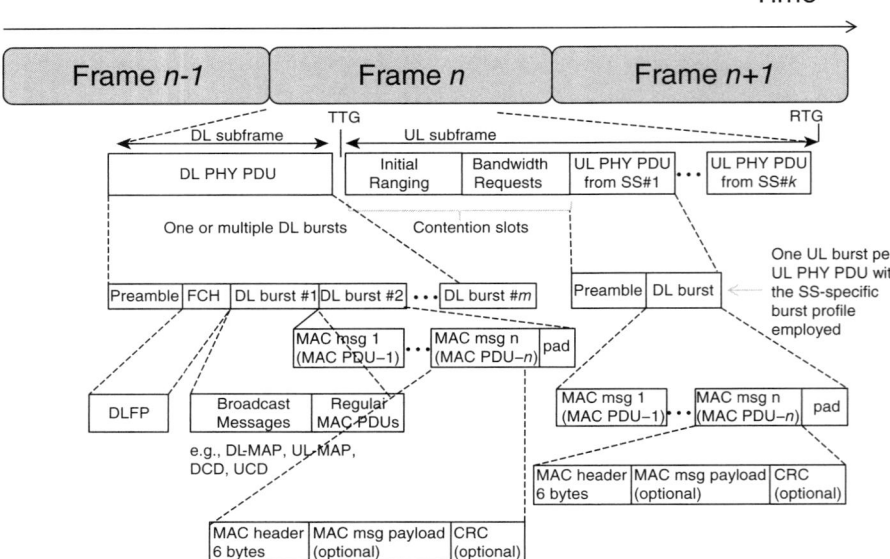

Figure 5.24   An example of a TDD OFDM frame structure. Adapted from Reference [45]. TTG, transmit/receive transition gap; RTG, receive/transmit transition gap; PDU, protocol data unit; FCH, frame control header; DLFP, downlink frame prefix; DL-MAP, downlink map message; UL-MAP, uplink map message; DCD, downlink channel descriptor; UCD, uplink channel descriptor; CRC, cyclic redundancy check.

5.25 illustrate examples of TDD and FDD frame structures, respectively. Each frame carries uplink and downlink subframes (PHY PDUs), guard intervals, and gaps. A downlink subframe is composed of only one downlink (DL) PHY PDU. An uplink (UL) subframe consists of contention intervals scheduled for initial ranging and various bandwidth requests and one or multiple UL PHY PDUs, each transmitted from a different SS. The allowed frame durations range between 2.5 and 20 ms.

When comparing Figure 5.24 and Figure 5.25, the FDD frame structure is quite similar to that of TDD with only two notable exceptions: (1) base station transmit/receive transition gap (TTG) and base station receive/transmit transition gap (RTG) are omitted and (2) the DL and UL subframes occur at the same time but on two different frequencies. Each DL PHY PDU starts with a preamble, which is used for PHY synchronization. The presence of preamble on the downlink allows new SSs to accurately track and adjust to the bases station's signal. On the uplink, the preamble is needed for initial ranging. In general, preambles are used for fine-tuning channel estimation as well as time and frequency estimation.

Channel Encoding   The OFDM PHY channel encoding always occurs before the inverse fast Fourier transform (IFFT) operation and is accomplished in four

Time

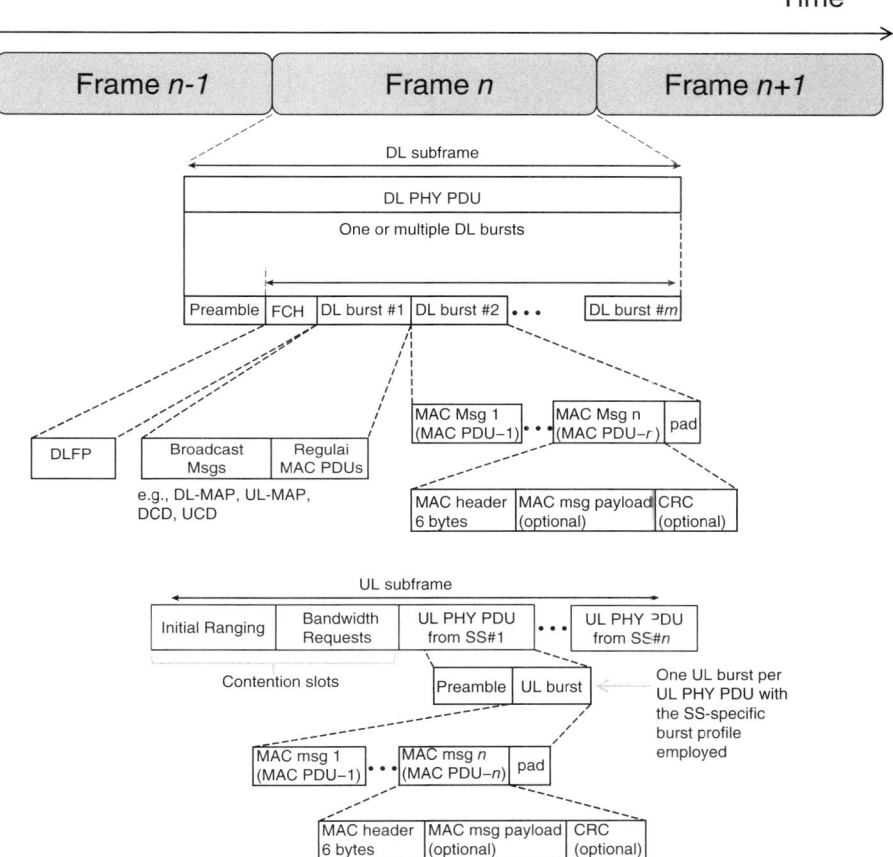

Figure 5.25   An example of an FDD OFDM frame structure. Adapted from Reference [45]. TTG, transmit/receive transition gap; RTG, receive/transmit transition gap; PDU, protocol data unit; FCH, frame control header; DLFP, downlink frame prefix; DL-MAP, downlink map message; UL-MAP, uplink map message; DCD, downlink channel descriptor; UCD, uplink channel descriptor; CRC, cyclic redundancy check.

distinct stages: randomization, forward error correction (FEC), interleaving, and modulation.

Randomization   Data randomization serves several purposes:

1. To minimize repeat receiver errors resulting from retransmission of data sequences with high peak-to-average power ratios (PAPRs)

2. To lessen the possibility of transmission of an unmodulated carrier

3. To ensure adequate numbers of bit transitions for clock recovery support

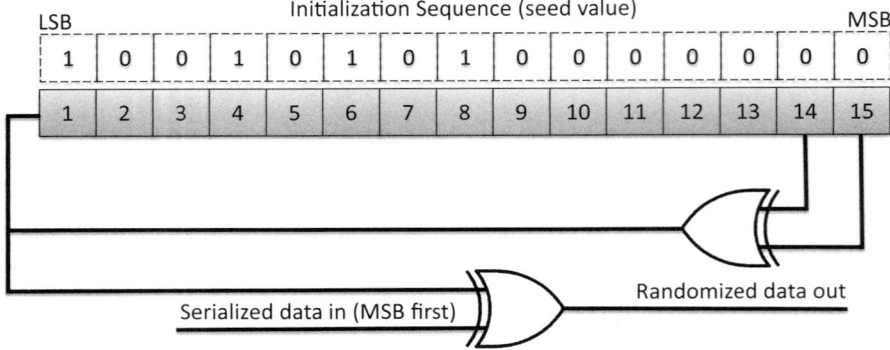

Figure 5.26   PRBS generator of $1 + X^{14} + X^{15}$ for data randomization. Adapted from Reference [45].

4. To provide PHY encryption in order to prevent an unauthorized receiver from decoding the data.

Each data burst on the uplink and downlink must undergo the data randomization process. During this process, a block of data, that is, subchannels in frequency domain and OFDM symbols in time domain, is transmitted through a pseudorandom binary sequence (PRBS) generator with $1 + X^{14} + X^{15}$ polynomial as shown in Figure 5.26 [45]. Each data byte is fed sequentially into the randomizer, MSB first. The randomizer sequence is used only on information bits, thus preambles are never randomized. The randomization bits are calculated from a specified seed value and are combined in an XOR operation with the serialized bit stream of each burst.

Forward Error Correction   Forward error correction (FEC) in OFDM PHY is accomplished via one of these three methods: (1) Reed–Solomon outer coding concatenated with a rate-compatible convolutional inner coding (RS-CC), (2) block turbo coding (BTC), and (3) convolutional turbo coding. RS-CC is mandatory and must be supported on both the uplink and downlink. Although BTC and CTC yield a better coding gain (approximately 2–3 dB better) than RS-CC, their support in OFDM PHY is optional. It is their significant implementation complexity that made them less attractive as a mandatory requirement when compared to RS-CC [95].

Reed–Solomon Coding Concatenated with Convolutional Coding   The Reed–Solomon concatenated with convolutional coding (RS-CC) encoding is performed with a systematic RS ($N = 255$, $K = 239$, $T = 8$) code over Galois field (GF) ($2^8 = 256$), where:

N is the number of overall bytes after encoding,

K is the number of data bytes before encoding, and

T is the number of data bytes which can be corrected.

The following polynomials are used for the systematic code:

Code Generator polynomial:

$$g(x) = (x + \lambda^0)(x + \lambda^1)(x + \lambda^2) \ldots (x + \lambda^{2T-1}), \lambda = 02_{HEX}. \quad (5.2)$$

Field Generator polynomial:

$$p(x) = x^8 + x^4 + x^3 + x^2 + 1. \quad (5.3)$$

This code can be shortened and punctured to ensure variability within block sizes and error correction capabilities. Each RS block is then followed by the binary convolutional encoder with native rate of 1/2, a constraint length of 7, and the generator polynomials codes of $G_1 = 171_{OCT}$ and $G_2 = 133_{OCT}$ to derive its two code bits [45].

The randomized data burst to be encoded is appended by a single tail byte (of zeros) to ensure zero-tail termination of the CC encoder. In the RS encoder, the redundant (parity) bits for each data block are sent before the data block itself, ensuring that the tail byte is passed at the end of the allocation. Note that the CC encoder is utilized over the entire data burst. To avoid performance issues due to the RS codes' weakness for small blocks, OFDM PHY does not employ the RS encoding when subchannelization is applied [95]. In addition, OFDM PHY bypasses RS encoder when BPSK modulation is employed. Table 5.9 [45] lists OFDM PHY mandatory channel coding per modulation along with allowed data block sizes.

Block Turbo Coding   Block turbo coding (BTC) encodes the data in a block form, where the block is initially filled row by row and then column by column. This method of encoding is fundamentally different from RS-CC, where the data are manipulated in one long string.

Essentially, BTC is the product of two simple component codes: binary extended Hamming codes or parity check codes. OFDM PHY defines the set of approved codes summarized in Table 5.10 [45].

Table 5.9   Required Channel Coding Parameters per Modulation Type

Modulation type	Uncoded block size (bytes)	Coded block size (bytes)	Overall coding rate	RS code	CC code rate
BPSK	12	24	1/2	(12,12,0)	1/2
QPSK	24	48	1/2	(32,24,4)	2/3
QPSK	36	48	3/4	(40,36,2)	5/6
16-QAM	48	96	1/2	(64,48,8)	2/3
16-QAM	72	96	3/4	(80,72,4)	5/6
64-QAM	96	144	2/3	(108,96,6)	3/4
64-QAM	108	144	3/4	(120,108,6)	5/6

Reprinted from Reference [45].

Table 5.10 OFDM PHY BTC Component Codes

Component code (n, k)	Code type
(64,57)	Extended Hamming code
(32,26)	Extended Hamming code
(16,11)	Extended Hamming code
(64,63)	Parity check code
(32,31)	Parity check code
(16,15)	Parity check code
(8,7)	Parity check code

Adapted from Reference [45].

Table 5.11 OFDM PHY Hamming Code Generator Polynomials

n'	k'	Generator polynomial
7	7	X^3+X^1+1
15	11	X^4+X^1+1
31	26	X^5+X^2+1
63	57	X^6+X+1

Reprinted from Reference [45].

The generator polynomials for the Hamming codes are provided in Table 5.11 [45]. Extended Hamming codes are created by adding an overall even parity check bit at the end of each code word.

In BTC, the component codes are used in a block form, which is illustrated in Figure 5.27 [45].

The k_x information bits in the rows get encoded into n_x bits via the component block (n_x, k_x) code specified for the respective composite code. After the encoding of the rows is complete, the columns are encoded with the block code (n_y, k_y). The check bits of the first code are also encoded. The overall block size of a product code is $n = n_x \times n_y$ with the total number of information bits $k = k_x \times k_y$ and the code rate $R = R_x \times R_y$, where $R_i = k_i/n_i$, $i = x, y$. The Hamming distance of the product code is $d = d_x \times d_y$. Data bit ordering in the composite BTC matrix defines the first bit in the first row as the LSB, and the last data bit in the last data row as the MSB. The actual transmission of the data block over the channel occurs linearly, with all bits of the first row transmitted from left to right followed by the second row, and so on and so forth.

BTCs may also be shortened by removing symbols from the BTC array to match a required packet size (see Figure 5.28 [45]). This is done by not filling the first I_x rows, I_y columns, and B bits and by zero-filling Q bits until the next byte boundary.

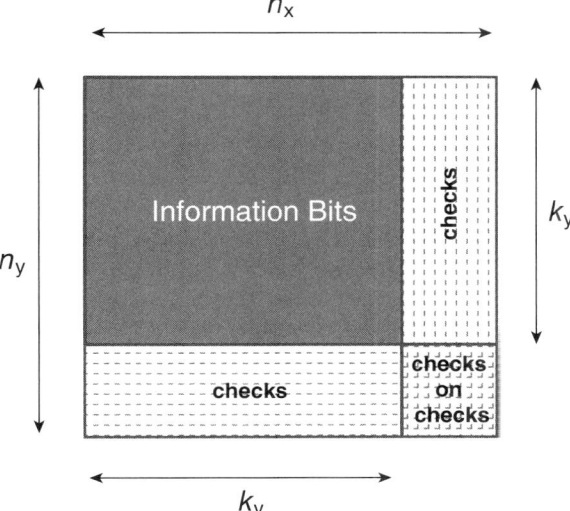

Figure 5.27   BTC structure. Adapted from Reference [45].

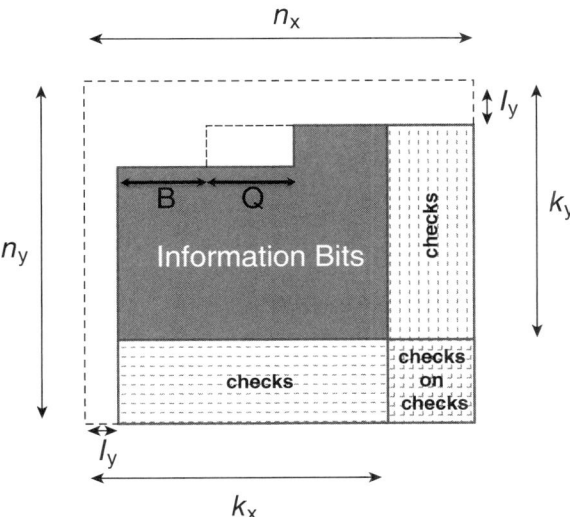

Figure 5.28   Shortened BTC structure. Adapted from Reference [45].

Table 5.12   Optional BTC Channel Coding Parameters per Modulation Type

Modulation	Uncoded block size (bytes)	Coded block size (bytes)	Overall approximate code rate	Efficiency (bits/s/ Hz)	Constituent codes (n_x, k_x) (n_y, k_y)	Code parameters
QPSK	23	48	1/2	1.0	(32,26) (16,11)	$I_x = 4, I_y = 2,$ $B = 8,$ $Q = 6$
QPSK	35	48	3/4	1.5	(32,26) (16,15)	$I_x = 0, I_y = 4,$ $B = 0,$ $Q = 6$
16-QAM	58	96	3/5	2.4	(32,26) (32,26)	$I_x = 0, I_y = 8,$ $B = 0,$ $Q = 4$
16-QAM	77	96	4/5	3.3	(64,57) (16,15)	$I_x = 7, I_y = 2,$ $B = 30,$ $Q = 4$
64-QAM	96	144	2/3	3.8	(64,63) (32,26)	$I_x = 3,$ $I_y = 13,$ $B = 7,$ $Q = 5$
64-QAM	120	144	5/6	5.0	(32,31) (64,57)	$I_x = 13,$ $I_y = 3,$ $B = 7,$ $Q = 5$

Reprinted from Reference [45].

As a result, an encoding is fully described by its two constituent codes and the parameters I_x, I_y, B, and Q [95]. A shortened BTC has a block size of the product code $n = (n_x - I_x)(n_y - I_y) - B$ with the information length $k = (k_x - I_x)(k_y - I_y) - B - Q$. Consequently, the code rate R is given by:

$$R = \frac{(k_x - I_x)(k_y - I_y) - B - Q}{(n_x - I_x)(n_y - I_y) - B}. \tag{5.4}$$

Table 5.12 [45] provides block sizes, code rates, channel efficiency, and code parameters for various modulation and coding schemes as specified in the OFDM PHY description.

When subchannelization is employed, the coding block size is limited to blocks of 96 bits in length or more. According to Reference [95], the performance of the BTC for an allocation of only a few subchannels was never thoroughly examined during the developmental stage of the subchannelization option. Thus, the use of the small block size might prove to be problematic in terms of the system performance.

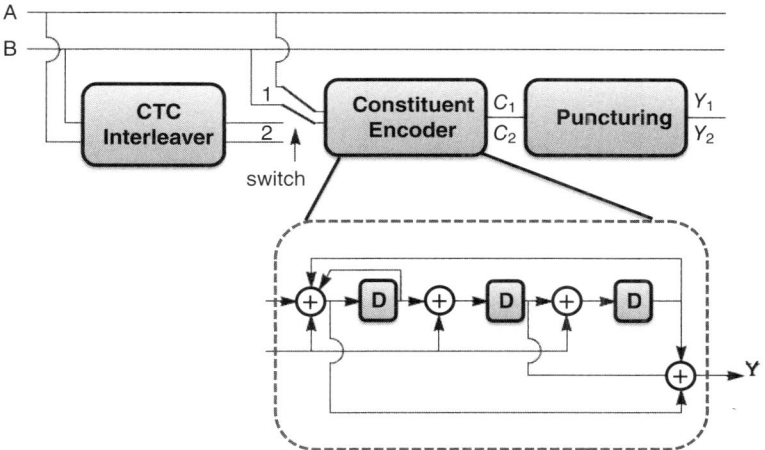

Figure 5.29   CTC encoder. Adapted from Reference [45].

Convolutional Turbo Coding   The CTC encoder is depicted in Figure 5.29 [45]. The data bits to be encoded are alternately fed into A and B, MSB first, with the switch in the position 1 as shown in Figure 5.29. The encoder is supplied with blocks of k bits or N couples ($k = 2 \times N$ bits). For all frame sizes, k is a multiple of 8 and N is a multiple of 4. N is bound by $8 \leq N/4 \leq 1024$. When subchannelization is used, the coding block size must be at least 48 bits in length, but no more than 1024. Moreover, k cannot be a multiple of 7. The polynomials defining the connections can be described in octal and symbol notations as follows:

- For the feedback branch (symbolic notation): $D^3 + D + 1$
- For the Y parity bit: $D^3 + D^2 + 1$

The encoding process is executed twice: one with the memory elements initialized to 0b000 with the switch in position 1 and another with the memory elements initialized with a value that was calculated by passing the first value of the first run through a lookup table with the switch in position 2 [95]. The final CTC-encoded stream consists of a sequence of alternated A and B bits, followed by punctured parity bits from the first encoding (when the switch was in position 1), then followed by punctured parity bits from the second encoding (switch position 2) [95].

Table 5.13 provides CTC block sizes, code rates, channel efficiency, and code parameters for various modulation and coding schemes. In Table 5.13, N_{sub} denotes the number of subchannels for the allocation in which the encoded data is to be transmitted.

Interleaving   The interleaver in OFDM PHY is defined by a two-step permutation. In the first step, the adjacent coded bits are mapped onto nonadjacent subcarriers. In the second step, the adjacent coded bits are mapped alternately onto less or

Table 5.13 Optional CTC Channel Coding Parameters per Modulation Type

Modulation	N	Overall code rate	Interleaver parameter P_0
QPSK	$6N_{sub}$	1/2	7
QPSK	$8N_{sub}$	2/3	11
QPSK	$9N_{sub}$	3/4	17
16-QAM	$12N_{sub}$	1/2	11
16-QAM	$18N_{sub}$	3/4	13
64-QAM	$24N_{sub}$	2/3	17
64-QAM	$27N_{sub}$	3/4	17

Reprinted from Reference [45].

Table 5.14 Bit Interleaver Sizes for Various Modulation and Coding Schemes

	Default (16 subchannels)	8 subchannels	4 subchannels	2 subchannels	1 subchannel
			N_{cbps}		
BPSK	192	96	48	24	12
QPSK	384	192	96	48	24
16-QAM	768	384	192	96	48
64-QAM	1152	576	288	144	72

Reprinted from Reference [45].

more significant bits of the constellation, thus avoiding long runs of low-reliability bits. All encoded data bits are interleaved with a block size corresponding to the number of coded bits in all allocated subchannels within an OFDM symbol, N_{cbps}. In other words, there is only one interleaving block per OFDM symbol for each transmitter [95]. Table 5.14 [45] lists the allowed bit interleaver sizes for various modulation and coding schemes.

Modulation   The OFDM PHY mandates support of BPSK, Gray-mapped 4-QAM (QPSK), and 16-QAM constellations. The use of Gray-mapped 64-QAM constellation is optional for license-exempt bands. The constellations of supported modulations are depicted in Figure 5.30 [45]. Each constellation is normalized by multiplying the constellation point with the indicated factor c to achieve equal average power. For each modulation type, b_0 denotes the LSB.

Pilot subcarriers are modulated with a BPSK signal. The values to be used are obtained from feeding fixed initialization sequences (one for UL and one for DL) through the PRBS generator with the polynomial $X^{11}+X^9+1$ clocked with the OFDM symbol rate (see Figure 5.31 [45]). The values of the individual pilot tones for an

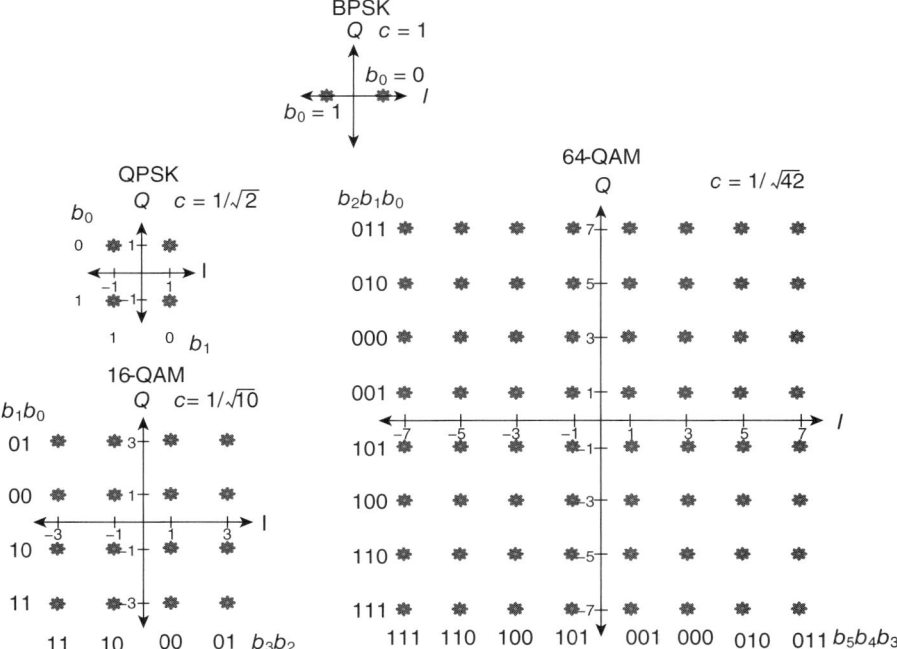

Figure 5.30 OFDM PHY: BPSK, QPSK, 16-QAM, and 64-QAM constellations. Adapted from Reference [45].

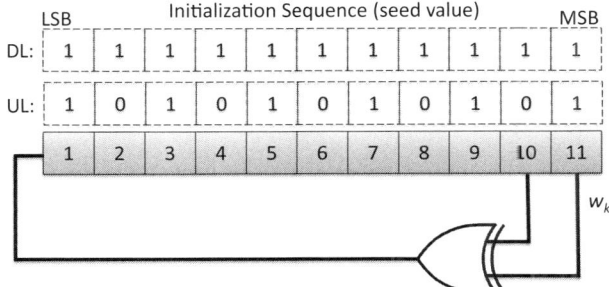

Figure 5.31 PRBS generator for pilot modulation. Adapted from Reference [45].

OFDM symbol k are derived from w_k and defined as being equal or negative of the BPSK-modulated PRBS generator's output [95]. For each pilot (indicated by frequency offset index), the BPSK modulation is defined as follows [45]:

Downlink:

$$c_{-88} = c_{-38} = c_{63} = c_{88} = 1 - 2w_k \quad \text{and} \quad c_{-63} = c_{-13} = c_{13} = c_{38} = 1 - 2\overline{w}_k. \quad (5.5)$$

Table 5.15   Sample OFDM PHY Data Rates across Various Modulation Types

Assumptions	
Channel bandwidth	3.5 MHz
FFT size	256
Oversampling	8/7
Downlink-to-uplink ratio	3 : 1
Duplexing	TDD
Frame size	5 ms
OFDM guard interval overhead	12.5%
Partial usage of subchannels (PUSC)	Yes

	PHY data rate (Mbps)	
Modulation type, CC code rate	Downlink	Uplink
BPSK, 1/2	0.946	0.326
QPSK, 1/2	1.882	0.653
QPSK, 3/4	2.822	0.979
16-QAM, 1/2	3.763	1.306
16-QAM, 3/4	5.645	1.958
64-QAM, 1/2	5.645	1.958
64-QAM, 2/3	7.526	2.611
64-QAM, 3/4	8.467	2.938
64-QAM, 5/6	9.408	3.264

Reprinted from Reference [91].
The values shown are from Reference [91] and represent the cumulative PHY data rate that is shared among all users within the sector (TDD case). To derive these rates, the authors in Reference [91] also assumed that all usable OFDM data symbols were available for the user traffic with the exception of one symbol reserved for the downlink frame overhead.

Uplink:

$$c_{-88} = c_{-38} = c_{13} = c_{38} = c_{63} = c_{88} = 1 - 2w_k \quad \text{and} \quad c_{-63} = c_{-13} = 1 - 2\overline{w}_k. \quad (5.6)$$

Physical Layer Data Rates   OFDM PHY data rate performance is highly dependent upon the chosen operating parameters such as channel bandwidth, coding and modulation scheme, oversampling rate, OFDM guard time, and subchannel allocation [91]. Table 5.15 [91] lists the sample OFDM PHY data rates for the 3.5-MHz channel BW across various modulation types and convolutional code rates.

Use of Multiple-Input, Multiple-Output (MIMO) Technology in Fixed WiMAX   OFDM PHY profile supports the use of transmit (Tx) diversity on the downlink, which is optional. Tx diversity is a form of MIMO and is accomplished

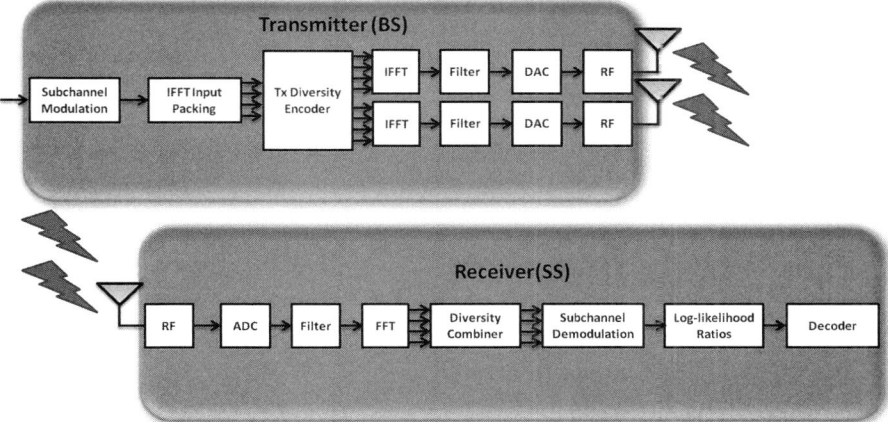

Figure 5.32   STC in OFDM PHY. Adapted from Reference [45].

through the use of space–time coding (STC) based on Alamouti's algorithm. Optional STC can be used on the downlink to provide higher-order spatial Tx diversity. Since two Tx antennas are employed on the BS side and only one receive antenna on the SS side, this scheme requires multiple-input, single-output (MISO) channel estimation. Decoding is performed in a manner similar to maximum ratio combining. Figure 5.32 [45] illustrates how STC is achieved in OFDM PHY [45].

On the BS side, both antennas transmit two different OFDM data symbols at the same time. Transmission is performed twice to decode and to get second-order diversity. During the first transmission, antenna 0 transmits complex symbol s_0 and antenna 1 transmits complex symbol s_1. During the second transmission, antenna 0 transmits $-s_1^*$ and antenna 1 transmits $-s_0^*$. STC is applied independently on each carrier with respect to pilot carrier positions [88]. Figure 5.33 [45] depicts an example of the STC usage for only one pilot subcarrier −88 [45].

Control Mechanisms   OFDM PHY provides BS with four control mechanisms to better manage the SS's behavior: synchronization, ranging, BW request, and power control.

Network Synchronization   OFDM PHY recommends (but not requires) for all BSs to be time synchronized to a common timing signal (both TDD and FDD realizations). In the event of the loss of the network timing signal, BSs will continue their operations and will automatically resynchronize to the network timing signal as soon as it is recovered. The synchronizing reference of a 1-pps timing pulse is recommended along with a 10-MHz frequency reference. Typically, these signals would be provided by the GPS receiver [45].

Ranging   Recall from our earlier discussion about initial network access that "ranging" is the process of acquiring the correct timing offset and power adjustments

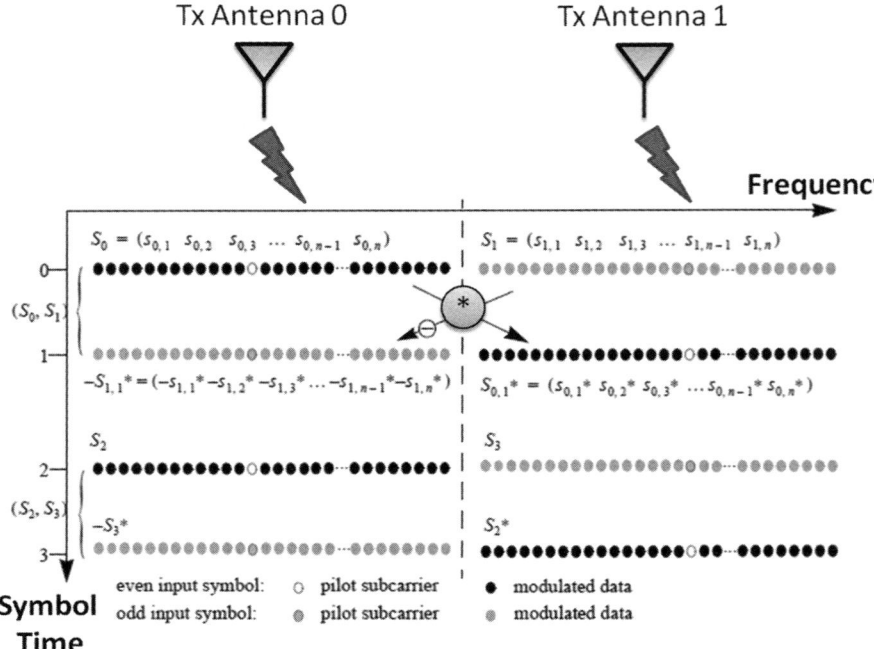

Figure 5.33   An example of STC Usage with OFDM PHY. Adapted from Reference [45].

in such way that the SS's transmissions are aligned with the BS receive frame for OFDM PHY (Fixed WiMAX), and are received within the appropriate reception thresholds. The WiMAX technology employs two types of ranging: initial ranging and periodic ranging. During initial ranging, the SS sends a brief ranging request message that allows the BS to send back a ranging response message with the amount of timing offset that the SS must use once it starts transmitting. Initial ranging is the process of time aligning with a communication system *before* establishing a communication session. The initial ranging process involves synchronization with an incoming transmission channel, sending a request at a particular time interval relative to the received channel and obtaining a response that allows for the time synchronization with the device or system.

Periodic ranging, in contrast, is the process of continuous time alignment with a communication system during a communication session. The periodic ranging process involves maintaining synchronization with an incoming transmission channel, periodically transmitting messages (e.g., transmitting user data) and receiving time alignment information from the other device or system.

The periodic ranging uses the regular uplink burst. Since periodic ranging is predominantly a MAC process, PHY does not make any special provisions in terms of waveform creation or scheduling. Initial ranging is different in that regard as it uses a contention-based protocol with a nonzero probability of collision and requires a long preamble [45, 95]. Initial ranging is performed during two phases of operation: (1) during (re)registration and when synchronization is lost, and (2) during transmission on a periodic basis.

Bandwidth Requests   Bandwidth can be requested in a number of ways: either via polling (see the section "Bandwidth Negotiation and Allocation"), piggybacking, or one of the two contention-based methods. From the PHY's perspective, the latter two are of interest. These two methods are *full contention* and *focused contention* [95]. Implementation of focused contention is optional in OFDM PHY.

Full contention can be used with full bandwidth allocations as well as with subchannelized allocations. When full bandwidth is employed, the request is constrained to two OFDM symbols reserved for short preamble and one data symbol carrying the actual BW request. In cases of subchannelized allocations, the width (in subchannels) and length (in OFDM symbols) is reported in the uplink channel descriptor (UCD), a MAC message describing PHY characteristics of the uplink. SS can make multiple BW requests within a single allocation since the BS can identify each SS by its unique basic CID during the BW request [95].

When optional focused contention is implemented, the SS randomly chooses a contention channel index out of the 48 available. This information is then mapped to four unique data subcarriers, which are arranged to form three sets of contention channels within each subchannel. After that, the SS transmits one of eight BPSK contention codes. For the subsequent OFDM symbol, the SS transmits again the same code on the same four subcarriers across all other subcarriers with the contention element minus 1 and inverted phase. As a result, the BS can now identify the SS by the contention channel index and contention code before issuing an uplink allocation.

Power Control   In OFDM PHY power control is specified only in the uplink, as it is needed to bring the received power density from a given SS to a desired level. The IEEE 802.16 specification defines the received power density as a total power received from a given SS divided by the number of active subcarriers. When subchannelization is not employed, the number of active subcarriers is equal for all the subscribers and the power control algorithm must bring the total received power from a given SS to the desired level. The BS must be able to provide accurate power measurements of the received burst signal. This value can then be compared against a reference level, and the resulting error can be fed back to the SS in a calibration message coming from the MAC. The power control algorithm must support power attenuation due to distance loss or power fluctuations at a rate of at least 30 dB/s. The exact algorithm implementation is vendor specific. The total power control range is composed of both the fixed portion and the portion that is automatically controlled by feedback. The power control algorithm must also take into account the interaction of the RF power amplifier with different burst profiles. For instance, when changing from one burst profile to another, margins should be maintained to prevent saturation of the power amplifier and to prevent violation of emissions masks [95].

OFDM PHY supports two types of uplink power control modes: closed-loop power control (mandatory) and open-loop power control (optional).

In closed-loop power control, when subchannelization is applied in the uplink the SS maintains the same transmitted power density unless the maximum power level is reached. In other words, when the number of active subchannels allocated to a user is reduced, the total transmitted power is reduced proportionally by the SS and vice versa.

In open-loop power control, which is optional in OFDM PHY, the power per subcarrier is adjusted depending on the link quality. The open-loop power control further divides into two types: passive and active. In passive open-loop power control, the SS does not participate in the power adjustment process, whereas in active open-loop power control, the SS can adjust power value that falls within a predefined range.

5.1.4 System Profiles

To ensure broad and swift acceptance of WiMAX technologies throughout the world, WiMAX Forum defines interoperable system profiles suitable for common licensed and unlicensed bands used around the globe. A WiMAX system profile is a set of defined network parameters such as frequency band of operation, channel bandwidth, and duplexing method. The current fixed profiles are defined for both TDD and FDD schemes and have channel bandwidths of 3.5, 5, 7, and 10 MHz. As of the last quarter of 2011, seven system profiles were outlined for Fixed WiMAX, which is based on the OFDM PHY specification.

5.2 USAGE

One of the most undeniable measures of the success of a technology is in the number of deployments it experiences. In 2011 WiMAX Forum tracked 583 WiMAX network deployments, either already in service or in the process of being deployed [93]. The total number of countries with WiMAX deployments reached 150 as of May 2011 [93]. Deployments by region are listed in Table 5.16 (derived using data from Reference [93]).

By the end of 2011, the marketplace offered 260 WiMAX Forum Certified™ products, 62 of which were base stations and the rest (198) were subscriber and mobile stations. Additionally, there were 26 base station vendors along with 51 SS/MS vendors proffering WiMAX Forum Certified products. Some of the current base station vendors are: Airspan Networks, Alcatel-Lucent, Alvarion, Axxcelera, Cisco Systems, E.T. Industries, Harris Stratex, Huawei Technologies, Institute for Information Industry, Motorola, NEC, Nokia Siemens Networks, Nortel, PointRed Telecom Pvt Ltd, POSDATA, Proxim Wireless Corporation, Redline Communications, Runcom Technologies Ltd., Samsung, Selex Communications, Sequans Communications, Soma Networks, SR Telecom, WiNetworks, and ZTE Corporation. Current SS/MS vendors tracked by the WiMAX Forum are: Acer Incorporated, Airspan Networks, Alcatel-Lucent, Alvarion, ApaceWave Technologies, Axxcelera Broadband Wireless, Beceem, Cisco Systems, C-motech Co. Ltd., D-Link, E.T. Industries,

Table 5.16 WiMAX Deployments by Region

Region	Countries	Deployments
CALA (Central America and Latin America)	33	120
Africa	43	117
North America (United States/Canada)	2	56
Asia Pacific	23	98
Eastern Europe	21	86
Western Europe	18	77
Middle East	10	29

Fujitsu, GCT Semiconductor, Inc., Gemtek Technology Co. Ltd, Gigaset Communications, Harris Stratex, Huawei Technologies, Infomark, Intel Corporation, Interbro, Japan Radio, Kouziro Co Ltd, Kyocera, Lenovo, MediaTek, Micro-Star Int'l co., LTD, Mitsumi Electric Co. ,Ltd., Modacom Co., Ltd., Motorola, NEC, Oki Electric Industry Co., Ltd., Onkyo, Panasonic, PointRed Telecom, POSDATA, Redline Communications, Runcom Technologies Ltd., Samsung, Selex Communications, Seowon Intech Co., Ltd., Sequans Communications, Siemens AG, Sony, SR Telecom, Toshiba, Ubee Interactive, Wavesat Inc., ZTE Corporation, and ZyXEL Communications, Inc. [93].

5.3 EVOLUTION

The evolution of BWA as it relates to WiMAX technology can be traced back to the desire to find a feasible alternative to traditional wired broadband infrastructures. As it was previously mentioned, WiMAX technology has evolved through four stages:

1. Narrowband WLL systems
2. First-generation LOS broadband systems
3. Second-generation non-line-of-sight (NLOS) broadband systems
4. Standards-based broadband wireless systems

The fourth stage is defined by the development of the IEEE 802.16 standard that had already gone through numerous revisions to address market's ever-changing needs for broadband access and mobility. See Table 5.2 for a detailed description of each revision.

Figure 5.34 illustrates various stages of BWA evolution in the context of the WiMAX technology.

5.4 WIMAX TRANSITION TO CELLULAR TECHNOLOGY

As WiMAX technology transitions into the realm of cellular networks, it faces many challenges, including difficulty of OFDMA implementation due to its complexity,

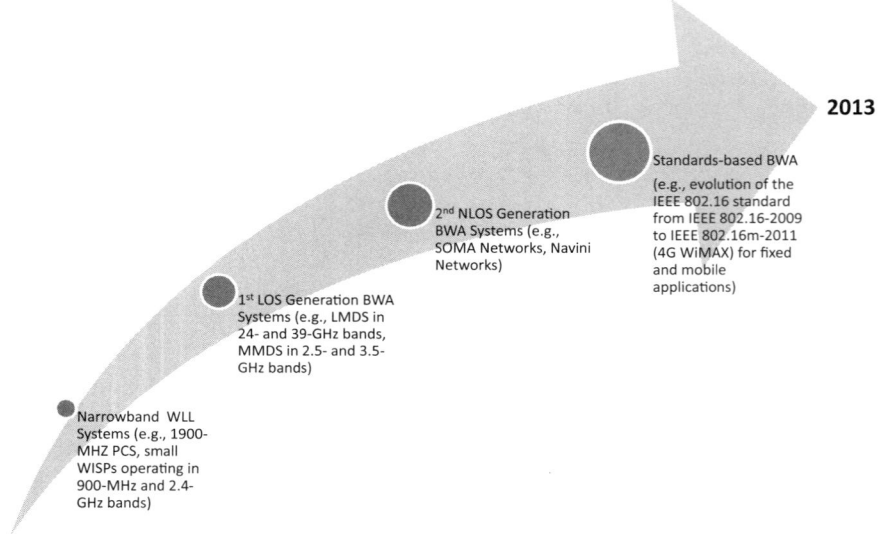

2013

Standards-based BWA (e.g., evolution of the IEEE 802.16 standard from IEEE 802.16-2009 to IEEE 802.16m-2011 (4G WiMAX) for fixed and mobile applications)

2nd NLOS Generation BWA Systems (e.g., SOMA Networks, Navini Networks)

1st LOS Generation BWA Systems (e.g., LMDS in 24- and 39-GHz bands, MMDS in 2.5- and 3.5-GHz bands)

Narrowband WLL Systems (e.g., 1900-MHZ PCS, small WISPs operating in 900-MHz and 2.4-GHz bands)

1990s

Figure 5.34   WiMAX evolution. Adapted from Reference [45]. WLL, wireless local loop; PCS, personal communications services; WISP, wireless Internet service provider; LMDS, local multipoint distribution systems; MMDS, multichannel multipoint distribution systems; BWA, broadband wireless access; 4G, fourth generation.

intercell interference at the cell edges due to WiMAX's nontraditional utilization of frequency reuse methods, and the need for development of more linear, more efficient (and thus more expensive) power amplifiers for WiMAX mobile nodes, as well as higher-performance synthesizers to support WiMAX's subchannelization. WiMAX networks must also be able to seamlessly integrate into existing cellular structures, without sacrificing connectivity and user experience. Of course, amalgamating WiMAX technology with the core network, BS, and MS is an incredibly difficult feat because WiMAX's modulation and coding schemes are vastly different from those of the legacy cellular systems.

RECOMMENDED ADDITIONAL READING

If the reader is interested in further study of Fixed WIMAX technology, the following resources are recommended:

> J.G. ANDREWS, A. GHOSH, and R. MUHAMED, *Fundamentals of WiMAX: Understanding Broadband Wireless Networking*, Prentice Hall, New York, 2007 [91].

> T. COOKLEV, *Wireless Communication Standards: A Study of IEEE 802.11™, 802.15™, and 802.16™*, IEEE Press/Standards Information Network, New York, 2004 [88].

C. EKLUND, R.B. MARKS, S. PONNUSWAMY, and N.J.M. van WAES, *Wireless-MAN: Inside the IEEE 802.16 Standard for Wireless Metropolitan Area Networks*, IEEE, New York, 2006 [95].

All three books are outstanding resources for those trying to understand the fundamentals of Fixed WiMAX technology. Please bear in mind that since these books were published a few years ago, some of the information might have become outdated due to ongoing technology evolution. However, since Fixed WiMAX is not necessarily the current focus of the IEEE 802.16 WG, it is believed that these texts will give the reader a relatively up-to-date treatment of the technology.

To download the IEEE 802.16-2009 specification, please visit the following Web site: http://standards.ieee.org/getieee802/download/802.16-2009.pdf.

Chapter 6

Second-Generation (2G) Cellular Communications

Second-generation (2G) cellular networks marked the beginning of digital mobile cellular communications. A combination of factors gave rise to the development of these technologies. First, the success of the original analog cellular networks, retro-actively dubbed first generation (1G), proved that the newly developed cellular industry was indeed a profitable sector of wireless radio communications. Second, the spectral inefficiency of 1G networks combined with a growing subscriber base meant that the existing cellular networks were physically constrained by the available spectrum—meaning that, by the late 1980s, 1G networks were quickly approaching the saturation point at which technology limitations would prevent network providers from supplying the user capacity to meet growing market demand. The 2G cellular standards generally approached the need for increased spectral efficiency by using time division multiple access (TDMA) methods to multiplex multiple users onto a single carrier frequency. In addition to increased spectral efficiency, the switch to digital transmission technology brought along other benefits. Advances in signal processing and compression coupled with powerful error control coding techniques allowed for improved speech quality in harsh channel conditions. Improved digital speech codecs eventually led to landline-quality voice, which was a key improvement that precipitated the trend of using the mobile phone as a replacement for—rather than a supplement to—the landline phone in the home. Digital transmission also allowed for improved security through the use of encryption and digital security algorithms. This was a tremendous advantage in digital cellular networks, as eaves-dropping and phone cloning (i.e., reproducing another subscriber's credentials, usually captured over the air, for the purposes of illegal service access without charge) were major drawbacks of 1G networks that proved very costly to cellular network operators in the 1990s. Advances in semiconductor technology also made it possible for pocket-sized mobile handsets to replace their brick-sized predecessors. All of these improvements, combined with more affordable prices due to the growing

Wireless Networking: Understanding Internetworking Challenges, First Edition.
Jack L. Burbank, Julia Andrusenko, Jared S. Everett, and William T.M. Kasch.
© 2013 the Institute of Electrical and Electronics Engineers, Inc. Published 2013 by John Wiley & Sons, Inc.

subscriber base and economies of scale, combined to make cellular services more accessible and fueled the exponential growth of cellular networks in the 1990s. In short, the creation of 2G cellular networks marked the coming of age of the cellular industry.

6.1 HISTORICAL PERSPECTIVES

The introduction of 2G cellular networks came at a very interesting time in the history of the Internet. These cellular technologies were designed primarily in the late 1980s and early 1990s, before the privatization of the Internet and invention of the World Wide Web. At that time, the internetworking of computers and computer networks for the purposes of communication and online services was still an immature technology. The use of computer networks for communication and dissemination of information was primarily limited to universities, government agencies, and businesses. Although some basic commercial services intended for home users did exist, the personal computer and the Internet were not yet a primary tool for communication as part of everyday life the way the telephone was. As a result, 2G cellular networks were designed and optimized to support voice telephony services. The provisions for data communications in the original 2G cellular standards were either minimal or nonexistent. By the time these technologies were brought to market, however, a very different story was about to unfold.

The various 2G cellular technologies saw initial network deployments between 1991 and 1995. By this time, all the conditions were in place for the Internet revolution to begin. The privatization of the Internet had started in the early 1990s, and on April 30, 1995, the U.S. government formally terminated its control over the Internet infrastructure, allowing commercialization. The emergence of the World Wide Web and Web browsers created an entirely new use for the Internet and helped to transform it from a research tool to a popular medium. In the midst of the rollout of 2G cellular networks, the disruptive technology of the Internet caused a fundamental paradigm shift in the communications world. The growing popularity of data services such as e-mail, Web browsing, and file transfer created a new cellular data market in the mid-1990s. The existing provisions for data services based on circuit switching in the 2G cellular networks were simply insufficient, and enhancements were needed to keep up with growing market demand. This resulted in the development of so-called *2.5G* technologies, which were primarily packet-switched data enhancements designed to bridge the gap between the limited data capabilities of 2G networks and the not-yet-released third-generation (3G) networks that were designed from the beginning to support mobile networking. What's truly interesting about the evolution of 2G cellular networks during the 1990s is that they began as technologies designed and optimized for one application (i.e., circuit-switched voice) but were later adapted to a completely different application entirely (i.e., packet-switched data services). So began the mobile Internet.

The first 2.5G packet-switched data networks were deployed around the late 1990s. They offered an advantage to network operators over emerging 3G cellular technologies in that the operators could leverage their existing 2G networks, which represented a sizable investment, rather than replace it entirely. Meanwhile, a gap had opened up between the data rates offered in the fixed infrastructure and those of cellular networks. While enhanced 2.5G cellular networks were able to offer data rates comparable to that of a dial-up modem, many Internet users were becoming used to broadband Internet connections in the fixed-line world through technologies such as the cable modems and DSL. Even new technologies such as Wi-Fi offered wireless Internet connections at significantly higher data rates. It would not be until the introduction of fourth-generation (4G) networks that the cellular industry would finally close the performance gap with wireless networking technologies, providing an equivalent user experience. In the meantime, the main competitive advantage of the cellular networks, however, was mobility—as even Wi-Fi offered only nomadic access.

6.2 OVERVIEW OF 2G TECHNOLOGIES

Following suite with their 1G analog predecessors, a number of mutually incompatible 2G technologies were developed and deployed in many regions around the world in the early to mid-1990s. The exception to this was the GSM standard developed in Europe, which is a unique example from this era of multinational collaboration and political compromise, resulting in a compatible system capable of spanning many national boundaries. The story of how this came to pass is an intriguing one that is discussed in the next section of this chapter. The cellular market of North America, on the other hand, stands in stark contrast to the collaborative approach adopted in Europe at the time. Based on the Federal Communications Commission (FCC) free-market approach of letting standards be developed by industry and allowing network operators to choose from the available offerings, a number of mutually incompatible digital cellular technologies were developed. These include the Integrated Digital Enhanced Network (iDEN), IS-136, and IS-95.

The iDEN standard is a proprietary technology developed by Motorola, which includes dispatch capabilities to support push-to-talk (PTT) services. IS-136 is the North American TDMA standard that developed as a backward-compatible digital extension of the Advanced Mobile Phone Service (AMPS) standard, the dominant 1G analog standard in North America. IS-136 is also referred to as Digital AMPS (D-AMPS), North American TDMA, or simply "TDMA" within the North American telecommunications context. The third major North American 2G standard is IS-95, commercialized under the name cdmaOne. This standard was developed by Qualcomm and is fundamentally different from all other 2G standards in that it utilizes a code division multiple access (CDMA) scheme based on direct-sequence spread spectrum (DSSS) techniques, whereas the other 2G cellular standards all used TDMA to multiplex multiple users onto a single carrier frequency. The CDMA technique introduced by IS-95 would prove to have a significant impact on the genesis of 3G cellular technologies. Although IS-95 is no longer widely deployed today, it is nonetheless a technology of historic significance as the first cellular

Table 6.1   Summary of 2G Cellular Technologies

Standard	Region of origin	First deployed	Access scheme	Carrier spacing	Number of TDMA time slots
GSM	Europe	1991	TDMA	200 кHz	8
iDEN	United States	1993	TDMA	25 kHz	6
IS-136 (D-AMPS)	United States	1993	TDMA	30 kHz	6
IS-95 (cdmaOne)	United States	1995	CDMA	1.25 MHz	–
PDC	Japan	1993	TDMA	25 kHz	3
PHS	Japan	1995	TDMA-TDD	300 kHz	4 Tx/4 Rx

technology to used CDMA. Furthermore, it formed the basis for the 3G standard CDMA2000 1xRTT, which is backward compatible with IS-95. CDMA2000 has played an important role in the cellular ecosystem, particularly in North America, and is discussed in further detail in Chapter 7.

Finally, two 2G cellular standards were developed in Japan: Personal Digital Cellular (PDC) and Personal Handy-phone System (PHS). PDC is a standard developed by what is now the Association of Radio Industries and Business (ARIB) standards body and deployed solely in Japan. PHS is a microcellular technology that is similar to a cordless telephone with added cell handover capabilities. An interesting aspect of PHS is that it uses TDMA in a time division duplexed configuration, where four time slots are assigned to the uplink and four are assigned to the downlink on a single carrier frequency. This design more closely resembles Digital Enhanced Cordless Telecommunications (DECT), the European cordless telephone technology, than the other 2G cellular systems. PHS has limited coverage and functionality as compared with other cellular technologies, and the small cell size is effective in high-density urban settings but does not translate well to low-density rural environments. PHS systems have seen deployment in parts of Asia and Brazil. The 2G cellular technologies previously described are summarized in Table 6.1.

6.3   2G DEPLOYMENTS

Today, the primary 2G cellular standard that remains widely deployed is GSM. While global GSM subscriber statistics continue to increase, the number of connections supported by other 2G technologies is on the decline. This is due in part to migration of network operators to other technologies. For example, most IS-136 operators have now migrated to GSM due to the widespread popularity of GSM combined with the similarity between the two standards. Most IS-95 operators have now migrated to the backward-compatible CDMA2000 1xRTT, leading to a steady decline of IS-95 systems. According to GSMA Wireless Intelligence, 2G cellular subscriptions made up roughly 75% of all global cellular connections as of July 2011. Of that 75%, however, GSM subscriptions accounted for over 99.3%, whereas all other 2G cellular subscriptions made up less than 0.7% combined—that is to say, non-GSM 2G connections comprised only 0.5% of the total global cellular subscriptions. In terms of absolute numbers, however, that still equates to over 25 million subscribers. Although numbers are dwindling, the second most widely deployed 2G technology as of July 2011 was iDEN, with roughly 20 million subscribers.

6.4 CHAPTER OVERVIEW

The remainder of this chapter focuses on providing a detailed technical overview of GSM and its evolutionary enhancements, as these technologies continue to play an important role in the modern mobile Internet. Due to the historic significance of GSM, this section includes an extended discussion of the history of GSM, from the original concept in 1982 up through the evolution to GERAN, realized in Release 5 first published in 2002, which aligns the GSM/EDGE Radio Access Network with the concepts of 3G Universal Mobile Telecommunications System (UMTS). Since the vast majority of today's GSM deployments include the General Packet Radio Service (GPRS) and EDGE enhancements, the technology overview focuses on the aspects of a GSM/EDGE-based deployment with emphasis on data services for mobile networking. We also discuss the next major step in the evolution of GSM networks—the 3rd Generation Partnership Project (3GPP) Release 7 enhancements commonly known as *EDGE Evolution*. Due to the similarities between GSM and the other 2G TDMA systems, as well as the much greater importance of GSM within both the historic and current cellular market, the other 2G TDMA systems are not addressed in further detail.

6.5 AN INTRODUCTION TO GSM

The Global System for Mobile Communications is by far the most successful 2G cellular technology and the first cellular technology to achieve market acceptance on a global scale. With the proper mobile equipment, subscribers can receive seamless connectivity in over 200 countries and territories spanning every continent of the world. In this respect, GSM may be considered one of the greatest achievements of the twentieth century, both technologically and politically, and a landmark in the history of wireless communication. The development of GSM began in 1982, when the European Conference of Postal and Telecommunications Administrations (CEPT)[1] created the Groupe Spécial Mobile[2] to develop a Pan-European cellular

---

[1] The CEPT acronym derives from the French, "Conférence européenne des administrations des postes et des telecommunications."

[2] English translation: "Special Mobile Group."

technology for use in the 900 MHz frequency band. This decision was forward looking for its time, as the 1G analog market was just emerging and the true market potential for cellular systems was not yet fully known. The GSM standard was originally frozen and published in 1990 by the European Telecommunications Standards Institute (ETSI), with initial deployment in Finland in 1991 and commercial deployments in most other CEPT member countries in 1992. In 2000, the technical work of maintaining and evolving the GSM standard was transferred to the 3GPP organization. More than a decade after the initial commercial launch of 3G cellular networks, GSM remains the dominant cellular technology worldwide, surpassing the 4 billion subscriptions mark in 2010. The continued success of GSM is due in part to the effort of its creators to make it "future proof" by establishing an open evolution of the standard, whereby new features are added to the standard through a series of backward-compatible releases. The open evolution of GSM has allowed the standard to remain relevant as time and technology have progressed.

Today, GSM operation has been defined for a number of frequency bands to address the demands of a global market. The four widely deployed bands are the 900 and 1800 MHz bands in Europe, Asia Pacific, and Africa and the 850 and 1900 MHz bands in North America and parts of Central and South America. GSM uses a combination of time division multiple access (TDMA) and frequency division multiple access (FDMA) to divide physical resources among multiple users. It is deployed in paired spectrum using frequency division duplexing (FDD) between uplink and downlink channels. It is a narrowband system with 200 kHz carrier spacing. Each radiofrequency (RF) carrier is divided in time into TDMA frames with eight time slots each. This allows up to eight full-rate traffic channels to be multiplexed onto a single RF carrier. To mitigate the negative effects of frequency selective fading due to multipath propagation in a mobile environment, GSM defines optional frequency hopping. The modulation used by GSM is Gaussian minimum-shift keying (GMSK), although enhancements to the standard use higher-order modulation such as eight-phase-shift keying (8PSK) to achieve higher data rates. The scope of the GSM standard encompasses not only the air interface, but also the architecture and protocols of the core network. This was done to ensure network compatibility for international roaming. The fact that GSM defines a full system architecture distinguishes it from other 2G technologies, as it was common practice at the time to specify only the air interface.

GSM was originally conceived as a circuit-switched system optimized for voice. It was designed to support a similar set of services as the Integrated Services Digital Network (ISDN), including facsimile (fax), low-rate circuit-switched data, and a rich set of supplementary services. The Short Message Service (SMS) is another novel contribution of GSM that has become extremely popular in the cellular world. As the disruptive technology of the Internet grew in popularity in the early 1990s, it soon became apparent that a packet-switched enhancement to GSM was needed. The General Packet Radio Service (GPRS) enhancement first defined in Release 97 addresses this need. GPRS adds a packet-switched domain for data services that is compatible with the GSM air interface architecture but utilizes its own signaling and traffic channel architecture as well as a parallel core network architecture. A key aspect of GPRS is that it makes more efficient use of the radio resources by using statistical multiplexing to share the physical channels of the air interface among multiple users on an as-needed basis. Furthermore, downlink and uplink resources are assigned independently in GPRS to support asymmetric data transfer. This approach is well suited to the bursty nature of packet data services, such as mobile Web browsing. Modern GSM networks nearly always include GPRS.

The open evolution of GSM has included numerous enhancements focused on improved data services for GSM/GPRS networks. The EDGE enhancement, first defined in Release 99, provides roughly a threefold increase in peak data rates. The application of the EDGE enhancement to GPRS data services is known as enhanced GPRS (EGPRS). EGPRS introduces PHY and RLC/MAC layer improvements to the air interface, including higher-order modulation (8PSK), link adaptation through adaptive modulation and coding, and hybrid automatic repeat request (HARQ). Further data enhancements have been defined in Release 7 that are often referred to collectively as "EDGE Evolution." These include increased spectral efficiency, downlink dual carrier (DLDC), mobile station (MS) receiver diversity (MSRD), and reduced latency. Increased spectral efficiency is achieved through the introduction of EGPRS phase 2 (EGPRS2): an expanded set of modulation and coding schemes (MCSs) for link adaption, including higher-order modulation (16-point and 32-point quadrature amplitude modulation [16-QAM, 32-QAM]) and turbo coding, and an optional higher modulating symbol rate. EDGE is now deployed in the vast majority of GSM networks worldwide. However, EDGE Evolution has not yet reached wide-spread market penetration as of this writing.

6.5.1   History of GSM

The widespread success of GSM has earned it a place of prominence in cellular history that will last well beyond the useful life of the system. The design decisions made during the creation of GSM were strongly influenced by both technical and political factors. GSM has left a lasting legacy for cellular networks, as the engineering concepts developed for GSM have influenced all subsequent standards in the "GSM Family" (i.e., GSM, UMTS, Long-Term Evolution [LTE]) and in many ways helped to shape the cellular networking world as we know it. In order to provide insight as to how and why GSM has taken its current form, special care is taken in elaborating the history of GSM in this section. For a more comprehensive description of GSM history, the interested reader is referred to the "Further Reading" section later in this chapter. An overall timeline of significant milestones in the history of GSM are summarized in Table 6.2.

6.5.2   Early History of GSM

The motivation for the development of GSM can be traced back to the International Telecommunications Union (ITU) World Administrative Radio Conference (WARC) in 1979, when a decision was made to allocate a block of RF spectrum in the

Table 6.2   Timeline of GSM Milestones

Year	Event
1982	CEPT forms the Groupe Spécial Mobile and recommends the reservation of frequencies in the 900 MHz band for a future Pan-European cellular system.
1986	Eight radio system proposals evaluated by CNET in Paris.
1987	Basic parameters of the GSM standard defined.
	European Council issues directive reserving 900 MHz band for GSM in member states.
	GSM Memorandum of Understanding signed by 14 operators in 13 European countries.
1988	Completion of the first set of detailed GSM specifications for infrastructure purposes.
1989	Standardization work transferred from CEPT to ETSI.
	CEPT Groupe Spécial Mobile becomes ETSI Technical Committee GSM.
1990	GSM Phase 1 specification frozen in ETSI Technical Committee GSM.
	GSM adaptation work begins for the DCS 1800 band to meet U.K. market demand.
1991	First GSM call made by Radiolinja in Finland.
	ETSI Technical Committee GSM begins UMTS specification activities and is renamed the Special Mobile Group (SMG).
1992	First commercial GSM networks come into service in seven European countries.
	First hand-portable cell phones become commercially available.
	First international roaming agreements between Telecom Finland and Vodafone (United Kingdom).
	First SMS sent by Vodafone (United Kingdom).
1993	GSM surpasses 1 million subscriptions.
	Telstra Australia becomes the first non-European operator to adopt GSM and sign MoU.
	ETSI Technical Committee SMG agrees to objectives and methodology for an open evolution of GSM to be implemented as Phase 2+.
	First DCS1800 network opened in the United Kingdom.
1994	Data capabilities launched in GSM networks.
1995	GSM surpasses 10 million subscriptions.
	GSM Phase 2 specification frozen in ETSI Technical Committee SMG.
	First North American PCS1900 network opened in the United States.
1996	First GSM networks in Russia and China launched.
1997	GSM Release 96 specification frozen in ETSI Technical Committee SMG—first Phase 2+ release, defines HSCSD enhancement.
	First tri-band GSM mobile devices become commercially available.
	GSM networks commercially deployed in 100 countries worldwide.
1998	GSM Release 97 specification frozen in ETSI Technical Committee SMG— defines GPRS enhancement.
	GSM surpasses 100 million subscriptions.
1999	GSM Release 98 specification frozen in ETSI Technical Committee SMG.

(Continued)

Table 6.2   *(Continued)*

Year	Event
2000	GSM Release 99 specification frozen in ETSI Technical Committee SMG—defines EDGE enhancement.
	GSM standardization work transferred from ETSI to 3GPP.
	First commercial GPRS networks launched.
	First GPRS handsets become commercially available.
2001	3GPP Release 4 specification frozen.
2002	3GPP Release 5 specification frozen—GERAN architecture compatible with 3G core network.
2003	First commercial EDGE networks launched.
2004	GSM surpasses 1 billion subscriptions.
2005	3GPP Release 6 specification frozen.
2006	GSM surpasses 2 billion subscriptions.
2007	3GPP Release 7 specification frozen—defines EDGE Evolution enhancements.
2008	3GPP Release 8 specification frozen.
	GSM surpasses 3 billion subscriptions.
2009	3GPP Release 9 specification frozen—VAMOS enhancement for increased voice capacity.
2010	GSM surpasses 4 billion subscriptions.
2011	3GPP Release 10 specification frozen.
2012	3GPP Release 11 specification frozen.

900 MHz band to be used for land mobile cellular communication systems in Europe. At the time, the European telecommunications market was badly fragmented. This trend carried into the emerging cellular market, as the 1G analog cellular technologies were initially deployed in parts of Europe in the early 1980s. Each country deployed its own cellular technology, which was generally incompatible with those of other countries with few exceptions. The cellular networks in most European countries were state-run monopolies controlled by the national Post, Telephone and Telegraph Administrations, which used the telecommunications industry as a tool for promoting their domestic manufacturing and operating industry. As a direct consequence, international roaming was limited and no single system reached the economies of scale necessary to provide inexpensive cellular phones. This entire system may seem foreign and self-limiting today, however one must consider that the globalization of the 1990s and 2000s—a globalization that technologies such as GSM helped precipitate—had not yet occurred [51, pp. 11–12].

In 1982, a decision was made by CEPT to reserve the newly available 900 MHz band for use by a new Pan-European cellular system. This decision was made in an attempt to prevent existing market fragmentation from carrying over into the 2G of cellular systems. Given the state of technology at that time, spectrum above 1 GHz was considered unsuitable for cellular communications. Since no other sufficiently wide spectrum was available below 1 GHz, the 900 MHz band was seen as the last chance to build a Pan-European system in the twentieth century [51, p. 12]. By the end of the year, CEPT established a study group, the Groupe Spécial Mobile, comprising delegates from 11 member nations.[3] The purpose of the study group was to define a system of open interfaces that would fulfill a number of requirements. The system should allow international roaming throughout Europe, offering the same services as the public switched telephone and data networks, without requiring significant modification to the fixed national telephone networks [51, p. 14]. The system should support services other than speech. However, there was much uncertainty as to what those services might be by the time the system reached deployment in the 1990s: a modular system structure using the same philosophy as for the ISDN and Open Standards Interconnect (OSI) should be applied. Furthermore, in order to be competitive with existing 1G cellular systems, the new system should improve over 1G systems in the following respects [99]:

- Spectral efficiency
- Speech quality
- Cost of equipment (mobiles and network infrastructure)
- Viability of handheld mobiles (in contrast to large vehicle-borne phones of the time)
- Flexibility to introduce new services
- Voice security (encryption)

The requirements of the new system were indeed ambitious. The prospect of Pan-European international roaming was in itself a sufficiently formidable political challenge, to the extent that some delegates supported the use of well-established analog technology to make the standardization process achievable [51, p. 13]. The risk associated with the timely implementation of the new GSM system would become a key technical driving factor.

The critical issue of the first 5 years of GSM technical work was to agree upon the basic parameters of the air interface [51, p. 20]. This demonstrates the central importance of the air interface to the GSM standard. Eight different radio system proposals were submitted by CEPT member administrations, and in 1986 a comparative evaluation was performed in Paris by CNET,[4] the research wing of French Telecom [51, p. 20]. In 1987, the basic GSM parameters were decided at a pivotal meeting of the GSM group in Funchal, Madeira, Portugal. These parameters included

---

[3] Denmark, Finland, France, Germany, Italy, The Netherlands, Norway, Portugal, Sweden, Switzerland, and the United Kingdom.

[4] CNET: Centre National d'Études des Télécommunications (English translation: National Center for Telecommunication Studies).

the access method, channel bandwidth, modulation, and speech codec. Based on the results of the Paris trials, a digital system using narrowband TDMA was chosen, as this radio system was shown to be most well suited to achieving the minimum system requirements discussed previously [51, p. 314]. It is interesting to note, however, that the basic parameters of the air interface were not finalized until 3 months later. This was due to the fact that French and German administrations supported a wideband TDMA scheme that they had developed and their delegates were not permitted to accept a compromise on access scheme. A significant amount of technical and political discussion took place before a compromise was eventually reached in which France and Germany accepted the basic parameters under the condition that they be "enhanced in the areas of modulation and coding" [51, p. 22]. This included a change of modulation scheme from adaptive differential phase modulation (ADPM), originally specified in the Madeira agreement, to GMSK. Although there are advantages to using GMSK over ADPM—for example, GMSK does not include redundancy, allowing all redundancy to be used for channel coding [51, p. 315]—this change fundamentally represents a political compromise, as the use of ADPM would have made a proprietary form of delay equalization developed by Alcatel SEL incompatible with the new GSM system. The significance of this decision is that it clearly illustrates the impact of political compromise on the early technical development of GSM. This sets GSM apart from other 2G cellular technologies, such as IS-95 and IS-136 developed in North America, for example. The creation of the GSM standard was based on the unanimous agreement of technical delegations from many independent states, each with their own national interests to uphold.

Another key milestone in the development of GSM occurred in 1987 with the signing of the GSM Memorandum of Understanding (MoU). The GSM MoU was an agreement signed by 14 network operators in 13 European countries, stating that they would support the new GSM technology and agree to launch the service in 1991. At the time, GSM was seen as a risky business endeavor for both network operators and suppliers. The MoU played a key role to mitigate that risk by securing network operator support and synchronizing system rollout. These aspects ensured that international roaming—a key feature of the new GSM system—would be commercially realized soon after initial deployment. These aspects also ensured that adequate economies of scale would emerge in a timely fashion. In this way, the GSM MoU helped to galvanize the European supply industry into investing the large sums of research and development (R&D) capital needed to make equipment based on the GSM standard a reality. The realization of GSM hand-portable devices, for example, required a tremendous R&D effort on the part of the semiconductor industry in order to create the new chipsets necessary to shrink cell phone sizes and reduce power consumption. The signatories of the GSM MoU worked together to help bring GSM to market in a timely fashion and made critical business and economic contributions to the development of GSM, such as the economic (rather than technical) realization of international roaming. The organization which grew out of the GSM MoU is now known as the GSM Association (GSMA) and claims membership of nearly 800 network operators globally . The GSMA continues to support the interests of mobile network operators using the GSM family of standards.

In summary, the early success of GSM was realized through four key contributing factors. First, political cooperation was required to preserve the 900 MHz frequency band for development of a Pan-European standard. This included political interaction and compromise at the highest levels of government to reach unanimous agreement on the basic parameters of the technology. Second was the technical standardization effort, which was necessary to develop a robust technology platform that would allow flexibility for future enhancement. The third factor was the network operator commitment established through the GSM MoU. Finally, the fourth factor was the industrialization of the supply industry, which included considerable R&D investment in areas of core network and mobile device development.

6.5.3　Overview of Phase 1, Phase 2, and Phase 2+

As the target commercial launch date of 1991 approached, it became apparent that further technical work would be required beyond that date. For this reason, the base GSM standard was released in two phases. The Phase 1 standard was frozen and published in 1990, allowing vendors to start developing equipment for initial commercial launch. By this time, the GSM technical specification work had been transferred from CEPT to ETSI, who would maintain and develop the GSM standard throughout the 1990s. The Phase 1 standard included basic voice services, almost all data services within the original GSM scope (i.e., circuit-switched data at various rates up to 9.6 kbps), facsimile (fax) service, SMS, a subset of supplementary services, and optional slow frequency hopping. It should be noted, however, that many of these services were not activated until some time after the initial commercial launch of GSM. True to the GSM MoU, agreement, the first GSM Phase 1 network was launched by Radiolinja in Finland in 1991. However, due to type approval issues that delayed the availability of mobile handsets, GSM did not see widespread commercial deployment until 1992 [51, p. 69]. The remaining work within the original technical scope of the GSM base standard was added in the Phase 2 specification originally published in 1995. The main focus of the Phase 2 specification work included maintenance and optimization of the Phase 1 standard and the development of a mature set of test specifications, in particular those used for type approval of mobile handsets. There were, however, a few key enhancements that were added to the Phase 2 standard. These included the enhanced full-rate (EFR) speech codec for landline-quality speech, fully harmonized support for GSM operation in the Digital Cellular Service (DCS) 1800 MHz frequency band, and improved 3-V Subscriber Identity Module (SIM) technology for lower battery consumption. It was originally anticipated that the new development work on GSM would end for the most part with the publication of the Phase 2 specification. In keeping with the paradigm established by the 1G cellular technologies, standards work beyond that time would primarily be limited to maintenance and incremental improvement of the GSM specification.

One of the main shortcomings of the GSM Phase 2 standard was its data service. It was limited to 9.6 kbps and operated in a circuit-switched mode only. The data rates supported by GSM actually compared favorably with those available in wired networks when GSM was being developed in the late 1980s. As the 1990s progressed, however, a data rate gap began to open between the services of the cellular world and the quickly increasing data rates of wired networks. Furthermore, the circuit-switched approach of GSM required the reservation of dedicated radio resources for the duration of a data session, making it poorly suited to the bursty nature of packet data. Despite the best efforts of its creators to make GSM future proof, the growing importance of the Internet and demand for an improved wireless data service would ultimately limit the useful life of GSM.

A turning point came in 1993, when representatives from Nokia tabled a short proposal entitled, "GSM in a Future Competitive Environment" [51, p. 74]. At the time, the idea that GSM standards work would continue under the name "Phase 2+" already existed, however the purpose of this work was maintenance and incremental improvement to the GSM standard. The proposal, however, added an entirely new dimension to Phase 2+. Rather than wait for 3G to arrive and make GSM obsolete, it proposed an open evolution of the standard to include, for instance, improved speech coding algorithms and higher rate data services—features that, at the time, were thought to be solely the territory of the new 3G UMTS system under development. In short, it would allow GSM to evolve to a competitive 2.5G standard and beyond. In 1993, the ETSI Technical Committee SMG agreed on the objectives and methodology of the open evolution of GSM, to be implemented as Phase 2+. The first Phase 2+ standard, Release 96 (R96), was frozen early in 1997. Since then, new releases of the GSM standard have been frozen approximately every 1–2 years. Each new Phase 2+ release is backward compatible and includes new features and enhancements. Notable data service enhancements of Phase 2+ include GPRS (R96) and EDGE (R99). The decision to establish an open evolution of the GSM standard beyond its original 2G functionality is a critical success factor that has extended the longevity of the GSM standard and allowed it to remain one of the fastest growing cellular technologies decades after its initial deployment. It is also historically significant, as it created a new paradigm of continuous evolution within the cellular industry that has been applied to all subsequent technologies in the GSM lineage— namely UMTS and LTE.

6.5.4   GERAN and the Transfer to 3GPP

In June 2000, shortly following the completion of Release 99, the GSM specification work was transferred to 3GPP and ETSI Technical Committee SMG was closed. Around the same time, the term "GERAN" (GSM/EDGE Radio Access Network) was introduced to the GSM vocabulary. GERAN refers to the evolution of GSM toward a 3G radio access network that is fully compatible with the UMTS core network, while maintaining full backward compatibility with earlier GSM releases. The motivation for the GSM evolution to GERAN is rooted in the fact that new system and architectural concepts are employed in UMTS. In particular, UMTS is designed to keep mobile-specific functions out of the core network. This results in a fundamentally different functional split between the radio access network and the core network than that of GSM. The GERAN architecture incorporates two key changes: the adoption of the RAN/CN functional split of the 3G UMTS protocol architecture and the addition of a new 3G core network interface (GERAN Iu mode). These changes allow GERAN to support a similar set of services to UTRAN,[5] thus facilitating internetworking by allowing easy transition of services between the two radio access technologies. The harmonization of GERAN and UTRAN allows network operators to leverage existing GSM/EDGE infrastructure in conjuncture with 3G and beyond deployments as part of a UMTS multiradio network with common connectivity to a single evolved core network. The evolution to GERAN was first realized in 3GPP Release 5 (Rel-5), published in 2002. Although the term "Phase 2+" is not used in 3GPP releases, the concept of open evolution of GSM—and now GERAN—continues in new releases of the 3GPP standard. The requirement for backward compatibility in new GERAN releases is also preserved.

6.5.5 Overview of GSM Releases

As previously discussed, the open evolution of GSM is characterized by a series of backward-compatible releases of the technical standard. The specific capabilities of a particular GSM network depend on which release of the standard it implements. Since the start of Phase 2+, new releases have historically been published approximately every 1–2 years and include new features and capabilities. As of 2012, 12 Phase 2+ releases had been published: Release 1996 (R96), R97, R98, R99, Release 4 (Rel-4), Rel-5, Rel-6, Rel-7, Rel-8, Rel-9, Rel-10, and Rel-11. The change in release notation corresponds to the transfer of the GSM standard development work from ETSI to 3GPP. In Rel-5 and beyond, the term "GERAN" is used to refer to the evolved radio access technology based on the GSM/EDGE air interface as previously described. A nonexhaustive list of key features introduced in each release of the GSM standard, as well as year of initial publication, are summarized in Table 6.3. For 3GPP Rel-4 and beyond, only features related to GSM/EDGE are listed.

6.6 GSM TECHNOLOGY OVERVIEW

This section provides a technical overview of the essential concepts, architecture, protocols, and procedures of a GSM cellular network. Recall that the wide majority of modern GSM deployments incorporate both the GPRS and EDGE enhancements. For the present discussion, a network infrastructure that supports both GPRS and EGPRS capabilities defined in Release 99 and beyond is assumed. The term "GPRS" will be used throughout this section to refer to both GPRS and EGPRS unless otherwise noted. EGPRS architecture and operation is nearly identical to that of GPRS, with the exception of the PHY and RLC layer enhancements. These enhancements

---

[5] UTRAN, Universal Terrestrial Radio Access Network, is the term used for the 3G UMTS radio access network.

Table 6.3 Summary of Key Features in GSM Releases

Release	Year	Key features
Phase 1	1990	• Basic telephony using (old) full-rate (FR) speech codec at 13 kbps (no longer widely deployed) • Basic circuit-switched data rates up to 9.6 kbps • Short Message Service (SMS) point-to-point or via cell broadcast • Partial set of supplementary services (e.g., call barring, call forwarding, call barring while roaming) • Support for facsimile (fax) communication • Security through authentication, confidentiality, and ciphering • International roaming • Support for GSM 900 MHz band operation only
Phase 2	1995	• Improved speech quality using enhanced full-rate (EFR) speech codec at 12.2 kbps (widely deployed in pre-R98 systems) • Half-rate (HR) speech codec at 5.6 kbps to increase maximum number of users (never commercialized) • SMS enhancements (e.g., concatenation, replacement) • Full set of supplementary services (e.g., call line identification, multiparty calls) • Support for DCS 1800 MHz band operation, including interworking with GSM 900 band and multiband operation by a single operator • Improved 3-V SIM technology • Base station subsystem (BSS) repeaters
R96	1997	• Improved single-slot circuit-switched data services at 14.4 kbps • HSCSD: Enhanced circuit-switched data rates of up to 57.6 kbps for a Type 1 MS through slot aggregation (i.e., multislot configurations) • Definition of multislot classes (1–18) • SIM Application Toolkit: Standardized approach to implement applications on the SIM (e.g., banking, weather) • ASCI Phase 1: Voice enhancements for GSM railway systems allowing, for example, multiple simultaneous talkers and listeners • CAMEL Phase 1: Enables the definition of intelligent network services on top of existing GSM services such as call control-related functionality • SMS interworking extensions • Extension of GSM emergency numbers for regions outside Europe • Definition of wireless local loop (WLL) using GSM

R97	1998	• GPRS Phase 1: Enhancement to the GSM architecture to enable packet-switched data. Includes fundamental changes to the air interface and core network, creating a so-called *packet-switched domain* in parallel to existing 2G circuit-switched domain
		• Extended multislot classes (19–29) for GPRS
		• Basic GPRS encryption using the A5 algorithm
		• Security mechanisms for SIM Application Toolkit
		• SMS enhancements Phase 1
		• ASCI Phase 2
		• CAMEL Phase 2
R98	1999	• Adaptive multirate (AMR) speech codec: Enables speech codec adaptation to match link quality and cell capacity requirements. Eight codec modes are defined ranging from 4.75 to 12.2 kbps. (Widely deployed in both GSM and UMTS networks.)
		• Location services (LCS) in the CS domain: Mechanisms to support location of an MS using the GSM network
		• GPRS Phase 2
		• Support for GSM/GPRS in the PCS 1900 MHz band for use in North America
R99	2000	• EDGE: Air interface enhancement that enables higher data rates in good channel conditions through the use of higher-order modulation (i.e., 8PSK), adaptive modulation and coding, hybrid ARQ, and enhanced link quality measurements.
		• DTM: Support for parallel CS and PS domain connections on the same carrier (e.g., for simultaneous voice and data over GSM) to implement GPRS Class A MS
		• Enhanced measurement reporting to support multiband MS
		• Interworking between GSM and UTRAN for mobility between 2G and 3G networks, including handover for voice traffic
		• Enhanced security using the A5/3 algorithm for 64-bit 3G encryption
		• Support for three new bands of operation in the 850, 450, and 480 MHz spectrum
Rel-4	2001	• NACC: Reduces access time during cell reselection by allowing the MS to notify the cell to which it will reselect
		• Delayed Downlink TBF Release/Extended Uplink TBF mode: Improves latency and reduces signaling for data traffic by maintaining a layer 2 link when between the MS and the network when no data are exchanged
		• Extended DTM multislot class
		• Gb over IP: Definition of IP transport over the Gb interface between the BSC and the SGSN as an alternative to frame relay
		• GERAN/UTRAN interworking additions for low chip-rate TDD
		• Dynamic ARFCN mapping to allow flexibility for deployment in new spectrum to address niche markets
		• Support for the GSM 750 MHz band

(Continued)

Table 6.3 (*Continued*)

Release	Year	Key features
Rel-5	2002	• GERAN Iu Mode: Network architecture enhancement that allows GERAN (i.e., GSM/EDGE air interface and base station subsystem) to connect to a 3G core network via Iu interface
		• Wideband AMR (AMR-WB) speech codec: Enables improved speech quality for high definition (HD) voice using 8PSK modulated traffic channels (currently deployed but less common than AMR as of 2011)
		• AMR 8PSK Half-Rate: Enables increased voice traffic capacity through support for the AMR speech codec on half-rate 8PSK traffic channels
		• Extended multislot classes (30–45) for (E)GPRS
		• Enhanced Power Control
		• Further cell reselection enhancement through external NACC (eNACC)
		• Enhancements to the GERAN/Core Network interfaces (A and Gb)
		• Improved GERAN/UTRAN interworking
		• Location services (LCS) in the PS domain
Rel-6 [100]	2005	• PS Handover: Allows PS resources to be assigned to an MS in a target cell before the MS is handed over (e.g., to support high QoS applications)
		• MAC enhancements to enable improved multiplexing of data flows with different QoS by using multiple parallel TBFs.
		• Seamless support of streaming services in GERAN A/Gb mode
		• MBMS: Enables the network to send broadcast and multicast data to multiple MSs on the same radio resources
		• GAN: Internetworking enhancement that enables access to GSM services (via the A and Gb interfaces) through non-GERAN Internet access technologies (e.g., WLAN, Bluetooth, etc.) by tunneling upper layer protocols between the core network and the MS over IP
		• FLO: Allows flexible configuration of the GSM PHY at call setup, allowing optimized support of IMS services for the PS domain in GERAN Iu mode
		• DARP Phase 1: Improved MS reception performance through the use of single antenna interference cancellation (SAIC)
		• Increased FACCH and SACCH robustness through repetition and soft combining
		• DTM enhancements, including extended DTM (E)GPRS multislot classes
		• U-TDOA: Enhanced location service for both GSM and GPRS to support FCC Emergency 911 requirements in the United States
		• Addition of frequency bands to support TETRA Advanced Packet Services (TAPS) for professional use (e.g., transportation, public safety, and military sectors)

Rel-7 [101]	2007	• EDGE Evolution: Enhancements to EDGE that reduce the performance gap between 2G and 3G networks by increasing data rates and reducing latency in GERAN (includes DLDC, EGPRS2, RTTI, and FANR)
		• DLDC: Increases EDGE data rates through simultaneous assignment of two downlink RF carriers to a single MS
		• EGPRS2: Increases EDGE data rates through the addition of higher-order modulation (16-QAM, 32-QAM) and coding (turbo codes) schemes
		• Reduced latency through reduced TTI and fast Ack/Nack reporting at the RLC layer
		• DARP Phase 2: Enables MS receiver diversity using two antennas
		• DTM Handover: Support for concurrent handover of CS and PS resources
		• SIGTRAN: Protocol stack to support PSTN signaling over IP (e.g., on the A interface between BSS and MSC)
		• Support for conversational services (e.g., VoIP) in GERAN A/Gb mode via PS domain
		• PS Handover between GERAN/UTRAN mode and GAN mode
		• RLC nonpersistent mode
Rel-8 [102]	2008	• Interworking between GERAN and E-UTRAN for mobility between GSM and LTE networks in the GSM to LTE direction (PS domain only)
		• A over IP: Definition of IP transport over the A interface between the BSC and the MSC to facilitate full BSS over IP (when combined with Gb over IP, Rel-4)
		• GAN Iu mode: Providing generic access to the Iu interface based on GAN
		• U-TDOA Enhancements
Rel-9 [103]	2009	• VAMOS: Allows two full-rate voice traffic channels to share a single physical channel concurrently
		• Enhanced security using the A5/4 algorithm in the CS domain and GEA4 algorithm in the PS domain for 128-bit 3G encryption (standardization started Rel-6)
		• GERAN aspects of HNB/HeNB enhancements: Support for active mode mobility between GERAN and 3G or 4G femtocells (PS services only)
		• GERAN aspects of MSR: Base station enhancements to support multiple radio access technologies (i.e., GSM/EDGE, UMTS, LTE) within the same frequency band
		• HL for GERAN: Optimizes concurrent use of various network-based and mobile-based location technologies (e.g., LCS, U-TDOA, satellite based)
		• Standardized CBC-BSC Interface

(Continued)

Table 6.3 (*Continued*)

Release	Year	Key features
Rel-10 [104]	2011	• LCLS: Allows local calls generated and terminated by users served by the same BTS or BSC to be locally switched, minimizing the use of backhaul resources on the Abis and/or A interfaces (MSC signaling required) • MOCN by GERAN: Allows GERAN resources to be shared by multiple core networks for GSM spectrum refarming purposes • Enhancements to the Iur-g interface • MSR enhancements for noncontiguous spectrum deployments • Tightened link level performance requirements for single antenna MS (test ongoing)
Rel-11 [105]	2012	• SRVCC GERAN aspects: Seamless support for voice handover from GSM (CS domain) to LTE or HSPA • Service Identification for RRC Improvements: Allows the PS core network to share mobile data application/service type information with the BSC to improve radio resource management and minimize network congestion • MOCN by GERAN Phase 2 • Introduction of the ER-GSM band for GSM-R

ASCI, advanced speech call items; CAMEL, customized applications for mobile networks enhanced logic; CBC, cell broadcast center; DARP, Downlink Advanced Receiver Performance; DLDC, downlink dual carrier; DTM, dual-transfer mode; E-UTRAN, evolved UTRAN (4G air interface/radio access network used in LTE); FANR, fast ACK/NACK reporting; FLO, flexible layer one; GAN, generic access network; GPRS, General Packet Radio Service; HL, hybrid location; HNB/HeNB, Home Node B/Home eNode B (3GPP term for 3G/4G femtocell); HSCSD, high-speed circuit-switched data; IMS, IP multimedia subsystem; LCLS, local call local switch; MBMS, multimedia broadcast and multicast service; MOCN, multioperator core network; MSR, multistandard radio; NACC, network-assisted cell change; RTTI, reduced transfer time interval; SIGTRAN, signaling transport; SRVCC, single-radio voice call continuity; TBF, temporary block flow; TETRA, terrestrial trunked radio; UMTS, Universal Mobile Telecommunications System (3G successor to GSM); U-TDOA, Uplink Time Difference of Arrival; UTRAN, Universal Terrestrial Radio Access Network (3G air interface/radio access network); VAMOS, Voice services over Adaptive Multi-user channels on One Slot.

are addressed at the end of this section. The evolution of GSM to GERAN and EDGE Evolution is beyond the scope of this section, but is addressed in Section 6.12 later in this chapter.

Although the vast majority of today's networks and devices support EDGE, some do not. This is still the case, for example, in some low-cost mobile phones that may provide only GSM voice and/or GPRS data services. Furthermore, older devices may still be in use that were designed before EDGE, or even GPRS, became a standard feature—though the number of these devices will clearly decrease over time. It is important to understand that these devices will still work on an EDGE-capable GSM network; they just won't be able to utilize EDGE functionality. For this reason, it is important to understand the differences between GPRS, EDGE/EGPRS, and traditional circuit-switched GSM. These differences are discussed throughout this section.

6.6.1 Circuit-Switched and Packet-Switched Domains

The introduction of GPRS into GSM networks in the late 1990s and early 2000s has resulted in a duality in GSM networks of today. The result is a network architecture that is characterized by two parallel yet coexistent domains: the circuit-switched (CS) domain and the packet-switched (PS) domain. Strictly speaking, the GSM standard uses these terms with reference to the core network architecture [106]. However, the general concept can be extended throughout the entire system from the core network to the mobile device. Although GSM and GPRS share the same physical resources on the air interface using a similar physical layer architecture with only minor modifications, the overall architecture of the RAN for GPRS is fundamentally different from traditional GSM, including separate protocol stack, signaling channels, mobility management procedures, and concept of operation.

GPRS was designed to be an overlay for traditional circuit-switched GSM networks. The decision to design GPRS as an overlay—resulting in a parallel PS domain—was influenced by pressures from GSM network operators and equipment manufacturers to minimize the impact of the new packet-switched features on the existing infrastructure [51, p. 306]. After nearly a decade of technical work and the large capital investments required to bring GSM to market, it is no wonder they were reluctant to introduce major changes to a fielded technology that was finally paying back dividends in terms of exponential customer growth.

Today, a typical GSM network uses the CS domain for voice traffic and the PS domain for data traffic. The SMS may be transmitted in both domains at the option of the mobile device. It is worth mentioning that packet data services are defined in the CS domain, though no longer widely used. These include the original 9.6 kbps circuit-switched data (CSD), high-speed circuit-switched data (HSCSD) defined in Release 96, and enhanced circuit-switched data (ECSD) defined in Release 99. The latter is the combination of EDGE techniques with HSCSD. HSCSD was never widely deployed outside Europe, due in large part to the introduction of GPRS shortly thereafter. These circuit-switched data services are not widely used in today's wireless Internet, and are not discussed in further detail. The focus of this chapter will be on the services and operation of the PS domain using GPRS and EGPRS. The fundamentals of the CS domain operation with respect to voice and dual-transfer mode (DTM) (i.e., simultaneous voice and data operation on a single RF channel).

6.6.2 GSM Network Architecture

The GSM network architecture consists of three functional subsystems: the mobile station (MS), the base station subsystem (BSS), and the network switching subsystem (NSS). The MS refers to the user equipment necessary to access the services of the cellular network. The role of the BSS is to support communication with the MS over the radio interface, called the Um interface, and deliver user traffic to and from the wired network infrastructure. The role of the NSS is to deliver traffic between the MS and the public switched telephone network (PSTN) or a packet data network (PDN), such as the Internet or corporate intranet, based on Internet Protocol version 4 (IPv4) or Internet Protocol version 6 (IPv6). The NSS is also responsible for various functions such as user registration and authentication. The BSS is also commonly referred to as the radio access network (RAN), and the NSS is commonly referred to as the core network (CN). The terms RAN and CN are used more frequently in newer releases of the GSM standard based on 3GPP nomenclature (Rel-4 and beyond). The combination of the BSS and NSS, as well as an operation and maintenance subsystem (OMS) used by the network provider to monitor and maintain the network, form a public land mobile network (PLMN). In GSM nomenclature, a *PLMN* is a complete cellular network that is operated by either a recognized private operating agency (e.g., Vodafone, Orange, T-Mobile, AT&T) or, in some cases, a government agency for the purpose of providing land-based mobile communication services to the public as an extension of the fixed network infrastructure (i.e., PSTN, Internet) [106].

The basic network architecture of a traditional 2G GSM network is shown in Figure 6.1.

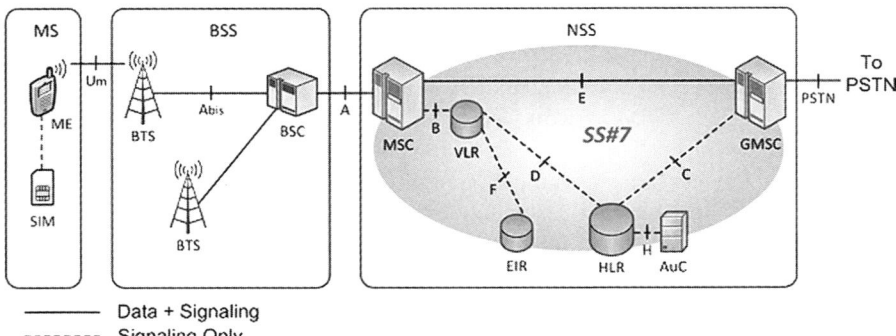

Data + Signaling
Signaling Only

Figure 6.1   GSM 2G network Architecture.

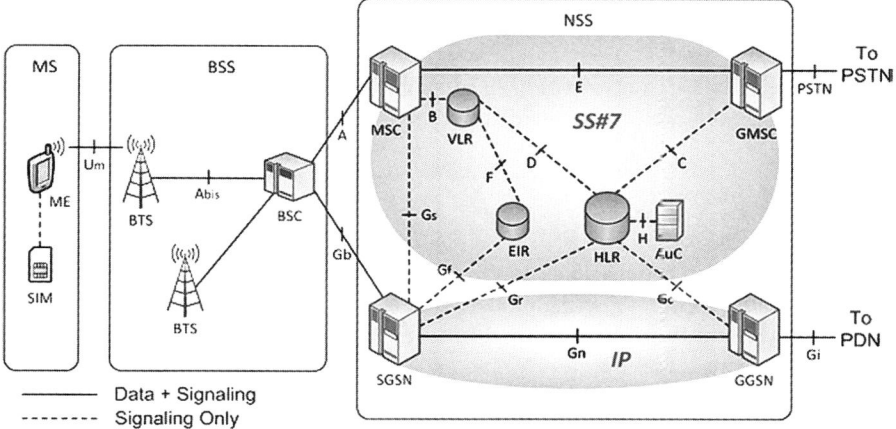

Figure 6.2   GSM/GPRS 2.5G network architecture.

It is characterized by a CS domain only. This architecture is indicative of the GSM networks deployed in the 1990s, however it is extremely uncommon today. The basic GSM/GPRS network architecture introduced in Release 97 is shown in Figure 6.2. It is characterized by the parallel CS and PS domains discussed previously that make up modern GSM networks. The network architecture shown in Figure 6.2 reflects that of GSM Release 97–Release 99. Further core network evolution has continued within 3GPP by means of a single evolved core network for GSM and UMTS; however, these architectures are generally considered 3G and beyond. The functional entities and interfaces of the GSM network are labeled in Figure 6.1 and Figure 6.2. Interfaces carry either user traffic and signaling (bold line) or signaling only (dashed line). Signaling transactions between entities of the NSS are performed using the Signaling System No. 7 (SS#7) network protocol in the CS domain[6] and Internet Protocol (IP) in PS domain. It is not a requirement that all functional entities of the GSM network architecture be implemented as separate equipment. Common examples of functional entities that may be integrated into a single device are the base transceiver station/base station controller (BTS/BSC), mobile services switching center/visitor location register (MSC/VLR), and the home location register/authentication center (HLR/AuC). In these cases, the manufacturer need not implement the interface between the two functional entities internally. The following sections provide an overview of each of the subsystems and functional entities of the GSM network.

---

[6] The SS#7 protocol is used for NSS signaling in the circuit-switched domain for all GSM releases up to Release 99. For core network architectures starting with Release 4, NSS signaling in the circuit-switched domain is a combination IP and SS#7. These core network architectures are considered 3G and beyond and are not addressed in detail in this chapter.

6.6.2.1   Mobile Station

The MS is the cellular equipment carried by the mobile user. It comprises two components: the mobile equipment (ME) and the Subscriber Identity Module (SIM).[7] Both components are required in order to form a functioning MS capable of network access.

6.6.2.2   Mobile Equipment

Mobile equipment (ME) refers to the GSM user device itself (e.g., mobile phone, Universal Serial Bus [USB] data modem, tablet computer). Every ME is uniquely identified by means of an International Mobile Equipment Identity (IMEI), a 15-digit hierarchical identification number that both identifies the ME and gives clues as to its origin (i.e., manufacturer, type approval date) [108, p. 26]. It is assigned by the equipment manufacturer and registered by the network operator in the equipment identity register (EIR).

6.6.2.3   Subscriber Identity Module

GSM distinguishes explicitly between subscriber mobility and equipment mobility by means of the Subscriber Identity Module (SIM). The SIM is the functional entity that stores subscriber-related information and implements the security functions pertaining to authentication and ciphering on the user side. The SIM stores a variety of permanent (i.e., fixed at subscription time) and temporary subscriber information. The following are examples of information stored on the SIM [109]:

- International mobile subscriber identity (IMSI)
- Security information (e.g., authentication key [Ki], authentication algorithm [A3], cipher key generation algorithm [A8], GPRS cipher key algorithm [A5], temporary cipher keys)
- Location information (e.g., temporary mobile subscriber identity [TMSI], packet TMSI [P-TMSI], current location, location update status)
- Network-specific data (e.g., list of carrier frequencies for cell reselection, list of forbidden PLMNs)
- Service-related data (e.g., charging information)

For security purposes, all MS processes that require the subscriber authentication key (Ki) or security algorithms are performed internally by the SIM to keep this information confidential [109, p. 9]. The SIM functions are implemented using a Universal Integrated Circuit Card (UICC), a removable plastic card bearing a silicon chip with both nonvolatile memory and a general-purpose processor [107].

---

[7] From 3GPP Release 5 onward, the SIM is replaced by the Universal Subscriber Identity Module (USIM), the SIM's evolved 3G counterpart, in all GSM MSs. The USIM is backward compatible with GSM Release 4 and prior and supports advanced features [107].

The UICC uses a standard interface, allowing it to be used interchangeably with different MEs. A key advantage of the SIM is that it allows the user to transfer easily between MEs while retaining the same phone number and subscriber data. The SIM concept is a major innovation of GSM. It has created a paradigm of subscriber mobility that continues to be used in UMTS and LTE, the 3G and 4G successors to GSM, in the form of the Universal Subscriber Identity Module (USIM) and has also been adapted to some CDMA2000 implementations in the form of the Removable User Identity Module (RUIM).

6.6.2.4 GPRS MS Classes

The ability of a GPRS handset to simultaneously access CS voice and PS data services is based on the capabilities of the ME. In order to address the various needs of different market segments, GPRS defines three MS classes corresponding to three modes of operation [110, pp. 17–19].

- *Class A.* The MS is able to support simultaneous operation of CS and PS services. This includes simultaneous attach, monitor, and invocation of CS and PS traffic.
- *Class B.* The MS is able to simultaneously attach and monitor both CS and PS services, but can only support one type of traffic at a time. A Class B MS is not required to monitor PS paging while on a voice call, but it must monitor CS paging during PS data traffic and alert the user with the option to suspend data traffic and accept the incoming call.
- *Class C.* The MS may only attach to CS or PS services at a given time, but not both simultaneously. The service that is not selected remains detached and unreachable unless a switch between domains is initiated.

In 3G UMTS networks, only Class A devices are allowed [110, p. 18]. Thus, for multimode GSM/UMTS devices, it is desirable to implement Class A mode of operation in GSM for continuity of services between 2G and 3G infrastructure. When GPRS was first introduced in Release 97, there was no coordination between the two service domains to support Class A mode of operation in GSM. Implementation was prohibitively complex and expensive, as it required two transceivers because the CS and PS domain services may operate on different frequencies. As a result, few such devices were initially developed. To simplify Class A mode of operation in GSM, dual-transfer mode (DTM) was defined starting in Release 99 [111]. DTM is a set of protocols that allow simultaneous access to CS and PS services using the same RF channel. Paging in DTM is similar to Class B mode of operation; however, the ability to move from a single traffic state (i.e., CS or PS traffic) to a dual-transfer state is supported. DTM support requires modifications in both the MS and the network. The capabilities associated with DTM have evolved since Release 99 to include, for example, extended multislot classes (Release 4, Release 6) and concurrent CS and PS handover (Release 7).

6.6.2.5 Multislot Class

Devices that support GPRS and EGPRS may achieve increased data rates by simultaneously transmitting and/or receiving on more than one of the eight available time slots within a single TDMA frame over the air interface if resources are available. The allocation of multiple channels in this manner is referred to as a multislot configuration. A multislot configuration occupies up to eight time slots with the same frequency parameters (i.e., RF channel or frequency hopping pattern), and may be asymmetric in the uplink and downlink. The maximum number of time slots an MS is capable of using at one time is defined by the multislot class of the MS, as summarized in Table 6.4[8] [112]. Multislot capable MSs are distinguished based on their ability to transmit and receive at the same time. A Type 1 MS is not required to transmit and receive at the same time. A Type 2 MS is required to be able to transmit and receive at the same time, which requires a more sophisticated RF frontend.

Three parameters are defined for each multislot class: the maximum number of downlink time slots the MS can receive (Rx), the maximum number of uplink time slots the MS can transmit (Tx), and the maximum total number of time slots that the MS can use for both transmission and reception combined (Sum). Note that the Sum parameter is less than the sum of the Rx and Tx parameters for some multislot classes, thereby adding an additional constraint to the multislot configurations supported by the MS. For multislot classes 13–29, no such constraint is imposed. For some higher-order multislot classes, the requirement of the MS to perform channel measurements and receive signaling channels are relaxed or modified to achieve high multislot configurations. In practice, the ability of an MS to actually utilize higher-order multislot configurations is determined by the multislot class of the MS, the multislot capabilities of the network, and the availability of sufficient radio resources.

6.6.2.6 Base Station Subsystem

The BSS, or GSM radio access network, is the system of base station equipment that creates the link between the NSS and the mobile station. It bridges the gap between the wired (i.e., core network) and wireless (i.e., air interface) network infrastructure. The BSS comprises two functional entities: the base transceiver station (BTS) and the base station controller (BSC).

Base Transceiver Station   The BTS is the part that we typically think of as the base station. It is the physical gateway between the wired and wireless portions of the GSM network, implementing the GSM radio link on the network side. The BTS consists of high-speed transmitter and receiver equipment, including antennas and amplifiers, that provides signaling and data channels to the MS over the air interface, called the Um interface in GSM. Each BTS forms one cell of coverage in the network, though it is common for multiple BTSs to be integrated into a single physi-

---

[8] Note that multislot classes 30–45 are defined in Release 5 and beyond but are included in this section for reference.

Table 6.4 MS Multislot Classes

Multislot class	MS type	Maximum number of slots			Release
		Rx	Tx	Sum	
1	1	1	1	2	R96
2	1	2	1	3	R96
3	1	2	2	3	R96
4	1	3	1	4	R96
5	1	2	2	4	R96
6	1	3	2	4	R96
7	1	3	3	4	R96
8	1	4	1	5	R96
9	1	3	2	5	R96
10	1	4	2	5	R96
11	1	4	3	5	R96
12	1	4	4	5	R96
13	2	3	3	N/A	R96
14	2	4	4	N/A	R96
15	2	5	5	N/A	R96
16	2	6	6	N/A	R96
17	2	7	7	N/A	R96
18	2	8	8	N/A	R96
19	1	6	2	N/A	R97
20	1	6	3	N/A	R97
21	1	6	4	N/A	R97
22	1	6	4	N/A	R97
23	1	6	6	N/A	R97
24	1	8	2	N/A	R97
25	1	8	3	N/A	R97
26	1	8	4	N/A	R97
27	1	8	4	N/A	R97
28	1	8	6	N/A	R97
29	1	8	8	N/A	R97
30	1	5	1	6	Rel-5
31	1	5	2	6	Rel-5
32	1	5	3	6	Rel-5
33	1	5	4	6	Rel-5
34	1	5	5	6	Rel-5
35	1	5	1	6	Rel-5
36	1	5	2	6	Rel-5
37	1	5	3	6	Rel-5
38	1	5	4	6	Rel-5

(*Continued*)

Table 6.4   (*Continued*)

Multislot class	MS type	Maximum number of slots			Release
		Rx	Tx	Sum	
39	1	5	5	6	Rel-5
40	1	6	1	7	Rel-5
41	1	6	2	7	Rel-5
42	1	6	3	7	Rel-5
43	1	6	4	7	Rel-5
44	1	6	5	7	Rel-5
45	1	6	6	7	Rel-5

cal tower to form a sectored base station. The portion of the BTS that implements transmit and receive functions is called the transceiver (TRX). The TRX implements most of the functions of GSM physical layer on the air interface (Um interface) on the BSS side. These include, for example, channel coding, encryption, frequency hopping, burst formatting, and modulation. The TRX is also responsible for taking channel measurements to report back to the BSC for radio resource management purposes. A single BTS my host up to 16 TRXs, however most practical BTS implementations host between one and four TRXs [113, pp. 19–21]. In the case of cell sectorization, each sector is itself a BTS hosting one or more TRXs.

Base Station Controller   The BSC is the switch responsible for combining user traffic in the network direction and distributing user traffic to the appropriate BTSs in the MS direction. The BSC can be considered the "brain" of the BSS. It is responsible for interpreting radio resource management from the MS, most of which is transmitted transparently through the BTS to the BSC. The BSC handles all traffic channel allocation, link supervision, and traffic channel release functions for all BTSs within the BSS. Each BSC typically controls several BTSs; however, this is not required. In practice, the split between the BSC and the BTS is implementation specific. They may be implemented separately or integrated into a single physical device. The BSC provides the single point of connection to the GSM circuit-switched core network over the A interface, and in turn communicates with the BTSs over the Abis interface.

Transcoding Rate Adaptation Unit   GSM transmits voice traffic over the air interface as a digitally encoding signal. The most common voice codecs used in GSM are the EFR codec, which encodes speech at 12.2 kbps, and the adaptive multirate (AMR) codec, which encodes speech at a variable rate of 12.2 kbps or less depending on channel conditions. However, voice is transmitted through the NSS using the ISDN standard format (ITU Telecommunication Standardization sector [ITU-T] A-law) at 64 kbps. The Transcoding Rate Adaptation Unit (TRAU) is the device that is used to adapt GSM-encoded speech to ISDN-encoded speech and vice versa in real time. The location of the TRAU is implementation specific. It may reside in the BTS, the BSC, or even on the MSC side of the A interface. The latter has an advantage to the network provider in that it substantially reduces the bandwidth required in the backhaul between the BSC and the NSS in the CS domain. If the TRAU is implemented outside of the BTS, it must receive in-band signaling from the BTS to allow synchronization and decoding. In this case, each GSM voice traffic channel utilizes a 16 kbps subchannel between the TRAU and the BTS. The TRAU is not considered a distinct functional entity within the GSM network architecture, but is rather a component of the BSS [114, pp. 123–125].

Packet Control Unit   Although GPRS reuses the basic structure of the air interface, new hardware is required within the BSS to implement the GPRS packet channels and the associated signaling and traffic. The device that does this is called the packet control unit (PCU). Like the TRAU, it is not considered a distinct functional entity within the GSM network architecture, but rather a component of the BSC. This is true even if the PCU is physically implemented remotely outside the BSC (e.g., at the BTS site or serving GPRS support node [SGSN] site) [115, pp. 151–152]. The physical resources of the air interface that are reserved for GPRS operation at any given time are assigned by the BSC to the PCU. On the air interface side, the PCU is responsible for radio resource management. On the network side, the PCU interfaces with the network switch responsible for delivering all PS domain traffic to the BSS (i.e., the SGSN; see Section 6.6.2.7) via the Gb interface.

Location Area   For the purpose of mobility management in the CS domain, the BSS is divided into groups of cells called location areas (LAs). As long as an MS in CS idle mode stays within a single LA, it may move freely without initiating a location update, thereby reducing the signaling burden on the MS and extending battery life. The number of cells contained within an LA is a matter for network planning and has a significant impact on the amount of signaling required within the BSS. As the size of the LA increases, the number of paging requests the MS must monitor increases because there are more subscribers within the LA. As the size of the LA decreases, the number of location update messages required for a mobile MS increases. Network optimization consists of striking a balance between these two conflicting parameters [116, p. 465]. In operational networks, typically 20–30 cells are grouped into an LA [117, p. 45]. A consequence of allowing the MS in idle state to roam freely within the LA is that the exact cell in which the MS is located is not known. For this reason, paging must be used to locate the MS in the event of an incoming call.

Routing Area   The routing area (RA) is the PS domain counterpart of the LA which is used for mobility management in GPRS. When a GPRS-attached MS is in the standby state, it will only update its location with the network when it roams into a new RA. In general, an LA is divided into one or more RAs, making the RA a subset of the LA. For an MS that is currently registered with the network both in the CS and PS domains (i.e., GPRS Class A or Class B mode of operation), joint location update/RA update messages may be sent to the network when appropriate to minimize signaling overhead.

6.6.2.7 Network Switching Subsystem

The network switching subsystem (NSS), or GSM core network, provides a wired connection between the BSS and the outside world. Although the focus of our discussion is on the wireless link, a basic understanding of core network design is helpful in the discussion of internetworking challenges in later chapters. The functional entities of the core network include the home location register (HLR), authentication center (AuC), and equipment identity register (EIR), which support signaling functions for both domains; the mobile services switching center (MSC), gateway MSC (GMSC), and visitor location register (VLR) in the CS domain; and the serving GPRS support node (SGSN) and gateway GPRS support node (GGSN) in the PS domain. Each of these functional entities is described briefly in the succeeding sections.

Home Location Register   The home location register (HLR) is a centralized database in charge of the management of mobile subscribers. It maintains subscriber profiles for all subscribers authorized to use the network. A GSM PLMN may contain one or multiple HLRs depending on the number of mobile subscribers, the capacity of the equipment, and the organization of the network [106, p. 11]. The HLR stores details for every SIM indexed by the IMSI and maps these entries to the telephone number (i.e., the Mobile Subscriber ISDN Number [MSISDN] address) of each MS. The HLR stores information about any active IP addresses assigned to an MS and informs the GGSN whether it is allowed to dynamically allocate IP addresses to a particular MS. For each subscriber profile, the HLR also stores subscription information and location information. The latter is used for routing user traffic between the PLMN gateways and the appropriate MSC in the CS domain or SGSN in the PS domain.

Authentication Center   The authentication center (AuC) is the functional entity on the network side which stores confidential data for each subscriber used for the security procedures of authentication and encryption. The AuC stores a secret authentication key, Ki, for each registered subscriber. It also implements the security algorithms: A3 (authentication), A8 (ciphering key generation), and A5 (GPRS ciphering key generation). For security, the AuC communicates only with the HLR over the H interface and does not actively participate in the authentication process [106, p. 13]. Rather, the AuC generates the required authentication challenge/response data, which is then transmitted to the VLR or SGSN via the HLR [118, p. 20]. The authentication process relies on the confidentially of the authentication key that is known only to the AuC and the SIM and is never transmitted over the air.

Once a subscriber is successfully authenticated, the HLR is allowed to manage the SIM and a ciphering key is generated to encrypt all subsequent communication. If authentication fails, then the MS is denied service by the network. Due to its isolation from the rest of the network, the AuC is commonly integrated into the HLR in practical implementations.

Equipment Identity Register The equipment identity register (EIR) is a database whose core function is to maintain a list of IMEIs for all known GSM MEs. This list is used to recognize whether an MS that attempts to register to the network is using equipment that is obsolete, stolen, or nonfunctional and allow the network to deny or limit service to such devices. Each IMEI entry in the EIR may be classified as "white listed," "black listed," or "gray listed" according to the following definitions [90, p. 6]:

- *White List.* A register of all equipment identities that are permitted for use.
- *Black List.* A register of all equipment identities that are barred from use. This list may include equipment that has been stolen or used for other nefarious purposes, and is periodically exchanged among network operators.
- *Gray List.* A register of equipment that is not necessarily barred from use, but are tracked by the network for evaluation and other purposes. This may include equipment that is malfunctioning or obsolete.

With the exception of the white list, the use of these IMEI registers is at the discretion of the operator. If an MS has an IMEI that is black listed or unknown to the network (i.e., not on the white list), network access is terminated with the exception of emergency calls [119, p. 6].

Mobile Services Switching Center The mobile services switching center (MSC) is the primary service delivery node within the CS domain, responsible for handing voice calls, SMS text messages, and other circuit-switched services. It performs all the switching and signaling functions necessary to support circuit-switched services for all MSs located within its geographical area, and is analogous to a telephone exchange in the wired PSTN. The main difference is that the MSC must implement additional functions necessary to support mobility in a cellular network. The primary functions of the MSC are traffic switching (i.e., establishment of traffic paths from PSTN to MSs), mobility management (i.e., location registration and handover), and allocation of radio resources [106, p. 13]. The MSC provides the single point of connection between the NSS and the BSS in the CS domain via the A interface [106, p. 14]. A typical PLMN configuration includes multiple MSCs, each of which interfaces with several BSSs, in order to obtain radio coverage over a large geographic area [106, p. 9].

Gateway MSC Incoming calls delivered to the PLMN from another network (e.g., the PSTN, other PLMNs) are first routed through a functional entity called the gateway MSC (GMSC). The core function of the GMSC is to provide a gateway to the PLMN in the CS domain. When a call is delivered to the PLMN from another network, the GSMC interrogates the HLR for location information and then routes the call to the appropriate MSC where the MS is located [106, p. 14]. The decision of which MSCs can act as a GMSC is left to the operator, though in practice there is typically only one designated GMSC per PLMN. From an implementation standpoint, the GMSC may be identical to an MSC (i.e., capable of serving one or more BSSs in addition to its gateway function) or, alternatively, it may be a high-capacity MSC designed solely to carry out the GMSC function (i.e., the A interface is not implemented).

Visitor Location Register   The visitor location register (VLR) is a database containing temporary information for the subscribers who have roamed into the geographic area it serves. A VLR may be in charge of one or several MSC areas. However, in practice, a one-to-one association between VLRs and MSCs is common. When an MS roams into a new LA, it registers its identity and location with the MSC, which then transfers this information to the VLR. If the MS is not yet registered, the VLR obtains subscriber information from the HLR to facilitate the proper handling of CS domain traffic and assigns a temporary MS identity (TMSI) to the mobile, which it will use to communicate confidentially with the MS while it is within the area of service of the VLR. Data stored in the VLR include the IMSI, MSISDN, Mobile Station Roaming Number (MSRN), TMSI (see Section 6.6.3 for definitions), and the last known location of the MS. The VLR also stores a small amount of PS domain information pertaining to the current SGSN where the MS is registered, if applicable [106].

Serving GPRS Support Node   The SGSN is the primary service delivery node in the PS domain, responsible for handling data traffic and SMS text messages sent via packet-switched traffic channels. The SGSN is the PS domain counterpart to the MSC; except instead of being a circuit switch, the SGSN is a router designed to support GPRS-specific protocols for packet-switched communication with the PCU and the GGSN. The SGSN is the functional entity responsible for session management on the network side, which includes the management of packet data protocol (PDP) contexts. In GPRS, PDP context activation is required before an MS is able to send or receive any data traffic in the PS domain. This operation sets up the routing tables of the PS domain nodes and makes the MS known to the corresponding gateway node, the GGSN, allowing interworking with the wired Internet or other PDN (e.g., corporate network) [115]. The PDP context activation is similar to dialing in a voice call in the CS domain, except that it operates in a connectionless state and no resources are reserved. The PDP context itself is essentially a record that makes it possible to route IP packets through the GPRS network and specifies certain GPRS-specific quality of service (QoS) parameters. Routing, SM, and PDP context procedures are discussed in further detail in Section 6.9 later in this chapter. The SGSN is also responsible for MM and the network-termination point for ciphering over the air interface in the PS domain. The SGSN has an integrated location register which stores the following information for each attached MS: subscriber information (i.e., IMSI, P-TMSI, current IP addresses[es]), location information (i.e., last registered RA), and information pertaining to any active PDP contexts (i.e., GGSN IP address, QoS profile) [106]. The optional Gs interface is defined between the SGSN and the MSC/VLR to support coordination of services between the CS and PS domains (e.g., combined LA/RA updates, coordinated MS paging).

Gateway GPRS Support Node   Similar to the SGSN, the GGSN is the PS domain counterpart to the GMSC. The GGSN is also a router with GPRS enhancements. The GPRS enhancements required in the GGSN are significantly less than those of the SGSN, as the only GPRS-specific protocol is the GPRS tunneling protocol (GTP), which is used for tunneling IP packets through the CN from the GGSN to the SGSN [115]. The GGSN serves a number of key functions. On the cellular network side, it serves as an anchor through which all IP packets are routed regardless of the mobility of the MS. From the perspective of the external PDN (i.e., the Internet or a corporate network), the GGSN hides the mobility of the user and provides a fixed gateway for all IP addresses in the PLMN. The GGSN is also the functional entity responsible for dynamic assignment of IP addresses during PDP context activation. The GGSN generally includes firewall and packet-filtering mechanisms as well.

6.6.3   Identifiers and Addresses

Several identifiers are utilized in GSM that serve to identify mobile subscribers, ME, and various geographic regions within a PLMN. These identifiers are generally hierarchical in nature, as this facilitates identification, routing, and localization within the global mobile infrastructure. The identification plan for mobile subscribers and allocation of MS roaming numbers conforms to the ITU-T Recommendations E.212 and E.213 [120, p. 7; 121, 122]. The most significant GSM identifiers are summarized in the succeeding sections. In each case, it is indicated whether the identifier or address is used in the CS domain, the PS domain, or both.

6.6.3.1   International Mobile Subscriber Identity

Every GSM subscriber is uniquely identified by means of the international mobile subscriber identity (IMSI), which is the primary subscriber identifier in both the CS and PS domain. The IMSI is completely independent from the subscriber's MSISDN number (i.e., phone number), making it possible to assign multiple phone numbers to the same subscriber and preserving subscriber confidentiality (i.e., a subscriber may distribute his phone number without divulging his subscriber identity). The IMSI consists of a maximum of 15 numeric digits. The IMSI is hierarchical and comprises three parts:

- *Mobile Country Code (MCC).* A three-digit code that uniquely identifies the home country of the subscription. In some cases, a country with a large number of subscribers may be assigned multiple MCCs (e.g., 310–316 for the United States, 460–461 for China). These numbers are administered by the ITU-T.
- *Mobile Network Code (MNC).* A two- or three-digit code that identifies the home PLMN of the subscriber.
- *Mobile Subscriber Identification Number (MSIN).* A number that uniquely identifies the subscriber within the PLMN. MSIN allocation is the responsibility of the network provider.

The hierarchical structure of the IMSI makes it easy to identify the home PLMN of a subscriber, as only the first six digits need to be analyzed. It also makes it possible to uniquely identify a subscriber within their home country using only a subset of the IMSI, the combination of MNC and MSIN, which is referred to as the national mobile subscriber identity (NMSI) [120].

6.6.3.2 Temporary Mobile Subscriber Identity

For the sake of subscriber confidentiality, the IMSI is broadcast over the air only when the subscriber initially registers to the network. This may occur at startup or if the subscriber roams into a new PLMN. Once a subscriber is authenticated, the serving VLR assigns the subscriber a temporary identifier, the temporary mobile subscriber identity (TMSI), which is used for all further communication over the air interface related to CS domain services. A specific TMSI is assigned only while the MS resides within a VLR region and is not propagated beyond the VLR within the CN. The VLR serves the function of translating the TMSI to its corresponding IMSI when communicating with the HLR. Since it has only local significance, the precise structure and coding of the TMSI is at the discretion of the VLR manufacturer and the network operator. However, in general, the TMSI is implemented as a 32-bit number [114, p. 49; 120].

6.6.3.3 Packet Temporary Mobile Subscriber Identity

The packet temporary mobile subscriber identity (P-TMSI) is the PS domain equivalent of a TMSI. It is assigned to the MS by the SGSN and used for subscriber confidentiality during communication over the air interface. The P-TMSI has the same 32-bit structure as the TMSI except that the first two bits are always set to 11. A consequence of this is that the VLR-assigned TMSI address space is limited to those starting with 00, 01, or 10. Like the TMSI, the P-TMSI has only local significance within the service area of a single SGSN. A new P-TMSI is assigned when the MS moves to a new RA. The SGSN may also choose to assign a new P-TMSI at any time [108, 115].

6.6.3.4 Mobile Subscriber ISDN Number

The actual phone number used within the PSTN to dial an MS is called the MSISDN.[9] The MSISDN is the primary MS address in the CS domain from the perspective of the outside world. The MSISDN follows a hierarchical structure consisting of a country code (CC),[10] national destination code (NDC), and subscriber number (SN). The combination of NDC and SN forms the national (significant) mobile number, which may be used for domestic dialing [120, p. 11]. This structure follows the international numbering plan set forth by ITU-T recommendation E.164, thus allowing the subscriber to call any number within the PSTN, ISDN, or any other PLMN and vice versa. When a call is initiated from the PSTN to an MS, the MSISDN is used to route the call to the HLR associated with the MS. Outside the PLMN, the MSISDN is the only identifier used to route voice calls to the MS [120, p. 11].

6.6.3.5 Mobile Subscriber Roaming Number

For routing purposes in the CS domain, a temporary location-depending ISDN number is assigned to the MS by the VLR. This number is the Mobile Subscriber Roaming Number (MSRN), and it follows the same structure as the MSISDN (i.e., CC + NDC + SN), where the CC and NDC reflect the location (i.e., country and region) of the VLR. The remaining SN is assigned by the VLR either when the MS registers to the network or when requested by the HLR for the purpose of setting up a connection for an incoming call.

6.6.3.6 IP Address

In the PS domain, the IP address is the primary address used by the outside world to identify the MS. The IP address of the MS may also be referred to as the PDP address, which reflects the fact that GPRS was originally designed to support other network protocols as well (e.g., X.25) [123]. IP addresses may be assigned to the MS statically or dynamically from the address space of the PLMN (i.e., *transparent* access), or it may be assigned from the address space of an external PDN, for example, in case the user is connecting to a corporate network (i.e., *nontransparent* access). In the common case of transparent Internet access, the address space of the PLMN forms a subnet that appears to external PDNs as a typical IP network. In general, GPRS supports both IPv4 and IPv6 addresses. An MS may be assigned multiple IP addresses concurrently, which may be a combination of IPv4 and IPv6 addresses.

---

[9] The specific meaning of the acronym has come to be defined in several ways depending on the standards body. In the original revisions of the GSM specification published within ETSI, it was defined as "Mobile Station International ISDN Number" (ETSI spec 03.03 v3.6.0 GSM Ph 1). Within 3GPP, it is defined as "Mobile Subscriber ISDN Number" (Ref: 3GPP TR 21.905 v10.3.0 3GPP Vocabulary Rel 10). Within the ITU, it is sometimes called the "Mobile International ISDN Number." In all cases, the underlying meaning is the same.

[10] The CC used in the MSISDN is different from the MCC used in the IMSI

6.6.3.7 Tunnel End Point Identifier

The tunnel end point identifier (TEID) is the identifier used by the GPRS tunneling protocol (GTP) for tunneling of user traffic and control data through the core network in the PS domain. The TEID is stored in the GGSN upon PDP context activation and is used to forward data between the GGSN and the SGSN. The TEID may also be used after an inter-SGSN routing area update to transfer IP packets between the old SGSN and the new SGSN as part of the mobility management procedure. The TEID has significance only within the GTP, and there is always a one-to-one correspondence between TEID and PDP context [115]. The term TEID is first defined with GTP version 1 (GTPv1) defined in Release 99 and beyond, and serves a similar function as the tunnel identifier (TID) defined in GTP version 0 (GTPv0) defined in Release 97 [115, 124].

6.6.3.8 Temporary Logical Link Identifier

The temporary logical link identifier (TLLI) is a unique identifier that is used to temporarily identify the MS across the air interface for the duration of a PDP session. The TLLI is a 32-bit value that identifies the logical link established between the MS and the SGSN over the Um and Gb interfaces in the PS domain. Since all active PDP contexts associated with an MS share a single logical link, there is a one-to-one correspondence between the TLLI and the IMSI within a single RA that is known only to the MS and the SGSN. In general, it is the TLLI—not the P-TMSI—that is used to confidentially identify the MS at the air interface. The value of the TLLI is derived from the P-TMSI and the current location of the MS. Three different types of TLLI may be derived depending on the scenario. The MS derives a local TLLI if it has a valid P-TMSI for the current RA. A foreign TLLI is derived if the MS has a P-TMSI from a different RA. A random TLLI is created in the absence of a P-TMSI. The foreign or random TLLI may be used until a valid P-TMSI is assigned for the current RA.

6.6.3.9 Network Layer Service Access Point Identifier

The network layer service access point identifier (NSAPI) is used to differentiate between different active PDP contexts maintained by the MS. More specifically, it marks the service access point (SAP) between the GPRS subnetwork-dependent convergence protocol (SNDCP) sublayer and the network layer (i.e., IP) above. The combination of the NSAPI and the TLLI unambiguously identifies a specific PDP context within a routing area. The NSAPI/TLLI combination is used for routing purposes between the SGSN and the MS during a PDP session.

6.6.3.10 International Mobile Equipment Identity

Recall that GSM explicitly distinguishes between the subscriber and the mobile equipment. Thus, an identifier is needed for the ME in addition to the IMSI, which is used to identify the subscriber. Every ME is uniquely identified by means of the International Mobile Equipment Identity (IMEI), a 15-digit number assigned by the equipment manufacturer. The precise structure of the IMEI has changed over time. Since April 1, 2004, the IMEI has been composed of the following three parts [125, p. 8]:

- *Type Allocation Code (TAC)*. An eight-digit code that identifies the ME type (i.e., model and build of the device) and provides information as the manufacturer and reporting body that allocated the TAC.

- *Serial Number*. A six-digit serial number assigned by the manufacturer that uniquely identifies each ME within a TAC.

- *Check Digit (CD)*. A one-digit checksum computed from the other 14 digits using the Luhn formula [108, pp. 71–72]. The purpose of the CD is to help guard against manually tampering with IMEI values in stolen MEs. Of course, this method works only against lazy criminals who do not know how to read the GSM standard.

The IMEI is requested when the MS registers to the network and is used to verify that the ME is approved for service (i.e., has not been black listed as stolen or malfunctioning). The hierarchical structure of the IMEI is helpful in limiting the size of the EIR database. For example, all mobile equipment of a specific model and build from a mobile equipment manufacturer will be assigned a single TAC. If all of these devices are white listed, only the eight-digit TAC need be recorded in the EIR, rather than the full list of IMEIs for every device. The IMEI structure previously described is used in UMTS and LTE, the 3G and 4G successors to GSM. Multimode MEs supporting 2G, 3G, and/or 4G are assigned a single IMEI [125, p. 8].

6.6.3.11 Location Area Identifier and Cell Identifier

Every LA in a GSM network is addressed by means of the location area identifier (LAI). The LAI is used for routing and localization purposes in the CS domain. It is also broadcast at regular intervals over the air interface on the broadcast control channel (BCCH) and utilized by the MS to determine when it has entered a new LA and must initiate a location update [114, p. 49]. The structure of the LAI is fundamentally similar to the IMSI in that it uses the same MCC and MNC combination to uniquely identify the PLMN to which the LA belongs. The structure of the LAI, therefore, is composed of three parts:

- *Mobile Country Code (MCC)*. A three-digit code, as defined for the IMSI.

- *Mobile Network Code (MNC)*. A two- or three-digit code, as defined for the IMSI.

- *Location Area Code (LAC)*. A 16-bit address that uniquely identifies the LA within the PLMN [120].

In addition to the LAI, every cell within an LA is assigned a unique 16-bit address called the cell identifier (CI). The concatenation of the LAI and the CI (i.e.,

LAI + CI) form a global identifier called the cell global identifier (CGI), which uniquely addresses every cell in the global GSM infrastructure.

6.6.3.12 Routing Area Identifier

Similar to the LAI, every RA in a GSM network is addressed by means of a routing area identifier (RAI), which is broadcast as system information and used for routing and localization in the PS domain. Since the geographic area covered by a RA is the subset of a single LA (i.e., an LA consists of one or more RAs), the RAI is formed by the concatenation of the LAI with an 8-bit routing area code (RAC) that uniquely identifies the RA within the LA (i.e., RAI = MCC + MNC + LAC + RAC).

6.6.3.13 Example Call Routing in CS Domain

Figure 6.3 shows an example of how the CS domain identifiers described earlier are used to route an incoming voice call through the network to the MS. First, the MSISDN is used by the PSTN to route the call to the GMSC, which then queries the HLR for the appropriate MSRN. The GMSC then uses the MSRN to route the call to the appropriate MSC. At the MSC, the VLR is queried using the MSRN to determine the local identity of the MS, defined by the TMSI, and the approximate location of the MS based on the last updated LAI. The TMSI is then used by the network to page the mobile, which is localized to a specific group of cells using the LAI. The mobile then responds to the network page with the TMSI to request resources from the serving BTS to begin call setup. Notice that the TMSI is used to identify the MS over the air interface rather than the IMSI to maintain subscriber confidentiality. In this example, it is assumed that the MS has already successfully authenticated with the network using the IMSI and a TMSI has been assigned by the serving VLR.

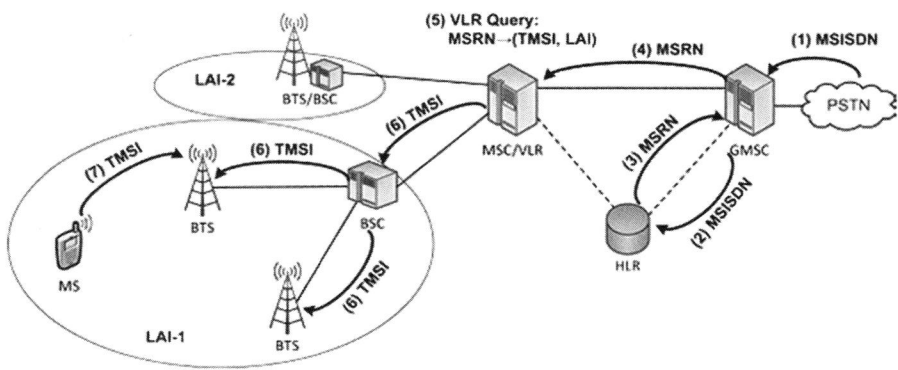

Figure 6.3 Incoming call routing example.

It will be beneficial to introduce a few fundamental GPRS concepts before delving into a PS domain addressing example. For an example of how the PS domain identifiers described earlier are used to route incoming (i.e., mobile terminated) and outgoing (i.e., mobile originated) IP packets, see Section 6.9.5 later in this chapter.

6.6.4   Air Interface Overall Structure

The air interface constitutes the radio link between the MS and the BTS within a GSM network. In GSM, this interface is defined as the Um interface, derived from "mobile U interface," as it is the mobile counterpart of the U interface in the ISDN architecture. The majority of the protocols and signaling that are unique to GSM reside in the air interface, as the protocol architecture beyond the BTS are primarily based upon ISDN signaling (i.e., SS#7) in the CS domain and IP in the PS domain. When many people talk about GSM, for example, that it is a narrowband TDMA technology, that it uses a 200 kHz channel bandwidth, or that it utilizes GMSK modulation, they are actually speaking of the GSM air interface. The design of the air interface is one of the most important aspects of any cellular system and has broad-reaching implications. The implementation of the air interface largely governs the data rate that a cellular technology can support. It also impacts the spectral efficiency of a system (i.e., number of users per unit of bandwidth), cell dimensioning (i.e., maximum cell size), flexibility of frequency allocation, spectral reuse, robustness of the radio link to bit errors, amount of signaling overhead, and the complexity and associated cost of MSs and BTSs, just to give a few examples. When one considers the impact of these design decisions, it is no wonder that it took the GSM technical committee over 5 years (from 1982 to 1987) and political involvement at the highest levels of government in many European countries to agree on the basic parameters of the GSM air interface.

In GSM, the fundamental problem of how to partition the physical resources of the radio link for multiple access is addressed through a combination of frequency and time division multiplexing. In this sense, GSM uses a combination of FDMA and TDMA. Frequency is divided into RF channels with 200 kHz carrier spacing. Time is partitioned into time slots, approximately 577 μs in duration, and grouped into TDMA frames with eight time slots per frame. This partitioning in both frequency and time results in a two-dimensional channel structure as shown in Figure 6.4. While all the RF channels supported by a single BTS are synchronized in time, GSM does not require time synchronization between BTSs. In addition to TDMA and FDMA, which are implemented as fixed assignment protocols, GSM also makes use of random access based on slotted ALOHA on the random access channel (RACH). The RACH is mapped onto the frequency–time structure described earlier and discussed further in subsequent sections.

The narrowband TDMA structure of the GSM air interface has a number of advantages. For example, TDMA allows multiple physical channels, divided in time, to be multiplexed onto a single carrier frequency. This increases capacity over the FDMA scheme of 1G analog systems by creating eight physical channels for each

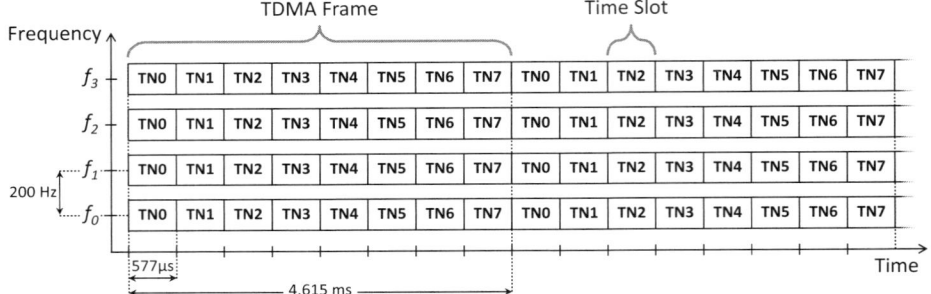

Figure 6.4   Overall radio resource structure of GSM.

carrier. Since the MS transmits and receives in bursts, this scheme also makes it possible to monitor other frequency channels during the time between bursts, which can be exploited in the case of handovers. Since the MS does not transmit all the time, this also allows for seamless handover between BTSs. The narrow bandwidth of GSM allows for flexible spectrum usage. This makes GSM scalable to a variety of deployment scenarios, such as high-density usage (more RF channels) in urban areas and low-density usage (fewer RF channels) in rural areas.

Today, TDMA may seem relatively simple compared with other multiple access schemes, such as CDMA or orthogonal frequency division multiple access (OFDMA) that are used in 3G and 4G cellular technologies. However, when it was first introduced to the GSM Working Party 2 (WP2) in the mid-1980s, TDMA represented a radical paradigm shift from the original 1G cellular technologies, which were restricted to a "one carrier per active user" FDMA scheme due to the use of analog modulation [51, p. 312]. At that time, the very idea that multiple users could coexist on the same carrier frequency without interference would have been an intriguing new idea to many experienced radio engineers. Of course, today we recognize that this is just one of the many benefits of using digital, rather than analog, modulation.

6.6.4.1 *Frequency Domain*

GSM is deployed in paired spectrum and uses frequency division duplexing (FDD) to separate the uplink and the downlink. This allows for simultaneous transmission and reception of downlink and uplink channels by the TRX of the BTS. As stated previously, frequency is divided into RF channels with 200 kHz carrier spacing. More precisely, the term "RF channel" in GSM refers to a pair of carriers, one each for the uplink and the downlink. Each RF channel is assigned an identifier, called the absolute RF channel number (ARFCN). The subset of all available RF channels that is assigned to a single cell is called the cell allocation (CA). By definition, a CA comprises one or more RF channel. In practice, however, each cell is typically allocated multiple RF channels. One RF channel is used to carry synchronization

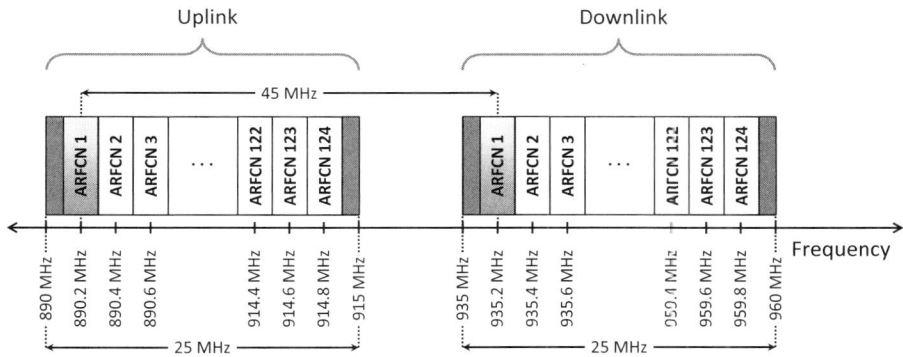

Figure 6.5   Frequency structure of the GSM 900 band.

information (i.e., frequency correction channel [FCCH] and synchronization channel [SCH]) and the broadcast control channel (BCCH) for the cell. This RF channel is designated the BCCH carrier and must be constantly transmitted by the BTS [112, p. 17].

When the original Phase 1 specification was frozen in 1990, GSM was only specified for operation in a single frequency band in the 900 MHz range, called the GSM 900 band. Since then, GSM has been adapted to many other frequency bands of operation. To illustrate the division of frequency resources in GSM, we use the GSM 900 band as an example. The overall structure of the GSM 900 band is shown in Figure 6.5. The frequency allocation for uplink and downlink channels is 25 MHz each. Each RF channel is formed by a duplex pair of carriers, one each for the uplink and the downlink, that are separated by a fixed offset called the duplex frequency. For GSM 900, the duplex frequency is 45 MHz. Since the carrier spacing in GSM is 200 kHz, the GSM 900 band is theoretically capable of supporting 125 RF channels. However, to avoid interference with adjacent bands, a guard band of 100 kHz is added to the upper and lower edges of the band. As a result, the GSM 900 band supports 124 duplex channels.

Today, there are four predominant bands of operation in which GSM is deployed: GSM 850, GSM 900, DCS 1800, and Personal Communications Service (PCS) 1900. Additionally, an extended GSM 900 band (E-GSM 900) has been specified to accommodate additional capacity in the 900 MHz range. This has lead to the original 124-channel GSM 900 band being referred to as the *primary* GSM 900 band (P-GSM 900). Each of these bands is summarized in Table 6.5. Although the details of each band vary (e.g., duplex frequency, number of channels), the basic structure follows that of the GSM 900 band described earlier. In addition to these four bands listed earlier, GSM has been adapted to many other frequencies of operation, primarily to address niche markets. As of Release 10 of the specification, 13 frequency bands have been defined for GSM operation. A detailed list of these frequency bands can be found in 3GPP Technical Specification 45.005.

Table 6.5   Summary of Main GSM Frequency Bands

Band name	Uplink frequencies (MHz)	Downlink frequencies (MHz)	Duplex frequency (MHz)	Number of RF channels	ARFCN range
GSM 850	824–849	869–894	45	124	128–251
P-GSM 900	890–915	935–960	45	124	1–124
E-GSM 900	880–915	925–960	45	174	0–124, 975–1023
DCS 1800	1710–1785	1805–1880	95	374	512–885
PCS 1900	1850–1910	1930–1990	80	299	512–810

6.6.4.2 Time Domain

Each RF channel is partitioned in time into time slots, approximately 577 μs in duration. The time slot is the fundamental physical resource unit of the GSM radio link. Eight consecutive time slots are grouped together to form a TDMA frame, approximately 4.62 ms in duration. Each time slot is assigned a number, from 0 through 7, which repeats modulo 8 called a time slot number (TN). Information is transmitted over a time slot using GMSK or 8PSK modulation[11] at a gross symbol rate of approximately 270.8 ksps, resulting in a modulated symbol period of roughly 3.7 μs. This means that one time slot corresponds to 156.25 symbol durations. Time slots are accessed by means of bursts, a period of RF carrier modulated by a bit stream. Different burst formats are defined depending on the type of information being transmitted. All bursts include a guard time to allow the transmitter to ramp its power up and down at the beginning and end of the burst. In general, the BTS is not required to ramp power up and down between bursts when transmitting on consecutive time slots. This is also true for an MS utilizing a multislot configuration. Due to the guard time, the useful duration of a burst is reduced. The relationship between TDMA frames, time slots, and bursts is shown in Figure 6.6. This example depicts the transmission of a so-called *normal burst*, used for transmission of most logical channels, using GMSK modulation. Since GMSK modulation is used, each modulation symbol duration corresponds to a single bit. The bottom portion of the figure depicts the transmitted power level versus time for the RF burst. The center of the figure shows how the normal burst format—explained later in Section 6.7.3—maps to the time mask for the normal burst. Note that the normal burst format includes 148 bits (not including guard period), however the useful duration of the normal burst is 147 symbol durations. This is because the first useful duration is defined as starting halfway through the first tail bit.

---

[11] In the case of EDGE, user traffic may be transmitted using 8PSK modulation in favorable channel conditions. The fundamental concept remains the same as GMSK, except that each symbol represents 3 bits instead of 1 bit. See Section 6.11 later in this chapter for further discussion.

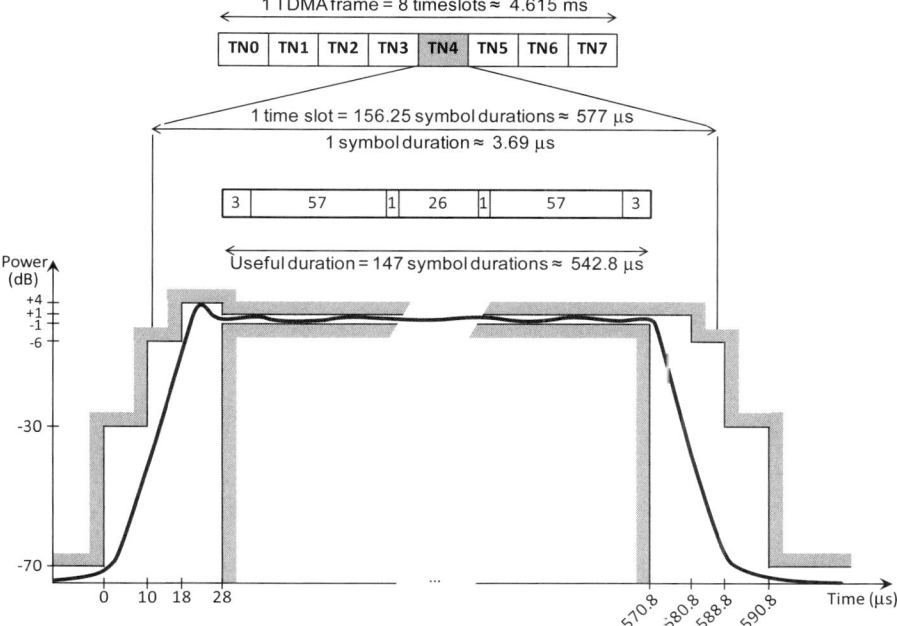

Figure 6.6   Time structure of GSM.

6.6.4.3   Time Synchronization between Uplink and Downlink

At the BTS, the TDMA frames of the uplink are delayed by three time slots with respect to the downlink. This is done in order to allow the same TN to be used on the uplink and the downlink while avoiding the requirement of the MS to transmit and receive simultaneously. By doing so, the RF front end of the MS does not require a duplex unit, which, in turn, allows the mobile to be manufactured as a less expensive and more compact device. Although this is not a major concern for modern manufacturing standards, this design decision reflects the effort made by the designers of GSM to make a standard that would support affordable, hand-portable mobile phones. Within the historical context of the 1980s, when the GSM air interface was designed, this was a significant change from the bulky, often vehicle-borne mobile phones of the time. The timing offset between uplink and downlink channels is particularly important for voice traffic in the CS domain, where both uplink and downlink resources are reserved for the mobile at time of call setup.

6.6.4.4   Timing Advance

The time synchronization of the uplink and the downlink is defined with respect to the BTS. However, the distance between the MS and the BTS, and the corresponding

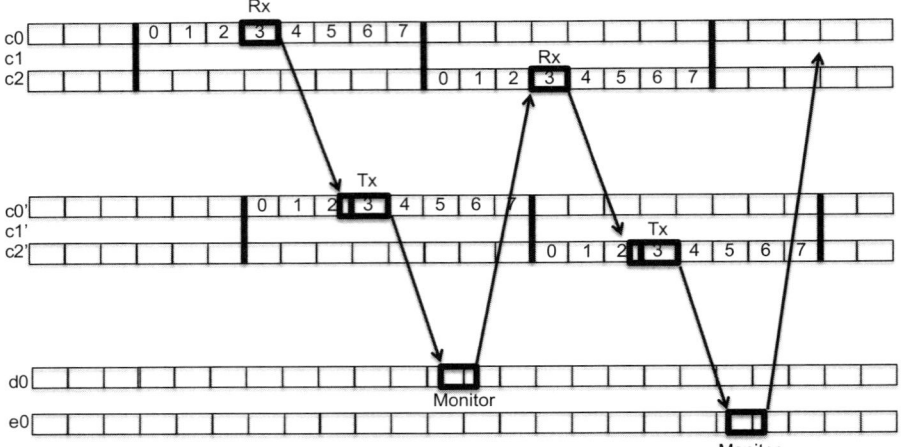

Figure 6.7   Transmission and reception from an MS perspective. Adapted from Reference [112].

propagation delay, can vary considerably due to mobility. This poses a problem, not only from the standpoint of synchronizing the uplink and the downlink, but also for ensuring that uplink bursts from different MSs are received at the BTS in the appropriate time slot and do not cross into the next time slot, thereby corrupting data in both time slots. These collisions are avoided through a process called adaptive frame alignment, whereby the BTS measures the propagation delay of the MS and instructs it to advance its timing alignment using a parameter called the timing advance (TA). From the perspective of the MS, this means that the delay between the TDMA frames of the downlink and the uplink is actually three time slot durations minus the TA. An example of the time synchronization from the perspective of the MS is shown in Figure 6.7 [112]. This example assumes that the MS a full-rate traffic channel with frequency hopping on time slot 3. The MS cycles between receiving on the downlink, transmitting, and monitoring the signal quality of adjacent cells. Note that there is no time synchronization between adjacent cells: each BTS has an independent clock.

6.6.4.5 Frequency Hopping

GSM defines a procedure for optional frequency hopping. Frequency hopping is a form of spectral diversity that can be used to compensate for frequency-selective fading caused by multipath propagation. It essentially averages the signal-to-noise ratios (SNRs) of multiple RF channels so that, in theory, no MS is assigned a low-quality RF channel all of the time. In general, at least four hopping frequencies are required to significantly improve performance. The frequency hopping technique used in GSM is said to be "slow frequency hopping" because the RF channel used for transmission changes with every TDMA frame, resulting in a hopping rate of approximately 217 hops per second. This basic procedure is illustrated in Figure 6.7. The exact hopping pattern used for a physical channel is defined by three parameters: the mobile allocation (MA), the hopping sequence generator number (HSN), and the mobile allocation index offset (MAIO). The MA lists the RF channels used, the HSN defines the hopping sequence generator, and the MAIO combined with the current frame number (FN) determines the current position within the hopping sequence. From this perspective, the case of no frequency hopping is actually a special case in which the MA contains a single ARFCN. The decision whether to activate frequency hopping is left to the network and may be decided on a per-cell basis. Since an MS must be capable of functioning in a cell that uses frequency hopping, all MSs are required to implement it.

6.6.5 GSM Physical and Logical Channels

6.6.5.1 *Physical versus Logical Channels*

An important concept in understanding the GSM PHY is the distinction between physical channels and logical channels. A physical channel is formally defined as a sequence of TDMA frames, a timeslot number (TN), and a frequency hopping sequence. A physical channel, therefore, refers to a subset of the physical resources that are used to transmit modulated data through the air. A logical channel, by contrast, refers to a specific type of information that is carried over one or more physical channels. Logical channels may be defined in the downlink direction (i.e., from BTS to MS), the uplink direction (i.e., from MS to BTS), or both. Logical channels are mapped onto physical channels by the network and must be managed (i.e., set up, maintained, torn down). There is not a one-to-one mapping between physical and logical channels, as multiple logical channels may be multiplexed onto a single physical channel. This is particularly true of signaling information sent over the air interface. By making the distinction between physical and logical channels, it is possible to discuss physical characteristics of the PHY that are independent of the type of data being transmitted. Conversely, it is possible to discuss the transmission of certain types of control and user data over the air interface without delving into the details of the time and frequency resources that carry them. The logical channels form the service access points (SAPs) between the PHY and layer 2, with the exception of the frequency correction channel (FCCH) and the synchronization channel (SCH), which exist only at the PHY. Furthermore, the mapping of logical channels from layer 2 onto physical channels is implemented within the PHY. This section describes the various logical channels and then discusses the basic concepts involved in mapping logical channels onto physical channels.

6.6.5.2 *Definition of Logical Channels*

Logical channels are classified based on the type of information they carry. In a broad sense, GSM logical channels, both those supporting CS domain services and PS domain services, can be divided into traffic channels and control channels. Traffic channels are used to carry user data exclusively, while control channels carry the various types of signaling between the MS and BSS. In this sense, traffic channels belong to the *user plane* while control channels belong to the *control plane*. The classification of traffic channels is generally quite simple. The classification of control channels, however, is more complex due to the significant signaling overhead involved in a cellular network. A taxonomy of GSM logical channels is provided in Table 6.6 [112, 126]. For the sake of discussion, the GSM logical channels can be divided into three categories: traffic channels, control channels, and packet data logical channels. This subdivision reflects the fact that GPRS uses a unique structure for mapping of logical channels to physical channels. Each logical channel is defined in the following discussion.

6.6.5.3 Traffic Channels

In GSM networks, the term *traffic channel* is often used to refer specifically to a circuit-switched traffic channel. When used in this sense, the abbreviation TCH is used. The TCH carries the user payload in the CS domain (e.g., voice traffic). Two types of traffic channels are defined: full-rate and half-rate. A full-rate traffic channel (TCH/F) fully utilizes one physical channel providing a gross data rate of 22.8 kbps. A half-rate channel (TCH/H) utilizes a physical channel only half of the time, providing a gross data rate of 11.4 kbps. The advantage of the half-rate channel is that it allows a single physical channel to be allocated to two subscribers simultaneously at a lower data rate. Traffic channels can be further subdivided based on the type of bearer service they support (e.g., EFR speech [TCH/EFS], adaptive full-rate speech [TCH/AFS]). A full list can be found in 3GPP TS 45.002 [112].

6.6.5.4 Control Channels

Cellular systems based on fixed assignment multiple access schemes generally require a large amount of signaling overhead in order to function. Signaling is required, for example, for synchronization, registration to the network, control of the radio link, request and assignment of radio resources, and localization of the MS within the PLMN. As shown in Table 6.6, GSM defines a number of different control channels to carry signaling and synchronization information over the air interface. These control channels can be generally divided into three categories: broadcast channels (BCHs), common control channel (CCCHs), and dedicated control channel (DCCHs). The broadcast channels (i.e., BCCH, FCCH, SCH) are unidirectional channels broadcast by the BTS and monitored by the MSs in the cell. The common control channels (i.e., PCH, RACH, AGCH, NCH) are primarily used for access management functions (e.g., paging of the MS, channel request by the MS, channel assignment by the network) and shared by all MSs in a cell. In combination, these four logical channels are referred to collectively as a CCCH. The dedicated control channels (i.e., SDCCH, SACCH, FACCH) are bidirectional control channels used for point-to-point signaling between the BSS and a single MS [112]. Note that these

Table 6.6 Taxonomy of GSM Logical Channels

General category	Specific category	Channel name	Function	Direction
Traffic channels	Traffic (TCH)	TCH/F	Full-rate traffic channel	BSS ↔ MS
		TCH/H	Half-rate traffic channel	BSS ↔ MS
Control channels	Broadcast	FCCH	Frequency correction	BSS → MS
		SCH	Synchronization	BSS → MS
		BCCH	Broadcast control	BSS → MS
	Common control (CCCH)	PCH	Paging	BSS → MS
		RACH	Random access	BSS ← MS
		AGCH	Access grant	BSS → MS
		NCH	Notification (voice group)	BSS → MS
	Dedicated control (DCCH)	SDCCH	Stand-alone dedicated control	BSS ↔ MS
		SACCH	Slow associated control	BSS ↔ MS
		FACCH	Fast associated control	BSS ↔ MS
	SMS-CB	CBCH	SMS cell broadcast	BSS → MS
Packet data logical channels	Packet data traffic (PDTCH)	PDTCH/D	Downlink packet data traffic	BSS → MS
		PDTCH/U	Uplink packet data traffic	BSS ← MS
	Packet broadcast	PBCCH	Packet broadcast control	BSS → MS
	Packet common control (PCCCH)	PPCH	Packet paging	BSS → MS
		PRACH	Packet random access	BSS ← MS
		PAGCH	Packet access grant	BSS → MS
		PNCH	Packet notification (IP multicast)	BSS → MS
	Packet dedicated control (PDCCH)	PACCH	Packet associated control	BSS ↔ MS
		PTCCH/D	Packet timing advance control (downlink)	BSS → MS
		PTCCH/U	Packet timing advance control (uplink)	BSS ← MS

control channels support various signaling functions of the air interface that are required to support both CS domain and PS domain services.

The specific functions of each of these control channels are summarized as follows:

- *Frequency Correction Channel (FCCH).* The FCCH carries information for frequency synchronization of the MS. By transmitting an unmodulated carrier at a fixed frequency offset from the BCCH carrier, the FCCH provides a sort of beacon that the MS can use to correct its frequency synchronization. In essence, the FCCH can be more accurately thought of as a physical layer signal, as it does not carry any binary data. The FCCH is used only by the PHY and is not visible to the upper layers.

- *Synchronization Channel (SCH).* The SCH carries time synchronization information and two parameters: the reduced frame number (RFN) and the BTS identity code (BSIC). These parameters are used for frame synchronization and identification of the BTS, respectively. The SCH is used only by the PHY and is not visible to the upper layers.

- *Broadcast Control Channel (BCCH).* Once the MS has synchronized to the system, it still needs information on the configuration of the network in order to register. This information is broadcast on the BCCH in a series of System Information messages, which contain information such as the number of ARFCNs in the current cell assignment (CA), the configuration of the CCCHs, and whether the BTS supports GPRS, as well as basic information about adjacent cells. The BCCH is always broadcast on the first frequency of the CA, which is called the BCCH carrier.

- *Paging Channel (PCH).* The PCH is a downlink channel used by the network to page the MS (e.g., for an incoming call). It is analogous to a loudspeaker paging system in an office building that can be used to locate an employee (e.g., to notify them that a package has arrived).

- *Random Access Channel (RACH).* The RACH is an uplink channel that is used by the MS to request a dedicated control channel (SDCCH) for point-to-point signaling with the network. It is used, for example, to initiate an outgoing call. It is the only uplink control channel that can be accessed by the MS without dedicated resources assigned. The RACH is unique in GSM, as it is the only channel that uses a random access scheme rather than fixed assignment. The random access scheme of the RACH is based upon the principle of slotted ALOHA.

- *Access Grant Channel (AGCH).* The AGCH channel is a downlink channel that is used to assign an SDCCH or a TCH to an MS in response to a channel request on the RACH.

- *Notification Channel (NCH).* The NCH channel is a downlink channel that is used to notify the MS of incoming voice group and voice broadcast calls.

- *Stand-Alone Dedicated Control Channel (SDCCH).* The SDCCH is used for dedicated signaling with a single MS when there is no active connection in the user plane. It is "stand-alone" in the sense that it is not associated with the existence of a TCH. The SDCCH is requested by the MS via the RACH, assigned via the AGCH, and released at the end of the signaling transaction. Examples of transactions which use the SDCCH include MS location update and connection setup.

- *Slow Associated Control Channel (SACCH).* The SACCH is used for the periodic transmission of control information during a connection, such as adaptive frame control (i.e., timing advance), adaptive power control, and channel quality measurements. A SACCH is always assigned in association with a TCH or an SDCCH.

- *Fast Associated Control Channel (FACCH).* The FACCH is a control channel that is used for occasional fast signaling associated with a TCH, for example at connection setup or during handover. The FACCH is obtained through preemptive dynamic multiplexing, whereby the physical channel of a TCH is temporarily "stolen" for signaling at the expense of user data.

- *Cell Broadcast Channel (CBCH).* The CBCH is defined on the downlink to support the short message service cell broadcast (SMSCB), which differs from the normal short message service (SMS) in that it can be used to broadcast massages to many MSs within a cell (e.g., to distribute emergency information or, perhaps more commonly, advertisements from the network provider). The CBCH carries both signaling and user payload. The CBCH is optional and, if used, uses the same physical channel as the SDCCH [112]

6.6.5.5 Packet Data Logical Channels

An additional set of logical channels is defined to support GPRS operation over the air interface. The categories of packet data logical channels are similar to those previously described and include traffic channels (i.e., PDTCH) and control channels. The control channels are similarly subdivided into broadcast (i.e., PBCCH), common control (i.e., PCCCH), and dedicated control (i.e., PDCCH) channels. However, the use of the PBCCH and PCCCH channels is optional. If the PBCCH or PCCCH are not allocated, the corresponding signaling for PS operation is carried by the BCCH or the CCCH, respectively. The specific functions of the PDCHs are summarized as follows:

- *Packet Broadcast Control Channel (PBCCH).* Broadcasts parameters used by the MS for PS domain operation. BCCH information is reproduced on the PBCCH such that GPRS Class A and Class B mobiles in GPRS attached mode need only monitor the PBCCH. The PBCCH is optional and its existence is indicated on the BCCH.

- *Packet Common Control Channel (PCCCH).* Analogous to the CCCH, the PCCCH comprises a PPCH, PRACH, PAGCH, and PNCH. If allocated, a PCCCH may transmit information for CS operation as well (e.g., paging a GPRS Class A or Class B MS in GPRS attached mode for incoming voice call).
  - ○ *Packet Paging Channel (PPCH).* Downlink channel used to page MSs; analogous to the PCH.
  - ○ *Packet Random Access Channel (PRACH).* Uplink channel used to request assignment of one or more PDTCHs; analogous to the RACH.
  - ○ *Packet Access Grant Channel (PAGCH).* Downlink channel used to allocate one or more PDTCHs; analogous to the AGCH.

○ *Packet Notification Channel (PNCH).* Downlink channel used to notify MSs operating in point-to-multipoint mode, for example, for IP multicast; analogous to the NCH.

- *Packet Associated Control Channel (PACCH).* The PACCH is a dedicated bidirectional signaling channel associated with an active PDTCH; analogous to the SACCH.

- *Packet Timing Advance Control Channel Uplink (PTCCH/U).* The PTCCH/U is an uplink channel used by the MS to transmit random access bursts (ABs) to the BTS for estimation of timing advance during packet transfer mode (i.e., MS is transmitting on one or more PDTCH/Us). One PTCCH/U is utilized by a single MS to avoid collision.

- *Packet Timing Advance Control Channel Downlink (PTCCH/D).* The PTCCH/D is a downlink channel used by the BTS to transmit timing advance updates to an MS in packet transfer mode. One PTCHCH/D is paired with several PTCCH/Us.

6.6.5.6 Mapping of Logical Channels to Physical Channels

GSM offers significant flexibility in the mapping of logical channels to physical channels. This makes it adaptive to a variety of deployment scenarios and configurable for network planning and optimization based on anticipated volume of user traffic. It also allows for dynamic allocation of resources. For example, if data traffic is high but voice traffic is low, the network may choose to allocate more physical channels to carry packet data logical channels. The GSM air interface was designed this way to make it equally suited to deployment in downtown London as to the Italian countryside, and all points in between. It was also done to make GSM "future proof," by allowing it to adapt to deployment scenarios that may have been unforeseen to GSM's creators [51, p. 548]. As a direct consequence, there is not a one-to-one mapping between physical channels and logical channels. There are, however, a certain set of rules that must be followed.

6.6.5.7 Channel Mapping in Frequency

Channel mapping comprises two components: mapping in frequency and mapping in time. In frequency, logical channel mapping is a function of the frame number, the frequency allocation of the BTS (i.e., the cell allocation [CA]), the frequency allocation of MS (i.e., the mobile allocation [MA]), and the frequency hopping sequence described previously. One carrier frequency of the CA is designated the BCCH carrier. This may be any single carrier within the CA, as no particular ordering with respect to carrier frequency is stipulated by the standard [112]. The BCCH carrier is unique as it is used to carry the FCCH, SCH, and BCCH. These logical channels are always carried on TN0 of the BCCH carrier, for which no frequency hopping is allowed. The BCCH carrier forms a beacon, allowing the MS to register to the network and make power measurements for the purpose of cell selection and reselection. For this reason, the BTS is required to transmit on every time slot of the BCCH carrier, using dummy bursts if necessary to fill time slots that are not used to carry any other user traffic or signaling information.

6.6.5.8 Channel Mapping in Time

In the time domain, logical channel mapping is based on a set of repeating, hierarchical structures that form logical groupings of TDMA frames. The organization of TDMA frame structures is illustrated in Figure 6.8 [127]. Note that each time structure shown is cyclic. In addition to creating a framework for the mapping of logical channels, this frame structure was chosen for a combination of time-synchronization and cryptographic reasons. From the bottom up, the GSM frame structure is defined as follows: TDMA frames are grouped into multiframes, of which there are three types:

- *51-Multiframe.* Comprised of 51 TDMA frames with a duration of 235.4 ms, this is used to transmit most types of control channels, including BCCH, CCCH (i.e., PCH, RACH, AGCH, NCH), SDCCH (and associated SACCH), and CBCH (optional).
- *26-Multiframe.* Comprised of 26 TDMA frames with a duration of 120 ms, this is used to transmit traffic channels in CS domain (TCH) and associated control channels (i.e., SACCH and FACCH).
- *52-Multiframe.* Comprised of 52 TDMA frames with a duration of 240 ms, this is used exclusively to transmit packet data logical channels of all types. The 52-multiframe, which occupies the space of two 26-multiframes, was added with the introduction of GPRS as a way to integrate the packet data service into the existing GSM air interface. As will be discussed later, the manner in which GPRS uses the physical resources of the 52-multiframe are fundamentally different than the 51-multiframe and 26-multiframe.

Since 51 and 26 are coprimes, these two types of multiframes will synchronize every $51 \times 26 = 1326$ TDMA frames. This structure is referred to as a superframe and has a short periodicity of 6 seconds and 120 ms. The structure of the superframe, as the superposition of two coprimes, is designed to assist in time synchronization.

The top level of the hierarchy is the hyperframe, which comprises 2048 superframes and has a long periodicity of almost 3 hours and 30 minutes. Each TDMA frame within a hyperframe is assigned a unique frame number. The frame number is cyclic and repeats with the long period of the hyperframe. The need for a hyperframe naturally arises from the encryption process, which uses the FN as an input parameter. The long duration of the hyperframe, which is substantially longer than a superframe, makes it more difficult for hackers to intercept a data stream.

In order to understand the time structure hierarchy of GSM, it is important to keep in mind that the multiframe structure is defined with respect to *one time slot* of a TDMA frame, not all the time slots, as Figure 6.8 may seem to imply. For example, the physical channel corresponding to time slot 0 of the TDMA frame may

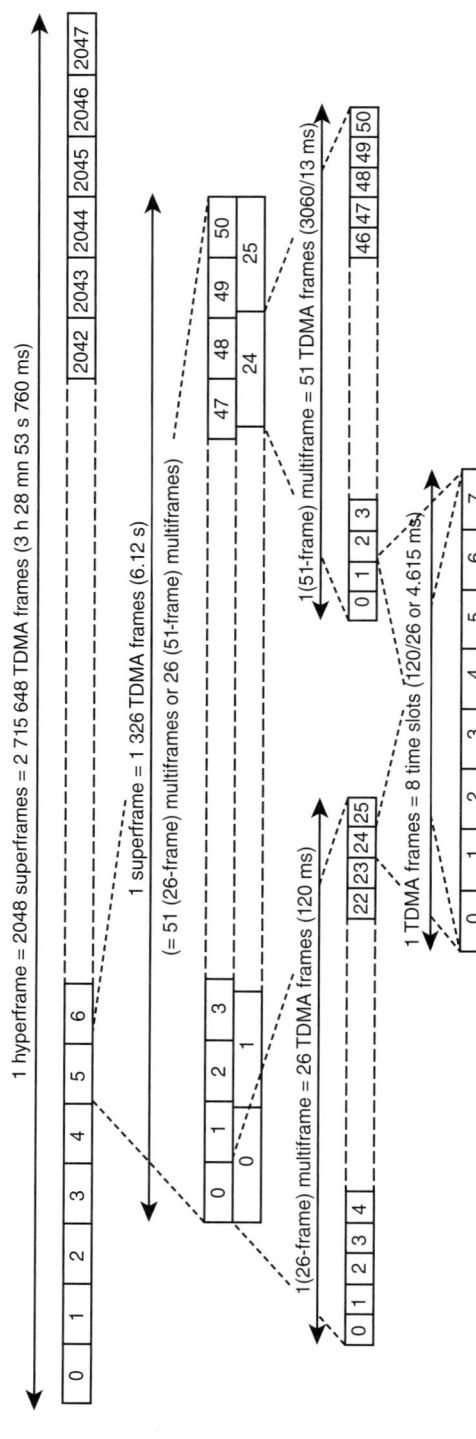

Figure 6.8 Hierarchy of GSM time-domain frame structures. Reprinted from Reference [127].

be used to carry a combination of control channels (e.g., FCCH, FCCH, SCH, CCCH, and SDCCH). Therefore, time slot 0 of successive TDMA frames will be organized into 51-multiframes. At the same time, the physical channels corresponding to time slots 1 through 4 of the TDMA frame may be used to carry four full-rate TCHs. In this case, time slots 1 through 4 will each carry a TCH/F that is organized into 26-multiframes. Finally, time slots 5 through 7 may be allocated for GPRS use. They will therefore carry all required packet data logical channels, both signaling and traffic, which are organized into 52-multiframes.

The multiframe structures provide a framework for channel mapping in the time domain, but additional rules are required to define how the different logical channels may be mapped to the multiframe structure. In general, each physical channel (i.e., one time slot on success frames) is not used for a single type of logical channel all of the time, but rather is shared by various logical channels which are transmitted at various intervals. Restrictions are placed on which combinations of logical channels may be combined onto a single physical channel. The permitted channel combinations (CC) that may be multiplexed onto a single physical channel are summarized in Table 6.7, where shaded areas represent legal channel combination-logical channel pairings. The bottom row indicates the multiframe type used for each CC of the corresponding column. The detailed list of rules used for mapping logical channels to physical channels is beyond the scope of this discussion. The interested reader may find these details in 3GPP technical specifications 45.002 and 43.064.

Table 6.7 Permitted GSM/GPRS Channel Combinations

Logical channels	Channel combinations										
	CC1	CC2	CC3	CC4	CC5	CC6	CC7	CC8	CC9	CC10	CC11
TCH/F	▓										
TCH/H		▓									
TCH/H			▓								
FCCH				▓	▓	▓	▓				
SCH				▓	▓	▓	▓				
BCCH				▓	▓	▓	▓				
CCCH				▓	▓	▓	▓				
SDCCH					▓	▓	▓				
SACCH	▓	▓	▓		▓	▓	▓				
FACCH	▓	▓	▓								
PDTCH								▓	▓	▓	▓
PBCCH								▓	▓		▓
PCCCH								▓	▓	▓	
PACCH								▓	▓	▓	▓
PTCCH								▓	▓	▓	▓
Multiframe type	26	26	26	51	51	51	51	52	52	52	52

6.6.5.9 Channel Mapping of Packet Data Channels

Notice that the packet data logical channels are not mixed with the other logical channel types. This is due to the fact that GPRS was introduced as an overlay to the existing circuit-switched GSM infrastructure. Although the GPRS physical layer architecture on the air interface is similar to and compatible with that of the original GSM architecture, the manner in which GPRS utilizes the physical resources of the air interface is distinct from traditional circuit-switched GSM and optimized for packet-switched services. Any physical channel that has been allocated to carry packet data logical channels is referred to as a PDCH. The PDCH resources are managed by the PCU in the BSS and carry only packet data logical channels. A key concept in mapping of packet logical channels to PDCHs in GSM/GPRS networks is that the physical resources of the air interface are dynamically allocated based on the current traffic demands of the cell. If demand for voice services is low, any unused channels may be allocated as PDCHs. However, CS domain resources take precedence due to the time-sensitive nature of voice service. This means that if demand for voice calls suddenly rises, a PDCH may be reallocated to carry CS traffic (i.e., TCH). In this case, each MS using the PDCH is notified via the PACCH and reassigned to a different PDCH if available. This concept is called *capacity on demand*.

PDCHs are organized using 52-multiframes. The overall structure of the 52-multiframe is divided into 12 *radio blocks*, each of which comprises four frames, along with two PTCCH frames and two idle frames. This frame structure is illustrated in Figure 6.9 [128]. Recall that the multiframe structure is defined with reference to a single time slot. Therefore, a radio block is physically realized as four radio bursts distributed over four consecutive TDMA frames, where each burst is transmitted on the same timeslot number in each frame.

Recall that the use of the PBCCH and PCCCH is optional in GPRS. When data traffic is low or if the network architecture demands it, broadcast and common control signaling required for GPRS operation may be transmitted on the BCCH and CCCH. In this case, all GPRS-attached MSs in the cell camp on the CCCH, which is used for all channel access signaling. When demand for data traffic increases or if extra resources are available, the PCCCH and (optionally) the PBCCH may be allocated in the cell. In this case, all GPRS-attached MSs camp on the PCCCH and the CCCH is not used. As demand for data services continues to increase, more PCCCHs may be allocated if resources are available and removed if demand wanes.

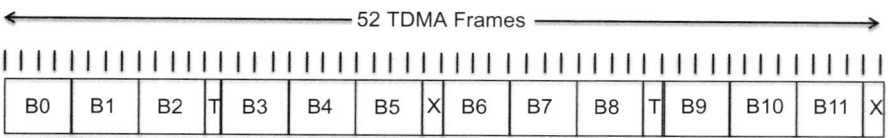

Figure 6.9   Organization of 52-multiframe structure for packet data channels. Reprinted from Reference [128]. X, idle frame; T, frame used for PTCCH; B0–B11, radio blocks.

If the PCCCH is used, the mapping of logical channels to PDCHs utilizes a *master–slave* relationship. One or more PDCHs that carry the PCCCH are designated master channels (i.e., CC8, CC9, and CC11 from Table 6.7). These channels carry packet control channels and, optionally, PDTCH(s). The remaining PDCHs, designated slave channels, carry only packet data traffic (PDTCH) and associated signaling (PACCH, PTCCH). The MS then uses master channels in packet idle mode and transfers data on master and/or slave channels.

6.7 GSM PHYSICAL LAYER

The GSM physical layer constitutes layer 1 of the OSI reference model and realizes all functions necessary for transmission of logical channels over the physical resources of the air interface. Although different mechanisms and protocol architecture are used at the air interface to realize circuit-switched services in GSM and packet-switched services based on GPRS/EGPRS, the configuration of the transmission chain used at the physical layer is fundamentally similar for both. Therefore, an understanding of the GSM physical layer is essential to understanding GPRS and EGPRS as well. This section briefly describes the GSM physical layer. Physical layer modifications and enhancements employed in GPRS and EGPRS are discussed later in this chapter in Sections 6.9.7 and 6.11, respectively.

6.7.1 PHY Reference Configuration

The GSM physical layer reference architecture is shown in Figure 6.10. This block diagram illustrates the reference configuration of the transmission chain used to transmit logical channels over physical channels. On the transmission side, the diagram begins with information bits, which may be user data (e.g., digitally encoded speech) or signaling (e.g., information carried on a BCCH, CCCH, or DCCH). The information bits first undergo channel coding for error detection and correction. This includes two codes, an outer block code, and an inner convolutional code, as well

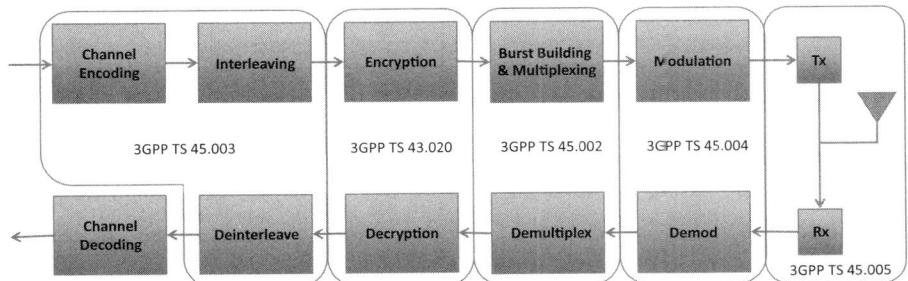

Figure 6.10   GSM physical layer reference configuration.

as interleaving. Next, the channel-coded bits undergo encryption.[12] The encrypted bits are then divided up and placed into a burst structure depending on the type of information that is being carried for transmission over a single TDMA time slot. The bursts are then modulated and transmitted over the channel. On the receive side, the received signals undergo an inverse process in order to recover the original information bits. The 3GPP specification numbers corresponding to each block of the reference diagram are indicated. This numbering scheme applies to Release 4 and beyond.

6.7.2 Channel Coding and Interleaving

GSM uses a combination of block coding, convolutional coding, and interleaving for the detection and correction of bit errors-caused channel impairments over the radio link. This process corresponds to the first two blocks of the transmission chain reference architecture shown in Figure 6.10. Although each logical channel has its own unique coding and interleaving scheme, the basic organization is shared to allow for a single unified decoder structure. The generic channel coding structure is shown in Figure 6.11. It comprises two serially concatenated codes followed by an interleaver. The outer block code is used purely for error detection. A cyclic redundancy check (CRC) is used to carry user data on the TCH. The majority of the control channels, however, use an extremely powerful fire code to ensure very low probability of an erroneous control message being passed to the upper layers. Specifically, the fire code used in GSM has a probability of 2–40 that it will fail to detect a block containing an error [114]. All logical channels use a rate 1/2 convolutional code for forward error correction (FEC).

In most cases, the output of the channel encoder results in coded blocks of length 456 bits, which is exactly four times the number of payload bits carried by a normal burst. These coded blocks can then be broken up and mapped onto the payload

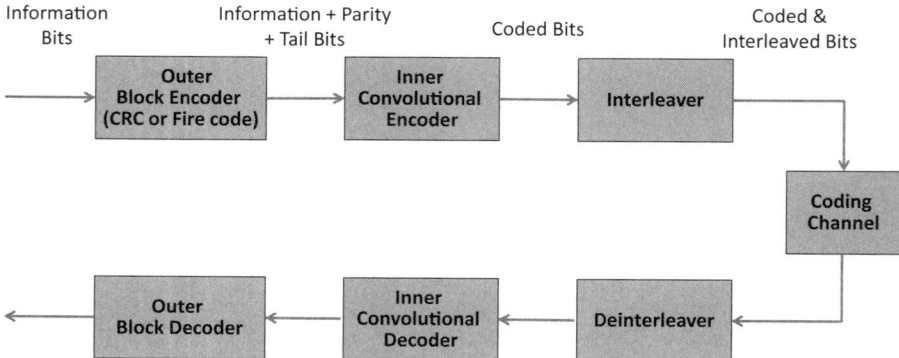

Figure 6.11   GSM channel coding architecture.

---

[12] GSM encryption is discussed later in Section 6.10.

Table 6.8 GSM Channel Coding Parameters

Logical channel type	Block code	Block code bits per block			Convolutional code rate	Encoded bits per block	Interleaver depth
		Data	Parity	Tail			
TCH/EFS	CRC	244				456	8
Class I bits		170	15	4	1/2	378	
Class II bits		74	4	0	–	78	
FACCH/F (full rate)	Fire	184	40	4	1/2	456	8
FACCH/H (half-rate)	Fire	184	40	4	1/2	456	6
SDCCH, SACCH, BCCH, NCH, AGCH, PCH, CBCH	Fire	184	40	4	1/2	456	4
RACH	CRC	8	6	4	1/2	36	1
SCH	CRC	25	10	4	1/2	78	1

portion of the bursts. The exceptions to the 456-bit encoded block length rule are the RACH and SCH, which are encoded into blocks that match the payload size of the access burst (AB) and synchronization burst (SB), respectively. Channel coding parameters for different logical channels are summarized in Table 6.8. The TCHs use different coding parameters depending on the bearer service they support. As an example, EFR speech (TCH/EFS) is shown. It should be noted that coding of speech data utilizes a hierarchical approach. The bits that are most critical for reconstruction of speech at the receiver are called Class I bits, while those that are less important for our perception of speech quality are called Class II bits. This allows for more efficient use of radio resources by applying less channel coding to Class II bits. A full list of channel coding parameters for various types of traffic channels can be found in 3GPP Technical Standard 45.001, clause 7.

6.7.3 Burst Building

A burst is a period of RF carrier modulated by a bit stream. It represents the physical contents of a time slot. The process of burst building described in this section corresponds to the "burst building and multiplexing" block of the reference architecture shown in Figure 6.10. GSM defines five different burst formats. The burst format used depends on the logical channel it carries. For this reason, the formal definition of a physical channel does not include a burst format, since there is not a one-to-one

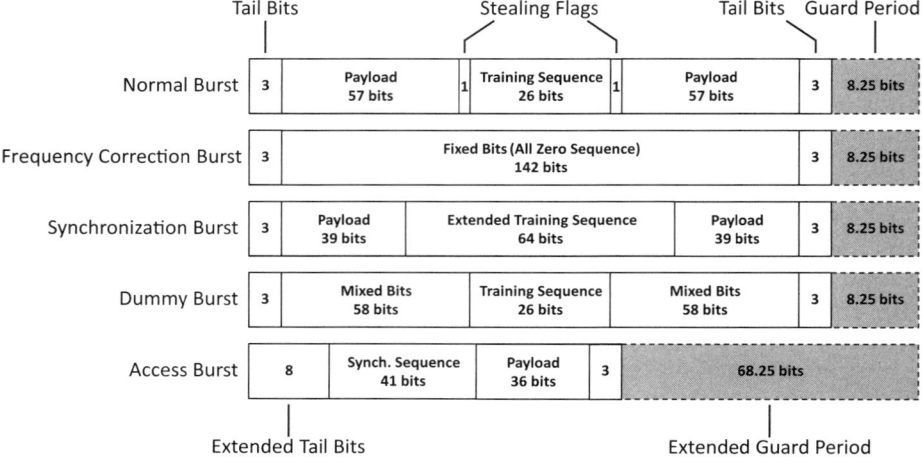

Figure 6.12   GSM burst formats. Adapted from Reference [112].

mapping between physical channels and the type of logical channels. The five different burst formats are shown in Figure 6.12 [112] and described as follows:

- *Normal Burst (NB).* As the name implies, the NB is the most common burst. It is used to carry the payload of all TCHs and most control channels, with the exception of the FCCH, SCH, and RACH, which use their own specialized bursts. At the beginning and end of the NB are three tail bits. These bits do not carry data, but are included to compensate for potential timing inaccuracies [129, p. 15] or switching transient leakage into the beginning or end of the burst due to RF power ramping [130, pp. 29–30]. The tail bits are also used by the demodulator to identify the start of the burst [114, p. 68]. In the center of the burst is a 26-bit training sequence that is used for channel quality estimation and Viterbi equalization to combat intersymbol interference (ISI) caused by multipath propagation in the mobile environment. The training sequence is chosen from a set of eight sequences that are known to both the MS and the BTS [112]. The payload of the NB comprises 114 channel-encoded, encrypted bits carried in two 57-bit blocks. A single bit on either side of the training sequence are used as stealing flags, which indicate whether the burst is used for traffic or signaling (e.g., FACCH signaling) [114, p. 68]. The remaining portion of the time slot forms a guard period.

- *Frequency Correction Burst (FB).* The FB is used for frequency synchronization of the MS. It has the same duration and guard period as the NB, however it consists of a series of 148 zeros (counting tail bits). This results in an unmodulated carrier at a fixed frequency offset of 67.7 kHz above the BCCH carrier frequency. The repeated transmission of the FB forms the FCCH [127, p. 15].

- **Synchronization Burst (SB).** The SB is used for time synchronization of the MS. It features an extended training sequence and carries the RFN and BSIC. Since the MS uses the SB during initial synchronization with the BTS, a single known training sequence is defined for the SB. The repeated transmission of the SB forms the SCH [127, p. 15].

- **Dummy Burst (DB).** The BTS must transmit a burst on every time slot in the downlink of the BCCH carrier in order for the MSs to allow MSs to perform power measurements [112, p. 55]. For this purpose, the DB is defined for transmission when no other burst is sent. It is composed of a predefined set of mixed bits and is transmitted on the BCCH carrier only [112, p. 28]

- **Access Burst (AB).** The AP is transmitted by the MS in the uplink for random access on the RACH. It has a fundamentally different structure from the other bursts due to its specialized function. Because an MS transmitting on the RACH has not yet been assigned a timing advance, the AB features an extended guard period of 68.25 bit periods to avoid collision with the adjacent time slot. This guard period dictates the maximum propagation delay that can be tolerated within a single cell, which allows for a maximum cell radius of approximately 35 km [127, p. 15].

In general, the guard period of 8.25 bit periods is specified to allow time for RF ramping of the transmitter's power amplifier without creating interference for adjacent time slots (with the exception of the AB). This is particularly applicable to the MS, which typically only transmits on one physical channel in the uplink at any given time. However, since the BTS generally transmits on many physical channels in the downlink consecutively, it would be inefficient to try to ramp power up and down between consecutive bursts. Therefore, the BTS is only required to perform power ramping in the downlink for unused time slots [112, p. 29].

6.7.4 Modulation

GSM uses Gaussian minimum-shift keying (GMSK) modulation to transmit bits over the radio link. The details of GMSK were described previously in the "Bluetooth" section of Chapter 3. The choice of GMSK modulation in GSM is an interesting one, as it is one of the few basic parameters of the GSM air interface that was chosen largely for political reasons—it is the result of a compromise made to acquire the political approval of France and Germany—rather than by pure technical consensus of the GSM plenary group. GMSK is a constant envelope modulation scheme that has low adjacent channel interference, making it well suited to the cellular radio application. The implementation of GMSK used in GSM utilizes differential encoding and a linear filter parameter BT = 0.3, where B is the 3 dB bandwidth of the filter and T is the bit duration.

In addition to GMSK, higher-order modulation schemes have been added in subsequent GSM enhancements to improve spectral efficiency and increase peak data rates over the GSM air interface. In EDGE (R99 and beyond), 8PSK is introduced along with adaptive modulation and coding. Further higher-order modulation has been added as part of the Rel-7 EDGE Evolution enhancement called EGPRS2. These include QPSK, 16-QAM, and 32-QAM. Another form of modulation called adaptive QPSK (AQPSK) has been introduced in Rel-9 as part of the VAMOS enhancement, which allows two users to use the same physical channel simultaneously for circuit-switched voice traffic. EDGE, EDGE evolution, and VAMOS enhancements are all discussed in greater detail later in this chapter.

6.7.5   Additional PHY Functions

Additional functions of the GSM physical layer are briefly summarized in the following list. These include mechanisms required for the physical establishment, control, and maintenance of radio links over the air interface.

- *Power Control.* The radiated power levels of the MS and (optionally) the BTS are adjusted over time to ensure link quality while minimizing output power. This is done for efficiency and to reduce interference levels in the network. GSM power control is defined in 2 dB increments.

- *Synchronization.* The MS must synchronize to the network both in frequency and time—including time slot and frame number synchronization. This is performed using the FCCH and SCH described previously, the generation and reception of which are defined at the physical layer.

- *Cell Selection and Reselection.* At power up, the MS performs measurements to determine the best available cell and then reads the BCCH so that it may register to the network. In idle mode (i.e., no radio resources allocated), the MS performs measurements of adjacent BCCHs to determine when it should listen to a different cell. These processes are called cell selection and cell reselection and are controlled by the MS. When an idle MS selects a cell and listens for its TMSI on the PCH, this is called *cell camping*. Link maintenance mechanisms are defined at the physical layer to ensure that a mobile camped on a cell can reliably communicate on the uplink and downlink.

- *Handover.* In dedicated mode (i.e., radio resources are allocated, for example, for an active call), handover is performed to maintain call quality. Handover may be performed between two cells (i.e., intercell handover) or between two channels belonging to a single cell (i.e., intracell handover). Handover in GSM is always controlled by the network, but is based on channel quality measurement reports that are communicated to the BTS on a continual basis by the MS via the SACCH.

6.8   GSM SIGNALING AT THE AIR INTERFACE

The GSM layered protocol architecture in the CS domain can be subdivided into the user plane and the control plane. The user plane is relatively simple, as is to be

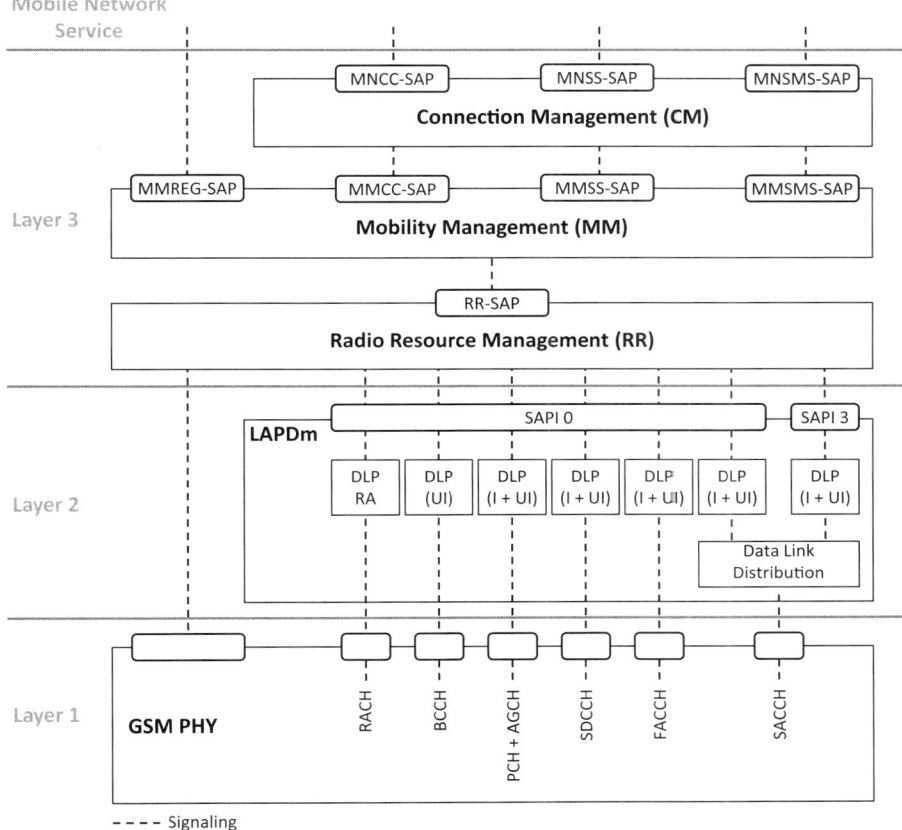

Figure 6.13   GSM layered control plane architecture in the MS.

expected for a circuit-switched service. Consider the transmission of voice traffic as the most common example. On the MS side, the voice traffic is exchanged directly between the GSM speech codec and the physical layer in the form of digitally encoded speech frames. These frames are passed through the physical layer transmission chain described previously and received at the BTS. Within the BSS, the TRAU converts speech frames between a GSM-specific speech format used on the air interface and an ISDN standard format used in the core network. The control plane, however, is a little more complex, as it must support all the signaling mechanisms required for the transmission of user data over the air interface. These mechanisms include registration, authentication, call establishment, call maintenance, and mobility handling. The layered architecture of the GSM control plane on the MS side is illustrated in Figure 6.13. GSM signaling at the air interface is divided into layer 2 and layer 3, which are briefly described in the remainder of this section.

6.8.1   GSM Signaling Layer 2: LAPDm

LAPDm is a GSM-specific protocol that constitutes the data link layer (i.e., layer 2) of the GSM control plane at the air interface. The layer 2 signaling terminates in the BTS on the network. The main purpose of LAPDm is to convey layer 3 signaling information across the air interface. The LAPDm protocol is based primarily on two data link protocols from fixed networks: the High-Level Data Link Control (HDLC) [131] protocol defined by the ISO and the Link Access Protocol Channel D (LAPD) protocol [132] defined for the ISDN. As the name implies, LAPDm is derived from the LAPD protocol but has been optimized for the specific characteristics of the wireless radio link in a mobile network.

LAPDm supports two modes of operation: acknowledged mode and unacknowledged mode. In acknowledged mode, layer 3 information is sent in numbered Information (I) frames between peer entities. The I frames are acknowledged by the receiver and passed in order to the upper layers. Flow control and error recovery mechanisms are also defined in acknowledged mode. In unacknowledged mode, layer 3 information is sent in Unnumbered Information (UI) frames between peer entities. The UI frames are not acknowledged by the receiver. Furthermore, flow control and error recovery mechanisms are not defined n unacknowledged mode. In HDLC terminology, acknowledged mode of operation is only supported once Asynchronous Balanced Mode (ABM) has been established between two peer entities. The BCCH, PCH, and AGCH logical channels are always transmitted in unacknowledged mode. The SDCCH, SACCH, and FACCH logical channels support both modes of operation.

The LAPDm defines two SAPs, identified by the so-called service access point identifier (SAPI), to provide services to layer 3 above. SAPI equal to 0 is used to carry nearly all layer 3 signaling (e.g., radio resource management, mobility management, call control). However, a second SAP corresponding to SAPI equals 3 is used to carry SMS messages, as these messages are lower priority. Note that each layer 2 SAP is associated with one or more data link connection end points. Each data link connection is controlled by an instance of the data link procedure and is uniquely identified by the combination of SAPI and type of control channel, the combination of which is referred to as the data link connection identifier (DLCI).

6.8.2   GSM Signaling Layer 3

GSM layer 3 signaling at the air interface is divided into three sublayers: radio resource management (RR), mobility management (MM), and call management (CM).

6.8.2.1   Radio Resource Management

The radio resource management (RR) sublayer is responsible for functions related to the establishment, maintenance, and release of radio resources. This includes, for example, paging procedures, radio link measurement reporting, and handover mechanisms. The RR sublayer is also responsible for generation and monitoring of the BCCH. The RR sublayer terminates on the network side in the BSC. A SAP is defined between the RR and the PHY that allows for the control of physical resources. The behavior of the RR depends on the current RR operating mode. For GSM CS services, two RR operating modes are defined.

- ***Idle Mode.*** In this mode, the MS has not been allocated any dedicated resources for CS services and no RR connection exists The MS chooses the cell with the highest quality signal and monitors the BCCH and CCCH from this cell, ensuring that uplink and downlink communication is possible. This procedure is called *cell camping*. Cell reselection is controlled by the MS and does not require notification of the network unless the target cell is in a new LA.

- ***Dedicated Mode.*** In this mode, the MS has been allocated dedicated radio resources for CS service (e.g., voice) and an RR connection exists. This includes two dedicated logical channels. One channel is either a TCH or an SDCCH, and the other channel is an SACCH. Since dedicated resources are required, movement to a new target cell requires handover procedures. Handover in GSM is always initiated by the network and assisted by the MS.

The RR mode state model is shown in the left part of Figure 6.14. The right part of the figure summarizes the corresponding operations of the GSM physical layer in the MS for each RR operating mode [133]. There are five states defined in the physical layer of the MS. The NULL state is when the equipment is powered off. The SEARCHING BCH state is when the PHY is searching for the best BCCH. The BCH state is when the MS is camped on a cell and listening to the BCCH and CCCH. The TUNING DCH state is when the physical layer seizes a dedicated physical channel. Finally, the DCH state is when the physical layer is operating on a dedicated physical channel.

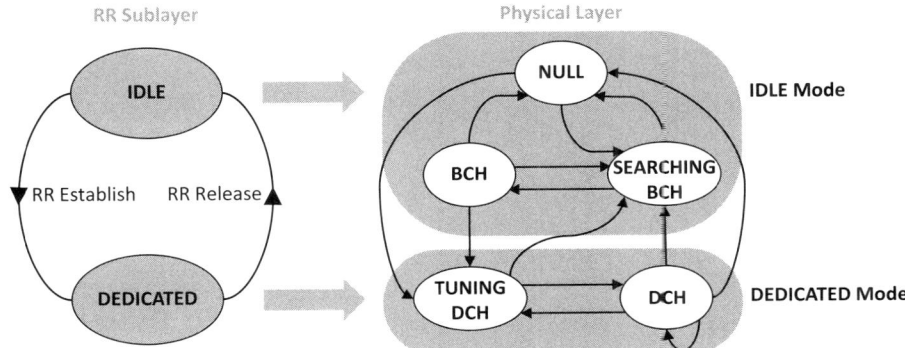

Figure 6.14   GSM RR state model and corresponding physical layer states in the MS.

6.8.2.2 Mobility Management

The mobility management (MM) sublayer is responsible for associated to user mobility, including registration (i.e., attach/detach procedures), authentication, TMSI (re) allocation, and location updating. The MM sublayer terminates on the network side in the MSC. The MM sublayer defines four SAPs to provide services to the upper layers. The MMREG-SAP is defined to allow the upper layers to initiate the registration procedure, known as IMSI attach, which is used to register to MS with the network for CS service. The remaining three SAPs provide an interface with each of the three entities of the CM sublayer.

6.8.2.3 Call Management

The call management (CM) sublayer is responsible for call handling, supplementary services, and the transmission of SMS messages. The CM sublayer terminates on the network side in the MSC. The CM sublayer includes three entities: call control (CC), supplementary service (SS), and short message service (SMS). Each of these entities has an associated SAP to provide its services to the upper layers.

6.9 GPRS OVERVIEW

This section addresses main concepts in the operation of GPRS networks. GPRS is designed to provide the user with PS connectivity to public networks, namely the Internet, as well as private networks, such as corporate intranets. The logical architecture of a GPRS network was previously introduced in the "GSM Network Architecture" section (Section 6.6.2). Table 6.9 provides a mapping of the various functions of a GPRS network to the logical entities, where shaded gray areas represent which logical entities play a role in various network functions and the black areas represent the sum (logical OR) of the gray boxes.

6.9.1 GPRS Attach

Before the MS can access the PS domain services, it must perform a GPRS Attach, which can be thought of as "logging on" to the GPRS system. The GPRS attach serves two purposes. First, it notifies the SGSN that the MS is present and desires access to the GPRS services. Second, it indicates the location of the MS to the network. After the GPRS Attach procedure, the SGSN is aware of the current RA and cell of the MS and records this information in the mobility management (MM) context associated with the MS. At the air interface, the GPRS Attach procedure is controlled by the GPRS mobility management (GMM) sublayer.

The GPRS Attach Procedure is illustrated in Figure 6.15 [115]. This procedure could take place, for example, when the MS is powered on. For this example, it is assumed that the MS has at some point previously attached to the network (i.e., this is not the first power on). Furthermore, it is assumed that the SGSN responsible for

Table 6.9 Mapping of GPRS Functions to Logical Entities

Function	MS	BSS	SGSN	GGSN	HLR
Network access control	■	■	■	■	■
Registration					▓
Authentication and authorization	▓		▓		▓
Admission control	▓		▓		
Message screening				▓	
Packet terminal adaptation	▓				
Charging data collection			▓	▓	
Packet routing and transfer	■	■	■	■	■
Relay	▓	▓	▓	▓	
Routing	▓	▓	▓	▓	
Address translation and mapping			▓	▓	
Encapsulation	▓		▓	▓	
Tunneling			▓	▓	
Compression	▓		▓		
Ciphering	▓		▓		▓
Mobility management	■		■	■	
Logical link management:	■		■		
Logical link establishment	▓		▓		
Logical link maintenance	▓		▓		
Logical link release	▓		▓		
Radio resource management	■	■			

the current routing area (i.e., New SGSN) is different from the SGSN that assigned the last valid P-TMSI (i.e., Old SGSN). The GPRS Attach Procedure is initiated when the MS sends an Attach Request to the New SGSN. The Attach Request provides key information to the SGSN, including:

- **MS Identity.** Typically the last valid P-TMSI and the old RAI in which it was assigned. Note that the old RAI must be sent because the P-TMSI has only local significance. Alternatively, if the MS does not have a P-TMSI and old RAI stored in its nonvolatile memory (e.g., this is the first power on), then the IMSI is sent and the SGSN proceeds directly to the authentication procedure.

- **Attach Type.** An indication of whether this is a GPRS Attach only, a GPRS Attach while already attached to CS domain services (i.e., IMSI Attached), or a combined GPRS/IMSI Attach

- **MS Capabilities.** This includes the so-called MS Classmark—which provides information such as the GPRS class (i.e., Class A, B, or C), multislot class, and DTM support—as well as the discontinuous reception (DRX) capabilities of the MS.

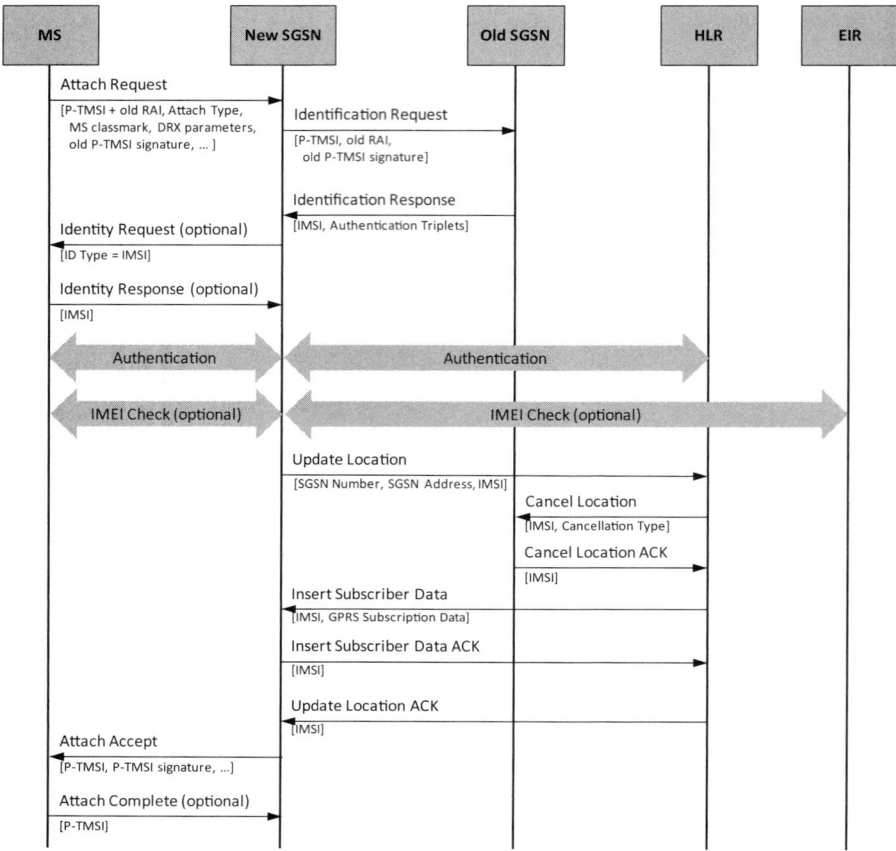

Figure 6.15 GPRS attach procedure. Adapted from Reference [115].

- **P-TMSI Signature.** The P-TMSI signature is a 32-bit value derived from the P-TMSI that may be assigned to the MS by the SGSN. Its use is optional and is decided by the network. The purpose of the P-TMSI signature if it is used is to provide added security, as a new P-TMSI signature is sent to the MS with every successful GPRS Attach, GPRS Detach, and RA Update. If the MS replies with an incorrect P-TMSI signature, the SGSN may choose to initiate an authentication procedure. Like the P-TMSI, the P-TMSI signature has only local significance within an SGSN area.

- **Location Information.** The BSS begins tracking the RAI and CI corresponding to the MSs current location.

Upon receipt of the Attach Request, the New SGSN uses the old RAI to determine the SGSN that assigned the last valid P-TMSI (i.e., Old SGSN) and attempts to determine the IMSI and related security information for the MS. If the New SGSN is unable to retrieve the subscriber identity from the Old SGSN for any reason, then the MS must transmit its IMSI unciphered for identification purposes. The network may then perform authentication and IMEI check procedures. Authentication is not mandatory if the Identification Request (from New SGSN to Old SGSN) was successful. If the SGSN number has changed (i.e., the New SGSN is different from the Old SGSN), then signaling is required in CN to update location information in the HLR, remove the MM context information from the Old SGSN, and validate that the presence of the MS in the New SGSN does not violate subscription restrictions. If the GPRS Attach procedure is successful, the New SGSN allocates a P-TMSI with local significance in the current RA and sends an Attach Accept message (from SGSN to MS). Recognizing a change in P-TMSI, the MS completes the procedure by sending an Attach Complete message (from MS to SGSN), verifying that it has changed its P-TMSI.

In the event that the New SGSN is the same as the Old SGSN (i.e., the MS has not moved to a new SGSN area since the last GPRS Detach), the GPRS Attach Procedure is much simpler. The SGSN will recognize the local P-TMSI, assuming no network error has occurred, and modify the associated MM context. In this case, the only signaling is the Attach Request (from MS to SGSN) followed by Attach Accept (from SGSN to MS). The MS is not required to respond with an Attach Complete message in this case, as the P-TMSI has not changed.

The GPRS Attach procedure is similar to the IMSI Attach procedure used to access CS domain services. A key difference, however, is that an MS that has performed GPRS Attach still cannot communicate with the outside world until it performs the so-called PDP context activation. If the optional Gs interface is supported between the SGSN and MSC/VLR in the CN, then it is possible for the MS to initiate a combined GPRS/IMSI attach procedure (i.e., attach to both CS and PS services) with a single Attach Request to the SGSN in order to save radio resources. In this case, the procedure in Figure 6.15 would be modified to include CN signaling immediately following the Update Location ACK (from HLR to New SGSN). The new TMSI assigned by the MSC/VLR is sent along with the P-TMSI in the Attach Accept (from SGSN to MS).

When the GPRS Attach procedure is initiated, the MS initially identifies itself at the RLC/MAC layer using a Foreign TLLI derived from the last valid P-TMSI in order to establish a radio link with the BSS. If a valid P-TMSI is not available, the MS generates a Random TLLI, which can be identified as such by the BSS based on the address space of the Random TLLI. Once a new P-TMSI is allocated within the current RA, a Local TLLI is derived from the P-TMSI to identify the MS in subsequent transmissions.

6.9.2 GPRS Mobility Management State Model

After a successful GPRS Attach, GPRS MM contexts are created in the MS and the SGSN. The GPRS Mobility Management (GMM) state model is illustrated in Figure 6.16 [115] for both the MS and the SGSN. In both cases, the GMM state model is characterized by three states:

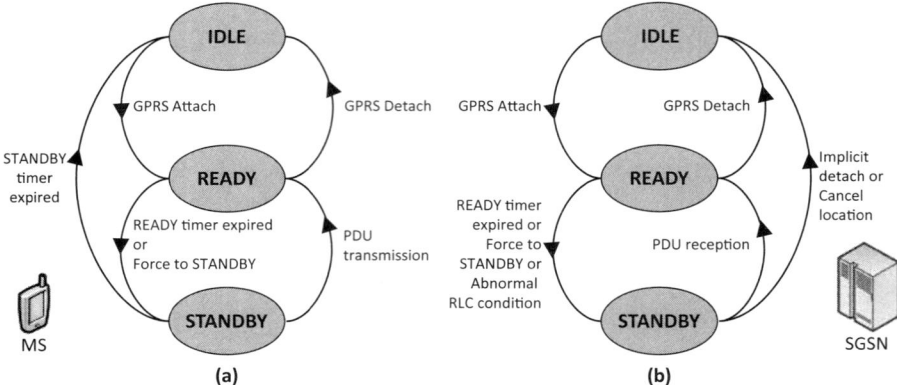

Figure 6.16   GPRS mobility management state model for (a) the MS and (b) the SGSN. Adapted from Reference [115].

- **IDLE.**   The MS is not GPRS Attached and is not known to the GPRS network. The MM contexts do not hold any valid location information about the MS. The MS is not reachable for paging, and data transmission to and from the MS is not possible.

- **STANDBY.**   The MS is attached to the GPRS network and MM contexts have been established in both the MS and the SGSN, which store the current RAI. The MS may receive pages for incoming data or signaling information from the network side; however, data reception and transmission is not possible (though transmission may be initiated by the MS as will be described later). In this state, the MS may initiate the activation or deactivation of *PDP contexts* (these will be discussed shortly).

- **READY.**   The MS is both attached to the GPRS network and able to send and receive packet data (i.e., IP packets) and signaling information. In this state, the SGSN MM context is extended to include the current cell identity, namely the CGI, where the MS is located. To maintain this information, the MS performs mobility management procedures upon cell reselection. Since the exact location of the MS is known by the SGSN to the cell level, paging is not required in this state. The READY state may be entered from the IDLE state when the MS performs a GPRS Attach, typically followed by a PDP context activation. The READY state may be entered from the STANDBY state when the MS initiates a packet transmission (possibly in response to a paging message from the network). The MS remains in the READY state between transmissions, even when no radio resources are allocated, until the READY timer expires.

The definition of two GPRS Attached states, namely READY and STANDBY, allows the MS to remain logged on to the GPRS network for long periods of time while minimizing the need for mobility management signaling during long periods of inactivity. This approach is efficient in the sense that it minimizes the utilization of radio resources for signaling overhead and improves battery life through the use of discontinuous reception (DRX) in STANDBY. Typically the MS performs cell selection and reselection locally in both states, although cell reselection may optionally be controlled by the network in READY state. In READY state, the MS must update its location with the network whenever it performs cell reselection. Since the CGI is included in the base station system GPRS protocol (BSSGP) header of all data packets transmitted from the BSS to the SGSN, no additional signaling is required as long as the MS has user data packets to transmit after cell reselection in the READY state. However, in the STANDBY state, when there are no user data packets to transmit, the MS may move freely between cells and need only update its location in the network when it moves into a new RA.

6.9.3 PDP Context

After the MS has performed a GPRS Attach and entered the READY state, there is one more step that must be executed before it is able to send and receive user data packets: PDP context activation. The *packet data protocol (PDP) context* describes the characteristics of the data session and is stored in the MS, the SGSN, and GGSN. The PDP context makes the MS known to the appropriate GGSN, allowing the exchange of information with the appropriate external PDN. Once the PDP context has been activated, the MS is visible to the outside world. Although most data sessions use IP, it is interesting to note that the name *PDP context* has been chosen rather than *IP context*. This reflects the fact that GPRS has been designed to be flexible in its ability to support arbitrary packet data protocols, including future protocols that are yet to be invented. Initially, GPRS was designed to support internetworking with IP and X.25 networks [134]. However, as the popularity of IP grew dramatically and X.25 became obsolete, the requirement to support X.25 was removed from the GPRS standard. From R99 onward, GPRS supports IPv4, IPv6, and also the point-to-point protocol (PPP) [115].

Each PDP context includes the following session information:

- *PDP Type.* This parameter indicates the type of packets carried over the GPRS bearer service, for example, IPv4, IPv6, or PPP.

- *PDP Address.* This parameter indicates the IP address of the MS, which will either be an IPv4 address or an IPv6 address depending on PDP Type.

- *Access Point Name (APN).* This parameter indicates the APN to be used for internetworking with the external PDN. The APN identifies the specific GGSN to be used and, if the GGSN supports internetworking with multiple external PDNs, distinguishes which PDN interface to use. The APN may optionally indicate a specific service to be offered as well.

- *QoS Profile.* This parameter indicates the QoS attributes that characterize the data session. The QoS Profile is negotiated between the MS and the SGSN based on the desired QoS of the MS and the radio resources available in the

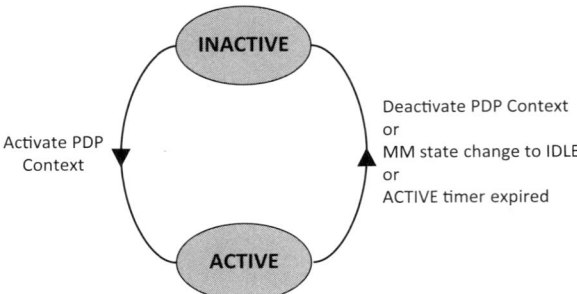

Figure 6.17 PDP state model.

network. The PDP contexts saved in the MS and the SGSN maintain a record of both the requested and negotiated QoS Profiles.

- *Routing Information.* When active, the PDP context also includes routing information necessary for the exchange of user data (e.g., IP packets) between the MS and the GGSN. These parameters include the NSAPI and, in the CN entities, the TEID.

The MS may maintain multiple PDP contexts that may include a combination of different PDP types, QoS profiles, and IP addresses and may access multiple PDNs. Each PDP context exists in one of two states, ACTIVE or INACTIVE, as characterized by the PDP state model in Figure 6.17. PDP contexts can be activated or deactivated by the MS only when it is in MM state READY or STANDBY. All PDP contexts belonging to an MS are associated with a single MM context in the network entities. If the MM state transitions to IDLE for any reason, all PDP contexts transition to the INACTIVE state and communication with the MS is not possible. If the MS uses a dynamic IP address, the address is allocated by the GGSN upon PDP context activation. Likewise, the MS loses any dynamically allocated IP addresses when all PDP contexts associated with that IP address are deactivated. Each PDP context also includes an ACTIVE timer that is reset every time packets are transmitted or received by the MS. If the timer expires, this will also trigger a transition to the INACTIVE state. The duration of the ACTIVE timer is controlled by the network and is operator specific; however, a value of around 1 hour is typical [135].

The PDP context activation procedure is illustrated in Figure 6.18. The MS first sends an Activate PDP Context Request message to the SGSN. This message specifies PDP context parameters described previously, including the QoS requested by the MS. If a dynamic IP address is to be used, the PDP Address parameter is left empty. After performing security functions, the SGSN validates the request against the subscription records for the MS. The SGSN then uses the APN to determine the GGSN address, creates a TEID associated with the PDP context for CN routing purposes, and sends a Create PDP Context Request to the GGSN. Once the PDP context has been created in the GGSN, the SGSN sends an Activate PDP Context

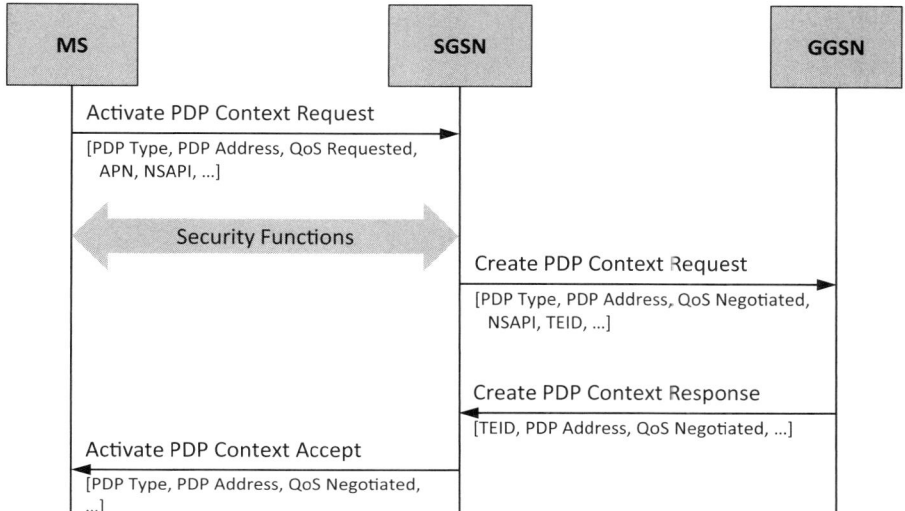

Figure 6.18 PDP context activation procedure.

Accept message to the MS that includes the negotiated QoS and the dynamically allocated IP address, if any.

Once a PDP context has been activated, the MS may initiate the activation of a Secondary PDP context that reuses the parameters of an existing PDP context, including IP address and APN, while specifying a different QoS profile. The secondary PDP context activation procedure is similar to the PDP Context Activation Procedure described earlier, except that the subscription checking, APN selection, and host configuration functions are excluded. The secondary PDP context may be useful, for example, to establish multiple Internet connections for different applications demanding different QoS (e.g., e-mail, Web browsing, and streaming media) while maintaining a single IP address. This results in more efficient use of the address space for dynamically allocated IP addresses. In addition to the secondary PDP context activation procedure, other procedures are defined for the modification and deactivation of PDP contexts [115]. At the air interface, PDP context-related procedures belong to the GPRS session management (SM) sublayer.

6.9.4 Quality of Service Profile

Quality of service (QoS) in GPRS is characterized by means of the QoS profile. As mentioned previously, each PDP context has an associated QoS profile. The QoS profile is considered to be a single parameter with multiple attributes. With the definition of 3G UMTS in R99, a completely new set of QoS attributes were defined. As part of the evolution of the GSM standard, GPRS R99 QoS attributes are equivalent to UMTS QoS attributes. This means that EGPRS systems use UMTS QoS

Table 6.10   GPRS R97/R98 QoS Delay Classes

Delay class	Delay (maximum values)			
	Packet size: 128 bytes		Packet size: 1024 bytes	
	Mean transfer delay (s)	95 percentile delay (s)	Mean transfer delay (s)	95 percentile delay (s)
1	<0.5	<1.5	<2	<7
2	<5	<25	<15	<75
3	<50	<250	<75	<375
4	Best effort	Best effort	Best effort	Best effort

Adapted from Reference [123].

attributes, rather than the QoS attributes defined in earlier GPRS releases (i.e., R97 and R98). This change affects both the CN signaling (i.e., change from GTPv0 to GTPv1) as well as the SM sublayer of the (E)GPRS air interface. Since UMTS QoS attributes are a 3G topic, the GPRS R97/R98 QoS attributes are presented here even though the majority of this chapter assumes a R99 baseline. The UMTS QoS attributes used in GPRS R99 will be discussed in the UMTS section of Chapter 7. The mapping between R97/R98 QoS attributes and R99 QoS attributes is defined in 3GPP TS 23.107 [136]. The GPRS R97/R98 standard defines four QoS attributes [124]:

- *Precedence Class.* This defines the relative importance of maintaining the service commitment in the event of network congestion. Three precedence classes are defined: high priority, normal priority, and low priority.

- *Delay Class.* This defines the expected delay in the end-to-end transmission of SDUs (e.g., IP packets) through the system between the MS and the Gi interface. GPRS is not a store-and-forward service, so all delay is due to transmission characteristics of the system, including radio resource signaling, transmission over the air, and tunneling through the PS core network. Four delay classes are defined, which are characterized by the maximum predicted mean delay and 95 percentile delay values indicated in Table 6.10 [123]. Here the 95-percentile delay value is the maximum delay predicted for 95% of all packets.

- *Reliability Class.* This defines the probability of reliable transmission in terms of residual error rate probabilities for the following cases: lost packet, out-of-sequence packet, duplicate packet, and corrupted packet. Five reliability classes are defined for real-time and non-real-time traffic with varying error and data-loss sensitivity. The reliability class dictates the mode of operation (i.e., acknowledged or unacknowledged) for the logical link control (LLC) and RLC layers at the air interface and the GTP layer in the core

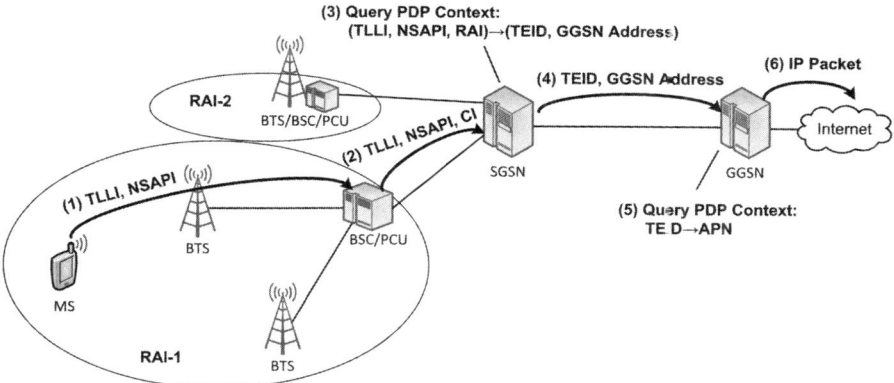

Figure 6.19   GPRS mobile originated IP packet routing example.

network. The reliability class also dictates whether L_C data protection is used.

- **Throughput Classes.** This defines the expected bandwidth required for the PDP context in terms of both peak and mean throughput. Nine peak throughput classes are defined, ranging from 8 kbps to 2 Mbps, and 18 mean throughput classes are defined, ranging from 100 bits/h to 50 Mbits/h. An additional mean throughput class is defined for best effort.

6.9.5   Routing, Tunneling, and Encapsulation

GPRS has two encapsulation functions that are used to transparently transport network PDUs (e.g., IP packets) between the MS and the external PDNs. One encapsulation function is defined between the MS and the SGSN for transmission over the air interface. The other is defined between the SGSN and the GGSN for transmission through PS domain core network. The former process is illustrated in Figure 6.19. The latter process uses GTP to perform tunneling of IP packets over a User Datagram Protocol (UDP)/IP protocol stack through the PLMN backbone network that connects the GSNs.

Based on the concepts described in this section, it is now possible to describe how IP packets are routed through the GPRS network. Recall that the PS domain identifiers were introduced previously in Section 6.6.3. Figure 6.19 shows a simple example of how these identifiers are used to route an outgoing (i.e., mobile originated) IP packet from the MS to the Gi reference point. For this example, let us assume that that the external PDN is the Internet and that the MS is requesting content from an external host, such as a Web server. Let us also assume that the MS has already activated a PDP context and that the MS is in either the READY or STANDBY state. The specific state is irrelevant, as the MM state of the phone will change to READY as soon as it begins the transfer. The following steps correspond to the numbering in Figure 6.19:

1. First, an IP packet is formed in the network layer of the MS, which will be routed transparently through the GPRS subnetwork. The IP packet contains a normal IP header, including the IP source address (i.e., IP address of the MS) and the IP destination address (i.e., IP address of the Web server). The IP packet undergoes the encapsulation process shown in Figure 6.26 for transmission over the air interface. Within the GPRS subnetwork, the TLLI and NSAPI are used to uniquely identify the source of the IP packet and the associated PDP context.

2. When LLC frames carrying the packet are forwarded from the BSC to the SGSN, the CGI is also sent to the SGSN so that it may perform cell-level tracking of the MS location while the MS is in the READY state.

3. At the SGSN, the IP packet is reassembled by the SNDCP function and passed to the relay function of the SGSN. Based on the TLLI/NSAPI pair, the SGSN is able to query the PDP context to determine where to forward the packet within the PS domain core network. This information includes the address of the GGSN within the PLMN backbone network and the TEID associated with the PDP context.

4. The SGSN relay function transfers the IP packet to the appropriate GTP tunnel identified by the TEID. GTP is used for encapsulation and tunneling. The GGSN address is used to route the packet through the IP-based PLMN backbone network to the appropriate gateway node.

5. Based on the TEID, the GGSN queries the associated PDP contexts to determine which access point name (APN) to use. The APN identifies the connection to a specific external PDN (e.g., the Internet).

6. Finally, the GGSN relays the IP packet over the Gi reference point to the appropriate external PDN.

As an extension of the previous outgoing routing example, Figure 6.20 shows a simple example of how these identifiers are used to route an incoming (i.e., mobile terminated) IP packet. For this example, let us assume that the MS is now in the process of receiving a series of IP packets as part of a file download from an external host, such as a Web server. Once again, we assume that the MS has an active PDP context and, for simplicity, that the MS is in the READY state.

1. First, the IP packet arrives at the GGSN (i.e., the Gi reference point) with the IP address of the MS in the destination field.

2. The GGSN then queries its stored database of PDP contexts to determine whether an active PDP context exists for IP destination address. The query is successful, so the GGSN determines the TEID and SGSN address from the PDP context.

3. Next, the GGSN uses GTP to tunnel the IP packet through the PLMN backbone network to the SGSN that is currently serving the MS.

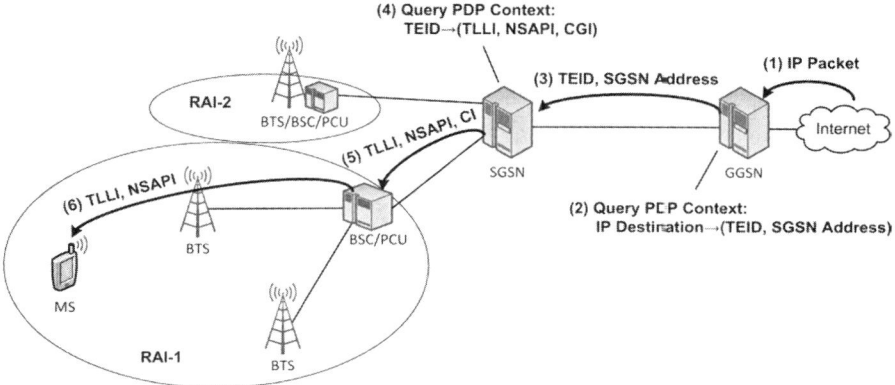

Figure 6.20 GPRS mobile terminated IP packet routing example.

4. The SGSN likewise performs a PDP context query to determine the temporary local identity of the MS, defined by the TLLI, and the NSAPI associated with the specific PDP context. If the MS is in READY state, then the SGSN has been tracking the current cell location of the MS and no paging is required.

5. The SGSN uses the CGI stored in the PDP context to route the IP packet to the BSC that is currently serving the MS. At this point, the destination of the IP packet is identified using the TLLI and NSAPI pair.

6. The PCU of the BSC allocates the appropriate radio resources if necessary to send the IP packet transparently to the MS based on the TLLI and CI. Finally, the NSAPI is used at the MS to associate the IP packet with the correct PDP context (e.g., to distinguish between multiple PDP contexts that may be simultaneously active), and the IP packet is delivered to its final destination: the network layer entity in the MS.

For the mobile terminated routing example previously discussed, the MS was assumed to be in the READY state. However, it is also possible that the MS could be in the STANDBY state. This might have happened, for example, if there was a long delay between the time that MS requested the file and the time when the first packet was received from the external server, causing the READY timer in the MS to expire. In this case, a paging message would be sent on the PPCH of every cell in the routing area. After the appropriate air interface signaling, the MS would respond by sending any arbitrary LLC frame to the network, indicating its location and causing a transition to the READY state. When the LLC frame is forwarded from the BSC to the SGSN, the CGI is also sent to the SGSN so that it may begin cell-level tracking of the MS location while the MS is in READY state. From this point, the example would continue with steps 5 and 6 as described previously.

In both routing examples, we have focused on a single SGSN and a single GGSN that are involved in the routing of IP packets between the radio access network and the external PDN. However, a typical GPRS network consists of multiple SGSNs and GGSN, where each SGSN provides an interface with part of the radio access network and each GGSN provides an access point to one or more PDNs. The PS domain backbone network therefore forms a private, IP-based network that connects the various GPRS support nodes.

6.9.6   Radio Resource Operating Modes

At the air interface, the GPRS RR management function is characterized by three modes: Packet Idle mode, Packet Transfer mode, and Dual-Transfer mode. The latter of these is only applicable for MSs supporting DTM. Each of these modes is described as follows [137]:

- *Packet Idle Mode.* In this mode, no radio resources are allocated to the MS and no data are being transmitted or received by the MS. At this RLC/MAC layer, no temporary block flow (TBF) exists, however the upper layers may initiate transmission of an LLC PDU, which will implicitly trigger the establishment of a TBF and transition to Packet Transfer mode. In this mode, the MS is camped on a cell and monitors the PBCCH and the relevant paging subchannel on the PPCH for the paging group to which the MS belongs. If no PCCCH is present in the cell at a given time, the MS listens to the BCCH and to the relevant paging subchannel on the PCH.

- *Packet Transfer Mode.* In this mode, the MS has been allocated radio resources on one or more PDCHs. Resources may be assigned on the downlink, uplink, or both. At the RLC/MAC layer, one or more TBFs exist. When performing cell reselection, the MS must leave this mode and return to Packet Idle mode before changing cells, at which time it will read the new cell system information (i.e., BCCH and [optionally] PBCCH) before resuming Packet Transfer mode in the new cell.

- *Dual-Transfer Mode.* In this mode, the MS has been allocated radio resources for both CS and PS services (e.g., concurrent voice and data). The MS performs all the tasks of Dedicated mode for CS RR management, plus the upper layers may trigger the release of all packet resources (transition to Dedicated mode) or the release of all RR resources (transition to Idle mode/Packet Idle mode). When performing handover, the MS must leave this mode and return to Dedicated mode before changing cells, at which time it may read the system information via messages sent on the SACCH and resume Dual-Transfer mode in the new cell.[13]

The transitions between RR modes depend on the GPRS MS class. For Class A and B MSs, both the CS and PS RR operating modes must be considered, as the MS capabilities include both voice and data services. Figure 6.21 [137] shows the

---

[13] In Rel-7 and beyond, DTM handover has been added to support concurrent handover of CS and PS domain resources.

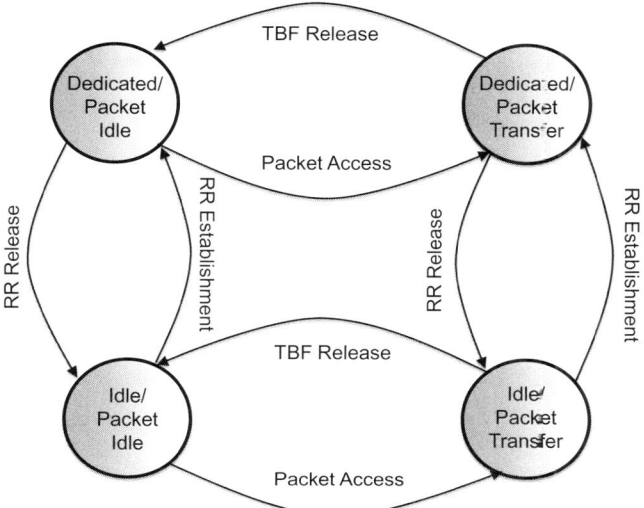

Figure 6.21   RR state model for Class A MS (non-DTM). Adapted from Reference [137].

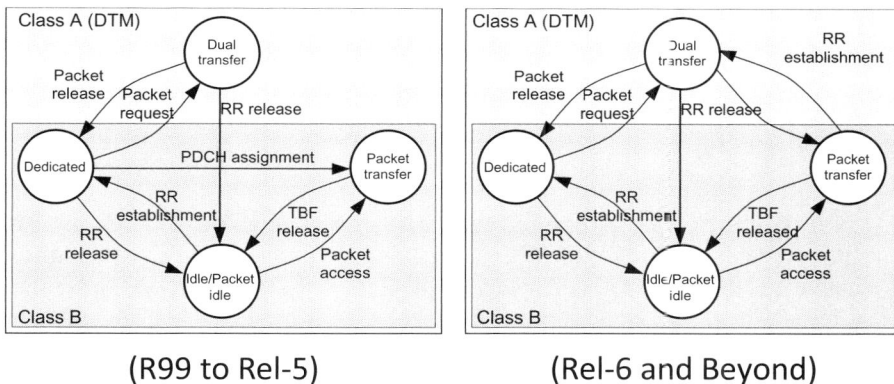

(R99 to Rel-5)   **(Rel-6 and Beyond)**

Figure 6.22   RR state model for Class A MS (DTM) and Class B MS. Adapted from Reference [137].

state model for a non-DTM Class A MS, which is essentially the combination of two state machines with two RR states each. Figure 6.22 [137] shows the state model for Class A MS using DTM and Class B MS. Note that the Class A DTM state model is essentially the Class B state model augmented by the DTM and corresponding state transitions. Furthermore, it is worth noting that the Class A DTM and Class B state models have been modified starting in Rel-6 to accommodate direct transitions between Packet Transfer mode and DTM for enhanced QoS. Finally, Figure 6.23 [137] shows the state model for Class C MS. Since a Class C MS can attach to only one service domain at a time, this state model consists of two disjoint state machines.

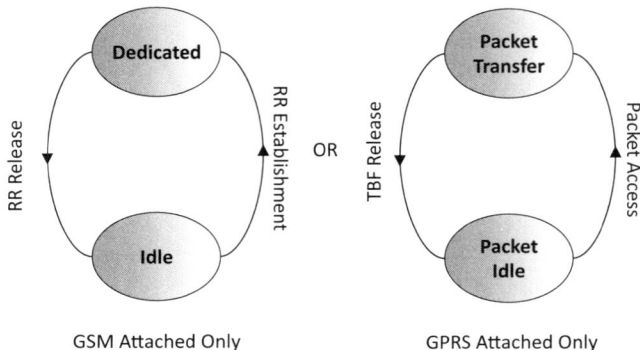

GSM Attached Only   GPRS Attached Only

Figure 6.23   RR state model for Class C MS. Adapted from Reference [137].

Table 6.11   Relationship between RR operating modes and MM states (non-DTM)

RR operating mode (BSS)	Packet transfer mode	Measurement report reception	No state	No state
RR operating mode (MS)	Packet transfer mode	Packet idle mode		Packet idle mode
GPRS MM state (network and MS)	Ready	Ready		Standby

Adapted from Reference [137].

Finally, it should be recognized that there is a relationship between the RR operating modes and the GPRS MM states described previously. This relationship is summarized in Table 6.11 and Table 6.12 [137] for non-DTM and DTM-supported MSs, respectively. By definition, the MS must be in Packet Idle mode when it is in MM STANDBY state. In the MM READY state, the MS may either be in Packet Transfer mode or Packet Idle mode.

6.9.7   GPRS Protocol Architecture

The GPRS protocol architecture can be divided into a user plane and the control plane. The user plane carries data and associated signaling (e.g., headers, flow control information) while the control plane carries signaling messages required to support the operations of the user plane (e.g., GPRS Attach procedures, PDP context activation). Both planes use the same set of lower layer protocols to transmit information over the air interface, namely the PHY, RLC/MAC, and LLC layers. The user plane protocol architecture is illustrated in Figure 6.24 and extends from the MS to the GGSN. The protocols that are specific to GPRS are highlighted in white, while the protocols that are outside the scope of GPRS are shaded in grey. The GPRS

Table 6.12 Relationship between RR operating modes and MM states (DTM supported)

RR operating mode (BSS)	Dual-transfer mode	Dedicated mode	Packet transfer mode	Measurement report reception	No state	Dedicated mode	No state
RR operating mode (MS)				CS idle and packet idle			CS idle and packet idle
GPRS MM state (network and MS)	Ready					Standby	

Adapted from Reference [137].

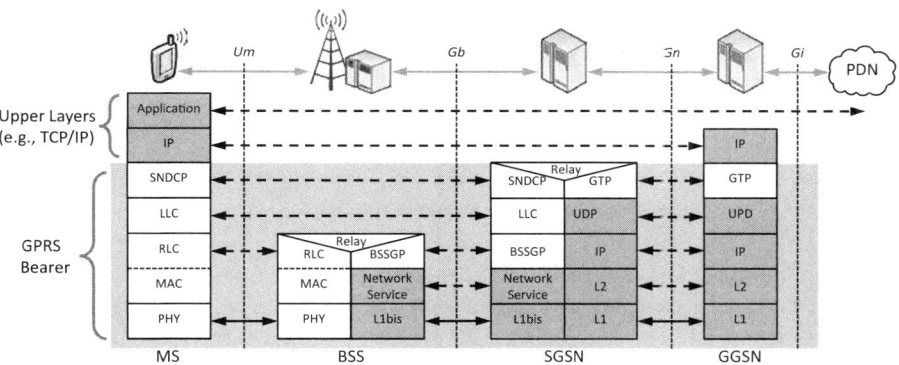

Figure 6.24 GPRS user plane protocol architecture. SNDCP, subnetwork-dependent convergence protocol; BSSGP, base station system GPRS protocol; GTP, GPRS tunneling protocol.

network effectively provides a GPRS bearer that transports IP packets between the GGSN and the MS.

The control plane protocol architecture in the radio access network is shown in Figure 6.25. Once again, the GPRS-specific protocols are highlighted in white and other protocols are shaded grey. The control plane layered architecture used between GSNs, namely on the Gn (SGSN-GGSN) and Gp (SGSN-SGSN) interfaces, is also shown. The full control plane architecture also includes the definition of layered protocol stacks at each of the interfaces of the core network in the PS domain. This includes the Gr (SGSN-HLR), Gs (SGSN-MSC/VLR), and Gf (SGSN-EIR) interfaces. A full treatment of the core network protocol architecture in the PS domain is defined in 3GPP TS 23.060 [115].

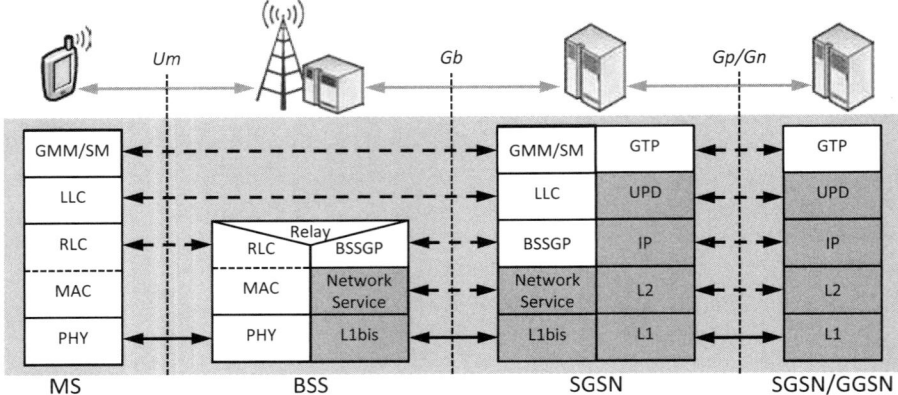

Figure 6.25   GPRS control plane protocol architecture. GMM, GPRS mobility management; SM, session management.

6.9.8 Segmentation and Encapsulation

IP packets are transmitted over the GPRS air interface through a process of segmentation and encapsulation. The purpose of this process is to condition the IP packets into a form that is suitable for transmission over the air using the physical structure of the existing GSM physical layer. The GSM PHY was designed and optimized for the transmission of voice. In order to transmit an IP packet efficiently over the existing radio interface, it must be compressed, segmented, and encapsulated as it passes from the network layer down to the GPRS physical layer. At the receiver, an inverse process is performed to reassemble the IP packet and either deliver it to the network layer in the MS or pass the packet along for transmission through the PS domain core network. The process of encapsulation and segmentation over the protocol layers of the GPRS user plane at the air interface is illustrated in Figure 6.26.

6.9.9 GPRS PHY Layer

A key objective of GPRS is to provide a packet radio service that reuses the GSM physical layer to the greatest extent possible. As you would expect, the characteristics of the GPRS PHY are very similar to that of GSM, though some modifications have been made to optimize for data transmission. Although the same physical resources are used (i.e., time slots), a special multiframe structure is defined for GPRS as described previously in Section 6.6.5.6 and radio resources are allocated on the basis of radio blocks consisting of four bursts on a given time slot over four consecutive frames. Furthermore, a new channel coding scheme has been defined for GPRS and the ciphering function has been moved from the physical layer to the LLC layer. The transmission chain used at the GPRS physical layer is similar to that

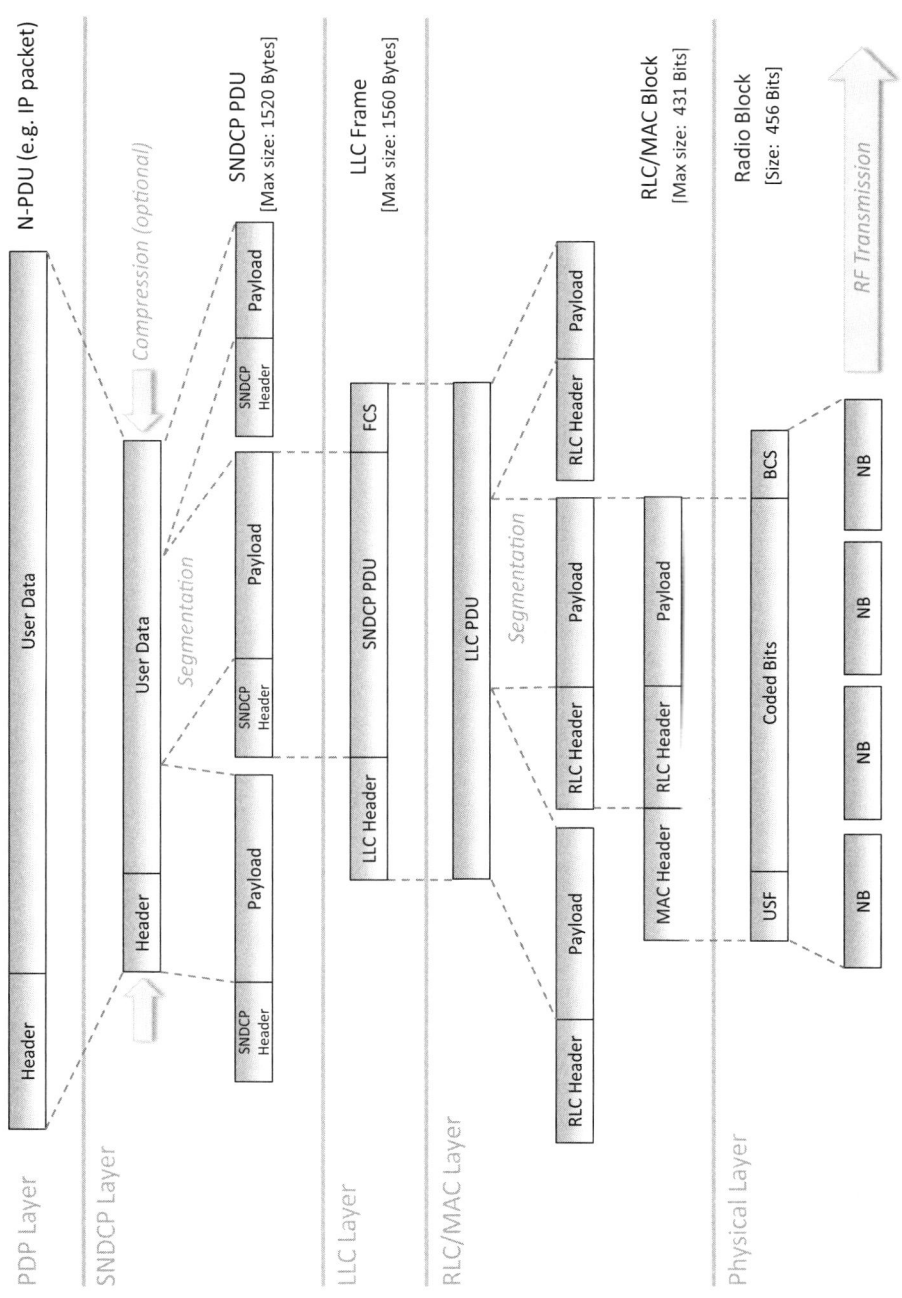

Figure 6.26 Segmentation and encapsulation process across GPRS protocol layers in the MS.

329

of GSM shown previously in Figure 6.10, with the encryption and decryption blocks removed. The GPRS PHY is logically divided into two sublayers in the standard: the physical RF layer and the physical link layer.

6.9.9.1   GPRS Physical RF Layer

The physical RF layer is responsible for the transmission and reception of bit streams over the physical resources. The generation of the bit sequence to be transmitted, as well as the interpretation of received bit sequences, is performed by the physical link layer. In the context of Figure 6.10, the physical RF layer is responsible for the blocks corresponding to modulation, transmission and reception, and demodulation. For GPRS (i.e., not including EDGE), these functions are identical to those defined for the GSM PHY.

6.9.9.2   GPRS Physical Link Layer

The purpose of the physical link layer is to convey information over radio interface using the services of the physical RF layer, thereby creating a physical link between the MS and the BTS for the transmission of RLC/MAC blocks. This layer includes functions necessary to maintain communication capability over the radio interface. The service primitives provided by the physical link layer to the RLC/MAC layer above are summarized in Table 6.13 [137], where shaded regions indicate which functions are performed by the various primitive types. The definition of each primitive type is summarized as follows [138]:

- *Request.* This is used when a higher layer is requesting a service from the next lower layer.
- *Indication.* This is used by a layer providing a service to notify the next higher layer of activities related to the request primitive type of the peer.
- *Response.* This is used to acknowledge receipt of an indication primitive type from the next lower layer.
- *Confirm.* This is used by the layer providing a requested service to confirm that an activity has been completed, either successfully or unsuccessfully.

In terms of transmission chain reference configuration of Figure 6.10, the physical link layer is responsible for the blocks corresponding to channel coding, interleaving, and burst building and multiplexing—as well as the corresponding inverse functions at the receiver. Note that this does not include ciphering, which is not performed at the GPRS PHY. The burst building function of GPRS is fundamentally similar to GSM, as the same burst structures are used. Specifically, the normal burst (NB) is used for all packet data logical channels except for PRACH and PTCCH/U, which use the access burst (AB).

The channel coding and interleaving scheme defined for GPRS have been modified from GSM to optimize for data. Here we must once again distinguish between

Table 6.13 Service Primitives Provided by the Physical Link Layer in GPRS

Name	Request	Indication	Response	Confirm	Comments
PH-DATA	▓	▓			Used to pass message units containing frames used for RLC/MAC layer respective peer-to-peer communications to and from the physical layer.
PH-RANDOM ACCESS	▓			▓	Used to request and confirm (in the MS) the sending of a random access frame and to indicate (in the network) the arrival of a random access frame.
PH-CONNECT		▓			Used to indicate that the physical connection on the packet data physical channel has been established.
PH-READY-TO-SEND	▓				Used by the physical layer to trigger, if applicable, piggybacking, the start of timer for the RLC/MAC layer, and the forwarding of a data unit to the physical layer
PH-EMPTY-FRAME	▓				Used by the RLC/MAC layer to indicate that no frame has to be transmitted after receiving the PH-READY-TO-SEND primitive

Adapted from Reference [137].

GPRS and EGPRS, as the latter utilizes additional coding schemes and a link adaptation mechanism described later in this chapter. GPRS defines four channel coding schemes for the transmission of user data and associated signaling over PDTCHs, as summarized in Table 6.14. CS-1 corresponds to the most robust FEC coding for reliability in exchange for decreased throughput. CS-4 corresponds to no FEC—with the exception of uplink state flag (USF)[14] precoding—to achieve maximum throughput in favorable channel conditions. For all four coding schemes, an outer block

---

[14] The USF is a 3-bit field in the MAC header described later in Section 6.9.10 .

Table 6.14 GPRS Channel Coding Schemes

Scheme	Overall code rate	Precoded USF size (bits)	Block code	Block code bits per block			Coded bits[b]	Punctured bits	Data rate per time slot (kbps)
				Data[a]	BCS	Tail			
CS-1	0.40	3	Fire	184	40	4	456	0	9.05
CS-2	0.59	6	CRC	274	16	4	588	132	13.4
CS-3	0.69	6	CRC	318	16	4	676	220	15.6
CS-4	0.95	12	CRC	440	16	0	456	0	21.4

[a]Input bits of the inner block code, including precoded USF bits and remainder of the RLC/MAC block (payload).
[b]Number of coded bits at the output of the convolutional encoder prior to puncturing.

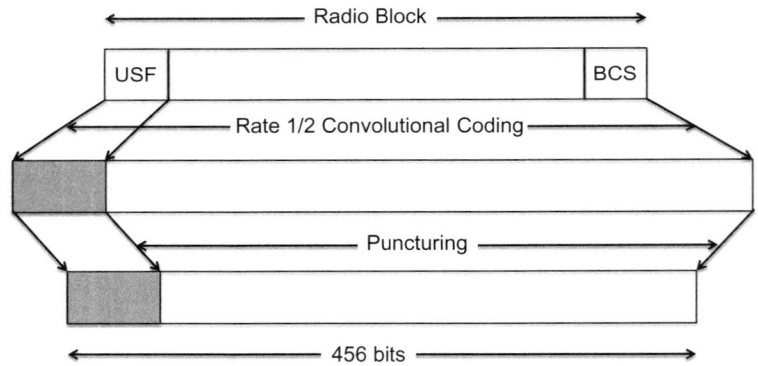

Figure 6.27 GPRS radio block structure for CS-1 to CS-3. Adapted from Reference [137].

code (i.e., CRC or Fire code) is applied first for error detection, generating a block check sequence (BCS) appended to the end of the RLC/MAC block. For CS-1 to CS-3, an inner rate 1/2 convolutional code is applied next for FEC. Finally, puncturing is applied in the case of CS-2 and CS-3. The overall process is summarized in Figure 6.27 and Figure 6.28 [137]. Note that in all cases, the final coded block size is 456 bits after any possible puncturing. This corresponds to the number of payload bits carried by four normal bursts, which make up one radio block.

Packet data logical channels used for signaling only use coding scheme CS-1. These include PACCH, PBCCH, PAGCH, PPCH, PNCH, and PTCCH/D. This ensures the robust transmission of control information. The PRACH and PTCCH/U channels use the coding scheme defined previously for RACH, as they are carried by the access burst rather than the normal burst.

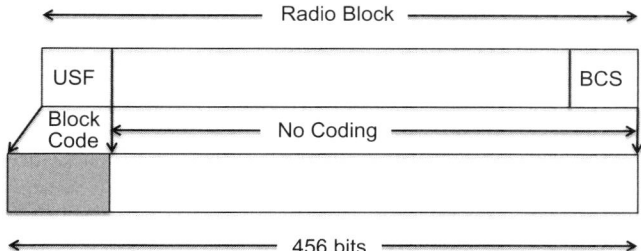

Figure 6.28   GPRS radio block structure for CS-4 (no FEC). Adapted from Reference [137].

Additional functions of the GPRS PHY closely parallel those of GSM, including synchronization procedures, monitoring and evaluation of radio link quality, cell selection/reselection, and power control. However, some of these functions have been modified to accommodate the connectionless operation of the GPRS air interface (e.g., cell reselection, timing advance, etc.). In particular, the physical layer functions required to maintain physical communication between the MS and the network must be modified based on the fact that handover is traditionally not defined during packet transfer mode, but rather the cell reselection process is controlled by the MS. The physical link layer is also responsible for discontinuous reception (DRX) procedures used to conserve battery power in the MS.

6.9.10   RLC/MAC Layer

The main purpose of the RLC/MAC layer is to establish a reliable link between the MS and the BSS. The MAC sublayer defines the procedures for the transmission of data and control messages over PDCHs, including the statistical multiplexing of users onto PDTCHs. The associated logical channels are PBCCH, PCCCH, PACCH, and PDTCH. The RLC sublayer accepts LLC PDUs from the layer above—which may contain user data, GMM/SM control messages, or SMS—and segments them into RLC data blocks which, after the addition of a MAC header, are of an appropriate size to be transmitted over one radio block at the PHY. The RLC sublayer also includes a selective automatic repeat request (ARQ) function and generates RLC/MAC control messages, such as ACK/NACK, uplink resource request/assignment, paging request, and packet system information (PSI) messages of the PBCCH. The RLC/MAC layer terminates on the network side in the BSC, or more specifically the PCU. The RLC/MAC layer service primitives are summarized in Table 6.15 [137], where shaded regions indicate which functions are performed by the various RLC/MAC layer primitive types.

6.9.10.1   Differences between GPRS and EGPRS RLC Function

It is important to note that the RLC function of EGPRS is fundamentally different than that of GPRS in two key ways. First, mechanisms must be included

Table 6.15 Service Primitives Provided by the RLC/MAC Layer in GPRS

Name	Request	Indication	Response	Confirm	Comments
RLC/ MAC-DATA					Used for the transfer of upper layer PDUs. Acknowledged mode of operation in RLC is used. The upper layer shall be able to request high transmission quality via a primitive parameter.
RLC/ MAC-UNITDATA					Used for the transfer of upper layer PDUs. Unacknowledged mode of operation in RLC is used.
RLC/ MAC-STATUS					Used to indicate that an error has occurred on the radio interface. The cause for the failure is indicated.

Adapted from Reference [137].

to accommodate the adaptive modulation and coding schemes supported by the EGPRS PHY. Second, EGPRS makes use of a Type II Hybrid ARQ scheme (HARQ) scheme using incremental redundancy for retransmissions. These differences require the definition of different RLC/MAC block structures. As a result, the RLC layers of GPRS and EGPRS are incompatible in the sense that an RLC/MAC layer connection must be defined as either GPRS or EGPRS at connection setup. The present section focuses specially on the RLC function of GPRS. EDGE enhancements are addressed in further detail in Section 6.11, "EDGE Enhancements."

6.9.10.2 Temporary Block Flow

The connection between two peer RLC/MAC entities that is used to transmit user data is called a *temporary block flow* (TBF). Formally, the TBF is defined as a physical connection used by two RR peer entities to support the unidirectional transfer of LLC PDUs on PDCHs [139]. Note the word "unidirectional" in that sentence and recall that GPRS radio resources are allocated to MSs asymmetrically in the downlink and uplink. This means that an MS simultaneously engaged in uplink and downlink data transfer will utilize two TBFs—one each for the uplink and downlink. A TBF is initiated when an LLC PDU is received at the RLC/MAC entity on the transmit side and released when the final queued LLC PDU is reassembled at the

RLC/MAC entity on the receive side. The TBF is temporary and is maintained only for the duration of the data transfer. In the case of acknowledged mode, the TBF is maintained until all ACK messages are received. A TBF may operate in either GPRS or EGPRS TBF mode, which is set by the network.

Each TBF is identified by means of a temporary flow identity (TFI). The TFI is a 5-bit identity assigned to the TBF by the network in the radio resource assignment message that precedes the transfer of LLC frames over the air interface. Since the TFI is unique among concurrent TBFs in a given direction (i.e., uplink or downlink), the TFI serves as a temporary, unique identifier for the MS at the RLC/MAC layer. The TFI is included in every RLC header and RLC control message (e.g., ACK/NACK) belonging to the TBF.

6.9.10.3 Medium Access Control Function

The MAC function is responsible for providing efficient multiplexing of data and control signaling onto the radio resources allocated to the GPRS in the uplink and the downlink. When considering the MAC function of GPRS, it is important to consider that access control of the shared medium (i.e., radio resources) in cellular networks is based on fixed assignment by a central authority, namely the network. On the downlink, multiplexing is controlled by a scheduling mechanism in the PCU. On the uplink, multiplexing is controlled by radio resource allocation to individual MSs in response to a service request. The exception, of course, is the random access channel (i.e., PRACH or RACH, depending on network configuration) that exists in all-cellular systems to allow the MS to request allocation of radio resources for mobile originated packet transfer on the uplink. In GPRS, the PRACH uses a slotted ALOHA-based reservation protocol. The MAC function is responsible contention resolution between channel access attempts on the PRACH, including collision detection and recovery. Finally, the MAC function performs scheduling of access attempts and queuing of packet accesses for mobile terminated packet transfer, as well as priority handling.

In GPRS, each PDCH can be considered as a shared medium that is utilized by multiple MSs. The manner in which the MS determines when it may transmit on the uplink is defined by the medium access mode. The MAC function supports four medium access modes [139]:

- *Dynamic Allocation.* In this mode, uplink radio resources are allocated dynamically through the use of the uplink state flag (USF), a 3-bit value that is assigned to the MS by the network during TBF establishment. USF is transmitted in the MAC header at the beginning of every RLC/MAC block on the downlink to indicate which MS can transmit on the following uplink radio block of that particular PDCH. That is to say, the USF transmitted on radio block B0 of PDCH downlink allocates uplink radio resources for radio block B1 of the associated PDCH uplink. Therefore, all mobile stations sharing an uplink PDCH must constantly monitor the downlink for the USF. When an MS reads its USF value in the Ddownlink, it recognizes that it may transmit in the subsequent radio block on the uplink of the same PDCH. The number of uplink radio blocks that an MS may utilize upon receipt of its USF is defined by the *USF granularity*, which is a parameter set by the network and may be set to either one radio block or four radio blocks. Ultimately, dynamic allocation mode allows the network to dynamically allocate uplink radio resources on a block-by-block basis.

- *Extended Dynamic Allocation.* This mode is similar to dynamic allocation (i.e., characterized by the use of a USF) with enhancements to accommodate multislot operation in the uplink. In the dynamic allocation mode described earlier, an MS transmitting on multiple slots in the uplink is required to monitor the USF independently on each time slot. As a result, for each allocated uplink time slot, the MS multislot class must support one uplink time slot (i.e., for transmission) and one downlink time slot (i.e., to monitor USF) simultaneously—quickly pushing the MS requirements into higher-order multislot classes. In extended dynamic allocation mode, the requirement to monitor the USF during multislot operation is relaxed in the following way: During block periods when it is not transmitting, the MS monitors the USF on all assigned PDCHs in the normal way. When the MS detects its USF value on one PDCH, it is allocated uplink resources in the next radio block on that PDCH and all PDCHs on higher numbered time slots. Additionally, the MS is not required to monitor the USF on these higher numbered time slots during transmission. To illustrate with an example, say that the MS was assigned to PDCHs on time slots TN0, TN2, and TN3 during TBF establishment. The MS then detects its USF on TN0. In the next block period, the MS will receive on TN0 to monitor the USF and transmit on TN0, TN2, and TN3. Now the MS detects its USF on TN2. In the next block period, the MS will receive on TN0 and TN2 to monitor the USF and transmit on TN2 and TN3. In both cases, the multislot class of the MS is required to support a total of four transmit and receive time slots.

- *Fixed Allocation.* In this mode, a fixed uplink resource allocation is communicated to the MS at TBF establishment. The fixed allocation consists of a start frame, slot assignment, and block assignment bitmap indicating which blocks to use per time slot. If additional resources are required, the MS may request them in one of the assigned uplink blocks. This triggers the allocation of a new TFI and a new fixed allocation is sent to the MS on the PACCH. The USF is not used in this mode and the MS is free to transmit on allocated uplink radio resources without monitoring the downlink. Therefore, dynamic allocation on a block-by-block basis is not possible in fixed allocation mode.

- *Exclusive Allocation.* In this mode, an MS is granted the exclusive right to transmit on the uplink radio resources of a particular PDCH for the duration of the TBF. The exclusive allocation includes neither a USF nor a block assignment bitmap. This mode applies only MSs operating in dual-transfer mode.

All MSs are required to support both dynamic and fixed allocation modes, while the network is required to support at least one of these two allocation modes. Support for extended dynamic allocation and exclusive allocation modes are optional in both the MS and the network.

Statistical multiplexing of multiple MSs on the downlink is a much simpler matter, as it is controlled by a scheduling mechanism on the network side and does not require contention arbitration. When the downlink TBF is established, the MS receives a TFI and a list of PDCHs that will be used for downlink transfer in the Packet Downlink Assignment message on the PCCCH. The MS then monitors the downlink of the appropriate PDCHs for its TFI, which is transmitted in the RLC header of every RLC/MAC block in the downlink.

The number of downlink TBFs that can be multiplexed onto a single PDCH is limited by the size of the TFI. The TFI is a 5-bit value, therefore a maximum of 32 MSs can be statistically multiplexed onto a single downlink PDCH. However, in the case of dynamic allocation mode, the number of uplink TBFs that can be multiplexed onto a single PDCH is limited by the size of the USF. The USF is a 3-bit value, therefore the number of MSs that can be statistically multiplexed onto a single uplink PDCH is 8. Furthermore, if the PDCH carries a PCCCH, then USF flag 111 is reserved for PRACH transmission—which further reduces the maximum number of concurrent uplink TBFs to 7.

6.9.10.4 *Radio Link Control Function*

The RLC function provides an interface between the MAC function and the LLC layer above. A key responsibility of the RLC function is the segmentation and reassembly of LLC PDUs. When an LLC PDU is received from the upper layers, it is segmented into one or more RLC data units. The RLC data unit is the payload of an RLC data block. The length of the RLC data unit is between 20 and 50 bytes depending on the coding scheme used at the PHY.[15] The RLC function, therefore, is also responsible for link adaptation on the PDTCH as it decides which coding scheme to use. After segmentation, a block sequence number (BSN) is included in the RLC header to allow for in-order reassembly of the RLC data blocks at the receiver. Since the maximum LLC PDU length is 1560 bytes, between 1 and 78 RLC data blocks are required to transfer a single LLC PDU over the air interface. If a segmented LLC PDU does not fill an integer number of RLC data units, the beginning of the next queued LLC PDU may be appended to the end of the final RLC data unit, otherwise zero padding is used. When an RLC data unit contains data from two or more LLC PDUs, this is signaled using an optional extension byte with the length indicator (LI) field in the RLC header to denote the boundary between two LLC PDUs.

Another key responsibility of the RLC is the selective ARQ functionality for retransmission. The RLC ARQ function supports two modes of operation [139]:

---

[15] RLC data unit size may be less if optional RLC header fields are used.

- *Acknowledged Mode.* The RLC function uses retransmission of lost or corrupted RLC data blocks to achieve high reliability. Retransmission is based on a sliding window and the use of ACK/NACK control messages.

- *Unacknowledged Mode.* The RLC function does not make use of retransmission of lost or corrupted RLC data blocks, except during the TBF release procedure at the end of the transfer.

The ARQ transmit window size for GPRS is 64 blocks. This means that at most 64 RLC blocks may be transmitted without the reception of an ACK/NACK message. Another key aspect of the GPRS RLC function is the countdown value. In the uplink direction, the MS assists the ARQ function and the scheduling of radio resources by transmitting a countdown value (CV) in the MAC header of every RLC/MAC block. The CV indicates to the network the BSN of the final RLC data block that will be sent in the current uplink TBF. The MS initiates the release of an uplink TBF by starting a countdown of the final 15 RLC data blocks. The final RLC data block is indicated with CV equal to 0. Regardless of operating mode, the uplink TBF release process involves a handshaking procedure. First, the network replies to the Final ACK/NACK message—denoted by setting the final block indicator (FBI) bit of the RLC header to value 1 and allocates uplink radio resources on the PACCH by means of the relative reserved block period (RRBP) field. The MS then waits for the radio block indicated by the RRBP field, transmits a Packet Control Acknowledgment message on the PACCH, and releases the TBF.

6.9.10.5   RLC/MAC Block Structures

There are primarily two types of RLC/MAC block: RLC data blocks, which carry a payload of (possibly segmented) LLC PDU data, and RLC/MAC control blocks. The general structure of RLC data blocks includes a MAC header, an RLC header, RLC data (payload), and spare bits if necessary. The spare bits are zeros that are padded at the end of the RLC data block to accommodate the code rate of the error control coding at the physical link layer. The number of spare bits is 0, 3, or 7 depending on the coding scheme, as shown in Table 6.16. The general structure of RLC/MAC control blocks includes a MAC header, optional RLC header (DLdown-

Table 6.16   RLC Data Block Sizes

Channel coding scheme	RLC data block size without spare bits (bytes)	Number of spare bits (bits)	Final RLC data block size (bytes)
CS-1	22	0	22
CS-2	32	7	32 + 7 bits
CS-3	38	3	38 + 3 bits
CS-4	52	7	52 + 7 bits

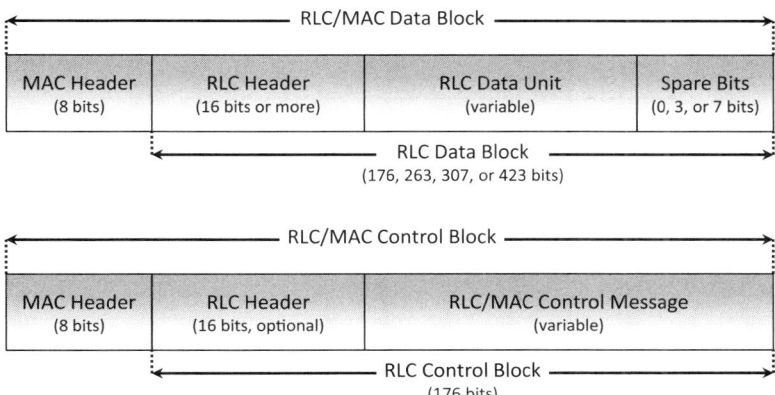

Figure 6.29   General RLC/MAC block structures for data and control blocks.

link only), and a control message generated by the RLC function. The general structure of RLC/MAC blocks is illustrated in Figure 6.29.

The specific contents of the MAC and RLC headers are different for the uplink and downlink. The four possible RLC/MAC block structures are illustrated in Figure 6.30 [139]. The definitions of each of the MAC and RLC header fields are summarized in Table 6.17 and Table 6.18, respectively. These RLC/MAC block structures apply to GPRS only. A separate set of RLC/MAC block structures are defined for EGPRS as a function of the modulation and coding scheme used at the PHY, as defined in 3GPP TS 04.60 (R99) and 3GPP TS 44.060 (Rel-4 and beyond).

6.9.11 GPRS LLC Layer

The LLC layer provides a highly reliable data link between the MS and the SGSN with optional ciphering. It is the lowest layer of the GPRS protocol architecture to span both the Um and Gb interfaces and carries user plane and control plane data, both of which are adapted to LLC PDUs and passed to the lower layers (i.e., the RLC/MAC on the MS side or BSSGP on the SGSN side). The responsibilities of the LLC include the detection and recovery of lost LLC PDUs, optional retransmission using ARQ, flow control, and ciphering. The frame formats used by the LLC layer are based on the LAPD protocol [132] defined by the ITU for the ISDN and the Radio Link Protocol (RLP) [140] defined for IS-95. The LLC procedures are modeled after the well-known HDLC protocol [131], which is an ISO standard. In these respects, the LLC protocol has similarities to the LAPDm protocol used for GSM layer 2 signaling at the air interface.

The LLC layer has two modes of operation:

- ***Acknowledged Mode.*** In this mode, SDUs received from the upper layer are transmitted in numbered Information (I) frames. The transmission of I frames

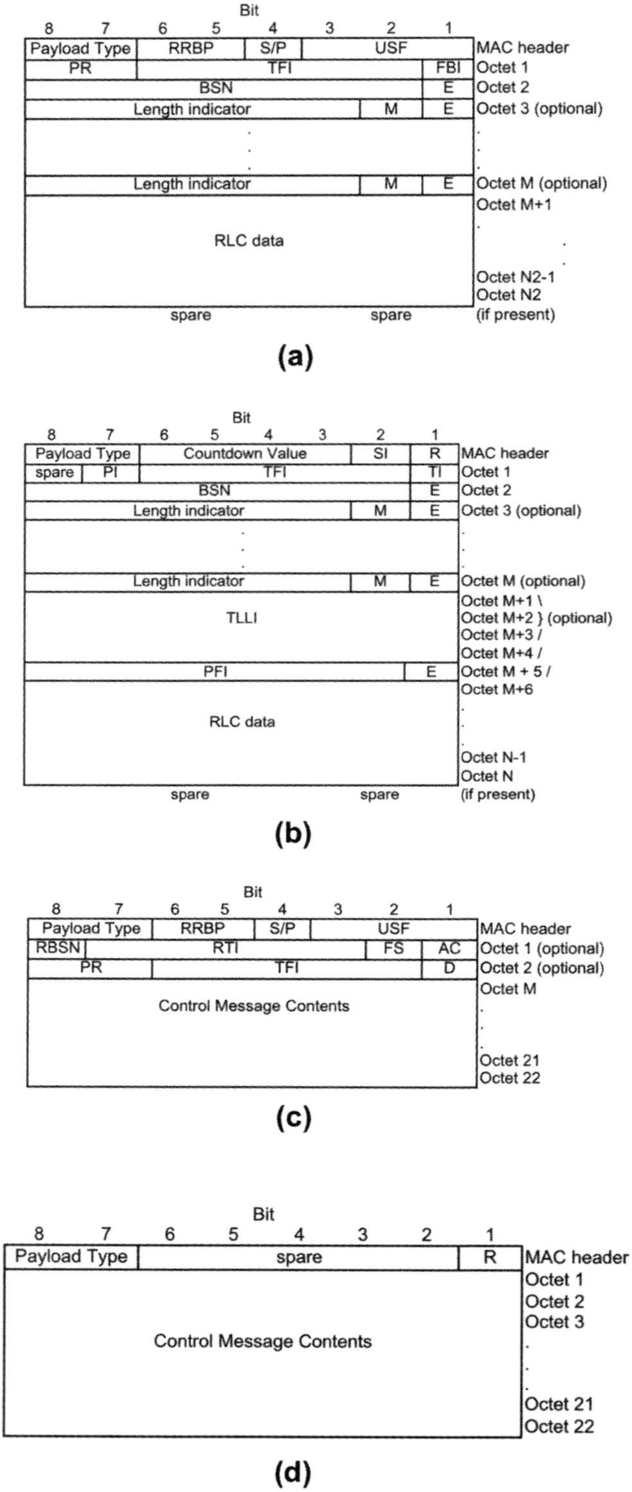

(a) Downlink RLC data block

Bit

8	7	6	5	4	3	2	1	
Payload Type		RRBP		S/P		USF		MAC header
PR		TFI					FBI	Octet 1
BSN							E	Octet 2
Length indicator						M	E	Octet 3 (optional)
				:				
Length indicator						M	E	Octet M (optional)
								Octet M+1
			RLC data					
								Octet N2-1
								Octet N2
spare				spare				(if present)

(a)

(b) Uplink RLC data block

Bit

8	7	6	5	4	3	2	1	
Payload Type		Countdown Value				SI	R	MAC header
spare	PI	TFI					TI	Octet 1
BSN							E	Octet 2
Length indicator						M	E	Octet 3 (optional)
				:				
Length indicator						M	E	Octet M (optional)
								Octet M+1 \
			TLLI					Octet M+2 } (optional)
								Octet M+3 /
								Octet M+4 /
PFI							E	Octet M + 5 /
								Octet M+6
			RLC data					
								Octet N-1
								Octet N
spare				spare				(if present)

(b)

(c) Downlink RLC/MAC control block

Bit

8	7	6	5	4	3	2	1	
Payload Type		RRBP		S/P		USF		MAC header
RBSN		RTI				FS	AC	Octet 1 (optional)
PR		TFI					D	Octet 2 (optional)
								Octet M
		Control Message Contents						
								Octet 21
								Octet 22

(c)

(d) Uplink RLC/MAC control block

Bit

8	7	6	5	4	3	2	1	
Payload Type		spare					R	MAC header
								Octet 1
								Octet 2
								Octet 3
		Control Message Contents						
								Octet 21
								Octet 22

(d)

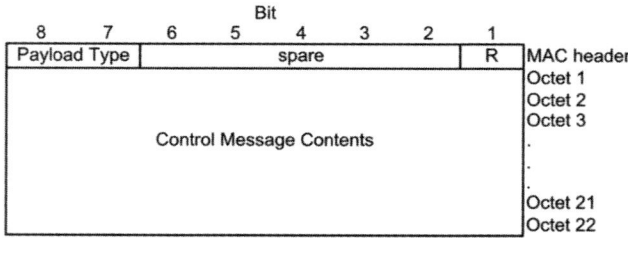

Figure 6.30 RLC/MAC block structures. (a) Downlink RLC data block; (b) uplink RLC data block; (c) downlink RLC/MAC control block; (d) uplink RLC/MAC control block. Reprinted from Reference [139].

Table 6.17 MAC Header Fields

Field	Field name	Size (bits)	Data/ control	UL/ DL	Definitions
USF	Uplink state flag	3	Both	DL	Used for allocation of uplink radio resources
S/P	Supplementary/ polling bit	1	Both	DL	Indicates whether RRBP field is valid
RRBP	Relative reserved block period	2	Both	DL	Used for allocation of PACCH uplink radio resources, for example, for TBF release handshaking
Payload type	–	2	Both	Both	Indicates whether payload is an RLC data block or control block and whether optional RLC header octets are included
R	Retry bit	1	Both	UL	Indicates whether the MS sent more than one PRACH/RACH channel request in the most recent channel access
SI	Stall indicator bit	1	Data	UL	Indicates whether the RLC transmit window can advance or not
CV	Countdown value	4	Data	UL	Indicates the number of RLC data blocks remaining at the end of an uplink TBF

between peer LLC entities is acknowledged and in-order delivery is guaranteed. Error recovery and reordering procedures based on retransmission (ARQ) are defined. The use of error detection (CRC) and flow control procedures are also defined.

- **Unacknowledged Mode.** In this mode, SDUs received from the upper layer are transmitted in numbered Unconfirmed Information (UI) frames. The transmission of UI frames between peer LLC entities is not acknowledged and in-order delivery is not guaranteed. Error recovery, reordering, and flow control procedures are not defined. The use of error detection (CRC) is defined, but the scope of error detection depends on the unacknowledged transport mode used. In unacknowledged mode, two transport modes are supported: protected and unprotected. With *protected transport mode*, error detection is applied to both the header and information fields and UI frames are discarded when errors are detected in the information field. With *unprotected transport mode*, error detection is applied to the header only and errors within the information field may be delivered to the upper layers.

These modes of operation are independent of the RLC modes of operation. The specific LLC operating mode that is used for the transmission of user data is a

Table 6.18 RLC Header Fields

Field	Field name	Size (bits)	Data/ Control	UL/ DL	Definition
FBI	Final block indicator bit	1	Data	DL	Signals the final RLC data block of a downlink TBF
TFI	Temporary flow identity	5	Both	Both	Identifies the TBF to which the RLC/ MAC block belongs
PR	Power reduction	2	Both	DL	Used for downlink power control
E	Extension bit	1	Data	Both	Indicates the presence of an optional extension octet in the RLC header
BSN	Block sequence number	7	Data	Both	Used to number RLC data blocks for in-order reassembly
M	More bit	1	Data	Both	Used to delimit LLC PDUs within an RLC data block (with E and LI fields)
Length indicator	–	6	Data	Both	Used to delimit LLC PDUs within the RLC data block (follows M bit)
TI	TLLI indicator bit	1	Data	UL	Indicates the presence of the optional TLLI field in the RLC header
PI	PFI indicator bit	1	Data	UL	Indicates the presence of the optional PFI field
PFI	Packet flow identifier	7	Data	UL	Used to support optional packet flow control procedures
TLLI	Temporary logical link identifier	32	Data	UL	Optional field containing the TLLI
AC	Address control bit	1	Control	DL	Indicates the presence of the optional TFI/D octet in the RLC header
FS	Final segment bit	1	Control	DL	Indicates the final segment of an RLC/MAC control message
RTI	Radio transaction identifier	5	Control	DL	Used to group the downlink control blocks that make up an RLC/MAC control message
RBSN	Reduced block sequence number	1	Control	DL	Used to number RLC/MAC control blocks for in-order reassembly
D	Direction bit	1	Control	DL	Indicates the direction of the TBF identified by the TFI field

function of the QoS profile defined in the PDP context for that transmission. Control and SMS frames are always transferred in unacknowledged mode.

The overall structure of the LLC layer is illustrated in Figure 6.31. LLC layer SAPs are defined to provide service to three upper layers: SNDCP, GMM, and SMS. The LLC-GMM SAP includes a signaling part to control the logical link management entity (LLME) and a data transfer part to carry GMM/SM control messages. The LLC-SNDCP interface is divided into four SAPs, each corresponding to a different QoS. Each SAP of the LLC has an associated logical link entity that is responsible for data transfer, flow control, and error detection for the data transfer associated with a particular SAP. As described previously in Section 6.6.3, the logical link between one SGSN and one MS is uniquely identified by means of a TLLI. The TLLI is a temporary identifier derived from the P-TMSI that uniquely identifies the MS within a particular RA. Within the LLC layer, the SAP and corresponding LLE are identified by the SAPI in both the MS and the SGSN. The combination of these two identities (SAPI + TLLI) is called the DLCI and it uniquely identifies the logical data link between two LLEs.

The general structure of an LLC frame is shown in Figure 6.32. The address field includes the SAPI and identifies the corresponding DLCI. Note that the address

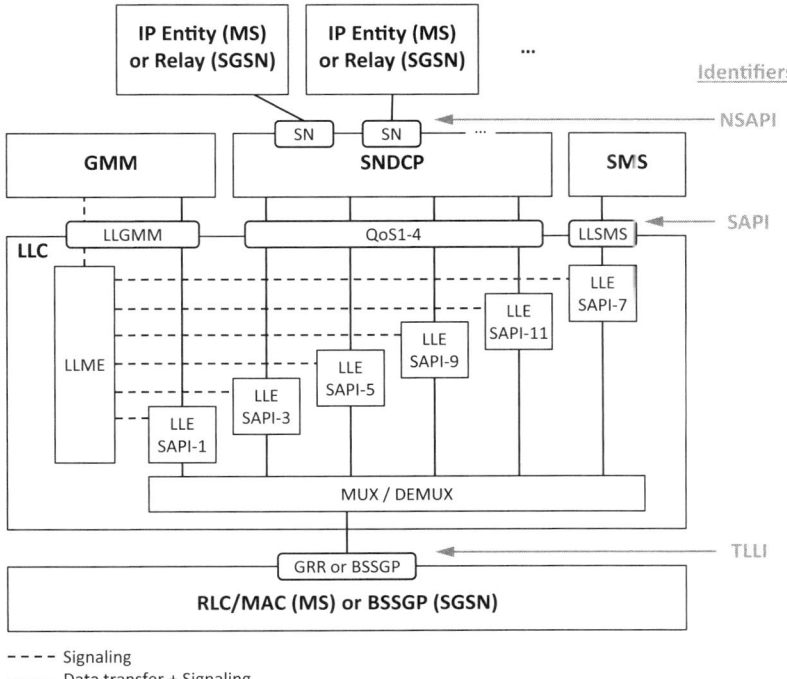

Figure 6.31   Structure of GPRS LLC layer.

Figure 6.32 GPRS LLC frame structure.

field in the LLC frame header does not include the TLLI. Rather, the TLLI is added by the next lower layer—either RLC or BSSGP—as part of that layer's header when necessary for disambiguation. The LLC header also includes a control field that ranges in size from 1 to 36 bytes and may include a variable length bitmap for selective acknowledgment (SACK) purposes. The information field carries the payload received from an upper layer entity (e.g., SNDCP PDU). The LLC operates under the principle of bit transparency, which means the bits of the information field are inserted into the LLC frame unaltered and variable length frames are supported. The maximum length allowed for the information field is a variable that may be set by the LLC, and is generally defined independently for each SAPI and frame type (i.e., I frame, UI frame). However, the maximum length supported is 1520 bytes. With the addition of LLC header and check bits, this means that the largest possible LLC frame length in GPRS is 1560 bytes.

A 3-byte frame check sequence (FCS) of parity bits is added at the end of every frame for error detection. The LLC layer performs frame-level error detection using a 24-bit CRC with the following generator polynomial:

$$G(x) = x^{24} + x^{23} + x^{21} + x^{20} + x^{19} + x^{17} + x^{16} + x^{15} + x^{13} + x^8 + x^7 + x^5 + x^4 + x^2 + 1.$$

In the case of unacknowledged unprotected mode, the CRC calculation is performed over the LLC and SNDCP headers only. The CRC calculation is performed immediately before ciphering (if used) on the transmit side, such that the entire frame is encrypted including the FCS.

6.9.12 GPRS SNDCP Layer

The SNDCP layer is the uppermost layer of the GPRS user plane at the air interface. This layer is responsible for the efficient and reliable transmission of N-PDUs (e.g., IP packets) over the Um and Gb interfaces. The main functions of the SNDCP layer are multiplexing, compression, segmentation, and the associated inverse functions. An essential aspect of the SNDCP layer is that it provides protocol transparency to the network layer entities above. This means that it abstracts the details of the GPRS air interface away from the upper layers, allowing the network layer protocol to operate the same way it would over a wired data link layer.

The SNDCP function multiplexes N-PDUs received from multiple network layer PDP entities, possibly using different protocols (e.g., IPv4, IPv6), onto the four LLC SAPs (i.e., from QoS1 to QoS4) based on the required QoS. In this sense, the

SNDCP offers a single point of convergence for packet data received from one or more layer 3 entities. Each network layer entity is identified by means of an NSAPI, which is saved in the associated PDP context.

To support the efficient use of radio resources, N-PDUs are first compressed. Two compression processes are defined, both of which are optional. The first is protocol-specific header compression. Common examples include Transmission Control Protocol/Internet Protocol (TCP/IP) header compression as specified in Request for Comments (RFC) 1144 and UDP/IP header compression as specified in RFC 2507 [141]. The second compression process is data compression, which applies to the entire N-PDU, including the possibly compressed protocol header. Compressed N-PDUs then undergo the segmentation process. This process ensures that the SNDCP PDUs formed after encapsulation conform to the maximum information field size defined by the LLC layer.

6.9.13   GPRS GMM/SM Layers

GPRS Mobility Management (GMM) and Session Management (SM) are the layer 3 signaling entities in the GPRS control plane. GMM is responsible for attach and detach procedures and associated mobility management states described previously in Section 6.9. GMM is also responsible for security functions (e.g., authentication), allocation of P-TMSI, assignment of TLLI (i.e., derived from the P-TMSI), and location management functions (e.g., cell update, RA update, combined RA/LA update). SM is responsible for the management of PDP contexts (i.e., activation, modification, deactivation). For coordination purposes, a SAP is defined between SNDCP in the user plane and SM in the control plane (i.e., SNSM-SAP). In an MS supporting both CS and PS services, GMM is part of the MM sublayer and coordinates with the MM entity defined for GSM layer 3 signaling. SM is also part of the CM sublayer, and becomes a fourth entity alongside CC, SS, and SMS.

6.10   GSM SECURITY ASPECTS

The security features of GSM focus primarily on preventing unauthorized users from gaining unpaid access to network services and protecting user data from passive eavesdropping. To these ends, GSM supports the following security mechanisms in both the CS and PS domains [142]:

- Subscriber identity confidentiality
- Subscriber identity authentication
- Ciphering of user data and associated signaling

6.10.1   Confidentiality

The GSM and GPRS signaling procedures have been designed to minimize as much as possible the number of times the user must transmit his IMSI over the air interface unencrypted. This is done primarily through the use of temporary identifiers, including the TMSI for CS domain services and the P-TMSI, TLLI, and TFI for PS domain services. Subscriber confidentiality makes it difficult to collect sensitive subscriber information for nefarious purposes, such as mobile phone cloning. The use of temporary subscriber identifiers also makes it more difficult to perform unauthorized user tracking through the network, as these identifiers change over time.

6.10.2   Authentication

GSM networks use authentication procedures to verify the identity of an MS to prevent unauthorized access to network services. It should be noted that GSM only supports authentication of the subscriber identity by the network, but does not support authentication of the network by the MS. The reason for this is because, at the time when GSM was designed, cell phone cloning was a major issue for network operators that cost the cellular industry a substantial amount of money. The authentication mechanism is designed primarily to address this problem. GSM authentication uses a challenge-and-response approach based on a secret 128-bit authentication key, Ki. The authentication key is assigned at time of subscription and is permanently stored in the SIM on the MS side and the AuC on the network side. All security functions are based on the secrecy of this key, which is never transmitted over the air. In the AuC, challenge–response pairs are generated using the GSM authentication algorithm, A3. The challenge is a 128-bit random number (RAND). Using algorithm A3, a 32-bit signature response (SRES) may be calculated that requires prior knowledge of the subscriber's secret key. In this sense, SRES is a digital signature that, in principle, can only be generated by the subscriber's SIM. The challenge–response values, RAND and SRES, may be sent back and forth over the air interface without compromising the integrity of the secret key, Ki. Furthermore, since a new challenge is sent each time, this prevents the impersonation of a legitimate MS by playing back an earlier SRES. The same authentication algorithm and authentication key, Ki, are used for both GSM and GPRS.

6.10.3   Ciphering

GSM provides confidentiality of user data at the air interface through ciphering, which includes the processes of encryption on the transmit side and decryption on the receive side. For GSM CS services, ciphering is implemented at the physical layer and terminates at the BTS. GSM ciphering provides confidentiality of user data and dedicated signaling transmitted on the TCH, SACCH, FACCH, and SDCCH. For PS services of GPRS, ciphering is implemented at the LLC layer—not the physical layer—and terminates in the SGSN. The reason for this is to simplify encryption key management, since GPRS operates in connectionless mode with cell reselection generally performed by the MS. By comparison, in GSM ciphering, key information is sent to the target BTS by the network when handover occurs. GPRS ciphering, when used, provides confidentiality of user data and GMM/SM signaling carried in LLC frames.

GSM ciphering is performed as a two-part process and involves the ciphering key generation algorithm, A8, and the ciphering algorithm, A5. The first part of the process is ciphering key generation. First, a 128-bit RAND is generated by the AuC. This is actually the same RAND generated for the authentication procedure. Using algorithm A8, a 64-bit ciphering key, Kc, is generated in the AuC. This key is saved in the HLR as part of an authentication and ciphering triplet (Kc, RAND, SRES) until it is requested by an MSC/VLR or SGSN. The algorithm A8 is also stored in the SIM, which may use RAND to generate a ciphering key identical to the one stored in the network. The second part of the process is the ciphering of data transmitted over the air. The ciphering key is transferred to the ME and the BTS, in the case of CS service, where encryption and decryption are performed using a bit-wise exclusive OR (XOR) scheme known as a *stream cipher* [142]. At the transmitter, each bit of the data stream undergoes an XOR operation with a ciphering bit stream to encrypt the data. At the receiver, decryption is performed using exactly the same method—the received data stream undergoes an XOR operation with the same ciphering bit stream. The ciphering bit stream is generated using algorithm A5. In GPRS, ciphering is performed in essentially the same way. The ciphering key generation process is identical, using algorithm A8 and the 64-bit ciphering key. However, a special GPRS encryption algorithm (GEA) is used for algorithm A5 that has been optimized for LLC frame transfer.

6.11 EDGE ENHANCEMENTS

EDGE was specified in Release 99 and provides improvements to the GSM/GPRS air interface designed to boost peak data rates, throughput, coverage, and capacity. This is accomplished by increasing the spectral efficiency of a single time slot. Although EDGE also defines improvements to the CS domain (ECSD), our focus will be on enhancements to the PS domain—known as enhanced GPRS (EGPRS). The main features of EGPRS include the addition of higher-order 8PSK modulation for roughly a threefold increase in peak data rate compared with GMSK modulation, link adaptation through adaptive coding and modulation schemes, and incremental redundancy (Type II HARQ) for improved RLC performance as compared with the Type I ARQ scheme of GPRS. These improvements have a significant impact on the PHY and RLC/MAC layer protocols. Since EDGE is an optional enhancement, an EDGE MS must be capable of operating in both GPRS and EGPRS modes.

6.11.1 Adaptive Modulation and Coding

Basic link adaption has already been specified in GPRS, which defines four coding schemes with different code rates to optimize the transmission of user data based on the current channel quality. EGPRS provides enhanced link adaption capabilities through the use of higher-order 8PSK modulation and additional modulation and

Table 6.19 EGPRS Modulation and Coding Schemes

Scheme	Modulation	Overall code rate	IR puncturing schemes	RLC data unit per radio block (bytes)	Data rate per time slot (kbps)	MCS Family
MCS-1	GMSK	0.53	2	22	8.8	C
MCS-2	GMSK	0.66	2	28	11.2	B
MCS-3	GMSK	0.85	3	37/31 + 37	14.8/13.6	A/A′
MCS-4	GMSK	1	3	44	17.6	C
MCS-5	8PSK	0.37	2	56	22.4	B
MCS-6	8PSK	0.49	2	74/68	29.6/27.2	A/A′
MCS-7	8PSK	0.76	3	56 × 2	44.8	B
MCS-8	8PSK	0.92	3	68 × 2	54.4	A′
MCS-9	8PSK	1	3	74 × 2	59.2	A

Adapted from References [137, 139].

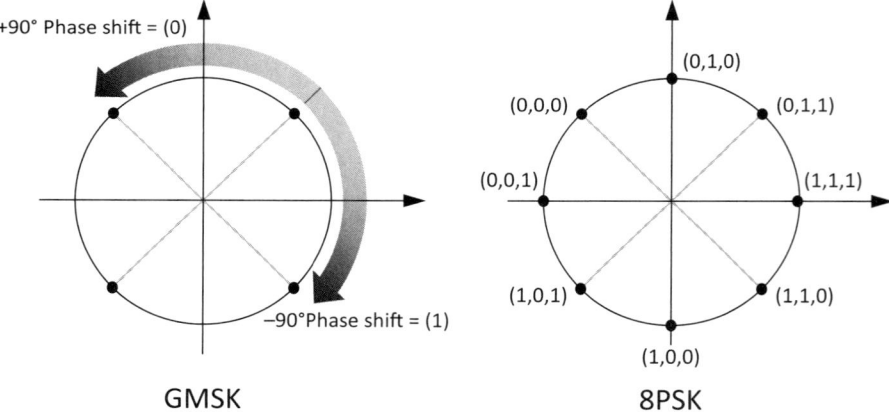

Figure 6.33 EGPRS modulation schemes.

coding schemes (MCSs). Since the purpose of these enhancements is to increase user data rates over the GPRS air interface, these enhancements apply specifically to the RLC data blocks of the PDTCH.

EGPRS defines nine MCSs, which are summarized in Table 6.19 [137, 139]. The first four schemes, MCS-1 to MSC-4, use standard GMSK modulation while the remaining five schemes, MCS-5 to MCS-9, use 8PSK modulation. Since 8PSK modulation carries 3 bits per symbol, a gray code is used for MCS-5 to MCS-9 to map triplets of encoded radio block bits onto 8PSK symbols, as shown in Figure 6.33. The coding scheme for all nine MCSs utilizes a block check sequence (BCS)

Table 6.20 EGPRS Modulation and Coding Scheme Families

Family	Associated MCS	Basic payload unit (bytes)
A	MCS-9, MCS-6, MCS-3	37
A′	MCS-8, (MCS-6, MCS-3)[a]	34
B	MCS-7, MCS-5, MCS-2	28
C	MCS-4, MCS-1	22

[a]For Family A′, six zero padding bytes are used when switching from MCS-8 to MCS-6 or MCS-3 to achieve the 34-byte basic payload unit.

for error detection, followed by a rate 1/3 convolutional code that is punctured to achieve the desired code rate. In all cases, the MAC header is coded separately from the remainder of the radio block to ensure strong header protection. In particular, the USF is precoded separately using a block code for extra robustness. The encoded USF takes up 12 symbols for each MCS—which means the 3-bit USF is encoded using 12 bits for GMSK and 36 bits for 8PSK. This applies additional protection when using 8PSK, as this modulation scheme is more susceptible to sudden decreases in channel quality. The remainder of the MAC header is protected with a header check sequence (HCS) consisting of an 8-bit CRC, followed by a rate 1/3 convolutional code with MCS-specific puncturing.

For retransmission purposes, the EGPRS MCSs are divided into four *families*, which are summarized in Table 6.20 (derived from information in Reference [137]). An MCS family is a group of two or three different MCSs that utilize the same basic payload unit. The unencoded size of the RLC data block (i.e., payload) that carried on a single radio block is an integer multiple (1, 2, or 4) of the same basic payload unit for all MCSs belonging to the same family. This concept is illustrated in Figure 6.34 [137] for each coding family. Note that Family A′ is formed using padded versions of MCS-6 and MCS-3 for lower data rates. When four basic payload units are transmitted using MSC-7 to MSC-9, the data rate per time slot is high enough that two RLC data blocks are carried on a single radio block.

The grouping of MSCs into families facilitates link adaptation by allowing erroneous RLC/MAC blocks to be retransmitted with a lower-order MCS on an integer number of radio blocks. For example, if Family A is used and if channel quality permits, the transmitter may start by using MCS-9 to transmit radio blocks carrying a payload of 148 bytes each (i.e., two 74-byte RLC data blocks). Over time, the link quality may be degraded due to increased noise and interference levels, causing a radio block not to be received. In this case, the transmitter may adapt to the current link conditions and retransmit the previous erroneous radio block using MCS-6—the next more robust MCS in Family A. In this case, the retransmitted RLC data is divided up and transmitted using two radio blocks carrying a payload of 74 bytes each. The data rate per time slot is divided in half in exchange for an increase in robustness. Now say that the link quality is degraded even further, causing another radio block error. The transmitter may further adapt to the degraded

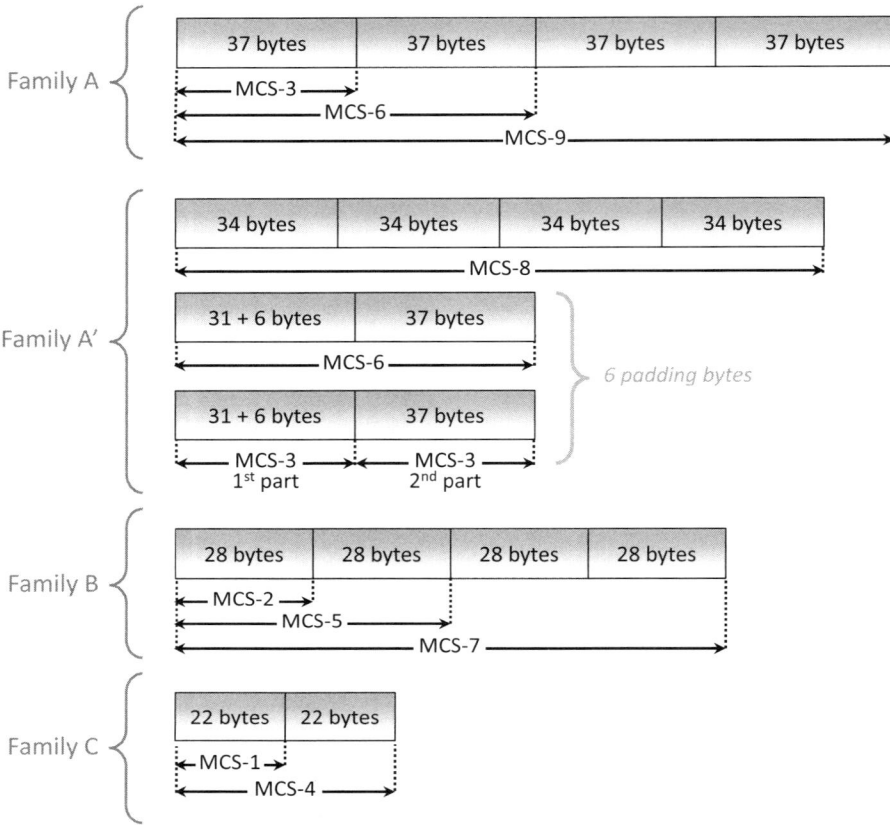

Figure 6.34   Payload sizes for EGPRS modulation and coding scheme families. Adapted from Reference [137].

channel condition by retransmitting an erroneous radio block using MCS-3. The retransmitted RLC data is once again divided in half and retransmitted using two radio blocks, this time with a payload of 37 bytes each—or one times the basic payload unit for Family A. The data rate per time slot is again divided in half in exchange for an increase in robustness, this time achieved by switching to GMSK modulation.

Clearly, a link adaptation mechanism is required for dynamic selection of MCS based on the current channel conditions. In EGPRS, a new parameter called bit error probability (BEP) is introduced to characterize the radio conditions on a burst-by-burst basis and determine which MCS should be used. The BEP is essentially a measure of radio link quality. Radio link quality measurements are performed in the downlink by the MS and in the uplink by the BTS, and communicated back to the transmitter through measurement reports.

6.11.2   Incremental Redundancy

Another key enhancement of EGPRS is the addition of a Type II HARQ scheme with incremental redundancy (IR), which is defined as part of the RLC layer [137]. When operating in IR mode, retransmissions of uncorrectable RLC data blocks use different puncturing schemes such that different coded bits are sent with each transmission. At the receiver, uncorrectable received blocks are not discarded. Instead, soft information from multiple, differently punctured received blocks is combined when decoding an RLC data block on subsequent attempts. In essence, this causes the number of coded bits sent across the channel to increase with each retransmission, thereby decreasing the code rate of the encoded data block at the decoder and increasing the amount of information at the receiver. This mechanism can significantly increase link performance as compared with the selective Type I ARQ mechanism of standard GPRS, in which uncorrectable data blocks are discarded. The EGPRS IR mechanism defines either two or three puncturing schemes for each MCS, as summarized in Table 6.17.

Incremental redundancy provides an additional mechanism to efficiently cope with variations in link quality without requiring the MS to switch to an MCS with a lower data rate. In terms of average throughput over time, it is generally better to retransmit a few erroneous data blocks than to use a slower data rate if channel conditions permit. EGPRS supports the use of incremental redundancy in combination with link adaptation as part of an overall link quality control scheme [143].

6.12   GSM EVOLUTION

This section describes key enhancements to the GSM standard beyond the introduction of EDGE in R99. Focus is placed on enhancements that significantly impact the air interface.

6.12.1   GERAN (Rel-5)

The GSM/EDGE Radio Access Network (GERAN) architecture is the evolution of GSM toward the concepts of 3G UMTS networks, supporting a similar set of services to that of 3G UMTS and allowing for a single converged core network evolution. The GERAN architecture was first defined in Rel-5, with the level of modifications comparable to that required for GPRS in R97 [143]. The motivation and historic aspects behind the definition of the GERAN architecture were discussed at the beginning of this chapter in Section 6.5.4. The main concept behind GERAN was to define a new, backward-compatible network architecture that allows network operators to leverage their existing radio access network infrastructure (e.g., towers, TRX hardware, BTS equipment) based on the 200 kHz TDMA radio interface of GSM/EDGE—now referred to as GERAN—as part of an enhanced 2G/3G multiradio network, including GERAN, UTRAN,[16] and a single converged core network. This new paradigm requires the adoption of the Iu interface for GERAN, which is supported by a new operating mode and protocol architecture.

The GERAN reference architecture is shown in Figure 6.35 [144]. GERAN may connect to the core network using the traditional A/Gb interfaces of GSM, the new Iu interface of UMTS, or a combination of these three interfaces. Based on this broad definition, the term GERAN may be used to include legacy GSM/EDGE networks which do not support the Iu interface. However, the term "GERAN" is really a part of the 3GPP Rel-5 and beyond vocabulary. The GERAN architecture also defines the optional Iur-g interface for signaling between two BSCs or between a BSC and a radio network controller (RNC) within the 3G infrastructure. This interface is based on the Iur interface, which may connect two RNCs in the UMTS architecture. The Iur-g interface may only be used in support of an MS operating in Iu mode.

An MS connected to GERAN operates in either A/Gb mode or Iu mode. The A/Gb mode uses the protocol architecture of earlier releases of GSM/GPRS/EDGE at the air interface and makes use of the A and Gb interfaces to connect to the core network. The A/Gb mode may be implemented in the network for backward compatibility with pre-Rel-5 terminals; however, a Rel-5 MS is required to support both A/Gb and Iu modes. The Iu mode, which is new to Rel-5, uses a new protocol architecture based on UMTS at the air interface and makes use of the Iu interface to connect to the core network. The Iu mode is used when a Rel-5 or beyond terminal

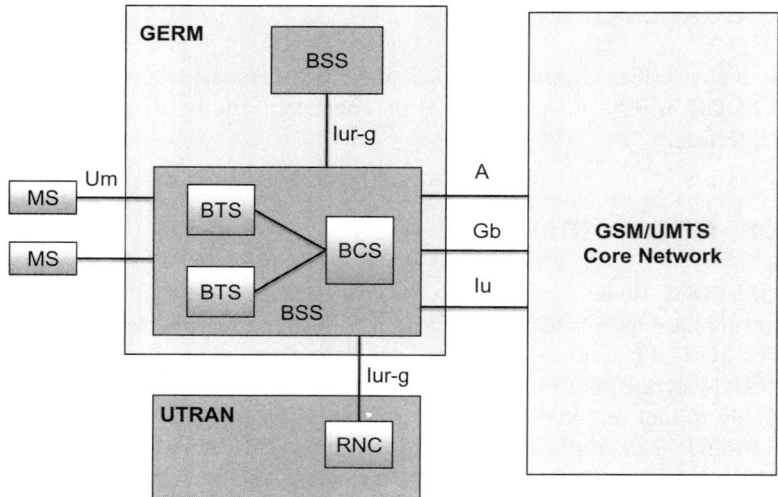

Figure 6.35   GERAN Reference Architecture. Adapted from Reference [144].

---

[16] Universal Terrestrial Radio Access Network (UTRAN) is the 3G UMTS radio access network, which is discussed in detail in the next chapter.

Table 6.21   Comparison of Network Functional Split in GERAN Modes

Network function	A/Gb mode	Iu mode
Ciphering	CN	RAN
Compression (IP header and payload)	CN	RAN
Termination of layer 2 protocols	RAN/CN	RAN
Termination of layer 3 signaling protocols	RAN/CN	RAN
Buffer management	CN	RAN
Radio resource handling and cell level mobility	RAN/CN	RAN

Adapted from Reference [143].

connects to a GERAN network supporting the Iu interface. The protocol architecture defined for GERAN Iu mode essentially uses the UMTS protocols at the air interface, except that the MAC and PHY layers are those of GSM/EDGE. At the Iu interface, the protocol architecture is identical to that of UMTS. A key difference between the Iu mode protocol architecture and the A/Gb mode protocol architecture is the functional split between the RAN and the core network. Table 6.21 [143] presents a comparison of the functional split of the two modes. By adopting an architecture based on UMTS, GERAN Iu mode allows for improved interworking between GERAN and UTRAN and provides greater continuity of service outside the 3G coverage area.

6.12.2 EDGE Evolution (Rel-7)

EDGE Evolution refers to a set of further enhancements to the GSM air interface defined in Release 7 under the umbrella term "evolved GERAN." The main driver for the development of EDGE Evolution is the widespread deployment of High-Speed Packet Access (HSPA), which has resulted in a large gap in terms of user experience for mobile Internet services between the 2G (i.e., GSM/EDGE) and 3G (i.e., Wideband Code Division Multiple Access [WCDMA]/HSPA) coverage areas. EDGE Evolution seeks to leverage existing GERAN infrastructure to provide a better minimum user experience for data services in multiradio networks, where GERAN is used to provide continuity of service over a broader geographic footprint than the 3G and 4G coverage areas alone.

EDGE Evolution increases peak data rates through the use of DLDC and EGPRS2—the latter of which improves upon EGPRS link adaptation capabilities. In addition to improved data rates, EDGE Evolution provides reduced latency through reduced transfer time interval (RTTI) and fast ACK/NACK reporting (FANR). It also includes mobile station receiver diversity (MSRD), which enables MS receiver diversity using two antennas for improved reception. All of this results in a significant improvement to the overall user experience for mobile Internet access over GERAN. Since Rel-7 EDGE Evolution is actually a set of enhancements, it provides an evolutionary pathway that may be rolled out in stages. It is anticipated that network operators will begin by deploying DLDC, followed by EGPRS2 [145]. As with EDGE, these enhancements can generally be deployed as a software upgrade, reducing cost and impact on the existing infrastructure. Each EDGE Evolution enhancement is now briefly described.

6.12.2.1 EGPRS2

EGPRS2 is the next step in EDGE link adaptation, which complements EGPRS and is compatible with it. Key features of EGPRS2 include higher-order modulation schemes, turbo coding, and higher symbol rate burst transmission—all of which are used hierarchically to adapt to current link conditions. EGPRS2 is divided into two separate levels based on MS capabilities: EGPRS2-A and EGPRS2-B. Both levels include new MCSs in both the uplink and downlink; however, EGPRS2-B schemes make use of the new higher modulating symbol rate of 325 ksymbols/s. Both levels make use of a rate 1/3 turbo code on the downlink and use the traditional rate 1/3 convolutional code of EGPRS on the uplink. As with EGPRS, EGPRS2 makes use of incremental redundancy for retransmissions. Either two or three puncturing schemes are defined for every EGPRS2-A and EGPRS2-B MCS. The definition of two different levels of EGPRS2 was done to provide manufacturers with an evolutionary pathway, allowing them to implement the EGPRS enhancements in two phases, thus facilitating early adoption.

Unlike EGPRS, the new MCSs defined in EGPRS2 are defined separately for the downlink and the uplink. A network supporting EGPRS2 may choose to implement only a subset of these MSCs. EGPRS2-A defines seven new MCSs in the downlink, named DAS-5 to DAS-12, and five in the uplink, named UAS-7 to UAS-8. These are used in conjuncture with a subset of the MCSs defined for EGPRS, namely MCS-1 to MCS-4 in the downlink and MCS-1 to MCS-6 in the uplink. The parameters for each MCS used in EGPRS2-A for both the downlink and the uplink are summarized in Table 6.22. MCSs that are new to EGPRS2-A are highlighted in white, while those that are reused from EGPRS are shaded in gray. EGPRS2-B defines seven additional MCSs in both the downlink and the uplink, named DBS-5 to DBS-12 and UBS-5 to UBS-12, respectively. All of these schemes utilize a new higher symbol rate of 325 ksymbols/s to transmit more symbols per burst, thus boosting data rates. EGPRS2-B also utilizes MCSs from both EGPRS and EGPRS2-A to achieve a full range of data rates for different link qualities. The parameters for each MCS used in EGPRS2-B are summarized in Table 6.23. Once again, MCSs that are new to EGPRS2-B are highlighted in white, while those that are reused from EGPRS and EGPRS2-A are shaded in gray. In addition to the MCSs shown, EGPRS2-A may optionally use MCS-7 and MCS-8 from EGPRS. EGPRS2-B may optionally use MCS-6 to MCS-9 from EGPRS and DAS-5 and DAS-8 as well as padded versions of DAS-10 and DAS-12 from EGPRS2-A.

The concept of MCS families developed for EGPRS has been reused for EGPRS2 (see Section 6.11). Each MSC within a family transmits one or more RLC data blocks that are an integer multiple of the same basic payload unit. The grouping

Table 6.22 EGPRS2-A Modulation and Coding Schemes

Direction	Scheme	Modulation	Symbol rate (ksymbols/s)	Channel code	Overall code rate	RLC data unit per radio block (bytes)	Data rate per time slot (kbps)	MCS Family
DL	MCS-1	GMSK	270.8	CC	0.53	22	8.8	C
	MCS-2	GMSK	270.8	CC	0.66	28/26 + 28	11.2/10.9	B/B'
	MCS-3	GMSK	270.8	CC	0.85	31 + 37	13.6	A'
	MCS-4	GMSK	270.8	CC	1	44	17.6	C
	DAS-5	8PSK	270.8	TC	0.37	56	22.4	B
	DAS-6	8PSK	270.8	TC	0.45	68	27.2	A'
	DAS-7	8PSK	270.8	TC	0.54	82	32.8	B'
	DAS-8	16-QAM	270.8	TC	0.56	56 × 2	44.8	B
	DAS-9	16-QAM	270.8	TC	0.68	68 × 2	54.4	A'
	DAS-10	32-QAM	270.8	TC	0.64	82 × 2	65.6	B'
	DAS-11	32-QAM	270.8	TC	0.80	68 × 3	81.6	A'
	DAS-12	32-QAM	270.8	TC	0.96	82 × 3	98.4	B'
UL	MCS-1	GMSK	270.8	CC	0.53	22	8.8	C
	MCS-2	GMSK	270.8	CC	0.66	28	11.2	B
	MCS-3	GMSK	270.8	CC	0.85	37/27 + 37	14.8/12.8	A/A"
	MCS-4	GMSK	270.8	CC	1	44	17.6	C
	MCS-5	8PSK	270.8	CC	0.37	56	22.4	B
	MCS-6	8PSK	270.8	CC	0.49	74/64	29.6/25.6	A/A"
	UAS-7	16-QAM	270.8	CC	0.55	56 × 2	44.8	B
	UAS-8	16-QAM	270.8	CC	0.62	64 × 2	51.2	A"
	UAS-9	16-QAM	270.8	CC	0.71	74 × 2	59.2	A
	UAS-10	16-QAM	270.8	CC	0.84	56 × 3	67.2	B
	UAS-11	16-QAM	270.8	CC	0.95	64 × 3	76.8	A"

CC, Convolutional Code; TC, Turbo Code.

Table 6.23 EGPRS2-B Modulation and Coding Schemes

Direction	Scheme	Modulation	Symbol Rate (ksymbols/s)	Channel Code	Overall code rate	RLC data unit per radio block (bytes)	Data rate per time slot (kbps)	MCS Family
DL	MCS-1	GMSK	270.8	CC	0.53	22	8.8	C
	MCS-2	GMSK	270.8	CC	0.66	28	11.2	B
	MCS-3	GMSK	270.8	CC	0.85	37/31 + 37	14.8/13.6	A/A'
	MCS-4	GMSK	270.8	CC	1	44	17.6	C
	DBS-5	QPSK	325	TC	0.49	56	22.4	B
	DAS-6	8PSK	270.8	TC	0.45	68	27.2	A'
	DBS-6	QPSK	325	TC	0.63	74	29.6	A
	DBS-7	16-QAM	325	TC	0.47	56 × 2	44.8	B
	DAS-9	16-QAM	270.8	TC	0.68	68 × 2	54.4	A'
	DBS-8	16-QAM	325	TC	0.60	74 × 2	59.2	A
	DBS-9	16-QAM	325	TC	0.71	56 × 3	67.2	B
	DAS-11	32-QAM	270.8	TC	0.80	68 × 3	81.6	A'
	DBS-10	32-QAM	325	TC	0.72	74 × 3	88.8	A
	DBS-11	32-QAM	325	TC	0.91	68 × 4	108.8	A'
	DBS-12	32-QAM	325	TC	0.98	74 × 4	118.4	A
UL	MCS-1	GMSK	270.8	CC	0.53	22	8.8	C
	MCS-2	GMSK	270.8	CC	0.66	28	11.2	B
	MCS-3	GMSK	270.8	CC	0.85	37/31 + 37	14.8/13.6	A/A'
	MCS-4	GMSK	270.8	CC	1	44	17.6	C
	UBS-5	QPSK	325	CC	0.47	56	22.4	B
	UBS-6	QPSK	325	CC	0.62	74/68	29.6/27.2	A/A'
	UBS-7	16-QAM	325	CC	0.46	56 × 2	44.8	B
	UBS-8	16-QAM	325	CC	0.60	74 × 2/68 × 2	59.2/54.4	A/A'
	UBS-9	16-QAM	325	CC	0.70	56 × 3	67.2	B
	UBS-10	32-QAM	325	CC	0.71	74 × 3/68 × 3	88.8/81.6	A/A'
	UBS-11	32-QAM	325	CC	0.89	68 × 4	108.8	A'
	UBS-12	32-QAM	325	CC	0.96	74 × 4	118.4	A

of EGPRS2 MSCs into families facilitates link adaptation by allowing erroneous radio blocks to be subdivided and retransmitted using a lower-order MSC that is more robust but less bandwidth efficient. EGPRS2 uses the same MSC families (i.e., A, B, C, and A') and basic payload unit summarized previously in Table 6.19. However, EGPRS2-A makes use of two additional families, denoted B' (downlink only) and A'' (uplink only), which use a basic payload size of 82 and 64 bytes, respectively. Similar to Family A', Family B' and A'' are modified versions of existing families, as they use padded members of Family B or Family A when changing from high-data-rate MCSs to a more robust, low-data-rate MCS.

6.12.2.2 *Downlink Dual Carrier*

Downlink dual carrier (DLDC) is a key feature of Rel-7 that allows a single MS to simultaneously receive on two RF channels belonging to the same serving cell. Radio resources are allocated by the BTS such that the MS receives simultaneously on two carrier frequencies for a given TDMA time slot. An example is illustrated in Figure 6.36, where the time slots allocated to a single MS supporting DLDC for a single radio block period are shaded in dark gray. For this example, the MS receives on nine time slots in the downlink and transmits on one time slot in the uplink. The maximum number of downlink PDCHs that may be allocated depends on the multislot class of the MS. For example, an MS supporting any of multislot classes 40–45 may be allocated as many as 12 downlink PDCHs spread over 6 time slots within a single TDMA frame in DLDC mode [146]. When used in conjuncture with EGPRS, this yields a theoretical peak data rate of 710.4 kbps (i.e., 12×59.2 kbps). When used in conjuncture with EGPRS2-B, this yields a theoretical peak data rate of 1.42 Mbps (i.e., 12×118.4 kbps). When operating in DLDC mode, the MS must report channel quality measurements on each downlink carrier channel. DLDC may be used in conjuncture with other common GSM features, such as frequency hopping and DTM.

Although DLDC only supports dual-carrier configuration in the downlink, there are benefits that may be realized in the uplink as well. If the number of downlink

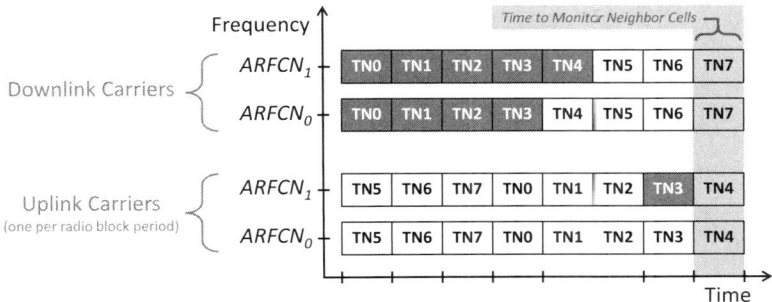

Figure 6.36 Example radio resource allocation for a single MS supporting DLDC.

PDCHs remains fixed, then the number of time slots needed to support the downlink may be decreased by a factor of 2. This approach frees up time slots to accommodate increased bandwidth on the uplink without sacrificing downlink data rates. Another benefit to the uplink is that radio blocks may be allocated on either of the uplink carriers associated with the dual downlink carriers, just not at the same time (i.e., not during a single radio block period). This provides increased flexibility for dynamic allocation of uplink radio resources.

Downlink dual carrier does not require any hardware upgrade on the part of the network, since user data on each individual carrier is still transmitted normally at the physical layer. Only a software upgrade is required with respect to radio resource management. The main impact of DLDC will be on the MS, which requires a new receiver architecture to support simultaneous reception on two carrier frequencies. Potential MS receiver architectures that could be used include: (1) separate narrowband receiver chains, using either the same antenna or separate antennas, and (2) single wideband receiver chain [147].

6.12.2.3 Latency Reduction

While DLDC and EGPRS2 may inherently decrease overall latency by allowing more RLC/MAC blocks to be transmitted within a single frame, Rel-7 includes two additional features that are specifically focused on latency reduction at the air interface: Reduced Transmission Time Interval (RTTI) and Fast ACK/NACK Reporting (FANR).

The RTTI feature reduces latency by reducing the amount of time necessary to transmit a single radio block over the radio link by a factor of two. In legacy GPRS/EGPRS networks, one radio block is divided into four normal bursts, which are then transmitted on a single time slot over the span of four consecutive TDMA frames. This configuration results in an average transmission time interval (TTI) of 20 ms per radio block,[17] which is now referred to as basic TTI (BTTI) configuration. The RTTI feature allows an alternative RTTI configuration to be used. In this configuration, two consecutive time slots of a TDMA frame are grouped together to form *PDCH-pair*. The four bursts of a radio block are then transmitted on the PDCH-pair over the span of two consecutive TDMA frames (i.e., two bursts per TDMA frame). The RTTI configuration, therefore, results in a reduced TTI of 10 ms per radio block.

The FANR feature takes a different approach to latency reduction and focuses on reduction in delay caused by ACK/NACK control message signaling at the RLC layer. The fundamental mechanism for communicating ACK/NACK messages between RLC entities operating in acknowledged mode has remained essentially unchanged since R97 GPRS. After a number of RLC data blocks have been transmitted, the receiver creates an ACK/NACK bitmap to indicate the reception status of each block and sends it to the transmitter in an RLC control block. To minimize impact on radio resource availability for user traffic, these control messages are

---

[17] Recall that a 52-multiframe carries 12 radio blocks over a duration of 240 ms counting PTCCH and Idle frames in pre-Rel-7 networks (i.e., BTTI configuration).

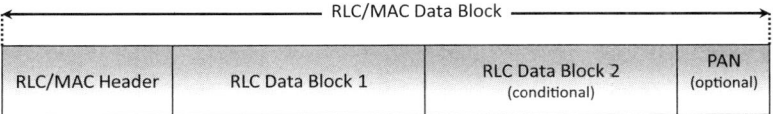

Figure 6.37   Example RLC/MAC block structure including PAN field for FANR.

generally kept to a minimum. In the event that an RLC data block is received in error, its retransmission is therefore delayed proportional to the wait time until the next ACK/NACK control message, thereby delaying in-order delivery of LLC PDUs to the layer above. FANR addresses this problem by allowing the piggybacking of ACK/NACK information within RLC/MAC blocks carrying user data. This is accomplished through the definition of the piggybacked ACK/NACK (PAN) field, which may be optionally appended at the end of an RLC/MAC block as shown in Figure 6.37. The benefits of FANR specifically apply to RLC acknowledged mode when a TBF exists in the opposite direction. For example, a downlink TBF in RLC acknowledged mode may utilize the PAN field for FANR on an existing uplink TBF. Because FANR impacts the structure of the RLC/MAC block, it requires modification to the modulation and coding schemes used at the physical layer. FANR-compatible MCS variants exist for all EGPRS and EGPRS2 schemes with the exception of MCS-4 and MCS-9.

6.12.2.4 Mobile Station Receiver Diversity

Mobile station receiver diversity (MSRD) is a downlink feature that utilizes an advanced receiver architecture based on two antennas to cancel interference. In addition to allowing the MS to operate in adverse interference environments, this improved receiver performance creates synergy when combined with EGPRS/ EGPRS2 link adaptation, allowing the MS to operate at higher data rates and improving spectral efficiency. MSRD is also referred to as Downlink Advanced Receiver Performance (DARP) Phase 2. It is a natural extension of the single antenna interference cancellation (SAIC) feature—also known as DARP Phase 1— specified in Rel-6. One driving factor for the development of MSRP for GSM/ EDGE is the definition of multiple-input, multiple-output (MIMO) techniques in HSPA+ and LTE-Advanced, which will increase the use of multiple antennas at the mobile device. MSRP allows multiradio devices that include these advanced receiver architectures to leverage this additional hardware for diversity gain over multipath fading channels when operating in GSM/EDGE mode.

6.12.3 VAMOS (Rel-9)

Voice services over Adaptive Multiuser Channels on One Slot (VAMOS) is an enhancement for GSM circuit-switched voice service first defined in Rel-9. VAMOS allows two full-rate voice traffic channels to share a single physical channel concurrently, thus improving the voice capacity of a single cell. VAMOS is the result of the Multi-User Reusing One Slot (MUROS) feasibility study concluded in Rel-8 and documented in 3GPP Technical Report (TR) 45.914 [148]. This study was primarily motivated by the tremendous increase in demand for mobile voice services that occurred in recent years in developing markets, such as China and India. As voice services become cheaper, operators face the challenge of more efficiently utilizing existing hardware and spectrum assets to maintain profitability. In densely populated cities where spectrum is scare and network load is high, an evolutionary option to increase voice capacity while minimizing impact on existing hardware, spectrum, and network planning is therefore desirable. VAMOS addresses this need by providing improved spectral efficiency at the air interface (i.e., voice traffic capacity may be increased by a factor of two per physical channel) and also improved BTS hardware efficiency (i.e., voice traffic capacity may be increased by a factor of two per TRX). It is expected that VAMOS can be implemented as a software upgrade for most BTS configurations, though backhaul upgrades may be required to support the increased voice traffic.

VAMOS is primarily an enhancement to the physical layer and radio resource management (RR) sublayer. On the downlink, multiple users are multiplexed onto a single time slot through the use of orthogonal subchannels (OSCs) that are created using a rotating hybrid quaternary complex symbol constellation called adaptive QPSK (AQPSK), as illustrated in Figure 6.38 [147]. The in-phase (I) and quadrature (Q) components of the AQPSK constellation are used to form the two OSCs. The constellation is adaptive in the sense that the parameter α, which represents the angle between the I and Q components, may vary over time. The constellation utilizes a special power control mechanism that supports the allocation of different power levels to each of the subchannels, defined by the subchannel power imbalance ratio (SCPIR). The SCPIR is given by the equation:

$$SCPIR(dB) = 20 \times \log_{10}(\tan \alpha) = 10 \times \log_{10}\left(\frac{Power(OSC\_B)}{Power(OSC\_A)}\right),$$

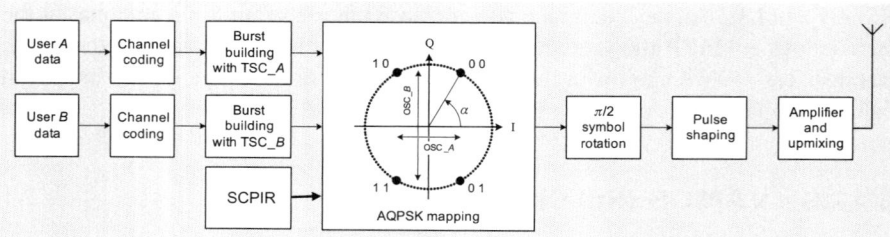

Figure 6.38 Baseband transmitter block diagram for VAMOS downlink. Adapted from Reference [147].

where OSC_A is the in-phase subchannel and OSC_B is the quadrature subchannel [149]. Furthermore, the value of α must be chosen such that:

$$|\text{SCPIR}| \leq 10 \text{ dB.}$$

To further minimize cochannel interference, a new set of training sequences have been defined for VAMOS, called training sequence code (TSC) set 2, to complement the existing set of eight training sequences defined for the normal burst. In particular, the training sequences in TSC set 2 have been selected to minimize the cross-correlation with the corresponding TSC in set 1. The MS operating on OSC_A is assigned a TSC from set 1, while the MS operating on OSC_B is assigned a TSC from set 2. In this way, only the second MS needs to support VAMOS, while the first MS may be a pre-Rel-9 legacy device. VAMOS also defines a legacy mode, in which two legacy mobiles are multiplexed using TSCs from set 1. However, this may require modification to existing cell planning, as the second TSC from set 1 may already be in use in nearby cells.

On the uplink, the physical layer is not changed, with the exception of the addition of TSC set 2. Each MS transmits at the same time using GMSK modulation and the same TSC as used in the downlink. From an MS perspective, no significant change is required at the physical layer as compared to a legacy GSM device. At the BTS receiver, however, a multiuser detection algorithm must be implemented, such as space–time interference rejection combining (STIRC), successive interference cancellation (SIC), or joint detection (JD) [147].

6.13 GSM USAGE

GSM was originally designed to provide digital mobile voice in a pocket-sized device that would allow the user to roam throughout all of Europe while maintaining continuity of service. It has certainly achieved that goal and more, with an expanded scope to include international roaming on a global scale. Today, GSM is the single most widely used technology for digital cellular voice service. GSM is also an important part of the wireless Internet. Enhanced data services realized through the addition of GPRS and EDGE are now widely deployed, and further evolutionary enhancements continue to be added.

By January 2012, GSM had achieved over 4.4 billion connections worldwide, accounting for roughly 73% of the total global cellular market according to the GSM Association (GSMA) Wireless Intelligence. Although the absolute number of GSM connections worldwide was still on the rise as of 2012, it is projected that these numbers will peak and—for the first time in the history of GSM—begin to decrease over the course of the next decade. This trend will be precipitated primarily by the maturation of 3G technologies and proliferation of new 4G technologies. However, after achieving over 4 billion connections spanning nearly every country around the globe, it is conceivable that GSM will remain in commercial use to some degree for decades to come.

The EDGE enhancement to GSM, first deployed in 2003, has become a mainstream technology. In fact, the Global Mobile Suppliers Association (GSA)[18] estimates that more than 80% of GPRS network operators have committed to using EDGE—including 545 GSM/EDGE networks in commercial service in 198 countries [150]. EDGE is an appealing upgrade option from the network operator's perspective because it is typically deployed as a software upgrade, making it possible to leverage existing GSM/GPRS infrastructure for increased data rates without significant capital investments. Although 3G UMTS networks have achieved widespread deployment over the past decade, they still are not as widely deployed in many countries as GSM in terms of geographic footprint. As 3G network coverage continues to expand in this regard, GSM/EDGE networks will continue to play an important role in providing continuity of service as the mobile Internet user leaves the 3G and 4G coverage areas. GSM/EDGE networks will also continue to play a key role in providing voice services in multiradio networks, thus freeing up 3G and 4G radio resources for high-speed data traffic. In terms of user devices, EDGE has become a standard handset feature in all but the lowest cost market segments. As of April 2011, it was being integrated into roughly 84% of HSPA mobile broadband devices, with a typical implementation supporting all four of the global GSM bands [150].

6.13.1 Global Spectrum Deployment

GSM quad-band phones have become very common, enabling users to make phone calls and access the Internet (for a fee) across most of the world. GSM is commercially deployed in four primary frequency bands. These bands occupy spectrum at 900, 1800, 850, and 1900 MHz as previously discussed. GSM global deployments operate in one or more of these frequency bands. These frequency bands are licensed spectrum and are therefore controlled by regional and international regulatory agencies. Figure 6.39 illustrates the global spectrum allocation for GSM network deployments. In most regions, GSM deployments utilize either a combination of GSM 900 and/or DCS 1800 spectrum or a combination of GSM 850 and/or PCS 1900 spectrum. These common spectrum combinations are largely due to historic reasons. The GSM 900 and DCS 1800 bands were originally developed for deployment in Europe, while the GSM 850 and PCS 1900 bands were originally developed for the North American market. Today, the 900 and 1800 MHz bands are used throughout Europe, Africa, the Middle East, Asia Pacific, Australia, and New Zealand. The 850 and 1900 MHz bands are used throughout North America and parts of Latin America. Throughout Central America, South America, and the Caribbean, a combination of

---

[18] The GSA is an advocacy group for the mobile supply industry related to GSM, UMTS, and LTE technology—including network infrastructure, mobile device, and semiconductor manufacturers. Their Web site is a good resource for market data pertaining to 3GPP-related technologies: http://www.gsacom.com.

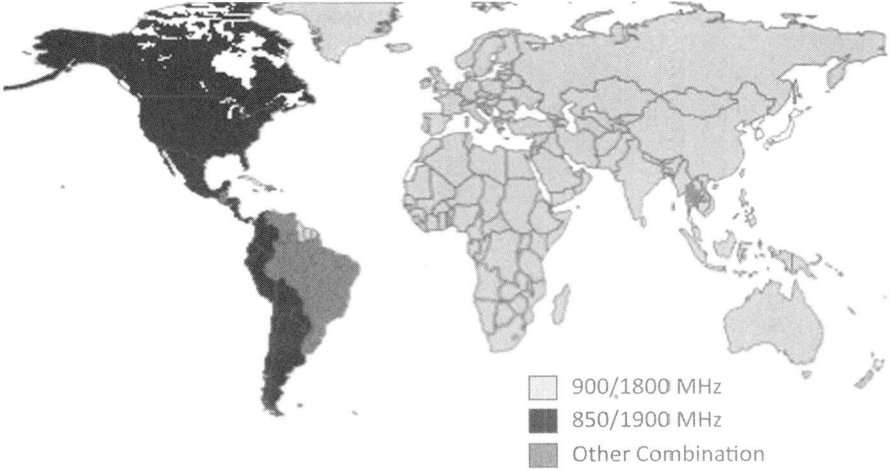

Figure 6.39 GSM global frequency allocations.

the four GSM frequency bands are used. While some countries conform to the conventional split between 900/1800 and 850/1900, many countries in this region now utilize a different combination of these frequency bands, often with deployments in three or four of these frequency bands coexisting within the same country. This is the case, for example, in Brazil, Guatemala, Jamaica, and Uruguay, to name a few. This situation is due in part to a combination of close proximity to North America combined with the presences of European overseas territories. As a result, the subject of spectrum allocation is a significant topic in the future cellular regulatory climate of this region. While GSM is deployed to some extent in nearly every country around the world today, coverage area and availability of services will vary from region to region. Figure 6.39 is not intended to imply that equal coverage exists in all parts of the shaded countries. Finally, it is worth noting that there are a few notable countries in which GSM has not been deployed, such as Japan and South Korea.

FURTHER GSM READING

GSM Standards documents are available for free through the 3GPP Web site (http://www.3gpp.org). Navigating the GSM standards documents can be intimidating at first due to the sheer number of standards documents available. However, with an understanding of the organization structure of 3GPP standards documents, access to such a large amount of information becomes a powerful tool. The 3GPP standards documents are organized into series based on general subjects. For example, the 45 series contains standards related to radio aspects of the GSM physical layer on the air interface. Each series is then subdivided into technical specification documents based on specific subjects; for example, modulation on the GSM air interface is

Table 6.24 GSM Technical Specifications for Further Reading

Topic	3GPP specification number (R99/ Rel-4 and beyond)	ETSI specification number (up to R99/Rel-4)
Network architecture	23.002	03.02
Numbering, addressing, and identification	23.003	03.03
Physical layer overview (radio aspects)	45.001	05.01
Physical layer overview (protocol aspects)	44.004	04.04
Enhanced full-rate (EFR) speech codec overview	46.051	06.51
GPRS overall description	23.060	03.60
GPRS radio interface overall description	43.064	03.64
GPRS RLC/MAC protocol	44.060	04.60
GPRS LLC protocol	44.064	04.64
GPRS SNDCP protocol	44.065	04.65
DTM	43.055	03.55
EDGE link adaptation	45.009	05.09
GERAN overall description	43.051	–

Technical Specification (TS) 45.004. Notice that the TS number takes the form *XX. YYY*, where *XX* specifies the series and *YYY* specifies the specific subject. On the Web page corresponding to TS 45.004, it is possible to access the version of TS 45.004 corresponding to each release of the specification. This makes it possible to see what specific features are included in different releases of the GSM standard; for example, Rel-10 is version 10.0.0. The GSM TSs are maintained and updated over time by the standards body. Notice that the version number takes the form *X.Y.Z*. For 3GPP TSs, where *X* corresponds to the release number, *Y* is incremented every time there is a substantive revision to the TS, and *Z* is incremented if there is an editorial revision to the TS. For archival purposes, all old versions are available on the Web site. However, to view the most up-to-date version of the standard, it is best to view the latest version number for the desired release.

As a starting point for further reading in the GSM standard, a number of important test specifications are listed in Table 6.24. This list includes a number of overview specifications pertaining to various topics of GSM, GPRS, EDGE, and GERAN. These documents provide a good launching point and, in many cases, provide references to the more detailed normative specifications that focus on more narrow topics. The numbering scheme previously described applies to 3GPP documents, which includes Rel-4 and beyond for TSs which apply only to GSM (i.e., series 41–55) or R99 and beyond for TSs which apply to both GSM and UMTS (i.e., series 21–28). For earlier releases, the corresponding ETSI specification number is provided as well.

In addition to the standard, there is a wide selection of books written on various aspects of GSM. Perhaps the most comprehensive book on the history and development of GSM from its inception until the year 2001 is *GSM and UMTS: The Creation of Global Mobile Communications*, edited by Friedhelm Hillebrand [51], which includes contributions from many of the individuals involved in GSM's development. It is highly recommended to those curious about why GSM was designed the way it was and the story of its development. *GSM Architecture, Protocols, and Services* by Eberspacher et al. [114] provides a comprehensive description of GSM operation in the circuit-switched domain. Finally, for a more up-to-date overview of GSM enhancements from Rel-7 to Rel-9, including a more in-depth description of Rel-7 EDGE Evolution, the authors recommend *GSM/EDGE: Evolution and Performance* by Säily et al. [147].

ACKNOWLEDGMENTS

Chapter 6 contains numerous figures and tables reproduced or adapted from 3GPP Technical Specifications. Each figure is individually referenced indicating the source 3GPP document. The authors would like to thank ETSI for kind permission to reproduce this information. Copyright 1999–2012. TSs and TRs are the property of ARIB, ATIS, CCSA, ETSI, TTA and TTC, who jointly own the copyright in them. They are subject to further modifications and are therefore provided "as is" for information purposes only. Further use is strictly prohibited.

Other figures and tables have been reproduced from John Wiley & Sons, Inc., sources, specifically Figures 6.3, 6.12, 6.33, and 6.39, along with Tables 6.6, 6.7, 6.8, and 6.21. Each figure is individually referenced indicating the source. This material is reproduced with permission of John Wiley & Sons, Inc.

Chapter 7

Third-Generation (3G) Cellular Communications

7.1 UNIVERSAL MOBILE TELECOMMUNICATIONS SYSTEM/WIDEBAND CODE DIVISION MULTIPLE ACCESS

The Universal Mobile Telecommunications System (UMTS) is a code division multiple access (CDMA)-based technology standardized by the 3rd Generation Partnership Program (3GPP), often marketed under the term "3GSM," that has experienced tremendous worldwide deployment. UMTS meets all of the requirements of the International Telecommunication Union (ITU) International Telecommunications-2000 (IMT-2000), a global standard for third-generation (3G) wireless mobile communications. UMTS maps out an evolutionary path from Global System for Mobile Communications (GSM) to achieve higher data rates and system capacity. While UMTS reuses the existing GSM core network, the UMTS Terrestrial Radio Access Network (UTRAN) introduces new radio access technologies (RATs), most prominently Wideband Code Division Multiple Access (WCDMA), which utilizes one 5-MHz channel for voice and data services. While initially offering data speeds up to 384 kbps, subsequent evolutionary upgrades to UMTS, such as High-Speed Downlink Packet Access (HSDPA) and High-Speed Uplink Packet Access (HSUPA), collectively known as High-Speed Packet Access (HSPA), offer much faster data speeds. After experiencing relatively slow deployment rates immediately after its conception, WCDMA-based UMTS networks have since amassed a significant number of deployments worldwide. According to the latest Global Mobile Suppliers Association (GSA) survey (January 20, 2012), there are over 424 commercial 3G WCDMA operators operating in 165 countries, serving over 822.4 million subscribers, including 469 million HSPA subscribers [151].

7.1.1 W-CDMA Historical Background

3G cellular communications systems are intended to provide a mobile wireless communications capability that will globally provide the ability to support a

Wireless Networking: Understanding Internetworking Challenges, First Edition.
Jack L. Burbank, Julia Andrusenko, Jared S. Everett, and William T.M. Kasch.
© 2013 the Institute of Electrical and Electronics Engineers, Inc. Published 2013 by John Wiley & Sons, Inc.

wide range of services to users, including voice, messaging, paging, and broadband data. The ITU began the process of 3G standardization with its IMT-2000 initiative. The IMT-2000 standard loosely specifies requirements for the 3G cellular network.

There are five standards in existence that satisfy all of the IMT-2000 requirements and are approved as 3G technologies by the ITU: UMTS, Code Division Multiple Access 2000 (CDMA2000), GSM Enhanced Data Rates for GSM Evolution (EDGE), Digital Enhanced Cordless Telecommunications (DECT), and Mobile Worldwide Interoperability for Microwave Access (WiMAX). Interestingly, while GSM EDGE, DECT, and Mobile WiMAX standards do meet all of the IMT-2000 requirements, they are not typically marketed as 3G, and employ completely different technologies.

3G standardization originally began within the 3GPP. The 3GPP was established in December 1998 as a collaboration between multiple regional telecommunications standards bodies: the Association of Radio Industries and Business (ARIB) in Japan, the Telecommunication Technology Committee (TTC) in Japan, the Alliance for Telecommunications Industry Solutions (ATIS) in the United States, the China Communications Standards Association (CCSA) in China, the Telecommunications Technology Association (TTA) in Korea, and the European Telecommunications Standards Institute (ETSI). Together, these standards bodies comprise the Organizational Partners for 3GPP. The 3GPP Project Agreement, signed by all the organizational partners, states that they shall cooperate in producing "globally applicable" technical specifications and reports for a 3G mobile system based primarily on GSM CN and the RATs they support, such as UMTS, which is based on a WCDMA air interface. The 3GPP was established primarily for preparation, approval, and maintenance of technical specifications and reports for 3G networks based on the GSM core structure. Furthermore, 3GPP is not considered a legal entity.

From the onset, it was apparent that 3GPP would not produce a meaningful technology given the rift that quickly formed between two groups of participant companies (representing current service providers and vendor equipment). This is due, in part, because cellular providers in the United States and elsewhere have failed to settle on a uniform second-generation (2G) mobile standard. In Europe, the uniform standard is GSM. However, in the United States and other countries, a mixture of systems currently exists, including GSM, the older time division multiple access (TDMA) Interim Standard (IS)-136, and IS-95 (the 2G CDMA cellular technology). GSM is similar to IS-136 in that it uses a TDMA scheme for multiple access, but the two are incompatible systems. However, CDMA is fundamentally different from GSM and IS-136 standards. Because of the investment by carriers in choosing one of these standards over another, if one single 2G standard was chosen over the other as the basis for a 3G migration, carriers not already on the base 2G standard would incur significant costs to upgrade their networks. Consequently, one group within 3GPP wanted to evolve existing GSM networks into the eventual 3G solution (the group that represents current 3GPP). The other group wanted to evolve existing IS-95 CDMA cellular networks (commonly referred to as CDMA One) into the eventual 3G solution.

For this reason, a second consortium was established. The 3rd Generation Partnership Program 2 (3GPP2) was formed to allow both consortiums to independently develop IMT-2000-compliant technologies. The 3GPP2 was established in December 1998 as a collaboration between multiple regional telecommunications standards bodies: the ARIB in Japan, the CCSA in China, the Telecommunications Industry Association (TIA) in North America, the TTA in Korea, and the TTC in Japan. Together, these standards bodies comprise the Organizational Partners for 3GPP2. Moreover, Market Representation Partners include the CDMA Development Group, the Internet Protocol version 6 (IPv6) Forum, and the International 450 Association. These Market Representation Partners offer market advice and a consensus view on market requirements. 3GPP2 was established primarily for preparation, approval, and maintenance of technical specifications and reports for 3G networks based on the cdma2000CN structure and cdma2000 air interface. Like 3GPP, 3GPP2 is not considered a legal entity.

The resulting technologies from the 3GPP and 3GPP2 groups are UMTS and CDMA2000, respectively, although both technologies rely on a CDMA-based air interface: WCDMA in the case of UMTS and CDMA2000 (commonly referred to as CDMA2000 1x).

UMTS supports high data rates and multiple services with individual quality-of-service (QoS) requirements. UMTS is capable of high-speed packet data by utilizing HSDPA and HSUPA technologies. In HDSPA, for instance, the maximum downlink (DL) PS data rate is 14.4 Mbps, while the maximum data rate in the original UMTS Release 99 is 384 kbps.

The development of 3G systems began in Europe as GSM was taking its first steps toward global success. 3G meant a system that was not only optimized for speech but also possessed a high service flexibility, high throughput, and capacity for high-speed data. It was anticipated that such a system would be required at the turn of the millennium. A simplified UMTS/WCDMA development history is represented in Figure 7.1 [152].

Several potential multiple access concepts for 3G, including advanced time division multiple access (A-TDMA) and code division testbed (CoDiT, based on Direct Sequence CDMA [DS-CDMA]), were examined in the RACE II project as part of the third European Union (EU) frame program for Pan-European research collaboration, which began in 1992. CoDiT's objective was to define a wideband (5 MHz per carrier) DS-CDMA-based radio access for 3G and to design a testbed capable of demonstrating its characteristics [153].

The first feasible WCDMA concept for cellular networks was created after the CoDiT testbed work concluded successfully in 1994, with a few field demonstrations [153].

The follow-up Future Radio Wideband Multiple Access Systems (FRAMES) project began in 1995. FRAMES was developed as part of the fourth EU frame program named Advanced Communication Technologies and Services and intended to put forward the European candidate 3G proposal. A number of schemes were developed under FRAMES, with many inheriting their basic principles from the earlier RACE II projects [153].

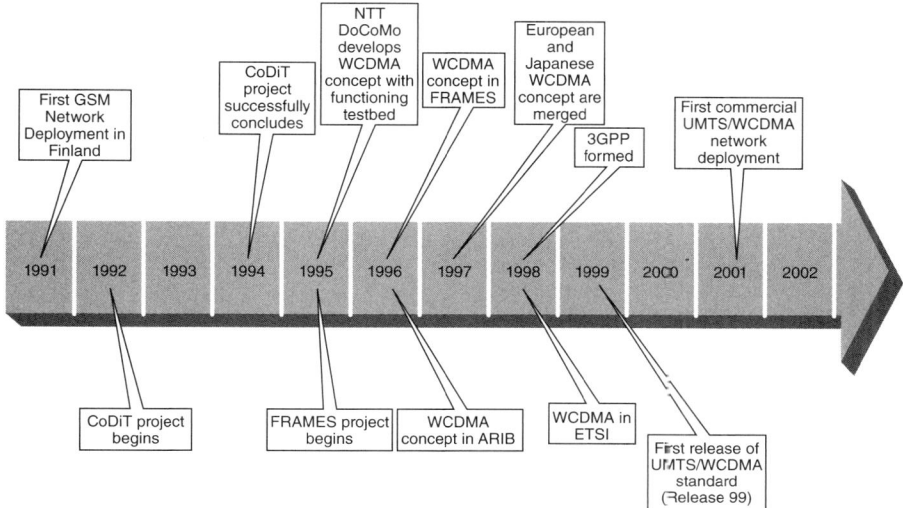

Figure 7.1   UMTS/WCDMA development. NTT, Nippon Telegraph, and Telephone.

In the beginning of 1997, the ETSI formed five concept groups where the 3G development continued. Three of the concepts came from FRAMES [153]:

- Alpha or WCDMA, originally from FRAMES
- Beta or orthogonal frequency division multiple access (OFDMA)
- Gamma or wideband TDMA (W-TDMA), originally from FRAMES
- Delta or time division–code division multiple access (TD-CDMA), originally from FRAMES
- Epsilon or opportunity-driven multiple access (ODMA), a concept later considered a complement to the others

Meanwhile, in Japan, by 1995, NTT DoCoMo had developed a WCDMA 3G concept, along with a functioning testbed. It was one of the proposals to ARIB and was preliminarily chosen as the Japanese 3G proposal in late 1996 [153].

The European and Japanese WCDMA concepts, already mature with respect to most of the basic building blocks of the radio part, were merged in 1997, and in January 1998, ETSI preliminarily selected the merged WCDMA as its proposal for a 3G Radio Access Network (RAN). WCDMA was to use the GSM/General Packet Radio Service (GPRS) CN, thus minimizing the need for new investments for the myriad existing GSM network operators. This and the subsequent founding of the 3GPP in December 1998, which took on the responsibility for all WCDMA and GSM standardization, ensured the global success of the UMTS/WCDMA standard [153].

7.1.2   W-CDMA Evolution

Figure 7.2 depicts the evolutionary path for wireless mobile broadband communications in terms of 3GPP UMTS and WCDMA releases [153].

Table 7.1 summarizes the key UMTS releases produced by 3GPP.

All UMTS specifications can be found in the central 3GPP document repository, publicly available from the 3GPP File Transfer Protocol (FTP) site: ftp://ftp.3gpp.org/Specs/latest. 3GPP specifications are typically segmented by layers. Many of the key UMTS specifications are summarized in Table 7.2.

Revisions of these specifications are published every 3 months as enhancements, and corrections are suggested by 3GPP members. When a current revision is believed to be correct and complete, that document (or set of documents) is frozen and designated as the next release. Once a release has been published, only corrections are allowed (not enhancements).

7.1.3   UMTS/WCDMA Technology Overview

7.1.3.1   Relationship between UMTS and GSM

A UMTS mobile uses WCDMA technology, a different RAT from that of the GSM, GPRS, and EDGE. It reuses most of the higher-layer software from a GSM phase 2+ mobile, which means that mobility management (MM), GPRS mobility management (GMM), call control (CC), Session Management (SM), and short message service (SMS) remain the same. The Universal Subscriber Identity Module (USIM) is based on the Subscriber Identity Module (SIM). Figure 7.3 illustrates how UMTS evolved from GSM/GPRS.

In general, a UMTS mobile device is also a GSM mobile device. In fact, the authors are not aware of any significant market penetration of UMTS-only devices. Rather, mobiles will seamlessly hand off between the GSM/EDGE Radio Access Network (GERAN) and UTRAN.

7.1.3.2   UMTS versus W-CDMA

The WCDMA air interface (Universal Terrestrial Radio Access–Frequency Division Duplex [UTRA-FDD]) has grown synonymous with UMTS because this is the only air interface that has been significantly developed or deployed. Even in this book, we will often use the terms "WCDMA" and "UMTS" interchangeably. However, in the strictest sense, the term "WCDMA" refers only to the air interface technology within the UTRAN, whereas "UMTS" refers to the system as a whole.

The UMTS structure is depicted in Figure 7.4 [154].

UTRA-FDD is the radio access system that utilizes WCDMA protocols.

Universal Terrestrial Radio Access–Time Division Duplex (UTRA-TDD) specifies two physical layers, one that operates at 3.84 Mcps and another one that operates at 1.28 Mcps. The physical layer operating at 1.28 Mcps is also called low chip rate TDD (LCR-TDD) or time division–synchronous code division multiple access

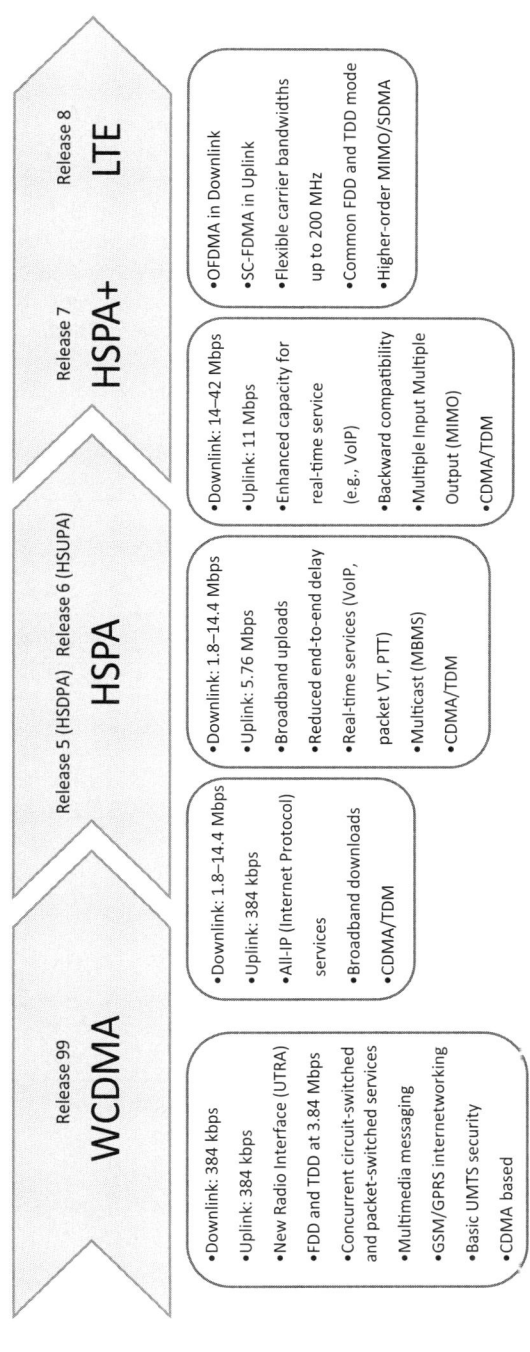

Figure 7.2 UMTS/WCDMA evolutionary path. VoIP, Voice over Internet Protocol; VT, Video Telephony; PTT, push-to-talk; MBMS, multimedia broadcast multicast service; TDM, time division multiplexing; OFDMA, orthogonal frequency division multiple access; SDMA, space division multiple access; HSPA+, HSPA evolved.

371

Table 7.1 Key UMTS Releases

Release	Description
Release 99	Specified the first CDMA-based UMTS 3G networks.
Release 4	Introduced circuit-switched core network split feature and some other minor enhancements.
Release 5	Introduced Internet Protocol Multimedia Subsystem (IMS), HSDPA that allowed broadband services on the DL, and other minor enhancements.
Release 6	Incorporated operation with Wireless Local Area Networks (WLANs) and added HSUPA (enables broadband upload and services), MBMS, and improvements to IMS such as Push-to-Talk over Cellular (PoC), video conferencing, messaging, and so on
Release 7	Introduced enhancements to HSPA+, QoS, and improvements to real-time applications, such as VoIP.
Release 8	Introduced E-UTRA or LTE based on OFDMA, All-IP Network (or System Architecture Evolution [SAE]) and femtocells operation.

Table 7.2 UMTS Specification Summary

Topic	Specification series number
RF performance	25.1xx
Physical layer	25.2xx
Layer 2 and layer 3	25.3xx
UTRAN	25.4xx
Nonaccess stratum (NAS) Layer (CC, SS, SMS, MM)	22.xxx, 23.xxx, 24.xxx
Packet-switched data service	22.060, 23.060
Circuit-switched data service	23.910
Voice service	26.xxx
Universal Subscriber Identity Module (USIM)	31.xxx
User equipment (UE) conformance	34.xxx
LTE	36.xxx

CC, circuit-switched call control protocol; SS, supplementary services; MM, mobility management.

(TD-SCDMA). The physical layer with 3.84 Mcps is also called high chip rate TDD (HCR-TDD) or TD-CDMA.

The GSM/GPRS Access Network is the radio access system that uses GSM and GPRS protocols. The user equipment could be a mobile, a fixed station, or a data terminal.

UMTS encompasses the entire mobile communication system, including the two CNs, circuit switched (CS) and packet switched (PS), and up to three RANs.

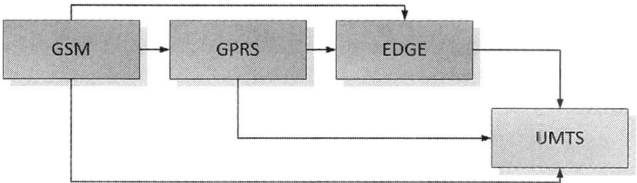

Figure 7.3   UMTS evolution from GSM/GPRS.

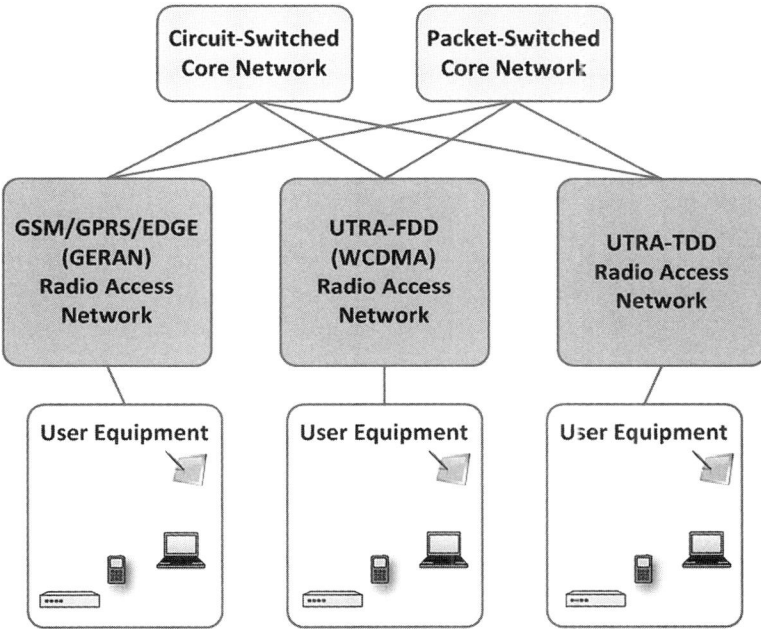

Figure 7.4   UMTS structure.

The UTRAN also allows for the use of different technologies such as WCDMA FDD, also known as UMTS-FDD, and TD-SCDMA time division duplexing (TDD), also known as UMTS-TDD.

The main focus of this chapter falls on the UTRA-FDD (WCDMA) access network because this implementation is more pervasive worldwide than the UTRA-TDD access network. As such, the discussion of UTRA-TDD access network is not included here.

7.1.3.3   UMTS Network Architecture Overview

A UMTS network is composed of three major subsystems:

User Equipment. This may be mobile, a fixed station, a data terminal, and so on. This includes USIM, an application residing on a "smartcard" used for accessing services provided by cellular networks where the application can register with the appropriate security. USIM contains all of the user's subscription information.

Access Network. This includes all of the radio equipment necessary for accessing the network. It may be UTRAN or GERAN.

Core Network. This includes all of the switching and routing capability for connecting to the Public Switched Telephone Network (PSTN) (CS calls) or a packet data network (PS calls) for mobility and subscriber location management and for authentication services.

In WCDMA, the functionality of the CN equipment is, in essence, unaffected when compared to that of a GSM/GPRS system. UTRAN, however, requires a new interface, and the access network and UE are entirely new. CN nodes can be 3G only, 2G only, or 3G and 2G capable. Most of the current deployments implement separate CN nodes for UTRAN and GERAN access.

The UMTS network architecture is shown in Figure 7.5.

User Equipment   User equipment (UE) is a device employed by a user to access network services (e.g., a cellular handset). The UE is divided into two logical parts:

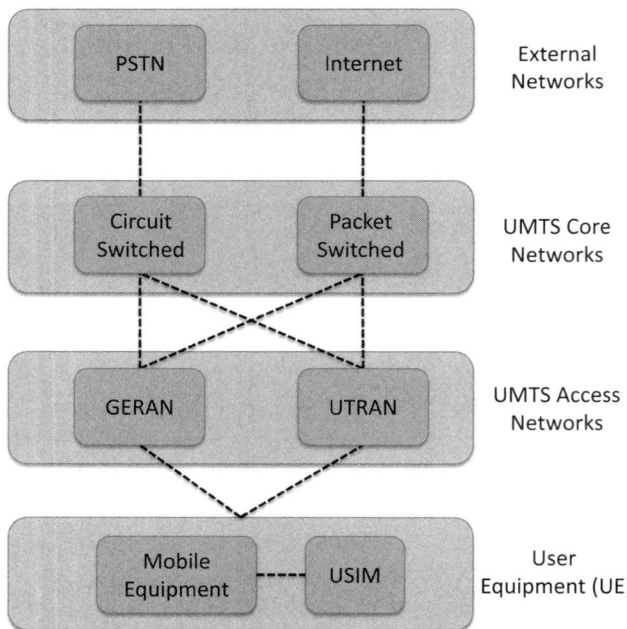

Figure 7.5   UMTS network topology. PSTN, Public Switched Telephone Network.

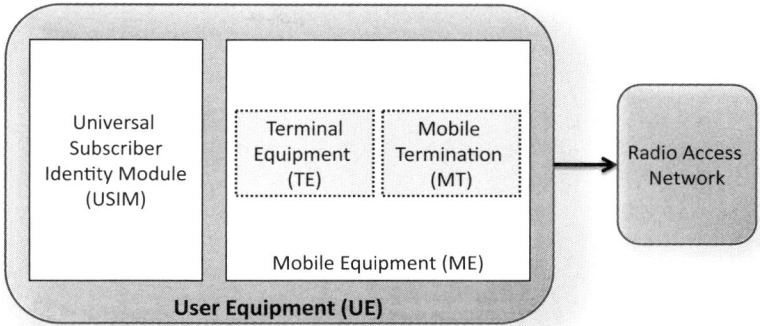

Figure 7.6   Logical structure of user equipment. Reprinted from Reference [155].

mobile equipment (ME) and USIM. The logical structure of the UE is shown in Figure 7.6 [155].

The ME, or the mobile handset, is manufactured by equipment vendors. The ME is further divided into mobile termination (MT) and terminal equipment (TE) functional groups. The MT is responsible for radio transmission termination, authentication, and MM. The TE manages the hardware (e.g., speaker, microphones, video cameras, and user display) and hosts user applications such as Web browsing.

The USIM, the equivalent of a SIM in GSM-only handsets, is a removable plastic card that stores a variety of information, including network-specific data, such as a list of carrier frequencies, various device and user identification numbers (e.g., international mobile subscriber identity [IMSI]), and authentication keys. The USIM also stores short messages, charging information, and telephone book data. The USIM allows for the separation of user mobility from equipment mobility; a user can transfer a USIM between MEs, and the change of MEs is not perceptible by the network (i.e., it looks like the same phone to the network).

Universal Terrestrial Radio Access Network   The UTRAN consists of one or more radio network subsystems (RNSs). Each RNS consists of a radio network controller (RNC) and one or more Node Bs (base stations, BSs). Each Node B controls one or more cells and provides the WCDMA radio link to the user equipment (e.g., cell phone). This interface is called Uu.

The interface between Node B and RNC is called Iub. The Iur interface, when present, allows soft handover to take place between cells connected to different RNCs. In such cases, one of the RNCs acts as the serving RNC, while the other is designated as the drift RNC. The serving RNC maintains the Iu connection to the core network and executes the selection and outer loop power control functions. The drift RNC transfers the frames exchanged over the Iur interface to the user equipment via one or more Node Bs.

The UTRAN structure is shown in Figure 7.7.

A Node B, analogous to a base transceiver station (BTS) in GSM networks, is the gateway between the wired and wireless portions of the network. The BTS

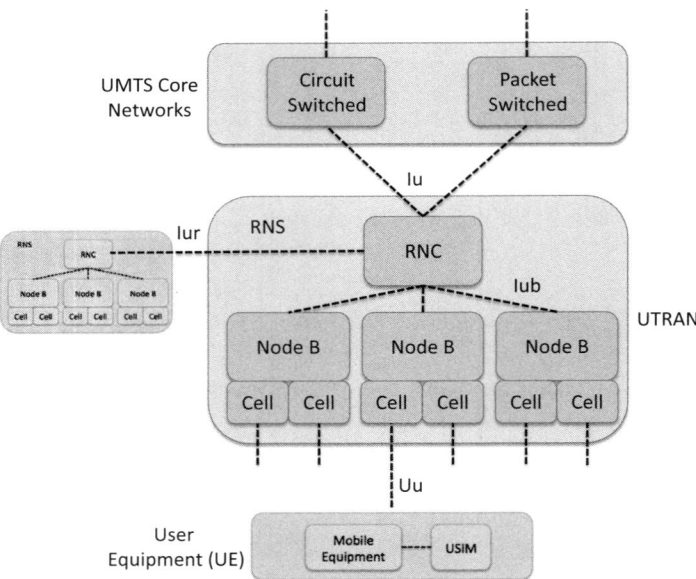

Figure 7.7 UTRAN structure.

consists of a high-speed transmitter and receiver and provides signaling and data channels to mobile stations (MSs). There is generally one Node B per cell of UMTS coverage.

The RNC, analogous to a base station controller (BSC) in GSM networks, generally controls multiple Node Bs and has several core functions, including radio resource management (e.g., assigning and releasing channels for all MSs within its area of control, handover).

Core Network The UMTS core network is based upon the GSM with GPRS core network. The purpose behind WCDMA design was to create a technology link to evolve 2G GSM networks to 3G services. WCDMA, therefore, was specifically designed to be interconnected with the GPRS core network. To introduce the UMTS core network, the original GSM 2G core network should be first reviewed. For more information about GSM 2G technology, please refer to Chapter 6. Let us take another look at the original GSM 2G network architecture (Figure 7.8).

Recall that the GSM 2G network architecture consists of two components: a base station subsystem (BSS) and a network subsystem (NSS). The NSS is also often referred to as the CS core network. The role of the BSS is to support transmissions from MSs and deliver wireless transmissions to the wired network infrastructure. The role of the NSS is to deliver MS traffic as presented by the BSS to the PSTN or vice versa. The NSS is also responsible for various infrastructural functions, such as user registration and user authentication. The BSS and NSS, combined with an operation subsystem (OSS), form a public land mobile network (PLMN).

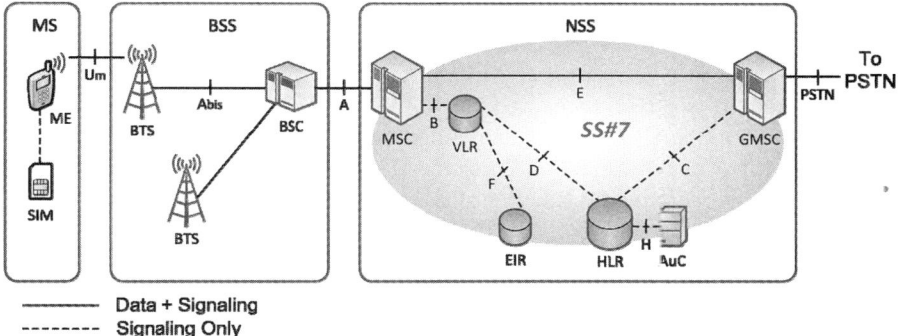

Figure 7.8   GSM 2G network architecture.

The following subsections provide an overview of each component in this architecture.

Mobile Switching Center   The mobile switching center (MSC) is the switching node of the network and is completely analogous to a circuit switch in the wired PSTN. The MSC is the primary service delivery node for GSM, responsible for handling voice calls, SMS text messages, and other services. The MSC sets up and releases the end-to-end connections, handles mobile and handover requirements during the call, and takes care of charging and billing. The primary functions of the MSC are traffic switching (i.e., establishment of data paths from PSTN to MSs) and mobility support (supports location registration of subscribers as they move about the network). There are generally multiple MSCs in the cellular network.

Gateway MSC   The gateway mobile switching center (GMSC) is identical to an MSC, except it connects the cellular network to the PSTN; the GMSC is the entry point into the PLMN. There is usually only one GMSC per cellular network, and it routes connections to the local MSC. The GMSC determines in which visited MSC the subscriber who is being called is currently located. All mobile-to-mobile and PSTN-to-mobile calls are routed through a GMSC.

Home Location Register/Visitor Location Register/Equipment Identity Register   The home location register (HLR) is a centralized database that maintains user profiles for all users authorized to use the network and supports connection routing by exchanging information with MSCs. The HLR stores details of every SIM/USIM card indexed by the IMSI, which is a unique identifier of the SIM/USIM. The HLR also maps these entries to the Mobile Subscriber Integrated Services Digital Network (MSISDN) of each MS, which is the telephone number used by the phone to make and receive calls. The type of information stored in the HLR for each IMSI/MSISDN entry includes services that the subscriber has requested or has been given, the current location of a subscriber, and GPRS settings to allow the subscriber to access packet services.

The visitor location register (VLR) is a temporary database containing information for the subscribers who have roamed into an area served by that VLR. Each VLR is typically associated with an MSC (often integrated directly into the MSC), and each Node B is served by exactly one VLR. The HLR or the MS has received the data stored in the VLR. These data include the subscriber's MSISDN, the subscriber's IMSI and authentication data, the GSM services that the subscriber is allowed to access, and the HLR address of the subscriber.

The equipment identity register (EIR) is a database whose main function is to maintain a list of International Mobile Equipment Identities (IMEIs) for all known GSM MEs.

This list is used to recognize whether an MS that attempts to register to the network is using equipment that is obsolete, stolen, or nonfunctional and allows the network to deny or limit service to such devices.

Authentication Center   The authentication center (AuC) provides support for authenticating each SIM/USIM card that attempts to connect to the GSM core network (typically when the MS is powered on). Once successfully authenticated, the HLR is allowed to manage the SIM/USIM and an authentication key is generated to encrypt all subsequent wireless communications. If authentication fails, then no services are available for that (SIM/USIM and PLMN) combination. The AuC does not actively participate in the authentication process; rather, the AuC generates data known as triplets that the MSC uses during the authentication procedure. The authentication process relies on a shared secret between the AuC and the SIM that is never transmitted over the air. This shared secret is combined with the IMSI to produce a challenge/response for identification purposes and an encryption key for over-the-air communications. A new triplet is requested by the MSC for a particular IMSI after each authentication process so that the same challenge/response is not used twice for a particular mobile. The AuC is typically collocated with the HLR, although this is not necessary.

GSM with GPRS Network Architecture   The network architecture depicted in Figure 7.9 is focused on supporting CS voice communications. Recall from Chapter 6 that, as GSM evolved, the network architecture also evolved, primarily to support data communications in addition to voice traffic. Figure 7.9 shows the network architecture defined in Release 97 (the GSM specification immediately preceding UMTS) to support GPRS data services. This is the core network architecture adopted by UMTS.

In the architecture of Figure 7.9 there are two core networks: a CS core network and a PS core network. In this network architecture, the serving GPRS support node (SGSN) is completely analogous to the MSC in the CS core network and the gateway GPRS support node (GGSN) is completely analogous to the GMSC in the CS core network. However, instead of switches supporting voice communications, the SGSN and GGSN are routers supporting PS data communications. The SGSN routes incoming and outgoing packets addressed to/from data subscribers located within the geographical area served by the SGSN. The GGSN serves as the interface to

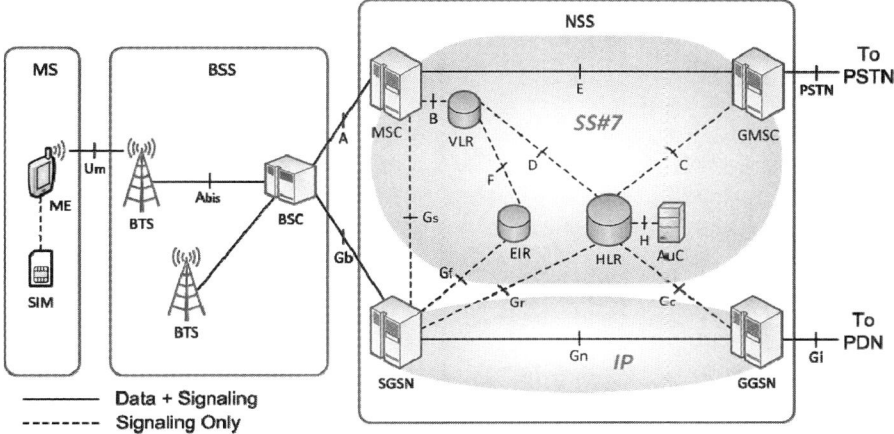

Figure 7.9   GSM with GPRS network architecture.

external Internet Protocol (IP) networks, which view the GGSN as an IP router serving all IP addresses in the PLMN. The GGSN stores SGSN addresses and the users within its area and tunnels protocol data packets to/from the SGSN. The GGSN generally includes firewall and packet-filtering mechanisms. In the GSM/GPRS core network architecture, voice calls are still supported by the CS core network (i.e., MSC and GMSC); data calls are supported by the PS core network (SGSN and GGSN).

Identifiers   There are several identification parameters utilized in the GSM/UMTS network:

IMSI. A unique 15-digit value assigned by a service provider and is a combination of a home country code, home GSM network code, mobile subscriber identification, and national mobile subscriber ID (i.e., phone number). This value is meant to identify the subscriber.

Temporary IMSI (TMSI). This is a 32-bit number assigned by the VLR to uniquely identify an MS within a VLR's area of responsibility. This value is meant to identify the subscriber.

Mobile Subscriber ISDN Number (MSISDN). This is the "real telephone number" of the MS and is centrally stored in the HLR. An MS can have several MSISDNs, depending on the SIM. The MSISDN consists of a country code (CC), national destination code (NDC), and subscriber number (SN).

International Mobile Station Equipment Identification (IMEI). This is a unique 15-digit value assigned by the equipment manufacturer and is a combination of a type approval code, final assembly code, and serial number. This number is meant to identify the handset. This number is registered by the network operator and centrally stored in an EIR. (The PLMN's OSS consists of an EIR and AuC.)

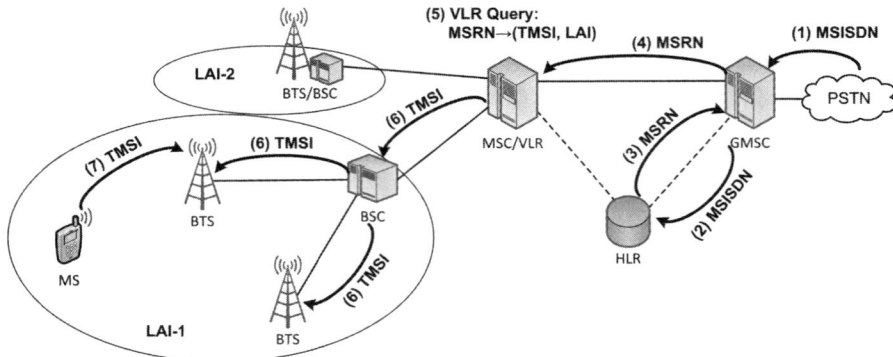

Figure 7.10   Identifiers and call routing.

Mobile Station Roaming Number (MSRN). This is a temporary location-dependent value based on the MSISDN and is the value by which calls are routed to an MS. It is assigned to an MS by the VLR and is done at registration or when the HLR requests it for setting up a connection for an incoming call. Its structure is the same as that of the MSISDN.

Location Area Identifier (LAI). The LAI is a number uniquely associated with a VLR and identifies a location area (LA). The LAI is regularly broadcast by the BTS on a broadcast channel.

Cell Identifier (CI). With an LA, individual cells are uniquely identified with a CI. The combination of the LAI and CI is that cell's global identity.

Figure 7.10 illustrates how these various identifiers are used in the call routing process of a GSM network. From the perspective of the outside world, the MSISDN is the addressing mechanism of a mobile subscriber. Within the CS core network, the MSRN is the primary addressing mechanism. Within the wireless radio subsystem (RSS) portion of the GSM network, the TMSI is the primary addressing mechanism. This architecture is identical in the UMTS paradigm.

7.1.4   UMTS Key Characteristics

Recall that UMTS is a CDMA-based technology that complies with the IMT-2000 standard. A 3G mobile wireless solution, UMTS evolves from GSM to provide high data service and higher system capacity. As seen from Section 7.1.3.3, UMTS reuses the GSM with the GPRS core network but implements a different air interface called WCDMA. This reuse of the GSM core network allows for easy handovers between the UTRAN and GERAN.

Overall, the UMTS network architecture is virtually identical to that of GSM (with many components renamed but serving largely the same function).

The main focus of this section is to highlight key characteristics of the UMTS system.

7.1.4.1 UMTS Signaling Architecture

The UMTS signaling protocol stack is divided into access stratum (AS) and nonaccess stratum (NAS). The NAS architecture evolved from the GSM upper layers and includes the following:

Connection Management. This contains sublayers responsible for CS services: call control (call setup and release), supplementary services (call forwarding and three-way calling), and SMS and PS services: session management and SMS.

Mobility Management. This deals with location updating/authentication for CS calls.

GPRS Mobility Management. This is responsible for location updating/ authentication for PS calls.

The UMTS NAS layer has remained unchanged from GSM. The AS architecture is new for WCDMA and is the main focus of this section. Figure 7.11 illustrates the protocol stack for the UMTS signaling between the UE and CN.

Access Stratum   The AS consists of the following layers

Radio resource control (RRC) (layer 3)

Packet data convergence protocol (PDCP) (layer 2)

Broadcast/multicast control (BMC) (layer 2)

Radio link control (RLC) (layer 2)

Media access control (MAC) (layer 2)

Physical layer (PHY) (layer 1).

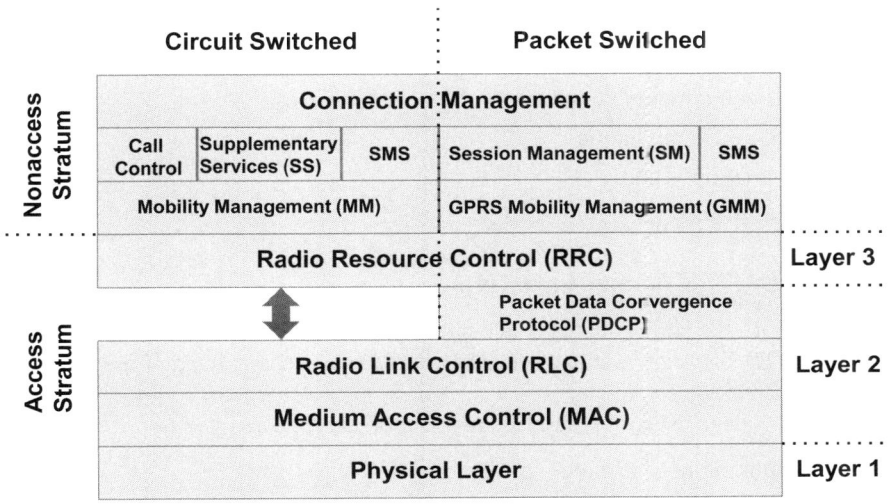

Figure 7.11   UMTS signaling protocol stack.

Data flow between layers is represented by the following:

Radio bearers carry signaling between peer RRC entities or carry user data between peer application layer entities.

Logical channel carries signaling and user data between peer RLC entities.

Transport channel carries signaling and user data between peer MAC entities.

Physical channels carry signaling and user data over the radio link.

RRC (layer 3) performs the following functions:

AS control

Paging and notification

Measurement control and reporting

RRC connection management

Radio bearer management

Broadcasts system information

Layer 2 is responsible for the following:

RLC: Segmentation, reassembly, concatenation, and padding

RLC: Retransmission control and flow control

RLC: Delivery assurance

MAC: Mapping logical channels to transport channels

MAC: Priority handling of data flows

MAC: Traffic volume measurements

PHY (or layer 1) performs the following:

Forward error correction (FEC) encoding/decoding

Interleaving and multiplexing of transport channels

Power weighting and combining of physical channels

Modulation/demodulation

Spreading/dispreading

Radiofrequency (RF) processing

7.1.4.2 UMTS Channel Structure

UMTS employs a sophisticated channel structure to interconnect the various layers of the UMTS signaling architecture. The UMTS channel architecture is depicted in Figure 7.12 [153].

UMTS channels can be described as one of the following:

Downlink, uplink, or both

Common (i.e., shared) or dedicated

Logical, transport, or physical

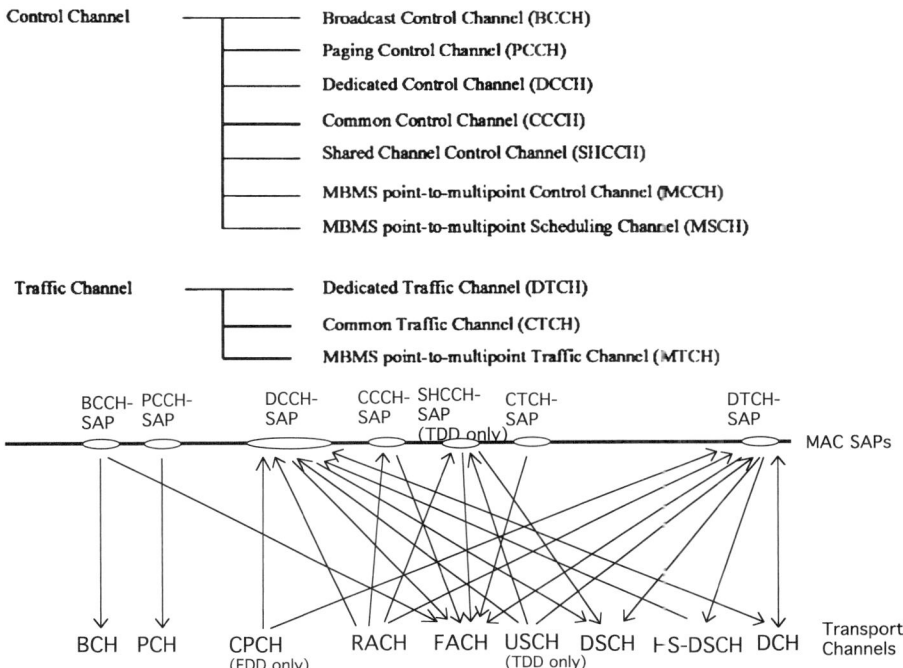

Figure 7.12   UMTS radio channel mappings. (a) Mapping channel type (i.e., bearers) to logical channels; (b) Mapping logical channels to transport channels (as seen by UTRAN). SAP, service access point. Reprinted from Reference [154]. © 2011. 3GPP™ TSs and TFs are the property of ARIB, ATIS, CCSA, ETSI, TTA and TTC, who jointly own the copyright in them. They are subject to further modifications and are therefore provided "as is" for information purposes only. Further use is strictly prohibited.

Brief definitions for all UMTS channel types are provided next:

Downlink channel: This is transmitted by UTRAN and received by UE

Uplink channel: This is transmitted by UE and received by UTRAN

Common: This delivers information to or from multiple UEs

Dedicated: This delivers information to or from a single UE

Logical: This is defined by the type of information transferred, for example, user data or signaling

Transport: This is defined by how data are transferred over the air interface, for example, multiplexing of logical channels

Physical: This defined by physical mappings and attributes employed to transfer data over the air interface, for example, spreading rate

Each of the "channels" in Figure 7.12 can be thought of as the end-to-end channel as seen by the next highest layer. For example, the channel as seen by the RLC is referred to as a logical channel. The channel as seen by the MAC layer is a transport channel. The channel as seen by the application is referred to as a radio bearer. Each particular channel type can be mapped to one or more lower-layer channels. Not all mappings are defined at the same time for a given UE. Multiple instantiations of some mappings may occur simultaneously. The common pilot channel (CPICH), synchronization channel (SCH), dedicated physical control channel (DPCCH), acquisition indicator channel (AICH), and paging indicator channel (PICH) exist only in the physical layer context and do not carry upper layer signaling or user data.

Transparent mode (TM), unacknowledged mode (UM), and acknowledged mode (AM) represent the modes in which RLC can be configured for a logical channel. The channels shown in Figure 7.12 are described in Table 7.3 [154].

The CPICH is a particularly important channel to the WCDMA air interface. The CPICH is broadcast by Node Bs using spreading code 0 with a spreading factor of 256. The CPICH is broadcast at a constant power level with a known bit sequence. The CPICH is utilized by mobile devices for phase and power channel estimations. While not required for operation, the CPICH does dramatically improve reliability of the UMTS system.

7.1.4.3   UMTS Services

An end-to-end (user-to-user) network service may have a certain QoS provided to the user. The user ultimately decides whether the provided QoS is acceptable or not.

UMTS offers different bit rates for different environments. The typical non-HSDPA bit rates are 144 kbps for rural outdoor environments, 384 kbps for urban outdoor environments, and approximately 2 Mbps for indoor and short-range outdoor environments. When HSDPA is included, the offered bit rate could be higher than 10 Mbps [154].

UMTS network can provide traffic-specific services with different QoS classes. The user traffic is typically divided into four classes and is summarized in Table 7.4 [154].

7.1.5   WCDMA Physical Layer

A UMTS physical channel consists of a WCDMA waveform utilizing FDD 5-MHz paired channels. This section discusses the formation of WCDMA physical channels and the processing applied at the physical layer level.

7.1.5.1   Channelization

The physical layer of the WCDMA air interface employs orthogonal variable spreading factor (OVSF) codes on the uplink and downlink to spread data streams to the chip rate of 3.84 Mcps, and then modulated into a quadrature phase-shift keying (QPSK) waveform. All OVSF codes are orthogonal to each other at a given spreading factor (SF) and form a tree that allows for multiple SFs to be used. Variable SFs allow support for users at different data rates and address the flexibility needs of 3G.

Table 7.3 UMTS Channel Descriptions

Downlink		
Type	List of channels	Function(s)
Broadcast	Broadcast Control Channel (BCCH) Broadcast Channel (BCH) Primary Common Control Physical Channel (PCCPCH)	Identification (cell ID, neighbor info, frame number, system ID)
Paging	Paging Channel (PCH) Secondary Common Control Physical Channel (SCCPCH) Paging Control Channel (PCCH)	Used to provide paging requests to notify subscriber of incoming call requests
Forward Access	Forward Access Channel (FACH) SCCPCH, AICH, DCCH, DTCH, FACH	When the subscriber equipment is without a dedicated channel, this is used to exchange less frequent periodic messages between the subscriber and the cell system. However, when the subscriber equipment is active, it will provide the dedicated paths for control and user traffic.
Uplink		
Random Access	Common Control Channel (CCCH) Physical Random Access Channel (PRACH) Random Access Channel (RACH) Dedicated Control Channel (DCCH) Dedicated Traffic Channel (DTCH)	The subscriber uses this channel to request access to the network when it does not otherwise have a channel assignment. In the case where channel assignment is already present, it is used to provide a dedicated path for both control and user traffic.
Downlink and Uplink		
Dedicated Physical Channel	Dedicated Physical Data Channel (DPDCH) Dedicated Physical Control Channel (DPCCH)	Dynamically assigned channels available to the subscriber when transmitting/receiving voice and/or data.

Table 7.4 UMTS Traffic Classes

Traffic class	Function	Time delay tolerance
Conversational	Voice, videophone, video games	Less than 1 second
Streaming	Multimedia, video on demand, webcasts	Approximately 1 second
Interactive	Web browsing, network gaming, database access	Less than 10 seconds
Background	E-mail, SMS, downloading	Longer than 10 seconds

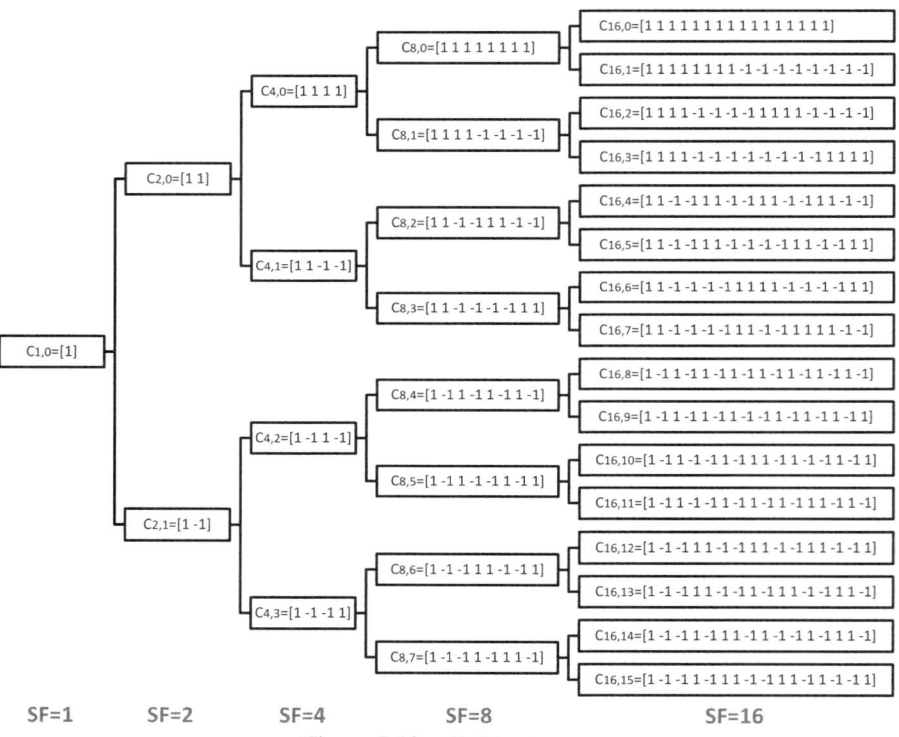

Figure 7.13 OVSF code tree.

Certain branches in the tree are prohibited for use to minimize cross-talk between users [153].

Figure 7.13 depicts the OVSF code tree, where C_{SFk} is described as

C: the OVSF code

SF: spreading factor

k: code number, where $0 \geq k \geq SF - 1$.

The two branches leading out of any given C are labeled by the sequences $[C\ C]$ and $[C\ -C]$.

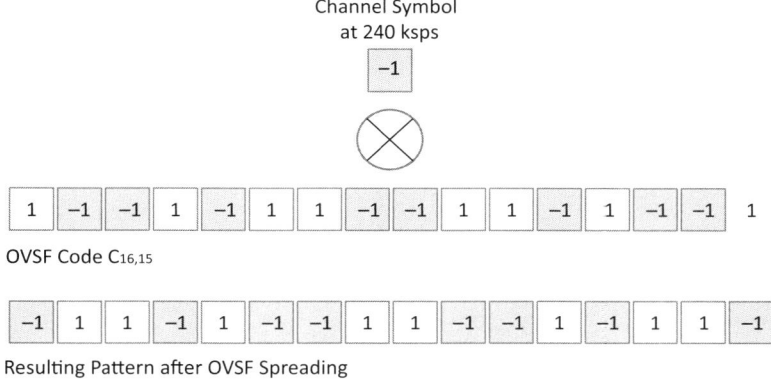

Channel Symbol
at 240 ksps

| −1 |

OVSF Code C$_{16,15}$

Resulting Pattern after OVSF Spreading

Chip Rate = Symbol Rate x SF = 240 ksps x 16 chips = 3.84 Mcps

Figure 7.14   Example of spreading with OVSF code.

SFs are related to data rates. A lower SF translates into a higher data rate because there are fewer chips per symbol. More detail is available through Reference [157].

In orthogonal spreading, each encoded symbol is multiplied with all the chips of a particular OVSF code. Figure 7.14 demonstrates how a symbol of value "−1" is orthogonally spread with OVSF code $C_{16,5}$, thus resulting in a 16-chip representation of the symbol. Each symbol in the data stream is now spread by one full repetition of an OVSF code.

7.1.5.2 Scrambling

While the OVSF codes are used for channelization, there are several spreading codes utilized within the WCDMA air interface. Scrambling codes are used on top of the channelization (OVSF) codes to separate the signals coming from different cells in the downlink and the signals coming from different users in the uplink. UMTS uses gold codes for scrambling. Gold codes simulate a random noise process via pseudorandom noise (PN) sequences and possess good cross-correlation properties necessary for separating cells and users.

In WCDMA, the gold code sequence is truncated to 38,400 chips to match the 10-ms radio frame. For the downlink, each cell has its own primary scrambling code (PSC). There are 512 PSCs. Secondary codes (7680) are used only when there are no more OVSF codes available for scrambling. The code length of $2^{18} - 1$ is also truncated to 38,400 chips.

Downlink PSCs have low cross-correlation among each other, regardless of the time offset between any two scrambling codes, which allows for the asynchronous cell deployment and the use of the secondary scrambling codes.

On the uplink, there are 2^{24} (16,777,216) codes. When a dedicated channel is assigned, the UE is signaled which uplink scrambling code to use. The code length

Table 7.5   Type of Spreading Codes Used in UMTS

UMTS Code	Purpose	Code Type	Length	Number of Distinct Codes
Downlink Channelization Codes	Handset separation	OVSF	4 to 512 chips (Dependent on channel type)	4 to 256
Uplink Channelization Codes	Physical channel separation	OVSF	4 to 512 chips (Dependent on channel type)	4 to 256
Uplink Scrambling Codes	Each handset is assigned a unique code that is used atop channelization codes for signal separation within physical channels. May either be long or short scrambling codes, depending on type of logical channel type.	Gold	Short: 256 chips Long: 38400 chips	16,777,216
Downlink Scrambling Codes	Used atop channelization codes for signal separation. Basis for PSCs and SSCs.	Gold	38400 chips	512 PSCs 7680 SSCs (15 SSCs for each PSC)

of $2^{25} - 1$ is truncated to 38,400 chips. Each UE is assigned a unique scrambling code, which channelizes the users.

Table 7.5 summarizes the various UMTS spreading codes.

7.1.5.3 *Asynchronous Network Operation*

UMTS is an asynchronous system. The gold codes used to differentiate cells are not synchronized to one another, removing the need for global synchronization. This is different from, say, the cdma2000 system, where all cells must be synchronized to a single time source; typically, Global Positioning System (GPS) timing is employed. This requirement for asynchronous operation became fundamentally important because it led to an autonomous and easily deployable system attractive to operators around the globe. This requirement also presented several new technological challenges to WCDMA, including cell reselection and handover, which are described next [153].

Table 7.6   UMTS-FDD Frequency Bands

Frequency band name	Region	Band number	Uplink range (MHz)	Downlink range (MHz)
UMTS-2100	Multiple	I	1920–1980	2110–2170
UMTS-2600	Multiple	VII	2500–2570	2620–2690
UMTS-1800	Europe	III	1710–1785	1805–1880
UMTS-900	Europe	VIII	880–915	925–960
UMTS-1900	Americas	II	1850–1910	1930–1990
UMTS-1700-2100	Americas	IV	1710–1755	2110–2155
UMTS-850	Americas	V	824–849	869–894
UMTS-800	Japan	VI	830–840	875–885
UMTS-1700	Japan	IX	1749.9–1784.9	1844.9–1879.9

7.1.5.4   UMTS Frequency Allocations

Different frequency bands have been assigned to UMTS in different regions and/or countries. Table 7.6 lists UMTS-FDD frequency allocations for various regions [155, 156].

The nominal UMTS channel spacing is 5 MHz, which can be adjusted to optimize performance in a particular deployment scenario. The channel raster is 200 kHz. Hence, the center frequency has to be an integer multiple of 200 kHz. The carrier frequency is represented using the UTRA absolute radio frequency channel number (UARFCN), where f_{center} = UARFCN*200 kHz [154].

7.1.5.5   Cell Reselection

Cell reselection can be defined as the process of selecting a new cell when the UE is not operating on a dedicated channel. The UE chooses a new cell autonomously without necessitating any intervention from UTRAN. UTRAN, however, supplies parameters in the system information messages that affect the UE's cell reselection choice.

There are three types of cell reselection: intrafrequency, interfrequency, and inter-RAT. Their definitions are provided next.

Intrafrequency cell reselection occurs between cells on the same RF. Because the neighboring cells use the same frequency, the UE can measure other cells' signal strength while utilizing the same RF channel in the present cell.

Interfrequency cell reselection happens between cells on different RFs. In order to measure the signal strength of an interfrequency neighboring cell, the UE has to tune to the neighboring cell's frequency.

Inter-RAT cell reselection occurs between cells that use different RATs.

7.1.5.6 Handovers

Handover is the process of adding or removing cells while the UE is communicating on a dedicated channel. The UE helps the process by measuring the signal strength of the neighboring cells and reporting it to the UTRAN. The UTRAN then decides when to perform a handover.

Handovers fall into three categories: soft, softer, and hard. Soft and softer handover are characterized by "make before break." The connection to the new cell is established before the connection to the previous cell is broken. Hard handovers are defined by "break before make." The connection to the previous cell is severed before the connection to the new cell is established.

Handovers are also categorized as intrafrequency, interfrequency, or inter-RAT (see cell reselection definitions).

Compressed Mode From the beginning, almost all WCDMA mobile devices were expected to be dual mode (to support GSM) because the initial WCDMA coverage was not expected to be as widespread as that of GSM. The handover of ongoing connections between WCDMA and GSM was essential for a seamless early deployment of WCDMA [153].

In order to perform a secure handover from WCDMA to GSM, the UE needs to monitor signal strengths on potential GSM target BSs in the area, even during an active call on the WCDMA system. This is time problematic because the UE is constantly busy with sending and receiving on a given WCDMA frequency. A technology dubbed "Compressed Mode," which was originally developed in the CoDiT project, was adopted in WCDMA to address this time issue [153].

Time for monitoring is made available to the UE by, for instance, changing the SF and thus "compressing" the information to be received and transmitted in a shorter time span, leaving idle time for measurements on other frequencies as well as other systems. Once these measurements are reported to UTRAN, a decision on the WCDMA to GSM (inter-RAT) handover can then be made. Compressed mode is also used for interfrequency handovers within the WCDMA system.

The compressed mode works as follows:

(a) Transmission and reception are halted for a short time to generate a gap.

(b) The gap is created via compression of the information in a radio frame in time.

(c) The transmission time can be reduced via SF or higher-layer scheduling.

(d) UTRAN issues the command to the UE to configure and activate compressed code.

7.1.5.7 Spreading Factor Reduction

Figure 7.15 illustrates compressed mode transmission using SF reduction. The SF of the compressed radio frames is divided by 2, allowing the same number of bits to be sent in a shorter amount of time. The instantaneous transmit power is increased

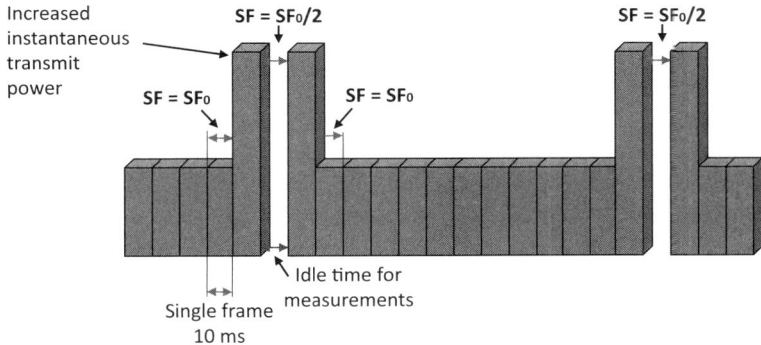

Figure 7.15   Compressed mode transmission—spreading factor reduction. Derived from Reference [329]. © 2006. 3GPP™ TSs and TRs are the property of ARIB, ATIS, CCSA, ETSI, TTA and TTC, who jointly own the copyright in them. They are subject to further modificat ons and are therefore provided "as is" for information purposes only. Further use is strictly prohibited.

to ensure that the frame quality remains unaffected by the reduced processing gain. This method is used when the SF > 4 and is applicable to uplink and downlink.

7.1.5.8 Higher-Layer Signaling

The upper layers can create a time gap during a data call by placing restrictions on the allowed transport format combinations. This method works well for the packet data, which is naturally bursty, but is not well suited for voice or CS data and is applicable to uplink and downlink.

7.1.5.9 Asynchronous Handover

As previously mentioned, WCDMA is deployed asynchronously, which means that each Node B has system frame timings that start at different time instants and that the system frame numbers are different. The system frame numbers are carried on the broadcast (transport) channel (BCH), which are transmitted on the primary common control physical channel (PCCPCH). Because each Node B has its own clock, the timing can also drift between Node Bs.

By contrast, in a synchronous deployment (e.g., IS-95 standard and cdma2000), the downlink frames transmitted by all the Node Bs in a network are aligned to within a few microseconds, which can be accomplished by using GPS receivers or via self-synchronizing procedures.

7.1.5.10 Uplink Power Control

The "near–far problem" resulting from using DS-CDMA in cellular systems was already known in the late 1970s [153]. One UE may be 10 km away from the Node B, while another UE may be only a few hundred meters away. UEs can experience

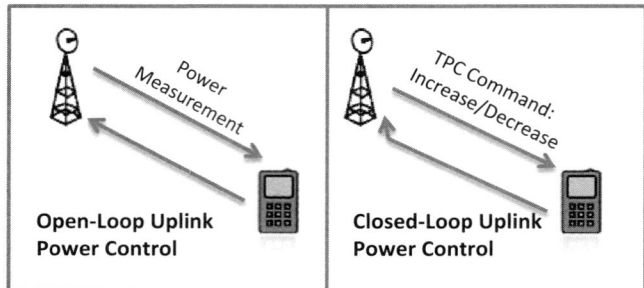

Figure 7.16 Two principles for uplink power control: open loop versus closed loop. TPC, transmit power control.

huge differences in path losses due to their varying distance from the Node B and multipath environments. Power control ensures that the signal sent from the UE that is near the Node B is relatively low compared with the signals sent from the UE that is far from the Node B.

Figure 7.16 illustrates two principles for uplink power control. Open loop means that the mobile device (UE) sets its transmission power based on the measurements of the received signal power, while closed loop means the mobile device sets its transmission power based on explicit commands received from the network.

For initial power settings, the UE uses an open-loop power control. The UE receives signaled parameters via RRC and makes channel measurements. The open-loop uplink power control is utilized for physical random access channel (PRACH) and dedicated physical control channel (DPCH) transmissions.

WCDMA implements two types of closed-loop uplink power control: inner loop and outer loop. Closed inner-loop power control, also known as fast power control, has an update rate of 1500 Hz, which means that the transmitter receives commands from the receiver 1500 times per second to decrease or increase its power.

Closed outer-loop power control, also known as slow power control, is an algorithm that finds the right signal-to-interference ratio (SIR) based on the 1% block error rate (BLER) requirement for a good voice QoS.

7.1.5.11 Downlink Power Control

For downlink power control, WCDMA utilizes closed-loop estimation (inner and outer).

In downlink closed outer-loop power control, the UE measures the BLER over a number of frames and increases or decreases the SIR target. Based on the SIR target and the SIR estimate, the UE commands the UTRAN to adjust the power of its dedicated channel accordingly.

Downlink closed inner-loop power control runs at a 1500- or 500-Hz update rate. The algorithm executes quickly to respond to rapidly changing channel conditions.

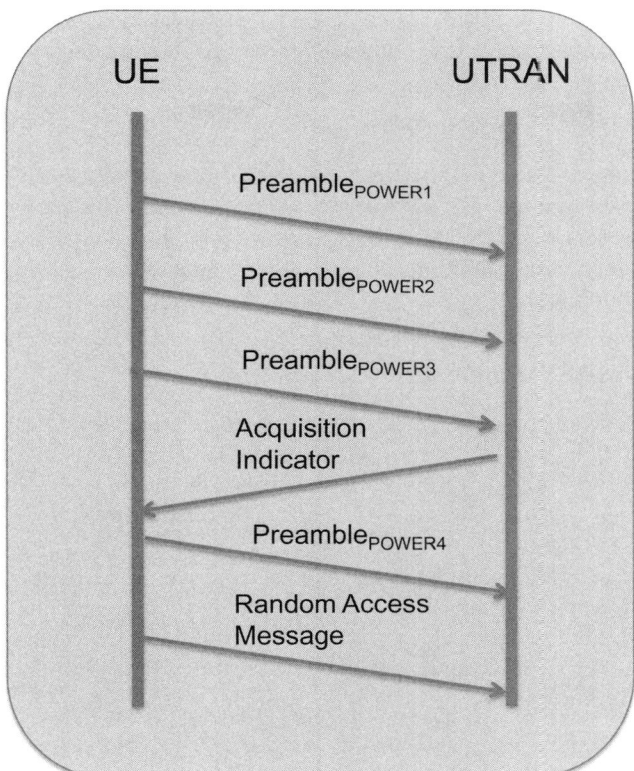

Figure 7.17 RACH power ramping procedure.

7.1.5.12 *Random Access*

WCDMA realizes its random access via the random access channel (RACH) (transport channel). Random access is an algorithm that the UE executes in order to gain initial access to the system.

To minimize signaling and required transmission power, WCDMA adopted a solution in which no explicit code signaling is required. All RACH-related codes are derived from given rules using only the downlink, that is, the transmission from the network, scrambling code, and a signature randomly selected by the UE.

Figure 7.17 illustrates how WCDMA preambles with randomized signatures are sent uplink with increasing power in a "ramping" procedure. When the power is sufficient, the network transmits an acquisition indicator to request transmission of the actual message from the mobile terminal. The MAC-directed ramping procedure is very fast and, when combined with the signature randomization, leads to an efficient, high-capacity RACH.

In contrast, the cdma2000 standard utilizes random access parameters derived from the UE's serial number and is based on time randomization instead of signature randomization, as in WCDMA. cdma2000, for instance, transmits and retransmits the entire random access message until it is received by the network.

7.1.5.13 Coherent Uplink Detection

WCDMA employs coherent detection on uplink and downlink based on the use of pilot symbols or common pilot. While already used on the IS-95 downlink, the use of the coherent detection on the uplink is a new technology that allows data reception under bad channel conditions and can increase the system capacity or the data rate on the uplink [156].

7.1.5.14 Data Rate Indication

When the physical layer receives data streams with a variable data rate, the problem arises on how to inform the receiver of the data rate currently used. The receiver is incapable of processing the received signal without that knowledge. To address this issue, WCDMA introduced the transport format combination indicator (TFCI), which adds specific information with a fixed and known coding to the transmitted signal, informing the receiver about the instantaneous data rate. TFCI notifies the receiver about all the transport formats employed in the currently used transport channels.

The WCDMA standard typically uses TFCI but allows for the blind rate detection, a trial-and-error procedure used by the receiver to blindly guess what transport formats are currently used.

7.1.5.15 Speech Codec

The WCDMA standard selected the 12.2-kbps adaptive multirate (AMR) codec (a speech compression/decompression algorithm) as its primary speech codec.

7.1.6 High-Speed Packet Access

The gap between digital subscriber line (DSL)-type broadband speeds and mobile broadband became narrower with the HSDPA delivering up to 14 Mbps in the downlink and the HSUPA returning 5.8 Mbps in the uplink. HSPA is specified in 3GPP Release 5 for downlink and Release 6 for uplink, although most have implemented HSDPA as it is described in Release 6. HSPA improvements in UMTS are achieved through the following:

New modulation (16-point quadrature amplitude modulation, 16-QAM) techniques

Reduced radio frame lengths

New functionalities within radio networks (including retransmissions between Node B and the RNC).

As a result, throughput is increased and latency is reduced (down to 100 ms or lower).

7.1.6.1 HSPA+

An evolution of HSPA was specified in Release 7, which added multiple-input, multiple-output (MIMO) antenna capability and 16-QAM (uplink)/64-point quadrature amplitude modulation (64-QAM) (downlink) modulation. Together with improvements in the RAN for continuous packet connectivity, HSPA+ will allow uplink speeds of 11 Mbps and downlink speeds of 42 Mbps within the Release 8 time frame [156].

7.1.6.2 HSDPA

The addition of HSDPA to an existing UMTS network does not require any new network entities; however, hardware and/or software changes may be required for each entity. The changes are concentrated in the UE, Node B, and RNC. Interface changes are concentrated on the Uu interface between UE and Node B and on the Iub interface between Node B and RNC.

UE and Node B. Hardware and software changes are required to support the new channels and functionality of HSDPA channels.

RNC. Requires software changes to support the new signaling messages for configuration and management of the HSDPA channels.

Uu Interface. Needs new signaling messages exchanged over existing signaling channels and new transport and physical channels to support high-speed operation.

Iub Interface. Needs a new frame protocol for sending high-speed user data from the RNC to the Node B.

The PS Iu remains unchanged but might experience bandwidth issues while supporting higher data rates to multiple users.

From a protocol perspective, the Release 5 specifications define a new sublayer of MAC called MAC-hs, which implements the MAC protocols and procedures for HSDPA. This sublayer operates at the Node B and the UE. The HSDPA channels cannot operate in soft handover because the MAC-hs sublayer of each Node B operates independently.

Figure 7.18 depicts the HSDPA protocol stack and where each layer's protocols are terminated.

HSDPA introduces three new downlink channels and one new uplink channel:

High-Speed Downlink Shared Channel (HS-DSCH). A downlink transport channel shared by a few UEs. The HS-DSCH channel is associated with one or several high-speed shared control channels (HS-SCCHs). It operates on a 2-ms transmission time interval (TTI).

HS-SCCH. A downlink physical channel for carrying downlink control information related to HS-DSCH transmission. The UE continuously monitors this channel to decide when to read data from the HS-DSCH and to determine the modulation scheme used on the assigned physical channel.

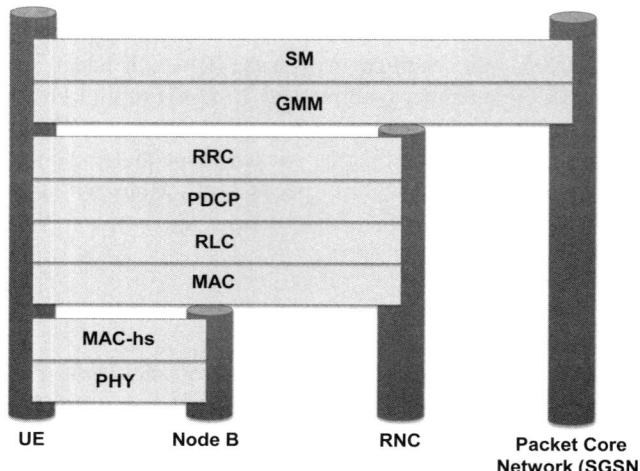

Figure 7.18 HSDPA protocol stack.

High-Speed Physical Downlink Shared Channel (HS-PDSCH). A downlink physical channel shared by a few UEs. This channel supports QPSK and 16-QAM and multicode transmission. It is allocated to a user at 2-ms intervals.

High-Speed Dedicated Physical Control Channel (HS-DPCCH). An uplink physical channel that carries feedback from the UE to the Node B's scheduling algorithm. The feedback contains a channel quality indicator (CQI) and a positive/negative acknowledgment (acknowledgement/nonacknowledgment, ACK/NAK) of a previous HS-DSCH transmission.

Only dedicated logical user DCHs can be mapped to HS-DSCH. When a dedicated traffic channel (DTCH) is mapped to HS-DSCH, only UM and AM can be used.

A UE in HSDPA mode has at least one dedicated channel (DCH/dedicated physical data channel [DPDCH]) allocated for the RRC and network application support (NAS) signaling, even if the UE is not capable of receiving the high-speed channels.

The HS-DPCCH is a physical layer control channel. Because HS-DPCCH does not carry any upper layer information, it has no logical or transport channel mapping.

7.1.6.3 HSUPA

HSUPA or FDD-enhanced uplink (also known as the enhanced uplink dedicated channel [E-DCH]) is specified in 3GPP Release 6. The main objective of HSUPA is to improve the performance of uplink dedicated transport channels, that is, to increase capacity and throughput and reduce delay. HSUPA's high performance is achieved via more efficient uplink scheduling at Node B and faster retransmission control [157].

HSUPA was designed based on the following Release 6 requirements (from Reference [157]):

The enhanced uplink feature shall aim at significantly improving user's throughput, delay, and/or capacity.

"The focus shall be on urban, suburban and rural deployment scenarios."

Full mobility shall be supported, including high-speed cases, but optimization should be performed for low-speed to medium-speed scenarios.

"The study shall investigate the possibilities to enhance the uplink performance on the dedicated transport channels in general, with priority to streaming, interactive and background services. Relevant QoS mechanisms shall allow the support of streaming, interactive and background PS services."

"It is highly desirable to keep the enhanced uplink as simple as possible. New techniques or group of techniques shall therefore provide significant incremental gain for an acceptable complexity. The value added per feature/technique should be considered in the evaluation. It is also desirable to avoid unnecessary options in the specification of the feature."

"The UE and network complexity shall be minimised for a given level of system performance."

"The impact on current releases in terms of both protocol and hardware perspectives shall be taken into account."

The enhanced uplink feature needs to be backward compatible with terminals from Release 99, Release 4, and Release 5. The enhanced uplink feature shall be compatible with HSDPA (when operated together) to achieve significant improvements in overall system performance. The deployment of the enhanced uplink feature shall be possible without any dependency on the deployment of the HSDPA feature. However, an HSUPA-supporting terminal must support HSDPA

General features of HSUPA include the following [157]:

Maximum data rate of 5.76 Mbps

Binary phase-shift keying (BPSK) modulation

No adaptive modulation

Multicode transmission

Spreading factor 2 or 4

10 and 2-ms TTI, but initially only 10-ms TTI to be used

Hybrid automatic repeat request (HARQ): rapid retransmissions of erroneously received data packets between UE and Node B

Fast packet scheduling in the uplink

Soft handover supported.

In order to support E-DCH, the following modifications to the existing nodes are required [157]:

UE. A new MAC entity (MAC-es/MAC-e) is added in the UE below MAC-d. MAC-es/MAC-e in the UE handles HARQ retransmissions, scheduling and

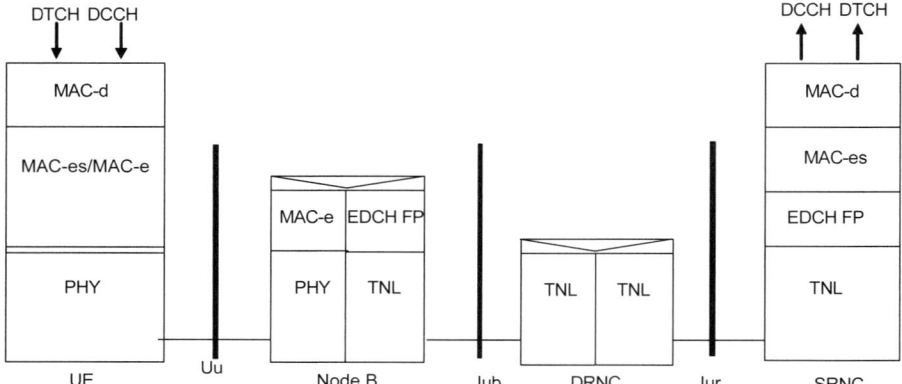

Figure 7.19   E-DCH protocol architecture. DCCH, dedicated control channel; DRNC, drift radio network controller; DTCH, dedicated traffic channel; EDCH FP, EDCH frame protocol; Iub, interface for RNC and Node B communication; Iur, interface for RNC-to-RNC communication; TNL, transport network layer; Uu, over-the-air interface between UE and Node B(s). Reprinted from Reference [160]. © 2006. 3GPP™ TSs and TRs are the property of ARIB, ATIS, CCSA, ETSI, TTA and TTC, who jointly own the copyright in them. They are subject to further modifications and are therefore provided "as is" for information purposes only. Further use is strictly prohibited.

MAC-e multiplexing, and E-DCH transport format combination (E-TFC) selection.

Node B. A new MAC-e is added in the Node B to handle HARQ retransmissions, scheduling, and MAC-e demultiplexing.

Serving Radio Network Controller (SRNC). A new MAC entity (MAC-es) is added in the SRNC to provide in-sequence delivery (reordering) and to handle combining of data from different Node Bs in case of soft handover.

The resulting E-DCH protocol architecture is shown in Figure 7.19 [157].

HSUPA Transport and Physical Channels   In the uplink, the following channels are added:

E-DCH. An uplink transport channel.

Enhanced Dedicated Physical Data Channel (E-DPDCH). An uplink physical channel that carries the user data bits to the network. (The bits from this channel end up being delivered up to MAC-e/es through the E-DCH transport channel.) This channel has a variable spreading factor and is variable in number. The UE selects the most appropriate configuration based on its current nonscheduled grant or scheduled serving grant and the amount of data it has waiting to be sent. The maximum HSUPA data rate is achieved with 2*SF2 codes plus 2*SF4 codes.

Enhanced Dedicated Physical Control Channel (E-DPCCH). An uplink physical channel that contains the PHY control information necessary for the BS to be able to demodulate and decode the variable E-DPDCH [158].

In the downlink, the following channels are added:

Enhanced Hybrid Indicator Channel (E-HICH). A downlink physical channel that signals the ACKs and NACKs to the UE.

Enhanced Access Grant Channel (E-AGCH). A downlink physical (shared) channel that signals absolute values for the serving grant. Each UE is assigned a unique enhanced radio network temporary identity (E-RNTI) at call setup, which is used to direct transmissions on the E-AGCH to specific devices.

Enhanced Relative Grant Channel (E-RGCH). A downlink physical channel that signals the incremental up/down/hold adjustments to the UE's serving grant.

Orthogonality is provided between the E-RGCH and E-HICH through the use of orthogonal 40-bit signatures. This is required because the E-RGCH and E-HICH share the same OVSF code space. These 40-bit signatures can be transmitted inverted or noninverted. The lack of a signature results in a 0 being assumed by the UE (this is also referred to as DTX). E-HICH and E-RGCH channels map the values of −1, 0, +1 to logical values of Up/Down/Hold or ACK/NACK. Given that 40 orthogonal signatures exist, it is possible to multiplex up to 20 UEs onto a single E-HICH/E-RGCH OVSF code.

HSUPA PHY Categories   Release 6 defines a set of HSUPA PHY categories that restrict the UE's data rate by specifying attributes such as the maximum number of E-DPDCHs, minimum E-DPDCH spread factor, 10-ms versus 2-ms TTI support, and so on [157]. Table 7.7 lists HSUPA PHY categories, as defined in Release 6 [330].

Fast Scheduling   HSUPA, unlike HSDPA, does not employ a shared channel for data transfer in the uplink. In WCDMA, each UE already uses a unique scrambling code in the uplink, which means that each UE already possesses a dedicated uplink connection to the network with plenty of code channel space in that connection. In the downlink, however, the Node B uses a single scrambling code before assigning different OVSF codes to different UEs. The shared resource in the uplink is compared relative to the interference level at Node B, which is managed by the network through the fast closed-loop power control. The presence of the UE's dedicated connection to the network in the uplink has a significant impact on the HSUPA design. As previously mentioned, HSUPA objectives were to support fast scheduling, which would allow the network to rapidly change UEs' transmission rates and to reduce the overall transmission delay. Transmission delay reduction is achieved via the HARQ algorithm. Because the primary shared resource on the uplink is the total power arriving at Node B, HSUPA performs its scheduling by directly controlling the maximum amount of power a UE can transmit at any given point in time.

The network has two methods for controlling the UE's transmit power on the E-DPDCH: by using a nonscheduled grant or a scheduled grant. In the nonscheduled grant, the network simply tells the UE the maximum block size it can transmit on

Table 7.7   HSUPA Physical Layer Categories

HSUPA Category	Maximum Number of HSUPA Codes Transmitted per Transport Block	Minimum Spreading Factor	Support for 10- and 2-ms TTI EDCH	Maximum Number of Bits of an E-DCH Transport Block Transmitted within a 10-ms TTI	Maximum Number of Bits of an E-DCH Transport Block Transmitted within a 2-ms TTI	Maximum Bit Rate (Mbps)
Category 1	1	SF4	10-ms TTI only	7110	–	0.73
Category 2	2	SF4	10-ms and 2-ms TTI	14,484	2798	1.46
Category 3	2	SF4	10-ms TTI Only	14,484	–	1.46
Category 4	2	SF2	10-ms and 2-ms TTI	20,000	5772	2.92
Category 5	2	SF2	10-ms TTI Only	20,000	–	2.00
Category 6	4	SF2	10-ms and 2-ms TTI	20,000	11,484	5.76

the E-DCH during a TTI. This type of grant is best suited for constant-rate delay-sensitive application such as Voice over IP (VoIP).

In the scheduled grant, the UE maintains a serving grant, which is updated based on information received from the network. The serving grant directly specifies the maximum power the UE can use on the E-DPDCH in the current TTI. There are two ways for the network to control the UE's serving grant. The first is through an absolute grant, transmitted on the shared E-AGCH downlink channel, which signals a specific absolute number for the serving grant. The other way is through relative grants, transmitted using the downlink E-RGCH channels that incrementally adjust a UE's serving grant up or down from its current value.

Rapid Retransmit   The rapid retransmit objective of HSUFA is achieved through the HARQ algorithm that runs inside the Node B. Many stop-and-wait HARQ processes operate in parallel, four for 10-ms TTI and eight for 2-ms TTI. Every time the UE transmits, the receiving HARQ process in the network attempts to decode the data block. If successful, the BS transmits an ACK to the UE over the E-HICH channel and that HARQ process in the UE will move onto the next data block. If the decoding fails, the Node B transmits a NACK to the UE on the E-HICH. If the maximum number of retransmissions has not been reached, the UE retransmits the data block again. The UE will use chase combining (transmission of the exact same bits again) or incremental redundancy (transmission of a different set of bits), depending on how the RRC layer configured the link at call setup.

MAC Layer   As previously mentioned, HSUPA adds a new transport channel, the E-DCH. The E-DCH connects to the new MAC sublayer, MAC-e/es. In the UE, MAC-e/es are considered one single sublayer; however, on the network side, MAC-e and MAC-es are considerate separate.

In the UE, the combined MAC-e/es sublayer executes HARQ processes, selects the uplink data rate based on the current serving grant, and provides status reporting.

On the network side, the MAC-e sublayer contains HARQ processes, some demultiplexing functionality, and a fast scheduling algorithm. MAC-es is responsible for reordering and combining and also disassembly of MAC-es protocol data units (PDUs) into individual MAC-d PDUs.

Broadcast and Multicast Services   Multimedia broadcast multicast service (MBMS) supports multicast and broadcast services such as mobile television (TV). With MBMS utilization, the same content is transmitted to multiple users in a unidirectional fashion, usually by multiple cells, to maximize the coverage area where the service is officered.

When a user is at the cell's edge, the signal from the neighboring BS can be used to improve the received signal quality. This is a form of a soft handover that leads to a significant increase in data rates and improves broadcast services' coverage.

In HSPA, a very strong emphasis was placed on supporting MBMS on the same carrier as other traffic (voice and data) to provide the operator with maximum flexibility. In cdma2000/Evolution–Data Optimized (EV-DO), for instance, a separate carrier is required, which means that a separate receiver is needed in the mobile device [153].

7.1.7   W-CDMA Usage

UMTS was first deployed in 2003 under the label of "Release 99," followed by HSDPA's first deployment in 2006 and HSUPA's first deployment in 2007. However, by the end of 2007, there were 166 commercial HSDPA network deployments in 75 countries, with additional 38 networks committed to deployment. The first commercial launch of HSUPA was in early 2007, and 24 networks had launched by the end of that year [158]. According to a recent GSA survey (January 20, 2012), there are over 424 commercial 3G WCDMA operators operating in 165 countries, serving over 822.4 million subscribers, including 469 million HSPA subscribers. In addition, 100% of WCDMA operators have upgraded to HSPA technology. Out of 424 commercial HSPA operators, 304 now support peak downlink data rates of at least 7.2 Mbps, and, as of 2010, HSPA+ has officially entered the mainstream with 211 HSPA+ network deployment commitments, and 152 HSPA+ and 49 42-Mbps DC-HSPA+ networks commercially launched.

7.1.8 Recommended Additional UMTS Reading

The GSA maintains the http://www.gsacom.com Web site, which is a great, single information resource targeted to the industry. GSA's Web site includes market surveys, technology and subscriptions updates, white papers, GSM/EDGE-WCDMA/ HSPA/HSPA+ network deployments information, devices availability, applications and services, operator case studies, success stories, and latest developments in the evolution to Long-Term Evolution/Service Architecture Evolution (LTE/SAE) network operator commitments. The authors also recommend the following references to the reader:

- H. Holma and A. Toskala, *WCDMA for UMTS: HSPA Evolution and LTE*, 4th Edition, John Wiley & Sons, Ltd., New York, 2007 [156]
- S. Su, *The UMTS Air-Interface in RF Engineering*, McGraw-Hill, New York, 2007 [154].

7.2 MOBILE WiMAX

In Chapter 5, we discussed the evolution of wireless metropolitan area network (WMAN) systems with a particular focus on Fixed WiMAX technology. You may recall that the development of the IEEE 802.16 standard had already gone through numerous revisions to address the market's ever-changing needs for broadband access and mobility. In Chapter 5, we focused primarily on WiMAX fixed systems (orthogonal frequency-division multiplexing [OFDM] PHY). The objective of this section is to describe how the IEEE 802.16 standard addresses mobility through the introduction of a different air interface, OFDMA PHY. Everything else described in Chapter 5 in terms of the evolutionary steps, deployments, history, recommended additional resources, and so on, remain applicable to Mobile WiMAX systems as well. Hence, the objective of this section is to present an overview of key technological components relevant only to Mobile WiMAX systems (based on the IEEE 802.16 standard-2009) that had not been previously described. Another objective is to also highlight major differences between Fixed and Mobile WiMAX technologies.

Mobile WiMAX is a 3G cellular technology fully compliant with all of the IMT-2000 requirements specified by ITU, though it is fundamentally different from other 3G technologies such as UMTS and CDMA2000 (which use CDMA) and is not typically marketed as a "true" 3G system.

Mobile WiMAX technology was originally based upon the IEEE 802.16 standard-2005, released in February 2006. It is an amendment to the IEEE 802.16 standard-2004, which is the base IEEE 802.16 standard for fixed broadband wireless communications (Fixed WiMAX). Both standards were later superseded by the IEEE 802.16 standard-2009. Perhaps one of the most important aspects of the IEEE 802.16 standard-2005 was the addition of the support for mobility through introduction of soft and hard handovers between BSs. Some other improvements include addition of the new OFDMA PHY based on the concept of scalable OFDMA (S-OFDMA) to support scalable channel bandwidths from 1.25 to 20 MHz, advanced antenna diversity schemes and MIMO technologies, turbo coding and low-density parity check (LDPC) coding, downlink subchannelization, and HARQ. The following sections detail key features of the Mobile WiMAX technology.

7.2.1 WiMAX Network Reference Model

The IEEE 802.16-2009 standard only defines the PHY and MAC layers. As a result, vendors and operators have formed two additional working groups within the WiMAX Forum, the Network Working Group (NWG) and the Service Provider Working Group, to define standard network reference models (NRMs) for open internetwork interfaces in order to address issues associated with air interface interoperability, roaming, multivendor access networks, and intercompany billing. The NWG is focused on creating higher-level networking specifications for fixed, nomadic, portable, and mobile WiMAX systems beyond what is defined in the IEEE 802.16 standard. The Service Provider Working Group assists with writing the network requirements and prioritizes them to help drive the work of the NWG [159].

The Mobile WiMAX network architecture is based on an all-IP platform without any dependence on legacy circuit telephony. WiMAX Forum industry participants have identified a WiMAX NRM to serve as a logical representation of the network architecture. Figure 7.20 [160] depicts the WiMAX NRM containing functional entities and corresponding links (i.e., interfaces or reference points) [159]. This architecture has been developed for the purpose of providing amalgamated support of functionality necessary in a range of network deployment models and usage scenarios (fixed, nomadic, portable, simple mobility, full mobility) [159].

The WiMAX NRM consists of four logical parts:

1. *Mobile Stations (MSs).* This represents the subscriber's mobile devices such as cell phones, wireless laptops, personal digital assistants (PDAs), and so on.

2. *Network Access Provider (NAP).* This provides radio access functionality. Some of the NAP functions include access service network (ASN), 802.16

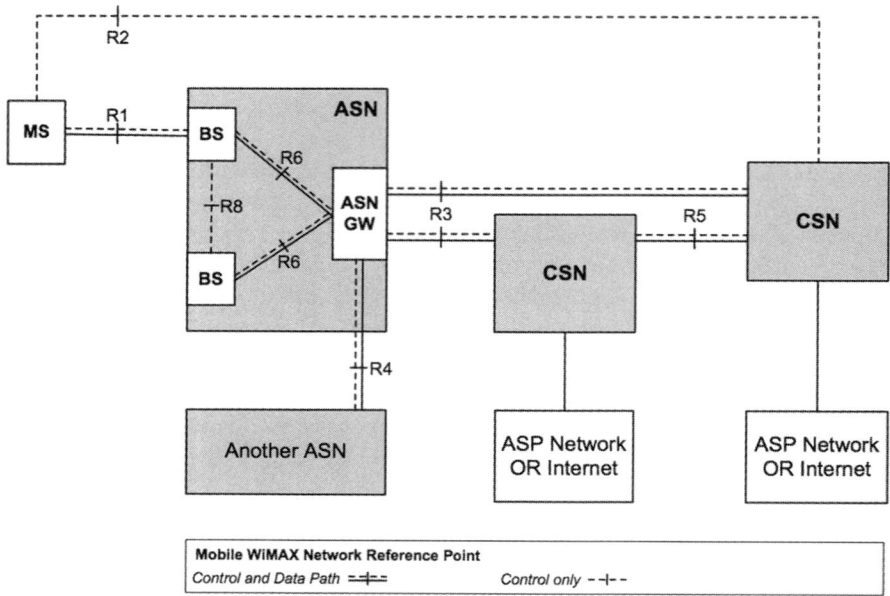

Figure 7.20 WiMAX network reference model. Reprinted from Reference [160].

interface with network entry and handover, ASN-GW (access service network gateway), BSs, foreign agent (FA), QoS and policy enforcement, and forwarding to a selected connectivity service network (CSN). A NAP may have contracts with multiple network service providers (NSPs).

3. *NSP.* This provides IP connectivity services. Some of the NSP functions include CSN, home agent (HA), visited and home authentication, authorization, and accounting (AAA) servers (visited AAA [VAAA] or home AAA [HAAA]), connectivity to the Internet, IP address management, AAA, and mobility and roaming between ASNs. An NSP may have a contract with another NSP and may also have contracts between multiple NAPs.

4. *Internet.* This delivers Internet content to a subscriber as well as connectivity to a NSP.

Reference points (e.g., R1 or R2) are conceptual links that connect two functional entities. Reference points represent a collection of protocols between peer entities (similar to an IP network interface). Interoperability is imposed through reference points without telling vendors how to implement the edges of those reference points. Table 7.8 defines each reference point (interface) shown in Figure 7.23 [160, 161].

Table 7.8   WiMAX Network Reference Model: Reference Points

Reference point	Description
R1	The interface between the MS and BS.
R2	Consists of protocols and procedures between the MS and CSN associated with authentication, services authorization and IP host configuration management.
R3	Consists of the set of control protocols between the ASN and the CSN to support AAA, policy enforcement and mobility management capabilities. It also includes the data path methods (e.g., tunneling) to transfer user data between the ASN and the CSN.
R4	The link between an ASN and another ASN. Consists of the set of control and data path protocols originating/terminating in an ASN-GW that coordinate MS mobility between ASNs and ASN-GWs. R4 is the only interoperable reference point between the ASN-GWs of one ASN or two different ASNs.
R5	The link between CSNs.
R6	Located within an ASN and represents a link between the BS and the ASN-GW.
R7	Located within the ASN-GW and represents internal communication within the gateway. Deprecated per latest standard [WMF10].
R8	Located within an ASN and represents a link between BSs to ensure fast and seamless handover.

Data derived from References [160, 161].

7.2.2   Differences between Fixed and Mobile WiMAX Technology

As mentioned earlier, the focus of this chapter is on Mobile WiMAX, so rather than providing a monolithic IEEE 802.16 standard-based technology overview, this chapter focuses on the key differences between Mobile WiMAX (802.16e-2005/802.16-2009) and Fixed WiMAX (802.16-2004/802.16-2009). This is largely motivated by the fact that there is an enormous amount of commonality between Fixed and Mobile WiMAX. This chapter assumes the reader has a basic knowledge of Fixed WiMAX, which is discussed in detail in Chapter 5.

Indeed, Fixed and Mobile WiMAX share great commonality in their base technology. Except for relatively minor differences, the MAC is virtually common between Fixed and Mobile WiMAX. However, there are several key differences worth discussing:

(a) New Orthogonal Frequency Division Multiple Access (OFDMA) PHY

(b) Advanced Antenna System (AAS) support

(c) New Forward error control (FEC)

(d) Hybrid Automatic Repeat Request (HARQ)

(e) Handover support (in the form of new MAC message types)

(f) New security mechanisms

(g) New optional MAC messages

(h) New bandwidth request mechanisms

(i) Modified bandwidth allocation mechanisms

(j) New device types and characteristics

(k) Sleep modes

(l) Time Division Duplexing (TDD) versus Frequency Division Duplexing (FDD)

(m) Channel bandwidth

(n) Mobile WIMAX frequency bands of operation

(o) Multicast and broadcast service (MBS)

7.2.3 OFDMA PHY of Mobile WiMAX

Perhaps the most significant difference between Fixed and Mobile WiMAX is the introduction of the OFDMA PHY. OFDMA is very similar to the OFDM PHY of Fixed WiMAX. To give a brief refresher on OFDM, the overall channel is subdivided into smaller "chunks," referred to as subcarriers. The high-rate data stream that is to be transmitted over the air interface is divided into smaller, lower-rate data streams that are then individually encoded and modulated, thus the FDM portion of OFDM. These subcarriers are separated in a way such that the center of a particular subcarrier falls at the first null of the adjacent channel, thus minimizing cochannel interference (CCI), the O portion of OFDM. The resulting OFDM spectrum is illustrated in Figure 7.21.

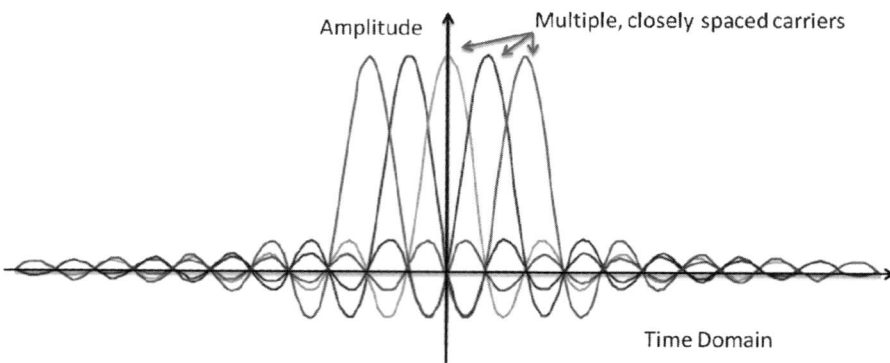

Figure 7.21   Illustration of OFDM spectrum.

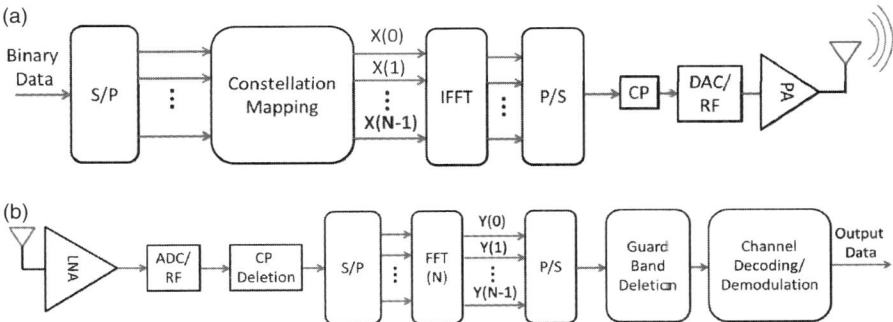

Figure 7.22   A high-level view of the end-to-end OFDM link. (a) Block diagram of a baseband OFDM transmitter; (b) block diagram of a baseband OFDM receiver. DAC, digital-to-analog converter; PA, power amplifier; LNA, low noise amplifier; ADC, analog-to-digital converter.

The potential advantages of such a multicarrier (MC) approach have been known for over 40 years [333] such as vastly improved robustness against multipath fading compared to single-carrier (SC) approaches. However MC approaches such as this proved overly complex for a long period of time. Indeed, an *n*-carrier approach would traditionally require *n* modulators, *n* oscillators, and so on. This proves overly complex when considering 256 carriers, 1024 carriers, 2048 carriers, and so on. This type of MC approach has only proved feasible in the past decade with advances in digital signal processing (DSP) capabilities. In the DSP paradigm, each subcarrier is digitally synthesized in the frequency domain. An inverse fast Fourier transform (IFFT) is then performed on the composite frequency-domain representation of all subcarriers to construct a composite time-domain waveform. At the receiver, a fast Fourier transform (FFT) is performed on the received composite time-domain waveform to produce the composite frequency-domain waveform. This frequency-domain waveform can then be separated into the individual subcarriers, demodulating and decoding these subcarrier streams before multiplexing them into the original high-rate data stream. This process is illustrated in Figure 7.22.

Note the addition of a cyclic prefix to each composite OFDM symbol, represented by the "CP" step in Figure 7.22a. Here, the last fractional portion of the OFDM symbol is replicated and added to the beginning of the original OFDM symbol. This cyclic prefix is intended to act as a guard time, can assume many values (1/4, 1/8, 1/16, 1/32—where this represents the fraction of the OFDM symbol that is replicated), and is usually chosen based on the root mean square (rms) Delay spread observed in the channel mitigates intersymbol interference (ISI). In the Fixed WiMAX OFDM 256-point FFT PHY, the OFDM symbol time varies between 11 and 128 μs. In the Mobile WiMAX OFDMA PHY, OFDM symbol time varies between 102 and 144 μs (102 μs is the envisioned dominant symbol time). Not all OFDM subcarriers are employed for data transmission. A subset of OFDM subcarriers is employed to carry pilot signals, which are used by receivers for synchronization and channel quality estimation. A subset of OFDM subcarriers is not used (i.e., null) and is employed to fit a particular OFDM scheme within a defined channel bandwidth.

In Fixed WiMAX, bandwidth is allocated to subscribers once every frame time, specifically telling the subscriber which portions of the downlink has data destined for the station and which portions of the uplink may be used by the subscriber to transmit data. The frame time can be one of multiple values (from 2.5 to 20 ms), and bandwidth allocations occur explicitly on a frame-by-frame basis. In Mobile WiMAX, the frame time can be chosen from a list of values ranging from 2 to 20 ms [45], though only the 5-ms radio frame is supported in the current Mobile WIMAX system profiles [162]. The key difference is the following:

(a) **Fixed WiMAX:** bandwidth is allocated in terms of time. A subscriber is allowed to utilize the channel for a given period of time, as defined by a number of OFDM symbols.

(b) **Mobile WiMAX:** bandwidth is allocated in terms of time and frequency. A subscriber is allowed to utilize the channel for a given period of time, as defined by a number of OFDM symbols, and for a given set of frequencies, as defined by a set of OFDM subcarriers.

As such, OFDMA acts as a multiple access mechanism—it is a logical channel defined by a period of time and a frequency range. This paradigm is reflected in the Mobile WiMAX TDD frame structure, as illustrated in Figure 7.23 [45].

Let us take a closer look at Figure 7.23. Each frame is divided into downlink (DL) and uplink (UL) subframes that are separated by Transmit/Receive and Receive/Transmit Transition Gaps (TTG and RTG, respectively) to prevent DL and UL

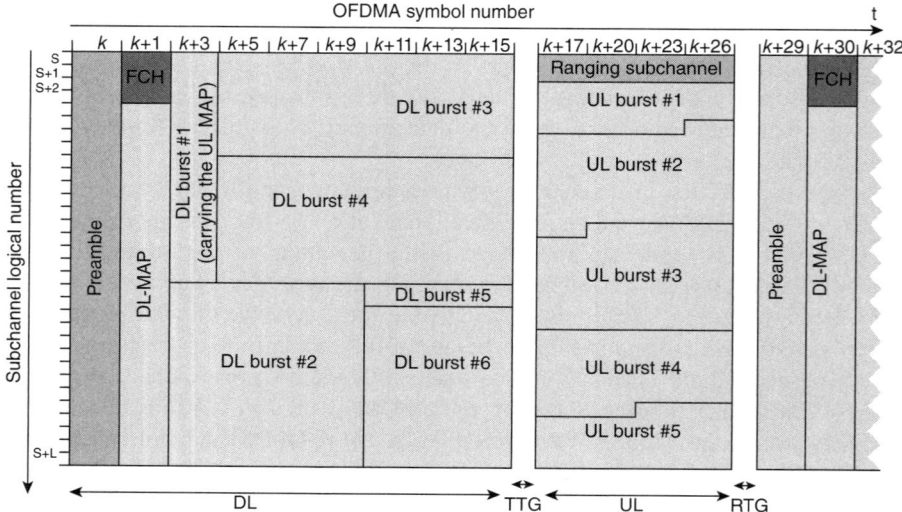

Figure 7.23   Mobile WiMAX TDD frame. Reprinted from Reference [45].

transmission collisions. The first OFDM symbol of the frame is reserved for the preamble, which is used for synchronization. The Frame Control Header (FCH) follows the preamble. It is responsible for carrying the frame configuration information about the MAP message length, coding scheme, and available subchannels. The DL-MAP and UL-MAP provide subchannel allocation and other control information for the DL and UL subframes, respectively. The UL ranging subchannel is allocated for MSs to perform closed-loop time, frequency, and power adjustment as well as bandwidth requests [159]. Bursts can either consist of all OFDM subcarriers, as in the case of the preamble shown in Figure 4.3, or a subset of all OFDM subcarriers (e.g., DL burst #2). In WiMAX terminology, this represents a full usage of subcarriers (FUSCs) and partial usage of subcarriers (PUSCs). Generally, portions of the UL and DL frame are identified as being used for PUSCs or FUSCs. These regions of usage are generally referred to as *zones*. A particular assignment of a group of subcarriers to a Mobile WiMAX subscriber is referred to as the creation of a *subchannel*. This is illustrated in Figure 7.24 [45].

It should be noted that while Figure 7.24 depicts subchannels as contiguous in frequency, this is generally not true. Subchannels generally represent a set of non-contiguous subcarriers.

Figure 7.25 and Figure 7.26 [45] illustrate the Mobile WiMAX FDD frame formats.

BSs of OFDMA FDD systems are specified to operate in full duplex mode, while SSs can operate in either full duplex (FDD) or half-duplex (H-FDD) mode. The FDD frame structures, shown in Figure 7.25 and Figure 7.26, support both FDD and H-FDD subscriber station (SS) types. Each frame structure is specified to support a coordinated transmission arrangement of two groups of H-FDD SSs (Group 1 and Group 2) that share the frame at distinct partitions of the frame. The frame structures shown in Figure 7.25 and Figure 7.26 support the concurrent operation of H-FDD and FDD SSs. The DL frame contains two subframes. DL Subframe 1 consists of a preamble symbol, a MAP region (MAP1), and data symbols (DL1). DL Subframe 2 is composed

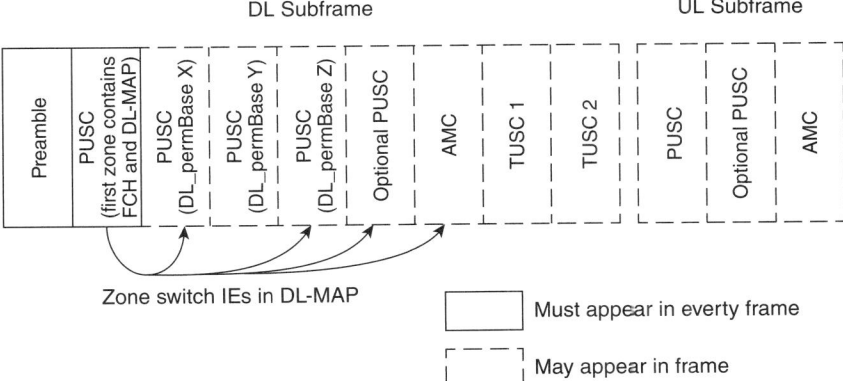

Figure 7.24   Example of a TDD frame with multiple zones. Reprinted from Reference [45].

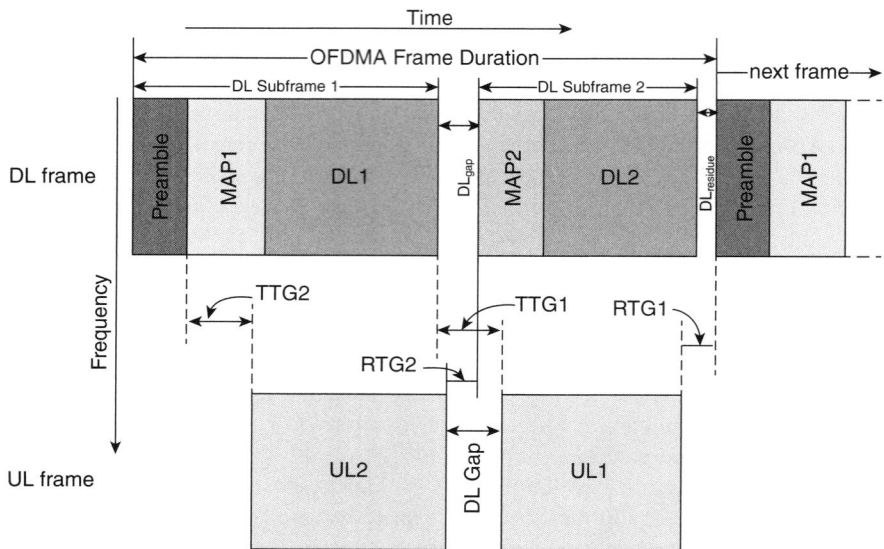

Figure 7.25 Generic OFDMA FDD frame structure (supporting H-FDD MS in two groups with residual at the end of the frame). Reprinted from Reference [45].

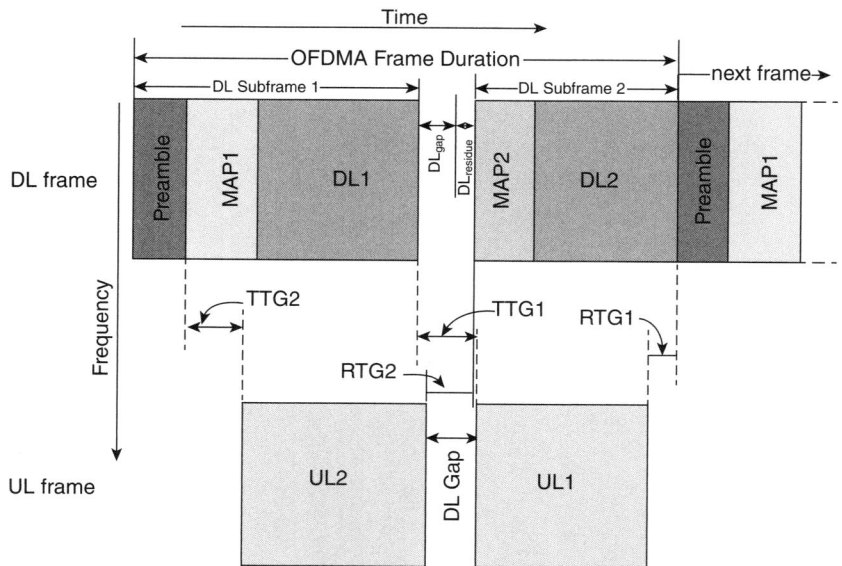

Figure 7.26 Generic OFDMA FDD frame structure (supporting H-FDD MS in two groups with residual between the downlink subframes). Reprinted from Reference [45].

of a MAP region (MAP2) and data symbols (DL2). The space between the two DL subframes is occupied by a gap DL_{gap} (see Figure 7.25), the size of which is an integer number of symbols (0, 1, 2, 3). Optionally, as shown in Figure 7.26, this gap may also include the residual frame time, $DL_{residue}$ (the frame duration minus the total time occupied by the frame symbols). The number of symbols in DL_{gap} and the location of $DL_{residue}$ is signaled in the downlink channel descriptor (DCD), in both DL subframes, using the "FDD DL gap" TLV (type/length/value). The BS cannot change the location and value of $DL_{residue}$ during operation. The UL frame contains two subframes, UL2 and UL1 (in this order). Figure 7.25 and Figure 7.26 illustrate the timing relationship of the UL subframes relative to the DL subframes. The four parameters TTG1, TTG2, RTG1, and RTG2 are stored in the DCD messages and should be large enough to accommodate the H-FDD SSs transmit receive switching time plus the round trip propagation delay [45].

Generally speaking, duplexing schemes aside, PUSCs have the obvious advantage of allowing the Mobile WiMAX BS much greater flexibility in its bandwidth allocation and scheduling algorithms. However, there are also significant advantages in the areas of range performance that allows for increased flexibility in frequency reuse planning. Range performance is achieved because the subscriber can focus its total available power into a fewer number of subcarriers, increasing the per-subcarrier power. This increases the robustness of the reverse link, which is typically the limiting factor in the bidirectional wireless link and also dramatically helps with near-cell-edge performance. The price for these advantages is decreased uplink data rate because of the decreased number of subcarriers.

As will be seen in subsequent sections, many other changes in Mobile WiMAX compared with Fixed WiMAX have been necessitated by OFDMA and PUSCs.

7.2.4 Advanced Antenna Systems

A key technology innovation included in Mobile WiMAX is its support of MIMO techniques and smart antenna systems. Here, multiple transmit and receive antennas separated in space are employed to provide numerous performance benefits including space–time diversity gains, spatial multiplexing gains, as well as adaptive smart antenna techniques. In space–time diversity coding, the same group of symbols are coded, interleaved, and distributed over multiple antennas. The symbols are typically transmitted in a different order on different antennas, providing time diversity as well as space diversity. This provides the ability to increase coverage and reliability. Using space–time diversity coding in a multipath environment can significantly increase received signal strength, increasing achievable link distance and/or link reliability.

In spatial multiplexing, different data streams are coded, interleaved, and distributed over multiple antennas. This provides the opportunity to significantly increase the achievable peak data rate. If N transmit and N receive antennas are employed (i.e., N-by-N MIMO), it is possible for the receiver to correlate the combined output of the N receive antennas to decode N different data streams because data from each of the N transmit antennas are received by N receive antennas.

Mobile WiMAX supports up to a 4-by-4 MIMO configuration. A typical WiMAX implementation combines MIMO techniques. A typical Mobile WiMAX BS will likely employ two transmit and two receive antennas per sector. Subscriber stations will likely employ two receive antennas but only a single transmit antenna. The Mobile WiMAX downlink will support both space–time coding and spatial multiplexing through adaptive switching. In unfavorable channel conditions, space–time coding will be employed to provide increased link robustness. In favorable channel conditions, spatial multiplexing will be employed to gain increased data rate. Nonstandardized link adaptation algorithms will determine the mode of operation at any point in time.

Smart antenna systems will also become an important part of the Mobile WiMAX technology portfolio. Here, Mobile WiMAX devices would have an array of multiple antenna elements. These antenna elements would be weighted to electronically form and steer narrow beamwidth radiation patterns. This adaptive beamforming capability would attempt to maximize receive signal strength while minimizing interference.

At this point it is important to mention that many current Mobile WiMAX products are single-input, single-output (SISO) systems employing only a single antenna, although there are already MIMO products available. Regardless, it is envisioned that future products will increasingly support MIMO and adaptive antenna concepts.

7.2.5   New FEC in Mobile WiMAX

The only forward error correction (FEC) mechanism that is mandatory in Fixed WiMAX is a concatenated Reed–Solomon–convolutional code (RS-CC) scheme. While convolutional turbo codes (CTCs) and block turbo coding (BTC) are also defined, they are optional approaches that have not been widely implemented in Fixed WiMAX devices.

The IEEE 802.16-2009 specification makes tail-biting convolutional codes (CC) the mandatory FEC scheme for Mobile WiMAX with the zero-tail CC, CTC, and BTC FEC approaches optional [45, 160].

7.2.6   Hybrid Automatic Repeat Request

HARQ is a promising approach that combines automatic repeat request (ARQ) with FEC such as convolutional or turbo codes. The transmitter sends a block of FEC-encoded data. If the decoder cannot recover the transmission, then the coded data blocks are stored at the receiver and retransmission is initiated. When an additional coded data block is received, both coded data blocks are combined and fed to the decoder. This approach adds incremental redundancy and hence improves the probability of recovering the data. The Mobile WiMAX HARQ approach employs a Stop-and-Wait protocol, where each transmitted message must be positively acknowledged before another message can be transmitted. Mobile WiMAX employs a synchronous acknowledgement (ACK) that is sent after a fixed delay. There is a dedicated PHY ACK/NACK available in the OFDMA uplink.

The IEEE 802.16-2009 specification supports the following two HARQ schemes:

(a) Chase combining

(b) Incremental redundancy

In chase combining, the exact same coded block is retransmitted. Here, the transmitted message (both original and subsequent replicas) consists of only a subset of the entire encoded data block in order to reduce overhead. In incremental redundancy, the transmitter generates several code blocks by subdividing the overall code block into smaller blocks. If the first code block cannot be decoded, the transmitter sends the second code block. The receiver combines these received blocks and feeds them into the decoder. The chase combining approach generally introduces higher overhead than the incremental redundancy approach. However, the chase combining approach is simpler to implement. For this reason, chase combining is the only approach required in Mobile WiMAX.

7.2.7 Mobile WiMAX Handover Support

There are two types of handover defined in Mobile WiMAX: break before make and make before break. In break-before-make handover, the mobile station (MS) first disconnects from its current BS before it then connects to and establishes service with the new MS. In cellular technologies, break before make is referred to as hard handover. In make-before-break approaches, the MS first establishes service and connects to the new BS before disconnecting from the previously serving BS. This is analogous to soft handover in cellular technologies. The advantage of break before make is that it is simpler to implement. The advantage of make before break is that it minimizes handover delay and packet loss. Because of its simplicity and the desire to enable low-cost subscriber units, break-before-make handover is the only mandatory handover method for Mobile WiMAX and is envisioned to be the dominant approach in fielded equipment.

7.2.8 Mobile WiMAX Security Mechanisms

The IEEE 802.16-2009 specification supports two Privacy Key Management (PKM) protocols: PKM version 1 (PKMv1) and PKM version 2 (PKMv2) with more enhanced features such as new key hierarchy, Advanced Encryption Standard–Cipher-based Message Authentication Code (AES-CMAC), Advanced Encryption Standard (AES) key wraps, and MBS. The PKM protocol allows for both mutual authentication and unilateral authentication (e.g., where the BS authenticates SS, but not vice versa). The PKM protocol uses either Extensible Authentication Protocol (EAP) or X.509 digital certificates together with Rivest, Shamir, and Adleman (RSA) public key encryption algorithm or a sequence starting with RSA-based authentication and followed by EAP authentication.

Within Fixed WiMAX, the support of PKM protocols is optional. Consequently, many Fixed WiMAX products do not employ any type of authentication mechanisms. Authentication is mandatory in Mobile WiMAX with PKMv2 EAP-based protocol with a mandatory AAA server component in the Mobile WiMAX infrastructure (CSN). This authentication is mutual. Support for PKMv2 is mandatory in Mobile WiMAX.

Furthermore, a much more comprehensive and secure approach to authentication and key distribution is employed in Mobile WiMAX. The encryption algorithms specified in Fixed WiMAX are Data Encryption Standard (DES) in cipher-block chaining (CBC) mode and AES in Counter with CBC-Message Authentication Code (CCM) mode. However, both of these are optional. Mobile WiMAX adds AES in Counter (CTR) mode (specified for protecting MBS) and AES in CBC mode. The cryptographic suite required for Mobile WiMAX certification is AES-CCM with 128-bit key, CCM Mode, and AES Key Wrap with 128-bit key. It should be noted that not all Mobile WiMAX traffic is encrypted. All data payloads and MAC subheaders are encrypted. Multicast frames may also be encrypted. However, many management frames are not encrypted, and the generic MAC header is not encrypted. Furthermore, the time division multiplexing (TDM) preamble, UL-MAP, and DL-MAP messages are not encrypted.

7.2.9 New MAC Messages

The basic generic MAC header remains largely unchanged between Fixed and Mobile WiMAX. However, Mobile WiMAX does extend the MAC subheader, providing for a variety of subheader types and additional uplink headers with no payload.

    (a) PHY channel report

    (b) Broadband wireless (BW) request with uplink transmit power report

    (c) BW request and carrier-to-interference plus noise ratio (CINR) report

    (d) BW request with uplink sleep control

    (e) Sequence number report

    (f) Channel quality indicator channel (CQICH) allocation request

Most of these are related to the mobile nature of Mobile WiMAX and the need for the mobile subscriber to report its changing channel quality conditions. It is important to note that these messages are in the form of MAC subheaders, which are actually part of the encrypted payload of the overall MAC frame (i.e., no payload refers to the absence of data, even if MAC subheaders are present within the payload portion of the MAC frame).

7.2.10 Bandwidth Request Mechanisms

WiMAX specifies two PHY-specific contention-based bandwidth requests: Focused Bandwidth Request for OFDM-based systems (Fixed WiMAX) and CDMA Bandwidth Requests for OFDMA-based systems (Mobile WiMAX). It should be pointed out that CDMA Bandwidth Requests were first specified in Fixed WiMAX. However, since the OFDMA PHY was optional, CDMA Bandwidth Requests were not widely adopted in Fixed WiMAX products. However, with the OFDMA PHY mandatory in Mobile WiMAX, CDMA Bandwidth Requests are the primary method for subscriber stations to request bandwidth from the WiMAX BS. In this approach, the ranging subchannel is employed to also serve bandwidth requests. The WiMAX BS defines the ranging subchannel, ranging codes to be used for ranging requests, and separate codes for bandwidth requests. An SS initiates a bandwidth request by putting a bandwidth request code on this ranging channel. The BS will then provide an uplink allocation so that the identified subscriber can then make a full bandwidth request. It is referred to as a CDMA bandwidth request because the contention codes are CDMA-like pseudonoise codes, resulting in robustness to collisions (the ranging subchannel is contention based and collisions are likely at times). For OFDMA, there are 256 contention codes of length 144 bytes, with the uplink channel descriptor (UCD) indicating which codes are allocated to initial ranging requests, periodic ranging, handover ranging, and bandwidth requests.

7.2.11 Bandwidth Allocation Mechanisms

Bandwidth allocation in Mobile WiMAX is very similar to Fixed WiMAX. The key difference between the two is the complexity introduced by the new OFDMA PHY of Mobile WiMAX. Because resources are not assignable to the OFDM subcarrier resolution, bandwidth allocation mechanisms must be adapted to support this new subcarrier-based allocation of resources. This equates to modifications in the UL-MAP and DL-MAP messages that are in every TDM frame. These modified UL-MAP and DL-MAP messages are required because MSs must now be told explicitly by the BS which subcarriers (both uplink and downlink) are allocated to them within the TDM frame.

Furthermore, the MAC scheduler controls MS's resource allocation that can vary from a single time slot to the entire frame, thus allowing for a very large dynamic range of throughput to a specific MS at any given time. Also, since UL-MAP and DL-MAP messages carry the resource allocation information at the beginning of each TDM frame, the MAC scheduler has an ability to change the resource allocation on a frame-by-frame basis to efficiently adapt to the traffic's bursty nature [159].

7.2.12 Mobile WiMAX Device Types and Classes

Perhaps the greatest difference between Fixed and Mobile WiMAX is the nature of the devices that will be utilizing the wireless network. Fixed WiMAX, as the name would suggest, was designed for fixed applications. Consequently, Fixed WiMAX SSs are often characterized by a relatively large size footprint, are not necessarily energy-efficient, and commonly employ directional antennas to connect to the Fixed WiMAX BS.

These same types of "cable modem equivalent" devices also exist in Mobile WiMAX. However, the trend clearly steers toward mobile platforms such as hand-held phones, PDAs, and laptop computers equipped with embedded Mobile WiMAX or Mobile WiMAX PC Card. Mobile WiMAX devices will increasingly have omni-directional antennas and have a very different RF design than Fixed WiMAX devices.

7.2.13   Sleep Modes in Mobile WiMAX

Mobile WiMAX has been developed with power consumption minimization as a clear design goal in order to enable devices that can provide long battery lives to consumers. Part of this approach is adaptive power control (power control is also employed in Fixed WiMAX). Another key feature for minimizing power consumption is the introduction of aggressive sleep modes into Mobile WiMAX. "Sleep modes" refer to the process of negotiating periods when the MS is not available on the air interface so that it can power down its transmitter/receiver to conserve energy. Here, an MS can be in one of two states, available or unavailable. Sleep modes are only optional for MSs, but mandatory for BSs. However, it is envisioned that a wide range of products will support sleep modes to varying extents. MSs that employ power-saving mechanisms will be assigned to 1 of 256 possible Power Saving Classes, where a Power Saving Class is a group of connections with common wake and sleep cycle priorities. Each Power Saving Class can be one of three Power Saving Class Types that are differentiated by the type of connection being served:

(a)   *Type I.* Best effort (BE) and non-real-time variable rate (NRT-VR)

(b)   *Type II.* Unsolicited grant service (UGS) and real-time variable rate (RT-VR)

(c)   *Type III.* Multicast and management.

It should be noted that only Power Saving Class Type I is required by WiMAX, with Types II and III being optional. An MS becomes a member of a Power Saving Class either via a request message to the BS or is assigned to a Power Savings Class by the BS. Once an MS belongs to a Power Saving Class, the BS will periodically send a traffic indication message to a Power Saving Class (via broadcast or multicast signaling). An MS belonging to Class Type I wakes up during its defined listening interval to look for this message, and then decides to either continue sleeping or remain awake based on the contents of this message. MSs belonging to other class types ignore this message and will make sleep and awake decisions without this information.

Mobile WiMAX also supports more drastic power saving mechanisms, placing MSs in "idle modes" when the MS is inactive for long periods of time. There is a page mechanism by which the BS can attempt to communicate with the idle MS, if necessary, to support incoming communications. In the idle mode, the MS is completely silent and may even move to new coverage sectors without signaling. Upon receipt of the broadcast page transmission or the need to originate data, the MS will respond in its current sector and only then initiate handover.

7.2.14   TDD versus FDD

The IEEE 802.16-2009 standard supports TDD and FDD (full and half-duplex) operations, and while Fixed and Mobile WiMAX system profiles implement both of them, TDD seems to be the preferred method of many vendors. This preference can be attributed to the following reasons:

- TDD uses half of the FDD spectrum, thus saving the bandwidth
- In TDD, the downlink/uplink ratio can be adjusted to efficiently support asymmetric downlink/uplink traffic, whereas with FDD, downlink and uplink always have fixed and, typically, equal downlink and uplink bandwidths
- TDD provides a better support of link adaptation, MIMO, and other closed-loop advanced antenna technologies due to its assurance of channel reciprocity
- Transceiver designs for TDD implementations are generally considered less complex and therefore less expensive.

Currently, there are 19 TDD (one is under development) and 12 FDD (one is a TDD/FDD mix) WiMAX Forum's Mobile WiMAX system profiles in existence [163], which are detailed in the next section.

7.2.15   Channel Bandwidths in Mobile WiMAX

Fixed WiMAX's 3.5 MHz is the most common bandwidth, primarily driven by the worldwide regulatory climate and the availability of spectrum in key markets and WiMAX Forum certification profiles. In Mobile WiMAX, the channel bandwidths can be 3.5, 5, 7, 8, 8.75, and 10 MHz. Table 7.9 (from the WiMAX Forum) summarizes key attributes of the current Mobile WIMAX certification profiles [163].

Table 7.10 lists Mobile WiMAX channel bandwidths with their respective FFT sizes [161].

7.2.16   Mobile WiMAX Frequency Bands of Operation

WiMAX spectrum is perhaps the most confusing issue surrounding WiMAX. WiMAX aims to define a global technology that can accommodate spectrum regulatory issues across all key markets. As such, it is anticipated that WiMAX will eventually operate in many frequency bands. To date, all certified products (both Fixed and Mobile WiMAX) operate in the following frequency bands [93]:

- 2.3 GHz (48 deployments)
- 2.5 GHz (112 deployments)
- 3.3 GHz (10 deployments)
- 3.5 GHz (308 deployments)
- 5+ GHz (21 deployments).

Table 7.9   Mobile WiMAX Release 1.0 and 1.5 Band Classes Groups

Band class group	Uplink MS transmit frequency range (MHz)	Downlink MS receive frequency range (MHz)	Channel bandwidth (MHz)	Duplex mode	WiMAX forum's air interface release
1.A	2300–2400	2300–2400	8.75	TDD	1.0
1.B	2300–2400	2300–2400	5 and 10	TDD	1.0
2.D	2305–2320, 2345–2360	2305–2320, 2345–2360	3.5, 5, and 10	TDD	1.0
2.E	2345–2360	2305–2320	2×3.5, 2×5, and 2×10	FDD	1.5
2.F	2345–2360	2305–2320	5 uplink, 10 downlink	FDD	1.5
3.A	2496–2690	2496–2690	5 and 10	TDD	1.0
3.B	2496–2572	2614–2690	2×5 and 2×10	FDD	1.5
4.A	3300–3400	3300–3400	5	TDD	1.0
4.B	3300–3400	3300–3400	7	TDD	1.0
4.C	3300–3400	3300–3400	10	TDD	1.0
5L.A	3400–3600	3400–3600	5	TDD	1.0
5L.B	3400–3600	3400–3600	7	TDD	1.0
5L.C	3400–3600	3400–3600	10	TDD	1.0
5.D	3400–3500	3500–3600	2×5, 2×7, and 2×10	FDD	1.5
5H.A	3600–3800	3600–3800	5	TDD	1.0
5H.B	3600–3800	3600–3800	7	TDD	1.0
5H.C	3600–3800	3600–3800	10	TDD	1.0
6.A	1710–1770	2110–2170	2×5 and 2×10	FDD	1.5
6.B	1920–1980	2110–2170	2×5 and 2×10 (20 MHz optional)	FDD	1.5
6.C	1710–1785	1805–1880	2×5 and 2×10	FDD	1.5
7.A	698–862	698–862	5, 7, and 10	TDD	1.0
7.B	776–787	746–757	2×5 and 2×10	FDD	1.5
7.C	788–793, 793–798	758–763, 763–768	2×5	FDD	1.5
7.D	788–798	758–768	2×10	FDD	1.5
7.E	698–862	698–862	5, 7, and 10 (TDD) 2×5 and 2×7 and 2×10 (FDD)	TDD/ FDD	1.5

Table 7.9   (*Continued*)

Band class group	Uplink MS transmit frequency range (MHz)	Downlink MS receive frequency range (MHz)	Channel bandwidth (MHz)	Duplex mode	WiMAX forum's air interface release
7.G	880–915	925–960	2×5 and 2×10	FDD	1.5
7.x[a]	730–770, 890–903, 915–950	730–770, 890–903, 915–950	5 and 10	TDD	1.5
8.A	1785–1805, 1880–1920, 1910–1930, 2010–2025, 1900–1920	1785–1805, 1880–1920, 1910–1930, 2010–2025, 1900–1920	5 and 10	TDD	1.5
8.G	1800–1830	1800–1830	5 and 10	TDD	1.0
9.D	170–202.5	170–202.5	5	TDD	1.0
10.A	5000–5150	5000–5150	5 and 10	TDD	1.0

Reproduced from Reference [166].

[a]The details of Band Class Group 7.x are under development pending finalization of Japanese regulations.

Table 7.10   Mobile WiMAX Channel Bandwidths and Associated FFT Sizes

Channel bandwidth (MHz)	FFT size
3.5	512
5	512
7	1024
8	1024
8.75	1024
10	1024

Reproduced from Reference [164].

It is currently envisioned that 3.5 GHz will remain the primary frequency band of operation for Fixed WiMAX. For Mobile WiMAX, it is envisioned that 2.5 GHz will be the dominant frequency because there are more favorable propagation characteristics at this band compared to higher bands, and several key proponents hold licenses in the 2.5-GHz frequency band (e.g., Sprint). One notable exception is the Wireless Broadband (WiBRO) network that has been deployed by Korea Telecom (KT) in South Korea that operates in the 2.3-GHz frequency band.

7.2.17   Multicast and Broadcast Service (MBS)

The Mobile WiMAX system supports MBS as an optional feature for the BS. MBS is designed to support a broadcast (i.e., point to multipoint) mode of operation, in contrast to the point-to-point connectivity typically supported by existing cellular networks. This type of service can be used for the broadcast TV type of application where, at the same time, several users under the same coverage area are connected to the same service (e.g., a TV channel). Some of the features of the WiMAX MBS include high data rates and coverage using a single frequency network (SFN) (based on the Digital Video Broadcasting–Handheld [DVB-H] SFN concept), a flexible allocation of RF channel resources, low MS power consumption, support for data-casting in addition to audio and video streams, and fast channel-switching.

In a multi-BS MBS system, several BSs in the same geographical locale transmit the same broadcast/multicast messages at the same time using the same frequency channel. These BSs actually belong to MBS_ZONE, a unique identifier transmitted from each BS of the set on the downlink channel descriptor (DCD) message. Note that a BS may belong to different MBS_ZONEs. A multi-BS MBS operation requires from the BSs in the same MBS_ZONE the following:

- Time synchronization (frame number and symbol level)
- Use of the same connection identifier (CID) for broadcast/multicast messages
- Use of the same security association (SA) for the encryption of the broadcast/ multicast messages

Use of a multi-BS MBS solution offers two major advantages over a single-BS MBS solution. First, the MSs successfully registered to a MBS service can receive MBS information from *any* BS of the MBS_ZONE without having to register to a specific BS of that zone (even MS previously in the Idle mode). Second, the MS receives MBS signals from multiple BSs simultaneously, which provides a macro-diversity gain and performance enhancement for the MBS signals.

7.2.18   Mobile WiMAX Summary

Instead of providing a monolithic overview of the entire Mobile WiMAX system, this chapter only focused on key differences between Mobile WiMAX and Fixed WiMAX. This was largely motivated by the fact that there is an enormous amount of commonality between Fixed and Mobile WiMAX technologies. This chapter assumed the reader has a basic knowledge of Fixed WiMAX, which is discussed in detail in Chapter 5.

7.3   CDMA2000

Communications networks based on CDMA technology continue to experience tremendous deployment success worldwide. In fact, there are now over 615 million CDMA subscribers worldwide (excluding UMTS-based systems) [164]. The CDMA air interface is used in both 2G and 3G networks. 2G CDMA standards include IS-95, IS-95A, and IS-95B specifications, collectively known as cdmaOne, while 3G CDMA standards serve as a foundation for 3G services and include CDMA2000 and UMTS. The CDMA2000 system is backward compatible with the cdmaOne technology and meets all of the IMT-2000 3G requirements. CDMA2000 not only utilizes the inherent advantages of CDMA technologies but also uses new enhancements such as OFDM, advanced control and signaling mechanisms, improved interference management techniques, end-to-end QoS, and MIMO to increase user data rates and quality of service as well as to drastically improve overall network capacity. CDMA2000 systems provide a number of services which include cellular, personal communications services (PCS), wireless local loop (WLL), and fixed wireless [164].

7.3.1 CDMA2000 Historical Background

A key reason for evolution to 3G networks was the introduction of connectivity to packet data networks over cellular infrastructures with the concurrent increase in voice capacity. Of course, this is vastly different from the original objective behind 2G systems to offer mobile CS voice and low date rate services.

ITU began the process of 3G standardization with its IMT-2000 initiative. The IMT-2000 standard loosely specifies requirements for the 3G cellular network. Currently, there are five standards in existence that satisfy all of the IMT-2000 requirements and are approved as 3G technologies by the ITU: CDMA2000, UMTS, GSM EDGE, DECT, and Mobile WiMAX, although only CDMA2000 and UMTS, both CDMA-based technologies, are typically marketed as true 3G systems.

You may recall from our earlier discussions that the reason behind the formation of the 3GPP2 was to allow both 3GPP and 3GPP2 consortiums to independently develop IMT-2000-compliant technologies, one based on GSM technology (3GPP) and another on cdmaOne (3GPP2).

The 3GPP2 was established in December 1998 as a collaboration between multiple regional telecommunications standards bodies: the ARIB in Japan, the CCSA in China, the TIA in North America, the TTA in Korea, and the TTC in Japan. Together, these standards bodies comprise the Organizational Partners for 3GPP2. Also, Market Representation Partners include the CDMA Development Group, the IPv6 Forum, and the International 450 Association. These Market Representation Partners offer market advice and a consensus view on market requirements. 3GPP2 was established primarily for preparation, approval, and maintenance of technical specifications and reports for 3G networks based on the CDMA2000 CN structure and CDMA2000 air interface. Like 3GPP, 3GPP2 is not considered a legal entity.

The resulting technologies from the 3GPP and 3GPP2 groups are UMTS (WCDMA) and CDMA2000, respectively.

7.3.2 CDMA2000 Evolution

The CDMA2000 is also known as CDMA2000 1xRTT (Radio Transmission Technology), the nomenclature intended to identify the version of CDMA2000 radio technology that operates in a pair of 1.25-MHz channels (1 times 1.25 MHz, as opposed to 3 times 1.25 MHz in 3xRTT). This 1xRTT air interface is denoted as IS-2000. The 3xRTT form of CDMA2000 utilizes a pair of 3.75-MHz channels to achieve increased data rates (1xRTT achieves data rates of 144 kbps). CDMA2000 3xRTT is also referred to as multicarrier (MC) CDMA and has not been deployed nor is it presently under development. In fact, the 3xRTT standard has been superseded by a two-phase strategy called CDMA2000 1xEV, where 1xEV stands for 1X evolution, or evolution using 1.25 MHz. Because of the relatively low data rates, the community often considers CDMA2000 1xRTT as a 2.5G technology (beyond 2G but not yet 3G). This has led to evolved versions of 1xRTT: 1xEV-DO (Evolution–Data Only or, lately, addressed in a community as Evolution–Data Optimized) and Evolution-Data/Voice (EV-DV).

The CDMA2000 1xEV-DO is an evolution of CDMA2000 1X with higher data rates (up to 2.5 Mbps downlink data rates and 154 kbps uplink data rates) and employs a TDM forward link. This air interface is denoted as IS-856. Revision A network was first deployed in the United States in October 2006, and in January 2010, Indonesia launched the world's first multicarrier broadband network using EV-DO Revision B [164].

The CDMA2000 1xEV-DV standard, also known as CDMA2000 Release D, was released in 2004 as a comprehensive standard aiming to support data and voice transmissions simultaneously. The standard was derived from proprietary Qualcomm technology. To date, Qualcomm owns many of the patents and intellectual property associated with this standard. CDMA2000 1xEV-DV supports downlink data rates of up to 3.1 Mbps and uplink data rates of up to 1.8 Mbps and can concurrently support operation of legacy 1X voice and data users. EV-DV development has been put on indefinite hold by Qualcomm due to lack of carrier interest. Verizon and Sprint, for instance, employ CDMA2000 1xEV-DO air interfaces. The prevailing belief is that 1xEV-DO provides a sufficient overlay above and beyond the CDMA2000 1X voice network to support data at high speeds without the need for deployment of 1xEV-DV. For this reason, CDMA2000 data networking is typically synonymous with CDMA2000 1xEV-DO, while voice traffic is supported by CDMA2000 1xRTT (CDMA2000 1X).

Figure 7.27 illustrates the CDMA air interface standards evolution [164–166].

As can be seen from Figure 7.27, the latest release for CDMA2000 1X (IS-2000) standard is 1X Advanced (Revision E), published by the 3GPP2 in 2009. 1X Advanced builds on the CDMA2000 1X technology platform. With the 1X Advanced technology, CDMA2000 1X operators can significantly increase their network's voice capacity by using various interference cancellation and radio link enhancements. Some of the improvements include BTS interference cancellation, improved power control, early frame termination, and smart blanking. These improvements can be integrated simultaneously or gradually, allowing operators to evolve their existing networks based on their individual market needs. The complete set of 1X Advanced enhancements can, in theory, quadruple the voice capacity of CDMA2000 1X systems in the same 1.25-MHz channel bandwidth. In addition, we can see from Figure 7.27 that the latest standard release for the EV-DO technology is DO Advanced

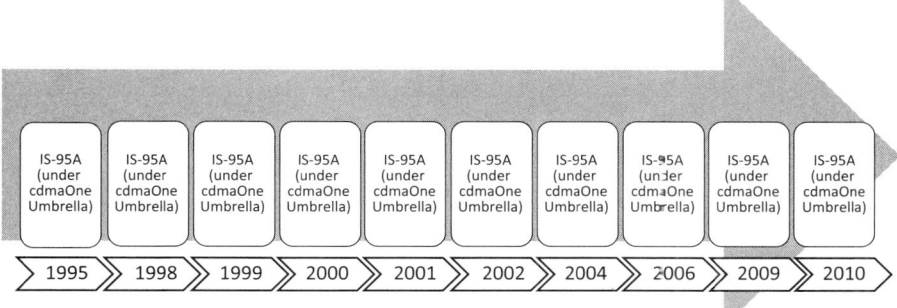

Figure 7.27   CDMA2000 air interface standards evolution.

(EV-DO Revision C), published by the 3GPP2 in 2010. According to the CDMA Development Group, the basic principle of DO Advanced technology is to make existing EV-DO networks and devices perform as efficiently as possible by exploiting unevenly loaded networks, and to do so without substantial changes to existing infrastructure. Interestingly enough, the Ultra Mobile Broadband (UMB) technology was supposed to be the fourth-generation (4G) successor to CDMA2000 system; however, in November 2008, Qualcomm announced it was ending development of UMB, favoring LTE instead [165].

Table 7.11 summarizes specified data rates of various CDMA2000 standards (where applicable) to show evolution toward higher data rates and larger network capacities [164, 165].

Table 7.12 (compiled from References [164, 167]) lists important historical milestones in the CDMA2000 technology development.

As we can see, the future of the CDMA2000 technology is far from being bleak. 1xEV-DO Revision B is gaining momentum with numerous operators and vendor commitments. Out of 334 CDMA operators, nine already launched Revision B networks with many more committed [164]. Amid 1X Advanced promising four times the 1xRTT voice capacity and DO Advanced maximizing EV-DO performance through existing assets, CDMA2000 is a perfect accompaniment to the next generation of cellular networks. As CDMA network operators gain access to wider spectrum due to augmentation with LTE, they will continue relying on 1X for voice services and EV-DO for ubiquitous data coverage [168].

In the following section of this chapter, we provide an overview of CDMA2000 1X and EV-DO technologies.

7.3.3 Technology Overview

To date, there are six 3GPP2-standardized CDMA2000 technologies in existence: CDMA2000 1X (1xRTT), 1xEV-DO Release 0, 1xEV-DO Revision A, 1xEV-DO Revision B, 1X Advanced, and DO Advanced. Some of these technologies have

Table 7.11 CDMA2000 Standards: Specified Data Rates

Technology	Common standard name	Year of publication	Features	Comments
CDMA2000 Release 0 (1xRTT)	IS-2000	1999	153.6 kbps for packet-switched data	Supports both voice and high-speed data; employs cdmaOne paging and access channels as well as new traffic channels
CDMA2000 Revision A	IS-2000 A	2000	Forward link (downlink)/ reverse link (uplink): 307.2 kbps (with one supplemental channel)	Supports simultaneous voice and high-speed data; implements new overhead channels
CDMA2000 1xEV-DO Release 0	IS-856/ TIA-856	2000	Forward link: 2.4 Mbps Reverse link: 153 kbps	Supports broadband data applications only
CDMA2000 Revision B	IS-2000 B	2001	N/A	Implements minor revisions such as additional multiplexing options and adds the Rescue channel
CDMA2000 Revision C (1xEV-DV)	IS-2000 C	2002	Forward link: 3.1 Mbps Reverse link: 307.2 kbps (with one supplemental channel)	Defines 1xEV-DV high-speed Forward link; has been put on hold due to lack of carrier interest
CDMA2000 Revision D (1xEV-DV)	IS-2000 D	2004	Forward link: 3.1 Mbps Reverse link: 1.8 Mbps	Defines 1xEV-DV high-speed Reverse link; Has been put on hold due to lack of carrier interest
CDMA2000 1xEV-DO Revision A	IS-856 A/ TIA-856-A	2004	Forward link: 3.1 Mbps Reverse link: 1.8 Mbps	Supports concurrent Voice over IP (VoIP) and advanced broadband data applications

Table 7.11   (*Continued*)

Technology	Common standard name	Year of publication	Features	Comments
CDMA2000 1xEV-DO Revision B	IS-856 B/ TIA-856-B	2006	Multicarrier EV/ DO Forward link: 9.3 Mbps Reverse link: 5.4 Mbps (5 MHz, FDD) Hardware upgrade Rev. B Forward link: 14.7 Mbps Reverse link: 5.4 Mbps (5 MHz, FDD)	Supports concurrent VoIP and advanced broadband data applications with much higher data rates than Rev. A
CDMA2000 1X Advanced (CDMA2000 1X Revision E)	IS-2000 E	2009	Four times increase in the voice capacity of CDMA2000 1X systems in the same 1.25-MHz channel; Forward link/ reverse link: 307 kbps	Includes a collection of enhancements that can theoretically increase voice capacity by up to 100 erlangs, or a factor of four over legacy CDMA2000 1X networks
CDMA2000 1xEV-DO Revision C (DO Advanced)	IS-856 C/ TIA-856-C	2010	Forward link: 32 Mbps Reverse link: 12.4 Mbps (4 × 1.25 MHz, FDD)	Allows an operator to dynamically allocate existing network capacity where and when it is needed the most

already been deployed, while others are either in the process of being deployed or have concrete operator commitments. Table 7.13 (derived from Reference [164]) summarizes key features of each CDMA2000 technology.

Subsequent sections provide a much more detailed overview of the two primary CDMA2000 air interfaces: CDMA2000 1x and EV-DO, including all of the enhancements described by the latest standard releases for each interface. However, before those air interfaces are considered in detail, Section 7.3.3.1 provides a general overview of a CDMA2000 network, demonstrating a context within which these air interfaces are employed. One goal of Section 7.3.3.1 is to provide a common lexicon that can be employed throughout the remainder of this chapter. Another goal of

Table 7.12 Notable Events in CDMA2000 Systems Development

Date	Event
October 1988	Conception of CDMA concept for commercial telecommunications applications
November 1989	Pac Tel Cellular, San Diego conducts first CDMA system demonstration
February 1990	NYNEX, New York City conducts first CDMA field trial demonstration
April 1993	South Korea adopts CDMA as its national cellular telephone system
July 1993	U.S. TIA adopts CDMA (IS-95A) as a North American digital cellular standard
December 1993	CDMA Development Group is founded
October 1994	Successful completion of CDMA networks' field tests in China
October 1994	South Korea unveils its first CDMA system
December 1995	World's first commercial launch of CDMA service by Hutchison Telecom, Hong Kong
December 1995	CDMA (IS-95A) standardized for U.S. PCS band (ANSI J-STD-008)
December 1996	More than 1 million CDMA subscribers worldwide
June 1997	IS-95B standard is completed (64 kbps data transmissions)
June 1997	cdmaOne™ trademark is launched by CDG for IS-95 CDMA
December 1997	cdmaOne service is commercially available in 100 U.S. cities
December 1997	7.8 million cdmaOne subscribers worldwide
March 1998	LG Telecom, South Korea launches first cdmaOne data service
December 1998	23 million cdmaOne subscribers in 30 countries worldwide
July 1999	CDMA2000 1X standard is completed and approved by ITU for publication
November 1999	ITU-R selects CDMA2000 1X as IMT-2000 (3G) MC-CDMA standard
December 1999	50.1 million cdmaOne subscribers worldwide (83 operators in 35 countries)
December 2000	World's first 3G CDMA2000 1X network is deployed (SK Telecom, LG Telecom, South Korea)
December 2000	80.4 million CDMA (cdmaOne and CDMA2000) subscribers worldwide
June 2001	CDMA2000 1xEV-DO is recognized by ITU-R as part of the 3G IMT-2000 standard
January 2002	World's first CDMA2000 1xEV-DO network launched (SK Telecom, South Korea)
May 2002	10 million commercial CDMA2000 subscribers worldwide
April 2004	3GPP2 approves CDMA2000 1xEV-DO Revision A specification
December 2004	240.2 million CDMA subscribers worldwide
July 2005	CDMA Certification Forum (CCF) is formed

Table 7.12   (*Continued*)

Date	Event
March 2006	TIA/EIA publishes CDMA2000 EV-DO Revision B standard
October 2006	World's first CDMA2000 1xEV-DO Rev. A network is launched (Sprint PCS, USA)
December 2006	373.5 million CDMA subscribers worldwide
December 2007	431.1 million CDMA subscribers worldwide, including 417.5 million CDMA2000 1X and 90.5 million EV-DO subscriptions
December 2008	464.7 million CDMA subscribers worldwide, including 454.9 million CDMA2000 and 112.4 million EV-DO subscriptions
August 2009	3GPP2 publishes 1X Advanced standard which quadruples CDMA2000 1X voice capacity
December 2009	528.1 million CDMA subscribers worldwide, including 524.2 million CDMA2000 and 141.3 million EV-DO subscriptions
January 2010	PT Smart Telecom, Indonesia launches world's first EV-DO Rev. B network
April 2010	3GPP2 publishes DO Advanced specification
March 2011	Sprint Nextel announces its plans to deploy 1X Advanced
April 2011	ZTE Corporation, China completes the industry's first CDMA2000 1x advanced interoperability testing in partnership with Qualcomm Incorporated
June 2011	KDDI, Japan announces its plans to deploy DO Advanced
June 2011	615 million CDMA subscribers worldwide

Data compiled from References [164, 167]).

Section 7.3.3.1 is to identify and describe the key components of a CDMA2000 network and how they are interconnected to form a CDMA2000-based network solution.

7.3.3.1 CDMA2000 Network Architecture

The CDMA2000 wireless network can be viewed as a collection of logical entities and the associated interfaces. Logical entities represent an architectural node where a function or group of functions is implemented. The links between the entities are referred to as "interfaces." An interface is an open specification of the functions and associated signaling that define the operational responsibility of the connected entities. An interface may alternatively be called a reference point if the two entities connected represent distinct physical devices.

The three primary components of a CDMA2000 network are MS, the RAN, and the core network (CN).

The CN is further divided into two components: the Packet CN, which deals with communication over the IP network, and the IS-41 CN, which interacts with the Public Switched Telephone Network (PSTN). Each component in the architec-

Table 7.13   CDMA2000 Technologies: Key Features

Technology	Features
CDMA2000 1X (IS-2000)	Voice Capacity: 33–40 simultaneous voice calls per sector in a single 1.25 MHz FDD channel; up to 55 simultaneous calls with implementation of a new codec (EVRC-B), additional Walsh codes, and handset interference cancellation High-Speed Data: max 153.6 kbps (on both Uplink and Downlink) with an average user data throughput of 80–100 kbps Average Latency: 250 msec node-to-node ping Applications: circuit-switched voice, short messaging service (SMS), ringtone downloads, multimedia messaging service (MMS), games, GPS-based location services, music and video downloads Backward Compatibility: backward compatible with 2G cdmaOne (IS-95A/B) systems and handsets
1xEV-DO Release 0	Broadband Data: up to 2.4 Mbps in the forward link and 153 kbps in the reverse link within a single 1.25 MHz FDD channel with an average data throughput of 300–700 kbps in the forward link and 70–90 kbps in the reverse link Average Latency: 110 msec node-to-node ping Applications: broadband data applications (e.g., broadband Internet, music downloads, 3D gaming, etc.) – serves as DSL substitute in many countries Backward Compatibility: compatible with CDMA2000 1X and cdmaOne systems
1xEV-DO Revision A	Advanced Broadband Data: up to 3.1 Mbps in the forward (down)link and 1.8 Mbps in the reverse (up)link within a 1.25 MHz FDD radio channel with an average data throughput of 600–1400 kbps in the forward link and 500–800 kbps in the reverse link in fully loaded networks; with MIMO EV-DO Rev. A networks can reach up to 1400–1700 kbps in the forward link and 970–2000 kbps in the reverse link Increased Capacity: Rev A. allows operators to support more users on both the forward and reverse link Symmetry: Offers a true synchronic broadband experience which is important for applications where users send packets of data as often as they receive them (e.g., receiving and sending email with attachments) Low Average Latency: under 50 msec node-to-node ping Advanced QoS: the prioritization and delivery of individual packets is based on the type of application or user profile All-IP: provides higher bandwidth efficiencies; supports advanced services such as VoIP, push-to-talk (PTT), push-to- media (PTM), video conferencing, etc. Backward Compatibility: backward compatible with Release 0, CDMA2000 1X and cdmaOne systems

Table 7.13   (*Continued*)

Technology	Features
EV-DO Revision B	Advanced Broadband Data: data rates are proportional to the number of carriers aggregated. The standard supports the aggregation of up to 15 channels in 20-MHz bandwidth, but the most common configuration is an aggregation of 3 carriers within a 5-MHz channel. Multicarrier EV-DO software upgrade (3-carrier configuration) delivers a peak data rate of 9.3 Mbps in the Downlink and 5.4 Mbps in the Uplink, and with a Rev. B hardware upgrade, the peak data rate in the Downlink increases to 14.7 Mbps.
Very Low Average Latency: under 35 milliseconds node-to- node ping
Advanced QoS: the prioritization and delivery of individual packets is based on the type of application or user profile
All-IP: provides higher bandwidth efficiencies: supports advanced services such as VoIP, push-to-X (talk, video, music, etc.), video conferencing, concurrent voice and multimedia services, online gaming, etc. Some applications also incorporate enhanced OFDM-based multicasting capabilities to enable the delivery of rich multimedia content.
Backward Compatibility: backward compatible with all the previous releases and cdmaOne systems
1X Advanced	A collection of enhancements that can theoretically quadruple voice capacity of the CDMA2000 1X network, or expand coverage by 70 percent. Supports uplink/downlink peak data rates of 317 kbps.
DO Advanced	Takes advantage of two basic characteristics of any mobile data traffic: (1) the mobile data traffic across the network is never uniformly distributed in space and time, and (2) many of the data sessions are relatively short in duration.
Provides network operators with flexibility to dynamically allocate existing network capacity where and when it is needed the most.
DO Advanced is comprised of three basic building blocks: Smart Networks (software upgrades), Enhanced Connection Management (software upgrades), and Advanced Devices (infrastructure- and standards-independent) to improve network capacity and user experience. Each of these blocks is described later on in the chapter.

Data derived from Reference [164].

ture contains a defining set of logical entities and interfaces. This is depicted in Figure 7.28.

Subsequent subsections further describe each of these functional components, along with the interfaces between them.

Mobile Station   The MS is a wireless terminal used by subscribers to access the PSTN or the IP network over a radio interface. MSs include portable units (e.g., handheld units), units installed in vehicles, and, somewhat paradoxically, fixed loca-

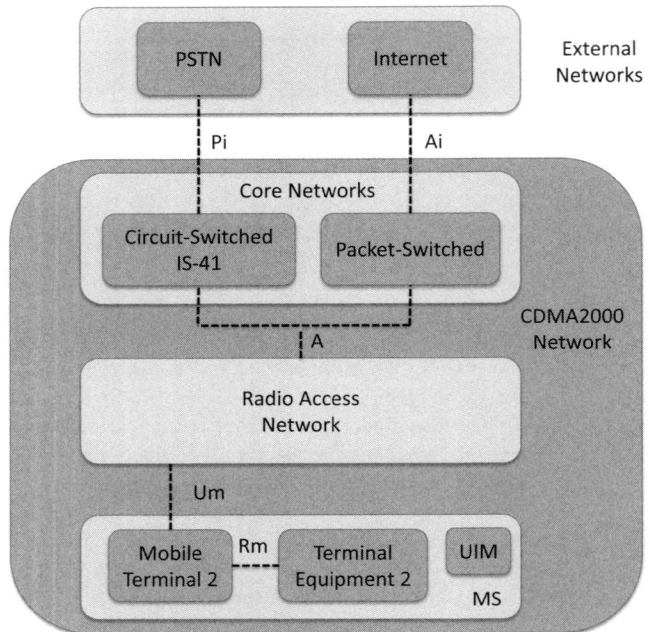

Figure 7.28   High-level reference model of a CDMA2000 wireless network.

Table 7.14   MS Internal and External Interfaces

Interface	Entities using the interface	Applicable standard
Rm	TE2-MT2	IS-707
Um	RAN-MS	IS-2000/IS-707

tion MSs [169]. The MS is the equipment used to terminate the radio path at the subscriber. The logical entities that make up the MS are the mobile terminal 2 (MT2), the terminal equipment 2 (TE2), and the User Identity Module (UIM). MT2 refers to a device used for voice communication (a phone) and TE2 refers to an external data processing device, such as a laptop computer. The UIM contains user subscription information. The MS components without the UIM are collectively known as the mobile equipment (ME). The UIM can be integrated into any ME or it may be removable.

MS Interfaces   Figure 7.28 depicts two interfaces employed by the MS. They are summarized in Table 7.14 (information derived from [169]).

Reference Point Rm   Reference Point Rm is the interface between the MT2 and TE2. For voice application, MT2 is sufficient and TE2 is unnecessary because, in

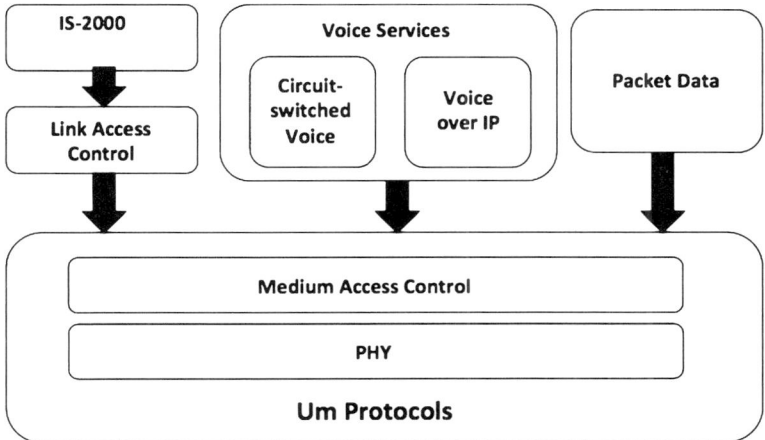

Figure 7.29   Um protocol layer structure.

addition to network connectivity, MT2 commonly provides CS voice service [170]. Also, in some configurations, MT2 may be solely sufficient for IP network communication, as well.

Reference Point Um   The air interface between the RAN and MS is known as Reference Point Um. The upper layer of the Um interface consists of user services, such as packet data service, VoIP, and CS voice and data, and of the IS-2000 layer 3 signaling protocol [171]. IS-2000 layer 3 supports broadcast of system information, mobility management, resource management, and establishment and maintenance of connections between the MS and the RAN. The layered protocol structure of the Um Reference Point is shown in Figure 7.29.

Before continuing, it is important to make the distinction between logical channels, which determine simply what data are transported, and physical channels, which represent the specific communication channels. Logical channels hide the upper layers from the characteristics of the physical channels [171].

The link access control (LAC) layer provides transport services over logical channels. The LAC is further divided into sublayers performing distinct functionality, such as authentication, integrity, and addressing.

The MAC layer maps the logical channels to physical channels and coordinates the use of physical resources [172]. The MAC also handles error-correction mechanisms for erroneously decoded data.

The physical layer provides transmitter and receiver functionality on physical channels. The physical layer protocol specifies channels' modulation characteristics and their numerology. In addition, the physical layer protocol specifies radio channels' related functions, such as link maintenance, power control, handoffs, rate control, and so on.

RAN   The RAN provides RF bearers between the CN and the MS for the transport of user data and nonaccess stream signaling, thus enabling MSs to access the service offered by the PSTN and Internet [173]. The RAN establishes, maintains, and terminates radio channels. The RAN also manages communication resources, allowing for mobility in communication. The two primary entities that make up the RAN are the BS and the packet control function (PCF), which is depicted in Figure 7.30 [169]. The BS includes two additional entities, the BSC and the BTS. The BSC provides control and management for the BTS. It also operates in conjunction with the PCF to perform signaling and to establish and tear down bearer channels. The BTS provides transmission capabilities across the Um reference point. Physically, the BTS represents radio devices and antenna equipment.

In Figure 7.30, all interfaces are typically wired interfaces while the Um interface between the BTS and MS is wireless.

RAN Interfaces   The RAN interfaces, which are depicted in Figure 7.30, are summarized in Table 7.15 (information derived from [169]).

Abis reference point (A7/A3)   The specification for exchange between the BSC and the BTS is defined by the Abis reference point. The reference point is divided into two subinterfaces: the A7 interface for signaling traffic and the A3 interface for signaling and user traffic. A7 traffic moves between source and target BSC, controlling allocation and release of radio resources on the target BSC. A3 signaling is used to establish and remove A3 user traffic connections and for the call-specific operational procedure. The A3/A7 protocol architecture is illustrated in Figure 7.31 and Figure 7.32 [169].

A8/A9 reference point   The A8/A9 reference point connects the BSC and PCF. There are several BSC entities per corresponding PCF. As with the Abis reference point, one interface, the A9, is responsible for signaling traffic while the A8 specifies the exchange of user data. The A8/A9 protocol architecture is depicted in Figure 7.33 [169].

A1/A2/A5 reference points   The A1, A2, and A5 interfaces represent the reference point between the BSC and the MSC, an entity in the CN. Signaling data flow over the A1 interface. The A2 interface is used to transfer pulse code modulation (PCM) voice over Digital Signaling Zero (DS0) channels (64 kbps channel), while A5 interface transports CS data. The application service layer of the A1 interface is called the base station application part (BSAP). The BSAP is further divided into two parts: base station management application part (BSMAP) and direct transfer application part (DTAP). The BSMAP messages are those processed by BSC and require actions on its part. DTAP messages are not processed by BSC; rather, they are piped through to and from the MSC or MS. This is depicted in Figure 7.34, Figure 7.35, and Figure 7.36 [169].

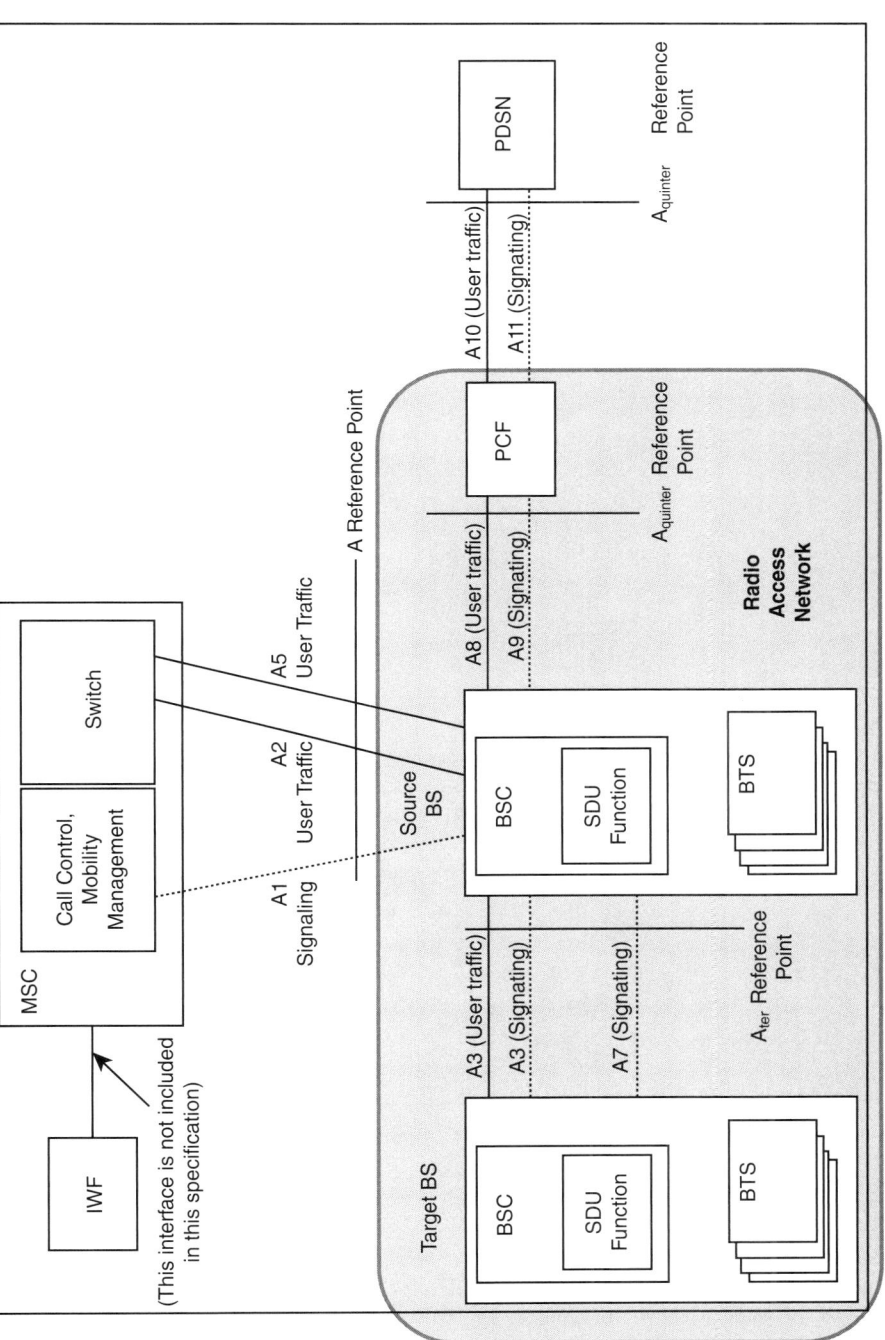

Figure 7.30 Radio access network reference model. Adapted from Reference [169]. This content, from the TIA-2001-C, "Interoperability Specification (IOS) for cdma2000® Access Network Interfaces Release C," is reproduced under written permission from Telecommunications Industry Association (www. tiaonline.org). All standards are subject to revision, and parties to agreements based on any Standard are encouraged to investigate the possibility of applying the most recent editions of the standards published by them.

Table 7.15   RAN Internal and External Interfaces

Interface	Entities using the interface	Applicable standard
Um	BSC/BTS-MS	IS-2000
Abis	BSC-BTS	IS-828
A3/A7	BSC-BSC/BTS	IS-2001
A8/A9	BSC-CF	IS-2001
A10/A11	PCF-PDSN	IS-2001
A1/A2/A5	BSC-MSC	IS-2001

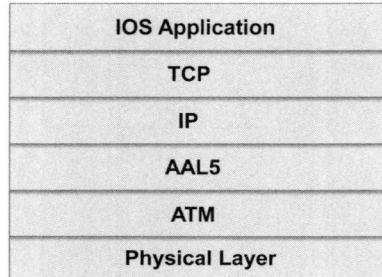

Figure 7.31   A3/A7 interface protocol architecture (control). IOS, Interoperability Specification (IOS); TCP, Transmission Control Protocol; AAL5, ATM Adaptation Layer 5; ATM, Asynchronous Transfer Mode. Reprinted from Reference [172]. This content, from the TIA-2001-C, "Interoperability Specification (IOS) for cdma2000® Access Network Interfaces Release C," is reproduced under written permission from Telecommunications Industry Association (www.tiaonline.org). All standards are subject to revision, and parties to agreements based on any Standard are encouraged to investigate the possibility of applying the most recent editions of the standards published by them.

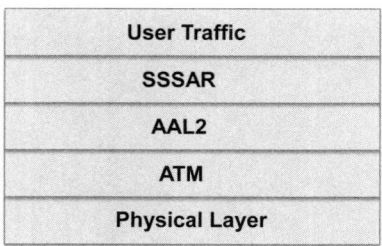

Figure 7.32   A3 interface protocol architecture (user data). SSSAR, Service Specific Segmentation and Reassembly; AAL2, ATM Adaptation Layer 2. Reprinted from Reference [172]. This content, from the TIA-2001-C, "Interoperability Specification (IOS) for cdma2000® Access Network Interfaces Release C," is reproduced under written permission from Telecommunications Industry Association (www.tiaonline.org). All standards are subject to revision, and parties to agreements based on any Standard are encouraged to investigate the possibility of applying the most recent editions of the standards published by them.

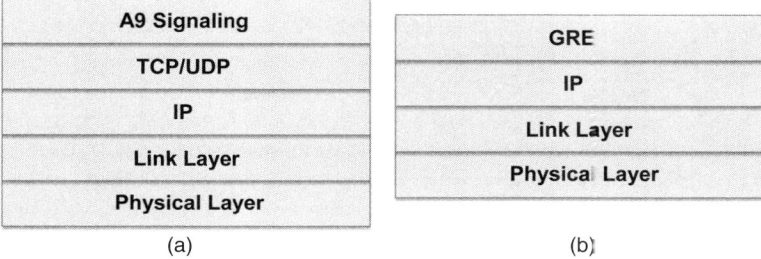

(a)   (b)

Figure 7.33 A8/A9 interface protocol architecture. (a) A9 signaling; (b) A8 user traffic. TCP/UDP, Transmission Control Protocol/User Datagram Protocol; GRE, Generic Routing Encapsulation. Reprinted from Reference [172]. This content, from the TIA-2001-C, "Interoperability Specification (IOS) for cdma2000® Access Network Interfaces Release C," is reproduced under written permission from Telecommunications Industry Association (www.tiaonline.org). All standards are subject to revision, and parties to agreements based on any Standard are encouraged to investigate the possibility of applying the most recent editions of the standards published by them.

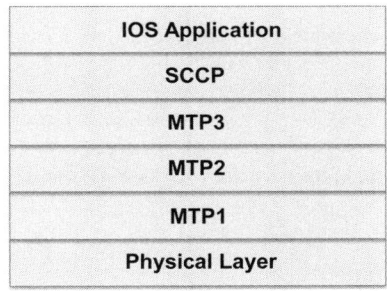

Figure 7.34 A1 protocol architecture (signaling). SCCP, Signaling Connection Control Part; MTP, Message Transfer Part. Reprinted from Reference [172]. This content, from the TIA-2001-C, "Interoperability Specification (IOS) for cdma2000® Access Network Interfaces Release C," is reproduced under written permission from Telecommunications Industry Association (www.tiaonline. org). All standards are subject to revision, and parties to agreements based on any Standard are encouraged to investigate the possibility of applying the most recent editions of the standards published by them.

7.3.3.2 Core Network

As mentioned earlier, the core network (CN) is divided into two major components. The IS-41 CN is the segment that manages CS data to and from the PSTN. The packet CN provides connectivity to the Internet. The following subsections present a review of these two components.

IS-41 CN The IS-41 CN reference model is shown in Figure 7.37 [174].

The primary functions of the IS-41 CN fall under two major categories: mobility management and call processing. Mobility management describes the group of func-

Figure 7.35 A2 protocol architecture (user traffic). Reprinted from Reference [172]. This content, from the TIA-2001-C, "Interoperability Specification (IOS) for cdma2000® Access Network Interfaces Release C," is reproduced under written permission from Telecommunications Industry Association (www.tiaonline.org). All standards are subject to revision, and parties to agreements based on any Standard are encouraged to investigate the possibility of applying the most recent editions of the standards published by them.

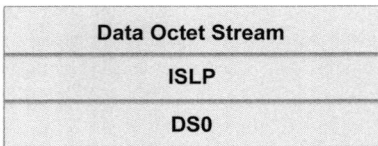

Figure 7.36 A5 protocol architecture (user traffic). ISLP, Inter System Link Protocol. Reprinted from Reference [172]. This content, from the TIA-2001-C, "Interoperability Specification (IOS) for cdma2000® Access Network Interfaces Release C," is reproduced under written permission from Telecommunications Industry Association (www.tiaonline.org). All standards are subject to revision, and parties to agreements based on any Standard are encouraged to investigate the possibility of applying the most recent editions of the standards published by them.

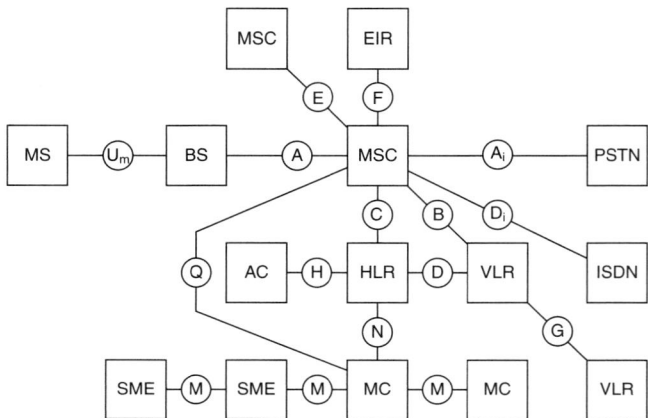

Figure 7.37 IS-41 CN reference model. Reprinted from Reference [177]. This content, from the ANSI/TIA/EIA-41-D, "Cellular Radiotelecommunications Intersystem Operations," is reproduced under written permission from Telecommunications Industry Association (www.tiaonline.org). All standards are subject to revision, and parties to agreements based on any Standard are encouraged to investigate the possibility of applying the most recent editions of the standards published by them.

tions that serve in maintaining service to the MS as it moves. More specifically, mobility management is composed of two types of functions: automatic roaming and authentication. Automatic roaming allows an MS to freely move outside the coverage area without any noticeable effect to the user. Authentication functions provide the ability to identify the MS from any location.

Call processing represents a group of functions that establish, maintain, and release calls to and from the MS. In the case of call establishment, different functions are necessary depending on whether the call is originated by the MS or terminated by the MS. For instance, MS-terminated calls require MS location and status functions while MS-originated calls do not. Other call-processing functions include obtaining routing information, paging, requesting the radio traffic channel setup, and applying call features such as call waiting, call forwarding, and three-way calling.

The logical entities that belong to the IS-41 CN are as follows:

MSC

Home location register (HLR)

Visitor location register (VLR)

Authentication center (AuC)

Message Center

Short Message Entity (SME)

Together, these entities provide the functions for mobility management and call processing.

MSC   The basic function of the MSC is to serve as a linkage between the RAN and PSTN. The MSC manages the CS traffic (usually voice) that flows back and forth between the PSTN and RAN. Typically, several networked MSCs divide the responsibility of serving an MS based on coverage area. In the case of traffic coming from the PSTN, the MSC routes the data to the MSC currently serving the destination MS. The MSC supports traditional call control signaling such as ISDN User Part (ISUP) to manage trunks toward the PSTN. Many of the call-processing functions are handled by the MSC. As for mobility management, the MSC is also responsible for intersystem handoffs that enable continuous service when the MS moves between cells belonging to different systems.

HLR   The HLR holds subscriber data for those users who reside in its assigned geographic area (or home network). The HLR stores data such as location, status, and identification information. The HLR is therefore used by the MSC to route calls from the PSTN to the home MSC of the destination MS. Specific information on the enabled and disabled features of the user subscription is also available in the HLR. An HLR is usually tightly coupled with one or more MSCs.

VLR   The VLR is very similar to the HLR, except that it contains data on the MSs that the associated MSC(s) is currently serving. Necessarily, the data stored in the VLR are transient due to the constant movement of MSs into and out of the coverage area. Although according to the reference model a VLR may serve multiple MSCs, it is most commonly collocated with the MSC, forming a single physical entity.

AuC The job of the AuC is to positively identify the connecting MS. The authentication process is achieved through encryption using authentication keys associated with the MS. The AuC is either associated with the HLR or physically a part of the HLR.

Message Center SMS is supported by the message center entity. Messages sent to and from MSs are routed through the message center. A significant feature of the message center is the ability to store messages for MSs that cannot be reached (due to an inactive state) for future delivery. Also, receipt verification may be sent to the message originator upon successful delivery. The message center performs signaling procedures to support MS location and status query.

SME The SME is a logical entity that can originate or terminate short messages or do both. The SME may be implemented independently in the HLR or on the MSC.

7.3.3.3 IS-41 CN Interfaces

The internal reference points in the IS-41 CN comply with a dual-layered protocol. The two layers are the application services layer and the data transfer service layer.

Ai and Di Reference Points The Ai and Di reference points represent the interfaces between the MSC and PSTN and the MSC and ISDN, respectively. The two interfaces are, in fact, the same. The Ai and Di interfaces use ISUP for signaling [175]. The only difference between the two is that the user data on the Ai interface are PCM voice, while the user data on the Di interface are unrestricted digital information.

Packet CN The Packet CN is the component that ultimately allows the MS to connect to the Internet. It interacts with the RAN's PCF entity on one side and the IP network on the other. The two major PCN entities are the PDSN and the AAA servers. AAA servers can be classified into three categories: HAAA, VAAA, and broker AAA (BAAA). One of the primary tasks that the PCN is responsible for is assigning the MS an IP address. The IP address assignment can be accomplished within two separate modes: simple IP and mobile IP. Usage of either of the two techniques has a large bearing on how the PCN functions.

7.3.3.4 Simple IP Access

With simple IP access, the access service provider dynamically assigns an IP address to the MS [175]. More importantly, the IP address assigned is always topologically correct with respect to the geographic area where the MS is currently located. Accordingly, new addresses are assigned as the MS moves across network areas. Simple IP access is depicted in Figure 7.38 [176].

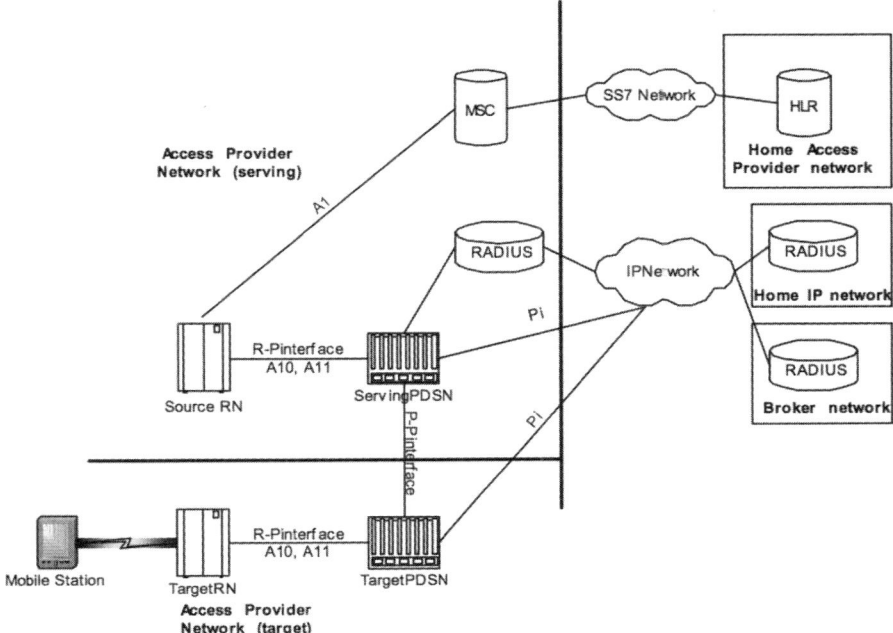

Figure 7.38   Packet CN reference model for simple IP access. Reprinted from Reference [176]. Copyrighted material. Reproduced and distributed by the authors under written permission of the Organizational Partners of the Third Generation Partnership Project 2 (3GPP2).

Mobile IP Access   Unlike the simple IP method, mobile IP access allows the MS to maintain an IP network address as it moves from one network to another. The MS's home network, not the access service provider network, assigns the IP address to the MS. Mobile IP access is depicted in Figure 7.39 [176].

Packet Data Serving Node   The role of the PDSN is to manage the point-to-point IP link with the MS. The PDSN performs all the functions provided by the service provider's network access server (NAS) [175]. It works closely with the AAA, passing it authentication credentials from the MS. The PDSN essentially acts as a router by forwarding packets to and from the attached IP network. When in simple IP access mode, the PDSN can either directly assign addresses to the MS or it will attain the IP address from the AAA and assist in the assignment. As for the mobile IP mode, the PDSN acts as a foreign agent (FA) in a home agent (HA)/FA scheme. In the scheme, the HA represents the connection point for the IP network, having an openly routable IP address. The HA forwards packets to the FA, which in turn forwards the data to the MS. The FA captures data originated at the MS and sends the data to the HA for final delivery. It is also the job of the FA (PDSN) to interact with the HAAA for the purpose of IP address assignment for the HA.

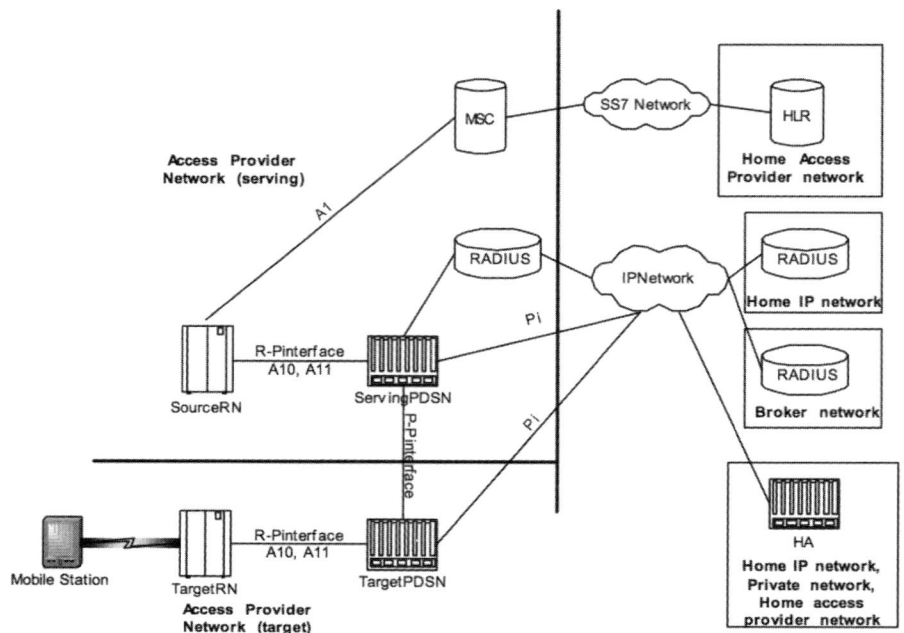

Figure 7.39   Packet CN reference model for mobile IP access. Reprinted from Reference [176]. Copyrighted material. Reproduced and distributed by the authors under written permission of the Organizational Partners of the Third Generation Partnership Project 2 (3GPP2).

AAA   The major purpose of the AAA is to identify and authorize the MS based on the credentials passed by the PDSN. Once again, the three types of AAA servers are VAAA, HAAA, and BAAA. The HAAA does the job of actually authenticating and authorizing. As mentioned earlier, the HAAA may also assign the address for the HA. The VAAA passes authentication requests from the PDSN to the HAAA and forwards authorization responses from the HAAA to the PDSN [177]. If the HAAA and VAAA are not bilaterally connected, the BAAA is necessary for forwarding requests and responses between the two.

Packet CN Interfaces   The Packet CN has three reference points associated with it: R-P, P-P, and Pi. The R-P reference point is the interface between the RAN and PDSN, the P-P reference point is the interface between two neighboring PDSNs, and the Pi reference point is the interface between the PDSN and external IP network.

R-P Reference Point   The PCF and the PDSN are connected by the R-P interface, which is also called the A10/A11 interface. The A11 interface is for signaling, while the A10 is for user data traffic. Besides signaling data, accounting information also passes over that A11 interface to the PDSN.

P-P Reference Point   Communication among PDSNs is provided thought the P-P reference point. More specifically, the P-P interface enables handoff from PDSN to PDSN as the MS moves between coverage areas. The handoff occurs quickly and without interruption of service for the MS. Signaling and user traffic flow over the interface. The signaling is used to set up, tear down, and refresh connections that carry data traffic [177].

Pi Reference Point   The Pi reference point serves exchange to and from the IP network. Separate PCNs are also connected through the Pi interface. The AAA servers pass their accounting and authentication user data through the interface to the PDSN. Depending on which entities are connected, the protocol over the interface varies.

7.3.3.5   CDMA2000 1X Air Interface

This section provides an overview of the CDMA2000 1X air interface, as defined by the IS-2000 specification (Revision A/Revision B). The focus here is to only highlight key concepts of the CDMA2000 1X technology.

CDMA2000 1X Layered Protocol Model   The IS-2000 specification follows the OSI reference model, defining layers 1, 2, and 3 of the protocol stack.

Layer 1   PHY defines air interface's physical channel structure, frequency, channel coding, modulation, and spreading schema.

Layer 2   Layer 2 consists of two sublayers: MAC sublayer and LAC sublayer. The MAC sublayer defines multiple access procedures for CS and PS users, data and signaling multiplexing, and QoS management. The LAC sublayer provides reliable delivery of SDUs to layer 3, construction of PDUs for transmission across the RF interface, segmentation and reassembly (SAR) of packets, security-based access control, and address control. The LAC sublayer is further subdivided into several sublayers:

The authentication sublayer is responsible for executing access control through global challenge authentication and/or message integrity validation.

The ARQ sublayer provides reliable delivery of SDUs to layer-3 peer entities.

The addressing sublayer is the response for address control to ensure delivery of PDUs based on addresses of MS'.

The utility sublayer is responsible for assembly and validating PDUs. This sublayer is used only in particular channel types: forward and reverse common signaling channels, dedicated signaling, and dedicated MAC channels (described later in this section).

The SAR sublayer is responsible for segmentation of encapsulated PDUs into fragments suitable for transfer by the MAC sublayer. Segmentation is implemented on the transmit side, necessitating reassembly at the receiver.

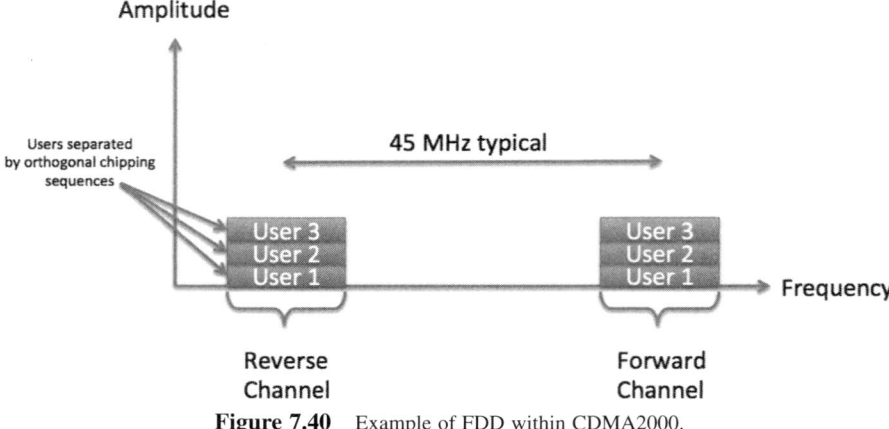

Figure 7.40 Example of FDD within CDMA2000.

Table 7.16 Specifications for CDMA2000 (Single Carrier)

Parameter	Value
Channel bandwidth	1.25 MHz
Frame length	5–20 ms
Spreading factors	4–256
Chipping rate	1.2288 Megachips per second (Mcps)

Layer 3 Layer 3 is responsible for upper layer signaling, data and voice services. Upper layer signaling executes message exchanges for radio resource control (handoff, channel configurations, power control, etc.), call control (setup, teardown, etc.), mobility (authentication, registration, etc.), and SMS.

CDMA2000 Physical Channelization The CDMA2000 network can either be configured to employ TDD or FDD. However, FDD is the predominant method deployed. Here, the forward and reverse channels are separated in frequency, as depicted in Figure 7.40.

For CDMA2000 in the 1X mode, key specifications are listed in Table 7.16. Recall that CDMA2000 supports two different channel configurations: Spreading Rate 1 (SR1) or 1X or 1xRTT and Spreading Rate 3(SR3) or 3X or 3XRTT or MC (multicarrier). The focus of this chapter is on 1X only.

Physical CDMA2000 channels are classified into two groups: dedicated and common channels. DPCHs offer a point-to-point connection, while common physical channels (CPCH) offer a point-to-multipoint connectivity. These channels are differentiated in frequency (forward/reverse) and by code (i.e., a particular channel has a particular spreading code associated with it). Table 7.17 and Table 7.18 summarize the types of channels within CDMA2000 1xRTT [183].

Table 7.17   CDMA2000 1xRTT Dedicated Physical Channel Types

Channel Name	Description
Forward fundamental channel (F-FCH)	Designed to transport dedicated data
Forward supplemental channel type (F-SCHT)	Allocated dynamically to meet a required data rate
Forward dedicated control channel (F-DCCH)	Transport of mobile-specific control information
Forward dedicated auxiliary pilot channel (F-DAPICH)[a]	Used with antenna beamforming
Reverse fundamental channel (R-FCH)	Analogous to F-FCH
Reverse supplemental channel type (R-SCHT)	Analogous to F-SCHT
Reverse dedicated control channel (R-DDCH)	Analogous to F-DDCH
Reverse pilot channel (R-PICH)	Provides the capabilities for coherent detection

[a]Optional.

Table 7.18   CDMA2000 Common Physical Channel Types

Channel name	Description
Forward pilot channel (F-PICH)	Provides capabilities for soft handoff and coherent detection
Forward common auxiliary pilot channel (F-CAPICH)	Provides capabilities for soft handoff and coherent detection
Forward common channel types (F-CCHT)	Paging channels, control channels, and sync channel
Reverse access channel (R-ACH)	Multiple access channel
Reverse common control channel (R-CCCH)	Transports control information

On the uplink, binary phase-shift keying (BPSK) data modulation and QPSK spreading modulation are employed. On the downlink, QPSK data modulation and spreading modulation are employed. As implied by Table 7.17 and Table 7.18, coherent detection is employed on both the uplink and downlink. To improve link robustness, there is a variety of forward error control (FEC) techniques employed, depending on configuration, including convolutional coding with Viterbi decoding and various rate turbo codes. The types of coding schemes employed in the various forms of CDMA2000 are depicted in Table 7.19, derived from [331].

Walsh Codes   CDMA2000 employs a set of orthogonal, binary codes called Walsh codes. Walsh codes are a form of spreading and are used to increase the

Table 7.19 CDMA2000 Forward Error Control Approaches

Spreading rate	Forward link			Reverse link		
	Radio configuration	Description	Data rates	Radio configuration	Description	Data rates
SR1 1 carrier 1.2288 Mcps	1	Required. IS-95B compatible, no cdma2000 coding features	9600, variable	1	Required. IS-95B compatible, no cdma2000 coding features	9600, variable
	2	Compatible with IS-95B RS2, no cdma2000 coding features	14400, variable	2	Compatible with IS-95B RS2, no cdma2000 coding features	14400, variable
	3	1/4 rate convolutional or turbo coding, base rate 9600	9600 153600	3	1/4 or 1/2 rate convolutional or turbo coding, 9600	9600 153600 307200
	4	1/2 rate convolutional or turbo coding, base rate 9600	9600 307200	4	1/4 rate convolutional or turbo coding, base rate 14400	14400 230400
	5	3/8 rate convolutional or turbo coding, base rate 14400	14400	5	Required, 1/4 or 1/3 rate convolutional coding, base rate 9600	9600 307200 614400
SR3 3 carrier 3.6864 Mcps	6	1/6 rate convolutional or turbo coding, base rate 9600	9600 307200	6	1/4 or 1/2 convolutional or turbo encoding	14400 460800 1036800
	7	Required, 1/3 rate convolutional or turbo coding, base rate 9600	9600 614400			
	8	1/4 or 1/3 rate convolutional or turbo coding, base rate 14400	14400 460800			
	9	1/2 or 1/3 rate convolutional or turbo coding, base rate 14400	14400 1036800			

data rate by a factor equal to the length of the Walsh code. In CDMA2000 1X, the supported lengths for Walsh codes are 4, 8, 16, 32, 64, and 128. Note that cdmaOne uses Walsh codes on the Reverse link only and does so in a manner completely different from that of CDMA2000 1X.

Pseudorandom Noise (PN) Sequences   CDMA2000 implements three PN codes on both the Forward and the Reverse links: two short PN codes 2^{15} bits long and one long PN code of length $2^{42} - 1$ bits, which repeats every 41 days. The purpose of PN codes is to scramble the data and make it look like noise.

7.3.3.6 CDMA2000 Handoffs

CDMA2000 supports three types of handoffs: idle handoffs access handoffs, and traffic handoffs.

Idle Handoffs   When the mobile is in the Idle state, it monitors the Forward link without actively exchanging data with the BS. The mobile measures the signal strength from nearby BSs to determine when to switch from one BS to another. The BS is never notified when MS performs an Idle handoff.

Access Handoffs   When in the System Access state, the MS can perform a handoff before the first probe is sent, between probes, or while waiting for a response to its Access message. Again, the BS is never notified when MS performs an Access handoff.

Traffic Handoffs   When the MS is in the Traffic Channel state, it is in an active voice or data connection with a BS. The BS ultimately decides when the MS should perform a handoff via a signaling message. CDMA2000 supports three types of Traffic handoffs: soft handoff, softer handoff, and hard handoff.

7.3.3.7 Power Control

CDMA2000 systems employ fast power control on both the forward and reverse links. Open-loop power control is used to address the near–far problem, whereas the closed-loop power control is used for mitigating the effects of fast fading.

7.3.3.8 Mobile State Machine

CDMA2000 specifies a call processing state machine for the MS. The four main states are defined as follows:

- *Initialization State.* After powering up, the MS performs System Determination to choose a CDMA system and processes the Pilot and Sync channels to acquire and synchronize with the CDMA system.

- *Idle State.* While in this state, the MS monitors the forward link paging channel (F-PCH) (or forward link broadcast control channel/forward link common control channel [F-BCCH/F-CCCH]) to receive overhead and paging messages from the BS.

- *System Access State.* The MS transmits messages to the BS on the reverse access channel (R-ACH, or reverse enhanced access channel [R-EACH]). The BS is always listening to these channels and responds to MS on either F-PCH or F-CCCH.

- *Traffic Channel State.* In this state, the BS and MS maintain their communication via dedicated Forward and Reverse Traffic Channels that carry user's voice and data (source: CDMA University).

7.3.3.9 CDMA2000 Band Classes

The CDMA2000 network defines several band classes reflecting different spectrum allocations around the world. Generally, all existing cellular and PCS bands in 800/900 and 1800/1900 MHz, as well as all IMT2000-recommended spectra, are supported. The following is a summary of the supported band classes in Release A of IS-2000:

Band Class 0. 800-MHz band

Band Class 1. 1900-MHz band

Band Class 2. Total Access Communications System (TACS) band

Band Class 3. Japanese Total Access Communications system(JTACS) band

Band Class 4. Korean 1800-MHz PCS band

Band Class 5. Nordic Mobile Telephony (NMT) 450-MHz band

Band Class 6. 1.9/2.1-GHz band

Band Class 7. 700-MHz band

Band Class 8. 1800-MHz band

Band Class 9. 900-MHz band

7.3.4 1X Advanced

1X Advanced includes a collection of enhancements that can theoretically quadruple the voice capacity of the legacy 1X networks. Some of the enhancements are based on the CDMA2000 1X Release E, while others are not. The CDMA2000 1X Advanced is implemented in the form of a software or channel card upgrade. The CDMA2000 1X Advanced enhancements include the following [178]:

- A new voice codec
- New interference cancellation techniques

- Mobile receive diversity (MRD)
- More efficient power control
- Smart blanking
- Frame early termination (FET)
- Quasi-orthogonal function (QOF)
- Simultaneous Voice and Data (SVDO).

Now, let us take a closer look at each enhancement.

7.3.4.1 A New Voice Codec

1X Advanced specifies the use of an advanced coding scheme (codec), called Enhanced Variable Rate Codec B (EVRC-B), to digitize voice communications before they are transmitted over the network. EVRC-B delivers higher voice capacity than Enhanced Variable Rate Codec (EVRC) while maintaining the same voice quality as EVRC [178].

7.3.4.2 New Interference Cancellation Techniques

1X Advanced offers few types of interference cancellation techniques, including quasi-linear interference cancellation (QLIC), advanced QLIC, and reverse link interference cancellation (RLIC) [178].

QLIC   QLIC reduces the interference in the Forward Link.

Advanced QLIC   CDMA2000 1X specifies 128 Walsh codes. Hence, in order to increase voice capacity, one would require additional Walsh codes. 1X Advanced makes use of QOFs to address the Forward link limitations imposed by Walsh codes. With that said, QOF is notorious for causing additional interference among users. To mitigate this, 1X Advanced uses advanced QLIC to improve the Forward link performance by not only canceling the interference from neighboring BSs but also canceling the additional interference caused by QOF [178].

RLIC   RLIC is a 1X Advanced technique for interference cancellation in the Reverse link, also known as BTS interference cancellation. RLIC can be also used for improving network coverage, beyond its primary purpose of increasing network capacity [178].

7.3.4.3 Mobile Receive Diversity

MRD utilizes two mobile device antennas to boost the quality of the received signal for a given transmit power level at a serving BS. As a result, the BS can reduce the amount of power necessary for the connection and use it to support additional voice/data sessions instead [178].

7.3.4.4 More Efficient Power Control

1X Advanced reduces the overhead allocated for the power control information exchanges between the BS and MS.

7.3.4.5 Smart Blanking

Smart blanking feature of 1X Advanced eliminates the transmission of background noise. Smart blanking of 1/8th rate frames effectively reduces Forward and Reverse links transmit (Tx) power used for background noise.

7.3.4.6 Frame Early Termination

If the FET feature is implemented in both the Forward and Reverse links, the BS (MS) is not required to transmit the entire frame granted that the MS (BS) has already successfully decoded the information and sent an ACK receipt. Much like EV-DO's HARQ, FET reduces the amount of transmit power over a period of time, thus increasing the overall network capacity [178].

7.3.4.7 Simultaneous Voice and Data (SVDO)

SVDO is a new device feature and is not standards based. In addition to 1X Advanced, it can significantly improve network performance without the need for infrastructural changes. Current CDMA2000 systems allow MS to receive 1X voice calls while in an active EV-DO data session. However, data connection is suspended if the MS remains in the voice call because separate receivers and transmitters are required (i.e., the MS would need two receivers and transmitters).

The new SVDO feature allows for concurrent voice and data sessions to be established using separate transmit and receive chains.

7.3.5   1X Advanced Future

To date, Qualcomm has outlined a strong chipset roadmap for 1X Advanced which will enable a wide range of devices supporting all market areas. Many operators have already announced their support for 1X Advanced. The first commercial 1X Advanced networks were deployed in 2012.

7.3.6   CDMA2000 EV-DO

7.3.6.1 1X EV-DO Revision A

The cdma2000 Release A (1xEV-DO) standard was developed by 3GPP2 to optimize performance for MSs that operate primarily in a "data" mode. The choice to develop this standard optimized for data only and not voice traffic stems from the view that QoS requirements vary significantly between voice and data traffic. This standard is also known as the high-rate packet data (HRPD) interface. For consistency, this subsection will refer to the standard as the 1xEV-DO interface.

The standard is based primarily on a proprietary technology developed by Qualcomm (as the case with virtually all CDMA-based technologies) and presented to the 3GPP2 forum for ratification. The forum developed enhancements to the standard and released it as the TIA/Electronic Industries Association (EIA) IS-856 standard in 2002. Because most of the applications envisioned to utilize the waveform are IP based and traffic loads are expected to be highly asymmetric, the 1xEV-DO interface is primarily focused on enhancing the Forward link. Key features of the 1xEV-DO interface include the following:

- A single high-speed packet bandwidth available to all active users
- Adaptive modulation techniques with peak data rates supported at up to 2.4 Mbps
- Physical layer hybrid ARQ that uses low-rate turbo codes with incremental redundancy
- Multiuser diversity support through short frames and fast scheduling
- Closed-loop rate control based on fast channel state feedback
- Site reselection diversity, also known as "virtual soft handoff"
- Optimal receive diversity located at the MT

The design of the reverse link in the 1xEV-DO interface is based primarily on Release 0 of IS-2000. Here, low data rates and power-controlled traffic channels are used for the transfer of data packets, code-division multiplexed.

1X EV-DO Air Interface Protocol Layers   The 1xEV-DO interface is based on a seven-layer protocol stack. While this interface is considered an "air" interface, the seven layers within the stack actually provide services similar to the seven-layer Open Systems Interconnection (OSI) communications reference model.

Each of the layers is summarized as follows:

- The physical layer defines air interface physical channel structure, frequency, channel coding, modulation, and spreading schema.
- The MAC layer defines procedures to control data rates, bandwidth allocations, and scheduling user traffic.
- The security layer provides encryption and authentication services.
- The connection layer defines signaling methods for establishing and maintaining the air interface connection.
- The session layer provides address management, data session configuration, and general management procedures.
- The stream layer allows distinct application streams to be multiplexed with signaling and user traffic over the same connection.

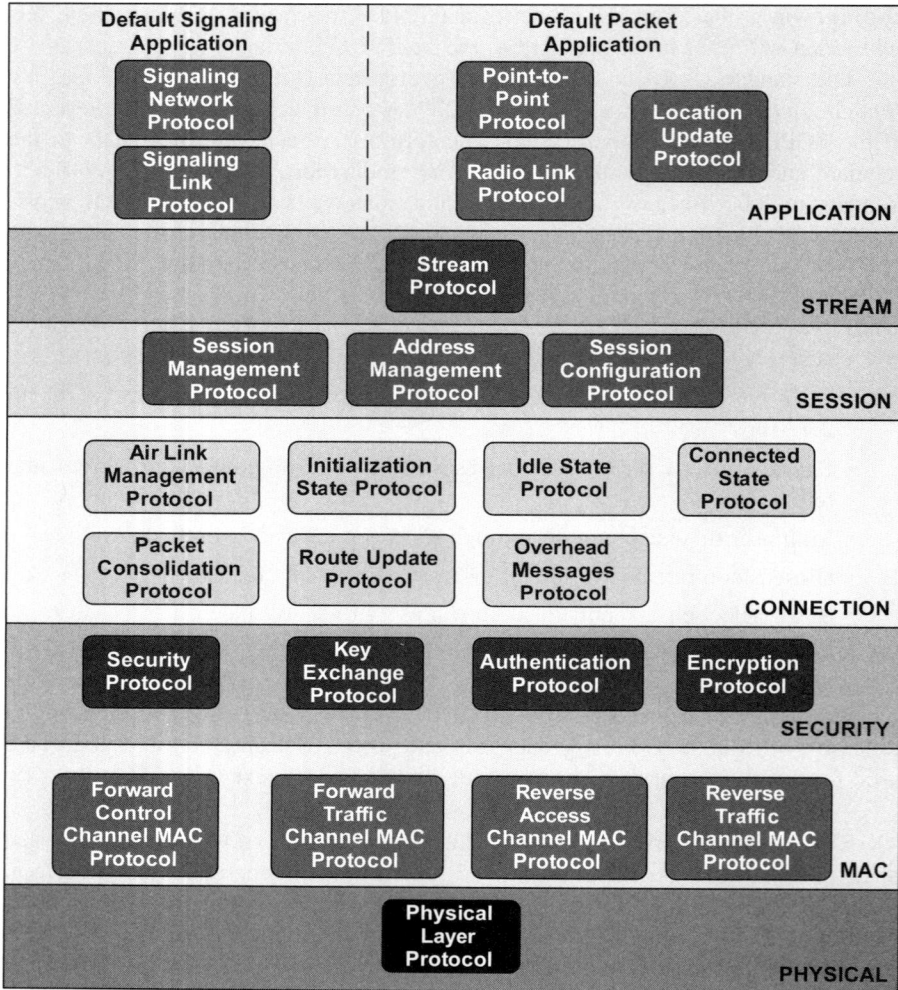

Figure 7.41 1xEV-DO layering structure and protocols. Reprinted from K. Etemad, *CDMA2000 Evolution: System Concepts and Design Principles*, Wiley-Interscience, Hoboken, NJ, 2004, with permission of John Wiley & Sons, Inc.

- The application layer refers to applications employed for carrying signaling and user traffic over the air—but does not refer to the user applications themselves.

Figure 7.41 [179] illustrates the seven-layer model for the 1xEV-DO interface, along with the default set of protocols used at each layer to provide the necessary functions.

Each of the protocols in the air interface specification is a self-contained object that has predefined messages, procedures, and interfaces with other protocols. This approach was envisioned to allow the flexibility of adding or modifying protocols to enhance the standard in later releases, if desired. The following sections describe each layer and its resident protocols in more detail.

Application Layer   Within the default-signaling application, there are two protocols: the signaling network protocol (SNP) and the signaling link protocol (SLP). Within the default packet application, there are three protocols: the point-to-point protocol (PPP), the radio link protocol (RLP), and the location update protocol (LUP).

SNP and SLP   SNP provides message transmission services for signaling messages, while SLP provides fragmentation methods and reliable and best-effort delivery mechanisms for signaling messages. When used in the context of the default-signaling application, SLP carries SNP packets.

PPP and RLP   The PPP protocol provides framing and multiprotocol support, while the RLP provides retransmission and duplicate detection for an octet-based data stream. When used in the context of the default packet application, RLP carries PPP packets.

LUP   The LUP is used to define the location update procedures and messages in support of mobility management for the default packet application.

Stream Layer   The stream protocol within the stream layer adds the stream header in the transmitting direction and also ensures packets are octet aligned. Furthermore, it removes the stream header in the receive direction and forwards packets to the destined application.

Session Layer   Within the session layer, there are three protocols: the session management protocol, the address management protocol, and the session configuration protocol.

Session Management Protocol   The session management protocol controls the activation and deactivation of individual sessions and their associated address management and session layer protocols.

Address Management Protocol   The address management protocol provides access terminal identifier (ATI) management. ATIs are defined to allow different types of user terminals.

Session Configuration Protocol   The session configuration protocol controls negotiation and configuration of protocols used in the session.

Connection Layer   The connection layer contains seven protocols. These are the air link management protocol (ALMP), initialization state protocol, idle state

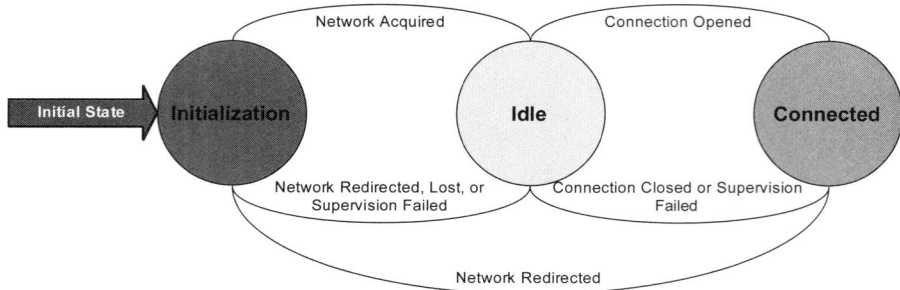

Figure 7.42 Access terminal (AT) state machine. Adapted from K. Etemad, *CDMA2000 Evolution: System Concepts and Design Principles*, Wiley-Interscience, Hoboken, NJ, 2004, with permission of John Wiley & Sons, Inc.

protocol, connected state protocol, route update protocol, overhead messages protocol, and packet consolidation protocol.

ALMP The ALMP protocol manages the overall state machine that terminals and the network follow during a connection. Depending on the state, the ALMP triggers the appropriate protocols. These states are shown in Figure 7.42 [179].

The access terminal (AT) may be in the initialization, idle, or connected states. Depending on its state, it follows a subset of the other protocols defined for the connection layer.

Initialization State Protocol The initialization state protocol defines the procedure followed when an AT acquires a network, as well as the procedure that the network follows to support network acquisition.

Idle State Protocol The idle state protocol defines the procedures that an AT and an access network follow when there are no connections.

Connected State Protocol The connected state protocol defines the procedures that an AT and an access network follow for an open connection.

Route Update Protocol The route update protocol provides a mechanism for maintaining a route between the AT and the access network.

Overhead Messages Protocol The overhead messages protocol provides broadcast messages that contain information mostly used by the connection layer protocols.

Packet Consolidation Protocol The packet consolidation protocol provides a transmit prioritization scheme and packet encapsulation at the connection layer.

Security Layer The security layer uses four protocols: the key exchange protocol, the authentication protocol, the encryption protocol, and the security protocol.

Key Exchange Protocol The key exchange protocol defines the method followed by the access network and AT to exchange security keys for encryption and authentication.

Authentication Protocol The authentication protocol provides the procedures followed by the access network and the AT for authentication of traffic.

Encryption Protocol The encryption protocol provides the method followed by the access network and the AT for encryption of traffic.

Security Protocol The security protocol provides the method used for generating a cryptosync used by the authentication and encryption protocols.

MAC Layer The MAC layer contains four protocols, each of which corresponds to a particular channel in the 1xEV-DO interface. These protocols include the access channel protocol, control channel MAC protocol, forward traffic channel MAC protocol, and reverse traffic channel MAC protocol.

Access Channel Protocol The access channel protocol defines timing, power levels, and transmission procedures on the reverse access channel during the initial system access by the MT.

Control Channel MAC Protocol The control channel MAC protocol defines the transmission rules and procedures to follow on the forward control channel, including scheduling of control channel packets. Also, the protocol defines how the AT acquires/monitors the system using the broadcast and common channel messaging on the forward control channel.

Forward Traffic Channel MAC Protocol The forward traffic channel MAC protocol defines the rules and methods that govern the scheduling packet transmission on the forward traffic channel based on the data rate control (DRC) command sent from ATs.

Reverse Traffic Channel MAC Protocol The reverse traffic channel MAC protocol defines rules and procedures that control the transmission of data, reverse rate indication (RRI), and soft handoff procedures.

Physical Layer The physical layer protocol provides the channel structure, power output, modulation specifications, and frequency for the forward and reverse links.

Channelization In the 1xEV-DO interface, TDM is used for channelization. Here, all power and code space entities are allocated to a single traffic channel that

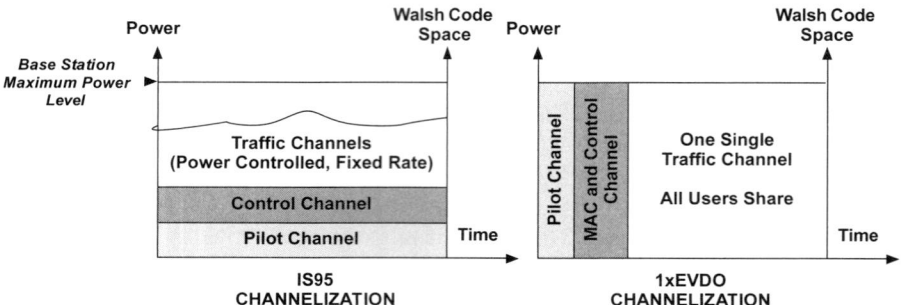

Figure 7.43 IS-95 versus 1xEV-DO channelization. Adapted from K. Etemad, *CDMA2000 Evolution: System Concepts and Design Principles*, Wiley-Interscience, Hoboken, NJ, 2004, with permission of John Wiley & Sons, Inc.

is time shared by all active users. In IS-95, each physical channel is transmitted the entire time using a portion of the total sector power and only one Walsh code. However, in 1xEV-DO, the forward physical channels are transmitted at full sector power using the entire Walsh code space but only occupying a discrete fraction of time. Figure 7.43 [179] depicts 1xEV-DO forward channel channelization, as compared with IS-95 channelization.

As with cdma2000 1xRTT, there are forward and reverse link channels for 1xEV-DO. The next subsections provide a summary of each of these channel types.

Forward Link Channels   All physical channels are time multiplexed on a single channel and transmitted at full power for all time for 1xEV-DO. Figure 7.44 [179] illustrates the 1xEV-DO forward link channel structure.

Pilot channel   The forward link pilot channel functions are the same as for those in IS-95 and cdma2000 1xRTT. It carries all zeros in its data stream. Power and timing of the pilot channel are used to determine the best BS available and to help ATs (i.e., mobile data clients) to acquire system timing and fast synchronization.

MAC channel   The MAC channel carries MAC layer information to control transmissions on the UL and also includes the following:

(a) *Reverse Activity Bit (RAB):* This has 1 bit per slot reserved that informs all ATs about the level of activity in the reverse link.

(b) *Reverse Power Control (RPC):* This is specific to each active user in a sector and provides closed-loop power control on active ATs.

(c) *DRCLock Channel:* This indicates to the AT if the network can receive the DRC sent by the AT. If set to zero, AT should stop pointing DRC at that sector.

Control channel   The control channel carries overhead broadcast information and user-specific common control information (e.g., a message assigning the traffic

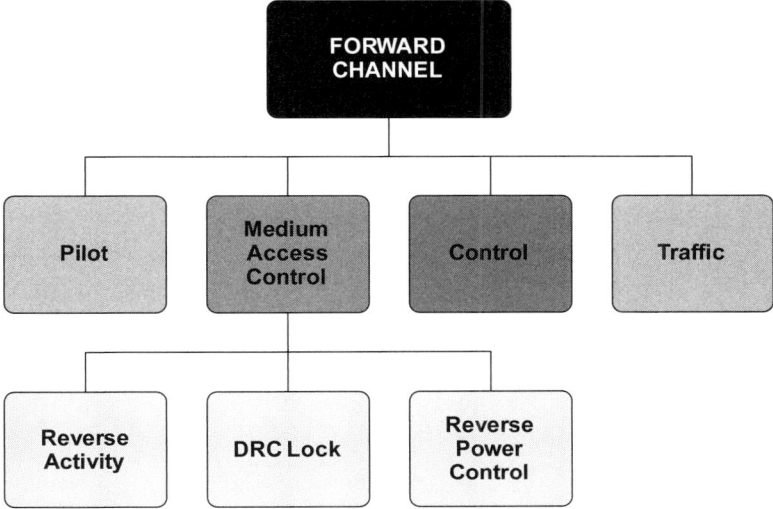

Figure 7.44 1xEV-DO forward link channel structure. Adapted from K. Etemad, *CDMA2000 Evolution: System Concepts and Design Principles*, Wiley-Interscience, Hoboken, NJ, 2004, with permission of John Wiley & Sons, Inc.

channel timeslots). Synchronous control packets are transmitted every 400 ms, while asynchronous control packets are transmitted when needed.

Traffic channel   The traffic channel is a variable-rate traffic bandwidth "pipe" time shared by many users. A packet data-scheduling algorithm is used to determine who accesses the bandwidth and when. User traffic data rates vary from 38.4 kbps to 2.45 Mbps, depending on modulation and coding scheme employed. Any time a physical packet frame is sent, a known preamble sequence is included for AT synchronization. The preamble also identifies whether or not the data are user traffic or control traffic and the destination user.

Reverse Link Channels   Physical channels in the reverse link for 1xEV-DO are code multiplexed and grouped either as access channel mode or traffic channel mode. Figure 7.45 [179] illustrates the reverse link channel structure.

Reverse access channel mode   The reverse access channel mode is used by the AT to initialize communications with the network or respond directly to an AT-directed message. There are two channels within the reverse access channel mode: the pilot channel and the data channel.

PILOT CHANNEL   The pilot channel is used to transmit the preamble and provide for time synchronization.

DATA CHANNEL   The data channel is used to carry any common control messages for the channel from the AT but only when the AT does not have an assigned channel

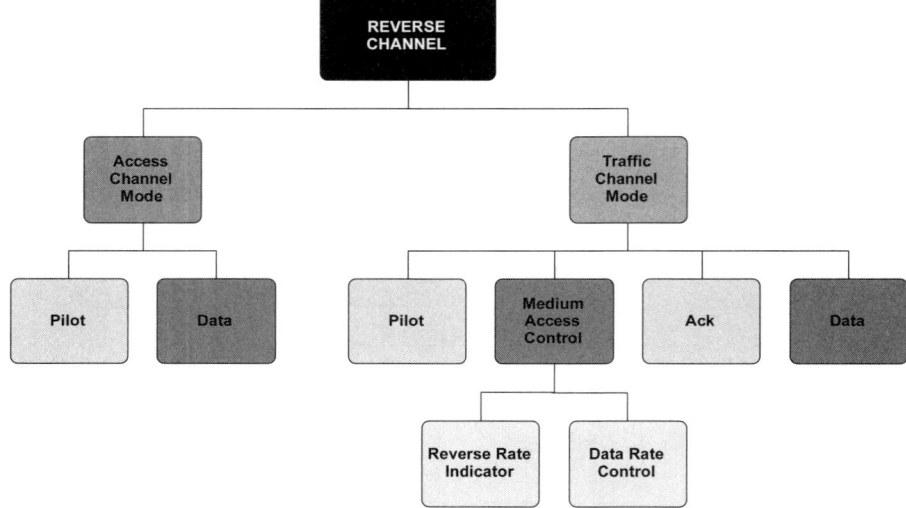

Figure 7.45   1xEV-DO reverse link channel structure. Adapted from K. Etemad, *CDMA2000 Evolution: System Concepts and Design Principles*, Wiley-Interscience, Hoboken, NJ, 2004, with permission of John Wiley & Sons, Inc.

(i.e., it is not in the connected state). The AT transmits the pilot and data channels simultaneously.

Reverse traffic channel mode   The reverse traffic channel mode is used by the AT to transmit user traffic or to transmit signaling information to the network. There are six channels within this mode: pilot, medium access, RRI, DRC, ACK, and data.

PILOT CHANNEL   The pilot channel in the reverse traffic channel mode is similar to the cdma2000 pilot channel. It provides time synchronization to the receiver so that the receiver can coherently demodulate symbols.

MEDIUM ACCESS CHANNEL   The medium access channel is a combination of two channels: the RRI and the DRC.

RRI CHANNEL   The RRI channel is used to indicate if the data channel is being transmitted on the reverse traffic channel and, if so, what data rate is being used.

DRC CHANNEL   The DRC channel is used by the AT to inform the network what the supportable forward traffic channel data rate is and its corresponding best serving sector.

ACK CHANNEL   The ACK channel is used by the AT to provide data packet acknowledgments to the access network from data received on the forward channel. The AT sends ACK or NACK for each physical layer slot received.

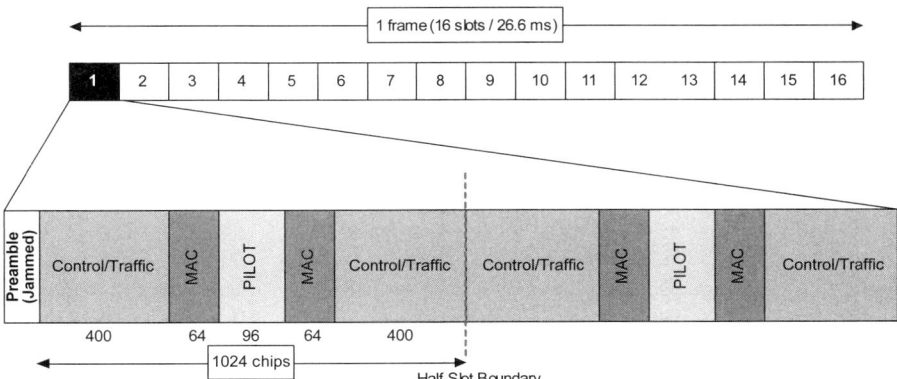

Figure 7.46   Forward physical channel structure. Adapted from K. Etemad, *CDMA2000 Evolution: System Concepts and Design Principles*, Wiley-Interscience, Hoboken, NJ, 2004, with permission of John Wiley & Sons, Inc.

DATA CHANNEL   The data channel is used to carry the user traffic and any dedicated control messages. Data rates for this channel vary from 9.6 up to 153.4 kbps.

Forward Physical Channel Structure   The forward physical channel structure consists of only one physical channel, TDM, among pilot, MAC, control, and traffic channels. It is divided into 26.67 ms frames, each of which is divided into 16 equal slots of 1.67 ms. Each of these slots is divided further into two half-slots, and each half-slot contains 1024 chips, resulting in 1.228 Mcps. Figure 7.46 [179] illustrates the physical channel structure for the forward channel. Here, control traffic within each half slot utilizes 800 chips divided into 400 chips on each side of the center of the half-slot. The center of each half-slot contains the MAC and pilot channels, each of which takes 128 and 96 chip resources, respectively. In an idle slot, the control and traffic channels do not contain any information; thus, only the MAC and pilot channels are transferred.

Within the control/traffic channels, the number of bits carried may differ from one half-slot to the next. This is a function of the selected modulation scheme and coding rate. Control channel slots are cycled every 256 slots. The network will send a broadcast and common channel message on the control channel slots for every 426.67 ms. Remaining slots are used for scheduled user data traffic. Figure 7.47 illustrates the control channel cycle, showing 8 slots allocated to control and 248 slots allocated to user data.

The traffic channel in the forward physical channel structure contains data encoded in blocks. Control channel data are sent at the 38.4 or 76.8-kbps data rate, while user traffic can be transmitted up to 2.45 Mbps. Table 7.20 [179] summarizes the data rates supported in this structure.

Spreading is achieved in the forward physical channel similar to IS-2000 (i.e., cdma2000 1xRTT). After orthogonal spreading, the combined time multiplexed forward channel sequence is quadrature spread. The sequence is a quadrature PN

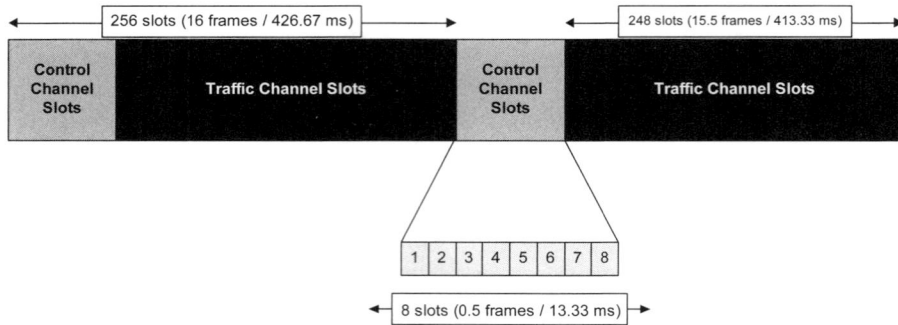

Figure 7.47 Control and traffic channel slots for the forward physical channel. Adapted from K. Etemad, *CDMA2000 Evolution: System Concepts and Design Principles*, Wiley-Interscience, Hoboken, NJ, 2004, with permission of John Wiley & Sons, Inc. Adapted from Reference [18].

Table 7.20 Modulation and Coding Schemes for Forward Traffic and Control Channels

| Data rate (kbps) | For every physical layer packet | | | | |
	Slots	Bits	Code rate	Modulation	TDM chips (preamble, pilot, MAC, data)
38.4	16	1024	1/5	QPSK	1024, 3072, 4096, 2456
76.8	8	1024	1/5	QPSK	512, 1536, 2048, 12288
153.6	4	1024	1/5	QPSK	256, 768, 1024, 6144
307.2	2	1024	1/5	QPSK	128, 384, 512, 3072
614.4	1	1024	1/3	QPSK	64, 192, 256, 1536
307.2	4	2048	1/3	QPSK	128, 768, 1024, 6272
614.4	2	2048	1/3	QPSK	64, 384, 512, 3136
1228.8	1	2048	1/3	QPSK	64, 192, 256, 1536
921.6	2	3072	1/3	8PSK	64, 384, 512, 3136
1843.2	1	3072	1/3	8PSK	64, 192, 256, 1536
1228.8	2	4096	1/3	16-QAM	64, 384, 512, 3136
2457.6	1	4096	1/3	16-QAM	64, 192, 256, 1536

Reprinted from K. Etemad, *CDMA2000 Evolution: System Concepts and Design Principles*, Wiley-Interscience, Hoboken, NJ, 2004, with permission of John Wiley & Sons, Inc.

sequence of length 32,768 chips, with a chip rate of 1.2288 Mcps and a repeat period of 26.67 ms.

Figure 7.48 [180] illustrates the Forward Channel Structure block diagram implementation.

As shown in the figure, forward traffic channel or control channel physical layer packets are fed into an encoder of rate 1/3 or 1/5, and the resulting sequence is multiplied by a scrambler to randomize the bits. The resulting sequence is then sent to a channel interleaver and then to a modulator to provide I and Q constellation points mappings to QPSK, eight-phase-shift keying (8PSK), or 16-QAM modulations. The resulting I and Q outputs are fed into a sequence repetition and symbol-puncturing device, and then demultiplexed to 16 different channels, each of which has a separate 16-ary Walsh cover associated with it, at a rate of 76.8 kilosymbols per second (ksps). Each of the Walsh-coded symbols of all the streams is summed together to form a single in-phase stream and a single quadrature stream at a total chip rate of 1.2288 Mcps. These chips are then TDM with the preamble, pilot channel, and MAC channel chips to form the resulting sequence of chips that undergoes the quadrature spreading operation. The quadrature spreading operation is the same as the reverse channel and is covered in more detail in the next section ("Reverse Physical Channel Structure"). Baseband filtering is also performed and is identical to the reverse channel; it is also covered in more detail in the subsequent section. Upconversion is performed on the quadrature-spread sequences for the I and Q channels to result in the final signal s(t).

Reverse Physical Channel Structure   The reverse physical channel structure for 1xEV-DO is less complex than the forward channel physical channel structure because most of the multiplexing between channel types is achieved in the code domain. However, it differs depending on the two mode types (access or traffic). Figure 7.49 and Figure 7.50 [179] illustrate the reverse physical traffic channel and reverse physical access channel structures, respectively.

In Figure 7.49, users x, y, and z may have frames allocated to them on an as-needed basis. Users with data are said to be utilizing the data channel within this TDM structure for the traffic channel. Table 7.21 [179] summarizes the reverse data channel modulation and coding parameters for the physical traffic channel.

The data rate is chosen by the AT based on algorithms defined by the network. Selected data rates are explicitly called out by the RRI field within the RRI channel.

In Figure 7.50, the reverse physical access channel structure transmits in-phase (I) and quadrature-phase (Q) components. The pilot channel is always transmitted along the I channel, while the data channel is transmitted along the Q channel. The data channel frame structure is equivalent to the data channel frame structure within the reverse traffic channel structure, but the data rate is limited to 9.6 kbps. The user may transmit frames of length 26.67 ms each. During the preamble transmission, only the pilot channel is transmitted on the I channel. During the access channel physical layer transmission, however, I and Q channels are transmitting the pilot and data channels simultaneously. The output power of the pilot channel during the preamble is equivalent to the sum of the output powers of the I and Q channels

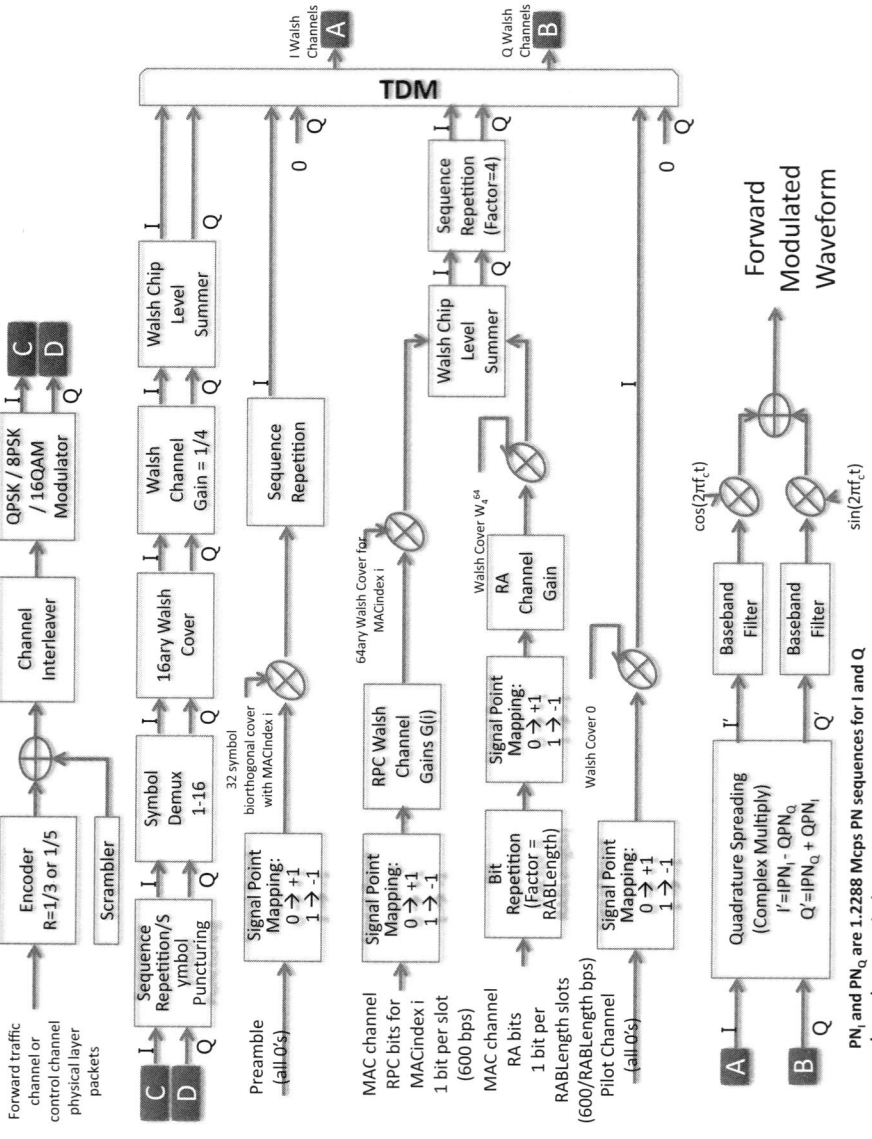

Figure 7.48 Forward channel structure block diagram implementation. Reprinted from Reference [180]. Copyrighted material. Reproduced and distributed by the authors under written permission of the Organizational Partners of the Third Generation Partnership Project 2 (3GPP2).

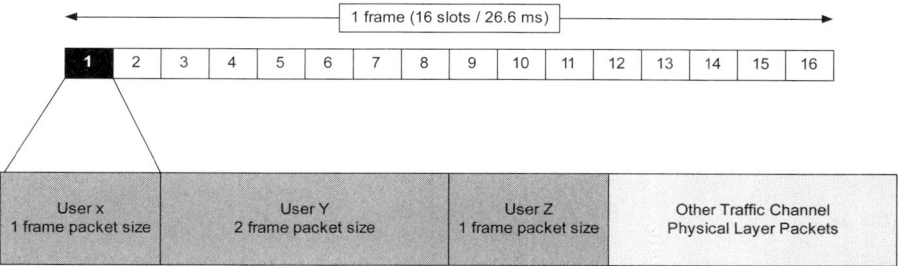

Figure 7.49   Reverse physical traffic channel structure. Reprinted from K. Etemad, *CDMA2000 Evolution: System Concepts and Design Principles*, Wiley-Interscience, Hoboken, NJ, 2004, with permission of John Wiley & Sons, Inc.

	Preamble Frame 1	Preamble Frame 2	Message Frame 3	Message Frame 4
I-Phase	Pilot Channel	Pilot Channel	Pilot Channel	Pilot Channel
Q-Phase	NOT TRANSMITTING		Data Channel	Data Channel

TIME

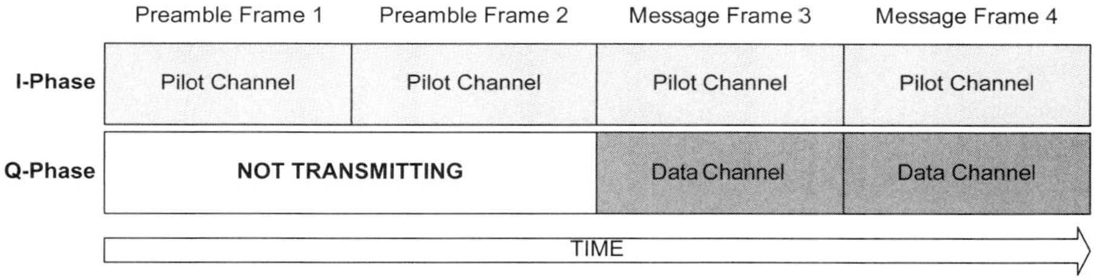

Figure 7.50   Reverse physical access channel structure. Reprinted from K. Etemad, *CDMA2000 Evolution: System Concepts and Design Principles*, Wiley-Interscience, Hoboken, NJ, 2004, with permission of John Wiley & Sons, Inc.

during the access channel frame times. These channels are spread by complex multiply PN sequences with a 1.2288-Mcps rate.

Figure 7.51 [180] illustrates the block diagram implementation of the reverse physical access channel. Here, the two channels (pilot and data) are shown. In the pilot channel, all zeros are generated as the bits and are mapped into +1s to be multiplied by the Walsh code sequence W_0^{16}. The resulting chipped sequence at a rate of 1.2288 Mcps is fed into the I channel input and complex multiplied to provide a complex spreading based on PN sequences (PN_I for the I channel and PN_Q for the Q channel). In Reference [31], PN_I and PN_Q are based on long codes known as U_I and U_Q multiplied by short PN sequences known as P_I and P_Q. These sequences utilize the same polynomial for generation but are made unique based upon which channel the AT is transmitting. The generating polynomial known as $p(x)$ for these sequences is shown next:

$$p(x) = x^{42} + x^{35} + x^{33} + x^{31} + x^{27} + x^{26} + x^{25} + x^{22} + x^2 + x^{19} + x^{18} +$$
$$x^{17} + x^{16} + x^{10} + x^7 + x^6 + x^5 + x^3 + x^2 + x + 1.$$

PN_I is simply P_I multiplied by U_I. The generating polynomial known as $P_I(x)$ is shown for P_I next:

Table 7.21   Modulation and Coding Schemes for Reverse Data Channel

Parameter	Data rate (kbps)				
	9.6	19.2	38.4	76.8	153.6
Reverse rate index	1	2	3	4	5
Bits per physical (PHY) layer packet	256	512	1024	2048	4096
PHY layer packet duration (ms)	26.67	26.67	26.67	26.67	26.67
Code rate	1/4	1/4	1/4	1/4	1/2
Symbols per PHY packet	1024	2048	4096	8192	8192
Code symbol rate (ksps)	38.4	76.8	153.6	307.2	307.2
Interleaved packet repetitions	8	4	2	1	1
Modulation symbol rate (ksps)	307.2	307.2	307.2	307.2	307.2
Modulation type	BPSK	BPSK	BPSK	BPSK	BPSK

Reprinted from K. Etemad, *CDMA2000 Evolution: System Concepts and Design Principles*, Wiley-Interscience, Hoboken, NJ, 2004, with permission of John Wiley & Sons, Inc.

$$P_I(x) = x^{15} + x^{13} + x^9 + x^8 + x^7 + x^5 + x^1.$$

The generating polynomial for P_Q is shown next:

$$P_Q(x) = x^{15} + x^{12} + x^{11} + x^{10} + x^6 + x^5 + x^4 + x^3 + 1.$$

Unlike PN_I, PN_Q is derived with a few more steps. First, P_Q and U_Q are multiplied together. Then, the resulting sequence is decimated by a factor of 2: The output of the decimator provides a constant output for two consecutive chips by deleting every other input value and repeating the previous input value in place of the deleted value. Then, the decimated output is multiplied by the Walsh cover sequence W_1^2.

Once the pilot chipped sequence is sent through the quadrature spreading phase, the resulting sequence passes through a baseband filter that allows all baseband signaling through up to 590 kHz with a 40-dB roll off and a stop band set at 740 kHz.

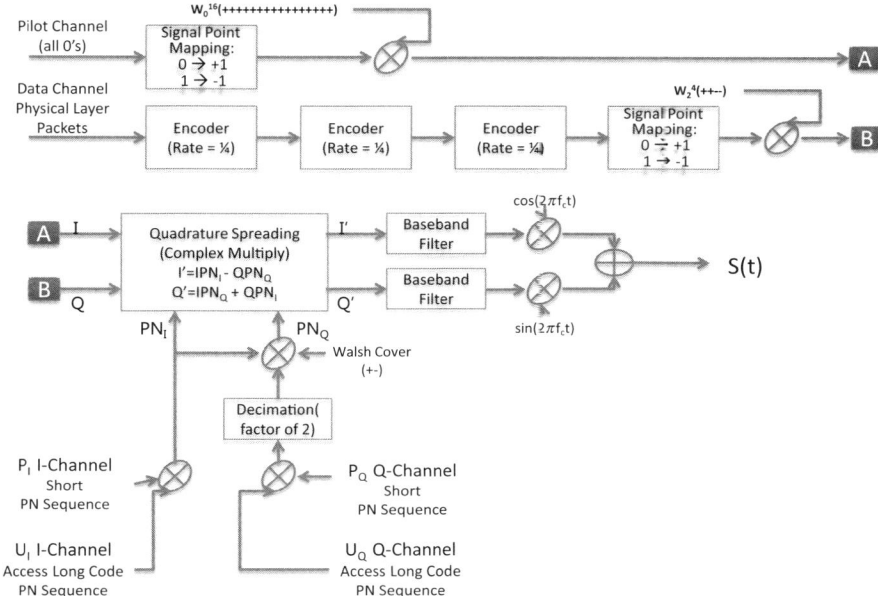

Figure 7.51   Block diagram of reverse physical access channel implementation. Reprinted from Reference [180]. Copyrighted material. Reproduced and distributed by the authors under written permission of the Organizational Partners of the Third Generation Partnership Project 2 (3GPP2).

Then, the filtered baseband is upconverted and summed together with the Q channel to result in the final signal s(t).

The data channel of the reverse physical access channel is encoded at a rate of 1/4, and then encounters channel interleaving, repetition of interleaved packets, and signal point mapping before it is multiplied by the Walsh code sequence W_2^4. The resulting sequence is adjusted for relative gain to the pilot sequence and complex multiplied by the PN sequences similar to the pilot. The resulting stream is then baseband filtered in the same fashion as the pilot channel and upconverted on the Q channel by multiplication of a sine wave at the desired frequency.

Figure 7.52 and Figure 7.53 [180] illustrate the Reverse Physical Traffic Channel Block Diagram Implementation, as specified in Reference [180].

Here, the pilot and ACK channels are mapped to the I channel, while the DRC and data channels are mapped to the Q channel. PN_I and PN_Q are equivalent to those in the reverse physical access channel, as are the baseband filters.

7.3.7   1XEV-DO Revision B

EV-DO Revision B is implemented in two phases. The first phase, also known as multicarrier EV-DO, is a software upgrade that leverages existing EV-DO Revision

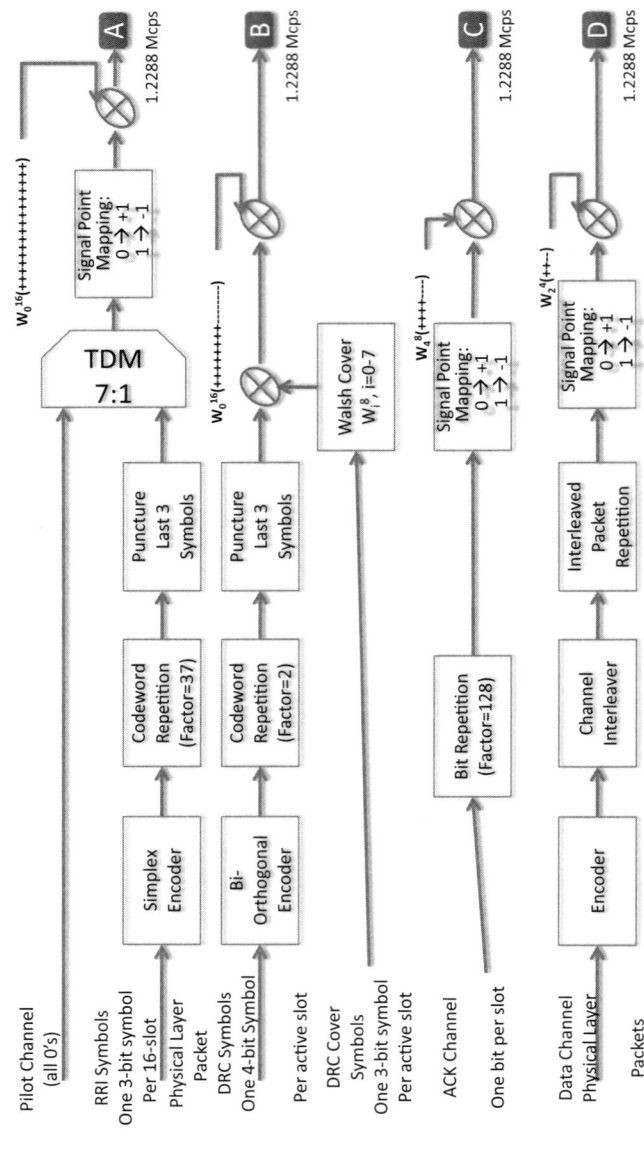

Figure 7.52 Block diagram of reverse physical traffic channel implementation (1 of 2). Reprinted from Reference [180]. Copyrighted material. Reproduced and distributed by the authors under written permission of the Organizational Partners of the Third Generation Partnership Project 2 (3GPP2).

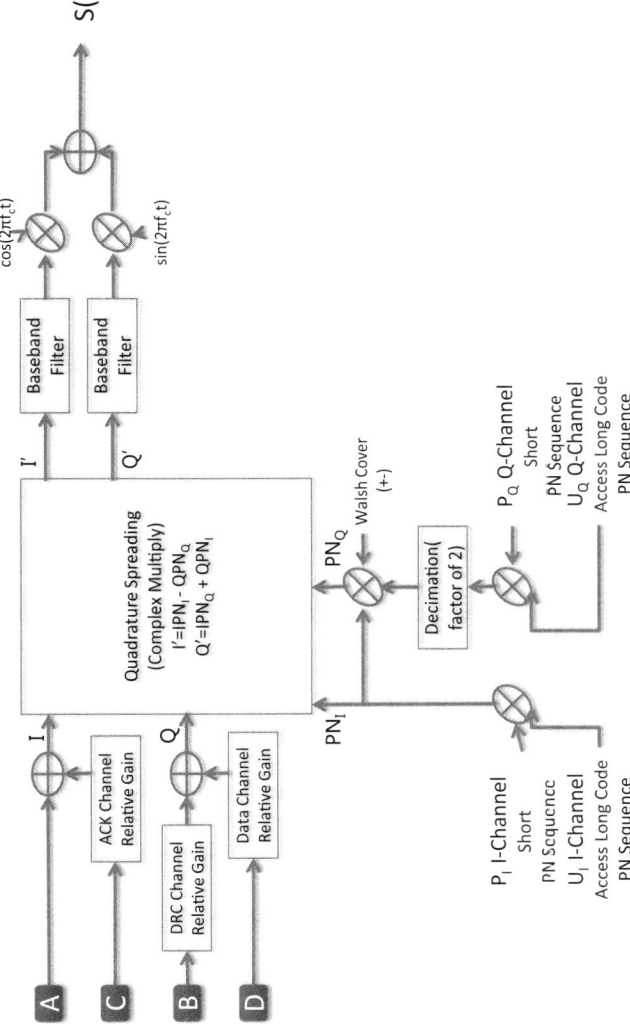

Figure 7.53 Block diagram of reverse physical traffic channel implementation (2 of 2). Reprinted from Reference [180]. Copyrighted material. Reproduced and distributed by the authors under written permission of the Organizational Partners of the Third Generation Partnership Project 2 (3GPP2).

A channel cards through the aggregation of multiple carriers to triple the data in the 5-MHz FDD spectrum or double the network capacity [178]. The increase in network capacity is achieved through logical combining of multiple EV-DO Revision A carriers. This phase maintains backward compatibility with legacy systems and can send unique data streams on different EV-DO Revision A carriers, thus increasing the throughput on both the Forward and Reverse links.

Multicarrier EV-DO combines up to three EV-DO Revision A carriers, thus achieving the Forward and Reverse link peak data rates of 9.3 and 5.4 Mbps, respectively, within the 5-MHz FDD spectrum (3x1.25 MHz carriers, plus guard band). Note that Revision A carriers do not need to be adjacent to each other [178].

The second phase of Revision B introduces even higher data rates through the introduction of 64-QAM modulation scheme, which increases the Forward link peak data rate to 14.7 Mbps (4.9 Mbps per 1.25-MHz carrier) with the Reverse link data rates remaining the same. The full upgrade to this phase requires a new channel card and new mobile devices [178].

In addition, Revision B provides significantly better coverage at the cell edge.

7.3.8   DO Advanced

DO Advanced improves overall network performance by exploiting unevenly loaded networks. It consists of three basic building blocks: Smart Networks, Enhanced Connection Management, and Advanced Devices.

7.3.8.1   Smart Networks

Smart Networks are essentially five unique features that can significantly increase the data capacity of an existing network with existing devices, without the need for additional infrastructure changes. They are:

- *Network Load Balancing.* This uses unexploited network capacity of lightly loaded neighboring cells.

- *Adaptive Frequency Reuse.* Also known as Demand Match Configuration, this adjusts the Tx power of the neighboring lightly loaded cells to reduce CCI and to improve overall network capacity and data rates.

- *Distributed Network Scheduler.* This is a multicarrier feature that prioritizes and allocates bandwidths to multiple users at the network level.

- *Single-Carrier Multilink.* With this feature, the EV-DO Revision B device (that can process two independent data streams simultaneously) can use the benefits of a multicarrier network even in a single-carrier handoff environment (same frequency can be used for both carriers serving a single mobile devices, as long as the carriers come from different cells or sectors). This feature improves cell-edge performance for multicarrier devices and provides better network load balancing.

- *Smart Carrier Management.* This can change the carrier assignment to the MS based on signal quality and the network load [178].

Deployment of Revision B is not mandatory for a lot of these features [178].

7.3.8.2 Enhanced Connection Management

Enhanced Connection Management in DO Advanced is responsible for increasing the number of connections, managing bursty applications, and improving user experience through parameter optimization, implementation enhancements, and Advanced Topology Networks (heterogeneous networks that include macrocells, picocells, microcells, and femtocells) [178].

7.3.8.3 Advanced Devices

Advanced Devices will include (1) enhanced equalizers for improved performance of bursty traffic and (2) Mobile Transmit Diversity (MTD) to achieve higher reverse link (uplink) capacity and data rates.

7.3.9 CDMA2000 USAGE

According to the CDMA Development Group, as of June 2011, there are 614,020,000 CDMA subscriptions (includes CDMA2000 1X, 1xEV-DO Rel. 0, and 1xEV-DO Rev. A) worldwide. Table 7.22 [181] totals the number of CDMA2000 subscriptions per geographic region.

ACKNOWLEDGMENT

Several figures found in this chapter have been reproduced with written permission from the Telecommunications Industry Association (TIA), 1320 N. Courthouse Road, Suite 200, Arlington, VA 22201; www.tiaonline.org.

Table 7.22   CDMA Subscriber Statistics: 4Q 2011

Region	Number of subscriptions	Percentage of overall (%)
Asia Pacific	374,300,000	59.9
North America	184,850,000	29.6
Caribbean and Latin America	25,700,000	4.1
Europe	7,780,000	1.2
Middle East	5,900,000	0.9
Africa	27,200,000	4.3
Total	625,730,000	100

Data from Reference [181].

RECOMMENDED ADDITIONAL READING

There are several additional sources of reading that we can refer the reader to. The following are some texts on the topic of CDMA2000 that we recommend:

K. ETEMAD, *CDMA2000 Evolution: System Concepts and Design Principles*, Wiley-Interscience, Hoboken, NJ, 2004 [179].

V. VANGHI, A. DAMNJANOVIC, and B. VOJCIC, *The CDMA2000 System for Mobile Communications*, Prentice Hall PTR, Upper Saddle River, NJ, 2004 [166].

Additionally, the CDMA Development Group is an excellent online resource for the latest information on the CDMA2000 technology, http://www.cdg.org.

Chapter 8

Fourth-Generation (4G) Cellular Communications

The aggressive evolution of the cellular networking landscape has rapidly brought us to the era of fourth-generation, "4G" cellular networks. New cellular technologies such as the 3rd Generation Partnership Project (3GPP) Long-Term Evolution (LTE)-Advanced and IEEE 802.16m promise an enhanced user experience with lower latency, higher throughput, and always-on connectivity—enabling a mobile broadband experience comparable to that of wireless local area network (WLAN) technology. With peak theoretical data rates of 1 Gbps or more, 4G cellular technologies are also fueling the convergence of mobile and fixed wireless broadband services. In recent years, the term "4G" has been far overused for marketing purposes, which has further blurred the already amorphous line between 4G and enhanced third-generation (3G) technologies. So, what constitutes a 4G cellular technology? One well-defined benchmark that is widely accepted in the technical community is the International Telecommunications Union (ITU) definition of performance requirements for International Mobile Telecommunications (IMT)-Advanced. From this perspective, there are currently only two technologies that fulfill the performance requirements of a 4G cellular technology: 3GPP LTE-Advanced and Wireless Metropolitan Area Network (WirelessMAN)-Advanced (based on the IEEE 802.16m standard). These technologies have a number of characteristics that collectively distinguish them from most other enhanced 3G networks. They define radio access technologies (RATs) based on orthogonal frequency division multiple access (OFDMA) and enhanced by multiple-input, multiple-output (MIMO) multiple antenna technology. They are all-Internet Protocol (IP) networks (AIPN), optimized for packet-switched (PS) data. They support wide bandwidth deployments of up to 40 MHz, combined with spectral efficiency up to 15 bps/Hz, resulting in high overall cell throughput as well as peak data rates of 1 Gbps or more in low mobility scenarios. Though some of these characteristics are now included in enhanced 3G technologies, in many cases they have been added after the fact as an evolutionary

Wireless Networking: Understanding Internetworking Challenges, First Edition.
Jack L. Burbank, Julia Andrusenko, Jared S. Everett, and William T.M. Kasch.
© 2013 the Institute of Electrical and Electronics Engineers, Inc. Published 2013 by John Wiley & Sons, Inc.

enhancement (e.g., MIMO enhancements to High-Speed Packet Access [HSPA]). In the case of 4G technologies, these characteristics have been chosen from the beginning as fundamental design requirements.

In this chapter, we focus on 3GPP LTE/LTE-Advanced, starting with the baseline standard defined in 3GPP Release 8 (Rel-8). Rel-8 LTE already meets many of the IMT-Advanced requirements, such as peak downlink spectral efficiency and low latency. Furthermore, it forms the basis for LTE-Advanced, which warrants its inclusion in this chapter. We then discuss the key enhancements defined in 3GPP Release 10 (Rel-10) for LTE-Advanced, which has been accepted into the IMT-Advanced family of standards as a fully compliant 4G cellular broadband technology.

8.1 LONG-TERM EVOLUTION

3GPP Long Term Evolution (LTE) is the highly anticipated next generation in the 3GPP family of technologies, which includes the Global System for Mobile Communications (GSM) and Universal Mobile Telecommunications System (UMTS) technology families described previously in Chapters 6 and 7. LTE is the predominant technology in the evolution to 4G cellular, mapping out an evolutionary path from GSM and UMTS that leads to higher data rates, reduced latency, and always-on IP connectivity for the delivery of cellular multimedia services. The name LTE derives from the fact that it was conceived as the long-term evolution of UMTS, designed to keep the 3GPP family of technologies competitive in the timeframe beyond 2015. To call LTE the long-term evolution of UMTS is a bit misleading in the sense that LTE is not a backward-compatible enhancement for UMTS systems, but rather a fundamentally new technology that has been fully optimized for packet data services without the constraint of backward compatibility. The system commonly referred to as LTE encompasses a new RAT complemented by an evolution of the nonradio network aspects, including a new core network called the evolved packet core (EPC). LTE and EPC together represent an important evolution in wireless communications with an emphasis placed on data networking. The first release of LTE, specified in 3GPP Rel-8, provides peak data rates of up to 300 Mbps in the downlink and 75 Mbps in the uplink. While these rates are impressive, it is important to keep in mind that LTE achieves other performance requirements that are equally as important for practical deployments—including significant improvements in latency, network capacity, spectral efficiency, and overall cell throughput. While Rel-8 LTE already meets some of the requirements for IMT-Advanced, it falls shy of full compliance, leading to it sometimes being referred to as "pre-4G." LTE forms the basis for Rel-10 LTE-Advanced, which is part of the open evolution of the LTE specification.

LTE introduces new RATs that utilize OFDMA in the downlink and a multiple access scheme called single-carrier frequency division multiple access (SC-FDMA) in the uplink, the latter of which combines discrete Fourier transform (DFT) precoding with an OFDM-like subcarrier mapping. SC-FDMA has many of the advantages of an OFDMA system, including frequency-domain scheduling of multiple users, but benefits from a reduced peak-to-average power ratio (PAPR), allowing for a less expensive transmitter design and reduced battery consumption in the user equipment. The use of MIMO antenna techniques for increased data rates and improved robustness is an integral component of the LTE air interface, providing an additional spatial dimension to the overall air interface. For maximum deployment flexibility with respect to spectrum allocation, LTE supports scalable channel bandwidths from 1.4 to 20 MHz. Furthermore, the specification of both frequency division duplexed (FDD) and time division duplexed (TDD) variants allow it to be deployed in either paired or unpaired spectrum. This makes it possible to deploy LTE in a large number of frequency bands.

A key characteristic of LTE is that it is an all-IP system. Unlike its predecessors, LTE does not have a circuit-switched domain to carry voice traffic. As a result, the entire system from the air interface to the core network has been optimized for IP packet traffic. To reduce latency, a flat network architecture is used, particularly in the radio access network (RAN) where all functionality has been condensed to a single functional node, the evolved Node B (eNB). The EPC system architecture has been designed from the start to support interworking with other RATs, both 3GPP and non-3GPP, for a seamless mobile experience. LTE thus provides an evolutionary pathway for network operators from various 3G technologies, including UMTS, code division multiple access 2000 (CDMA2000), and Mobile Worldwide Interoperability for Microwave Access (WiMAX).

This chapter provides a technical overview of the essential concepts, architecture, and protocols of an LTE cellular network. For the first portion of this chapter, a baseline system configuration based on 3GPP Rel-8 LTE is assumed unless otherwise noted. Further enhancements as part of Rel-10 LTE-Advanced are addressed in Section 8.2 later in this chapter.

8.1.1 History of LTE Development

An overall timeline of significant milestones in the history of LTE and LTE-Advanced are summarized in Table 8.1. The early development of LTE is discussed further in the following sections, including an introduction to LTE and System Architecture Evolution (SAE) standardization.

8.1.1.1 Origin of the LTE Study Item

The origins of LTE date back to November 2004, when the 3GPP TSG RAN[1] Future Evolution Workshop was held in Toronto, Canada to discuss potential technologies for the long-term evolution of the UMTS air interface and RAN architecture [182]. Proposals were presented by various 3GPP participants, including both network operators (e.g., NTT DoCoMo, Vodafone, TeliaSonera, China Mobile) and

---

[1] Technical services group for radio access networks.

Table 8.1 Timeline of LTE Milestones

Year	Event
2004	Work begins in 3GPP on the long-term evolution of UMTS; Start of the LTE Study Item.
2005	LTE target requirements defined in 3GPP TR 25.913; TSG RAN officially endorses OFDMA/SC-FDMA for LTE multiple access.
2006	Conclusion of LTE Study Item/Start of LTE Work Item.
2007	First 3GPP Workshop on LTE-Advanced.
2008	Start of LTE-Advanced Study Item; LTE-Advanced target requirements defined in 3GPP TR 36.913; 3GPP Release 8 specification frozen—first release of LTE.
2009	LTE-Advanced proposal submitted to ITU as candidate for IMT-Advanced; 3GPP Release 9 specification frozen; First commercial LTE deployments by TeliaSonera in Stockholm, Sweden and Oslo, Norway.
2011	3GPP Release 10 specification frozen—definition of LTE-Advanced; LTE-Advanced officially accepted by ITU to IMT-Advanced; First LTE-enabled smartphones commercially launched.

manufacturers (e.g., Qualcomm, Nokia, Ericsson). These proposals provided a variety of perspectives on what RATs should be used in the next-generation RAN and what the target performance of such system should be. Although a variety of technologies and opinions were presented, there are a few key points that appeared repeatedly and resonated throughout the workshop:

- Many participants supported an air interface based on *OFDM*, a technology that had matured significantly since the development of the UMTS air interface in the mid-1990s.

- Increased data rates and robustness through the use of *MIMO antenna systems* was proposed by a number of presenters.

- *Spectrum flexibility* to allow wideband deployments of up to 20 MHz bandwidth while still allowing for deployment in available spectra with bandwidths of less than 5 MHz.

- An air interface that has been optimized for the efficient transmission of *IP packet data*.

It is interesting to note that OFDM was previously considered as a possible air interface for UMTS during its early development in the mid-1990s prior to the decision to use Wideband Code Division Multiple Access (WCDMA) as the air interface [51]. There were a number of factors that lead to the decision not to use OFDM at that time. One factor was the lack of an appropriate uplink multiple access scheme, as OFDM was considered inappropriate as a cellular uplink technology largely due to the impact of its large PAPR on handset cost and power consumption. Another factor was that MIMO antenna technology—which is an essential part of MIMO-OFDM systems—was still an immature technology and somewhat cost prohibitive for implementation in handsets.

In December 2004, a proposal was approved by TSG RAN to begin a feasibility study on the development of a new air interface and RAN based on the technologies discussed at the RAN Future Evolution Workshop [183]. Although formally referred to as the "Feasibility Study on Evolved UTRA and UTRAN," this study item came to be known within the 3GPP as the LTE study item. The name "long-term evolution" originated simply from the fact that LTE was developed as the long-term evolution of the UMTS RAT, intended to keep the 3GPP family of standards competitive beyond the deployment of HSPA systems, over a timeframe of 10 years and beyond. It was not until later that the term LTE became the official name used by 3GPP to refer to the new system.

8.1.1.2 Early Performance Targets

One of the first points of business was to define the target performance for the new system. The LTE study item began around the same time that the final work was completed on 3GPP Rel-6, which defined the enhanced uplink (High-Speed Uplink Packet Access [HSUPA]) for UMTS. The high performance achieved by Rel-6 HSPA was considered to be sufficient to keep UMTS networks competitive for several years. Therefore, the objectives of the long-term evolution of 3GPP radio access needed to focus on improvements that would ensure competitiveness over a longer timeframe spanning from 2015 and beyond. This meant that the evolved Universal Terrestrial Radio Access (UTRA) and Universal Terrestrial Radio Access Network (UTRAN) must achieve significant performance improvements over the current system, targeted at supporting the next-generation cellular networks. The target performance requirements for the new LTE air interface and RAN were defined in 2005 in 3GPP Technical Report (TR) 25.913 [184]. Many of the performance targets are defined in terms of the required improvement over a baseline system configuration of Rel-6 HSPA. This baseline reference was convenient because, at the time, the Rel-6 HSPA system had just been completed and its performance was well defined. The requirements specified in Reference [184] were quite comprehensive. A few of the key requirements are summarized as follows:

- Downlink peak data rates of 100 Mbps (assuming two received antennas and a 20 MHz transmission bandwidth).
- Uplink peak data rates of 50 Mbps (assuming one transmit antenna and a 20 MHz transmission bandwidth).
- Latency reduction to include round-trip time of <10 ms and initial access delay time of <300 ms.
- Spectral efficiency increased two to four times greater than Rel-6 HSPA in a loaded network.

- Individual user throughput increased two to four times greater than Rel-6 HSPA for both average user and cell-edge user.
- Capacity increased to include at least 200 active users per cell in 5 MHz channel bandwidth and at least 400 active users in larger channel bandwidths.
- Flexible bandwidth allocations ranging from below 5 up to 20 MHz.
- Flexible coverage supporting cell ranges up to 5 km with full performance, 30 km with slightly degraded performance, and 100 km with degraded performance.
- Support for high mobility users moving at speeds of up to 350–500 km/h, depending on operating frequency.

The LTE study item resulted in a technical report assessing the proposed design for evolved UTRA (E-UTRA) and evolved UTRAN (E-UTRAN) in 3GPP TR 25.912 [185]. This feasibility study concluded that the proposed RAN and air interface based on OFDMA/SC-FDMA are capable of meeting or exceeding the target requirements set forth for the new system. In 2006, the LTE study item concluded and the LTE work item began to define the detailed E-UTRA and E-UTRAN design. It is worth noting that the LTE system as it was eventually defined in 3GPP Rel-8 meets all of these performance requirements and, in some cases, actually exceeds them. For example, the theoretical peak data rates of the final Rel-8 LTE system under the assumptions listed earlier are 150 Mbps in the downlink and 75 Mbps in the uplink. Furthermore, using the highest category of UE capabilities, a theoretical peak data rate of 300 Mbps can be achieved in the downlink using 4×4 MIMO spatial multiplexing.

8.1.1.3 The SAE Study Item

As the LTE study item progressed, it eventually became clear that the new RAT would need to be accompanied by a complementary evolution of the non-RAN aspects of the system, including perhaps most notably the core network architecture. This need spawned a study item in 3GPP TSG SA[2] under the name "System Architecture Evolution" (SAE), which was intended to address end-to-end system aspects not covered by the LTE study item, including the evolution of the core network. The SAE study item had already been preceded by a study item in TSG SA on the feasibility of an AIPN architecture, which would do away with the traditional split between the circuit-switched and packet-switched domains of the GSM/General Packet Radio Service (GPRS) and UMTS core network architectures in favor of a single packet-switched domain [186]. Concepts from the AIPN study item were taken into account in the SAE study item. The key results of the SAE study item are documented in 3GPP TR 23.882 [187] that, due to delays, was not finalized until 3GPP Rel-8.

---

[2] Technical services group for services and system aspects.

One of the most important work items to emerge from the SAE study item was the evolution of the 3GPP core network, called the EPC. The EPC is an AIPN in the sense that it comprises only a PS domain. The movement toward the AIPN architecture has been foreshadowed by developments in earlier 3GPP releases, such as the definition of an all-IP transport network and the IP Multimedia Subsystem (IMS) defined in Rel-5. However, the EPC takes the all-IP concept a step further by completely eliminating the circuit switched. Instead, all circuit-switched services, such as voice and short message service (SMS), are carried over IP (e.g., Voice over Internet Protocol [VoIP]). A services layer, such as IMS, may be used to control such services. Complementary to the PS-only EPC design is the concept of always-on IP connectivity, whereby the UE is assigned a basic IP connection when it registers to the network and that IP connection is maintained until the UE is powered off. This approach is different from GSM/GPRS and UMTS systems, where an IP session is only created upon request from the UE and is released when it is no longer required. The reason for this is that, in an all-IP system, the user cannot receive any services in the absence of an IP connection. Another key concept that has been identified as part of SAE is the ability to support mobility between heterogeneous access networks, including E-UTRAN, UTRAN, and GSM/EDGE Radio Access Network (GERAN) as well as non-3GPP RANs such as CDMA2000 Evolution–Data Optimized (EV-DO) or Wi-Fi. All of these heterogeneous access networks can be connected seamlessly through the EPC.

8.1.2 Summary of LTE Releases

3GPP Rel-8, first published in 2008, marked the initial release of the LTE standard. Within today's cellular market, this is the standard that is most commonly referred to as "LTE." 3GPP Release 9 (Rel-9) includes enhancements to the base LTE standard. In many cases, these enhancements include nonessential features that could not be completed in time for the Rel-8 publication and were postponed for future addition. A good example is the multimedia broadcast multicast service (MBMS). Many parts of the MBMS were already defined in Rel-8, such as the physical (PHY) layer design. However, the details of MBMS, including the higher-layer protocols, were not finalized until Rel-9.

3GPP Rel-10 marks a major enhancement to the LTE standard commonly known as LTE-Advanced. The set of enhancements in Rel-10 are aimed at evolving LTE to meet the performance requirements set forth by the ITU for IMT-Advanced compliance, which is the internationally recognized standard for 4G cellular communication. To meet these requirements, substantial improvements were made to achieve wider bandwidths, improved spectral efficiency, and higher peak data rates. Key air interface enhancements include increased bandwidths of up to 100 MHz through OFDM carrier aggregation and enhanced MIMO capabilities in both the downlink and the uplink. These MIMO enhancements notably include support for 8×8 MIMO spatial multiplexing (i.e., eight spatial streams) in the downlink as well the introduction of uplink MIMO spatial multiplexing with up to four transmit antennas (i.e., four spatial streams). Another key feature of Rel-10 LTE-Advanced is improved support for Heterogeneous Networks (HetNets), which allows dense network deployments with overlapping layers of coverage (e.g., macrocells overlapping with small cells, such as picocells, femtocells, or relay nodes [RNs]). Rel-10 LTE-Advanced was officially accepted by the ITU in 2011 as part of the IMT-Advanced family of standards, making it the first official 4G technology—at least in the ITU sense—to emerge from the 3GPP standards body.

Further enhancements to LTE-Advanced are specified in 3GPP Release 11, including carrier aggregation enhancements and coordinated multipoint operation (CoMP). The latter refers to the coordinated transmission from multiple cells to a single UE concurrently. Initial publication of Rel-11 was completed in 2013. 3GPP TSG RAN held a workshop in Slovenia in June 2012 to address requirements for the further evolution of the LTE RAN. The features to be included in Rel-12 are still being developed at the time of this writing. Initial publication of Rel-12 is anticipated for 2014, with initial deployment following a few years later and global adoption in the timeframe around 2020.

The new features introduced in each LTE release as well as year of initial publication are summarized in Table 8.2 [7]. While not exhaustive, this list includes the main features associated with each new release.

8.1.3   Introduction to 3GPP Terminology

The following terminology is used throughout the 3GPP technical specifications to refer to certain aspects of LTE/SAE systems. Their definition here will be helpful

Table 8.2   Summary of Key Features in LTE Releases

Release	Year	Key features
Rel-8	2008	• *LTE Initial release*: Introduction of the Evolved Packet System (EPS) including E-UTRA, E-UTRAN, and EPC • Self-organizing networks (SON): Concepts and requirements, eNB self-configuration, and Automatic Neighbor Relations (ANR) list management • Home eNB (i.e., LTE femtocell) • LTE FDD Repeaters
Rel-9	2009	• MBMS: Multimedia Broadcast/Multicast functionality in LTE, including MBSFN operation, for services such as mobile TV[a] • Dual-layer Beamforming (downlink) • SON enhancements • Enhancements for Home eNB (i.e., LTE femtocell) • Location services (LCS) over EPS • UE positioning over LTE • IMS emergency calls over EPS • Specification of Operating Bands 18–21 for FDD

Table 8.2   (*Continued*)

Release	Year	Key features
Rel-10	2011	• *LTE-Advanced*: Set of enhancements to LTE aimed at meeting the performance requirements of IMT-Advanced for 4G cellular compliance
		• Carrier aggregation (up to 100 MHz bandwidth)
		• Enhanced downlink MIMO spatial multiplexing up to eight spatial streams (8×8 MIMO)
		• Uplink MIMO spatial multiplexing up to four spatial streams (4×4 MIMO)
		• SON enhancements: Coverage and capacity optimization (CCO), mobility robustness optimization (MRO) enhancements, and mobility load balancing (MLB) enhancements
		• LTE relay nodes (Part 1)
		• Enhanced support for heterogeneous networks (HetNets)
		• MBMS enhancements
		• Mobility enhancements for Home eNB
		• Specification of Operating Bands 22–25 for FDD and Bands 41–43 for TDD
Rel-11	2012	• Carrier aggregation enhancements
		• Coordinated multipoint (CoMP)
		• Further SON enhancements
		• Carrier-based HetNet ICIC
		• NBPS: Network-based positioning support for LTE based on UTDOA using the uplink SRS (souncing reference signal)
		• LTE relay nodes (Part 2)
		• Network energy saving for E-UTRAN
		• Signaling and procedure for interference avoidance for in-device coexistence
		• Specification of Operating Bands 26–28 for FDD and Band 44 for TDD

*a*It is worth noting that many aspects of MBMS for LTE have already been defined in Rel-8. However, full support for MBMS, particularly with respect to the higher-layer protocol definitions, were not completed until Rel-9.

MBMS, multimedia broadcast/multicast service; MBSFN, multimedia broadcast over single frequency network; NBPS, network-based positioning support.

to avoid ambiguity within the following discussion, and also to assist the interested reader to interpret the 3GPP LTE standards documents.

- *Evolved Universal Terrestrial Radio Access Network (E-UTRAN).* The LTE RAN is formally named E-UTRAN, as it is an evolution of the UMTS RAN, UTRAN. E-UTRAN utilizes a flat architecture comprised of a single logical entity, the eNB, which combines functionality traditionally implemented in the Node B and radio network controller (RNC) entities of UTRAN.

- *Evolved Universal Terrestrial Radio Access (E-UTRA).* The LTE RAT utilizing OFDMA in the downlink and SC-FDMA in the uplink is formally named E-UTRA. This includes the LTE air interface, which is formally referred to as the *E-UTRAN Uu interface* and commonly abbreviated as *LTE-Uu.*

- *Evolved Packet Core (EPC).* The evolution of the 3GPP core network architecture resulting from the SAE work item is named the EPC. In essence, the EPC defines a new evolved packet-switched (PS) domain of the core network that is distinct from the GPRS-based PS domain used for legacy systems.

- *Evolved Packet System (EPS).* The combination of E-UTRAN and EPC form the EPS, which is what is commonly referred to as the LTE system architecture in casual parlance. The EPS is the full end-to-end system resulting from LTE and SAE.

8.1.4 System Architecture Evolution

In order to meet the ambitious requirements for reduced user plane latency, 3GPP SAE has adopted a flat network architecture with as few nodes as possible, particularly in the user plane. Although core network enhancements have been made with varying degrees with every 3GPP release, the series of nodes involved in data transmission in the PS domain remained essentially unchanged from the original definition of UMTS in R99 through Rel-6. The shift toward a flatter user plane architecture had already begun in 3GPP Rel-7 HSPA+, which defines a direct tunneling option to bypass the serving GPRS support node (SGSN) in the user plane and allows the integration of RNC functions into the Node B [188]. In Rel-8, the direct tunnel and integrated RNC/Node B features have become cornerstones of the EPS architecture. The progression of 3GPP system architectures from HSPA in Rel-6 to LTE in Rel-8 is illustrated in Figure 8.1 [188].

The EPS system architecture consists of three functional subsystems: the user equipment (UE), the radio access network (E-UTRAN), and the CN (EPC). The essential functionality of each of these subsystems has been preserved from UMTS. The key architectural changes as a result of the LTE/SAE development work have taken place in the E-UTRAN and EPC subsystems. The functional architecture of the UE has remained unchanged, although it should be noted that significant enhancements are required in terms of implementation to support the MIMO-OFDM-based RAT of LTE. The system architecture of EPS in terms of its functional entities and interfaces is shown in Figure 8.2. For simplicity, a nonroaming system supporting only E-UTRAN is shown. Additional interfaces for interworking with other 3GPP and non-3GPP RATs are discussed later in this section.

The EPS architecture is characterized by a single PS domain. Unlike its predecessors, there is no CS domain. This is evident from the architecture, as the user plane has a single functional entity for connectivity between E-UTRAN and the EPC, namely the serving gateway (S-GW). This is in contrast to the mobile switching center

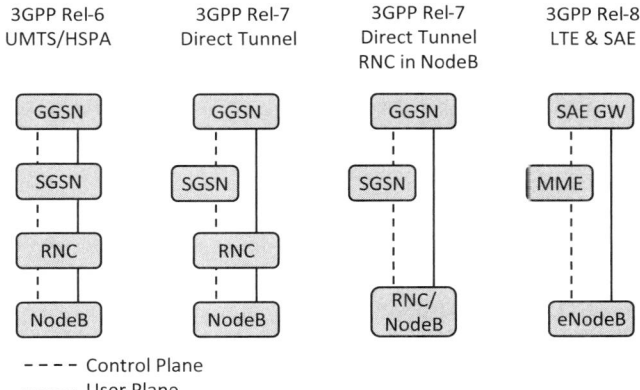

Figure 8.1   3GPP architecture evolution toward flat architecture. Reprinted from Reference [188].

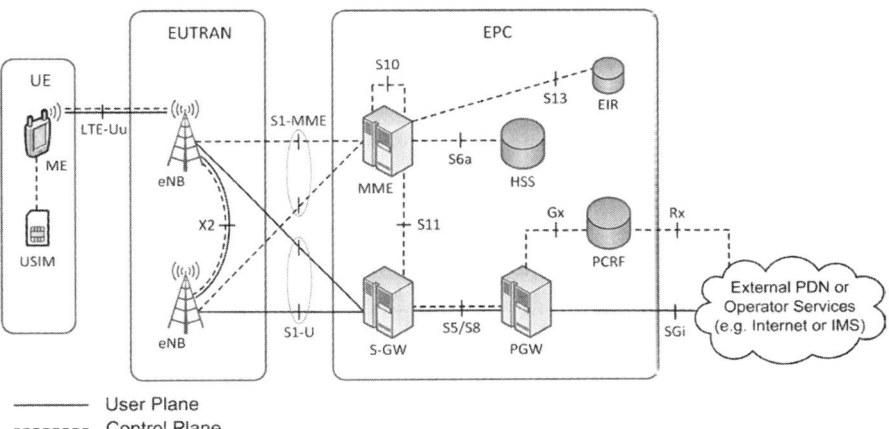

Figure 8.2   System architecture for EPS.

(MSC)/SGSN split of the Iu interface used in UMTS. All subscriber services, such as voice, data, and multimedia, are provided over IP. For this reason, the entire EPS system has been optimized for IP traffic. For this reason that LTE supports only IP services, LTE is often referred to as an "all-IP" network. Furthermore, the transport network used to connect the fixed infrastructure of the EPS is entirely based on IP. The domain of the EPS extends to the SGi reference point, which provides connectivity between the PDN gateway (P-GW) and a packet data network (PDN) and is analogous to the Gi interface of legacy 3GPP networks. The PDN may be an external IP network, such as the Internet or a corporate network, or it may be an additional CN infrastructure for the delivery of operator services, such as an IMS. In the latter case, the operator service could be, for example, a VoIP service to provide voice connectivity with other networks, such as the Public Switched Telephone Network (PSTN) or the CS domain of legacy cellular networks. The following sections provide an overview of each of the subsystems and functional entities of the EPS architecture.

8.1.5   User Equipment

As in UMTS, the term "user equipment" (UE) is used in LTE to refer to the cellular equipment utilized by the subscriber to access network services. The UE may be a smartphone or cellular data card, but it may also be an embedded device contained in a laptop or in machine-to-machine (M2M) equipment for supervisory control and data acquisition (SCADA) applications, for example. The functional architecture of an LTE UE is identical to that of UMTS. The UE comprises two parts: the mobile equipment (ME) and the Universal Subscriber Identity Module (USIM). The type of information stored on the USIM remains the same, but the specific fields have been expanded to include the EPS-specific addresses, identities, and security functions [189]. In practice, it has also become common to implement the USIM functionality in a smaller form-factor Universal Integrated Circuit Card (UICC) commonly referred to as a "micro-SIM."

8.1.5.1   LTE UE Categories

3GPP Rel-8 defines five UE categories for LTE, which are used to indicate the UL and DL radio access capabilities of the UE to the network. The definition of multiple UE categories is advantageous in allowing varying degrees of UE capability, complexity, and cost to address diverse market segments. The characteristics of each UE category are summarized in Table 8.3 [190]. Values for data rates and memory requirement for hybrid automatic repeat request (HARQ) processing are approximate. There is a direct relationship between UE category and the number of transport block (TB) bits that can be transmitted or received in a single transmission time interval (TTI), and by extension the peak data rate supported by the UE. In terms of MIMO spatial multiplexing support in the DL, capabilities range from no support for UE category 1 to 4×4 MIMO support (i.e., four layers) for UE category 5. In all cases, however, at least two receive antennas are required. In 3GPP Rel-10 and beyond, three additional UE categories have been defined to support higher-order MIMO configurations and higher data rates in both the UL and DL. These additional UE categories for LTE-Advanced are described in 3GPP TS 36.306 Rel-10 and beyond.

8.1.6   E-UTRAN

The E-UTRAN contains a single functional entity, the eNB, which serves one or more E-UTRAN cells. The eNB combines the functionality of the Node B and the RNC from 3G UMTS networks. Therefore there is no need for a separate node with

Table 8.3 Summary of LTE Rel-8 UE Categories

Capability	UE category				
	1	2	3	4	5
Maximum DL data rate (Mbps)	10	5C	100	150	300
Maximum UL data rate (Mbps)	5	25	50	50	75
Number of receive antennas	2	2	2	2	4
Number of supported layers for MIMO spatial multiplexing in DL	1	2	2	2	4
Support for 64-QAM in DL	Yes	Yes	Yes	Yes	Yes
Support for 64-QAM in UL	No	No	No	No	Yes
Layer 1 (L1) memory requirement for HARQ circular buffer (kB)	32	155	155	228	458
Total layer 2 (L2) buffer size (kB)	150	700	1400	1900	3500

base station controller type functionality. The adoption of this flat architecture results in a more access point-like network topology, similar to that used in Wi-Fi and WiMAX, which brings the user closer to the core network, thereby minimizing latency and improving efficiency. Essentially, the eNBs can be thought of as a group of peers, each of which participates in the overall optimization of the RAN. Neighboring eNBs are connected via the X2 interface, which may be used for signaling and to transfer user data in the case of a handover event. In this configuration, the E-UTRAN forms a mesh network of eNBs. Although the X2 interface is logically represented as a point-to-point link between eNBs in Figure 8.2, the physical realization need not be a point-to-point link. The implementation of the X2 is not mandatory, but it is generally required to take advantage of advanced network optimization techniques such as self-organizing networks (SON).

The eNB is connected to the EPC via the S1 interface, which is subdivided into two component interfaces to allow strict separation of control plane and user plane connectivity. The S1-mobility management entity (MME) interface connects the eNB to the MME in the control plane to carry signaling messages. The S1-U interface connects the eNB to the S-GW in the user plane to carry user IP traffic. E-UTRAN supports a many-to-many relationship between eNBs and these CN entities. This means that an eNB may be connected to one or more MME and one or more S-GW.

The functional split between E-UTRAN and EPC is illustrated in Figure 8.3 [191]. The eNB provides the point of termination on the network side of the so-called access stratum (AS) protocols, which are those air interface protocols related to radio access functionality. The responsibilities of the eNB include all functions related to radio resource management (RRM). These include [192]:

- radio bearer control and maintenance
- radio admission control (i.e., acceptance or rejection of requests for new radio bearers based on the overall resource situation of the E-UTRAN)

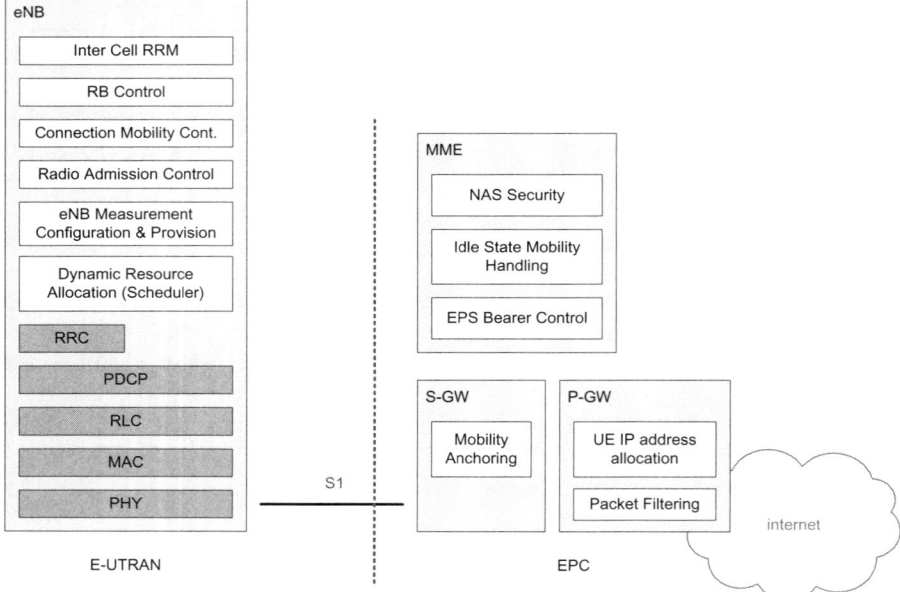

Figure 8.3 Functional split between E-UTRAN and EPC. Reprinted from Reference [191].

- connection mobility control (i.e., handover and configuration of idle mode cell reselection parameters)
- dynamic resource allocation (i.e., packet scheduling)
- intercell interference coordination (ICIC)
- load balancing (i.e., coordination between cells to redistribute traffic from highly loaded cells to underutilized cells to maintain quality of service [QoS])

To support these functions, the eNB is responsible for monitoring channel state information (CSI) that is required for mobility and scheduling based on current channel conditions. This includes measurement of the uplink radio channel and configuration of measurement reporting of the UE with respect to the downlink radio channel.

8.1.6.1 Tracking Area

For the purpose of mobility management, the E-UTRAN is divided into groups of cells called tracking areas (TAs). The TA is analogous to the routing area (RA) used in UMTS. When the UE is in Idle mode, it may camp on any cell within the current TA without signaling such changes to the network. If the radio environment is such that the UE must perform cell reselection into a new TA, for example due to mobility, then the UE initiates a TA update procedure to register its new location with the MME.

8.1.6.2 Pool Area

E-UTRAN utilizes the concept of pool areas that was defined previously for UTRAN in Rel-5. Since there is a many-to-many relationship between eNBs and core network nodes, each TA may be served by multiple MMEs and multiple S-GWs in parallel. A collection of MMEs or S-GWs that serve a common area is called an *MME pool* or *S-GW pool*, respectively. A collection of one or more TAs that is served by such a pool of CN nodes is called a *pool area*. In this configuration, the UEs within a single cell may be served by multiple CN nodes, thus increasing network efficiency through load sharing. The introduction of redundancy also minimizes the probability of loss of service due to a single point of failure. Each UE is served by one MME and one S-GW at a time, and it is the responsibility of the eNB to perform MME selection when no serving MME exists. The UE remains associated with the same MME until it moves to a different pool area.

8.1.7 Evolved Packet Core

The EPC comprises three main functional entities: the mobility management entity (MME), the serving gateway (S-GW), and the PDN gateway (P-GW). These are the three main elements that provide signaling support and IP connectivity to the UE. In practical networks, however, additional legacy CN entities are required for full network functionality. The most important among these are the Home Subscriber Server (HSS) and the Policy and Charging Rules Function (PCRF). Additionally, an Equipment Identity Register (EIR) may be used to verify UE International Mobile Station Equipment Identity (IMEI) values. Formally, these additional entities are common entities that are not specific to the EPC.

As with previous 3GPP technologies, the functional entities of the EPC may be implemented separately or integrated into one or a small number of nodes. Figure 8.4 illustrates four potential implementations of the MME, S-GW, and P-GW, each of which has its own benefits and disadvantages. The full split implementation provides the most flexibility but also requires the most number of core network nodes, which may increase infrastructure and operational costs. A common implementation scenario is the integrated gateway topology, in which the S-GW and P-GW have been integrated into a single node. This network topology is equivalent to the architecture shown previously in Figure 8.1, where the S-GW/P-GW form the so-called SAE Gateway. This network topology benefits from the separation of the control plane and the user plane, while at the same time minimizing the number of user plane nodes for minimum latency. Other potential network topologies include the legacy split (i.e., combined MME/S-GW—similar to SGSN in the GPRS architecture) and full integration. These options simplify the core network architecture through a reduced number of nodes; however, they also provide the least flexibility and do not benefit from the separation of user plane and control plane. In practice, the integrated gateway topology is expected to be most prevalent. Commercial offerings for integrated S-GW/P-GW devices are already available from vendors such as Ericsson, Huawei, and Nokia Siemens.

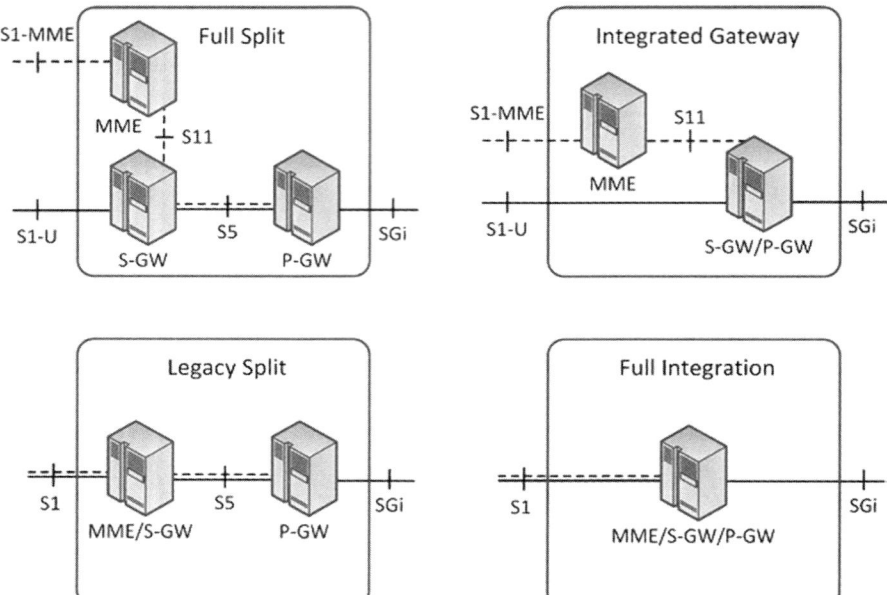

Figure 8.4 Possible implementations of EPC functional entities. Adapted from Reference [327].

8.1.7.1 Mobility Management Entity

The mobility management entity (MME) is the core network entity that performs signaling and control functions to manage the UE's connection to the EPC. The MME serves as the termination point on the network side for air interface protocols related to mobility and session management signaling. These protocols are referred to as the nonaccess stratum (NAS), as they do not directly relate to radio access functions of E-UTRAN. The responsibilities of the MME include mobility management functions (e.g., tracking and paging of the MS in Idle mode), session management functions (e.g., EPS bearer activation and management, selection of S-GW and P-GW), security functions (e.g., authentication, encryption, and integrity protection of NAS signaling, AS security control, security key establishment, and management), roaming functions (e.g., MME selection for inter-MME handover, handover to second-generation [2G GSM] and 3G UMTS core network elements), and allocation of temporary UE identities. The MME interfaces with the HSS via the S6a interface for subscription and authorization information. This information along with other UE-related state information is stored locally as the *UE context* to reduce overhead in the eNB and processing in UE during long periods of inactivity.

8.1.7.2 Serving Gateway

The serving gateway (S-GW) serves the user plane mobility anchor in the EPC, forwarding data from source eNB to target eNB during inter-eNB handover. The

S-GW also provides mobility anchor functionality for inter-RAT handovers to other 3GPP access networks (i.e., GERAN, UTRAN) for 2G or 3G interworking. The S-GW is primarily responsible for packet routing and forwarding between the eNBs and one or more P-GW, and plays only a minor role in control functions. In EPS connection management (ECM)-IDLE mode, the S-GW initiates the network-triggered service request procedure for incoming data and performs downlink packet buffering while the MME initiates paging of the UE until an E-UTRAN radio access bearer (E-RAB) is established. If the S-GW and P-GW are implemented separately, then they are connected via the S5 interface in the home network. In the roaming case where the S-GW and P-GW are in different networks, this interface is called the S8 interface but performs equivalent functionality. The S5/S8 interface may be implemented using a protocol stack based on either GPRS tunneling protocol (GTP) or Proxy Mobile IP (PMIP). PMIP is an Internet Engineering Task Force (IETF)-based network mobility management protocol that provides flexibility for network operators evolving from existing non-3GPP infrastructure, for example. When a PMIP-based S5/S8 interface is used, an additional interface is implemented between the S-GW and the PCRF called the Gxc interface. The responsibilities of the S-GW are then extended to include mapping between IP service data flows on S5/S8 and GTP tunnels on S1-U based on mapping information received from the PCRF.

8.1.7.3 PDN Gateway

The PDN gateway (P-GW) provides EPC connectivity with an operator services subsystem, such as the IMS, or directly to an external PDN, such as the Internet or an enterprise network. A key function of the P-GW is the allocation of IP addresses to UEs. If the address space of an external PDN is used (e.g., a corporate intranet), the P-GW tunnels packets between the EPS and the PDN. The Policy and Charging Enforcement Function (PCEF) of the Policy and Charging Control (PCC) architecture also resides in the P-GW, which includes IP service data flow detection, filtering, gating, QoS enforcement, and flow-based charging [193]. These functions are carried out based on policies received from the PCRF. The filtering functionality involves inspection of the packet contents (e.g., deep packet inspection) to filter the uplink and downlink packets of a single UE among multiple bearers supporting different QoS. The gating functionality involves allowing or blocking packets associated with an IP service data flow based on PCC rules. If the S5/S8 interface is based on GTP, the P-GW also performs mapping of IP service flows onto GTP tunnels associated with the EPS bearers. The P-GW also serves as the mobility anchor for interworking with non-3GPP access networks, such as CDMA2000, Wi-Fi, or WiMAX.

In general, the EPC may include multiple P-GWs, each of which provides access to one or more PDNs. Although the UE is served by a single S-GW, it may be connected to multiple P-GWs simultaneously, each supporting different EPS bearers.

8.1.7.4 Home Subscriber Server

The Home Subscriber Server (HSS) is a legacy node first defined in the UMTS core network architecture of 3GPP Rel-5. Although it is not unique to the EPC (i.e., as compared to legacy CS and GPRS PS domains), it is a required entity of a functional EPS network. The HSS combines legacy home location register (HLR) and authentication center (AuC) functions described previously in Chapter 6 and Chapter 7. The HSS stores a variety of user-related information including subscriber information (e.g., identities, QoS profiles subscribed, roaming restrictions), location information (e.g., address of current serving MME, inter-PLMN location information), and security information (e.g., identity key). Additionally, the HSS generates vectors for mutual authentication, ciphering, and integrity protection that are transmitted to the serving MME to support security procedures. The HSS may integrate heterogeneous information pertaining to multiple RATs and also plays a role in the IMS subsystem if deployed by the network operator.

8.1.7.5 Policy Charging and Rules Function

The Policy Charging and Rules Function (PCRF) is part of the PCC architecture first defined in 3GPP Rel-7. The PCRF is responsible for policy control decisions and flow-based charging control functions [193]. These are realized in the form of PCC rules that are sent to the PCEF in the P-GW whenever a new bearer setup is required. In this sense, the PCRF is responsible for control of the PCC functions that are enforced by the PCEF (e.g., filtering, gating, QoS enforcement). This responsibility includes provision of QoS class identifiers (QCIs) and bit rates in accordance with the user's subscription profile. The PCRF interfaces with the P-GW via the Gx interface and the Application Function of the IMS subsystem via the Rx interface if applicable. If PMIP-based S5/S8 interface is used, then the PCRF also interfaces with the S-GW via the Gxc interface. In practice, the PCRF is realized in the form of a server located in the operator's core network infrastructure [188].

8.1.7.6 Equipment Identity Register

The Equipment Identity Register (EIR) is a legacy database used for access control of MEs based on the IMEI of the device. It can be used, for example, to deny or limit service to devices that are obsolete, stolen, or nonfunctional. Communication between the MME and the EIR is supported via the S13 interface. For more information on the EIR, see the GSM section of Chapter 6.

8.1.8 Interworking with other RATs

To facilitate evolution toward a single all-IP 3GPP core network, the EPC architecture has been designed to support multiple heterogeneous access networks. Interworking support for both 3GPP and non-3GPP RATs has been a fundamental requirement from the very beginning. This section introduces the additional functional entities and interfaces defined to support mobility between E-UTRAN and other access networks. Two architectures are presented: one for interworking with 3GPP access networks (i.e., GERAN and UTRAN) and one for interworking with

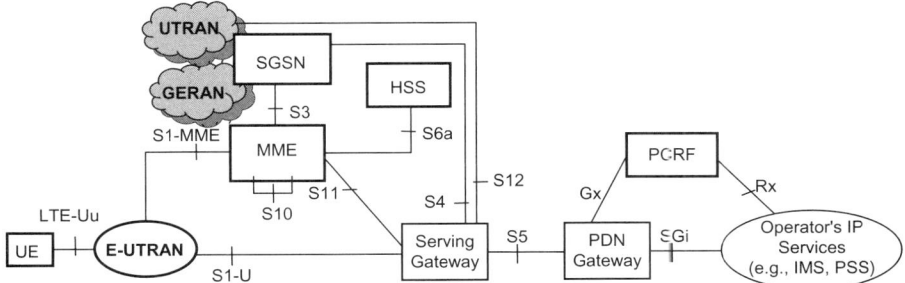

Figure 8.5   EPS system architecture for 3GPP access. Reprinted from Reference [194].

non-3GPP access networks (e.g., CDMA2000, Wi-Fi, or WiMAX). In both cases, the EPC provides connectivity to the external IP networks. In practice, the network operator may choose to implement any combination of 3GPP and non-3GPP access networks. For example, the operator network would include E-UTRAN, UTRAN, and GERAN for 2G/3G/4G 3GPP access, augmented by Wi-Fi hotspots in dense urban areas. Connectivity and interworking support across these heterogeneous networks may be provided through a common EPC core network, thereby simplifying operational costs.

8.1.8.1 3GPP Interworking

The EPS system architecture for 3GPP access is shown in Figure 8.5 [194]. For simplicity, the nonroaming case is shown. Interworking with GERAN and UTRAN is supported via a new functional entity called the *EPC SGSN*. The EPC SGSN supports all the functionality of a 2G/3G SGSN, but also supports some MME-like functionality, such as signaling for mobility between legacy 3GPP access networks and E-UTRAN and MME selection in the case of handover from GEREN/UTRAN to E-UTRAN. The EPC SGSN also supports S-GW/P-GW selection for UEs connected via GERAN or UTRAN. For this purpose, the EPC SGSN supports two new interfaces: the S3 interface to the MME and the S4 interface to the S-GW. A third EPC-specific interface, S12, is defined to support UTRAN direct tunneling. In this optional configuration, UTRAN connects to EPC via the SGSN in the control plane and directly to the S-GW via the S12 interface in the user plane. This configuration results in an architecture similar to that of EPS, where the SGSN performs the functions of the MME.

Although it is possible for UTRAN and GERAN access networks to utilize the EPC, it should be noted that there is no provision to use E-UTRAN with legacy core network architectures, such as the GPRS PS domain. When E-UTRAN is introduced to a network, it is mandatory to introduce the MME, S-GW, and P-GW functionality to the core network.

8.1.8.2 Non-3GPP Interworking

A number of mechanisms have been defined in the EPC to accommodate mobility between E-UTRAN and arbitrary non-3GPP IP access networks. The EPS system architecture for E-UTRAN interworking with generic non-3GPP access networks is shown in Figure 8.6 [195]. This architecture has been designed to support network-based mobility (e.g., PMIP) between E-UTRAN and arbitrary non-3GPP IP access networks, such as Wi-Fi, WiMAX, or potentially other access technologies not yet conceived. Non-3GPP IP access networks are divided into two categories: trusted and untrusted. A trusted network is one that can safely be assumed to execute 3GPP security functions such as authorization, authentication, and key agreement. By contrast, an untrusted network is not assumed to perform any functions other than delivery of IP packets [188]. The trust relationship of a non-3GPP access network is independent of the specific RAT. The architecture for untrusted non-3GPP inter-working is essentially an extension of the 3GPP Interworking with WLAN (I-WLAN) architecture first defined in Rel-6 [196].

In the case of untrusted non-3GPP access networks, EPS service is accessed through a functional entity called the enhanced packet data gateway (ePDG). The ePDG resides within the EPC and supports processes such as authorization, service selection, and subscription checking to determine whether 3GPP service may be granted to a UE via untrusted non-3GPP access. A secure IPSec tunnel is established between the UE and the ePDG via the SWu reference point. Functionality across SWu includes UE-initiated tunnel establishment, secure user data packet transmission, and support for fast teardown and update of IPSec tunnels during handover

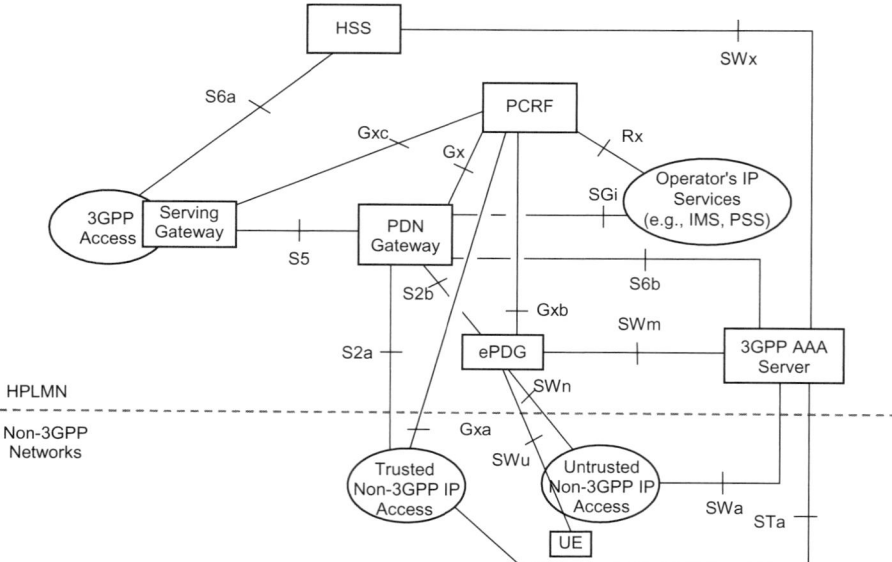

Figure 8.6 EPS system architecture for generic non-3GPP access. Reprinted from Reference [195].

between two untrusted non-3GPP IP accesses [195]. The ePDG in turn supports routing of user packets to and from the P-GW via the S2b interface. In the case of trusted non-3GPP access, the ePDG is not required and the non-3GPP access network interfaces directly with the P-GW via the S2a interface. In both trusted and untrusted cases, support for HSS-type functionality for non-3GPP access users is provided through the 3GPP Authentication, Authorization, and Accounting (AAA) Server. The 3GPP AAA Server retrieves subscriber profiles and authentication information from the HSS and supports signaling for authentication, authorization, and location management services. The 3GPP AAA Server coordinates closely with the HSS via the SWx interface to support mobility and service continuity across all access networks associated with the EPS.

In addition to network-based mobility, EPC also supports UE-based mobility using Dual-Stack Mobile IP (DSMIP). In this configuration, mobility is handled as an overlay function and the UE is logically connected with the P-GW (i.e., the mobility anchor) via the S2c interface.

Non-3GPP interworking based on the generic architecture is also referred to as "handover without optimization," as it has not been optimized for a specific access network. To achieve optimized performance and minimize handover gap time, specific architectures must be defined that have been optimized for interworking with a single access technology. This type of interworking is referred to as "handover with optimization." In 3GPP Rel-8, the only optimized architecture that has been defined is for interworking with CDMA2000 EV-DO,[3] as shown in Figure 8.7 [195].

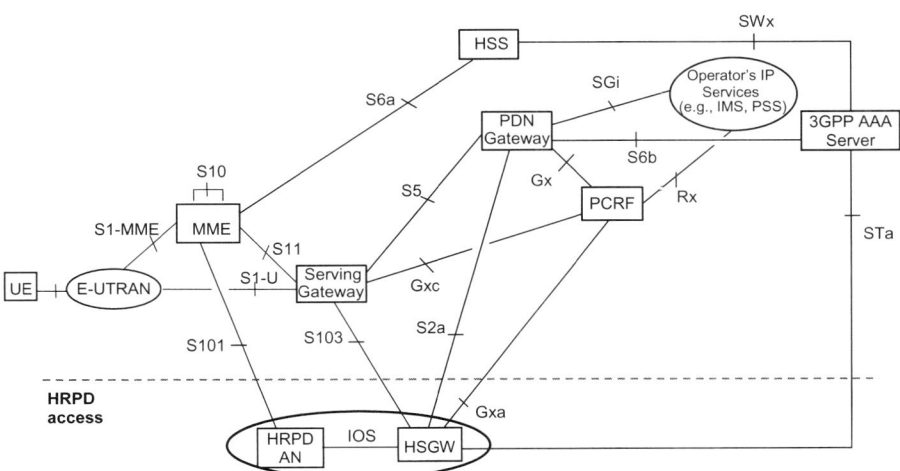

Figure 8.7 EPS system architecture for optimized interworking with CDMA2000 EV-DO. Reprinted from Reference [195].

---

[3] Note that CDMA2000 EV-DO is referred to in the 3GPP literature as CDMA2000 HRPD (High-Rate Packet Data), an alternate name for the technology.

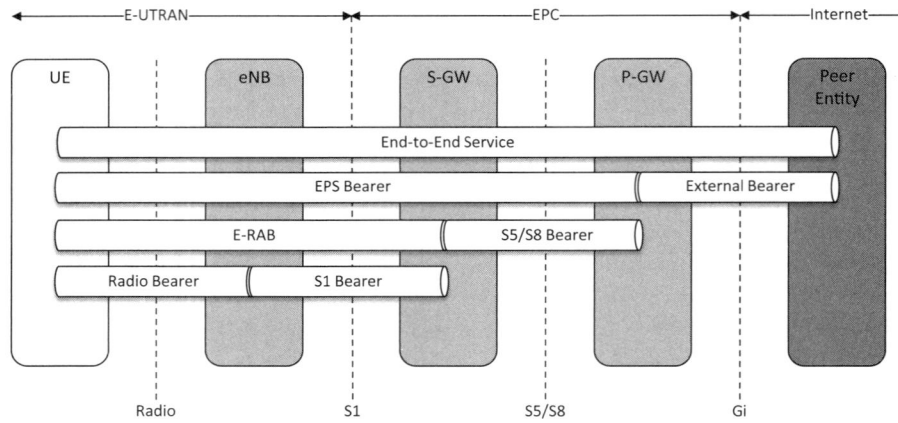

Figure 8.8   EPS bearer model for GTP-based S5/S8 interface. Reprinted from Reference [192].

Two additional interfaces are defined in this architecture to support optimization. The S101 interface provides direct connectivity between the CDMA2000 access network and the MME. The S103 interface provides direct connectivity between the HRPD Serving Gateway (HSGW) in the CDMA2000 infrastructure and the S-GW in EPC.

8.1.9   Bearer Model and QoS Concept

As an all-IP system, the EPS must support multiple concurrent IP-based services for each UE. For example, the user may have an active VoIP call while surfing the Web and downloading multimedia content all at the same time. As in this example, these services may have different QoS requirements (e.g., packet delay, packet loss, jitter). To support these services, the EPS uses the concept of bearers to transport IP traffic to and from the UE. This concept allows the IP flows of different applications to be handled differently by the EPS even though they may be transported between the same UE and PDN. The EPS bearer model is illustrated in Figure 8.8 [192]. For this figure and the following discussion, GTP-based S5/S8 is assumed.[4] The PDN connectivity service between the UE and SGi reference points is provided by an EPS bearer. An EPS bearer uniquely identifies traffic flows that receive a common QoS treatment. The EPS bearer is the level of granularity for QoS control in the EPS in the sense that each EPS bearer may support different QoS parameters. Furthermore, multiple IP traffic flows between a common UE and PDN, with a common QoS being aggregated onto a single EPS bearer.

---

[4] In the case when S5/S8 is based on PMIP, PDN connectivity service is provided by an EPS bearer between the UE and S-GW concatenated with IP connectivity between the S-GW and P-GW, as described in clause 4.10 in Reference [195].

Each bearer has an associated set of packet filters, collectively referred to as a traffic flow template (TFT), that are used to map IP traffic flows to the appropriate bearer. The TFT applies the concept of deep packet inspection and filters IP packets based on a variety of information, which may include source and destination IP addresses, port numbers, or other protocol information that may be used to distinguish multiple IP services associated with a single UE. An evaluation precedence index is assigned to each TFT to determine which packet filters are applied first when multiple bearers are established. In the UE, a UL TFT maps one or more IP traffic flow onto a bearer. Likewise, a DL TFT performs this mapping of IP traffic flows to bearers in the P-GW. This process is known as bearer binding. The EPS bearer is mapped onto lower layer bearers according to the hierarchy shown in Figure 8.8. Across each interface of the user plane, a bearer is established with its own identity and it is the responsibility of each network node to maintain the appropriate mapping between EPS bearers and lower layer bearers. The UE stores a one-to-one mapping between TFTs and radio bearers, which are used to transport packets of an EPS bearer between the UE and the eNB over the air interface. In the eNB, a one-to-one mapping is stored between a radio bearer and an S1 bearer. Likewise, the S-GW stores a one-to-one mapping between S1 bearers and S5/S8 bearers. Both the S1 and S5/S8 bearers are realized in the form of GTP tunnels and identified by a tunnel end point identifier (TEID). The concatenation of a radio bearer and an S1 bearer are also referred to as an E-RAB, which transports packets between the UE and the EPC through the E-UTRAN. The bearer mapping that occurs in each network node is illustrated in Figure 8.9 [194].

With respect to QoS, EPS bearers can be divided into two classes: guaranteed bit rate (GBR) and nonguaranteed bit rate (non-GBR). GBR bearers define a guaranteed minimum bit rate that will be supported for all IP traffic flows aggregated onto a single EPS bearer. For non-GBR bearers, no such guarantee is made. GBR bearers are general used for real-time service such as VoIP, while non-GBR bearers

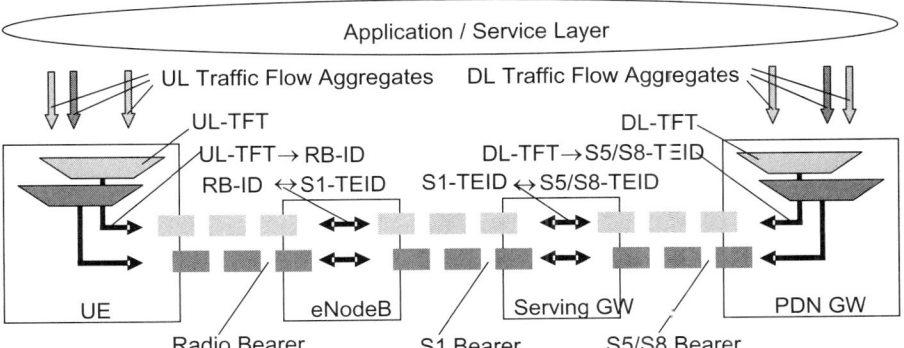

Figure 8.9 Example mapping of traffic flows to two unicast EPS bearers. Reprinted from Reference [194].

Table 8.4   Standardized QCI Characteristics for LTE [193]

QCI	Resource type	Priority	Packet delay budget (ms)	Packet loss rate	Example services
1		2	100	10^{-2}	Conversational voice (VoIP)
2		4	150	10^{-3}	Conversational video (video chat)
3	GBR	3	50	10^{-3}	Real-time gaming
4		5	300	10^{-6}	Nonconversational video (buffered streaming)
5		1	100	10^{-6}	IMS signaling
6		6	300	10^{-6}	Video (buffered streaming), TCP-based services
7	Non-GBR	7	100	10^{-3}	Interactive gaming, video (live streaming)
8		8	300	10^{-6}	TCP-based services (e.g.,
9		9	300	10^{-6}	Web browsing, email, FTP)

FTP, File Transfer Protocol.

may be used for non-real-time services such as Web browsing or file transfer. For GBR bearers, standard also defines a maximum bit rate (MBR) parameter that defines the maximum bit rate that may be supported on a GBR bearer when additional resources are available in E-UTRAN. However, for 3GPP Rel-8, the MBR is always equal to the GBR. In this sense, the GBR parameter not only defines the GBR, but also places a cap on the MBR achievable for that bearer.

Regardless of GBR/non-GBR status, each bearer is characterized by two additional QoS attributes: the QCI and the allocation and retention priority (ARP). The QCI defines attributes for three parameters: priority, packet delay budget, and packet loss rate. Furthermore, the QCI indicates whether a bearer is GBR or non-GBR. Nine standard QCI indexes are defined in the standard, as summarized in Table 8.4. The QCI provides a simple index that can be signaled between network nodes at bearer establishment to configure the node-specific parameters that control bearer-level packet forwarding treatment (e.g., scheduling weights, admission thresholds, queue management thresholds, link layer protocol configuration) [194]. This simplifies the QoS signaling between nodes, as the node-specific parameters are preconfigured for a given QCI and do not need to be explicitly signaled. The primary purpose of the ARP is to decide whether a bearer establishment or modification request needs to be rejected due to resource limitations during times of network congestion. The ARP does not affect packet forwarding parameters after the bearer is established, and the ARP is never signaled to the UE.

Always-on IP connectivity is a key characteristic of LTE. To support this feature, one EPS bearer is established when the UE first connects to a PDN. This EPS bearer is called the *default bearer*, and it remains established for as long as the

UE is connected to that PDN. The QoS parameters of the default bearer are assigned by the network based on subscription data stored in the HSS. The default bearer is always a non-GBR bearer. Additional bearers may be established to the same PDN to support different QoS. These bearers are called *dedicated bearers*. The distinction between default and dedicated bearers is transparent to E-UTRAN. The decision to establish or modify a dedicated bearer is always taken by EPC, never initiated by E-UTRAN, and the associated QoS parameters are signaled from the EPC. A direct consequence of the default bearer establishment procedure is that the UE is assigned an IP address immediately when it attaches to the network. This is a requirement for receiving service in the EPS, since it is all-IP and does not support a CS domain.

The bearer model and QoS concept used in the EPS is similar to that defined previously for UMTS, however a number of key improvements have been made. For example, the bearer model has been simplified and uses fewer layers than UMTS to avoid redundancy. Furthermore, the responsibility for deciding QoS parameters for a bearer now resides in the network. The UE may simply request a specific type of service and the EPC configures the appropriate QoS parameters to assign the required packet transfer characteristics to each network node. This simplifies the QoS configuration process as compared to UMTS, where the UE was responsible for setting the appropriate QoS attributes, resulting in a user-friendlier QoS concept for LTE.

8.1.10 Identifiers and Addresses

Many of the identifiers and addresses used in LTE are similar or identical to those of its 3GPP predecessors. This is to be expected, as the EPC architecture makes use of the HSS defined in the legacy 3GPP core network. The international mobile subscriber identity (IMSI) and IMEI identifiers have been preserved unchanged to identify the UE and ME, respectively. The PLMN identity, which comprises a 3-bit mobile country code (MCC) and a 2- or 3-bit mobile network code (MNC), has also been preserved and is incorporated into a number of new identifiers in LTE. A detailed description of these identifiers may be found in the GSM section of Chapter 6. Other legacy identifiers, such as the Mobile Subscriber ISDN Number (MSISDN) and Mobile Station Roaming Number (MSRN), are no longer used since there is no equivalent to the CS domain in LTE. The following is a list of key new identifiers and addresses defined in LTE [197], some of which are analogous to existing identifiers from UMTS.

- *Globally Unique MME Identifier (GUMMEI).* A permanent, unambiguous MME identifier formed by the concatenation of the PLMN identity and a 24-bit MME identifier that uniquely identifies the MME within the PLMN.

- *MME Temporary Mobile Subscriber Identity (M-TMSI).* A 32-bit temporary identity assigned to the UE by the MME for subscriber confidentiality. The M-TMSI is used for local identification of the UE, and is only unique when associated with a particular MME. The M-TMSI is analogous to the

Temporary Mobile Subscriber Identity (TMSI) or Packet TMSI (P-TMSI) of legacy 3GPP systems.

- *Globally Unique Temporary Identity (GUTI)*. A temporary UE identity formed by the concatenation of the GUMMEI and the M-TMSI, which unambiguously identifies the UE within the EPS. The GUTI provides both a temporary UE identity and also indicates where the identity was assigned. It conveys the information of both the P-TMSI and the RAI from legacy 3GPP systems.

- *SAE Temporary Mobile Subscriber Identity (S-TMSI)*. A shortened form of the GUTI that is used for more efficient radio signaling when the full GUMMEI is not required (e.g., for paging). It is a 40-bit identity formed from the concatenation of the last 8 bits of the GUMMEI and the M-TMSI. The key difference compared to the GUTI is that the S-TMSI uniquely identifies the UE within a specific MME pool area only.

- *Cell Radio Network Temporary Identity (C-RNTI)*. A 16-bit temporary UE identity assigned by the eNB to identify the UE's radio connection within a specific cell. It is used at the air interface to identify the radio resource control (RRC) connection of the UE for scheduling of uplink and downlink radio resources. The C-RNTI is used within E-UTRAN only and is not visible to the EPC.

- *Tracking Area Identity (TAI)*. An unambiguous, global TA identity formed by the concatenation of the PLMN identity (i.e., MCC + MNC) with a 16-bit tracking area code (TAC) that uniquely identifies the TA within the PLMN. The TAI is analogous to the RAI from legacy 3GPP systems.

- *E-UTRAN Cell Global Identifier (ECGI)*. Each cell in E-UTRAN is identified by an ECGI, formed by the concatenation of the PLMN identity and a 28-bit E-UTRAN cell identifier (ECI). The ECGI and ECI are fully analogous to the CGI and CI from legacy 3GPP systems.

Although not new to LTE, it is worth noting that the IP address is a critical UE identifier as it is the primary means of addressing the UE from outside the EPS since the MSISDN is no longer used. An IP address must be allocated to the UE before any service can be provided. IP address allocation is performed when the default bearer is established. Additional IP addresses may be subsequently allocated if UE connectivity is established with multiple PDNs. Therefore, the UE always maintains at least one IP address while attached to the LTE network. Although LTE supports both Internet Protocol version 4 (IPv4) and Internet Protocol version 6 (IPv6) addresses, it is expected that IPv6 addressing will be most common due to the large address space required in an all-IP system.

8.1.11   Protocol Architecture

The EPS user plane architecture is shown in Figure 8.10 [194] and extends from the UE to the P-GW. The protocols that are within the scope of the 3GPP are highlighted

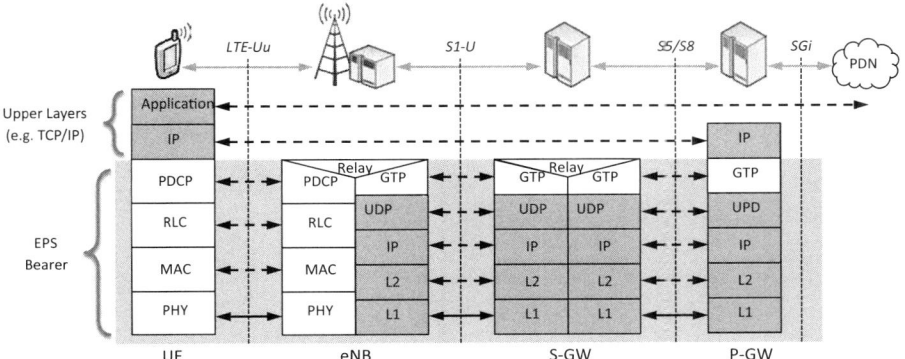

Figure 8.10   LTE user plane architecture. Reprinted from Reference [194].

in white, while those that are not are shaded in gray. Transmission of IP packets between the UE and E-UTRAN occurs at the LTE-Uu interface and utilizes the LTE layer 1 (i.e., LTE PHY) and layer 2 protocols. The LTE layer 2 is divided into three sublayers: medium access control (MAC), radio link control (RLC), and protocol dependent convergence protocol (PDCP). The MAC layer controls scheduling procedures for the radio channel, as well as HARQ functionality. The RLC layer provides segmentation/reassembly and logical link control over the air interface, which may include multiple simultaneous RLC links per UE. The PDCP layer provides protocol transparency to the higher layers, including in-sequence delivery of IP packets. It also performs robust header compression (ROHC) and ciphering of user data. Within the fixed network infrastructure, user data are tunneled over the S1 and S5/S8 interfaces using GTP tunnels.[5] The well-known GTP-UDP-IP stack is used, which was previously used in the GPRS PS domain of the GSM and UMTS core networks. A key difference is that a new version, GTPv2, has been defined in Rel-8 to support improved performance in the EPS. As an example, GTPv2 supports piggybacking of certain GTP control (GTP-C) messages to minimize connection setup time and user plane latency.

The EPS control plane architecture is shown in Figure 8.11 [194]. At the LTE air interface, the division of protocols into the access stratum (AS) and the nonaccess stratum (NAS) has been used as defined previously for UMTS. The AS includes those protocols that handle radio-specific functionality. The layer 1 and layer 2 protocols are identical to those of the user plane described previously, though certain functions such as ROHC are not required in the control plane. As an added security feature, an integrity protection function has been added to PDCP for control plane signaling. Additionally, the RRC protocol is included in the AS. The RRC is the primary protocol in charge of radio bearer setup and control plane signaling, with

---

[5] GTP-based S5/S8 is assumed. For PMIP-based S5/S8, a PMIP-over-IP protocol stack is used. The interested reader is directed to Reference [195] for further details.

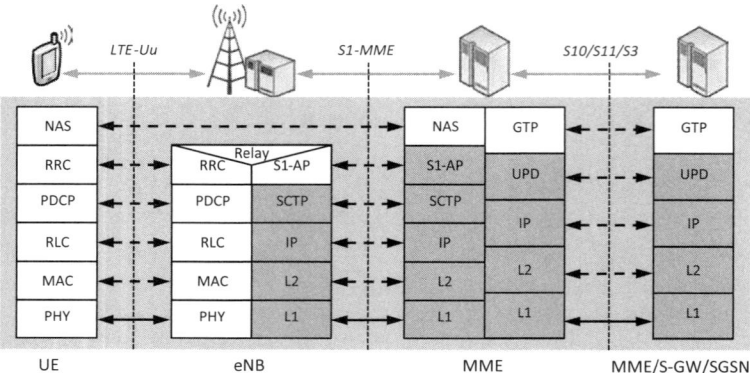

Figure 8.11 LTE control plane architecture. Adapted from Reference [194].

messages carried by signaling radio bearers (SRBs) over the air interface. The RRC forms layer 3 of the control plane architecture. Above the AS protocols is the NAS layer, which includes EPS mobility management (EMM) and EPS session management (ESM). The NAS layer terminates in the MME on the network side, and NAS messages are transported transparently through the E-UTRAN. The full control plane protocol architecture between all network entities is defined in Reference [194]. Many interfaces, including S10, S11, and S3, are based on the GTP-C part protocol as shown in Figure 8.11. In contrast, signaling between the MME and the HSS (S6a), or the MME and the EIR (S13), is based on the Diameter signaling protocol.

At the air interface, the protocol architectures of LTE and UMTS are similar, except that the PDCP sublayer is now used in both the user plane and control plane protocol stacks in LTE. However, a key distinction between LTE and UMTS protocols themselves must be made. While the names are the same, the LTE air interface protocols have been optimized for packet-switched operation only and are, therefore, not backward compatible with UMTS. In fact, this ability to optimize without the requirement of backward compatibility is one of the key advantages of defining LTE as a new standard rather than a part of the open evolution of UMTS. Furthermore, with all RAN functionality collapsed into a single node—the eNB—all of the radio protocols are terminated on the network side in the base station, as no separate RNC is defined.

8.1.12 Channel Architecture

Data flow between layers at the LTE air interface is represented by the concept of channels. The use of the word channel in this context may prove confusing to some readers. In this context, the term "channel" refers to a certain type of information that is transmitted between two peer protocol layers of communicating radios (i.e., the eNB and the UE). This is in contrast to the traditional meaning of the word in the communications discipline as the medium through which transmission occurs. The LTE channel architecture defines three categories of channels:

- *Logical Channels.* These carry signaling and user data between peer RLC entities in the UE and the eNB. The type of logical channel identifies *what type of information* is being transferred. The logical channels form the SAPs between the RLC and the MAC.

- *Transport Channels.* These carry signaling and user data between peer MAC entities in the UE and the eNB. The type of transport channel identifies *how* and with what characteristics data are transferred over the radio interface. The transport channels form the service access points (SAPs) between the MAC and the PHY.

- *Physical Channels.* These carry signaling and user data over the physical radio link between two peer PHY layer entities. The physical channels form the interface between the PHY layer and the physical medium.

Furthermore, the radio bearers described previously form the SAPs between layer 3, either IP for user plane or RRC for control plane, and layer 2. The RLC is then responsible for mapping between radio bearers and logical channels. A taxonomy of LTE channels, including each of the three types defined earlier, is provided in Table 8.5. The channel mapping between each channel type in the downlink and the uplink is illustrated in Figure 8.12 and Figure 8.13, respectively [198]. Each channel is briefly defined in the following sections.

8.1.12.1 *Logical Channels*

The classification of logical channels is divided into two groups based on the type of information transferred. Control channels are used to carry control plane signaling and traffic channels are used to carry user plane data. The logical channels are summarized as follows [192]:

- *Broadcast Control Channel (BCCH).* This carries broadcast system information (e.g., downlink transmission bandwidth, system frame number (SFN), PLMN identities, cell reselection information) in the form of the master information block (MIB) and a number of system information blocks (SIBs) (downlink only).

- *Paging Control Channel (PCCH).* This is used to page mobiles for incoming (i.e., mobile terminated) data transfers when the UE is in Idle mode and the exact location is not known to the network (downlink only).

- *Common Control Channel (CCCH).* This carries common control information between UEs and the network before an RRC connection is established.

- *Dedicated Control Channel (DCCH).* Carries dedicated (i.e., point-to-point) control information between a single UE and the network after an RRC connection has been established.

Table 8.5   Taxonomy of LTE Channels

Channel type	Channel	Full name	Direction	Control plane	User plane
Logical	BCCH	Broadcast control channel	eNB → UE	▓	
	CCCH	Common control channel	eNB ↔ UE	▓	
	DCCH	Dedicated control channel	eNB ↔ UE	▓	
	DTCH	Dedicated traffic channel	eNB ↔ UE		▓
	MCCH	Multicast control channel	eNB → UE	▓	
	MTCH	Multicast traffic channel	eNB → UE		▓
	PCCH	Paging control channel	eNB → UE	▓	
Transport	BCH	Broadcast channel	eNB → UE	▓	▓
	DL-SCH	Downlink shared channel	eNB → UE	▓	▓
	MCH	Multicast channel	eNB → UE	▓	▓
	PCH	Paging channel	eNB → UE	▓	▓
	RACH	Random access channel	eNB ← UE	▓	▓
	UL-SCH	Uplink shared channel	eNB ← UE	▓	▓
Physical	PBCH	Physical broadcast channel	eNB → UE	▓	▓
	PDSCH	Physical downlink shared channel	eNB → UE	▓	▓
	PMCH	Physical multicast channel	eNB → UE	▓	▓
	PRACH	Physical random access channel	eNB ← UE	▓	▓
	PUSCH	Physical uplink shared channel	eNB ← UE	▓	▓

Shaded areas indicate whether channels support the control plane, data plane, or both.

- *Multicast Control Channel (MCCH).* Carries broadcast (i.e., point-to-multipoint) control information for MBMS (downlink only).
- *Dedicated Traffic Channel (DTCH).* The primary logical channel in the user plane, it is a dedicated (i.e., point-to-point) channel used for the transfer of user data to a single UE. A DTCH may exist simultaneously in both the downlink and the uplink.

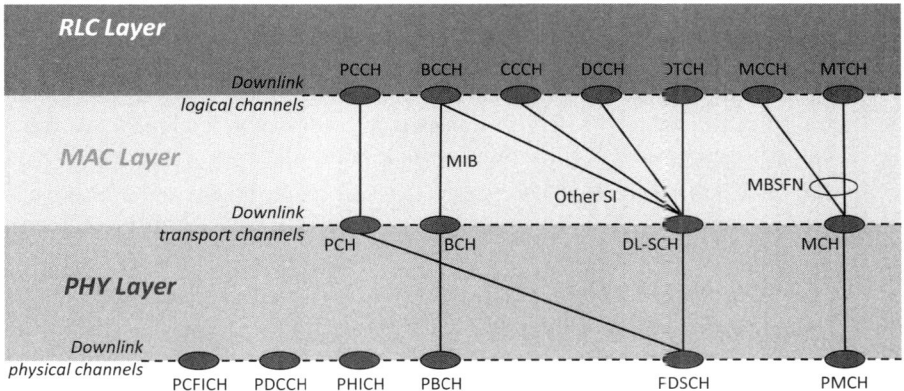

Figure 8.12   LTE downlink channel mapping. Reprinted from Reference [198].

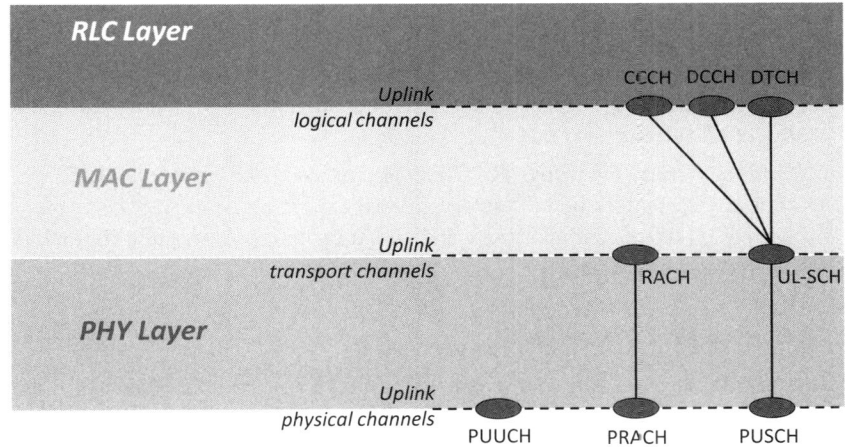

Figure 8.13   LTE uplink channel mapping. Reprinted from Reference [198].

- *Multicast Traffic Channel (MTCH).* A point-to-multipoint channel for MBMS traffic (downlink only).

8.1.12.2 *Transport Channels*

In contrast to logical channels, transport channels are always unidirectional and may be classified as either downlink or uplink transport channels. Multiple logical channels may be mapped to a single transport channel, meaning that the same transport characteristics are applied by the PHY layer information transfer service. A single transport channel may apply to a combination of control and traffic channels, as is the case for the DL-SCH and UL-SCH. The transport channels are summarized as follows [192]:

- *Broadcast Channel (BCH).* This is characterized by a fixed, predefined transport format used solely to transport the MIB; must be broadcast in the entire coverage area of the cell (downlink only).

- *Paging Channel (PCH).* This is used to transport the PCCH and is characterized by support for UE discontinuous reception (DRX); must be broadcast in the entire coverage area of the cell (downlink only).

- *Downlink Shared Channel (DL-SCH).* This is used to transport a number of logical channels including the CCCH, DCCH, DTCH, and the SIBs of the BCCH. Characterized by support for HARQ, link adaptation, dynamic and semistatic resource allocation, support for DRX, and the possibility to use MIMO techniques such as spatial multiplexing and beamforming (downlink only).

- *Multicast Channel (MCH).* This is used to transport MBMS multicast services; must be broadcast in the entire coverage area of the cell (downlink only).

- *Uplink Shared Channel (UL-SCH).* This is used to transport uplink signaling and traffic, including support for both dynamic and semistatic resource allocation (uplink only).

- *Random Access Channel (RACH).* This is used for uplink random access; random access may be contention free (e.g., during handover) or contention based (e.g., during initial access, tracking area update, or mobile-initiated data transfer) (uplink only).

8.1.12.3 Physical Channels

A common transmission approach has been employed at the LTE PHY layer, which minimizes the number of physical channels. A shared channel is defined in each of the downlink and the uplink. It is clear from Figure 8.12 and Figure 8.13 that nearly all logical channels are carried over the physical radio link by means of these shared channels, with the exception of broadcast and multicast channels. This approach provides flexibility to dynamically adapt allocation of physical resources between signaling and user traffic based on current cell load. Furthermore, a number of physical channels are defined solely to support the operation of the LTE PHY layer, which does not carry transport channels. Physical channels are always unidirectional and may be classified as either downlink or uplink physical channels. The physical channels are summarized as follows [199]:

- *Physical Broadcast Channel (PBCH).* This carries cell-specific system information from the BCH, referred to as the MIB, which includes the downlink transmission bandwidth, physical hybrid automatic repeat request (ARQ) indicator channel (PHICH) configuration, and SFN. The MIB is transmitted over the span of four consecutive frames for a TTI of 40 ms (downlink only).

- *Physical Downlink Shared Channel (PDSCH).* This carries control and user traffic from the DL-SCH and PCH, and comprises the majority of the physical time/frequency resources (downlink only).

- *Physical Downlink Control Channel (PDCCH).* This carries downlink control information (DCI) messages, which may include radio resource assignments and other control information for a UE or a group of UEs. Many PDCCHs may be transmitted in a subframe (downlink only).

- *Physical Control Format Indicator Channel (PCFICH).* This carries the control format indicator (CFI), which indicates the number of OFDM symbols used for transmission of downlink control channel information in a given subframe (downlink only).

- *Physical Hybrid ARQ Indicator Channel (PHICH).* This carries the HARQ acknowledgement/negative acknowledgement (ACK/NACK) messages to the UE in response to messages sent on the UL-SCH. This message is called the HARQ indicator (HI) (downlink only).

- *Physical Multicast Channel (PMCH).* This carries the MCH transport channel for MBMS services (downlink only).

- *Physical Uplink Control Channel (PUCCH).* This carries uplink control information (UCI) messages when no physical uplink shared channel (PUSCH) resources are allocated to the UE. To minimize the amount of radio resources required for signaling overhead, PUCCHs are code division multiplexed and can be shared by multiple UEs (uplink only).

- *Physical Uplink Shared Channel (PUSCH).* This carries dedicated traffic and control data, including UCI messages (uplink only).

- *Physical Random Access Channel (PRACH).* This is used to support random access procedures by through the transmission of RACH preambles. Unlike UMTS, the PRACH does not carry any user data or signaling in LTE, but only predefined preamble sequences (uplink only).

8.1.13 LTE Air Interface Overview

3GPP LTE defines a new cellular air interface that uses OFDMA in the downlink and SC-FDMA in the uplink. The LTE radio link is essentially divided into three dimensions: frequency, time, and space. This division is illustrated in Figure 8.14. The frequency dimension is divided into subcarriers with a 15 kHz spacing in normal operation. Spectrum flexibility allows for scalable total channel bandwidth from 1.4 to 20 MHz, a value which is generally fixed for a given network deployment. The time dimension is organized into 10-ms radio frames, which are further subdivided into 1-ms subframes and 0.5-ms slots. Slots are further subdivided into OFDM symbols[6] of duration 66.7 μs. A cyclic prefix (CP) is inserted between OFDM

---

[6] The 3GPP standard terminology refers to *OFDMA symbols* in the downlink and *SC-FDMA symbols* in the uplink. Here we use the general term *OFDM symbol* for simplicity when such a distinction is not required.

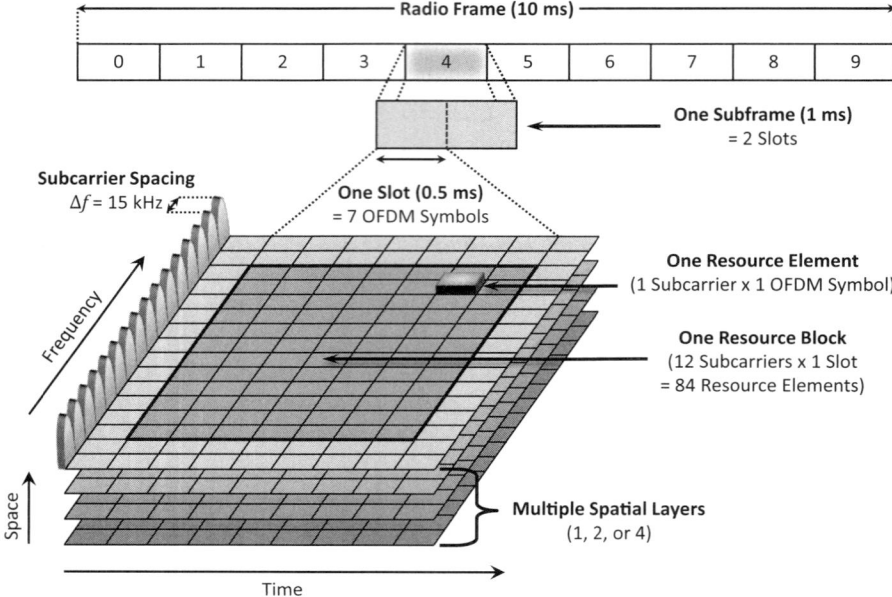

Figure 8.14 LTE radio resources divided into frequency, time, and spatial dimensions (normal cyclic prefix case).

symbols, effectively forming a guard period. LTE supports two CP durations, referred to as the normal CP and extended CP. The number of OFDM symbols per slot is 7 for the normal CP case or 6 for the extended CP case. The combination of the frequency and time division of the radio resources results in a time–frequency grid. The fundamental unit of this grid, comprising one subcarrier for one OFDM symbol duration, is called a *resource element* (RE). For the purpose of resource allocation, the REs are grouped into so-called *resource blocks* (RBs) that comprise 12 subcarriers for one 0.5-ms slot duration. RBs are allocated in pairs, resulting in a minimum TTI of 1 ms. The use of multiple transmit antennas in the downlink effectively creates a third spatial dimension, which is accessed by means of logical antenna ports. One resource grid is defined per antenna port, and each antenna port has its own reference signals (i.e., pilots) that are used for channel estimation and coherent demodulation at the receiver [200]. By providing radio access that can be dynamically allocated in frequency, time, and space, the transmission scheme is highly adaptive to changes in the state of the mobile radio channel, resulting in highly efficient use of the radio resources that approaches information-theoretic limits for data throughput. Key attributes of the LTE air interface are summarized in Table 8.6 and elaborated further in the following sections.

8.1.13.1 OFDMA versus SC-FDMA

The LTE air interface, like many modern wireless communication systems, is based on OFDM. In the downlink, LTE uses OFDMA, which is the adaptation of OFDM

Table 8.6   LTE Air Interface Key Attributes

Attribute	Description
Downlink access scheme	OFDMA
Uplink access scheme	SC-FDMA (DFTS-OFDM)
Duplex scheme	FDD or TDD
Channel bandwidth	1.4, 3, 5, 10, 15, or 20 MHz[*]
Subcarrier spacing	15 kHz
Radio frame length	10 ms—divided into 1 ms subframes
Minimum transmission time interval	1 ms (one subframe)
OFDM symbol length	66.7 μs + cyclic prefix length
Cyclic prefix length	5.2 μs/4.7 μs (normal)[**], 16.7 μs (extended)
Modulation	BPSK[***], QPSK, 16-QAM, 64-QAM
MIMO techniques	Spatial diversity (SFBC)
	Spatial multiplexing (SU-MIMO and MU-MIMO)
	Beamforming

[*]Fixed value for a particular network deployment, not dynamically reconfigurable.
[**]For normal CP case, a CP length of 5.2 μs is used only on the first OFDM symbol of each slot.
[***]BPSK modulation used for PHICH and PUCCH control channels only.

to a multiuser scheme. While the basic principles of OFDM have been known since the 1950s, the concept of OFDMA for cellular applications was first published by Reiners and Rohling in 1994 [201]. The fundamental concepts of OFDM and OFDMA, including the definition of cyclic prefix (CP) and baseband signal generation, are addressed in greater detail in Chapter 10. OFDMA has also been discussed previously in the context of Mobile WiMAX in Chapter 7. However, the specific way in which OFDMA is implemented in LTE in terms of parameterization (i.e., subcarrier spacing, cyclic prefix duration), physical channelization, and user resource allocation is different than in Mobile WiMAX, as we shall see in the following sections.

The LTE uplink uses a multiple access scheme called SC-FDMA, also referred to as DFT-spread OFDM (DFTS-OFDM). With early publications originating from the late 1990s [202], SC-FDMA is a more recent technology that is currently unique to LTE—as compared to the other wireless standards considered in this book—and warrants further discussion. It is important to understand the motivation for the use of SC-FDMA in the LTE uplink. OFDM in its fundamental form is sometimes viewed as unsuitable for use as an uplink transmission scheme because of its large PAPR and vulnerability to phase noise. SC-FDMA provides a single-carrier alternative, which is in many ways similar to OFDMA, but has a lower PAPR—or, more accurately, a lower cubic metric—which results in good power conversion efficiency in the power amplifier of the UE transmitter, which lowers the power consumption in the UE. The lower PAPR also reduces the linearity requirement of the power

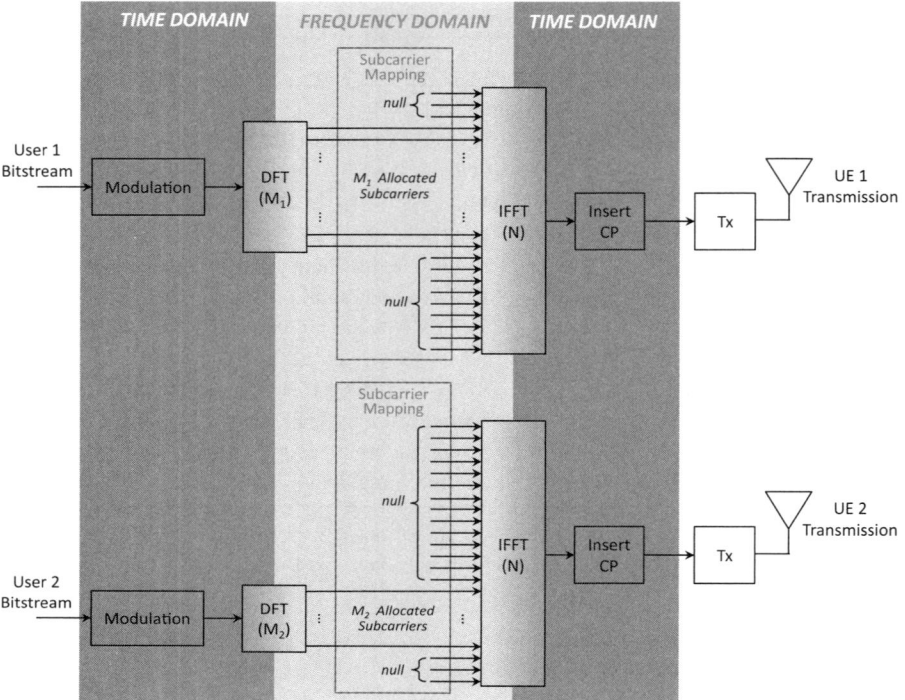

Figure 8.15   Example of an SC-FDMA transmitter block diagram (two UEs shown).

amplifier, thereby reducing the overall UE cost. One drawback to SC-FDMA is that it results in a more complex receiver design than that of OFDMA. This is due in part to the increase in equalizer complexity that results from the fact that there is essentially one CP for a block of SC-FDMA symbols, resulting in intersymbol interference (ISI) within one OFDM symbol duration of the time–frequency resource grid. However, this characteristic is well suited to an uplink technology, where additional complexity at the eNB in exchange for decreased cost and complexity at the UE is considered an acceptable trade-off.

An example of an SC-FDMA transmitter block diagram structure is shown in Figure 8.15 for two UEs. In this example, we assume that each UE has been allocated a certain number of subcarriers for transmission, denoted as M_1 for UE 1 and M_2 for UE 2. For the sake of discussion, we focus on User 1's transmission from UE 1. The generation of one SC-FDMA symbol begins with the modulation of M_1 data symbols in the time domain. Constellation mapping is performed according to the modulation scheme, for example, 2 bits per symbol for quadrature phase-shift keying (QPSK), 4 bits per symbol for 16-point quadrature amplitude modulation (16-QAM), and so on. Next, M_1 consecutive complex modulation symbols are collected and an M_1-point DFT is applied (i.e., DFT spreading). This operation converts the complex modulated data stream into the frequency domain, resulting in M_1 complex symbols each with its own amplitude and phase. From this point onward, signal generation is essentially performed in the same manner as OFDMA. The complex symbols of the DFT output are mapped to the M_1 subcarriers that have been allocated to UE 1 by the eNB for uplink transmission. In theory, these subcarriers could be localized (i.e., consecutive subcarriers) or distributed (i.e., nonconsecutive subcarriers). In E-UTRA, only localized subcarrier mapping is used for uplink data and control transmissions on the PUSCH and PUCCH. All subcarriers that are not allocated to UE 1 are null subcarriers and a zero value is assigned. Some of these subcarriers may correspond to transmission by other UEs, in this example UE 2. After subcarrier mapping, an N-point inverse fast Fourier transform (IFFT) is applied for conversion to the time domain, where N is dictated by the overall channel bandwidth of the E-UTRA carrier. Finally, a CP is added to the 66.7 μs SC-FDMA symbol, which is then sent for digital-to-analog conversion and radio frequency (RF) transmission.

At the eNB, the inverse process occurs for SC-FDMA reception. An example of an SC-FDMA receiver block diagram structure is shown in Figure 8.16. After the fast Fourier transform (FFT) block, the subcarriers corresponding to each UE are separated and frequency-domain equalization is applied. Next, an IDFT is applied (i.e., IDFT despreading) to each individual UE allocation to convert back to the time domain. For UE 1, the original M_1 modulated symbols have been recovered—with channel impairments—and demodulation may be performed to produce an estimate of the bit stream transmitted by User 1. Throughout the process, one key point that is important to notice is that there is always a one-to-one relationship between the number of modulated symbols (i.e., QPSK, 16-QAM) to be transmitted during the SC-FDMA symbol period and the number of subcarriers allocated for user transmission. Both of these values are always equal to the size of the initial DFT used for spreading—namely, M_1 in this example. Therefore, the DFT spreading size may vary

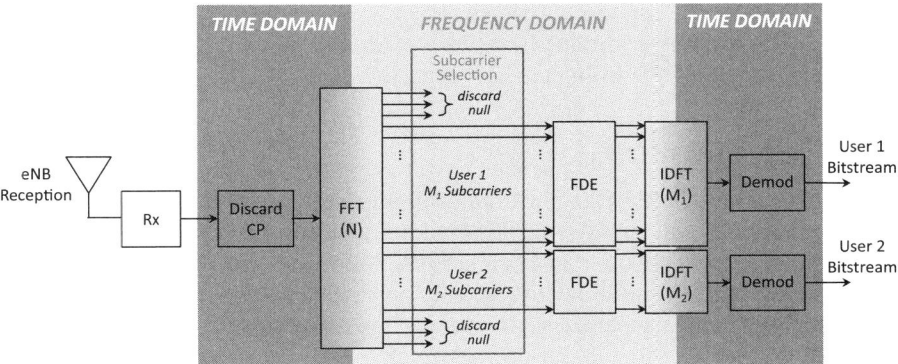

Figure 8.16   Example of an SC-FDMA receiver block diagram with frequency-domain equalization (FDE).

between users. In contrast, the IFFT/FFT size—namely, N in this example—used to produce the SC-FDMA symbols is a function of the E-UTRA carrier bandwidth and remains constant for all users at all times, just as in OFDMA transmission on the downlink.

Conceptually, SC-FDMA in this form can be thought of as a special form of single-carrier modulation that uses OFDM-like frequency-domain signal generation. The result is essentially a hybrid of single-carrier and OFDM schemes, which shares certain characteristics of both. From the single-carrier perspective, if we start with QPSK, then the transmitter output is equivalent to QPSK as a direct consequence of the DFT/IFFT combination. The addition of the subcarrier mapping is equivalent to a frequency shift, and the CP insertion is equivalent to adding a guard between each block of M symbols—but these operations do not change the fundamental properties of QPSK. The equivalent QPSK modulation scheme has a symbol period of $(1/M) \cdot 66.7$ μs[7] and span a bandwidth of $M \cdot 15$ kHz. Therefore, the equivalent QPSK data rate is variable according to the number of subcarriers in the UE allocation. However, the frequency-domain generation using DFT spreading imparts some of the benefits of an OFDM scheme that are not achieved by an equivalent time-domain-generated QPSK modulation scheme. In particular, the output of the DFT produces a single complex value per subcarrier, which remains constant for the full duration of the SC-FDMA symbol. For this reason, SC-FDMA is also referred to DFTS-OFDM. From the DFTS-OFDM perspective, the modulation scheme shares some of the advantages of OFDMA, such as long symbol durations and robustness to multipath fading. The OFDM-like frequency-domain signal generation using orthogonal subcarriers also means that guard bands are not required between RBs allocated to different users. This is an important characteristic of SC-FDMA that distinguishes it from other single-carrier multiple access schemes, such as time division multiple access (TDMA).

The similarity of SC-FDMA to OFDMA has the advantage that a fundamentally similar radio resource structure may be used in both the downlink and the uplink. The basic radio resource structure shown in Figure 8.14 applies equally to the SC-FDMA uplink as it does to the OFDMA downlink. The key exception is that Rel-8 LTE does not support multiple antenna transmission from a single UE in the uplink, which restricts the resource mapping to a single antenna port (i.e., one time–frequency resource grid) from the UE perspective.

8.1.13.2 Duplex Modes

In order to provide increased bandwidth flexibility to accommodate various deployment scenarios, LTE defines two duplex modes: frequency division duplexing (FDD) and time division duplexing (TDD). These duplex modes are illustrated in Figure 8.17. In FDD mode, the downlink and uplink transmit on separate frequencies. This approach requires paired spectrum, meaning that two frequency bands are required: one for uplink and one for downlink. In TDD mode, the downlink and

---

[7] Recall that one SC-FDMA symbol has a fixed duration of 66.7 μs prior to CP insertion.

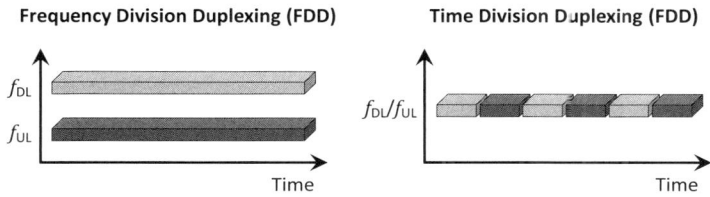

Figure 8.17   LTE FDD and TDD duplex modes.

uplink transmit on the same frequency but are nonoverlapping in time. The main advantage of TDD mode is that it may be deployed in unpaired spectrum, thus allowing the entire available bandwidth to be allocated to downlink or uplink transmission within a given subframe. This mode also offers a smooth evolutionary pathway from TDD-based 3G cellular technologies, including Mobile WiMAX and the UMTS-TDD variant based on the time division–synchronous code division multiple access (TD-SCDMA) air interface.

FDD mode is currently the most commonly deployed mode of operation for LTE and is anticipated to remain so for the foreseeable future. However, it is worth noting that a number of network operators have either deployed or begun trails for TDD mode—also referred to as *TD-LTE*—in various regions around the world. The protocols and procedures for FDD and TDD modes have been designed to be as similar as possible to encourage the development of multimode devices capable of operating on either type of network. This approach will also help TDD operators benefit from the same economies of scale enjoyed by the FDD market.

In addition to full-duplex FDD mode, LTE also supports a half-duplex FDD mode (HD-FDD) from the perspective of the UE. For a UE operating in HD-FDD mode, downlink and uplink transmissions do not occur simultaneously. The term half-duplex here applies only to the UE, as the base station must still support simultaneous downlink and uplink transmission for multiple UEs. The main benefit of this mode is that it reduces the complexity and cost of RF electronics by eliminating the need for duplex filters in the UE. This makes HD-FDD mode a potentially appealing option for applications in which cost is more important than data rate, such as embedded M2M modules.

8.1.13.3 *Frequency Domain Aspects*

Although the LTE air interface has high spectral efficiency, the peak data rates offered by LTE are achieved through wideband deployments of up to 20 MHz. As compared to 2G and 3G cellular technologies, this requires a large amount of spectrum which may not be possible in some regions, particularly for early deployments. In order to accommodate deployment in such regions and facilitate uptake in the short term until more spectrum can be made available (e.g., through refarming of existing 2G spectrum), the LTE air interface has been designed to provide a great deal of spectrum flexibility. This is done primarily in three ways. First, as mentioned previously, LTE defines support for deployment in both paired (FDD) and unpaired (TDD) spectra within a single RAT. Second, the LTE standard defines a large number of frequency bands to allow global deployment in a variety of different spectra. Third, LTE is designed to be bandwidth agnostic, allowing for channel bandwidths ranging from 1.4 to 20 MHz. We now describe these spectrum aspects of the LTE air interface in greater detail, including the bandwidth agnostic channel arrangement.

LTE Frequency Bands   Frequency bands supported for E-UTRA operation are summarized in Table 8.7 for paired spectrum (i.e., FDD mode) and Table 8.8 for unpaired spectrum (i.e., TDD mode) [203]. From a practical deployment perspective, the different LTE bands can generally be grouped into two categories: those that provide *coverage* and those that provide *capacity*. For coverage, lower frequencies are required for their propagation characteristics, which accommodate larger cell sizes and better building penetration for indoor coverage. For capacity, higher frequencies are generally used, as these frequency bands tend to include wider spectra and less spectrum fragmentation between different network operators. A common paradigm for a network operator seeking to provide comprehensive nationwide coverage is to use one band (e.g., 700 MHz) for broad coverage at a moderate data rate and a second band (e.g., 2600 MHz) for capacity/high data rates in dense urban areas where small cell sizes and indoor relays can be utilized. Deployment trends related to specific operating bands will be discussed further in Section 8.1.17 later in this chapter.

The original specification of LTE in 3GPP Rel-8 includes 15 FDD bands and 8 TDD bands, however additional bands have been added with each new release as indicated in Table 8.7 and Table 8.8 [204–206]. In order to allow UEs conforming to a particular 3GPP release to support operation in a frequency band from a later release, requirements for UEs supporting so-called release-independent frequency bands have been retroactively added to the standard. Consider, for example, E-UTRA band 19, which has been added in 3GPP Rel-9. In order to allow a UE conforming to 3GPP Rel-8 to operate in Band 19, it is necessary for the UE to conform to certain parts of the Rel-9 specifications, such as the RF performance and RRM requirements specific to Band 19. These requirements are specified in Reference [207]. The result is that new bands can be added to the LTE standard as new spectra become available without confining their use to a specific 3GPP release.

Channel Arrangement   In order to support multiple channel bandwidths, a scalable channel arrangement is required. OFDMA and SC-FDMA are well suited to this requirement, allowing scalable channel bandwidths to be achieved by changing the number of orthogonal subcarriers without requiring a change to fundamental parameters such as the 15 kHz subcarrier spacing or cyclic prefix dimension. The channel arrangement for one E-UTRA carrier is illustrated in Figure 8.18 [203]. For this particular example, a 5 MHz channel bandwidth is shown. However, the overall structure applies to all channel bandwidth configurations. There are different types

Table 8.7 E-UTRA Frequency Bands for FDD Mode

Initial release	Band number	Common name	Uplink (MHz)	Downlink (MHz)	Total spectrum (MHz)	Duplex frequency (MHz)	Band gap (MHz)
Rel-8	1	IMT 2.1 GHz	1920–1980	2110–2170	2×60	190	130
	2	PCS 1900 MHz	1850–1910	1930–1990	2×60	80	20
	3	DCS 1800 MHz	1710–1785	1805–1880	2×75	95	20
	4	AWS	1710–1755	2110–2155	2×45	400	355
	5	850 MHz	824–849	869–894	2×25	45	20
	6	(Not applicable; UTRA only)	830–840	875–885	2×10	35	35
	7	IMT extension 2.6 GHz	2500–2570	2620–2690	2×70	120	50
	8	GSM 900 MHz	880–915	925–960	2×35	45	10
	9	1700 MHz	1749.9–1784.9	1844.9–1879.9	2×35	95	60
	10	Extended AWS	1710–1770	2110–2170	2×60	400	340
	11	1500 MHz lower	1427.9–1447.9	1475.9–1495.9	2×20	48	23
	12	700 MHz US digital dividend (lower blocks A + B + C)	699–716	729–746	2×18	30	30
	13	700 MHz US digital dividend (upper block C)	777–787	746–756	2×10	−31	21
	14	700 MHz US digital dividend (public safety)	788–798	758–768	2×10	−30	20
				. . .			
	17	700 MHz US digital dividend (lower blocks B + C)	704–716	734–746	2×12	30	18
Rel-9	18	Japan 800 MHz lower	815–830	860–875	2×15	45	30
	19	Japan 800 MHz upper	830–845	875–890	2×15	45	30
	20	800 MHz European digital dividend	832–862	791–821	2×30	−41	71
	21	1500 MHz upper	1447.9–1462.9	1495.9–1510.9	2×15	48	33

(*Continued*)

Table 8.7   (*Continued*)

Initial release	Band number	Common name	Uplink (MHz)	Downlink (MHz)	Total spectrum (MHz)	Duplex frequency (MHz)	Band gap (MHz)
Rel-10	22	3.5 GHz	3410–3490	3510–3590	2×90	100	10
	23	2 GHz S-Band	2000–2020	2180–2200	2×20	180	160
	24	L-Band	1626.5–1660.5	1525–1559	2×34	−101.5	135.5
	25	PCS 1900 + G block	1850–1915	1930–1995	2×65	80	15
Rel-11	26	Extended 850 MHz	814–849	859–894	2×35	45	10
	27	850 MHz lower	807–824	852–869	2×17	45	28
	28	700 MHz APAC	703–748	758–803	2×45	55	10

AWS, Advanced Wireless Services.

Table 8.8   E-UTRA Frequency Bands for TDD Mode

Initial release	Band number	Common name	Uplink/ downlink (MHz)	Total spectrum (MHz)
Rel-8	33	TDD 1900 MHz	1900–1920	20
	34	TDD 2 GHz	2010–2025	15
	35	TDD 1900 MHz (PCS lower)	1850–1910	60
	36	TDD 1900 MHz (PCS upper)	1930–1990	60
	37	TDD 1900 MHz (PCS band gap)	1910–1930	20
	38	TDD 2.6 GHz	2570–2620	50
	39	China TDD 1.9 GHz	1880–1920	40
	40	China TDD 2.3 GHz	2300–2400	100
Rel-10	41	TDD 2.5 GHz	2496–2690	194
	42	TDD 3.4 GHz	3400–3600	200
	43	TDD 3.6 GHz	3600–3800	200
Rel-11	44	TDD 700 MHz APAC	703–803	100

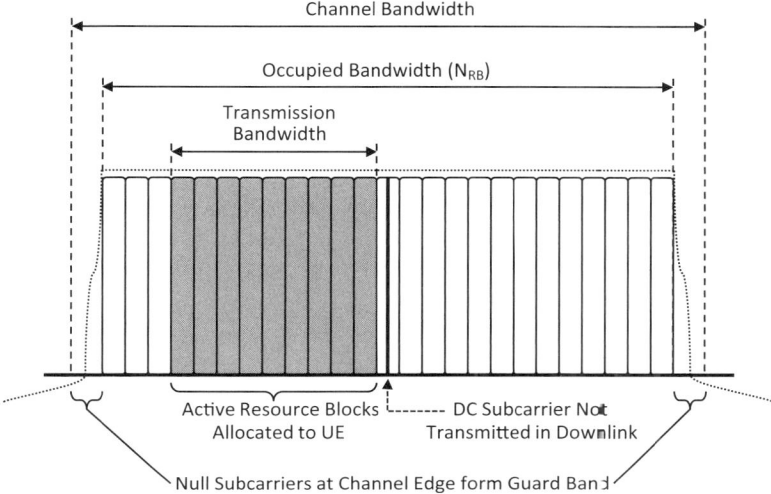

Figure 8.18   Channel arrangement for one E-UTRA carrier (example: 5 MHz channel bandwidth). Reprinted from Reference [203].

of subcarriers, which can generally be classified as data (i.e., user plane or control plane data), pilot, or null subcarriers. Null subcarriers are placed at the channel edges, forming guard bands to reduce interference with adjacent bands. Guard bands are generally a requirement in OFDM-based systems due to the large shoulders created by the sinc pulse shape of the subcarriers. In the downlink, a null subcarrier is also placed at the center subcarrier corresponding to DC at baseband. The DC subcarrier is not used because it may be subject to high interference, for example due to local oscillator leakage. In the uplink, the use of the DC subcarrier cannot be avoided due to the single-carrier nature of SC-FDMA. To minimize distortion effects, the uplink subcarriers are frequency shifted by one-half of a subcarrier spacing, such that two subcarriers straddle the DC subcarrier location. This results in a 7.5 kHz frequency offset in the uplink.

The remaining subcarriers which are not null form the occupied bandwidth of the E-UTRA carrier and are grouped into 180 kHz resource blocks (RBs) comprising 12 subcarriers each. Each channel bandwidth configuration occupies an integer number of RBs. The RBs include data subcarriers as well as pilot subcarriers, referred to in LTE as *reference signals*, which are distributed throughout the band. Unlike Mobile WiMAX, pilots are distributed in time as well as frequency—meaning they only occupy a specific subcarrier during certain OFDM symbol durations. The rest of the time, the same subcarriers may be used to carry data. The relationship between channel bandwidth, occupied bandwidth, number of RBs, and number of occupied subcarriers for each channel bandwidth configuration is summarized in Table 8.9.

Table 8.9   Summary of Channel Bandwidth Configurations

Channel bandwidth (MHz)	1.4	3	5	10	15	20
Occupied bandwidth (MHz)	1.08	2.7	4.5	9	13.5	18
Number of resource blocks (N_{RB})	6	15	25	50	75	100
Number of occupied subcarriers	72	180	300	600	900	1200
Nominal FFT size (N)	128	256	512	1024	1536	2048
Nominal sampling rate (MHz)	1.92	3.84	7.68	15.36	23.04	30.72
Relation of nominal sampling rate to UMTS chip rate (3.84 MHz)	0.5×	1×	2×	4×	6×	8×

Not all combinations of channel bandwidth configuration and frequency bands are supported by the standard. Supported combinations are summarized in Table 8.10. These combinations were defined based on input from the network operators regarding their intended network deployment configurations [188]. In general, wide channel bandwidth configurations are restricted by spectrum availability as well as other RF constraints. For example, the combination of a wide channel bandwidth configuration and a small band gap can result in UE self-interference. In some cases, the RF performance requirements of the UE may be relaxed to allow wider channel bandwidth configurations in certain bands. These bands are denoted by an asterisk (*) in Table 8.10 (from Reference [206]). Narrow channel bandwidth configurations, namely 1.4 and 3 MHz, are intended primarily for network migration pathways from legacy GSM and cdma2000 networks that use 200 kHz and 1.25 MHz channel bandwidths, respectively. These narrow channel bandwidth configurations also provide spectrum flexibility for deployment in regions where wider spectra are not available. To avoid spectrum fragmentation, these narrow channel bandwidth configurations are generally limited to narrow bands or bands that are expected to be used for 2G/3G migration scenarios. Key examples include bands 2, 3, 5, and 8—which correspond to the four main GSM bands (i.e., 1900, 1800, 850, and 900 MHz, respectively).

Frequency Reuse and Interference Coordination   LTE networks are designed to support a frequency reuse factor of 1 (i.e., full frequency reuse). This approach is spectrally efficient in the sense that it allows all of the available spectrum to be utilized by each cell with E-UTRAN. The concept of full frequency reuse was first introduced in 3G cellular networks, where orthogonal spreading codes are used to separate the channels transmitted by neighboring cells to minimize interference. In contrast to 3G systems, however, LTE does not utilize orthogonal spreading codes in this manner. Transmissions from neighboring cells may not be orthogonal and intercell interference may occur, resulting in a low signal-to-interference ratio (SIR). Consequently, alternate means must be used to mitigate the effects of intercell interference in E-UTRAN.

Table 8.10   Supported Channel Bandwidths by E-UTRA Frequency Band

E-UTRA band No.	Channel bandwidth (MHz)					
	1.4	3	5	10	15	20
1						
2					*	*
3					*	*
4						
5				*		
6				*		
7						*
8				*		
9					*	*
10						
11				*		
12			*	*		
13			*	*		
14			*	*		
17			*	*		
18				*	*	
19				*	*	
20				*	*	*
21				*	*	
22					*	*
23						
24						
25					*	*
26				*	*	
27				*		
28				*	*	*
33						
34						
35						
36						
37						
38						
39						
40						
41						
42						
43						
44						

*Denotes relaxed UE RF sensitivity requirements due to self-interference.

One method has been to design the physical control channels and physical signals to be inherently tolerant to high levels of interference, for example through the use of robust modulation and coding. A good example is the PCFICH, which utilizes rate 1/16 coding and QPSK modulation to transmit a 2-bit control field over 16 REs. The network can also be planned in such a way that some physical control channels and physical signals may be staggered in time and/or frequency between neighboring cells. This approach is commonly used for transmission of PBCH, reference signals (pilots), and synchronization signals in the E-UTRA downlink. Bit scrambling is also used at the E-UTRA PHY layer, which helps to distribute interference evenly by making neighbor cell transmissions appear more noise-like. Furthermore, multiple antenna transmit diversity using space frequency block coding (SFBC) may be used on the downlink to improve robustness in low SIR conditions.

Another key mechanism to mitigate intercell interference, especially with respect to cell-edge throughput performance, is the so-called *intercell interference coordination* (ICIC). The primary goal of ICIC is to keep interference levels in E-UTRAN at a manageable level through coordination of radio resource allocations within a localized geographic area [192]. ICIC is an inherently multicell RRM mechanism that is supported through standardized signaling over the optional X2 interface between eNBs. The standardized ICIC signaling messages can be generally divided into two categories: proactive and reactive. Proactive messages are sent by the eNB to its neighbors to announce how it plans to schedule users in the near future. Two proactive messages are defined. The uplink high interference indicator (HII) indicates which RBs will be allocated to high-interference-sensitivity users (i.e., cell-edge users) in the near future. The downlink relative narrowband transmit power (RNTP) serves a similar function for downlink ICIC by indicating whether downlink transmission power may exceed a certain threshold on a per-RB basis. In contrast, reactive messages are sent to neighbor eNBs to provide feedback on interference conditions based on past measurements. One reactive message is defined, which is the uplink interference overload indicator (OI) that indicates interference levels experienced on a per-RB basis [208].

Since there is no equivalent of a RNC in E-UTRAN, decentralized ICIC controlled by the eNBs is assumed. Although the 3GPP does not constrain implementation, ICIC schemes in Rel-8 and Rel-9 LTE are normally assumed to be frequency-based. A common approach is to divide the cell into two regions corresponding to the cell center and the cell edge. In general, the cell-center users will experience higher SIR and lower path loss, allowing a frequency reuse factor of 1 to be adopted. For the cell-edge users, a less aggressive frequency reuse scheme is adopted. Different schemes have been proposed for partitioning cell-center and cell-edge resources [209] as illustrated in Figure 8.19 [203]:

- *Fractional Frequency Reuse (FFR).* Certain RBs are reserved for cell-edge users such that these resources are nonoverlapping between neighboring cells, as shown in Figure 8.19c. This is essentially a hybrid approach where a frequency reuse factor 1 is used in the cell center and frequency reuse factor of 3 is used at the cell edge.

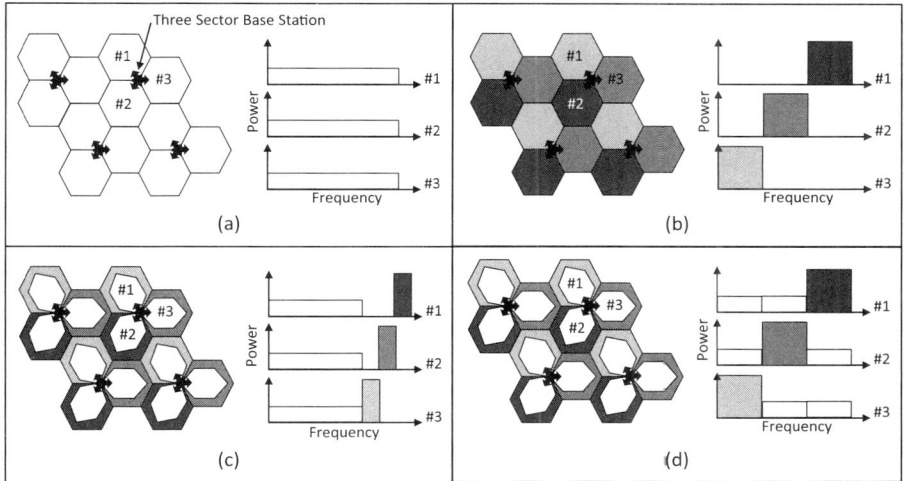

Figure 8.19   Example frequency reuse schemes in macrocellular network with trisector base stations. (a) Reuse factor of 1; (b) reuse factor of 3; (c) fractional frequency reuse (FFR); (d) soft frequency reuse (SFR). Reprinted from Reference [203].

- *Soft Frequency Reuse (SFR).* Power coordination techniques are used such that high-interference cell-edge users are scheduled on RBs corresponding to low-interference cell-center users in neighboring cells, as shown in Figure 8.19d.

Further time-based ICIC enhancements have been introduced in Rel-10 LTE-Advanced, which are specifically tailored to support heterogeneous network deployments, in which macrocells and small cells (e.g., operator-deployed picocells, user-deployed femtocells) may overlap geographically, forming a dense network topology. These enhanced ICIC (eICIC) mechanisms are addressed later in Section 8.2.4.

8.1.13.4   Time Domain Aspects

The LTE radio resources are organized in the time domain into radio frames, which are 10 ms in duration for both FDD and TDD modes. The LTE radio frame duration has been chosen for compatibility with UMTS, which also uses a 10 ms radio frame duration. The radio frame defines a repeating structure that forms the basis for mapping of physical channels to the radio resources of the air interface in time. Each radio frame is assigned an SFN, which repeats every 1024 frames, resulting in a periodicity of approximately 10 seconds. The SFN is necessary, for example, for the transmission of SIBs that span multiple radio frames (e.g., the MIB transmitted on the PBCH is transmitted over the course of four radio frames).

Radio Frame Structure Type 1 (FDD Mode)   LTE supports two radio frame structures: Type 1 for FDD mode and Type 2 for TDD mode. First we describe frame

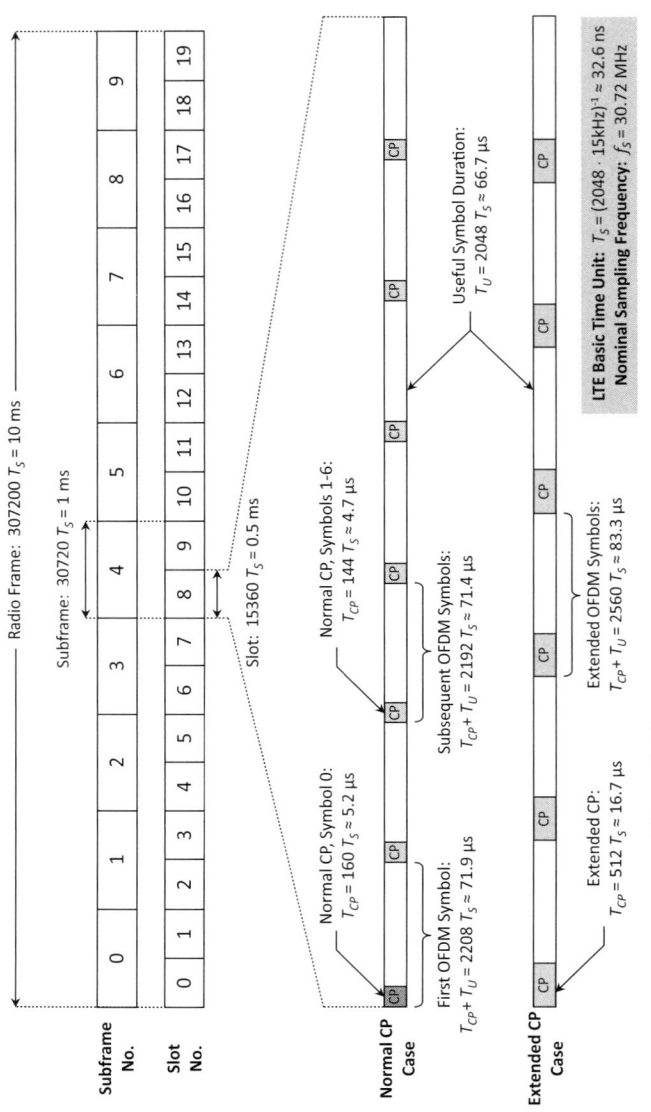

Figure 8.20 LTE radio frame structure type 1 for FDD mode.

structure type 1, which is shown in Figure 8.20. The 10-ms radio frame is subdivided into 10 subframes of 1 ms each. In the case of frame structure type 1, each subframe is further subdivided into two 0.5-ms slots. The number of OFDM symbols contained in a single slot depends on the CP length that is used. E-UTRA supports two CP lengths to accommodate different network deployment scenarios: a normal length of approximately 4.7 μs and an extended CP length of 16.7 μs. In the normal CP case, there are seven OFDM symbols per slot. In the extended CP case, the number of OFDM symbols per slot is reduced to six as a consequence of the additional overhead created by the extended CP. In FDD mode, this time structure applies to both the uplink and the downlink. The uplink and downlink radio frames are time synchronized at the eNB. Due to propagation delay, this requires the use of a timing advance for UE transmission on the uplink. The radio frame transmissions from different cells within E-UTRAN are not required by the standard to be synchronized. However, synchronized network operation may be optionally be implemented, for example to support ICICs.

Radio Frame Structure Type 2 (TDD Mode)   The radio frame structure for TDD mode has been designed to share a common overall parameterization with that of FDD mode described previously. In fact, the overall structure is essentially the same, including the definition of subframe, slot, OFDM symbol, and CP durations. The key difference is that frame structure type 2 must include special provisions for switching between downlink and uplink transmission. Such provisions are not required in FDD mode since the downlink and uplink are separated in frequency rather than time. The subframes of the TDD structure must be divided into those that are allocated for downlink transmission and those that are allocated for uplink transmission. A key feature of TD-LTE is that it accommodates asymmetric allocation of radio resources according to network demands. There are six uplink–downlink configurations, as shown in Figure 8.21. This allows for asymmetric allocation of resources: eight downlink subframes and one uplink subframe on the one extreme to two downlink subframes and six uplink subframes on the other. In practice, the uplink–downlink configuration is typically configured statically or semistatically within a given network deployment and is shared by neighboring cells to avoid interference between transmission directions [188].

The transition from uplink transmission to downlink transmission does not have a significant impact on the time structure, since all transmissions within the cell are time synchronized at the eNB. In contrast, the transition from downlink transmission to uplink transmission requires special treatment to account for propagation delay and UE timing advance. For this purpose, a special subframe is defined. The special subframe is composed of three fields: downlink pilot time slot (DwPTS), guard period (GP), and uplink pilot time slot (UpPTS). The precise duration of these fields is configurable to match the characteristics of the specific network deployment, such as cell size, antenna configurations, and coexistence with other TDD technologies. The DwPTS is reserved for downlink transmission. It essentially functions as a normal downlink subframe in which only the first portion of OFDM symbols is used. The UpPTS is reserved for uplink transmission and may be used for uplink channel

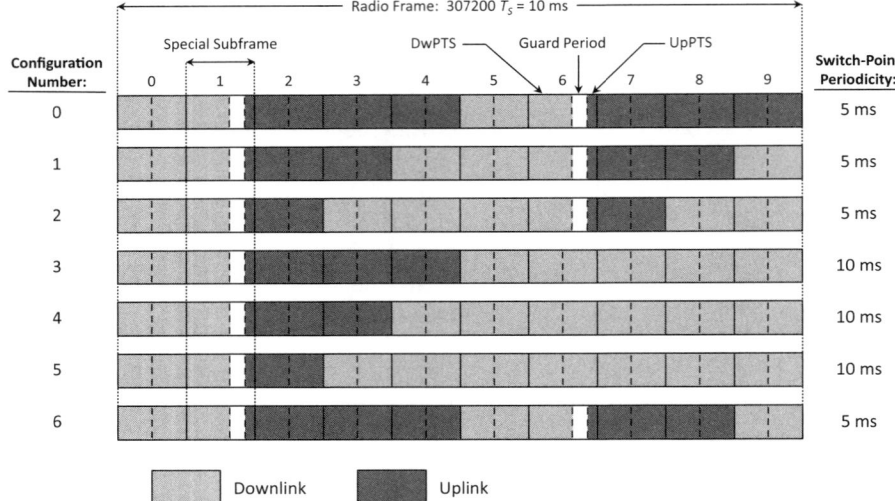

Figure 8.21 LTE radio frame structure type 2 for TDD.

sounding transmissions (i.e., sounding reference signal [SRS]) or for random access requests (i.e., PRACH). The uplink–downlink configurations shown in Figure 8.17 may include either a single special frame in subframe 1 for 10-ms switch-point periodicity or two special frames in subframes 1 and 6 for 5-ms switch-point periodicity. The latter are suitable for coexistence scenarios with 3G TD-SCDMA networks.

Cyclic Prefix Dimensioning An essential parameter in any OFDM-based system is the CP length. The CP provides a guard period between OFDM symbols, and its length determines the amount of channel delay spread that can be tolerated without experiencing ISI. A sufficiently long CP duration is particularly important in cellular systems, where large delay spreads due to multipath propagation are commonly experienced. Due to the orthogonality principle of OFDM systems, the useful duration of the OFDM symbol in E-UTRA is fixed at 66.7 μs as a consequence of the 15 kHz subcarrier spacing. The dimensioning of the CP length, therefore, represents a trade-off between spectral efficiency and robustness against ISI. A longer CP duration increases the amount of overhead in the form of redundant information, which decreases spectral efficiency. However, a longer CP duration also means tolerance for larger channel delay spreads, which has a direct impact on cell dimensioning. In order to accommodate a wide range of deployment scenarios— including both urban and rural, picocell and macrocell, and low mobility and high mobility—LTE defines two CP durations. The largest number of OFDM symbols with a 66.7 μs useful period that can fit into a 0.5 ms slot is seven. Therefore, the normal CP is designed to accommodate seven OFDM symbols per slot, resulting in a CP duration of approximately 4.7 μs. In most scenarios, the normal CP is sufficiently long to deal with ISI and represents a good trade-off between the metrics described earlier. However, in some cases such as rural deployments, where very large macrocells are used, an extended CP may be advantageous to combat large channel delay spreads. Finally, we note that for the normal CP case, the first OFDM symbol of each slot has a slightly longer CP duration than the subsequent six OFDM symbols, as shown in Figure 8.17. This is a consequence of the practical need to divide the 0.5-ms slot duration into seven symbol intervals while making each symbol an integer multiple of the basic time unit, which is described next.

Basic Time Unit   All time durations in the LTE specification are defined as multiples of the basic time unit, T_S, which is defined in Reference [200] as:

$$T_S = \frac{1}{2048 \cdot 15 \text{ kHz}} \approx 32.6 \text{ ns.}$$

This value is equivalent to the sample period for a transmitter/receiver implementation using a 2048-point FFT/IFFT and a corresponding sampling rate of 30.72 MHz. This implementation was shown in Table 8.9 to correspond to a 20 MHz channel bandwidth configuration. Therefore, the basic time unit T_S can be thought of as the smallest time unit of practical interest for LTE radio implementations. However, it was shown in Table 8.9 that a sampling rate of 30.72 MHz may be unnecessary for narrow channel bandwidth configurations. In these cases, a lower sampling rate and a correspondingly smaller FFT size may be preferable for efficient DSP implementation. Inspection of the OFDM symbol durations and CP durations in Figure 8.17 shows that the number of samples of duration T_S is always divisible by a factor of 16 for both the normal CP case and extended CP case. The reason for this is to support the scalable bandwidth of LTE deployments by accommodating all of the nominal FFT sizes summarized in Table 8.9.

For example, if the slot is resampled at a rate of 1.92 MHz—corresponding to a 1.4 MHz channel bandwidth configuration—then each of the CP durations and OFDM symbol durations will remain an integer number of the new sample period:

$$T_{S'} = \frac{1}{128 \cdot 15 \text{ kHz}} \approx 520.8 \text{ ns} = 16 \cdot T_S,$$

where $T_{S'}$ is the new sample period (i.e., for 1.92 MHz sample rate) and T_S is the basic time unit as defined previously. This relationship explains why the normal CP case defines a slightly longer CP duration (i.e., $160 \cdot T_S \approx 5.2$ μs) for the first OFDM symbol of each slot and a shorter CP duration (i.e., $144 \cdot T_S \approx 4.7$ μs) for the remaining OFDM symbols. This approach results in each CP duration being a multiple of 16 times the basic time unit.

8.1.13.5   Time–Frequency Grid

The subdivision of radio resources in both time and frequency in an OFDMA system leads to the definition of a time–frequency resource grid as shown in Figure 8.22

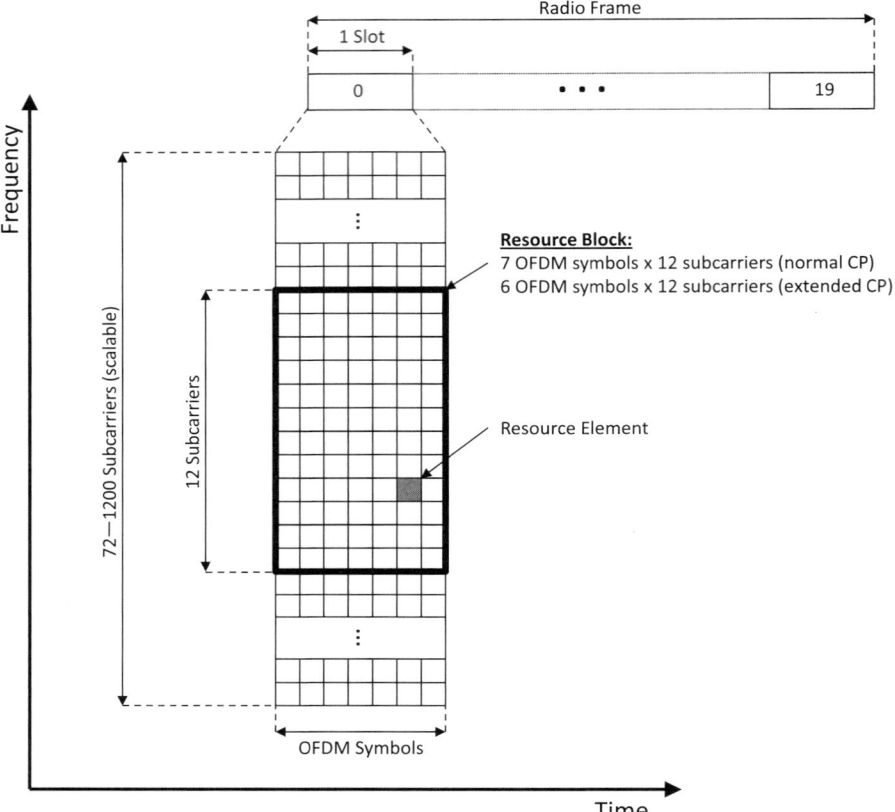

Figure 8.22   E-UTRA time–frequency resource grid. Reprinted from Reference [200].

[200] for the normal CP case. The fundamental unit of physical radio resource in E-UTRA is the *resource element* (RE), which consists of one subcarrier for a duration of one OFDM symbol. Each RE may be used to carry a data symbol or pilot symbol, or it may be unoccupied, for example, during times when cell traffic is low and not all radio resources are utilized. Data subcarriers may be modulated with QPSK, 16-QAM, or 64-point quadrature amplitude modulation (64-QAM)[8] modulation, allowing each RE to carry 2, 4, or 6 bits of coded data. Additionally, some REs are used to carry PHY layer signals that are mapped to specific REs in the time–frequency grid. Radio resources are grouped into units of 12 subcarriers, such that one unit of 12 consecutive subcarriers for a duration of one slot is termed a *resource block* (RB). An RB is thus comprised of 84 REs (i.e., 12×7 REs) when using the normal CP or 72 REs (i.e., 12×6 REs) when using the extended CP.

---

[8] In Rel-8, 64-QAM modulation is only supported in the downlink. Uplink 64-QAM has been added as a feature of LTE-Advanced starting in Rel-10.

The RB is an import unit in E-UTRA because it represents the basis of radio resource allocation for data transfer on both the uplink and the downlink. In theory, it is possible to allocate resources to individual users on a per-RE basis. However, it is impractical to allocate resources to users in this way, as the control overhead would be tremendous. In E-UTRA, resources are actually allocated to users on the basis of RB pairs, where one RB always falls in each 0.5-ms slot of a subframe. In the downlink, the first few OFDM symbols of each subframe are used to transmit control information, which includes user resource allocations for both the downlink and uplink. The network can therefore schedule a different set of users in each subframe.

The 3GPP specification defines two types of RBs: physical and virtual. A physical resource block (PRB) refers to the physical resources themselves, while a virtual resource block (VRB) is a logical mapping that is defined for scheduling purposes. Scheduling on the basis of VRBs may be used, for example, for optional frequency hopping. Frequency hopping may be configured either at the subframe boundary (i.e., intersubframe hopping) or at the slot boundary (i.e., intrasubframe hopping) depending on network configuration.

Frequency-Selective Scheduling   Broadband wireless radio links, and macrocellular links in particular, are subject to frequency-selective fading and interference, which varies on a per-UE basis as a function of both time and frequency. As a result, different portions of the spectrum will be more attractive for some UEs and less attractive for others. LTE takes advantage of these channel conditions through the use of frequency-domain scheduling with a short transmission time interval of 1 ms, which is on the order of the coherence time of typical cellular channels. Instantaneous channel quality measurements are performed by the UE and sent back to the eNB as often as every 1 ms, allowing for efficient scheduling on the downlink. In the uplink, a wideband SRS may be transmitted by the UE, allowing measurements to be performed at the eNB to estimate uplink channel quality. All of this information is considered by the scheduler in the eNB, which then allocates uplink and downlink radio resources to maximize overall system throughput while maintaining QoS requirements defined for each user. The fundamental concept of frequency-selective scheduling is illustrated in Figure 8.23 [210]. For simplicity, only two users are shown. The upper part of the figure shows the channel quality variations as a function of frequency and time for the two users. Where channel conditions are significantly better for user #1, those RBs are allocated to user #1. Similar allocations are made for user #2. In some cases, user #1 experiences deep nulls. It is generally more efficient to assign these RBs to user #2. In practice, scheduling is often based on a so-called proportional fair algorithm to ensure that a user with overall poor channel conditions is still allocated some amount of radio resources in order to maintain QoS requirements, even when other users experience better channel conditions.

8.1.13.6 *Space Domain Aspects*

The use of multiple antenna techniques in LTE effectively forms a third spatial dimension in the overall structure of the LTE air interface. As mentioned previously,

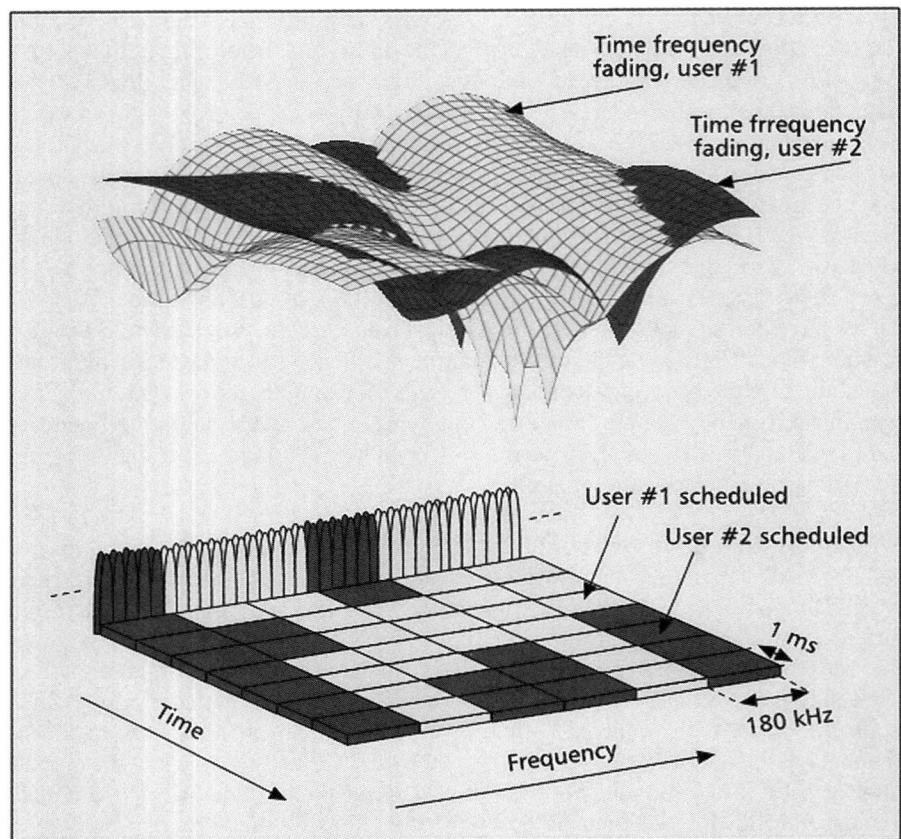

Figure 8.23   Channel-dependent scheduling based on channel quality variations in time and frequency. Reprinted from Reference [210].

the use of multiple transmit antennas only applies to the downlink transmission from the eNB in Rel-8 LTE.[9] In the spatial dimension, radio resources are organized on the basis of logical *antenna ports*. In the case of multiple antenna transmission, there is one time–frequency resource grid defined for each antenna port, resulting in a three-dimensional radio resource mapping. A list of antenna ports defined in LTE Releases 8 and 9 are summarized in Table 8.11. The eNB is required to support one, two, or four so-called cell-specific antenna ports such that antenna ports {0}, {0,1}, or {0,1,2,3} are defined. These antenna ports are used to support most multiple antenna transmission modes, which may include single antenna port

[9] Multiple uplink transmit antennas have been included as part of the enhanced MIMO operation in Rel-10 LTE-Advanced, as discussed later in this chapter.

Table 8.11 Summary of E-UTRA Downlink Antenna Ports in Rel-8 and Rel-9

Antenna port	Initial release	Description
0	Rel-8	Cell-specific transmission (always supported)
1	Rel-8	Cell-specific transmission (2 or 4 transmit antennas)
2	Rel-8	Cell-specific transmission (4 transmit antennas)
3	Rel-8	Cell-specific transmission (4 transmit antennas)
4	Rel-9[a]	MBSFN
5	Rel-8	Beamforming
6	Rel-9	UE positioning reference signals
7-8	Rel-9	Dual-layer beamforming

[a]Antenna port 4 has been defined for MBSFN in Rel-8 but not fully supported until Rel-9.

transmission on antenna port 0, two- and four-port spatial multiplexing, and two- and four-port transmit diversity, depending on network configuration.

A common pitfall in understanding multiple antenna techniques in LTE is to assume that the term "antenna port" is equivalent to a physical antenna at the transmitter or receiver—however, this is not necessarily true. An antenna port in LTE is actually a logical construct that represents a single transmission over common channel conditions. In reality, an antenna port may be mapped to one or more physical antenna elements. This is particularly true of antenna port 5, which is used for beamforming. An example mapping between logical antenna ports and physical antennas for a Rel-8 eNB supporting 4 × 4 MIMO spatial multiplexing and beamforming is shown in Figure 8.24. Each logical antenna port appears to the receiver as a single antenna, even though the transmitter side may include multiple antenna elements (e.g., for beamforming). Furthermore, each antenna port is uniquely identified by a training sequence, called a reference signal in LTE, which is used for channel estimation and coherent demodulation. In this way, multiple antenna port transmissions can be separated at the receiver, assuming an equivalent number of receive antennas exist.

8.1.14 Multiple Antenna Techniques in LTE

An essential characteristic of LTE that distinguishes it from all of its 3GPP predecessors is that it has been designed from the very beginning to support multiple antenna technology, commonly referred to as MIMO. In the previous section, we have already introduced the concept of antenna ports, which are used to realize the spatial dimension of the LTE air interface. In this section, we provide an overview of the various multiple antenna techniques that are supported in the E-UTRA downlink and uplink. Multiple antenna techniques serve a variety of purposes in LTE—which may include increasing the peak data rate of a single user, extending the effective coverage area of a cell, increasing the number of active users served at one time (i.e.,

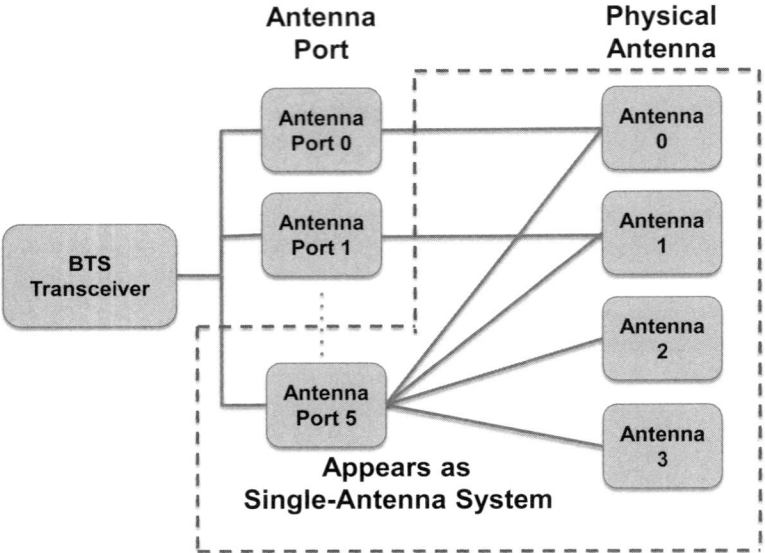

Figure 8.24   Example mapping of logical antenna ports to physical antennas for E-UTRA downlink transmission.

network capacity), or increasing the robustness of the communication link in the form of increased signal-to-interference-noise ratio (SINR). There are three fundamental categories of MIMO techniques that are employed in LTE: spatial diversity, spatial multiplexing, and beamforming. The fundamental concepts behind these technologies are described in Chapter 10. As we shall see, multiple antenna techniques are utilized differently for downlink transmission versus uplink transmission in accordance with the asymmetric capabilities of a base station transmitter (i.e., large antenna spacing, high power, high complexity, and cost) versus a typical UE transmitter (i.e., small antenna spacing, limited power, relatively low complexity, and cost). In general, MIMO techniques are concentrated in the downlink, where multiple antennas may be used for both transmission and reception. Though constrained to a single transmit antenna per UE, the Rel-8 LTE uplink design is compatible with so-called multiuser MIMO (MU-MIMO) spatial multiplexing techniques.

Before proceeding, it is worth making a special note on terminology. In the LTE community, the term "MIMO" is commonly used to refer specifically to spatial multiplexing techniques. In fact, this is how the term is used in the 3GPP specification. However in the greater wireless communications community—and throughout the rest of this book—the term "MIMO" is commonly used to refer to any communication system that utilizes multiple transmit antennas and multiple receive antennas to improve system performance, typically through some form of advanced signal processing techniques. The latter definition is much broader, encompassing spatial diversity and beamforming techniques in addition to spatial multiplexing.

The differences in the use of the term MIMO sometimes lead to ambiguities in the LTE literature. For consistency with other chapters, the term MIMO is used in this chapter in the general sense to refer to multiple antenna techniques unless otherwise noted. Every effort has been made to clearly delineate spatial multiplexing when referring to this specific form of MIMO.

8.1.14.1 Downlink Multiple Antenna Techniques

All downlink control channels are transmitted from the eNB using either single antenna port transmission (i.e., single input, single output [SISO]) or transmit diversity, depending on network configuration. These include the PCFICH, PDCCH, PHICH, PBCH, and PMCH, as well as well control signaling of the PHY and MAC layers. The type of transmission used is implicitly signaled to the UE based on the number of cell-specific reference signals present (i.e., the reference signals which identify antenna ports 0, 1, 2, and 3 defined previously). This approach is appropriate for control signaling because it does not depend on the current state of the channel between the eNB and any one UE. This simplifies the control transmission and makes it possible for the control channels to be received by all UEs within the coverage area (e.g., to receive broadcast information on the PBCH, DCI, and scheduling assignments on the PCCH).

In contrast, data transmission on the PDSCH may make use of a variety of MIMO techniques to increase performance—resulting in increased user data rates and/or cell throughput. In Rel-8 LTE, seven transmission modes are defined, each of which accommodates different multiple antenna processing of DL-SCH transport blocks for transmission on the PDSCH. In Rel-9 LTE, an eighth transmission mode has been defined to support the dual-layer beamforming enhancement. These transmission modes and the associated antenna ports are listed in Table 8.12 [211]. Adaptive multiple antenna transmission in the downlink is supported through feedback of CSI from the UE to the eNB. The UE takes measurements of the channel conditions and then reports these measurements back on a periodic or aperiodic basis to the eNB through over the PUCCH and/or PUSCH. In all cases, channel quality measurements are communicated to the eNB using the channel quality indicator (CQI), which is also used to set the adaptive modulation and coding scheme used for downlink transmission. For some spatial multiplexing techniques, additional CSI may be communicated using the rank indicator (RI) and precoding matrix indicator (PMI), the purpose of which is elaborated in the following list. With a basic understanding of uplink feedback reporting established, we now summarize the key multiple antenna transmission techniques associated with each transmission mode.

- *Single Antenna Port Transmission on Port 0 (Mode 1).* This mode corresponds to basic SISO antenna transmission using port 0. No MIMO precoding is used. This type of transmission is used by eNBs equipped with only one antenna per cell.

- *Transmit Diversity (Mode 2).* A single modulated symbol stream is mapped to two antenna ports using SFBC or four antenna ports using a combination

Table 8.12   Summary of PDSCH Transmission Modes and Associated Feedback Reporting for LTE Rel-8 and Rel-9

Mode	Associated transmission scheme	Ports	UE feedback CQI	RI	PMI
1	Single antenna transmission (SISO)	0			
2	Transmit diversity (SFBC or SFBC/FSTD)	{0,1}, {0,1,2,3}			
3	Open-loop spatial multiplexing	{0,1}, {0,1,2,3}			
4	Closed-loop spatial multiplexing	{0,1}, {0,1,2,3}			
5	Multiuser MIMO spatial multiplexing	{0,1}, {0,1,2,3}			
6	Single layer closed-loop spatial multiplexing	{0,1}, {0,1,2,3}			
7	Beamforming	5			
8	Dual-layer beamforming (Rel-9) and beyond)	{7,8}		Configurable	

SISO, single input, single output; SFBC, space frequency block code; FSTD, frequency-switched transmit diversity.

of SFBC with *frequency-shift time diversity* (FSTD). SFBC is similar to space–time block coding (STBC), also called Alamouti codes, in common MIMO terminology. However, the outputs of the block coding matrix operation are mapped to different antenna ports in frequency (i.e., consecutive data subcarriers within the same OFDM symbol time) rather than time. Most STBC codes operate under the assumption of block fading, meaning that the channels remain unchanged within the block time. Placing the subcarriers for SFBC close together in frequency has the benefit that this assumption can be made for SFBC as well. Only basic CQI reporting feedback is required from the UE for transmit diversity on PDSCH, similar to SISO transmission.

- *Open-Loop Spatial Multiplexing (Mode 3).* The modulated symbols resulting from one or two transport blocks are divided into multiple streams— referred to as *layers* in LTE. A precoding matrix is then used to map the layers to multiple antenna ports. In general, the number of layers must be less than or equal to the number of antenna ports supported at the eNB as well as the UE. The number of layers defines the *transmission rank* and is based on the perceived rank of the channel, which is measured by the UE and reported to the eNB through the RI. The precoding matrix is chosen from a predefined set of matrices, called a *codebook*, which are known to both the eNB and the UE. In the open-loop case, the precoding matrix is predefined and is not based on PMI feedback from the UE. For two antenna transmission, a fixed precoding matrix is used corresponding to codebook index 0. For four antenna transmission, all codebook entries are cycled among the subcarriers. Instead of adapting the precoding matrix to the spatial signature of the channel, a cyclic phase shift is applied to each of the layers during precoding. Each layer receives a different cyclic phase shift, which effectively creates artificial time dispersion that looks like multipath propagation to the UE. This approach is called *large-delay cyclic delay diversity* (CDD), and it effectively averages out the spatial variations in channel conditions experienced by the different layers [212]. UE reporting for open-loop spatial multiplexing includes CQI and RI feedback, but PMI reporting is not required—hence the name "open loop." This transmission mode is appropriate, for example, in the case of high UE mobility scenarios when the channel variations occur too quickly for accurate PMI feedback reporting. Since all the spatial layers are received by a single UE, this mode is sometimes referred to as open-loop *single-user MIMO* (SU-MIMO). It is one of multiple SU-MIMO modes supported.

- *Closed-Loop Spatial Multiplexing (Mode 4).* This is the transmission mode that allows LTE to reach its peak data rates by sending multiple parallel streams of data over the same time–frequency resources to a single user—hence it is also a form of SU-MIMO spatial multiplexing. This mode is similar to mode 3 described previously, in that one or more layers are precoded using a known codebook of precoding matrices. The key difference is that, in addition to CQI and RI reporting, the UE is required to measure the spatial signature of the channel and feedback this CSI to the eNB in the form of the PMI, which is a suggestion as to which precoding matrix the eNB should select from the codebook for the next transmission. Up to four spatial streams are supported in Rel-8 (i.e., 4×4 SU-MIMO), but this configuration requires a category 5 UE. A maximum of two transport blocks may be transmitted in a single TTI.

- *Multiuser MIMO (MU-MIMO) Spatial Multiplexing (Mode 5).* A variation on closed-loop spatial multiplexing in which multiple layers are assigned to different UEs. This approach allows multiple parallel streams to be transmitted to different users simultaneously using the same time–frequency resources. For this reason, MU-MIMO is also known as spatial division multiple access (SDMA). During MU-MIMO transmission, each user is limited to single layer transmission (i.e., rank = 1). Therefore, only CQI and PMI feedback reporting are required. Each UE is assigned a different precoding matrix from the codebook, which allows each UE to demodulate their own data stream while perceiving the other spatial stream as interference. The precoding matrices must be chosen by the eNB to maximize received power for each UE while minimizing mutual interference. MU-MIMO spatial multiplexing does not increase the data rate for a single user, but it may be used to increase overall cell capacity or optimize cell throughput.

- *Single-Layer Closed-loop Spatial Multiplexing (Mode 6).* Essentially a subset of mode 4 in which only a single layer may be used, corresponding to a transmission rank of one. The same codebook and UE CSI feedback reporting is used as in mode 4, except that the RI is not necessary. This mode requires less overhead signaling on both the downlink and the uplink. The codebook-based precoding may also be considered as a basic type of beam-forming. However, the predefined codebook means that only a fixed set of antenna beam patterns may be used (i.e., 4 patterns for 2 antenna ports or 16 patterns for 4 antenna ports).

- **Single-Antenna Port Transmission on Port 5 (Mode 7).** This mode is suitable for conventional beamforming using single layer transmission. The baseband processing and antenna port mapping is essentially the same as mode 1, except that antenna port 5 is used. The eNB transmitter then feeds the same signal with different weighting and phase shifts to each antenna element, allowing the wavefront to be spatially focused in the direction of the UE. This process is also referred to as non-codebook-based precoding. Special UE-specific reference signals are inserted, which receive the same complex weighting as the data transmission, allowing for coherent demodulation. The resulting transmission appears to the UE as though it were transmitted from a single antenna with increased SINR due to beamforming. The 3GPP specification does not define the method used to calculate complex weights used for beamforming. In practice, estimating these parameters is easier in TDD mode due to channel reciprocity in unpaired spectrum. Based on operator prioritization, mode 7 has been defined as a high-priority capability for TDD mode and a low-priority capability for FDD mode [188].

- **Dual-Layer Beamforming (Mode 8).** LTE Rel-9 defines transmission mode 8 for enhanced beamforming techniques. This mode combines spatial multi-plexing with beamforming to allow two-layer transmission using non-codebook-based precoding. Dual-layer beamforming supports both SU-MIMO and MU-MIMO cases, whereby the two layers may be transmitted to a single UE or multiple UEs. As with mode 7, this scheme is most suitable for TDD mode of operation.

To simplify the control signaling related to the various MIMO modes of operation, the UE is configured semistatically to one transmission mode by the RRC layer. There is not a strict one-to-one relationship between the transmission mode and multiple antenna technique used, as the earlier discussion may seem to imply. More accurately, the transmission mode defines a set of multiple antenna processing techniques that may be applied to the transport blocks processing of the DL-SCH transport channel. For transmission modes 3–8, transmit diversity is always supported as a fallback without implying a change in transmission mode. This approach allows for MIMO link adaptation to match changing channel conditions.

Finally, we note that further MIMO enhancements have been defined in Rel-10 as a key feature of LTE-Advanced. Enhanced downlink spatial multiplexing has been defined to support up to 8 spatial layers. Furthermore, these layers may be used for SU-MIMO or MU-MIMO spatial multiplexing scenarios, allowing a great deal of flexibility in comparison to Rel-8 MU-MIMO support, which is somewhat limited. A new transmission mode 9 is defined for this purpose. These enhancements are addressed later in this chapter in Section 8.2.

8.1.14.2 Uplink Multiple Antenna Techniques

In LTE Rel-8, only single-antenna transmission is supported by the UE. Closed-loop adaptive antenna selection transmit diversity is supported as an optional feature. This scheme allows the UE to select the transmit antenna with the best performance as a form of pseudo-transmit diversity, but the result is still a single-antenna transmission in the uplink at a given time. This capability is optional for the UE and must be signaled to the eNB as part of the UE capabilities.

MU-MIMO spatial multiplexing is supported on the uplink. In this scheme, multiple UEs are able to transmit on the same time–frequency resources in the uplink using single-antenna port transmission. The two UE transmissions will experience different channels and can be separated at the eNB using the uplink demodulation reference signals (DM-RS) transmitted from each UE. Uplink MU-MIMO is implicitly supported in Rel-8 LTE through the definition of orthogonal DM-RS sequences based on cyclically shifted Zadoff–Chu (ZC) sequences. By assigning a different DM-RS to each UE, the eNB is able to perform channel estimation and coherent demodulation for each transmission and separate the two data streams. Uplink MU-MIMO is therefore entirely a receiver-side technology, and is not defined in the 3GPP specification. Rather, it may optionally be implemented as a vendor-specific eNB feature.

8.1.15 E-UTRA Physical Layer

This section provides an overview of the E-UTRA PHY layer, which constitutes layer 1 of the Open System Interconnection (OSI) model on the LTE air interface. The PHY provides a data transport service to the higher layers, which is accessed by means of transport channels via the MAC sublayer. The data delivered by the MAC layer to the PHY layer, and vice versa, is called a *transport block*. Transport blocks are delivered once every TTI [213]. The main functions of the PHY layer include channel coding for error detection (i.e., cyclic redundancy checksum [CRC]) and error correction (i.e., forward error correction [FEC]); HARQ soft-combining; rate matching and multiplexing of codewords; bit scrambling; modulation and demodulation; MIMO antenna processing; and OFDM symbol mapping and signal generation. The PHY also performs link adaption in both the DL and the UL in the form of adaptive modulation and coding to achieve the maximum throughput based on perceived channel quality. The introduction to the PHY layer specifications is defined in 3GPP Technical Specification 36.201 [214], which is a good launching point for the reader interested in exploring the LTE PHY layer standards documents.

8.1.15.1 Summary of E-UTRA Physical Layer Signals

Section 8.1.12.3 presented an overview of E-UTRA physical channels, which are used to carry control information or data originating from a higher layer between peer PHY entities. In addition to the physical channels, a number of physical signals are defined which support the operation of the E-UTRA PHY layer. These signals are generated by the PHY and mapped to specific REs in the uplink and the downlink. In contrast to physical channels, the physical signals do not carry any data per se, although they can convey information such as the physical cell identity. One key role of physical signals is to provide a basis for channel measurements, for example to support channel-dependent scheduling, MIMO operation, or handover decisions. Different physical signals are defined in the downlink and uplink, as will be summarized later.

In the downlink, two types of physical signals are defined: reference signals and synchronization signals [200].

- *Downlink Reference Signals.* These are equivalent to what are commonly called pilot symbols or pilot tones in other OFDM systems. They consist of known symbols that are transmitted on specific REs embedded within the time–frequency resource grid. A reference signal (RS) is permanently mapped to each antenna port at the eNB and is used for channel estimation and coherent demodulation of the downlink channels by the UE. In Rel-8 LTE, two types of RSs are defined: cell specific and UE specific. The cell-specific RSs are common reference signals transmitted on antenna ports 0 through 3 (when present) that are used by all UEs within the cell. The cell-specific RSs are distributed across the entire channel bandwidth and form the basis for UE channel measurements, which are used to acquire CSI that is reported back to the eNB to support channel-dependent scheduling, MIMO operation, and handover decisions. The UE-specific RSs, in contrast, are used by a single UE to support coherent demodulation of user data transmitted on the PDSCH using transmission mode 7 (i.e., MIMO beamforming). The UE-specific RSs are transmitted on antenna port 5 and are only transmitted in RBs where MIMO beamforming is used. Additional types of downlink RSs are defined in subsequent 3GPP releases. These include the MBSFN-specific RS defined in Rel-9 to support MBMS operation and the CSI-RS defined in Rel-10 to support 8×8 MIMO spatial multiplexing for LTE-Advanced.
- *Synchronization Signals.* These signals are used to achieve initial synchronization during the cell selection process and are also used for synchronization during cell reselection and handover. Two synchronization signals are defined, called the primary synchronization signal (PSS) and the second synchronization signal (SSS). The combination of the PSS and the SSS inherently signal the physical cell ID, which takes on one of 504 possible values and is used to distinguish nearby cells within E-UTRAN.

In the uplink, two types of physical signals are defined: DM-RS and SRS. Both signals fall under the general category of uplink reference signals, but they are used for somewhat different purposes.

- **Demodulation Reference Signal (DM-RS).** These signals are used for channel estimation for coherent demodulation associated with UE uplink transmission on the PUSCH or PUCCH. The DM-RS sequences are based on prime length Zadoff–Chu (ZC) sequences that are cyclically extended to the appropriate length. The DM-RS are embedded in the UE uplink transmission, occupying one or more OFDM symbol and spanning the same bandwidth allocation as the associated PUSCH or PUCCH.

- **Sounding Reference Signal (SRS).** In contrast to the DM-RS, the SRS is not associated with PUSCH or PUCCH transmission from the UE. The primary function of the SRS is to allow the eNB to acquire CSI through uplink channel sounding. To support frequency-selective scheduling, the SRS may assume a very wide bandwidth that is configurable up to the full bandwidth of the E-UTRA carrier. During PUSCH transmission, the SRS may be used to support optional adaptive antenna selection transmit diversity in the uplink. The SRS may also be used to support other PHY layer procedures when no other uplink transmissions are configured. These include, for example, uplink timing advance estimation, link adaptation, and uplink power control. Depending on intended use, the UE may be configured by the eNB for single or periodic SRS transmission.

8.1.15.2 *Mapping of Downlink Physical Channels and Signals*

This section describes the general principles for mapping of physical channels and signals to the physical resources of the OFDMA downlink. The LTE PHY layer is characterized by the design principle of dynamically allocated, shared radio resources. The OFDMA interface is particularly well suited to this approach, as it allows for flexible resource allocation on the basis of RB pairs. Furthermore, the resource allocation among different users can be dynamically scheduled with a very fast 1-ms minimum TTI. The result is an air interface architecture that is well suited to the bursty nature of packet data transmission, for which LTE has been optimized. This approach stands in contrast to the PHY of legacy UMTS systems, which used dedicated channels to transmit user data.

The mapping of physical channels and signals to radio resources is called *resource element mapping*. In the downlink, this is a three-dimensional process. For the time–frequency resource grid corresponding to an antenna port, p, each RE is uniquely identified by an index pair (k, l). The frequency index k here is an integer value which identifies the subcarrier, starting at $k = 0$ for the lowest frequency subcarrier. Likewise, the time index l identifies the OFDM symbol, starting with $l = 0$ for the first OFDM symbol of each slot. The mapping of data streams to REs is generally performed by first incrementing the frequency index k across the allocated bandwidth and then incrementing the time index l, starting with the first slot of a subframe. Certain REs are reserved for control information (i.e., PDCCH), broadcast information (i.e., PBCH), and physical signals (i.e., reference signals, synchronization signals). The remainder may be used for PDSCH allocations. When network load is low, some REs may remain unused in a given subframe [200].

The detailed mapping of downlink physical channels and signals will depend on a number of variables associated with the specific network deployment, as well as the amount of control signaling required based on current cell load. Figure 8.25 [215] illustrates an example downlink mapping for a single radio frame in FDD mode of operation for a typical system configuration. This example assumes a 5 MHz channel bandwidth comprised of 25 RBs, a normal CP duration, and two

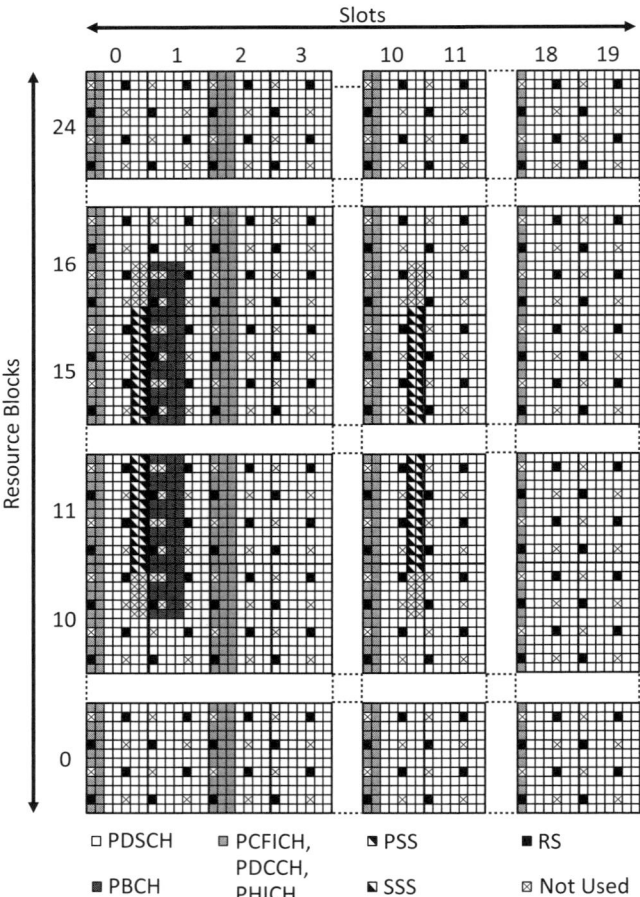

Figure 8.25 Example FDD downlink physical mapping on antenna port 0 with normal cyclic prefix and 5 MHz channel bandwidth. Reprinted from Reference [215].

cell-specific antenna ports. The location of the cell-specific reference signals indicates that this transmission is from antenna port 0 and that the cell has a physical cell ID of 1. For space, subframes 2 through 4 and 6 through 8 are not shown. These subframes assume the same general structure as subframes 1 and 9 shown. That is to say, the first one to three OFDM symbols are used to carry control channels (i.e., PDCCH, PCFICH, PHICH) and the remaining OFDM symbols carry the PDSCH, with the exception of those REs reserved for reference signals. Further information on the mapping of each downlink physical channel and signal is provided in the following sections.

Downlink Reference Signals   The cell-specific RS are transmitted in every RB of the downlink, with one RS transmitted from each of antenna ports 0 through 3 (as supported by the eNB). The cell-specific RS mapping takes on a sparse lattice structure distributed throughout the time–frequency resource grid. This approach is an efficient way to support channel estimation at the UE receiver in the sense that it minimizes the amount of control overhead. An example mapping of the cell-specific for one, two, and four antenna ports is illustrated in Figure 8.22 for the normal CP case. While the time-domain mapping of the RS is fixed, the frequency-domain mapping to subcarriers is a function of the physical cell ID. The entire lattice structure may be shifted by up to six subcarriers for a given cell, where the shift is determined by the physical cell ID modulo 6 [200]. This design minimizes the interference between cell-specific RS of adjacent cells in the network.

When one antenna port transmits an RS on a specific RE, all other antenna ports are quiet (i.e., a null subcarrier is mapped to this RE). This is required for the UE to be able to estimate the channels corresponding to each antenna port transmission. Furthermore, REs that are used to transmit the RS are reserved for this purpose, and may not be used to transmit any physical channel. From Figure 8.26, it is clear that additional cell-specific antenna ports, therefore, result in additional signaling overhead at the PHY layer, which reduces the physical resources available for data transmission on each individual antenna port. To minimize the added overhead associated with using four cell-specific antenna ports at the eNB, only two REs per RB are used for RS transmission on the third and fourth antenna ports (i.e., antenna ports 2 and 3).

The UE-specific RS is defined for MIMO beamforming transmission on antenna port 5. These signals are only embedded in RBs where transmission mode 7 is used for PDSCH transmission and are transmitted in addition to the cell-specific RS. The structure of the UE-specific RS for transmission mode 7 in Rel-8 LTE is shown in Figure 8.27 [200].

Synchronization Signals   The PSS and SSS are always mapped to the central 72 subcarriers of the downlink regardless of channel bandwidth. This frequency mapping corresponding to a bandwidth of 1.08 MHz, which is the smallest bandwidth supported by LTE. This approach makes it possible for the UE to locate and synchronize to the network without any prior knowledge of the cell configuration.

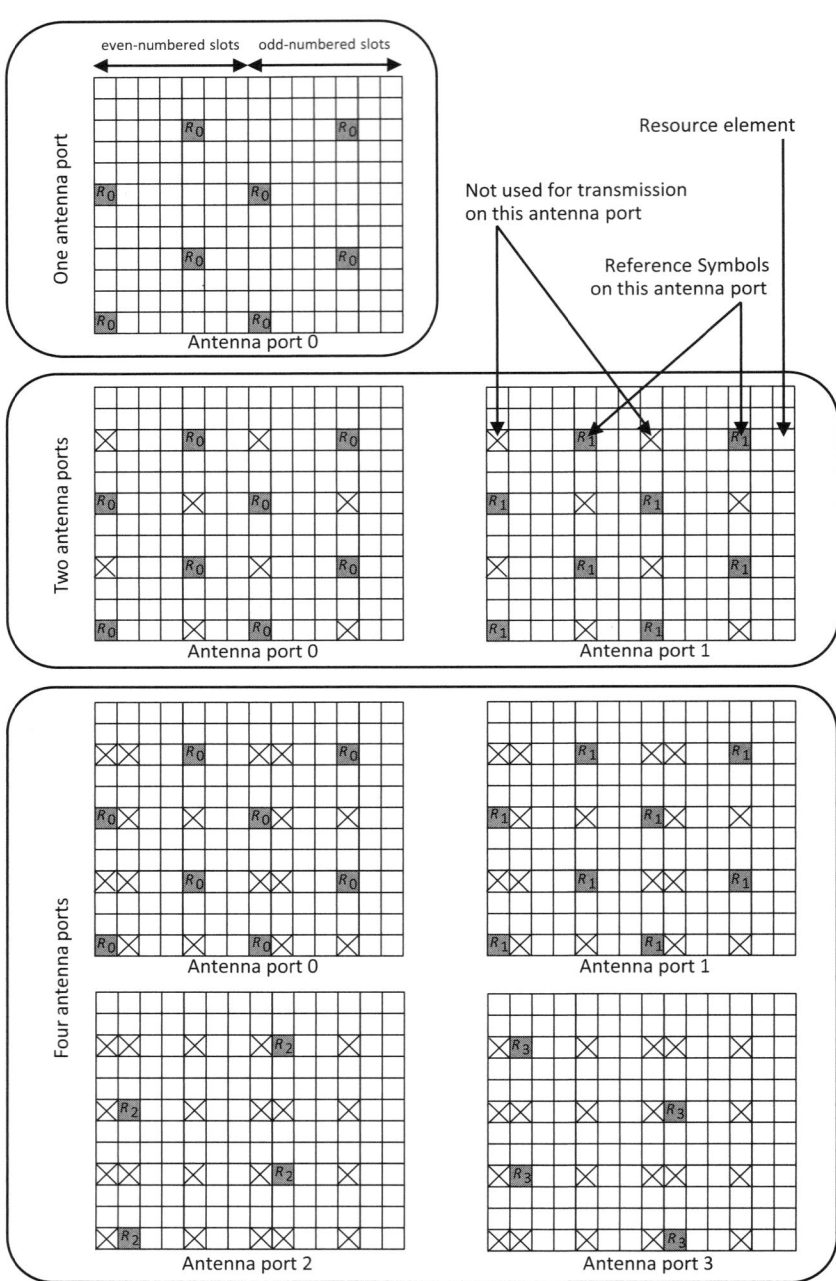

Figure 8.26 Example mapping of downlink cell-specific RS for 1, 2, and 4 antenna ports (normal CP case).

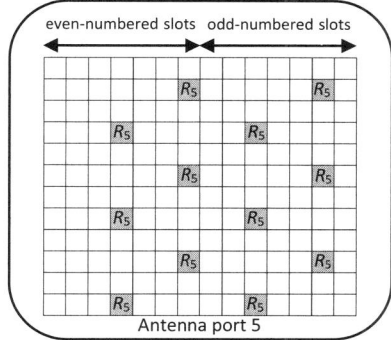

Figure 8.27   Example mapping of downlink UE-specific RS for antenna port 5.

Only 62 of the central 72 subcarriers are actually used to carry the synchronization signals, resulting in null subcarriers on five of the upper- and lowermost subcarriers [200]. This approach reduces the signal processing complexity by allowing the UE to use a 64-point IFFT, which increases the speed of the cell search procedure.

The specific location of the PSS and SSS in time depends on the CP length configured and the duplexing mode of the E-UTRA carrier. In the case of FDD mode, the PSS is located in the last OFDM symbol of the slot and is preceded by the SSS in the second to last OFDM symbol of the slot, as illustrated in Figure 8.21. In the case of TDD mode, the PSS is located the third OFDM symbol of subframes 1 and 6 and the SSS is located in the last OFDM symbol of slots 1 and 11 [200]. This scheme for PSS and SSS mapping implicitly signals to the UE the essential configuration information of the cell.

The specific structure of the PSS and SSS sequences also implicitly signals the physical cell ID. There are 168 unique SSS sequences, each of which represents a physical cell identity group. There are three unique PSS sequences, each of which indicates a unique identity within the physical cell identity group. The physical cell ID can be calculated as the sum of the PSS index plus six times the SSS index—resulting in 504 unique cell identities. In practice, the physical cell IDs will be assigned to cells that are geographically dispersed during network planning so that the cell is never able to receive a signal from two different cells with the same physical cell ID at the same time.

The PSS sequences are generated using a frequency-domain, length-63 Zadoff–Chu sequence based on the following equation:

$$ZC_u^{63}[k] = \exp\left[-j\frac{\pi u k (k+1)}{63}\right], \quad k = 0, 1, \ldots, 62,$$

where u is the ZC root index [200]. The selected roots for LTE are $u = 25, 29, 34$, which correspond to the three different PHY layer cell identities. Note that in LTE, the 32nd symbol (i.e., $ZC_u^{63}[31]$) is punctured due to the unused central DC subcarrier of the LTE downlink. The ZC sequence is then mapped to subcarriers in order of increasing index in the frequency domain. A key property of the ZC sequence is that it is a constant amplitude zero autocorrelation (CAZAC) sequence, meaning that its cyclic autocorrelation is an impulse at zero lags. However, since the PSS is punctured due to the unused DC subcarrier, a small amount of correlation can be observed at nonzero cyclic shifts of the cyclic autocorrelation. Furthermore, the cyclic cross correlation between the three LTE PSS sequences corresponding to the three different ZC sequence root index values are very low. The low cross correlation between the three LTE PSS sequences facilitates the synchronization and cell identification procedure. The PSS is suitable for noncoherent detection as a first step in achieving time synchronization. It provides a half-frame timing reference. However, this reference is ambiguous due to the fact that the same PSS sequence is used at the beginning and the middle of the radio frame.

The SSS sequences are generated using maximum length binary sequences called M-sequences, which are generated through cyclic shifts of a shift register. Each SSS is constructed by interleaving two length-31 M-sequences in the frequency domain [198]. The sequences are generated in such a way that the SSS is different at the beginning and middle of the radio frame, thereby resolving the ambiguity of the PSS acquisition step. After SSS acquisition, the UE can identify the start of the radio frame and begin decoding information transmitted on the PBCH.

Physical Broadcast Channel (PBCH)   The PBCH is transmitted once during each radio frame on the first four OFDM symbols of slot 1 for both FDD and TDD modes. Similar to the PSS and SSS, the PBCH is mapped in frequency to the central 72 subcarriers of the downlink regardless of channel bandwidth. However, unlike the synchronization signals, the PBCH utilizes the full 1.08 MHz bandwidth with the exception of those REs reserved for cell-specific RS. Regardless of the number of cell-specific antenna ports supported in a given cell, the resources used for cell-specific RSs for the four antenna port configuration are reserved and may not be used to transmit PBCH. As an example, consider the two antenna port case shown in Figure 8.21. Even though cell-specific antenna ports 2 and 3 are not used in this configuration, the REs where these cell-specific RS would be mapped remain unused on any antenna port. The PBCH is transmitted on all cell-specific antenna ports supported by the cell (i.e., antenna ports 0, {0,1}, or {0,1,2,3}) using transmit diversity when applicable.

Downlink Control Channels: PDCCH, PCFICH, and PHICH   The physical downlink control channels—namely the PDCCH, PCFICH, and PHICH—are carried on the first few OFDM symbols of every subframe and span the full system bandwidth. The exact number of OFDM symbols used to carry these control channels may vary from subframe to subframe. The number of OFDM symbols supported for this purpose depends on the channel bandwidth of the cell. In most cases, one to three symbols may be used for this purpose. However, for narrow bandwidth deployments with 10 RBs or less, two to four symbols may be used. The latter corresponds, for example, to the 1.4 MHz channel bandwidth configuration. A special RE grouping called a *resource element group* (REG) is defined for mapping of control channels. An REG consists of four usable REs corresponding to one OFDM symbol duration and consecutive subcarriers, excluding those reserved for RSs. The PDCCH, PCFICH, and PHICH are transmitted on the same set of antenna ports as the PBCH.

The sole purpose of the PCFICH is to carry a 2-bit field called the CFI. The CFI indicates the number of OFDM symbols used for control channels within a given subframe. Given the critical importance of the CFI, these 2 bits are encoded using a highly robust rate 1/16 block code and mapped to 16 symbols in the first OFDM symbol of every subframe. These symbols are grouped into four quadruplets and mapped to four REGs that are spread out across the channel bandwidth. The exact frequency mapping of the CPFICH depends on the physical cell ID, thereby reducing intercell interference [200].

The PHICH is responsible for carrying the HARQ indicator (HI), a 1-bit field that indicates whether the eNB has correctly received a transmission on the PUSCH (i.e., ACK/NACK). For uplink transmission on subframe n, the UE looks for a ACK/NAK on the PHICH in subframe $(n+4)$. The HI is encoded with a rate 1/3 repetition code and spread with one of eight 4-bit complex orthogonal Walsh sequences. The resulting 12 bits are mapped to 12 REs in 3 REGs within the control region. Due to the use of orthogonal spreading sequences, each PHICH group can carry up to eight PHICH channels when normal CP is used. The exact position of the PHICH depends on the cell ID.

The PDCCH is responsible for carrying DCI messages, which include information about scheduling assignments for both downlink and uplink, multiple antenna transmission configuration (e.g., precoding matrix used) for downlink, and power control commands for uplink. For security, each PDCCH is given a CRC that is scrambled with the RNTI associated with the intended recipient, which may be a UE-specific C-RNTI or a common SI-RNTI. The PDCCHs are organized on the basis of control channel elements (CCEs), where one CCE corresponds to nine REGs. A PDCCH may occupy between one and eight CCEs depending on the PDCCH format [200]. The PDCCHs are serially multiplexed and mapped to the remaining REs within the control region not occupied by the PCFICH, PHICH, or the DL-RS.

Physical Downlink Shared Channel   The remaining downlink physical resources that are not reserved for other purposes are used for PDSCH transmission, as shown in Figure 8.21. The UEs use scheduling information received on the PDCCH to determine which RBs to read in a given subframe. Three methods of PDSCH resource allocation are defined:

- *Type 0.* A simple bitmap is sent on the PDCCH, which indicates the resource block group (RBG) allocated to the scheduled UE. The RBG is a set of consecutive PRBs. No hopping is supported across slots within a subframe.

- *Type 1.* This is similar to the bitmap method of Type 0, but with associated DL bandwidth subsets. It is possible to define RBG sets with distributed resources over the bandwidth. No hopping is supported across slots within a subframe.

- *Type 2.* VRBs are used which are then mapped to PRBs. This approach provides flexibility and less signaling overhead than Types 0 and 1. This supports localized or distributed RB allocation and hopping across slots within a subframe for frequency diversity.

The resources of the PDSCH may be allocated to a single UE or to a group of UEs, as addressed by different types of radio network temporary identifiers (RNTIs) on the PDCCH. For example, a UE with an active RRC connection in the process of receiving a downlink transmission may be allocated PDSCH resources using either a cell RNTI (C-RNTI) or a semipersistent scheduling RNTI (SPS-RNTI), depending on the nature of the user data being received. Alternatively, if system information is being transmitted on the PDCCH for common control purposes, then the system information RNTI (SI-RNTI) is used [192].

8.1.16   Overview of E-UTRA Radio Protocols

The EPS protocol architecture has already been introduced in Section 8.1.11. We now take a closer look at the LTE-specific protocols of the E-UTRA air interface. The air interface protocols are divided into the user plane protocols, which are used to carry user data, and the control plane protocols, which are used to carry signaling between the UE and the network. The lower layer protocols are collectively called the access stratum (AS), and they are all terminated in the network side in a single node, the eNB. The PHY layer protocol has already been introduced, which makes up layer 1 of the protocol stack. The layer 2 functionality is divided among three sublayers: MAC, RLC, and packet data convergence protocol (PDCP). The overall structures of the layer 2 protocols are shown in Figure 8.28 for the downlink and Figure 8.29 for the uplink. In the control plane, the RRC makes up layer 3, which is the primary entity in charge of radio control functions. Above the RRC layer is the NAS protocol, which terminates in the network side at the MME and provides control signaling directly between the UE and the core network.

8.1.16.1   Medium Access Control Sublayer

The MAC sublayer forms the lower portion of layer 2 at the LTE air interface and is configured by the RRC layer above. It provides data transfer and radio resource allocation services to the upper layers. To provide these services, the MAC sublayer relies on the services provided by the PHY layer below, including lower layer signaling (e.g., HARQ feedback, scheduling requests) and channel measurements (e.g., CQI). The logical channels form the SAPs between the MAC and RLC sublayers. The MAC sublayer is responsible for mapping between logical channels and transport channels, as described previously in Section 8.1.12. The main functions of the MAC sublayer that support this service include [216]:

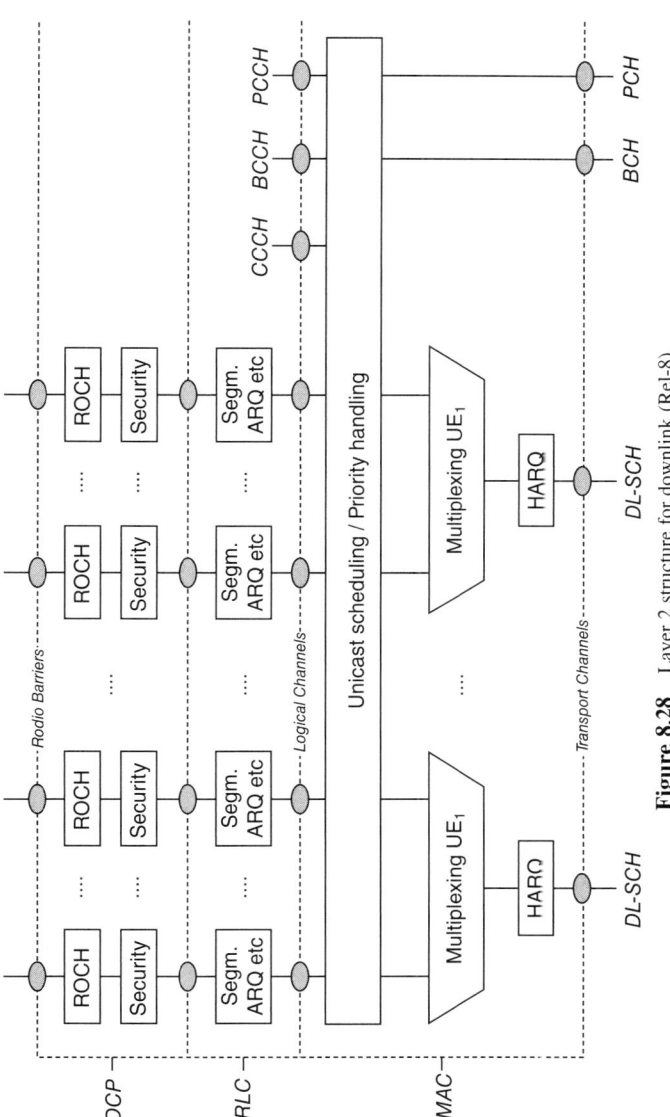

Figure 8.28 Layer 2 structure for downlink (Rel-8).

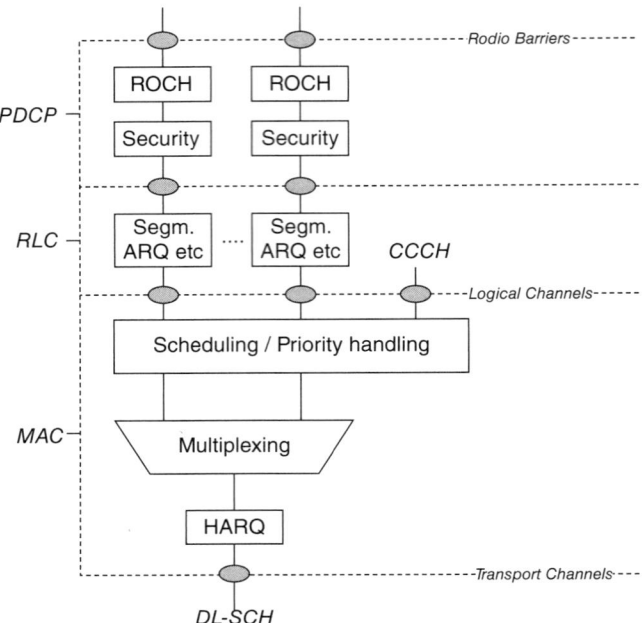

Figure 8.29 Layer 2 structure for uplink (Rel-8).

- Dynamic scheduling and scheduling information reporting.
- Priority handling of data flow associated with multiple UEs and between logical channels associated with a single UE.
- Multiplexing and demultiplexing of MAC SDUs from one or more logical channels onto/from TBs that are delivered to and received from the PHY layer via transport channels.
- Error correction through HARQ.
- Random access procedures which may be executed either in a contention-based process (e.g., for initial access) or non-contention-based process (e.g., for handover).
- Uplink timing maintenance, which encompasses the required procedures for mandatory periodic synchronization of UE with the cell.
- Transport format selection.

It is clear from this list of services that the MAC sublayer design is closely coupled to that of the PHY layer—particularly for functions such as scheduling and HARQ. There is one MAC entity per UE or eNB, which must multiplex and demultiplex packets to and from multiple radio bearers. Multiplexing and prioritization is performed to decide how much data are placed in a TB when radio resources are

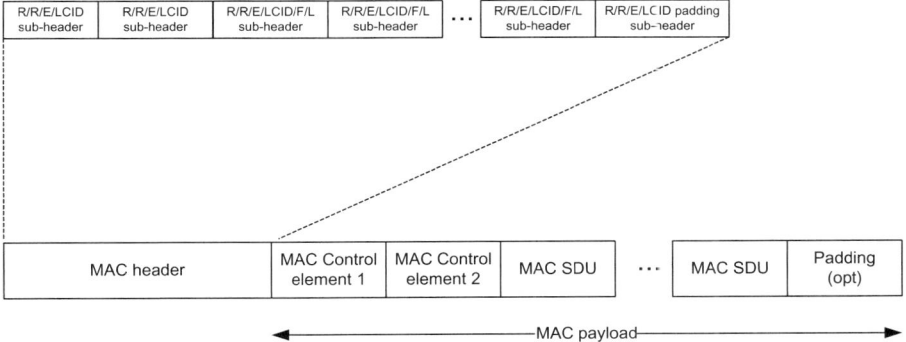

Figure 8.30   MAC PDU structure for DL-SCH and UL-SCH.

available. Demultiplexing reassembles packets and distributes them to the appropriate RLC entity. In LTE, scheduling is considered a MAC layer function and dynamic frequency-selective scheduling is supported based on scheduling information reported from the PHY. The MAC PDU structure used for the DL-SCH and UL-SCH is shown in Figure 8.30. Latency is reduced by allowing a "piggybacking" approach in which MAC control elements may be concatenated with payload from the layers above in a single transport block. The MAC protocol is defined in 3GPP Technical Specification 36.321 [216].

8.1.16.2 *Radio Link Control Sublayer*

The RLC layer is responsible for transferring upper layer PDUs provided by the RRC or PDCP. The RLC has three modes of operation that provide varying levels or service. Each RLC operating mode represents a different trade-off between reliability and latency.

- *Transparent Mode (TM).* As the name implies, this mode is essentially a pass through. None of the major RLC functions, such as ARQ or segmentation/concatenation, are applied in this mode. TM mode is used for common signaling channels, namely BCCH, PCCH, and CCH.

- *Unacknowledged Mode (UM).* In contrast to TM mode, UM mode is responsible for all RLC functions with the exception of ARQ functionality and resegmentation of RLC SDUs. UM mode provides an unreliable transport service that is appropriate for real-time applications, such as VoIP, where retransmissions are not required. UM mode may be used for DTCH as well as MBMS logical channels in Rel-9 and beyond.

- *Acknowledged Mode (AM).* AM mode offers a reliable transport service through the use of ARQ retransmissions. It also supports resegmentation of

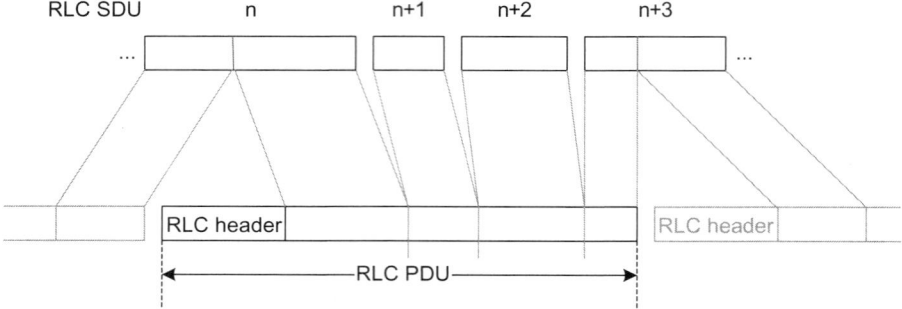

Figure 8.31   Segmentation and concatenation of RLC SDUs.

RLC SDUs. AM mode is always used for dedicated signaling on the DCCH and may optionally be used for user traffic on the DTCH.

The main functions of the RLC vary by mode of operation and include ARQ for error correction; concatenation, segmentation, and reassembly of RLC SDUs; in-sequence delivery of upper layer PDUs; duplicate detection; and protocol error detection and recovery. The RLC segments and concatenates packets from the PDCP to adapt packets to the needed size for the air interface as indicated by the MAC. This process is shown in Figure 8.31. When delivering packets to the PDCP, the RLC reassembles and reorders packets that may arrive out of sequence from the MAC. The RLC also provides error detection and correction through ARQ to handle residual errors not corrected at the MAC sublayer. There is only one RLC entity per radio bearer. The RLC protocol is defined in 3GPP Technical Specification 36.322 [217].

8.1.16.3   *Packed Data Convergence Protocol Sublayer*

The PDCP sublayer process IP packets received from the user applications. Header compression and decompression of user data is done using ROHC. In LTE, the PDCP also performs security functions over the air interface. These functions include ciphering as well as control plane integrity protection. During handovers, the PDCP handles retransmission and reordering to ensure lossless data transfer and in-sequence delivery of packets to the upper layers. There is one PDCP entity per radio bearer. The PDCP is also used for most SRBs associated with the RRC layer in the control plane. The PDCP protocol is defined in 3GPP Technical Specification 36.323 [218].

8.1.16.4   *Radio Resource Control Layer*

In the control plane, layer 3 is also included in the AS [328]. This layer is the RRC, and it is responsible for performing many of the functions previously implemented in the RNC of UMTS systems. The functions performed by the RRC are generally those associated with managing the access of UE to the network, and they include:

- Broadcast of system information related to the AS and NAS
- Paging
- Radio bearer management
- RRC connection management
- Security functions including key management
- Mobility functions including UE measurement reporting, handover, and cell selection
- QoS management
- Transfer of NAS signaling

The radio bearers associated with RRC signaling are referred to as SRBs. These radio bearers are only used for transmission of RRC and NAS messages. In the downlink, piggybacking of NAS messages is used for bearer establishment, modification, and release procedures. This approach reduces control plane latency associated with connection setup. Differences between RRC control signaling and MAC control signaling are summarized in Table 8.13 [192].

The system information from the RRC is divided into a number of different information blocks, which are passed to the lower layers through the BCCH. The *master information block* (MIB), transmitted on the BCH, has already been introduced. The MIB includes a limited number of the most essential cell parameters that are required to access the cell, for example, for network registration when the UE is powered on. In addition, a number of *system information blocks* (SIBs) are also defined. The SIBs are transmitted on the DL-SCH. The different types of information carried by the various SIBs are summarized in Table 8.14.

Table 8.13   Summary of Differences between MAC and RRC Control

	MAC control		RRC control
Control entity	MAC	MAC	RRC
Signaling	PDCCH	MAC control PDU	RRC message
Signaling reliability	$\sim 10^{-2}$ (no retransmission)	$\sim 10^{-3}$ (after HARQ)	$\sim 10^{-6}$ (after ARQ)
Control delay	Very short	Short	Longer
Extensibility	None or very limited	Limited	High
Security	No integrity protection No ciphering	No integrity protection No ciphering	Integrity protected ciphering

Table 8.14   Summary of Broadcast System Information

Block	Description
MIB	Essential cell information, including downlink channel bandwidth and system frame number
SIB 1	Cell selection information and scheduling of other SIBs
SIB 2	Radio resource configuration information common to all UEs
SIB 3	Common cell reselection information
SIB 4	Neighbor cell information for intra-frequency cell reselection
SIB 5	Interfrequency cell reselection information
SIB 6	Inter-RAT cell reselection information for UMTS
SIB 7	Inter-RAT cell reselection information for GSM
SIB 8	Inter-RAT cell reselection information for CDMA2000
SIB 9	Home eNB identity for femtocell deployments
SIB 10	Earthquake and Tsunami Warning System (ETWS) Primary Notification
SIB 11	Earthquake and Tsunami Warning System (ETWS) Secondary Notification

8.1.16.5 Nonaccess Stratum

The E-UTRA higher layer protocols run on top of the AS. In contrast to the AS, the higher-layer protocols communicate between the UE and the core network, using the eNB as a relay. In the control plane, the NAS protocol forms the highest layer of the protocol stack. It resides above the RRC layer and is terminated on the network side at the MME, thus providing signaling between the UE and the core network. The main functions of the NAS protocol are as follows:

(a) *Mobility Management.* Supports mobility of the UE. This includes informing the network of the present location of the UE, establishing and managing connections between the network and the UE, and providing user identity confidentiality. These functions are performed by the EMM sublayer.

(b) *Session Management.* Supports EPS bearer context handling in the UE and the MME. This includes establishment, maintenance, and release of EPS bearers. These functions are performed by the ESM sublayer.

The NAS procedures, especially the mobility management procedures, are based on and fundamentally similar to UMTS and GSM. An important improvement from UMTS is that the EPS allows concatenation of some NAS and AS procedures for bearer activation. This allows for faster connection establishment and speeds up idle-to-active transition, thus improving the control plane latency of LTE.

Security of NAS signaling between the UE and MME is an additional function of the NAS. NAS security protection involves the handling of EPS security contexts as well as integrity protection and ciphering of NAS messages. When a UE attaches with the network, a mutual authentication of the UE and the network is performed between the UE and the MME/HSS. This means that the network must authenticate the UE and the UE must authenticate the network. This authentication function establishes security keys that are used for encryption of the EPS bearers.

8.1.17 LTE Usage

The LTE ecosystem is developing so rapidly that it is impossible to provide an LTE market summary capable of remaining up to date for more than a few months. Therefore, in this section, we provide a snapshot of LTE deployment information up to 2012. A number of good resources are publicly available which provide regularly updated LTE market information online. The interested reader is directed to Reference [219].

8.1.17.1 Deployments

The first commercial LTE networks were launched in December 2009 by TeliaSonera, a Scandinavian network operator, in the capital cities of Stockholm, Sweden, and Oslo, Norway. Since then, LTE networks have been deployed in many countries around the world spanning every inhabited continent. Although LTE currently accounts for a small percentage of the global cellular marketplace in terms of number of connections, it is important to consider that LTE is still in the very early stages of deployment. As a point of comparison, consider the fact that in 2011—more than a decade after the initial commercial deployment of 3G—these 3G technologies accounted for just under 25% of all global connections while 2G technologies (primarily GSM) still accounted for nearly 75% of global connections. These numbers illustrate the point that, even in the fast-paced world of cellular, it takes a long time for a new technology to displace an older, massively deployed technology such as GSM. That being said, the LTE ecosystem has already demonstrated rapid growth in the relatively short time it has been commercially deployed. In fact, in terms of absolute number of connections after the first 2 years of service, LTE is officially the fastest growing cellular technology of all time. This is no doubt influenced by the fact that the cellular market is larger than it ever has been in the past.

8.1.17.2 Spectrum

Spectrum is a big topic in LTE deployment at the moment. Nearly 40 frequency bands of operation have been defined in the 3GPP standard to support LTE network deployments. It has been up to the network operators and the regional and national telecommunications authorities to determine what frequencies will be used in practice. Some predictions suggest a future in which LTE is deployed in many of the supported bands, possibly enabled by software-defined radio devices which can easily switch between bands. However, for the time being, most networks and devices are constrained by RF limitations to a limited number of bands. Some bands have been newly defined for use by emerging 4G cellular technologies, such as the 2.6 GHz IMT Extension band—defined by the ITU to support IMT-Advanced deployments—or the so-called Digital Dividend bands made available by the global switch from analog to digital television. LTE will also be deployed in frequency bands previously defined for 2G and 3G networks. These deployments are a combination of spectrum refarming (e.g., replacing existing 2G deployments with LTE) and deployment in spectra that was allocated for 2G/3G networks but never utilized in certain regions (e.g., 1800 MHz band).

The key LTE frequency bands in the current LTE ecosystem are briefly summarized in the following list. The corresponding 3GPP operating band numbers are provided. For a full description of downlink and uplink frequencies, see Table 8.7 and Table 8.8 in Section 8.1.13.3.

- *700 MHz Band.* This band is also known as the "Digital Dividend" band, because these frequencies have been made available by the global switch from analog to digital television. The timeline for this band to become available varies by country and has already occurred in many countries around the world. Since this band is tied to former TV spectrum, the exact frequencies vary by country and region. This is currently the primary band in terms of number of users, largely due to its use in the U.S. market. The low frequency has good propagation characteristics to provide large cell sizes for broad coverage in lower-density areas (e.g., rural deployment or macrocell compliment to high-frequency microcells). There are actually a number of operating bands defined by the 3GPP which operate near 700 MHz and are commonly referred to as "700 MHz Band." These bands include Bands 12, 13, 14, 17 in North America and Band 28 in Asia Pacific (APAC).

- *800 MHz Band.* This is the digital dividend band in Europe, which exists at different frequencies than North America. In most European countries, this band was made available by 2012 as part of the switch to digital television. This is a primary band of operation in Europe, and is commonly used in conjuncture with the 2600 MHz band. Corresponds to Band 20.

- *1800 MHz Band.* This is one of the main GSM bands used in the majority of the world outside North America. There has been a lot of industry push to refarm this spectrum for LTE use for a number of key reasons. It offers a large bandwidth (i.e., 75 MHz of paired spectrum), it is harmonized across most of the world, it is largely unfragmented (i.e., licenses are available for 10+ MHz of consecutive spectrum), and in many parts of the world, it is only partially utilized for existing cellular technologies. The 1800 MHz is poised to be the primary GSM cellular refarming band. Corresponds to Band 3.

- *AWS (1700/2100 MHz) Band.* This band is known as the Advanced Wireless Services (AWS) band. It includes a block of uplink frequencies in the 1700 MHz range and an associated block of downlink frequencies in the 2100 MHz range. It is used throughout the Americas. Corresponds to Band 4. An extended AWS band is also defined (Band 10).

- *2600 MHz Band.* This is the so-called IMT Extension band that has been defined by the ITU for use by enhanced 3G and 4G standards (e.g., HSPA, LTE, LTE-Advanced) to support the explosion in mobile broadband data service usage. Since this band has been established specifically to support wide-bandwidth advanced cellular technologies, there is plenty of spectrum to support up to 20 MHz LTE configurations, making this band ideal for high throughput cells. The trade-off, however, is decreased cell size and building penetration due to the high frequency propagation characteristics. Corresponds to Band 7.

- *TDD 2300 MHz Band.* This is one of two main bands supported for TD-LTE network deployments. The TDD 2300 MHz band is expected to be widely deployed in China by network operator China Mobile Corresponds to Band 40.

- *TDD 2600 MHz Band.* Along with the TDD 2300 MHz Band, this is the second band to see widespread adoption for TD-LTE network deployments. This band is defined in the band gap between the uplink and downlink portions of the 2600 MHz IMT Extension band described earlier for FDD mode. This band corresponds to Band 38, though a few devices support the overlapping Band 41 as well.

8.1.17.3 *Devices*

Device availability is a crucial component of any thriving technology ecosystem. LTE user devices have been designed for a variety of applications including mobile, nomadic, and fixed wireless usage scenarios. Correspondingly, these devices come in a number of different form factors. LTE Universal Serial Bus (USB) data modems were the first LTE devices to hit the market and have been commercially available in many parts of the world since 2010. These devices are small form-factor USB dongles, often with a built-in antenna, which are commonly used for nomadic wireless broadband Internet access for laptop computers. LTE-enabled smartphones, on the other hand, did not appear on the market until 2011. It can safely be said that 2011 was the year of the LTE smartphone—as nearly 50 LTE-enabled handsets were announced by end of year 2011 [220]. In general, these early smartphones supported LTE for data only, and used existing 2G and 3G circuit-switched networks to carry voice traffic (i.e., circuit-switched fallback).

As LTE network deployments and subscriber base have increased, so have the number of LTE consumer devices on the market. According to the Global Mobile Suppliers Association (GSA), there were over 400 LTE user devices announced by 67 manufacturers as of July 2012 [221]. The predominant form factor in terms of number of devices launched was the LTE router, accounting for 38% of all devices launched. This category actually encompasses two types of devices that serve similar functions. First, it refers to fixed wireless devices that use LTE for wireless broadband access as a wireless alternative to digital subscriber line (DSL) or cable, for example. Second, this category also refers to so-called LTE mobile hotspots, which tend to have smaller form factor and operate off a battery, allowing for portability. As a consequence of the smaller size, the mobile hotspot devices generally do not support wired connectivity via Ethernet but rather rely solely on WLAN connectivity. The second and third most common device form factors were LTE smartphones and LTE USB modems, respectively. Additional form factors included modules, tablet computers, notebook computers, personal computer (PC) cards, and femtocells. The term "module" here refers to an LTE modem that is designed for embedded applications. They may be integrated into a consumer device (e.g., tablet computer, notebook computer) or used in M2M applications. These early LTE devices, including all devices announced prior to 2011, conformed to UE category 3—as defined previously in Section 8.1.5.1. The first UE category 4 devices were announced at the GSMA Mobile World Congress in February 2012 and became commercially available by early 2013 [222].

8.1.17.4   TD-LTE Ecosystem

Although most LTE deployments use FDD mode, TD-LTE has been gaining industry momentum. TD-LTE has the potential to fill a number of roles in mobile and wireless broadband networks. One such role is the 4G successor to TD-SCDMA, the UMTS-TDD variant developed for the Chinese market. So far, TD-SCDMA has seen widespread deployment only in China by network operator China Mobile. When LTE was still in the early stages of development, it was a logical question to wonder whether TD-LTE would be confined to the Chinese market and possibly a few other niche markets around the world. As it turns out, TD-LTE has begun to see deployment on a global scale. In fact, the first network operator to commercially launch TD-LTE was the Polish operator Aero2, whose network supports both FDD and TDD modes as complementary access schemes. By mid-2012, nine commercial TD-LTE networks had been launched in eight countries, with numerous other network operators investing in predeployment or commercial trials [222].

As the TD-LTE ecosystem develops, TD-LTE is emerging as a global technology. In addition to providing an evolutionary pathway to TD-SCDMA, it also provides an evolutionary pathway for Mobile WiMAX. As an example, the U.S. network operator Clearwire has announced plans to deploy TD-LTE in the 2.5 GHz TDD band (band 41). TD-LTE is also seen as a complement to FDD deployments, and a small number of network operators have deployed both FDD and TDD modes. Such developments open new opportunities for global cellular roaming. An industry consortium called the Global TD-LTE Initiative has been formed by a number of key players in the cellular world to promote the development of the TD-LTE ecosystem. Their Web site (http://www.lte-tdd.org/) is a good resource for up-to-date news on the development of the TD-LTE ecosystem.

8.1.18   Further Reading

Table 8.15 summarizes some of the key 3GPP technical specification documents related to LTE/SAE. Although this list is not exhaustive, it provides the interested

Table 8.15   LTE Technical Specifications for Further Reading

Topic	3GPP specification number	Initial relevant release
EPS Network architecture, including interworking with 3GPP access networks	23.401	Rel-8
EPS network architecture for interworking with non-3GPP access networks	23.402	Rel-8
PHY layer overview (radio aspects)	36.201	Rel-8
E-UTRA and E-UTRAN overall description	36.300	Rel-8
E-UTRA MAC protocol	36.321	Rel-8
E-UTRA RLC protocol	36.322	Rel-8
E-UTRA PDCP protocol	36.323	Rel-8
E-UTRA RRC protocol	36.331	Rel-8
E-UTRAN NAS protocol	24.301	Rel-8
PCC architecture (including PCRF, PCEF, etc.)	23.203	Rel-8
E-UTRA Multiplexing and Channel Coding	36.212 [324]	Rel-8

reader with a good launching point for investigating various aspects of LTE systems in greater detail within the standard itself. The authors also recommend the following texts for a more comprehensive overview of LTE and LTE-Advanced.

C. Cox, *An Introduction to LTE: LTE, LTE-Advanced, SAE and 4G Mobile Communications*, John Wiley & Sons, Chichester, UK, 2012 [215].

H. Holma and A. Toskala, *LTE for UMTS: OFDMA and SC-FDMA Based Radio Access*, John Wiley & Sons, Chichester, UK, 2009 [188].

8.2   LTE-ADVANCED

LTE-Advanced represents a major milestone in the open evolution of the LTE air interface. First defined in 3GPP Rel-10 in 2011, LTE-Advanced achieves a number of performance enhancements over Rel-8/9 LTE, allowing it to meet or exceed all of the requirements set forth by the ITU for IMT-Advanced. Thus, LTE-Advanced is the first 3GPP RAT to officially bear the pedigree of 4G cellular technology in the ITU sense of the term. It is important to note that LTE-Advanced does not define a fundamentally new air interface technology, but rather it is an evolution of the LTE air interface described earlier in this chapter. This allows for backward compatibility between LTE-Advanced and earlier releases of LTE. This means that an LTE user device conforming to Rel-10 can operate on a Rel-8/9 network and vice versa. In this section, we provide an introduction to the key enhancements defined in 3GPP Rel-10 as part of LTE-Advanced—namely, carrier aggregation for wider bandwidths and increased bandwidth flexibility; enhanced MIMO multiple antenna transmission for improved spectral efficiency; relay nodes for expanded coverage; and enhanced support for heterogeneous network (HetNet) deployments.

8.2.1 Carrier Aggregation

The requirements for IMT-Advanced stipulate support for up to 40 MHz of bandwidth. Therefore, a means of supporting wider channel bandwidth configurations than that of Rel-8/9 LTE is an important requirement for LTE-Advanced. In theory, it would be possible to directly extend the bandwidth of LTE by simply aggregating more subcarriers to a single E-UTRA channel in order to define wider bandwidths than the 20 MHz maximum supported in earlier LTE releases. However, there are a couple disadvantages to this approach. First, this approach is not amenable to backward compatibility with Rel-8/9 UEs, largely due to the resource mapping structure of physical control channels. Furthermore, this approach limits deployments to the specific scenario where a contiguous spectrum block wider than 20 MHz is available.

To support bandwidth flexibility, as well as backward compatibility with Rel-8/9 devices, LTE-Advanced uses a technique called *carrier aggregation* to achieve wider bandwidth deployments. The basic concept is to create wider channel bandwidths by aggregating multiple E-UTRA carriers together, each called a *component carrier*. Each component carrier is essentially a normal LTE Rel-8/9-compliant carrier. In this way, LTE-Advanced devices conforming to Rel-10 or beyond may make use of the wider bandwidth associated with multiple component carriers through carrier aggregation. At the same time, older Rel-8/9 devices still see each component carrier as a normal LTE carrier and can therefore use one of the component carriers without requiring any modification to the device.

Rel-10 LTE advanced supports up to five component carriers allowing for a maximum aggregated channel bandwidth of up to 100 MHz. Each component carrier may be a different bandwidth. When carrier aggregation is configured, the UE only maintains one RRC connection through a single serving cell, called the primary cell. The component carrier associated with the primary cell is called the primary component carrier (PCC), which is typically used for uplink and downlink control signaling. The UE may then be configured by the eNB to support multiple secondary component carriers (SCCs), which are generally used for data transmission. To support this paradigm, so-called *cross-carrier scheduling* is defined, which allows the PDCCH from the one component carrier (e.g., the PCC) to allocate resources on other component carriers (e.g., one or more SCCs) [223]. A UE may be configured to use carrier aggregation on a different number of SCCs in the downlink than in the uplink to support asymmetric throughput scenarios. However, it is not possible for the UE to be configured to have more uplink SCCs than downlink SCCs. At the MAC layer, there is one independent HARQ entity associated with each component carrier. To reduce UE batter consumption, an activation/deactivation mechanism is defined at the MAC layer that allows SCC use to be quickly reconfigured [223]. Carrier aggregation in Rel-10 is solely a MAC and PHY layer enhancement. It is not visible to the layers above, except that higher data rates must be supported.

Carrier aggregation in LTE-Advanced is intended to do more than simply accommodate very wideband deployments. It also makes it possible for operators to utilize fragmented spectrum licensed for one or multiple frequency bands. Therefore, three distinct use cases have been defined, as illustrated in Figure 8.32 [224]. The simplest use case is intraband contiguous carrier aggregation. For fragmented spectrum scenarios, two cases have been defined: intraband noncontiguous and interband carrier aggregation. The latter can be considered as the most generalized scenario, where the carriers span multiple frequency bands and each component carrier may be contiguous or noncontiguous within a given band. These three scenarios may be dealt with in a common way at baseband [225]. However, RF performance requirements must be developed, not only for each of the three different use cases, but for each frequency band or combination of frequency bands. The focus on initial carrier aggregation work has been to support 20–40 MHz deployments and the intraband contiguous or interband use cases. Additional RF requirements are continually being developed as part of the ongoing standardization process based on input from the network operators [224].

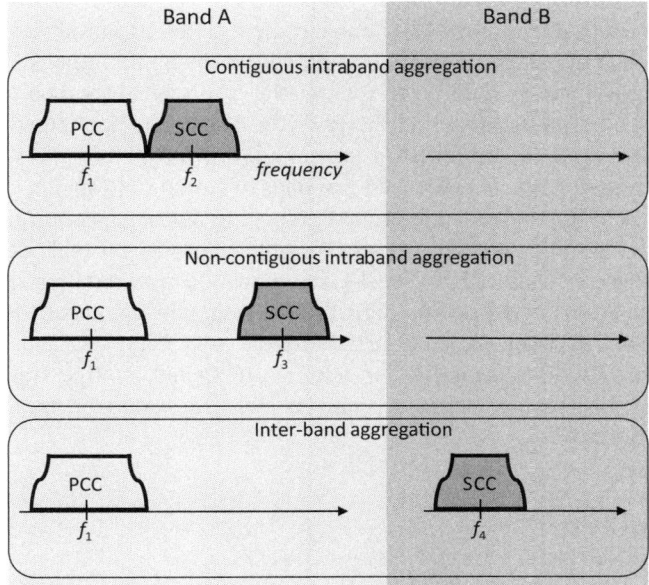

Figure 8.32   Carrier aggregation use cases.

8.2.2   MIMO Spatial Multiplexing Enhancements

Rel-10 LTE-Advanced achieves increased peak spectral efficiency in both the down-link and the uplink through the definition of higher-order MIMO spatial multiplexing schemes. In the downlink, up to eight spatial layers are supported allowing, for example, 8×8 MIMO spatial multiplexing scenarios. Rather than simply increasing the number of cell-specific RS, which would result in considerable overhead in the downlink for non-spatial-multiplexing scenarios, a new set of reference signals are defined. In Rel-8 LTE, the downlink reference signals are used both for coherent demodulation and as a power reference for channel quality measurements. In Rel-10, these functions are split between two sets of reference signals. UE-specific RSs are transmitted on antenna ports 7–14 to support coherent demodulation. These RSs are actually an extension of the UE-specific RS designed in Rel-9 to support dual-layer beamforming on antenna ports 7 and 8. Additionally, a new set of reference signals called *CSI reference signals* (CSI-RS) are transmitted on antenna ports 15–22 to support channel quality measurements. The CSI-RS are not transmitted in every subframe, but rather have a long periodicity that is configurable from twice per subframe to once every eight subframes, thereby minimizing overhead [199]. Eight-layer MIMO spatial multiplexing is supported by the new PDSCH transmission mode 9. In addition to SU-MIMO spatial multiplexing, enhanced MU-MIMO spatial multiplexing is supported for up to eight antenna transmission. In the MU-MIMO case, up to eight spatial layers may be divided among multiple users, for example four layers to UE #1 and four layers to UE #2. The specific mapping of the new UE-specific and CSI reference signals for eight antenna transmission is defined in 3GPP Technical Specification 36.211 [199].

In addition to downlink MIMO enhancements, Rel-10 introduces support for multiple antenna transmission from a single UE on the uplink. Specifically, closed-loop spatial multiplexing with up to four spatial layers is now supported for 4×4 SU-MIMO spatial multiplexing on the PUSCH. The UE is configured into one of two transmission modes for PUSCH transmission. Mode 1 corresponds to single antenna transmission, while mode 2 corresponds to closed-loop spatial multiplexing. All CSI required for rank adaptation and precoding matrix selection may be measured at the eNB. The eNB in turn specifies the transmission rank and precoding matrix to be used by the UE in the DCI as part of the scheduling grant. A new set of antenna ports are defined on the UE-side to support uplink MIMO operation, as summarized in Table 8.16 [199]. Different antenna ports are defined for PUSCH and PUCCH transmission. In addition to spatial multiplexing support on the PUSCH, open-loop transmit diversity is now supported on the PUCCH for improved robustness.

8.2.3   Relays

A new type of base station node has been introduced in Rel-10 called a *relay node* (RN). The purpose of the RN is to allow operators to improve and extend coverage

Table 8.16   Uplink Antenna Ports in Rel-10 LTE-Advanced

Antenna ports	Channels	Application
10	PUSCH, SRS	Single-antenna transmission
{20, 21}		Two-antenna closed-loop spatial multiplexing
{40, 41, 42, 43}		Four-antenna closed-loop spatial multiplexing
100	PUCCH	Single-antenna transmission
{200, 201}		Two-antenna open-loop transmit diversity

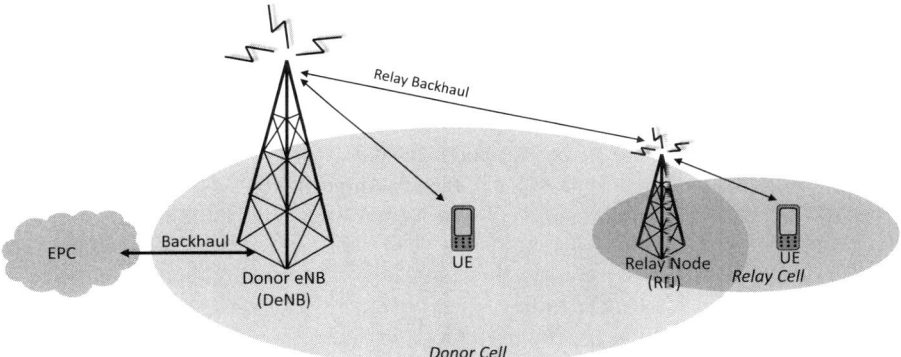

Figure 8.33   LTE-Advanced relay node topology.

by introducing low-power nodes without the requirement of a wired backhaul link
to the rest of the network [226]. The RN can be considered as a special type of eNB
that uses a wireless backhaul to connect to the rest of the RAN. As shown in Figure
8.33, the backhaul for the RN is provided wirelessly by a so-called Donor eNB
(DeNB) [225]. The relay backhaul link between the RN and the DeNB is a modified
version of the E-UTRA air interface, which has been defined as the new *Un interface.*
This interface serves the functionality normally associated with the S1 interface as
well as the X2 interface, which allows ICIC mechanisms to be utilized by the RN.
From the UE perspective, the RN appears as a normal eNB, providing network-side
termination of the E-UTRA radio protocols. The RN must also support a subset of
UE functionality in order to realize the wireless backhaul over the Un interface.
From a frequency planning perspective, two types of RNs are defined: out of band
and in band. For out-of-band operation, the backhaul link and the access link (i.e.,
RN-to-UE link) associated with the RN operate on two separate frequencies. While
this approach avoids intercell interference, it requires additional spectrum. A much
more attractive approach for network operators is in-band operation, in which the
backhaul and access links associated with the RN use the same frequency. Clearly,

in-band operation poses much greater design challenges. Therefore, mechanisms to support in-band relaying scenarios have been the focus of 3GPP work in this area.

8.2.4 Enhanced Support for Heterogeneous Network Deployments

With the deployment of 4G technology, the performance of the point-to-point cellular radio link is now approaching its information theoretical limits for spectral efficiency. In the face of exponentially increasing data consumption over cellular networks, new means must be found to improve the overall cell throughput and single user data rates that can be achieved in cellular networks. Wider bandwidths through carrier aggregation is one option, however this approach is often limited by economic, regulatory, and practical RF restrictions. One approach that has traditionally been used in macrocellular networks is to increase cell density through cell splitting or sectorization. However, as macrocellular networks become increasingly dense in urban areas, this approach is reaching its limits as well. Heterogenous network (HetNet) deployments offer a potentially attractive approach to increase network capacity and improve user experience through the introduction of small cells that may overlap with existing macrocells in both spectrum and coverage area.

Cellular networks are traditionally designed to cover a geographic area using a homogenous network of macrocells. The typical data rates that are experienced by the users in a practical macrocellular deployment scenario are generally much less than the peak theoretical data rate for a given technology. They can also vary widely, with users close to the base station experiencing much higher data rates than average users or those closer to the cell edge. In a HetNet deployment, the macrocellular network is complimented by small cells that are more access-point like, similar to a WLAN paradigm. The small cells may be used for hot-spot-type coverage in high usage areas or to provide coverage in dead spots. Essentially, HetNet deployments increase the practically achievable data rates and improve the overall user experience by bringing the base station closer to the user.

HetNet deployments are already supported by the LTE Rel-8 specification, which defines support for picocells, femtocells (i.e., Home eNBs), and frequency-based ICIC. In Rel-10, further emphasis has been placed on defining new mechanisms to support HetNet deployments. A number of different types of small cells are supported in LTE-Advanced, as summarized in Table 8.17. Additional approaches that have been proposed include the use of remote radio heads (RRHs) and massively distributed antenna architectures [228].

A simplified example of a HetNet deployment involving one macrocell and one picocell is shown in Figure 8.34 [225]. For this example, a cochannel deployment is assumed in which the macrocell and picocell operation are in shared spectrum The presence of overlapping high-power base stations (i.e., macrocell) and low-power base stations (i.e., small cell) generally puts the lower-power base station at a disadvantage. In traditional networks, cell association is based primarily on downlink received signal strength. However, for uplink transmission, the optimal cell

Table 8.17   Types of Small Cells Supported in Rel-10 LTE-Advanced

Type	Initial release	Typical output power[a]	Description
Picocell	Rel-8	250 mW–2 W[b]	Low-power eNB with wired S1 backhaul, typically deployed by the network operator in hotspot areas with omnidirectional antennas.
Femtocell	Rel-8	≥100 mW	Called *Home eNB* (HeNB) in LTE, typically deployed indoors by the end-user in a home or enterprise environment, may use a closed subscriber group (CSG) configuration to limit access
Relay	Rel-10	250 mW–2 W[b]	Similar to a picocell, but uses a wireless LTE backhaul via the Un interface. See Section 8.2.3

[a]Power levels based on Reference [227].
[b]Power levels assume outdoor deployment.

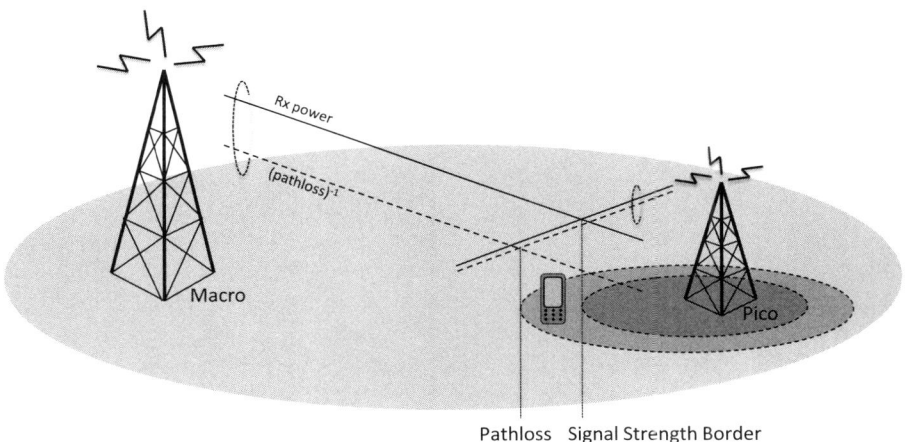

Figure 8.34   Heterogeneous network deployment example.

association should be based on the minimum path loss. To avoid underutilization of disadvantaged small cells, Rel-10 LTE-Advanced defines a mechanism to allow the eNB to intentionally bias the handover offset values to transfer more UEs to the picocell under certain conditions [198]. This mechanism essentially extends the coverage are of the picocell, and is often referred to as *cell range expansion*.

For a UE that has been transferred to the picocell using biased handover offset values, such as the UE shown in Figure 8.34, there is the potential for strong intercell interference on the downlink. This arises from the fact that the serving cell (i.e., the

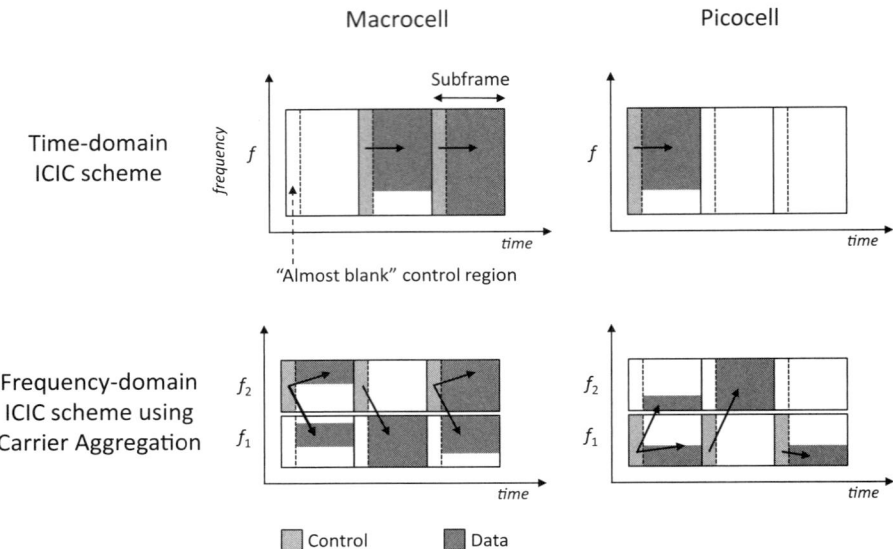

Figure 8.35   Enhanced ICIC schemes in LTE-Advanced.

picocell) is not the cell with the strongest received signal strength on the downlink. Therefore, some form of interference mitigation is required. For data transmission, this is not a problem because frequency-domain ICIC is already supported for PDSCH in Rel-8 LTE. This approach leverages X2 signaling and frequency-selective scheduling to avoid downlink interference between the macrocell and the picocell. One problem, however, is that Rel-8 ICIC does not apply to control signaling. For large cell range expansion scenarios, this can lead to radio link failure. Therefore, new eICIC mechanisms are defined in LTE-Advanced.

Rel-8 ICIC mechanisms can be characterized as frequency-domain based. In Rel-10, a new time-domain ICIC scheme is defined. This scheme is illustrated in the upper part of Figure 8.35 [225]. This scheme makes use of the so-called *almost-blank subframe* (ABS), which includes very minimal transmission from the eNB [198]. The macrocell transmits ABSs on certain subframes, which are then used by the picocell for transmission. The UE may also be configured to take channel measurements during the ABS. This approach effectively provides time separation for at least some of the control regions. From a backward compatibility perspective, this approach is beneficial because a Rel-8/9 UE may still take advantage of the full system bandwidth. However, these devices will not benefit from cell range expansion.

In addition to time-domain ICIC, interference mitigation can be achieved through the use of cross-carrier scheduling associated with carrier aggregation. This approach is illustrated in the lower part of Figure 8.35. In this approach, the available spectrum is divided into two E-UTRA carriers. One carrier serves as the PCC for the macrocell, while the other carrier serves as the PCC for the picocell. This effectively divides the control region at the start of the subframe into two channels, nonoverlapping in frequency, thereby reducing control channel interference. The carrier that is not used as the PCC may then be used solely for data transmission as an SCC. In this way, carrier aggregation cable UEs are still able to make use of the entire data region through the use of cross-channel scheduling. For Rel-8/9 UEs, this scheme will essentially appear as though the macrocell and picocell operate on separate E-UTRA carriers (i.e., not a cochannel deployment). Therefore, Rel-8/9 UEs will benefit from larger picocell coverage. However, this comes at the expense of bandwidth, as the Rel-8/9 UE can only access one of the component carriers at a time.

8.2.5 Further Reading

For the reader interested in exploring the LTE-Advanced specifications, all of the 3GPP specification numbers for LTE listed previously in Section 8.1.18 are applicable to LTE-Advanced as well. The exception is that LTE-Advanced features begin in Rel-10. Additionally, the study items that preceded the development of LTE-Advanced are listed in Table 8.18. The authors also recommend the following texts on LTE:

> H. HOLMA and A. TOSKALA, *LTE Advanced: 3GPP Solution for IMT-Advanced*, John Wiley & Sons, Chichester, UK, 2012 [320].

> S. SESIA, I. TOUFIK, and M. BAKER, *LTE The UMTS Long Term Evolution: From Theory to Practice*, 2nd Edition, John Wiley & Sons, New York, 2011 [198].

8.3 IEEE 802.16M

IEEE 802.16m, which is an amendment to the IEEE 802.16-2009 standard, was ratified as a technology standard in May 2011 [49]. Along with LTE, IEEE 802.16m is the other primary technology to have been officially declared as "4G." In October 2010, the ITU approved the "WirelessMAN-Advanced" air interface of IEEE

Table 8.18 Additional LTE-Advanced Technical Specifications for Further Reading

Topic	3GPP specification number	Initial relevant release
Feasibility study for LTE-advanced	36.912	Rel-9
Requirements for LTE-advanced	36.913	Rel-8
All LTE specifications as listed in Table 8.15	–	Rel-10
E-UTRA and E-UTRAN Overall Description	36.306 [325]	Rel-10

802.16m as an IMT-Advanced technology. This 4G-compliant WirelessMAN-Advanced air interface of 802.16m now forms the basis of WiMAX Release 2, the next-generation technology being developed by the WiMAX Forum.

We consider WiMAX Release 2 an important part of the wireless networking landscape. We do not, however, provide a detailed overview of WiMAX Release 2 or its base technology specification IEEE 802.16m. We hope to remedy this in future editions. Rather, we refer readers to our favorite references on the subject:

> S. AHMADI, *Mobile WiMAX: A Systems Approach to Understanding IEEE 802.16m Radio Access Technology*, Elsevier Inc., Burlington, MA, 2011.

> M. ERGEN, *Mobile Broadband Including WiMAX and LTE*, Springer Science and Business Media, New York, 2009.

ACKNOWLEDGMENTS

Chapter 8 contains numerous figures and tables reproduced or adapted from 3GPP Technical Specifications. Each figure is individually referenced indicating the source 3GPP document. The authors would like to thank ETSI for kind permission to reproduce this information. Copyright 1999–2012. TSs and TRs are the property of ARIB, ATIS, CCSA, ETSI, TTA and TTC, who jointly own the copyright in them. They are subject to further modifications and are therefore provided "as is" for information purposes only. Further use is strictly prohibited.

Other figures and tables have been reproduced from John Wiley & Sons, Inc. sources, specifically Figures 8.1, 8.12, 8.13, 8.20, 8.21, and 8.25 along with Tables 8.3, 8.12, and 8.16. Each figure is individually referenced indicating the source. This material is reproduced with permission of John Wiley & Sons, Inc.

Chapter 9

Mobile Internetworking

Wireless networks are often built and operated with host devices aiming to provide a wired-network experience. A simple example may reference a laptop using a Wi-Fi card to access the Internet. If the bandwidth is high enough and the number of users low, the user experience varies little compared with a wired Internet connection. However, the behavior of wireless networks in the context of the five-layer TCP/IP model may look drastically different compared with a wired network. This is because wireless networks face different challenges. Unlike a wired network, wireless networks must operate in the presence of potentially many interference sources and changing channel conditions. Wireless platforms can be mobile, resulting in changing network topologies, which severely challenges traditional network layer approaches. Furthermore, contention between users sharing the same wireless medium requires additional mechanisms to share the channel fairly. Most of today's wired networks do not share these constraints—they are essentially error free and available bandwidth is much higher. Switched Ethernet has essentially rendered the CSMA/CD mechanism irrelevant, as contention between Ethernet hosts now takes place on the backplane of the switch and not on the actual transmission medium. This does not mean that wired networks have unlimited bandwidth, however. Such networks do experience congestion conditions and must still share available bandwidth.

This chapter discusses some of the higher-layer challenges that are often encountered when attempting to provide the wired network experience to wireless devices.

9.1 WHAT IS MEANT BY MOBILE INTERNETWORKING?

Before we continue, it is important to clearly define what we mean when we say mobility, as this is a highly overused term to describe many attributes of a network.

Nodal Mobility. Network membership mobility (i.e., nodal mobility), but not subnetwork mobility (e.g., a network of mobile nodes, all moving relative to one another, but the point of attachment to the larger network such as the

Wireless Networking: Understanding Internetworking Challenges, First Edition.
Jack L. Burbank, Julia Andrusenko, Jared S. Everett, and William T.M. Kasch.
© 2013 the Institute of Electrical and Electronics Engineers, Inc. Published 2013 by John Wiley & Sons, Inc.

Internet is not moving—the network as a whole is not moving, at least as far as the Internet can tell).

Network Mobility. Mobility of a network in its entirety (not individual nodal mobility) (e.g., think of a collection of networked sensors on a vehicle. They are not moving relative to one another, but as a group they are moving relative to the rest of the network).

Mobile Ad hoc Network (MANET). Ad hoc networks (fixed or stationary) are those that can be formed in an immediate timeframe, without the need for preplanning or configuration. Rather, they form as needed. This addresses both network membership mobility and subnetwork mobility

The cases of nodal and network mobility are both predicated upon the concept of a fixed infrastructure. Only the generalized MANET makes no assumptions regarding infrastructure.

Thinking in terms of the key wireless networking technologies that we have considered in this book, the challenges they face all can be placed into one or more of these categories. It is obvious that cellular technologies such as GSM, UMTS, and CDMA2000 face nodal mobility challenges, as mobile handsets move about a cellular network. This is also true with advanced IEEE 802.11 networks, where users may roam between access points. Clearly, this challenge spans a wide range of wireless networking technologies.

In those cellular networks, base stations, the bridge between the wired and wireless domains, remain stationary. Also, IEEE 802.11 access points are also typically stationary. So at first glance one would think that network mobility is not problematic an issue. However, there is some degree of network mobility to be considered. There are numerous examples of cellular devices that also provide IEEE 802.11 access point capability to support multiple devices, utilizing the cellular access network as the Internet backhaul. However, this type of mobility is somewhat masked from the larger network (cellular wireless service provider or Internet in general) through the proxy nature of these devices. Nevertheless, there are still technical challenges surrounding providing seamless service to a group of users, and their potentially diverse set of applications, in a network that may be moving between points of attachment to the Internet.

The generalized MANET problem space is often considered slightly more academic in nature. However, there are numerous applications that we can think of where a MANET capability would prove useful. The obvious example is that of the military scenario, where there may not be existing infrastructure. Examples of the types of mobility scenarios that are generally considered by ad hoc network researchers can be found in Reference [229].

9.2 NETWORK LAYER CONSIDERATIONS

IP, and its supporting suite of protocols, has proven itself to be incredibly resilient over the past decades. The Internet has seen growth that none of its original creators could have foreseen. While many have predicted the failure of IP, it has, until now, proved all the naysayers wrong. However, it is clear that the widespread high-performance wireless networks of today are placing an incredible toll on the Internet, and IP itself. This section discusses some of the performance issues that wireless networks face and the problems posed for today's Internet architecture.

9.2.1 Addressing

IP inherent design is predicated upon fixed networks, not the mobile networks of today. This is clear through the fact that an IP address represents not only *who* a node is, but also *where* that node is. The network prefix portion of an IP address is a *locator* of a network device, where the remainder of the IP address is the *identifier* of the network device. It is the network prefix upon which the Internet's infrastructure (e.g., routers) acts to forward traffic from source to destination. So now imagine how this architecture reacts to the scenario where network clients are now free to move about, changing its point of attachment to the Internet. In this scenario, what does the IP address of that node mean? Does the device need to now obtain a new address every time it moves? While this may not seem like a difficult task, now imagine hundreds of millions of network devices moving, all requiring themselves to be renumbered periodically.

There have been several proposed solutions to this problem, all with strengths and weaknesses. However, the recurring theme of these solutions is to hide from the greater Internet the fact that these devices are mobile. An example of such a solution is Mobile IP (MIP).

9.2.1.1 Mobile IP

Mobile IP was developed within the IETF by the MIPv4 and MIPv6 Working Groups. MIPv4, defined by RFC 3344 [230], provided mobility support for IPv4. MIPv6, defined by RFC 3775 [231], provided mobility support for IPv6. These solutions were focused on addressing the problem of nodal mobility.

At the core of this protocol is the concept of each node having two IP addresses: a "home address" and a "care-of address." A mobile node maintains its identity represented by its home address (HoA). Regardless of its location, a node is reachable at its home address. This represents the node's "real" IP address, the address by which the rest of the world knows the node. When the node becomes mobile, a care-of address (CoA) represents the node's current point of attachment to the network. In order for another node (e.g., a correspondent node or core network as referred to in RFC 3775) to reach the mobile node (MN), a mechanism is needed to map the home address to a care-of address. The mechanism defined in RFC 3775 requires the assistance of a node within the home network of the MN.

A home agent (HA), located in the mobile node's home network, assists the mobile node. The HA performs proxy functionality for the mobile node on the home

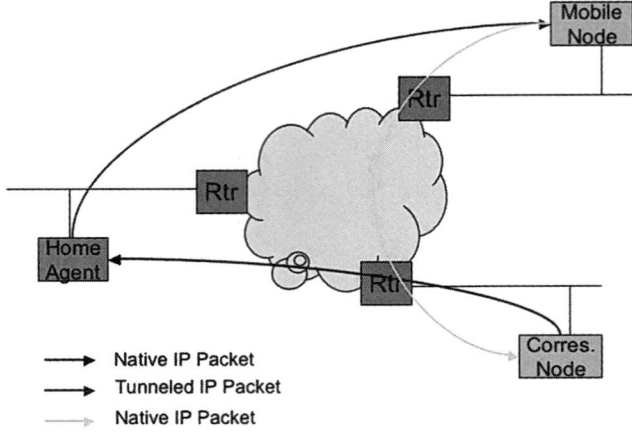

Figure 9.1   Overview of Mobile IP model.

network. This proxy functionality ensures that packets destined for the MN are properly directed to the MN. This is depicted in Figure 9.1.

Typically, the HA is a router or server maintained within the home network. As an example, consider if this type of approach was employed within a UMTS network. In this scenario, the SGSN would likely serve the role of the HA.

The capabilities needed by an HA are:

- Proxy Neighbor Advertisement messages for the MNs
- Process Binding Update messages from MNs
- Encapsulate data packets to the MNs

The mobile node signals its point of attachment to the HA by sending a binding update (BU) message which indicates its current care-of address. The BU provides sufficient information to the HA to allow data sent to the mobile node at its home address to be forwarded (encapsulated) to it at its care-of address. BU messages map the home address to the MN's current care-of address. Each time the MN changes its point of attachment, it signals the new mapping to the HA via the BU message. The BU process includes a rudimentary security mechanism, called the Return Routability check, to ensure that BUs are not spoofed.

The Return Routability check allows a correspondent node to be reasonably assured that the MN it wishes to communicate with is reachable at the care-of address reported in the received BU. The process contains four separate messages exchanged between the MN and the CN. The MN generates cryptographic *cookies* for use within the messages. The MN sends a HOME TEST INIT (HoTI) message (encapsulated through the HA) to the CN with a cookie. At the same time, the MN also transmits a CARE-OF TEST INIT (CoTI) message directly to the CN with a different cookie. The HA deencapsulates the HoTI and forwards it to the CN. After

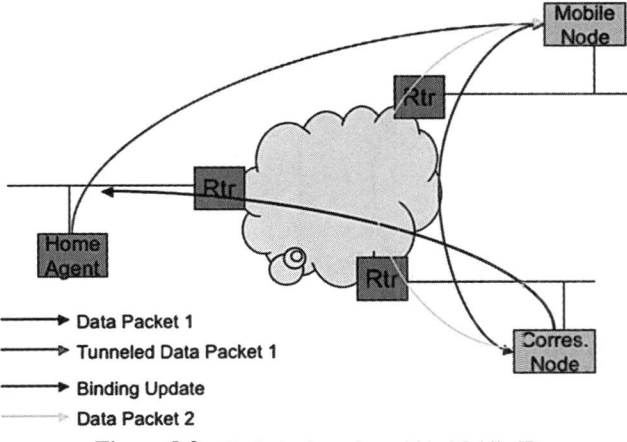

Data Packet 1
Tunneled Data Packet 1
Binding Update
Data Packet 2

Figure 9.2   Optimized routing within Mobile IP.

reception of the HoTI and the CoTI, the MN responds with a HOME TEST (HoT) message sent via the HA and a CARE-OF TEST (CoT) sent directly to the care-of address of the MN. These messages contain the cookies received from the MN and a keygen token generated based on the data in the CoTI and HoTI messages. The MN can then use the tokens to protect the BU sent to that CN.

When the mobile node receives data via the HA, it transmits a BU to the source of the data packet. This will allow the corresponding node, if it supports MIPv6, to utilize the Care-of Address for reaching the mobile node, thus bypassing the suboptimal triangular path through the mobile node's home network. This optimized mode of operation is depicted in Figure 9.2.

In order to efficiently forward data to MNs, IPv6 nodes will implement an *MIPv6 cache*. This cache maintains home address-to-care-of address mappings within the IPv6 stack of each node. Upon reception of a valid BU (and successful completion of the Return Routability check), the cache is updated to reflect the current home address/care-of address mapping. Applications running on the node will still refer to the MN by its home address, even though the data is routed to the care-of address. This transparent mechanism eliminates the need for applications to be aware of mobility issues.

9.2.1.2 Network Mobility

Network Mobility (NEMO) is a technology developed by the NEMO working group within the IETF to address the problem of network-level mobility, where an entire subnetwork's point of attachment to the network can change over time. This working group produced the NEMO specification as defined in RFC 3963 [232]. The assumptions embedded within this solution are that the subnetwork is a stub network (no transit traffic) and one (or more) mobile router(s) connect the subnetwork to the larger network.

The NEMO network is considered a leaf network within the larger topology. The leaf may have a single point-of-attachment to the larger network or it may be multi-homed. The goal is to provide a solution that hides the network mobility from the members of the network in order to reduce complexity.

The current basic NEMO specification, defined in RFC 3963, reuses the MIPv6 model. The mobile router (MR) that connects the leaf network to the larger network assumes the role of the MIPv6 mobile node. The MR acquires a care-of address from its point of attachment and uses that as a tunnel end point for communicating with the home agent located in the MR's home network. All traffic destined for or leaving the leaf network traverses the MR–HA tunnel. This is done to accommodate nodes within the NEMO network that are not aware of mobility (i.e., do not support MIPv6 extensions).

Even though the MR is a router, when it is attached to a foreign network, it does not perform router functionality on its egress interface (i.e., does not send Router Advertisements nor participates in any routing protocol exchanges). It appears as a host on the visited network. After it obtains its care-of address, the MR uses an MIPv6 BU message to signal the new location to the HA within the home network. The BU is extended to indicate NEMO routing capability and supported mobile prefixes to the HA. Once the HA acknowledges the BU, a bidirectional tunnel is established between the HA and the MR.

On the home network, the HA supporting an MR does not have to proxy Neighbor Discovery messages for all nodes in the NEMO. Rather, the prefixes advertised to the HA via the BU are injected into the routing system. The HA advertises itself as the shortest path to the mobile prefixes. When data packets destined to the mobile prefixes are received, the HA encapsulates them to the MR. The MR deencapsulates the packet and forwards it toward the destination within the NEMO.

9.2.2 Routing

It is generally well known that traditional Internet routing approaches do not work well in the generalized multihop wireless network (e.g., MANET). This holds true for both interior and exterior routing protocols. It is well known that traditional wired interior routing protocols such as open shortest path first (OSPF) can perform very poorly in highly dynamic unreliable wireless networks. There has also been previous literature that has made compelling arguments that traditional exterior routing approaches, that is, border gateway protocol (BGP), is not well suited for the generalized multihop wireless network [233]. The analysis presented in Reference [233] centered on two key arguments: (1) BGP is not able to detect and connect dynamically to arbitrary peers and (2) instability of links and peers because of mobility will require a prohibitive amount of routing information to be exchanged given the limited bandwidth of a wireless network [234]. This argument is illustrated in Figure 9.3 [233].

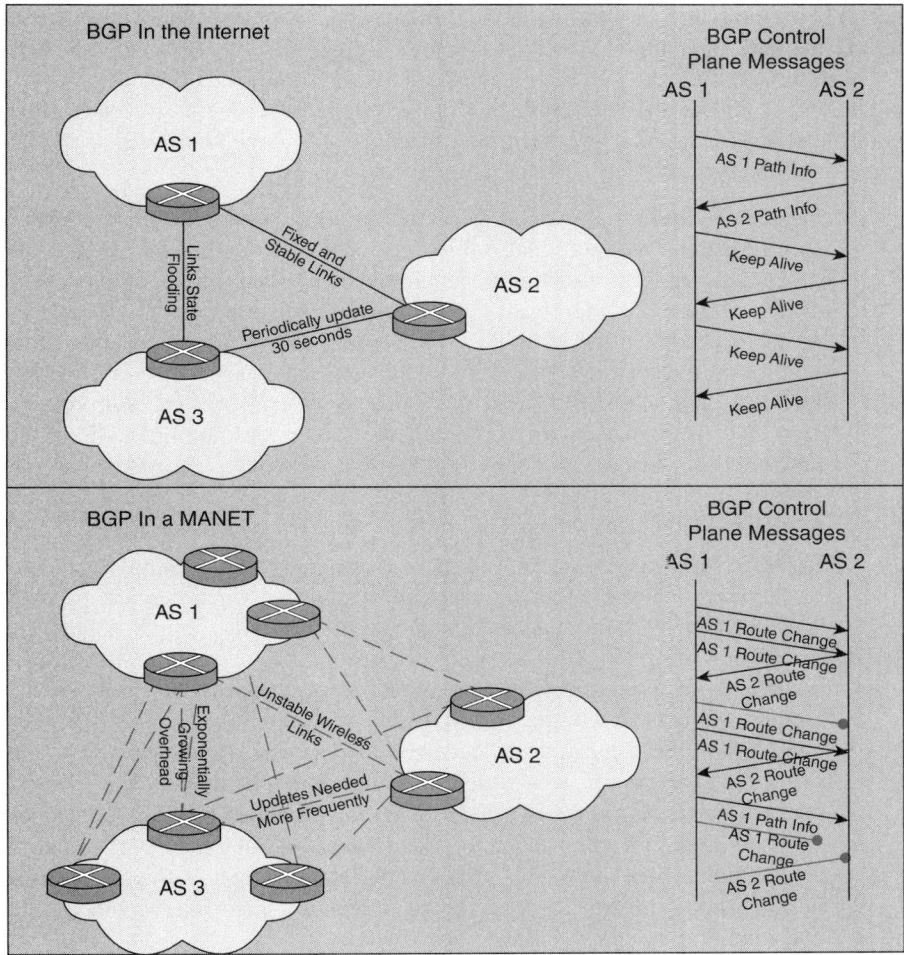

Figure 9.3   Issues surrounding BGP employment in a MANET. Reprinted from Reference [233].

In the case of exterior routing, there are no viable alternatives to BGP. This is problematic as wireless networks increasingly assume a more important role in the overall Internet landscape. However, the immense complexity of the Internet as a whole combined with the huge number of participants and administrators make developing a viable alternative in this space difficult.

In the case of interior routing, this is somewhat of a nonissue for many of the wireless technologies considered in this text. Many of the technologies presented in this text ultimately form a hub-and-spoke topology where all communications across the wireless air interface is single hop in nature. Here the "hub" is the access point or the base station and the "spoke" is the mobile device. This removes the need for interior routing. However, interior routing remains an issue for the case of the generalized multihop wireless network. An example of this scenario is an IEEE 802.11 network in an ad hoc mode of operation.

For the generalized multihop wireless network scenario, there have been a plethora of technical proposals and solutions developed over the past decade, from both the standards community and the academic community. Generally speaking, these solutions typically fall into one of three categories:

1. ***Proactive Routing Protocol.*** Proactive routing protocols operated in a similar manner as traditional routing protocols. State is maintained for the entire network, thus allowing for the immediate forwarding of data packets. The downside to this approach is the amount of network control traffic generated. The nodes in the MANET must continuously exchange routing messages to ensure that every node in the network has reachability information for all other nodes. Most proactive routing protocols are *link-state* routing protocols. They exchange information about all nodes and links within the network. The path calculation within a link-state protocol can take into consideration multiple variables that can affect the performance of the network (e.g., bandwidth, link delays, and traffic engineering metrics).

2. ***Reactive Routing Protocol.*** A reactive routing protocol does not keep state on the entire network. Rather, routes are discovered when they are needed. The forwarding state for each active route is maintained as long as data traffic is utilizing the path. After a period of nonuse, the forwarding state is discarded. By maintaining only active state, a reactive protocol reduces the amount of network control traffic on the wireless links. The cost of this approach is the setup latency occurred to discover routes when they are needed. This approach works well with highly mobile networks. Typically, reactive protocols utilize the *distance vector* model of routing. A distance vector protocol maintains a vector of distances, or hop counts, toward each destination. These protocols select paths by minimizing the number of network hops. In other words, the selection of a path does not take into consideration the quality of the network links.

3. ***Mesh Networking.*** The wireless network is designed so that all communications are one hop in nature, removing the need for layer 3 routing.

It is important to note that the last category is not a real solution to the problem. Rather, it is simply shifting the problem from layer 3 down to layer 2 (MAC layer).

The next several sections provide an overview of several of the interior routing solutions developed for the MANET problem space. For a much more exhaustive treatment of MANET routing protocols, we recommend Reference [235] to the reader.

9.2.2.1 Examples of IETF MANET Routing Protocols

The IETF's MANET working group, which resides within the Routing Area of the IETF (http://www.ietf.org/html.charters/manet-charter.html), is standardizing IP layer routing protocols aimed at dynamically forming, typically mobile, networks. These networks' characteristics differ dramatically from those of traditional wired networks.

The goal of the MANET working group is to develop lightweight, IP layer routing protocols that will provide connectivity across a wide range of mobility. Due to the wide range of deployment scenarios, the working group has explored a wide range of solutions, publishing several proactive and reactive approaches.

Two examples of reactive routing protocols published are dynamic source routing (DSR) [236] and ad hoc on-demand distance vector (AODV) [237]. Two proactive routing examples are optimized link-state routing (OLSR) [238] and topology-based reverse path forwarding (TBRPF) [239]. All of these protocols were published by the IETF as Experimental Standards. The goal of Experimental Standards is to use deployment experience from these protocols as input to the development of the Standards Track protocols. These protocols will be described more in depth in the following subsections.

The first targeted deployment scenario places the MANET at the edge of the fixed IP infrastructure. The network could be composed of all MNs or a hybrid collection of mobile and fixed routing platforms. In the second scenario, the MANET is an isolated network with no global connectivity to the larger network. Each scenario introduces characteristics that can have dramatic impacts on the routing protocol used.

The MANET approach has several limitations that can have adverse affects on the operation of a network. By reusing IP layer routing mechanisms, every node in the network must be constantly listening for routing messages. This prevents nodes from powering down in order to conserve battery life. Another drawback is the lack of hierarchy in the network. Nodes do not coordinate their IP addresses, so packets must be "flat routed" through the network, which limits the scaling factor. Reactive MANET routing protocols rely on hop counts to determine the best path through a network. This approach can lead to inefficient use of the RF spectrum since the shortest hop count path may be across a poor performing link. In contrast, proactive protocols define link metrics that could avoid this issue. The drawback there is the amount of control traffic that is generated as the rate of mobility increases. The proactive protocols may never reach convergence in highly mobile networks.

There are several areas of work that still need to be done in the MANET problem space. Many research projects have shown that MANET protocols that depend entirely on IP layer messages do not perform well in a highly dynamic environment. Ideally, there should be feedback utilized from layer 2 in order to optimize performance.

Dynamic Source Routing The DSR protocol [236] is a reactive MANET routing protocol based on the concept of source routes. DSR is composed of two main processes, Route Discovery and Route Maintenance. These two functions generate and maintain routing information for all destinations of interest. Since DSR is a reactive protocol, it is designed to support networks with high degrees of mobility. The protocol is limited in capability due to its dependence on source routing.

The DSR specification indicates the ability to support networks with approximately 200 nodes and a network diameter (longest multihop path) between 5 and 10 hops.

Route Discovery is the process within DSR that allows a sending node to determine a path to a target destination. When a node (*S*) needs a path to a destination (*D*), it begins the discovery process. This mechanism relies on two main messages, Route Requests and Route Replies.

The Route Request message is transmitted as a local broadcast message to all nodes within *S*'s transmission range. The target destination (*D*) is listed in the Route Request payload and not the IP destination address. Each node receiving the request determines whether it has a forwarding state for *D*. If so, the node can return a Route Reply message to *S* with the cached path to *D*. Otherwise, the node adds its address to the source route list in the message and transmits it to all its neighbors. When the Route Request reaches *D*, a Route Reply is returned. Multiple Route Requests can reach *D* due to multiple paths through the network.

The Route Reply message contains the source route list constructed within the Route Request message as it traversed the network. Since multiple paths may exist within the network, more than one Route Reply message may be returned to *S*. Multiple Route Replies allows nodes to maintain multiple paths to target destinations. The transmitter of a Route Reply has several options for sending the packet back to the source. It may initiate its own Route Request toward *S* and piggyback its Route Reply. Otherwise, *D* may simply take the source route built up in the Route Request and reverse it for its use. The latter is meant for networks where symmetrical routes are known to exist, while the former allows for asymmetrical routes and unidirectional links. The DSR discovery process is depicted in Figure 9.4.

Nodes can cache the forwarding state learned from multiple means. The most common method is the reception of a Route Reply for a destination that is locally

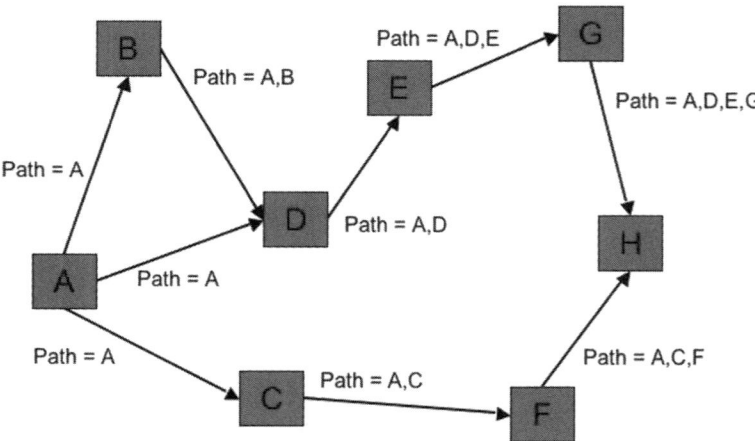

Figure 9.4 The DSR discovery process.

relevant. Cache information can also be updated or created based on Route Replies that are "overheard." Routes are also gleaned from data packets being forwarded.

Route Maintenance allows nodes within the network to learn of failures along network paths. Acknowledgments are employed to allow nodes to discover the failure of directly attached links. These acknowledgments can either be provided by the MAC layer (e.g., 802.11) or implemented as software-based DSR acknowledgments. When acknowledgments are not received or information is provided indicating the loss of a link, nodes generate Route Error messages to the source of the data packets affected by the failure. When a node receives a Route Error, alternate paths can be attempted or the communication aborted.

Once a path exists to D, data packets from S are augmented with additional data that indicates the source-routed path the data will take within the network. DSR requires the use of a fixed-size DSR header with 0 or more optional data fields. In IPv4, this is implemented as an IP Option that must immediately follow the base IPv4 header. For IPv6, the DSR header is implemented as a Hop-by-Hop Extension Header. Nodes along the source-routed path utilize the information in the DSR header to determine the next-hop node toward D.

Advantages of DSR   The ability to discover and maintain multiple paths to a destination gives DSR nodes the flexibility to recover from errors in a timely fashion.

Disadvantages of DSR   The main drawback to DSR is its basis in source routing. By including each hop's IP address, a large amount of a data packet can be consumed by overhead. As an example, if two nodes are communicating via DSR over IPv6 through five intermediate nodes, 80 bytes of the packet are consumed for listing the path.

Ad Hoc On-Demand Distance Vector   Another reactive MANET protocol is the AODV routing protocol [237]. AODV discovers and maintains a single path from a source to a destination as long as data packets are being transmitted. Based on distance vector techniques, AODV is not subject to the same overhead limitations as DSR. Since control information is not embedded in data packets, AODV scales to much larger networks (thousands of nodes) while still being able to support high degrees of mobility. The protocol itself is composed of three primary messages: Route Requests, Route Replies, and Route Errors.

A Route Request message is transmitted by a source when it does not have routing information for a target destination. Each RREQ is transmitted as a local broadcast to all neighbors. When a node receives an RREQ, it queries its forwarding cache to determine if it has a route to the target. If it has a route, a Route Reply can be sent to the requestor; otherwise, the RREQ is flooded to that node's neighbors. Nodes may see the same RREQ multiple times due to the flooding mechanism employed. Duplicate RREQs are discarded in order to minimize the potential for broadcast storms. Intermediate nodes use the information within the RREQ to create a forwarding state toward the originator of the RREQ. This allows for efficient forwarding of replies that utilize the reverse path.

The Route Reply (RREP) message is used to indicate the next hop to use to reach the destination. Recall that AODV is a distance vector protocol, which means that network paths from a source to a destination are not known. Rather, each node keeps track of only the next hop to use to reach a destination. The ultimate destination or an intermediate node originates an RREP back toward the RREQ originator using the next-hop information cached during the forwarding of the RREQ. As the RREP traverses the network, each hop along the path creates a forwarding entry toward the destination, updates the hop count field in the RREP, and forwards the RREP based on the cached state for the RREQ originator. Once the RREP reaches the final destination (the original source), the two-way path is established through the network. In the event of multiple RREPs being received, the hop count of each is taken into account in order to select a single path for use.

As the RREP message flows through the network, intermediate nodes create and/or update route table entries for the source and destination nodes. The key element of the route table management is the maintenance of *precursor lists*. These lists indicate dependencies imposed on route table entries by directly attached neighbors. When an RREP is received from D toward S, the next hop listed in the route to S is copied to the precursor list in the route to D. This mechanism allows for an intermediate node to identify all neighbors that need to be notified in the event of a failure of a next-hop node or link. The basic operation of the AODV protocol is depicted in Figure 9.5.

Route Error (RERR) messages are used to signal the failure of a link or a next hop. When a node detects a problem, it queries its route table for all entries utilizing the failed link. For each route match, if the precursor list is nonempty, an RERR is sent to the nodes listed. This notification will be propagated toward the sources

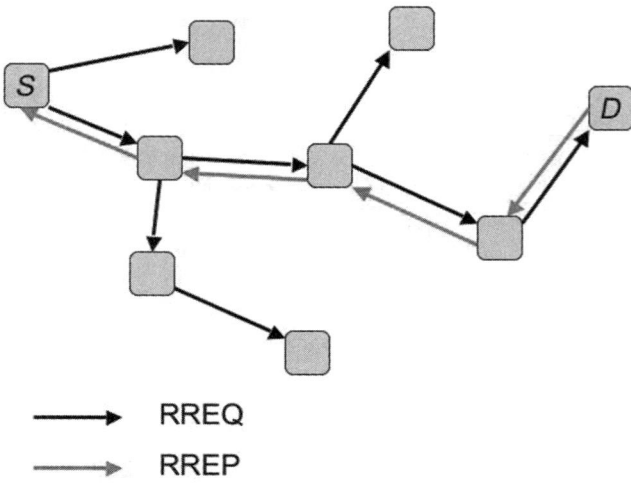

————▶ RREQ

————▶ RREP

Figure 9.5 Illustration of the AODV routing protocol.

dependent on the failed link. The reception of an RERR should trigger the invalidation of routes using the failed link and the transmission of RERRs toward any downstream-dependent nodes.

In order to avoid loops (cyclic paths preventing delivery to final destination), AODV employs sequence numbers. A node will maintain a local sequence number to indicate the age of information transmitted within AODV control messages. At startup, a node initializes its sequence number. A node increments its sequence number when:

- It transmits a new RREQ
- It generates an RREP when it is the destination

Each node maintains information from RREPs and RREQs only if the sequence number within the message indicates that it is the latest. This is accomplished by including the sequence number of other nodes in the route table entries for those nodes. If messages are received with sequence numbers older than what is installed in the route table, the message contents are ignored and not propagated. This mechanism prevents stale information from looping through the network and adversely affecting the forwarding state.

Advantages of AODV   The key benefit of AODV is its simplicity. It is possible to operate AODV on low-power devices without the overhead of DSR. AODV is also capable of scaling to larger network sizes (thousands of nodes) while still supporting a high degree of mobility (random waypoint model of mobility with nodes traveling ~10 m/s).

Disadvantages of AODV   The primary limitation of AODV is the distance vector mechanism. The dependency on selecting paths based solely on a hop count is limiting. In the case of wireless links, it can be disruptive due to the selection of weak paths (long transmission times like that of a satellite link) that have low hop counts.

Optimized Link State Routing   The OLSR routing protocol [238] is a proactive MANET protocol. Unlike the previous routing protocols discussed, OLSR maintains a topology database for the entire MANET. Such an approach works well for networks with a dense distribution of nodes and moderate amounts of mobility that do not cause wild fluctuations in the topology. OLSR operates like a traditional link-state routing protocol in that state is flooded throughout the network and nodes generate forwarding tables based on its local topology database. The flooding is accomplished using three key message types: HELLO, Multiple Interface Declaration (MID), and Topology Control (TC).

The OLSR HELLO message functions in much the same manner as hello messages in other routing protocols. These messages allow for the dynamic discovery of direct neighbors (HELLOs are not flooded through the network) within the routing domain. In the case of OLSR, this information is used to determine adjacent neighbors, link quality (e.g., symmetric vs. asymmetric), and Multipoint Relay (MPR) candidates. The HELLO messages are transmitted on a periodic basis on all interfaces participating in the OSLR protocol. Upon reception of a HELLO message, a node can modify its neighbors' adjacency status.

Once a node has identified its MPR candidate set, it selects an MPR set. This set of nodes allows the OLSR protocol to scale by limiting the number of nodes that flood data through the network. A node selects its MPR set by choosing the smallest set of nodes needed to reach all possible two-hop peers. Once a node selects its MPR set, it notifies these peers by changing their state within the HELLO messages being transmitted. An MPR is responsible for aggregating topology information from all of its *MPR selectors* and flooding that information throughout the network. This mechanism reduces the amount of flooded data through the entire network and limits the number of messages that non-MPR nodes need to transmit.

In addition to HELLO messages, all nodes also transmit MID messages. This message advertises the set of addresses in use by an OLSR node, the state of the interfaces associated with those addresses, and a *main address* that acts as a network ID for the node. The MID message contents are incorporated into the topology database of all recipients. MID messages allow for the efficient creation of a directed network graph used for route calculation. While the HELLO is limited to the neighbor–neighbor scope, the MID message is retransmitted by the MPRs in order to diffuse it to all portions of the network.

Once the MPRs have been selected, they take on the role of information distributor. Each MPR is responsible for generating TC messages within the MANET. The TC message, at a minimum, advertises the link state between the advertising MPR and all nodes that have selected the advertiser as its MPR. Those nodes are identified in the TC message by their main address.

When a node receives a TC message, it must determine if the data should be processed, forwarded, or dropped. Like AODV messages, the TC message contains a sequence number that is used to ensure the latest data are used. If a node receives a TC message with a larger sequence number, the data in that packet overrides data currently in the node's topology database. If the TC contains new data, the node must also flood the data. Otherwise, the TC message can be dropped.

Once the MID and TC messages have been propagated through the network, each node is responsible for calculating its view of the topology. The main addresses contained in MIDs are mapped to the link states advertised in the TCs to create a directed graph of the network. Then, as with other link state protocols, a shortest-path algorithm is run on the graph. Whenever local state changes (e.g., MPR set change) or new TC or MID messages are received, the route table must be recomputed.

Advantages of OLSR The primary advantage of OLSR is that there is no latency associated with route setup for new data flows. OLSR generally works well in networks with densely distributed nodes and only moderate amounts of mobility such that wild fluctuations in topology do not result.

Disadvantages of OLSR The need to distribute topology information across the entire network is a serious drawback for OLSR. In periods of high mobility (i.e., severe topology change), the behavior of a link-state protocol is nondeterministic. The possibility exists that convergence may never be reached, further degrading ability to forward packets.

Topology Dissemination Based on Reverse Path Forwarding Another proactive MANET routing protocol is the TBRPF protocol [239]. Like other link-state protocols, it generates a topology database within each node that reflects the state of the entire network. The major difference between other link-state protocols and TBRPF is the use of differential advertisements. The goal of this approach is to reduce the amount of control traffic transmitted. The TBRPF protocol is conceptually broken into two components: Neighbor Discovery and Route Management.

The TBRPF Neighbor Discovery (TND) module is responsible for maintaining state for all directly attached (one-hop) neighbors. The TND module detects adjacency state based on the exchange of *differential* HELLO messages. These messages only contain state changes detected by the transmitting node. This reduces the amount of information carried in the control messages. The HELLO messages report the perceived state change of the links between the sending node and all of its detected neighbors. The neighbor link state can be:

- *1-WAY.* Reachability is unidirectional
- *2-WAY.* Reachability is bidirectional
- *LOST.* No reachability

Each node maintains a neighbor table per interface. The HELLO messages transmitted on an interface only contains information derived from that interface's neighbor table. The HELLO messages are transmitted on a periodic basis (possibly with no data) in order to maintain reachability state and solicit adjacencies with new TBRPF peers.

The TBRPF Route Management (TRM) module is composed of two subcomponents. One subcomponent is responsible for topology discovery and management. The other performs route table computation based on the topology database. Together, they generate a node's view of the network in the form of a routing table.

The topology discovery component operates by advertising partial forwarding state to peers and receiving partial forwarding state from peers. Conceptually, this component must generate enough information that allows other nodes within the network to determine the most efficient path to any node. A node (X) generates a partial tree, rooted at X, and advertises it as its partial forwarding state. This tree comprises all of the two-hop paths that must pass through X. The set of nodes dependent upon X for these paths is X's Reported Node (RN) Set. It should be noted that this is analogous with X being selected an OLSR MPR for all members of the RN Set.

As the partial trees are advertised through the network, nodes update their topology tree. The updated tree is then processed using a modified Dijkstra's algorithm to create a set of shortest paths to each destination (node or network). A mechanism is included within the protocol to switch to a full, rather than partial, advertisement model. Such a step makes TBRPF perform in much the same manner as OSPF.

Advantages of TBRPF   TBRPF does allow for arbitrary metrics to be used for route calculation. This feature can avoid the pitfalls of using simple hop counts for path determination, as is the case in other MANET routing protocols such as AODV.

Disadvantages of TBRPF   TBRPF, like OLSR, is affected by the amount of mobility within the network. The protocol specification states that, on average, 200 nodes can be supported by TBRPF as long as the mobility is moderate. Another drawback to TBRPF is the relative complexity of the protocol in relation to other MANET protocols. Even though differential messaging reduces control traffic, the complex processing rules makes the protocol heavyweight.

9.2.2.2   Other MANET Routing Protocols

Of course, the protocols described in Section 9.2.1 are only a small portion of the proposed solutions in this problem space. Figure 9.6 provides a basic taxonomy of ad hoc routing protocols. We admit that almost all taxonomies are arbitrary and thus equally limited in utility. However, it seems intuitively obvious to us to divide the space of MANET routing based along proactive/reactive lines, as most MANET routing protocols can be classified as proactive, reactive, or a hybrid of both approaches. Note that the taxonomy of Figure 9.6 is also not close to being exhaustive. There are predictive routing techniques such as route-lifetime assessment-based routing (RABR) [240] and link life-based routing protocol (LBR) [241], power-aware routing techniques such as power-aware routing (PAR) [242], and geographic-based routing techniques such as location-aware (LAR) [243]. The reader is referred

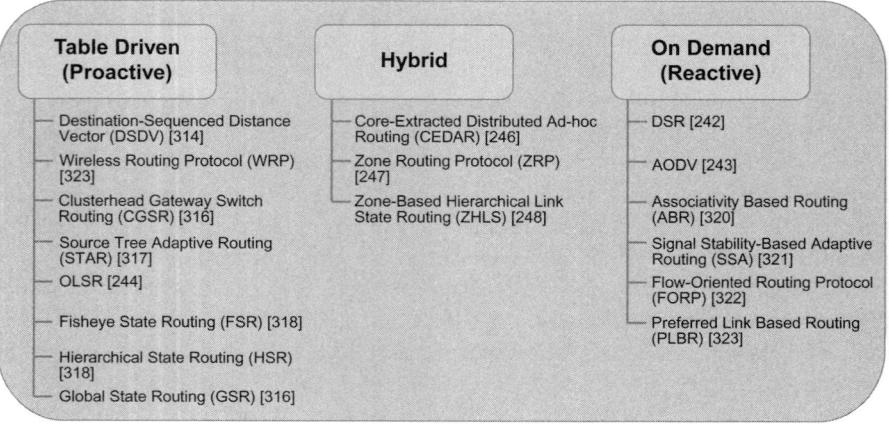

Figure 9.6   Taxonomy of ad hoc routing protocols.

to Reference [235] for a more thorough treatment of the many proposed ad hoc routing protocols.

9.2.2.3 Novel Routing Approaches

This section briefly discusses several particularly novel approaches to interior routing for ad hoc networks.

Epidemic Routing   Epidemic routing [244] is an adaptation of epidemic algorithms that have been applied for distributed database synchronization. The application of this technique to MANET routing uses the principle of transitivity of contacts in the ad hoc network. That is, messages are transferred, not necessarily by A coming into contact with B, but by A coming into contact with C, which comes into contact with D, and so on until some node in the chain comes into contact with B. In order to *guarantee delivery*, there must be an assumption that all nodes come into contact with each other eventually, but in order to get a *high probability of delivery*, this is not a necessary assumption.

The epidemic approach is interesting because of the lack of assumptions that it makes about the network. In particular, it does not assume that any two given nodes are connected at a given time; that is, it does not assume that there is ever an entire path that exists between the source and the destination. In addition, it does not assume that the sender knows where the destination is, or how to get there.

Epidemic routing is not strictly a routing protocol in the sense that OSPF is a routing protocol. It is more accurately a method of distributing information through an ad hoc network. There are no routing prefixes distributed. There are no routing tables maintained.

The basic operation of epidemic routing is quite simple. Every node in the network is a router and has the ability both to transmit and store messages. Each node has a finite memory in which to keep messages destined for other nodes. When two nodes meet, and have not met previously for some network-dependent period of time, they exchange information about what messages they are carrying. Each node checks the list against a similar list of the contents of its own buffer and computes the set difference. Each node transmits to the other the list of messages that it has not yet seen that the other node has. The nodes then send the missing messages to each other.

The nodes can decide whether to accept a message and whether to throw out a message that it already has to make room for another. It can also make policy decisions, such as whether to carry messages for particular nodes or sets of nodes. In addition to messages, the protocol can also accommodate acknowledgments that let not only the source and destination know that a message has been delivered, but also let the carrying nodes know that they can free up buffer space by discarding delivered messages. This is a feature of pair-wise communication. In broadcast communication, the number of acknowledgments becomes a scaling problem.

Simulation results for this approach are provided in Reference [244]. As one might expect, for a given area, number of nodes, and radio range, the larger radio range decreases latency and increases delivery rate. Hop count also plays a role in this performance. Very short transmission ranges tend to have much poorer performance, for example, six orders of magnitude difference in latency between the shortest and longest transmission radii. Buffer space also plays an important role: one can trade off delay against buffer size (since a message will exist as multiple copies in the network simultaneously), where smaller buffer sizes don't affect the eventual delivery rate, but do affect the latency. There appear to be critical points both in buffer size and in hop count, where a small change causes a large effect in performance metrics.

Probabilistic Routing Probabilistic routing is a modification of epidemic routing that relies on the observation that human movement is rarely random. There are regular movement patterns that can be exploited to improve the performance of epidemic-type algorithms. An example of this is the probabilistic routing protocol using history of encounters and transitivity (PROPHET) approach presented in Reference [245].

PROPHET uses as its primary metric an estimate of delivery probability, meaning the probability that if a message is passed to a contact, the contact will be able to deliver the message to the final destination. Each node keeps a vector of such estimates for itself—that is, the probability that it will make contact with each possible destination in the network, or the one-hop delivery probability. The nodes do not keep n-hop delivery probabilities.

PROPHET works in essentially the same way as epidemic routing. When two nodes make contact, they exchange message summary vectors but also exchange delivery probability vectors. Based on the information exchanged, the nodes decide to request messages from their peer. There are many possible strategies for forwarding messages to other nodes in this scheme. Trade-offs include the number of other nodes to which the message should be forwarded, the resources available (e.g., buffer space), latency, and the number of hops that the message may take.

PROPHET was simulated with two different scenarios in Reference [245]: A random waypoint model similar to what was used to simulate the epidemic routing and a scenario where movement is more directed. In the latter scenario, nodes move between a home community, a common gathering place, and multiple other communities, with high probabilities of going to the gathering place and back home again. In the simulations, the PROPHET nodes used the simple forwarding strategy of passing the message to another node if it has a higher delivery probability. However, it is unclear what network attachment information is used while making forwarding decisions. Much of this behavior appears to be implicitly assumed.

The results of the simulations of PROPHET are similar to the simulation results for epidemic routing for the scenario with the random waypoint mobility model. PROPHET performed slightly better than epidemic routing. For both, as radio range increases, delivery rates are better, latency decreases quite a bit, and the total number of messages forwarded increases. Also, as buffer sizes are increased, the delivery rates increase and latency also increases. For the more realistic scenario, with directed mobility, PROPHET performed substantially better than epidemic routing.

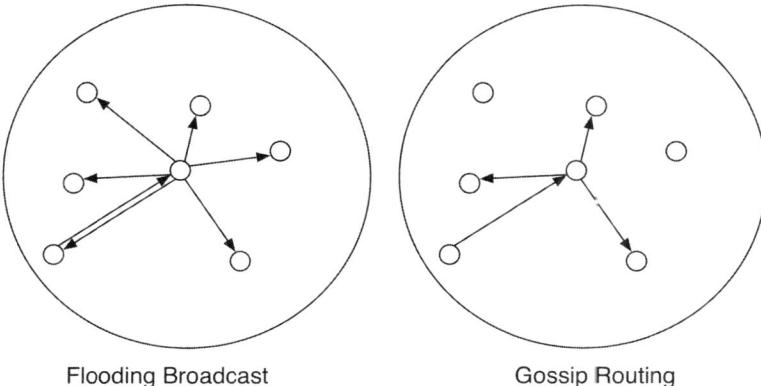

Flooding Broadcast   Gossip Routing

Figure 9.7   Flooding broadcast versus gossip routing.

The addition of the probability of delivery information gives PROPHET an advantage in latency and delivery rates, especially at larger buffer sizes.

Gossip Routing   The two previous protocols do not assume that an end-to-end path between source and destination ever exists. In the gossip routing algorithm, described in Reference [246], there is an assumption of an end-to-end path and the intent of the protocol is to find it.

The basic idea of the gossip protocol is simple. A node initiates a route request and sends the request to all its neighbors. For k hops from the origin of the request, nodes forward the message to their neighbors with probability 1. After k hops, nodes forward the request to each neighbor with probability p. If the node does not propagate the message (probability $1 - p$), it simply drops it. This is shown in Figure 9.7. Any duplicate of the message received by a node is dropped whether or not it was propagated in the first place. The goal of the protocol is, in large ad hoc networks, to reduce the number of routing messages (vs. the normal *flooding* algorithm) and consequently to increase the scalability of the network.

This algorithm produces varying behavior. Either almost all nodes receive the message, or almost none do. In general, near the edge of the network, the request is more likely to "die out" since nodes there have fewer neighbors, and because they don't benefit from a kind of reverse propagation where nodes that were "skipped" in the outward direction receive a request for the first time flowing back toward the origin.

The authors of the gossip routing algorithm propose a number of heuristics to mitigate the "die-out" behavior of the gossip protocol. One approach is to have a two-probability scheme, so that if a node has fewer than some threshold number of neighbors, it instructs the neighbors to gossip (forward the message) with a higher probability. This decreases the probability that all the neighbors will decide not to gossip, and on the average it increases the number of copies of the message that are forwarded into the rest of the network. A variation of this is if some neighbors of a node form a cut set of the network graph, it could instruct them to gossip with probability 1. Other proposed modifications include the use of zones within which proactive routing approaches would be used.

A version of the gossip protocol was added to an AODV implementation by the authors of Reference [246]. It was found that, in general, even for relatively small networks, it improved AODV performance somewhat in terms of latency, packets delivery ratio, and, more significantly, in terms of routing load. The simulations presented in Reference [246] use a random waypoint model and the results are given as a function of the pause time between node movements. As one might expect, as pause time increases, routing overhead and delay decrease because the network is stable for longer periods of time. The best improvement of delivery ratio when gossiping is added to AODV is about 8% [246]. The best improvement in routing overhead is about 25% [246].

9.3 TRANSPORT LAYER CONSIDERATIONS

In this section, we'll cover a discussion of the transport layer in detail, and explain how wireless network protocols affect the transport layer. In particular, we will focus on analyzing the Transmission Control Protocol and discuss some of the key technical features of TCP. We will also compare TCP's performance for both wired and wireless networks. Next, we will discuss a few different implementations of TCP and what features each variant has provided. Finally, we will discuss other transport protocols besides TCP, including why these protocols were developed.

9.3.1 Overview of TCP

The original TCP protocol is defined in Reference [247]. TCP is a connection-oriented, multiplexing protocol that rides on top of IP, ensuring that different data streams arrive in order and error free, with flow control algorithms that adjust flow rate based on network congestion conditions and other indications. Multiple variants of TCP exist that provide different treatment of some aspects of the protocol (such as flow control). This overview will cover a discussion of the motivation of the TCP protocol, its unique role within the Internet, and a basic overview of the congestion control mechanisms. One key assumption in the original design of TCP congestion control is the expectation that any segment loss is caused by congestion conditions in the network. For wired networks, this design element helps to reduce overall congestion and improve network conditions for all users. However, for wireless networks, significant segment loss can be caused by network congestion and/or bit errors from channel conditions. Yet, any segment loss detected by TCP is treated equally—and such behavior has consequences for wireless networks with hosts utilizing TCP.

In the 1970s and early 1980s, the U.S. Department of Defense (DoD) had sponsored efforts to develop a series of standardized protocol specifications for

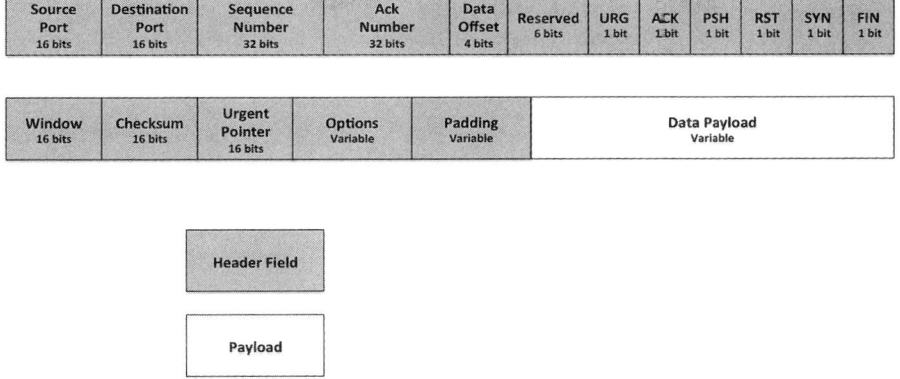

Figure 9.8   TCP protocol segment format with header. Reprinted from Reference [247].

military networks. The Undersecretary of Defense for Research and Engineering declared TCP as a basis for DoD-wide internetwork protocol standardization. In particular, military networks often experienced congestion and unreliable links—and TCP aimed to mitigate these conditions by providing various mitigating protocol functions. In TCP, data is sent in "segments," with each segment containing a set of data. Segments are delivered in order and acknowledged by TCP.

Figure 9.8 [247] illustrates the segment format of the TCP protocol. Note in the figure that shaded boxes denote header fields while the payload field is not shaded. Also note that the source port field denotes the start of the header, while the padding field denotes the end of the header.

The source and destination port fields, both 16 bits in length, indicate the source and destination ports of the hosts associated with the TCP session, respectively. These ports are analogous to the ports employed in the User Datagram Protocol (UDP) and, together with the source and destination host IP addresses as well as the protocol field (indicating TCP or UDP), form the quintuple identifier for a specific data flow. The sequence number is 32 bits. It is used to keep track of the segments transmitted, and enables the destination to keep segments in order before presenting data to the next higher layer. The acknowledgment (ACK) number is also 32 bits, and is used to indicate the next expected segment to be sent from the source. The ACK number is used in conjunction with the ACK flag (set equal to 1) when the destination responds explicitly with an acknowledgment of the last segment sent. The data offset field (4 bits) is the number of 32-bit words in the TCP header, and is used to indicate where the data payload begins. The reserved field (6 bits) has not been defined. The urgent pointer (URG) flag field (1 bit) indicates to check the urgent pointer field in the header when set to 1. The acknowledgement (ACK) flag field (1 bit) indicates to check the acknowledgment number field in the header when set to 1. The push function (PSH) flag field (1 bit) indicates the sender has requested all data to be sent by TCP at that point be transmitted immediately when set to 1.

The reset (RST) flag field (1 bit) indicates a request to reset the TCP connection when set to 1. The synchronize sequence numbers (SYN) flag field (1 bit) indicates the request to synchronize the sequence numbers between sender and receiver when set to 1. The finish (FIN) flag field (1 bit) indicates that the sender has no more data to send when set to 1. The window field (16 bits) indicates the number of data octets beginning with the one indicated in the acknowledgment number field, which the sender of this segment is willing to accept. Note that sender in this context does not indicate the primary source or destination of data, but is heavily predicated on the direction of flow. For instance, if one node (A) sends TCP data in segments to another node (B), node B will respond with TCP segments indicating the acknowledgment of the data segments being transmitted from A to B, and these segments sent from B to A in response will contain a window field that indicates the number of segments B is willing to accept. This is a key distinction of TCP and is used in conjunction with congestion control mechanisms that adjust the window based on perceived congestion conditions. The checksum field (16 bits) is used for data integrity, and is calculated across a hypothetical pseudoheader prefixed to the actual TCP header, the actual TCP header and text. More detail can be found in RFC793 [247]. The urgent pointer field (16 bits) conveys the present value of the urgent pointer as an increased offset to the sequence number in the segment. The options field (variable) must be an integer multiple of 8 bits in length. Options can be formatted in two ways: a single octet of "option-kind," or an octet of "option-kind," "option-length," and the actual "option-data" octets. TCP implementations must support all options. The padding field (variable) is used to ensure the TCP header ends and the data payload begins on a 32-bit boundary.

9.3.1.1   *TCP Congestion Control*

One primary feature of TCP is congestion control. RFC 5681 [248] documents a detailed set of TCP congestion control algorithms. In particular, there are four primary congestion control algorithms described: slow start, congestion avoidance, fast retransmit, and fast recovery. Figure 9.9 illustrates a general error-free flow exchange between a sender host and a receiver host using TCP.

Here, host A sends a SYN segment to start the connection establishment procedure. Once an acknowledgement (ACK) is sent back from host B acknowledging the SYN segment, host A responds with another ACK to signify the connection is established. Host A then sends segments to host B, and each time host B receives a segment it sends back an ACK. The ACKs are used in host A's congestion control algorithm to increase the number of segments sent at one time by increasing its sending congestion window (cwnd). The receiver also has an advertised receiver window (rwnd) that is used for flow control purposes. The minimum value of the set (cwnd, rwnd) is used to determine the number of segments transmitted at one time. For most implementations of TCP congestion control, the cwnd increases linearly each time an ACK is sent, which has the effect of increasing the window size exponentially as a function of the round trip time (RTT). RTT is generally defined as the time it takes

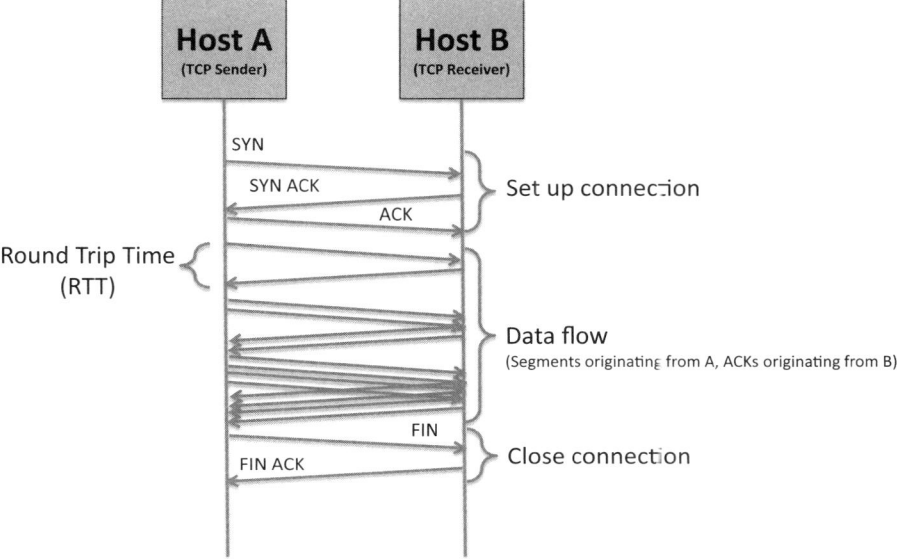

Figure 9.9   TCP flow exchange.

from initial sending of a segment until an acknowledgment of the segment is received at the sender. We will cover the cwnd in further detail later. Once all the data are sent, host A sends a finish (FIN) message to host B, indicating there is no more data to be sent, and host B responds with a FIN ACK to finalize connection closing.

The congestion control algorithms in TCP are divided into two primary states of operation: slow start and congestion avoidance. During a connection between two hosts, the cwnd is being adjusted according to the behavior of one of these two states. Figure 9.10 illustrates the general behavior of cwnd over time in both states assuming no errors or retransmissions. The transition point between the states is indicated by the slow start threshold (ssthresh), which indicates a distinct value at which the cwnd will grow linearly with RTT during the congestion avoidance phase, compared with the exponential growth in the slow start phase. Generally, this mechanism is referred to as additive increase/multiplicative-decrease (AIMD). TCP variants that support this mechanism of cwnd control include TCP Reno and TCP New Reno.

Note in the slow start phase that cwnd increases exponentially with each integer multiple of RTT. This is because cwnd increases by the sender's maximum segment size (SMSS) in bytes each time an ACK is successfully received at the transmitting host, sent from the receiving host. The transmitting host will send a number of segments equal to the value of cwnd/SMSS, and then wait for ACKs to come back to increase cwnd before sending more segments. Once cwnd reaches the value defined by ssthresh, growth then becomes linear with RTT instead of exponential; cwnd will

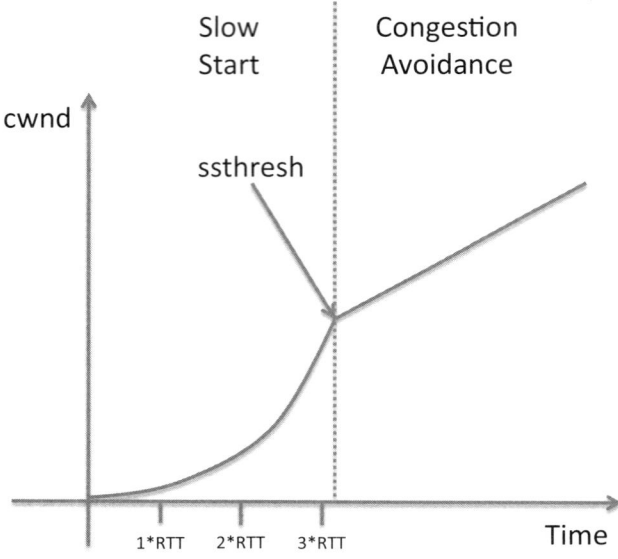

Figure 9.10   Behavior of cwnd over Time.

grow one SMSS per RTT. In a steady-state condition, cwnd will seesaw along a range of values; this is because TCP will always push the value of cwnd past the available capacity of the network.

When a TCP sender provides a segment to the IP layer below it, it starts a retransmission timer. If the sender does not successfully receive an ACK before the timer expires, then the segment is retransmitted. Key parameters that affect cwnd such as ssthresh are set when the retransmission timer expires to adjust the TCP flow of data into the network. In order to control the flow of segments into the network at a reasonable rate that minimizes congestion, it is important to consider physical parameters like the bandwidth of the link and the time it takes for data to travel from the sender to the receiver. The product of these two values, known as the bandwidth-delay product (BDP), is a key parameter that drives the maximum amount of data "in flight" in the network. Generally speaking, it is desirable to fill the network to the maximum capacity so that the amount of bits in flight is equal to 100% of the BDP. Any less would suggest inefficiencies in the protocols handling the data transmission and reduce the amount of user data rate available. This is illustrated in Figure 9.11.

In TCP especially, accurate BDP estimation is critical to minimize congestion and maximize efficiency of the underlying link. Consider Figure 9.11 as an example. If a TCP session sender can, at most, send 50% of the bits that the network could handle in flight, then it is essentially only using 50% of the network's available resources, because the receiver must acknowledge all segments. The time it takes

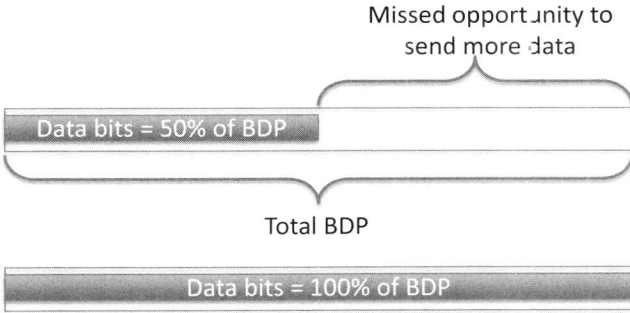

Figure 9.11 BDP considerations.

for an acknowledgment to come back from the receiver is the same regardless of the amount of data sent, up to the maximum BDP. Because of this, it is desirable to increase the amount of data sent in the network to 100% of EDP. The network can handle a total of BDP bits in flight from the sender to the receiver in any given instant in time, but if the maximum offered number of bits is limited by the sender to any value less than 100% of BDP, there is a missed opportunity to send more data, and thus a reduction in the TCP efficiency of the link. This problem is especially acute in the satellite communications domain, where BDP values are quite large, because of the increased propagation delay relative to terrestrial networks. TCP tries to account for this by using the slow start and congestion avoidance algorithms to increase the number of bits offered to the network while minimizing congestion and maximizing the number of bits in flight from the sender to the receiver.

When the retransmission timer expires and the segment has not been transmitted, TCP senders must adjust the ssthresh value according to the following equation:

$$\text{ssthresh} = \text{maximum}(\text{flightsize}/2, 2*\text{SMSS}),$$

where flightsize is the amount of outstanding data in the network. This effectively reduces the aggressiveness of the TCP sender in increasing cwnd (and thus increasing the offered traffic to the network) by lowering the threshold between slow start and congestion avoidance.

The fast retransmit and fast recovery mechanisms in TCP dictate what to do when segments are received out of order. RFC 5681 specifies that TCP receivers should send duplicate ACKs, acknowledging the last successfully received segment, if a segment with a sequence number is received that is higher than the next expected sequence number indicated in the ACK. Such events can be caused by lost segments, network reordering of segments, or replication of segments or ACKs in the network. When a TCP sender receives a duplicate ACK, it generally uses a fast-retransmit algorithm to detect and repair the loss. It waits for three duplicate ACKs to be successfully received before transmitting the missing segment, and does so without waiting for the retransmission timer to expire.

TCP mechanisms have been designed to effectively combat network congestion and adjust the offered traffic flow to the network based on measurements such as RTT and BDP. Such conditions are generally caused by retransmission timer expirations on effectively error-free links. For wireless systems, TCP may interpret such events as network congestion events, when they represent something else entirely.

In a wired network, TCP segments are likely to be lost during congestion events. Such events generally cause router buffers to overflow and drop IP packets, which cause the retransmission timer to expire in the TCP sender, who adjusts the rate of offered traffic to the network to reduce the drop rate and reduce network congestion. In a wireless network, all of these events can still happen, in addition to packet loss due to handoff or bit errors caused by adverse wireless channel conditions. TCP interprets the segment losses similarly, however, and because of that, performance can be adversely impacted.

Consider a simple example of a two-node terrestrial wireless network with a TCP sender host connected to one node and a TCP receiver host connected to the other. If the two wireless radios are well within the operating range, the bit error rate may be quite low, and TCP will generally perform reasonably well, similar to a wired network of equal bandwidth. Now consider that the two nodes are further away from each other, and as such, the bit error rate increases. In this example, the only network congestion that could be caused is from the two wireless nodes. Congestion is therefore only a function of how close to the maximum BDP the sender is operating when deciding how many segments to transfer at once into the network.

In the case where the nodes are far apart, bit errors cause segments to be dropped (Figure 9.12). Assuming the wireless network has no mechanism to retransmit lost frames that encapsulate the segments at the link layer, this results in the retransmission timer expiring or subsequent segments to be successfully received (and duplicate ACKs generated). In the case where the retransmission timer expires, the TCP sender will reduce ssthresh, reducing the flow of data to the network, and setting cwnd = SMSS (one segment). In the case where three duplicate ACKs are received before the retransmission timer expires, the TCP sender will reduce ssthresh and

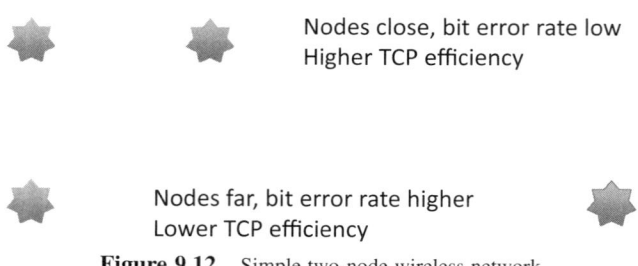

Figure 9.12 Simple two-node wireless network.

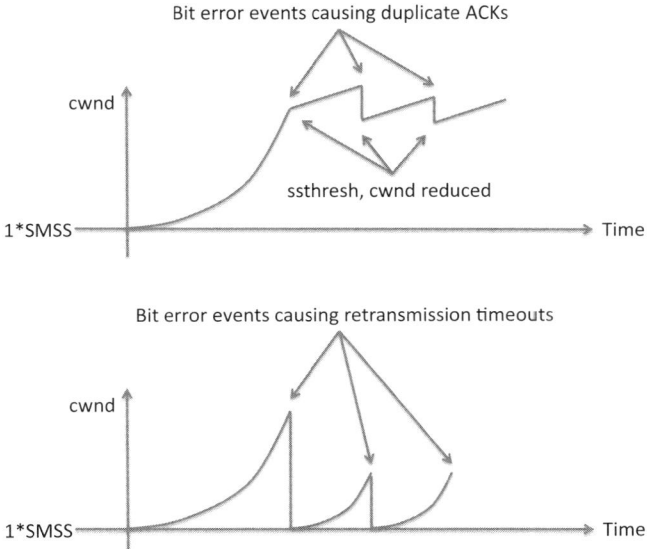

Figure 9.13   Behavior of cwnd based on bit errors for simple wireless network.

reduce the flow of data to the network, but will set cwnd to ssthresh upon successful reception of further ACKs of new data, suggesting the repair was successful. This improves the efficiency in the transient state where cwnd is increasing, but does so carefully by forcing the congestion control algorithm into the congestion avoidance phase where the cwnd increases by approximately one SMSS per RTT.

Therefore, in this example, what parameters should be considered when optimizing TCP? It seems reasonable to think that increasing the retransmission timeout value is a good thing—in this sense, we would observe that segment loss events would generally cause cwnd to drop down to ssthresh and then increase slowly. Is that better than dropping cwnd down to 1 SMSS each time a retransmission timer expires? Figure 9.13 illustrates the effect on cwnd caused by bit errors in these two cases. Note that the upper graph illustrates how the cwnd behaves linearly with time in the congestion avoidance phase upon a bit error event. Fast recovery and fast retransmit algorithms are being used to detect and correct any errors, while the cwnd stays within a range dictated by a minimum ssthresh value upon detecting an error. However, in the lower graph, the bit errors are causing TCP to expire a retransmission timer each time, reducing ssthresh, but also reducing cwnd to 1 SMSS. While this is a simple example, it is used to illustrate the different behavior of cwnd based on these two cases. In the case of a wireless network without congestion and with high bit error rates, it is generally more beneficial to operate in the congestion avoidance phase upon detecting an error than it is to reduce cwnd to 1 SMSS and exponentially increase. However, there may be some cases where it is indeed advantageous to operate in the slow start phase and set the retransmission timer lower, but it is important to study such cases carefully based on the expected network conditions. Modeling and simulation techniques to estimate the behavior of TCP and ultimately the goodput rates available should be used to explore the trade-offs adequately.

Many commercial wireless networking products based on technologies such as IEEE 802.11 or 802.16 use link layer strategies to remedy error-prone frames. In particular, techniques such as forward error correction (FEC) and automatic repeat request (ARQ) are utilized. These mechanisms aim to present data bits to the IP layer above error free; however, this is not without affect. In the case of FEC, additional delay is introduced into the link (albeit this may be very small as a percentage of the total delay) to account for the algorithm processing time in decoding the bit stream. In the case of ARQ, frames are retransmitted at the link layer when errors are detected, usually by a checksum calculation. These retransmitted frames may be presented up to the IP layer upon a successful checksum calculation, but this serves to increase the delay even further while increasing the probability that the IP packets arrive to a protocol like TCP out of order. A typical TCP implementation may simply use the three-ACK method to go back and retransmit the missing packets and reduce the value of cwnd accordingly, or sometimes, in the case of high error rate links, short retransmission timers may expire and cause cwnd to drop to a very low value, reducing TCP goodput substantially. However, other types of TCP implementations and options are available to improve performance for these types of systems.

9.3.2   TCP Variants and Options

This section covers some of the different TCP implementations and options developed to address some of the issues concerning TCP performance over unconventional links.

9.3.2.1   Selective Acknowledgments

RFC 2018 [249] specifies the TCP selective acknowledgment (SACK) option, which was originally developed as an answer to the problem of multiple missing/errored segments that could occur within a single RTT. Typical implementations of TCP acknowledge all data and rely on mechanisms such as fast retransmit/fast recovery to correct any errors in the segment stream. However, these mechanisms reduce efficiency especially in the case of large BDP links or where cwnd is large, because of the need for the sender to retransmit segments, starting with the segment indicated by the next expected sequence number in the duplicate ACK. This can occur even if the receiver has successfully received subsequent segments but needs only one segment to repair the stream. SACK options enable the receiver to selectively acknowledge those segments that were received successfully, enabling the sending TCP to only selectively retransmit those segments that were lost. This improves efficiency substantially over the conventional duplicate ACK method, especially for error-prone and high BDP links.

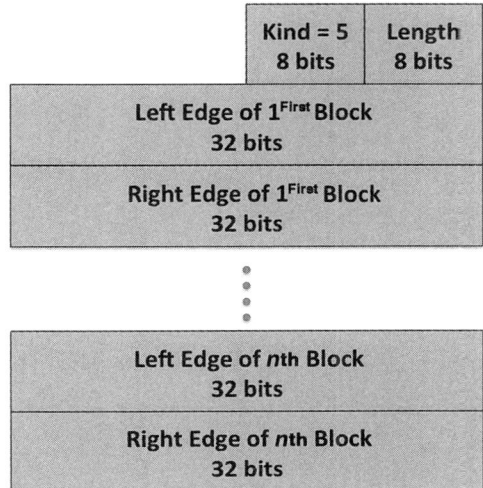

Figure 9.14 SACK option format.

The SACK-permitted option is defined in RFC 2018 as being sent within a SYN segment, of Kind = 4 and Length = 2. These fields are populated in the TCP options field of the TCP header. Any time a SACK is sent to acknowledge particular segments, it follows a format to define acknowledged "left" and "right" sides of a bit stream or series of bits in a stream to be acknowledged. Figure 9.14 illustrates the SACK option format. Here, a series of bits in a segment stream are acknowledged by the left and right edges of a piece of the total bit stream. Figure 9.15 illustrates this more clearly.

Note in Figure 9.15 that the received TCP segment stream has shaded areas that indicate successful reception of TCP segments defined by a left and right edge. The left and right edges correspond to the TCP sequence numbers that bound the data to be acknowledged. In the example shown, there are three distinct blocks: the first block is defined by a left and right edge, and the values of the sequence numbers defining the boundaries of this block are populated in the SACK option fields for the first block to be acknowledged. The second block to be acknowledged in this example is illustrated as well—while the missing stream data in the middle of the stream are not acknowledged. When the TCP sender receives this SACK, it will then know that the middle chunk of data was not acknowledged and will resend that data without having to resort to retransmission timeouts or substantial reductions in cwnd to account for the missing data. In a TCP implementation without SACK, all data up to the first right edge would be acknowledged, and then a retransmission timer would expire and/or a series of duplicate ACKs acknowledging the stream starting at the second left edge and beyond would cause a reduction in cwnd and potentially a fast retransmit operation. However, the SACK method allows TCP to continue to send and receive data in discontinuous pieces while keeping track of those pieces of the bit stream that are missing and selectively retransmitting as needed.

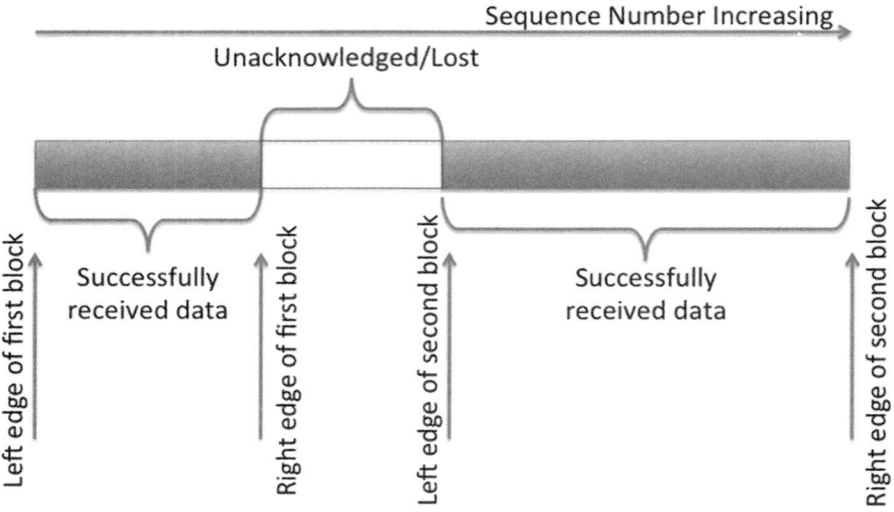

Figure 9.15   Selectively acknowledged data by TCP with SACK option.

9.3.2.2 *Hybla*

TCP Hybla was developed by researchers at the University of Bologna, Italy, specifically for SATCOM links on heterogeneous networks [250]. In particular, Hybla scales the behavior of the growth of cwnd in both slow start and congestion avoidance phases by way of a calculated ratio between the "target" RTT and the actual measured RTT. This is described in the following equation:

$$\rho = \frac{RTT_{actual}}{RTT_0},$$

where RTT_0 is set to a value representative of a particular link to be emulated and not necessarily the actual *RTT* of the link. When the ratio between the two values is one, cwnd grows similarly to typical TCP Reno or New Reno implementations. However, when the ratio is more than one, the window is scaled by the ratio. For instance, consider an $RTT_0 = 25$ ms and an $RTT_{actual} = 50$ ms. In this case, cwnd is scaled by 2, indicating that for every ACK received by the TCP sender, cwnd increases by 2*SMSS, instead of just 1*SMSS. This has the effect of doubling the offered traffic to the network compared to a typical TCP implementation. Furthermore, as RTT_{actual} increases even higher, the value of the cwnd increase scales accordingly. This can be valuable in cases where the increase in the actual RTT is caused by an increase in the number of hops in the network (and thus increase in network

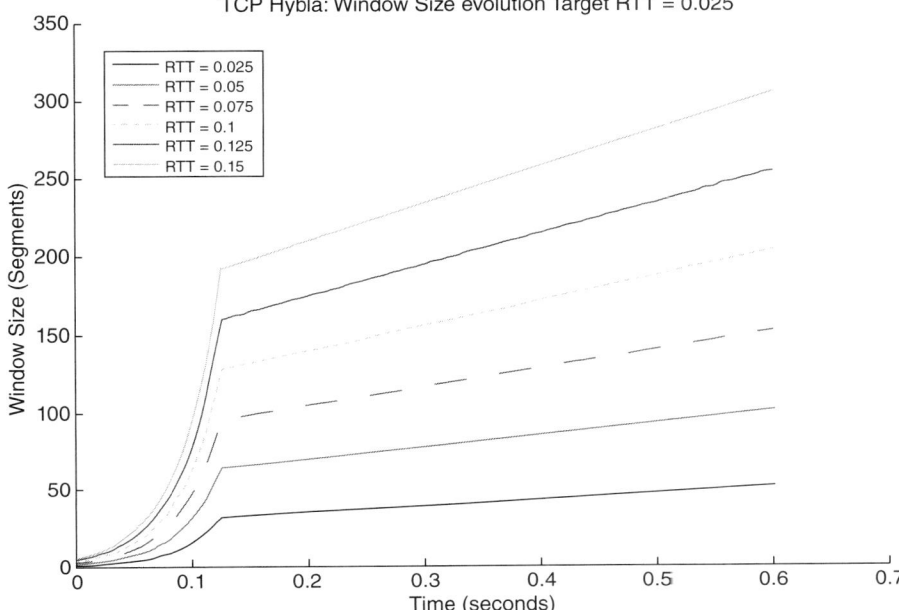

Figure 9.16 TCP Hybla scaling effect on cwnd.

capacity), but can be detrimental if the increased RTT is caused by congestion. In this sense, Hybla can quickly bring a network to its knees if it is not tuned properly; however, it can also improve the goodput ratio relative to other implementations. Figure 9.16 illustrates the scaling effect on cwnd for different values of RTT_{actual} and RTT_0.

9.3.2.3 Performance Enhancing Proxies

Performance enhancing proxies (PEPs) are generally used to split the TCP connection from source to destination host. In particular, these proxies often employ buffering strategies and immediately send back ACKs to the sender so the sender opens up the cwnd window as quickly as possible, maximizing the sender's goodput in the direction of the source PEP. At the destination side, the destination PEP handles all TCP semantics to the destination host, and presents the data to the destination host in a way that obscures the actual link behavior. Between the PEPs, mechanisms are in place to handle reliable transmission over the unconventional link, including selective acknowledgements and sometimes TCP variants that improve performance over such links (e.g., SATCOM). Figure 9.17 illustrates a split TCP PEP design.

Here, notice that the TCP sender sends SYN to initiate the connection to the receiver. However, PEP A receives this SYN and responds immediately with a SYN

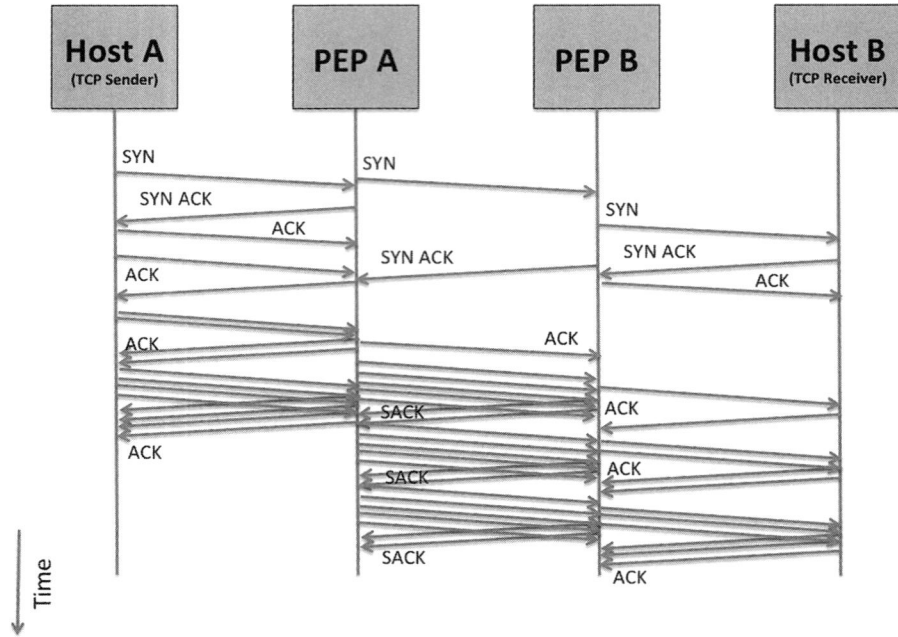

Figure 9.17 PEP design with split TCP connection.

ACK, fronting for the receiver. In the meantime, it also sends a SYN to PEP B to initiate a connection between the PEPs. PEP B then sends a SYN to the TCP receiver, fronting for the TCP sender. Both the TCP sender and receiver are led to believe they are operating on high bandwidth, low error rate links; in reality, the PEPs are fronting for each side and handling actual data transmission utilizing aggressive segment transmission strategies and SACKs to maximize efficiency. This is just one example of how a PEP could be implemented—there are many variants and algorithms that can be used in a split TCP scheme utilizing PEPs.

The authors in Reference [251] thoroughly examined the problem of TCP over wireless links in 1999. In particular, they compared the efficiency of TCP based on a variety of optimizations, including SACK, IETF-SMART, and TCP-aware LL strategies. Figure 9.18 [251] illustrates a relative comparison of some of the key results presented in the paper.

Here, four strategies were compared—the link layer (LL) strategy can be considered the baseline—note that there was no specific strategies employed to recognize and act on TCP-specific systems in this comparison, and the resultant efficiency was around 60% of the maximum. However, the LL-TCP-AWARE case specifically suppresses duplicate ACKs, reducing overhead on the network due to out-of-order packets, and results in a 90% efficiency for TCP. LL-SMART uses a SACK method

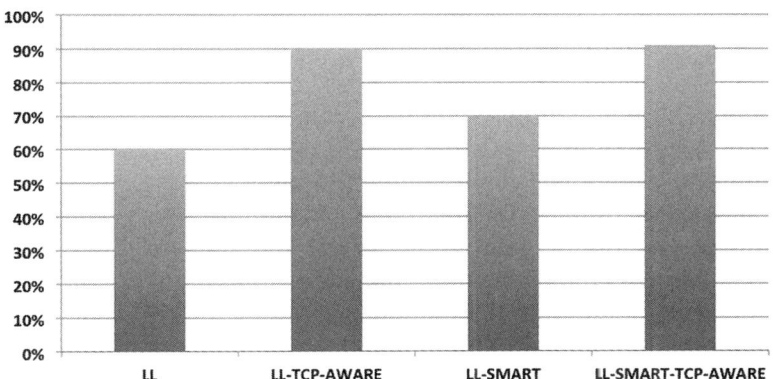

Figure 9.18   TCP efficiency. Reprinted from Reference [251].

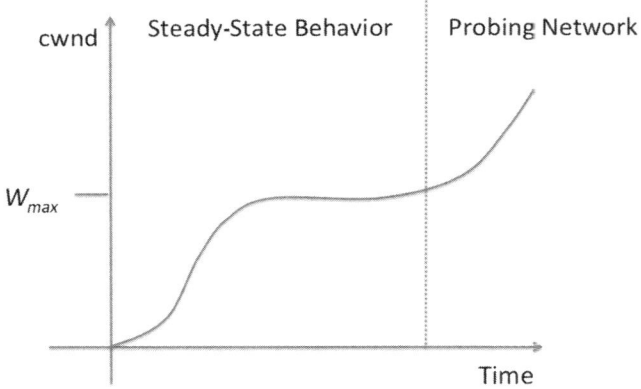

Figure 9.19   TCP CUBIC behavior of cwnd.

that improves efficiency relative to LL, around 70%. The LL-SMART-TCP-AWARE case compares similarly to the LL-TCP-AWARE case.

There are a couple of takeaways we should consider from the results presented in Reference [251]. One is that the potential gain in TCP efficiency based on employing some more strategies (SACK, link-layer awareness to suppress duplicate ACKs) can be substantial and can increase the user's effective data rate substantially. In Figure 9.18, the baseline efficiency of 60% compared with a maximum efficiency of 90% suggests a 50% increase in the user data rate for the same network! However, it is also important to note that any strategy should be considered carefully—additional complexity in the TCP and link layer stacks can perform marginally in cases where the error rates are lower. Newer versions of TCP stacks implemented generally try to adapt the behavior of cwnd growth based on the sensed RTT and its variance. For instance, TCP CUBIC [252] generally forces the cwnd curve to look much like a cubic function, flattening out at a programmable threshold value and then probing for an increase using the standard congestion avoidance mechanism. Figure 9.19 illustrates the behavior of cwnd in TCP CUBIC.

Note here that cwnd increases until some maximum W_{max}, at which point it remains near W_{max} for some period of time before probing the network to see if it can support more segments in flight. This gives other TCP sessions a chance to utilize the network in a fairer manner than other variants. Furthermore, backoff during the steady-state phase due to segment loss is not necessarily half, but rather driven by a configurable parameter to scale the backoff. Generally speaking, this factor is set below the typical value of half the original window, which results in a cwnd a bit higher than typical TCP after a segment loss. This can result in longer convergence times, but since W_{max} also throttles the increase of cwnd, this has shown improved network stability relative to congestion.

Chapter 10

Key Wireless Technology Trends: A Look at the Future

To understand the future wireless networking landscape, one needs to look only as far as academia. Over the past 20 years, academia has proven to be the engine of change within the wireless network industry. Figure 10.1 provides a somewhat tongue-in-cheek overview of this process.

This model has proven itself true time after time. As the wireless industry was perfecting the craft of building TDMA-based handsets, academia declared that CDMA was far superior. Eventually, the industry agreed and embraced CDMA. Then, after several years of study, academia declared that OFDM was far superior. Eventually, the industry agreed and embraced OFDM, and later OFDMA. Academia declared this new concept called MIMO was far superior to single-antenna systems. Eventually, the industry agreed and embraced MIMO. While this is a somewhat simplistic representation of these complex events, we hope you see the point we are trying to make. If you want to understand the technology of tomorrow, you can usually find your answer in the research papers of today.

The current driving factor for all technical change is the need for more data, the need for increased capacity to meet user demands. The smartphone and tablet revolution has fueled an insatiable appetite for bandwidth. And technology developers are struggling to keep up. To be more specific, the driving factor is to provide more data with no more, or even less, physical resources. Devices must be of lower power because consumers won't tolerate poor battery performance. Devices must be smaller and lighter because consumers won't tolerate bulky devices. Efficiency must improve, in terms of power, size, and bandwidth consumption.

There is no additional RF spectrum to be discovered. There is no getting around the physics of the problem. The only way to provide an increased data rate is to improve bandwidth efficiency. This leads to the need for improved modulation techniques. This leads to the need for more efficient MAC designs. This leads to the need for fundamental new approaches to old problems. Claude Shannon set the bar over 60 years ago. But clever scientists have found ways "around" Shannon's

Wireless Networking: Understanding Internetworking Challenges, First Edition.
Jack L. Burbank, Julia Andrusenko, Jared S. Everett, and William T.M. Kasch.
© 2013 the Institute of Electrical and Electronics Engineers, Inc. Published 2013 by John Wiley & Sons, Inc.

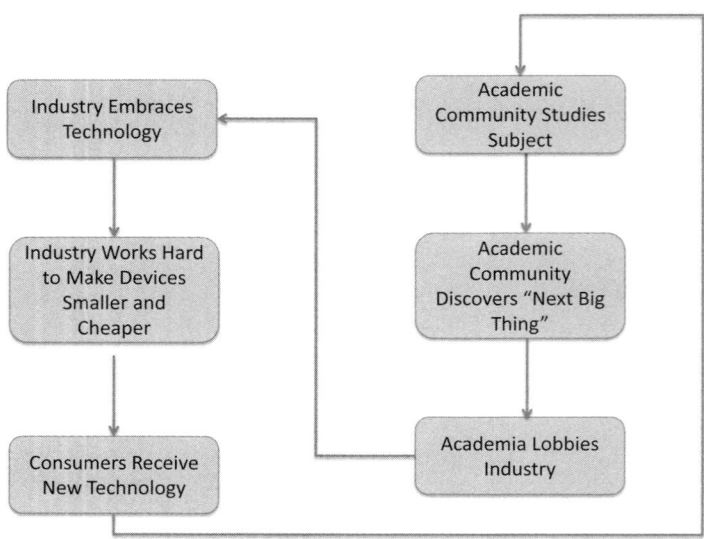

Figure 10.1 What drives the future commercial wireless networking landscape.

bandwidth efficiency limits by redefining what a channel means. We used to define a channel in terms of frequency in the era of analog communications. As we ushered in the digital era, we began to also think of a channel in terms of time (e.g., time slots). But over time, we found other fundamental aspects of a signal that we can manipulate to redefine a channel. We next discovered how to further define a channel in terms of power (e.g., CDMA). Last, we have discovered how to further define a channel in terms of space (e.g., MIMO). No one has proven Shannon wrong, nor is anyone likely to ever do so. We are simply discovering new ways to define the inputs into Shannon's arguments.

We can see clear trends across virtually all emerging wireless networking technologies. In fact, many communities are beginning to converge toward similar answers to many problems. We contend that we can examine those common themes across emerging technologies and understand how most, if not all, technologies will look over the next 5 years. One must look to academia to gain an understanding of what technologies might look like in 10 years. This chapter provides an overview of some of the key "themes" that we believe will dominate, or in some cases continue to dominate, the wireless networking landscape for the foreseeable future.

10.1 MIMO

MIMO is not a "future" technology. It has already arrived across virtually all emerging technologies. The IEEE 802.11 WLAN community has embraced MIMO. The IEEE 802.16 WMAN community has embraced MIMO. The 3GPP and 3GPP2 cellular communities have embraced MIMO. While MIMO has been on the scene only for a relatively short period of time, it is firmly entrenched in the landscape. The authors contend that MIMO will continue to be a driving force in future technologies for the next 10–20 years. This section provides a brief overview of MIMO. We recommend *Coding for MIMO Communication Systems* by Tolga Duman [253] as a wonderful MIMO reference.

10.1.1 Introduction to MIMO

The use of multiple antennas at both ends of the wireless link is not a novel concept. In fact, this is a natural causatum of more than four decades [254] of evolution within the realm of adaptive antenna technology. Recent technological advances have demonstrated that MIMO-enabled wireless systems can achieve significant increases in data throughput and link range without the need for additional bandwidth or increased transmit power. Because of these properties, MIMO technologies were integrated into various wireless standards such as IEEE 802.11n/ac (Wi-Fi), CDMA2000, W-CDMA, Long-Term Evolution (LTE), and WiMAX.

The history of MIMO can be traced back to the development of antenna arrays, primarily to form directed beams for radar applications [255]. The use of antenna arrays was an active research area during and after World War II in radar systems. In the 1970s, the introduction of digital signal processors made more sophisticated adaptive processing technologies possible. These techniques were vigorously developed for military applications. In the early 1990s, the community saw new proposals for using antennas to increase the capacity of wireless links. Roy and Ottersten in 1996 proposed the use of base station antennas to support multiple cochannel users. Paulraj and Kailath in 1994 proposed a technique for increasing the capacity of a wireless link using multiple antennas at both the transmitter and the receiver. These ideas, along with the fundamental research done at Bell Labs, began a new revolution in information and communications theories in the mid-1990s [255].

Figure 10.2 depicts a simplified version of the MIMO landscape. The previously independent concepts now feed into the general framework of MIMO technologies. The technologies are also supported by the theories about the channel and performance bounds. This development further benefits a number of commercial applications.

Today's theory-based MIMO technologies exist in three fundamental forms: spatial multiplexing (SM), spatial diversity, and beamforming with a few of evolving areas such as multiuser (MU) MIMO, cooperative MIMO, and microelectromechanical system (MEMS), an emergent area in adaptive beamforming.

Table 10.1 summarizes at a very high level all known commercial MIMO-enabled wireless standards and various forms of MIMO they support.

10.1.2 MIMO Technical Overview

10.1.2.1 MIMO Basics

The basic idea behind MIMO is one or more transmit antenna(s) communicating with one or more receive antenna(s). Various forms of MIMO exist, as shown in

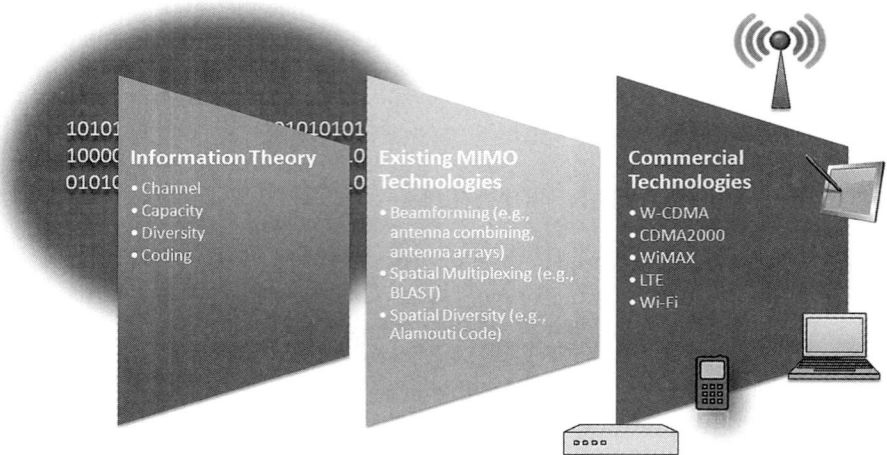

Figure 10.2   MIMO landscape. BLAST, Bell Laboratories Layered Space-Time.

Figure 10.3, such as single input, single output (SISO), single input, multiple output (SIMO), multiple input, single output (MISO), and, collectively, MIMO.

In the following few subsections we briefly discuss various MIMO forms.

10.1.2.2 Spatial Multiplexing

In spatial multiplexing, data streams are transferred simultaneously within one spectral channel of bandwidth. SM can significantly increase data throughput as the number of resolved data streams is increased. However, such improvement in throughput comes at a cost of increased system complexity (each stream requires a separate radiofrequency [RF] chain), which ultimately translates to higher implementation costs.

The basic idea of SM is illustrated in Figure 10.4. Note that in SM the number of receive antennas must be equal to or greater than the number of transmit antennas.

10.1.2.3 Spatial Diversity

Essentially, two types of spatial diversity exist: receive diversity and transmit diversity. Receive diversity demonstrates the ability to coherently resolve information from multiple signal paths using spatially separated *receive* antennas (see the SIMO illustration in Figure 10.3). Multipath signals are the reflected signals arriving at the receiver sometime after the line-of-sight (LOS) signal has been received. Typically, multipath is perceived to be the interference that negatively affects the receiver's ability to recover the intelligent information. Receive diversity MIMO, however, uses this opportunity to spatially resolve multipath signals, while providing the

Table 10.1   MIMO-Enabled Commercial Wireless Technologies

Technology	Beamforming	Spatial multiplexing	Spatial diversity	Cooperative MIMO/ MU-MIMO
CDMA2000 1X Advanced (3G)	No	No	Yes (UP: Mobile Rx diversity)	No
CDMA2000 1xEVDO Advanced (3G)	No	No	Yes (UP: Mobile Tx diversity)	No
UMTS/W-CDMA (up to HSPA+) (3G)	Yes (downlink)	Yes	Yes (DL: BS Tx diversity)	No
LTE (3.5G)	Yes (downlink)	Yes (SU-MIMO: 4 data streams on DL and 1 on UL)	Yes (DL: BS Tx diversity)	Yes (MU-MIMO for both UL and DL)
LTE-Advanced (4G)	Yes (downlink)	Yes (SU-MIMO: 8 data streams on DL and 4 on UL)	Yes (DL: BS Tx diversity)	Yes (MU-MIMO for both UL and DL)
Mobile WiMAX Release 1 (3G)	Yes (uplink and downlink)	Yes (up to 4×4 MIMO)	Yes (DL: BS Tx diversity; UL: Mobile Rx diversity)	Yes (uplink)
WiMAX-Advanced (Release 2) (4G)	Yes (uplink and downlink)	Yes (DL: up to 8 streams; UL: up to 4 streams)	Yes (DL: BS Tx diversity; UL: Mobile Rx diversity)	Yes (MU-MIMO for both UL and DL, Multi-BS MIMO)
802.11n (Wi-Fi)	Yes	Yes (up to 4×4 MIMO)	Yes	No
802.11ac (Gigabit Wi-Fi)	Yes	Yes (up to 8×8 MIMO)	Yes	Yes (MU-MIMO)

DL, downlink; UL, uplink; SU, single user; MU, multiuser.

diversity gain that contributes to the receiver's ability to recover intelligent information.

Now, let us consider the case where there are two transmit antennas sending the same data stream, and one receive antenna (see the MISO illustration in Figure 10.3). With this arrangement, there are two channels to utilize for transmission. Let us assume that the two transmit antennas are reasonably separated in space, that is, the channels are independently fading. Since the two channels are not likely to go into

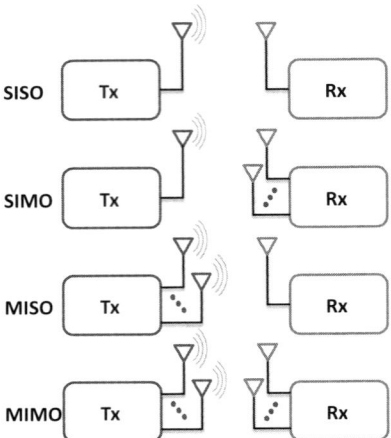

Figure 10.3   Various forms of MIMO.

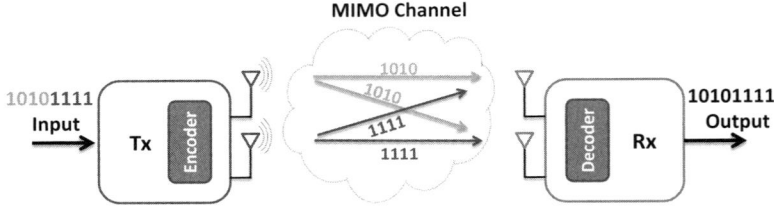

Figure 10.4   A 2 × 2 spatial multiplexing example.

deep fading at the same time, one might be able to use the "good channel" and thus avoid very low channel capacity. This is the basic idea of transmit diversity.

The transmit diversity can be achieved through the use of the space–time block codes (STBCs). The first and the simplest of all the STBCs is the Alamouti code, which is still widely used.

10.1.2.4 Beamforming

Fundamentally, there are two types of beamforming: transmit beamforming and receive beamforming/combining. Phased array transmit beamforming is used for focusing the energy to each receiver, whereas receive combining is used for boosting the reception of radio signals. Transmit beamforming offers a power gain due to the simultaneous use of the power amplifiers, up to the regulatory limits, of course. A rudimentary example of transmit beamforming is shown in Figure 10.5.

In general, beamforming increases the transmission range through the multipath mitigation and reduction in cochannel interference.

Figure 10.5   Transmit beamforming example.

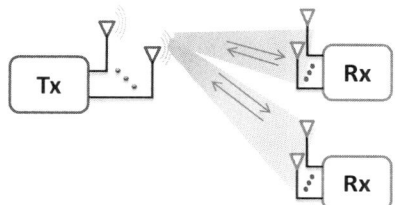

Figure 10.6   MU-MIMO example.

10.1.2.5 *Emerging MIMO Schemes*

Multiuser MIMO   MU-MIMO is a configuration where one node (e.g., a base station) has many antennas and communicates to multiple clients (e.g., mobile stations) at the same time and frequency. Signal processing is performed at the base station to minimize mutual interference among the links, while the client nodes operate independently (see Figure 10.6).

Cooperative MIMO   Cooperative MIMO has many possible configurations (see Figure 10.7).

The nodes involved are usually single-antenna transceivers, although multiantenna ones are not excluded. A number of nodes form clusters with high-speed interconnection. Nodes in a cluster act like the antennas in a multiantenna system. They jointly transmit and receive radio signals, achieving MIMO connections between the clusters.

MEMS for MIMO   Microelectromechanical system (MEMS) is a device (or a group of them) with a very small scale [256].[1] Typically, mechanically movable parts are fabricated using lithography or other nanofabrication technologies. These parts interact with electrical parts on the same device through electromagnetic fields. MEMS can serve as electrical and optical control devices, as well as sensors for mechanical conditions such as pressure and acceleration.

---

[1] All of the information presented in this subsection is based on the primary research conducted by Dr. Feng Ouyang, The Johns Hopkins University Applied Physics Laboratory, Laurel, MD.

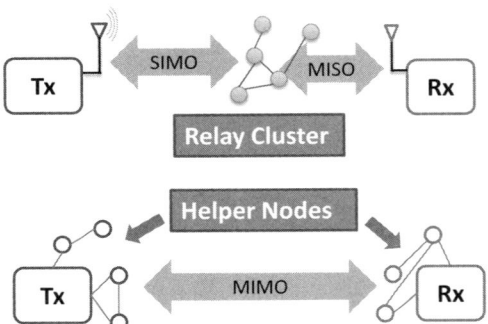

Figure 10.7   Cooperative MIMO example.

One application of MEMS in wireless systems is RF switches. An RF-MEMS switch consists of a movable beam on top of a substrate. When a voltage is applied to the device, the beam bends down to touch the substrate, closing the circuit. An RF-MEMS does not turn on or off at RF frequencies. However, its electronic characteristics are suitable for controlling RF circuits [257].

An RF-MEMS switch can be used in many ways. It can serve as a circuit switch in the RF chain, directing the antenna to transmission and receiving modules in a TDD system. It can also be part of the antenna, to form reconfigurable antennas. A reconfigurable antenna has multiple components that can be connected or disconnected to the feed. By controlling these connections, the system can control the frequency response, polarization, and directionality of the antenna [258–260].

The concept of reconfigurable antennas using RF-MEMS can be expanded to boost the performance of MIMO-enabled systems. The performance of a MIMO system heavily depends on the wireless channel, not only on attenuation but also on scattering properties. By manipulating the antenna characteristics, it may be possible to change the effective channel experienced by the communicating nodes and improve performance.

Typically, a reconfigurable antenna has multiple elements. In theory, we can consider each element as an antenna in the MIMO system and achieve full optimization by feeding them separate RF signals. However, this requires many RF chains, thus increasing system cost and complexity. In this view, a MIMO system with reconfigurable antennas (such as a MEMS-MIMO) is equivalent to grouping several "component antennas" into a reconfigurable antenna, sharing the same RF feed. Therefore, this is actually a compromise between performance and complexity. Such trade-off has been an important design issue for MIMO systems [261]. MEMS-MIMO introduces another dimension for optimization: changing the antenna patters of the grouped antenna. This further improves performance [262, 263].

In a spatially multiplexed MIMO system, ideally one would like to select antenna patterns on both sides. This can be done by measuring channel characteristics for each combination of antenna pattern selections. Selection can thus be made based on predicted channel capacity values. Depending on the scenarios and antenna designs, capacity gain for 2×2 MIMO systems range from 50% to 150% [259, 264, 265]. This selection method, in contrast, introduces extra tasks for the training process. The number of channel states to be measured equals to the number of possible antenna selection combinations. In the simplest case, a 2×2 MIMO system supporting two configurations for each antenna is used. For such system, a total of 16 channel measurements would need to be performed. This would increase training overhead in the protocol, particularly if the channel condition changes frequently (such as in a mobile application). Moreover, the selection decision reached by the receiver needs to be communicated to the transmitter, requiring further overhead and delay. A compromise can be made by selecting receiver antenna configurations only [266].

MEMS-MIMO can also be used for space–time coding systems [267]. In the conventional space–time coding systems, the diversity gain (the performance gain against channel fading) is limited by NM, where N and M are the number of antennas on the transmitter and receiver sides, respectively. In a MEMS-MIMO system, if each receiver antenna has P elements whose connections can be changed based on channel condition, the resulting diversity gain would be NMP. In other words, when the P elements are grouped into one reconfigurable antenna instead of P independent antennas, the diversity gain will not be degraded.

In addition to switching the connection of antenna elements, MEMS can switch circuit components as well. This means that circuit characteristics such as filter frequency responses or phase shifts can be changed in real time. MEMS can support the MIMO mode of beamforming by adjusting the phase difference between various antennas on the transmitter or receiver side [268, 269].

MEMS-MIMO can also be used for enhancing the conventional MIMO operation modes of a given wireless system. New modulation modes can also be implemented with MEMS switches. For example, a transmitter can have only one RF chain, which is switched to one of the multiple antennas. The identity of the antenna for a particular symbol carries an additional bit of data, and thus realizing the capacity gain of a MIMO system [270]. This method reduces the cost of the multiple RF chains required in a conventional MIMO system.

At present, MEMS-MIMO remains to be an active research topic. As discussed before, a MEMS-MIMO system needs additional training time and protocol support (if optimization at the transmitter is included). No major commercial wireless technology provides such support yet. However, MEMS-MIMO is an attractive option for proprietary systems that are sensitive to cost and size limitations.

10.2   MULTICARRIER MODULATION

This is another concept that is not in the "future" but rather in the "here and now." Multicarrier modulation (MCM) techniques such as OFDM currently dominate the commercial wireless networking landscape, having dominated virtually every sector

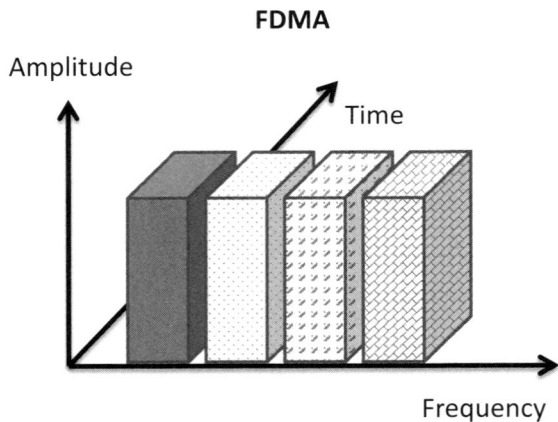

Figure 10.8   FDMA approach to multiple access in cellular systems.

of the industry. OFDM achieves high data rates by allocating a single data stream in a parallel fashion across multiple subcarriers. OFDM is widely deployed across various technologies such as Digital Video Broadcasting (DVB), Digital Audio Broadcasting (DAB), Asymmetric Digital Subscriber Line (ADSL), IEEE 802.11, IEEE 802.16, and LTE. It is believed by the authors that this trend is going to continue into the foreseeable future. This section provides a basic overview of the OFDM concept.

10.2.1   OFDM Background

Prior to OFDM, there were three primary cellular multiple access methods employed within cellular networks: frequency division multiple access (FDMA), time division multiple access (TDMA), and CDMA. In FDMA, the overall wireless bearer is divided in the frequency domain. A mobile user is assigned to a particular frequency channel, which the mobile user can utilize exclusively, as depicted in Figure 10.8. This was the basis of first-generation (1G) Advanced Mobile Phone System (AMPS) cellular systems.

In TDMA, channelization is provided by dividing the overall wireless bearer in the time domain and frequency domain. As such, a mobile user is assigned to a particular frequency channel and given a particular time assignment or time slot during which the mobile user has exclusive use of that frequency channel, as depicted in Figure 10.9. Usually, a mobile is assigned two time slots for the forward and reverse links (from base to mobile and from mobile to base, respectively). This was the basis of the International Standard (IS)-136 2G cellular standard and IS-54 1G cellular standard. Furthermore, this is the basis of the GSM technology standard, which is deployed worldwide.

TDMA

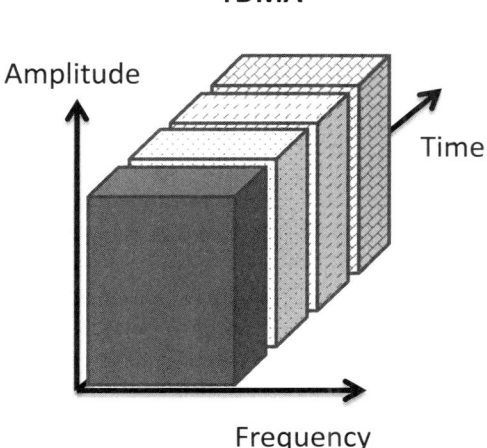

Figure 10.9   TDMA approach to multiple access in cellular systems.

During the evolution of 2G cellular technologies, the introduction of CDMA represented a revolutionary paradigm shift in the way cellular systems allocated system capacity. In CDMA, all mobile users occupy the same frequency channel at the same time, separated from each other by their unique modulation signatures, known as "spreading codes". This is illustrated in Figure 10.10.

CDMA technology operates on the premise of multiplying each user's data channel by a unique binary chipping sequence at a much faster rate than the symbol rate of the modulation used. This, in effect, spreads the spectrum of each user to cover a bandwidth much greater than the baseband signal bandwidth so that all users share the entire spectrum at the same time and with the same power. Interference and other channel impairments are minimized in this fashion because the chipping sequences are orthogonal (or nearly orthogonal) to one another in signal space. At the receiver, the orthogonal chipping sequence is applied to the received signal, mitigating signals other than the signal of interest. Both CDMA2000 and Wideband Code Division Multiple Access (W-CDMA) use the same fundamental concept of CDMA.

OFDM uses a large number of closely spaced narrowband carriers (Figure 10.11). In a conventional FDM-based system, the frequency spacing between carriers is chosen with a sufficient guard band to minimize interference. However, in OFDM, the carriers are packed much closer together to increase spectral efficiency by using a carrier spacing that is the inverse of the symbol or modulation rate. Moreover, each modulation symbol uses a simple rectangular pulse.

In OFDM, the frequency spacing between the subcarriers is chosen in a manner to maintain orthogonality between subcarriers, regardless of their close spacing. This is a key feature of OFDM that effectively mitigates (1) mutual interference

CDMA

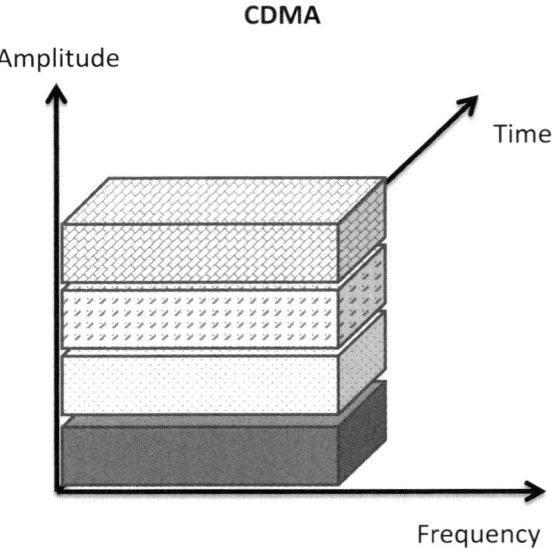

Figure 10.10 CDMA approach to multiple access in cellular systems.

OFDM

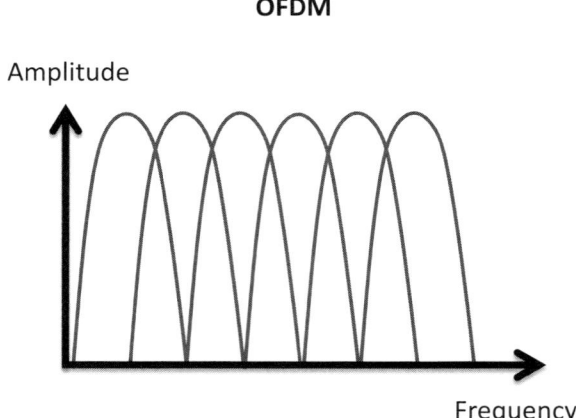

Figure 10.11 OFDM approach to multiple access.

between signals carried on different subcarriers (intercarrier interference) and (2) the guard band between subcarriers, a necessary feature in a conventional FDM system. This approach results in a relatively high achievable spectral efficiency, also eliminating the need for receiver filters for each subcarrier (as in a conventional FDM system).

10.2.2   OFDM Modulation

In the time domain, each OFDM subcarrier symbol is a rectangular pulse. A rectangular pulse of period T_u in the time domain corresponds to a sinc-shaped function in the frequency domain, with zero crossings every Δf or $1/T_u$ (Figure 10.12).

Since it is a fundamental principle of OFDM that each OFDM subcarrier symbol is a rectangular pulse, the subcarrier spacing equals the inverse of the symbol period (Eq. 10.1):

$$\text{Subcarrier Spacing} = 1/T_u. \tag{10.1}$$

Hence, for a specific subcarrier in the frequency domain, the zero crossings from adjacent subcarriers will coincide with its center frequency.

10.2.3   OFDM Orthogonality

In OFDM, adjacent subcarriers are separated in frequency by Δf or $1/T_u$. Having an identical period $T_u\Delta f$ or $1/T_u$ for each subcarrier ensures zero intercarrier interference at each subcarrier's center frequency, as shown in Figure 10.13.

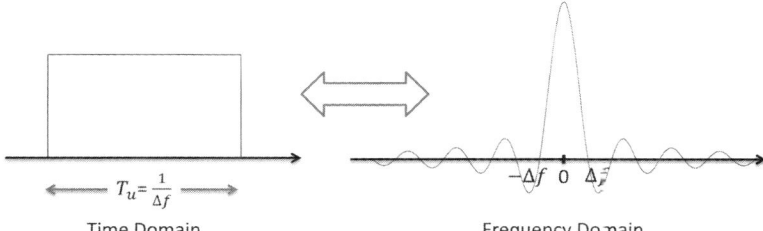

Time Domain   Frequency Domain

Figure 10.12   OFDM modulation.

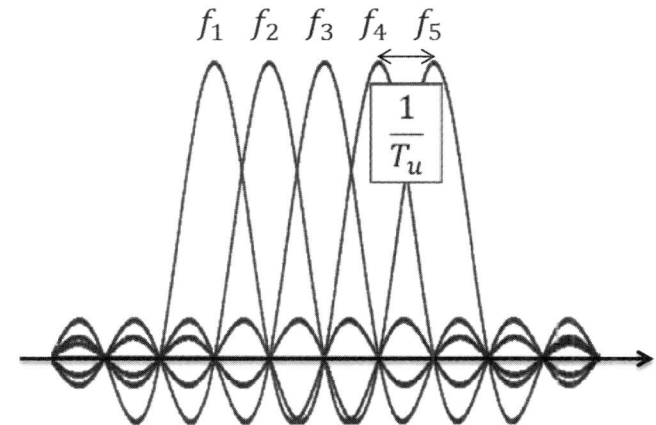

Figure 10.13   OFDM subcarriers with $1/T_u$ spacing.

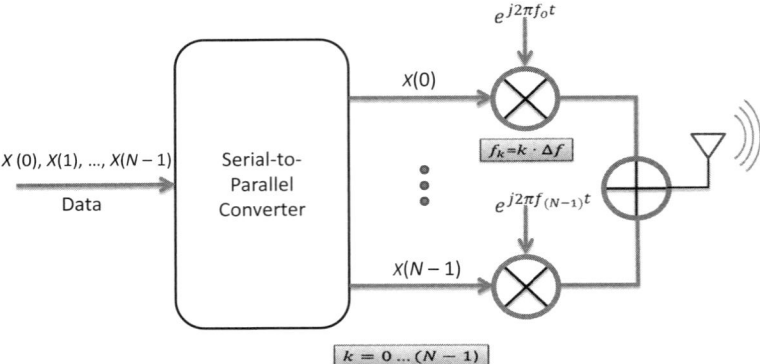

Figure 10.14   OFDM signal generation using banks of subcarrier oscillators.

Application of this criterion to two subcarriers in the time domain, $x_1(t)$ and $x_2(t)$, with center frequencies separated in this manner results in the following condition being met such that:

$$\int_{nT_u}^{(n+1)T_u} x_1(t)x_2^*(t)\,dt = 0, \tag{10.2}$$

during the nT_u to $(n + 1)T_u$ time period, where n is an integer.

OFDM signal generation, in its simplest form, is depicted in Figure 10.14. OFDM assigns a stream of data in parallel fashion to a number of subcarriers, where the subcarriers' orthogonality is ensured via the relationship between the subcarrier spacing and symbol period. Figure 10.14 depicts the use of banks of subcarrier oscillators in OFDM signal generation. A more efficient way of OFDM processing is via software using the fast Fourier transform (FFT) (discussed later) [271].

OFDM achieves an increased spectral efficiency by spacing adjacent subcarriers closely together. It is also quite resilient to multipath fading. OFDM sends a high-rate data stream over many low-rate subcarriers. This is different from single-carrier modulation schemes such as those employed in CDMA-based systems. In CDMA, the transmitted symbol rate increases proportionately with required user data rate. The symbol interval in these single-carrier (SC) modulation approaches becomes much shorter than the delay spread of the propagation environment, turning ISI into a significant performance problem.

10.2.4   OFDM Signal in Practice: Generating and Receiving

Until now we discussed OFDM transmitter implementation using banks of subcarrier oscillators. The potential advantages of such a multicarrier (MC) approach have

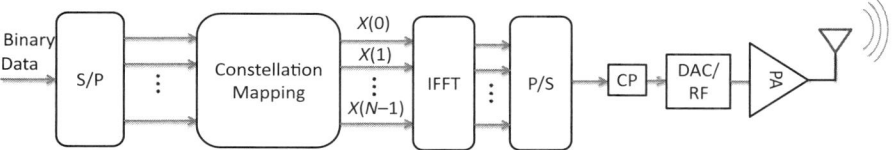

Figure 10.15   Block diagram of a baseband OFDM transmitter. DAC, digital-to-analog converter; PA, power amplifier.

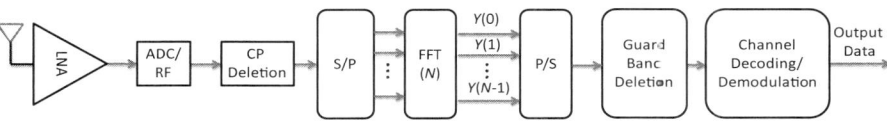

Figure 10.16   Block diagram of a baseband OFDM receiver. LNA, low noise amplifier; ADC, analog-to-digital converter.

been known for over 40 years, such as vastly improved robustness against multipath fading compared with SC approaches. However, these MC approaches proved overly complex for a long period of time. Indeed, an n-carrier approach would traditionally require n modulators, n oscillators, and so on, in a hardware-based implementation. This proves too complex when considering 256 carriers, 1024 carriers, 2048 carriers, and so on. This type of MC approach has only proved feasible in the past decade with advances in digital signal processing (DSP) capabilities. A more efficient way of OFDM processing is via software using the FFT algorithm on the receive end and the inverse fast Fourier transform (IFFT) algorithm on the transmitter side Figure 10.15 and Figure 10.16 illustrate digital implementation of a baseband OFDM transmitter and receiver, respectively.

Binary data is sent through the serial-to-parallel (S/P) converter. After that, the bits are mapped to complex digital symbols. The data stream is modulated such that a certain number of bits are represented by each modulation symbol. For instance, 16-QAM modulation has 4-bit modulation symbols, whereas 64-QAM has 6-bit ones. LTE, for example, supports QPSK, 16-QAM, and 64-QAM on both the downlink and uplink. No information is transmitted on guard subcarriers. The complex modulated symbols $X(k)$, where $k = 0,1,...,(N-1)$, are then mapped to the input of IFFT. A CP is added after the IFFT operation, where the resulting sequence is upconverted to RF and amplified as shown in Figure 10.15. At the receiver, the received signal is filtered, amplified, and downconverted from RF as shown in Figure 10.16. The CP samples are discarded before the FFT operation is performed on the sequence of received samples. These samples are then fed through the parallel-to-serial (P/S) converter. After that, the guard subcarrier information is deleted from the sequence. The remaining information is then decoded and demodulated.

10.2.5 Cyclic Prefix

In the absence of multipath, a received OFDM signal does not suffer from intersubcarrier interference and ISI. In a multipath fading environment, however, orthogonality between OFDM subcarriers can be partially lost due to the symbols received from reflected/delayed paths overlapping onto the following consecutive symbol. This issue is addressed with the CP, where the tail of each symbol is copied and pasted onto the front of the OFDM symbol, as shown in Figure 10.12. CP insertion increases the length of the symbol from $T_u T_u$ to $T_u + T_{CP}$, where T_{CP} is the length of the CP; this insertion effectively creates a guard time between symbols (Figure 10.17).

At the receiver, the integration is carried out over the original symbol time T_u. Properly selected CP ensures that any multipath components from a previous symbol fall within the effective guard time at the start of each symbol and can be discarded, thus avoiding intersymbol interference (ISI) and preserving orthogonality. Care must be taken when choosing the length of CP. A long CP can accommodate a large range of channel delay but at the expense of efficiency. A shorter CP is more efficient but increases the probability of signal distortion due to multipath.

In summary, the symbol period should be chosen such that the channel does not vary significantly over each period to preserve orthogonality and limit intersubcarrier interference. The carrier frequency should be as small as possible to maximize the symbol period for the CP overhead reduction.

The overall OFDM process is illustrated in Figure 10.18.

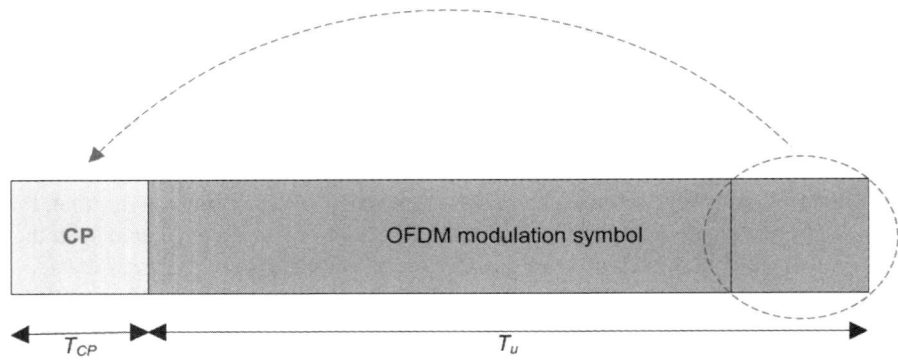

Figure 10.17   Cyclic prefix.

Figure 10.18   A High-level view of the end-to-end OFDM link.

10.2.6   Advantages and Disadvantages of OFDM

OFDM offers several key advantages:

(a) High spectral efficiency

(b) Robustness to multipath fading

(c) Easily scalable bandwidth allocation

In OFDM, a relatively high spectral efficiency can be achieved through the ability to closely space the subcarriers. In addition, carriers can be overlapped due to the orthogonality achieved by the frequency separation selection and symbol duration.

OFDM demonstrates a significant robustness in multipath fading environments by operating with narrowband channels that exhibit flat fading. The use of CPs effectively mitigates the effects of ISI due to delay spread.

OFDM-based systems are known for their flexibility. The use of multiple narrowband subcarriers allows the total bandwidth used to be easily scaled based on demand and spectrum availability.

OFDM also suffers from several disadvantages:

(a) High peak-to-average power ratio (PAPR)

(b) Sensitivity to frequency and timing errors

(c) Guard band requirements

PAPR is caused by drastic instantaneous power variations within the time-domain OFDM symbol (e.g., time-domain symbol of Figure 10.12). The modulation of each subcarrier in an OFDM system changes independently, resulting in the requirement of a power amplifier with a large range of linearity, which can lower efficiency and increase cost. This limits the utility of OFDM on the uplink of a cellular system because it can negatively impact the affordability of mobile subscriber equipment that can provide acceptable performance and battery life.

Imperfect frequency synchronization between the transmitter and receiver will cause loss of orthogonality between carriers, resulting in intersubcarrier interference. Since subcarriers are tightly packed, this becomes a critical issue. One way of dealing with this is through the use of reference (pilot) symbols, which are transmitted with a known periodicity. These pilot symbols have a sufficient density to interpolate for any subcarrier during any symbol time (see Figure 10.19).

In the frequency domain, the nature of a rectangular pulse as the OFDM modulation symbol results is a spectrum distribution that rolls off slowly in comparison to other multiple access techniques, hence the need for guard bands (Figure 10.20).

This generates a lower maximum number of carriers that can be reasonably utilized by a defined bandwidth. In LTE, the majority of OFDM carriers use approximately 90% of the assigned channel bandwidth.

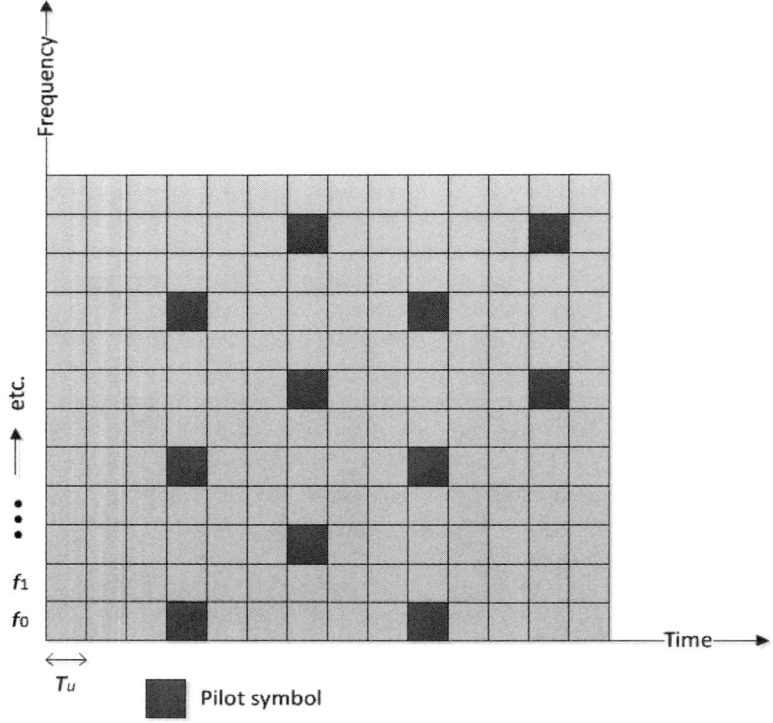

Figure 10.19   OFDM reference (pilot) symbols to help receiver with channel estimation.

10.2.7 OFDMA

OFDMA is a variant of OFDM. In the previous discussion, we assumed that a single user (SU) receives data on all the subcarriers at any given time. OFDMA, however, distributes subcarriers to different users at the same time, so that multiple users can be scheduled to receive data simultaneously. Subcarriers are usually allocated in contiguous groups for simplicity and to reduce the overhead of indicating which subcarriers have been allocated to each user. Figure 10.21 illustrates a general comparison between OFDM and OFDMA. Each pattern represents a different user. In a given time period, OFDMA lets users share the available bandwidth. OFDM supports QPSK, 16-QAM, and 64-QAM modulations.

OFDMA is employed by LTE on the downlink. When OFDMA is used in combination with TDMA, the resources are partitioned in the time-frequency plane, that is, groups of subcarriers for specific time duration. In LTE, such time-frequency blocks are the previously mentioned resource blocks (RBs) (see Chapter 8). Figure 10.21 illustrates such an OFDMA/TDMA mixed strategy used in LTE. RBs define contiguous groups of subcarriers and OFDM symbols.

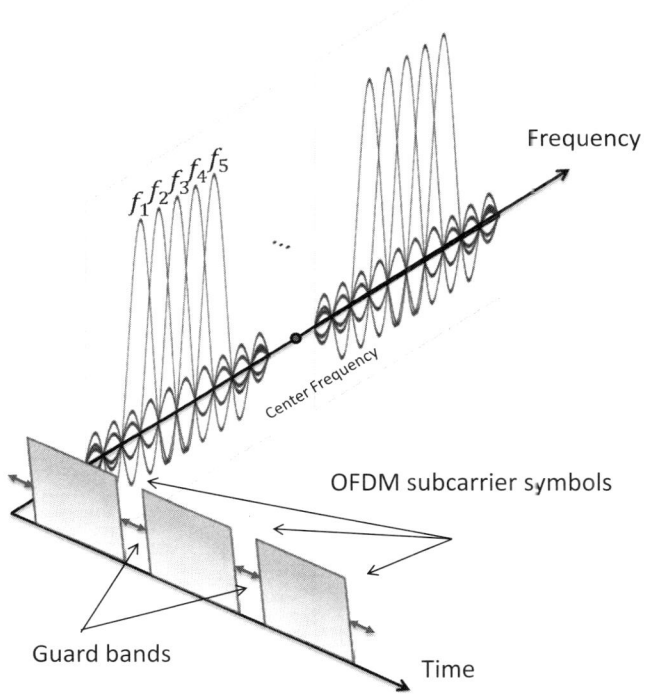

Figure 10.20   OFDM signal: slow spectrum roll-off.

10.3 COGNITIVE RADIO

Spectrum is a precious resource, and there is simply not enough to meet the needs of today and tomorrow's user base. This problem is exacerbated by the outdated way in which we manage spectrum. Regulatory agencies, such as the FCC in the United States, allocate spectrum for particular types of services that is then licensed to bidders for a fee. Those allocations and licenses are static in nature, which means that this spectrum is unavailable for use, even if those who own the rights to that spectrum do not use it. This has led to considerable inefficiency in spectrum utilization and has created an unnecessary shortage of spectrum. This issue has been temporarily alleviated by providing for the availability of spectrum for unlicensed usage and has fueled the global deployment of 802.11-based technology. However, these unlicensed frequency bands are becoming overpopulated and interference has grown to be a significant deployment constraint. All of these factors have led to the need to make dramatic changes in the spectrum regulatory process, as existing practices and policies are not capable of scaling with demand. This, in part, has led to the concept of cognitive radio.

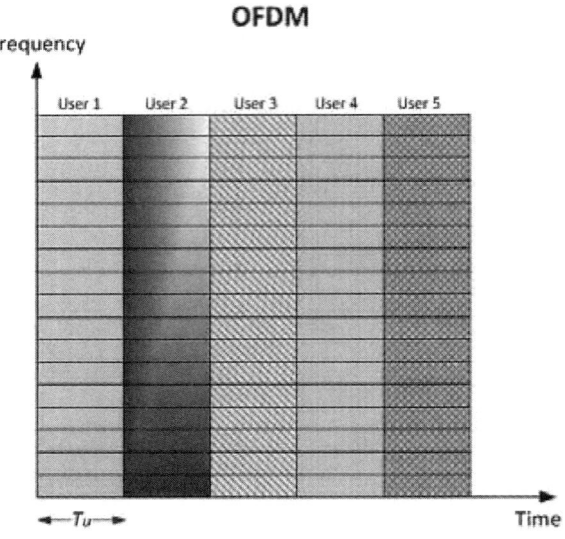

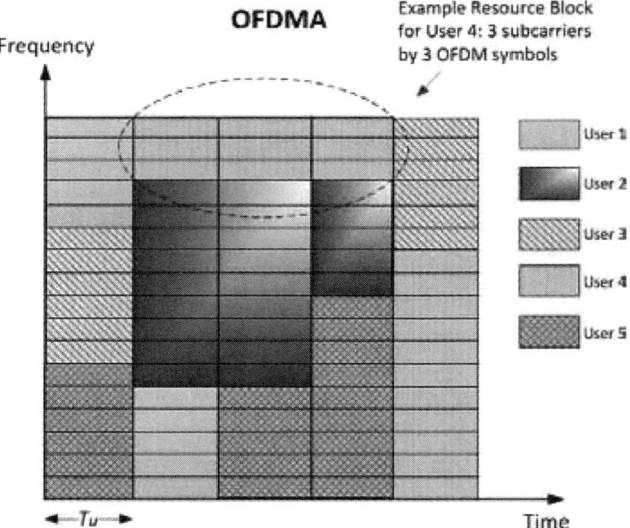

Figure 10.21   OFDM versus OFDMA.

Despite the growing body of work on the topic of cognitive radio and cognitive networking, there has yet to emerge a common understanding for what exactly is a cognitive radio. This includes the lack of well-understood cognitive networking terminology and definitions. cognitive radio is in itself a confusing concept that holds many potential meanings. For convenience, we restate some high-level attributes of cognitive radios that are generally agreed upon by the greater community:

1. *adapts* to its environment to meet requirements and goals

2. maintains *awareness* of surrounding and internal state

3. uses *reasoning* on ontology and/or observations to adjust adaptation goals

4. exhibits *learning* from past performance to recognize conditions and enable faster reaction times

5. *plans* to determine future decisions based on anticipated events

6. *collaborates* with other devices to build greater collective observations and knowledge

The FCC defines CR as a "radio system whose parameters are based on information in the environment external to the radio system." The National Telecommunications and Information Association (NTIA) has proposed to define CR as a "radio or system that senses its operational electromagnetic environment and can dynamically and autonomously adjust its radio operating parameters to modify system operations, such as maximize throughput, mitigate interference, facilitate interoperability, and access secondary markets."

IEEE 1900.1 defines a CR as:

a) a type of *Radio* in which communication systems are aware of their environment and internal state and can make decisions about their radio operating behavior based on that information and predefined objectives. The environmental information may or may not include location information related to communication systems.

b) Cognitive Radio, as defined in a), that utilizes *Software Defined Radio, Adaptive Radio*, and other technologies to automatically adjust its behavior or operations to achieve desired objectives.

The recent transition to digital television has, and will further result, in empty and unused bands of spectrum (white space), which have been allocated for broadcast TV channels, referred to as TV white space (TVWS). Furthermore, numerous regulatory authorities around the world are also actively redesignating former TV spectrum for use by commercial broadband providers on an unlicensed not-to-interfere basis. Making the best use of this available spectrum without at the same time infringing on the incumbents (e.g., TV service providers) has proven to be a significant technical and regulatory challenge and has slowed the development of such technologies. Thus, this has become the first large-scale application of cognitive radio concepts, with a particular focus on dynamic spectrum access (DSA) aspects of cognitive radio.

There are several standardization efforts aggressively working to develop dynamic spectrum access (DSA) solutions to provide broadband data services in these underutilized TV broadcast bands on a noninterference basis. These standardization efforts includes the much anticipated IEEE 802.22 Wireless Regional Area Network (WRAN) standard, the IEEE 802.11af Wireless Local Area Network (WLAN) TVWS standard, along with several areas. Regulatory developments of the past 2–3 years have dramatically influenced these development efforts.

10.3.1 Regulatory Positions

The first key regulation regarding TVWS usage came in November 2008 with FCC 08-260 [272]. This regulation was important because it paved the way for TVWS unlicensed usage, allowing both fixed and personal/portable broadband TV band devices (TVBDs). The frequency ranges that were opened for unlicensed usage span the VHF Band I (often referred to as "VLow"), VHF Band II (often referred to as "VHigh"), and UHF frequency bands:

- 54–72 MHz (TV channels 2–4)
- 76–88 MHz (TV channels 5 and 6)
- 174–216 MHz (TV channels 7–13)
- 470–512 MHz (TV channels 14–20)
- 512–692 MHz (TV channels 21–51)

Note that channels 14–20 are shared with Public Safety in certain urban areas.

Fixed devices are required to include geolocation capabilities and the ability to query geolocation databases to determine the allowed set of frequency channels based on current location. This is intended to provide protection to incumbent TV broadcasters, protecting a contour surrounding a particular broadcast market. The included geolocation capability is required to have an accuracy of ±50 m.

There are two types of portable devices defined by the FCC:

- *Mode I.* Under control of a device that employs geolocation and database access
- *Mode II.* Employs geolocation and the database access itself

The class of a device (i.e., fixed or portable) in turn defines various RF parameters, such as allowable channels of operation, power limits, and required spectrum-sensing sensitivity. These are summarized in Table 10.2.

The FCC regulation also contains stringent out-of-band emissions requirements, requiring that out-of-band emissions be at least 55 dB below the highest average power of the intended broadcast by the immediate adjacent channel. The requirement is even more stringent at the edges of particular channels (such as channels 36 and 38).

Table 10.2   Key RF Parameters for TVBDs from FCC-08-260

Type of device	Maximum EIRP	Allowed channels	Sensing requirement (dBm)
Fixed	1 W[a]	2, 5–36, 38–51	−114
Portable	100 mW[b]	21–36, 38–51	−114

[a]40 mW if operating on a channel adjacent to an incumbent TV broadcast.

[b]50 mW if device uses spectrum sensing only (no database query access).

Perhaps the most stringent, and certainly the most controversial, regulation in the original 2008 FCC report was the requirement for all TVBDs, both fixed and portable, to include a spectrum-sensing capability in order to protect wireless microphone signals that operate in licensed TV band. The locations of these wireless microphones are not known a priori (e.g., newsworthy event that draws field reporters to a location utilizing wireless microphones), so a database approach is not adequate. Therefore, all TBVDs are required to actively sense the spectrum in which it is operating to determine if an incumbent signal is present (per the sensing threshold requirements of Table 10.2). This ruling dramatically effected many development efforts and slowed the standardization and implementation of devices capable of utilizing TV white spaces.

A subsequent ruling in September 2010 by the FCC in FCC 10-174 [273] removed the requirement for sensing. Rather, two TV channels were reserved for use by wireless microphones to ensure they would not operate in the same channels as TVBDs. This completely changed the path of ongoing standardization efforts, leading toward rapid development and deployment. There are now numerous ongoing standardization efforts working to develop TVWS technologies, now moving forward very fast given the 2010 FCC relaxation of spectrum-sensing requirements on TVWS devices. The majority of these standardization efforts are now solely predicated upon geolocation database query approaches for incumbent protection. The FCC mandates that geolocation databases be maintained by independent third parties. The FCC will certify these databases.

We believe that the majority of the world's regulatory authorities will come to similar conclusions as the FCC, eventually relaxing spectrum-sensing requirements and emphasizing geolocation database approaches. We believe this is a historic event in the wireless networking landscape, perhaps of the same importance as the FCC ruling that made the 2.4 GHz ISM band available for unlicensed usage in the 1980s. We believe that this is such a historic event that we predict that future revisions of this text will focus as much on TVWS technologies as much as all other technologies.

One note we feel important to make here: DSA does not equal cognitive radio. A common misconception associated with cognitive radio is that cognitive techniques are limited to the radio layers of the protocol stack, such as the examples of DSA or adaptive modulation and coding (AMC). Although these are certainly useful examples in illustrating cognitive techniques, one can postulate cognitive techniques at each and every layer of the protocol stack. Table 10.3 lists example cognitive techniques throughout the protocol stack.

10.4 CROSS-LAYER RADIO

The OSI reference model is predicated upon independence between the layers in order to enable future technology insertion at a particular layer without the need for other technology replacements and decoupling the performance of a particular layer from the performance of the other layers. Unfortunately, this model is limiting. The

Table 10.3 Examples of Cognitive Techniques throughout the Protocol Stack

Layer	Technique	Awareness	Adaptation	Reasoning	Learning	Planning[a]	Collaboration
PHY and MAC	DSA	Spectrum sensing	Frequency adaptation	Dynamic threshold based on current EME	Dynamic threshold based on current and past EME	Preplanned change in operating frequency to accommodate known, future incumbents transmitters	Shares and uses EME information with and from neighbors for adaptation decisions
PHY and MAC	AMC	RSSI measurement	Modulation and coding adaptation	Dynamic RSSI threshold based on current RSSI	Dynamic RSSI threshold based on RSSI history	RSSI threshold is made a function of time of day to accommodate known RF propagation patterns	Shares and uses RSSI information with and from neighbors for adaptation decisions
Network	Adaptive hybrid mobile ad hoc network routing protocol	Network state	Changes routing protocol based on network state (e.g., switch between proactive and reactive protocols)	Dynamic tuning of network state criteria based upon network performance measurement	Dynamic tuning of network state criteria based on network performance history	Network preemptively changes routing protocol because of preplanned network topology changes (e.g., preplanned node exit)	Shares and uses network performance measurements with and from neighbors.
Transport	Adaptive Transmission Control Protocol (TCP)	Measures flow performance statistics	Adapts TCP parameters based on measured performance	Adapts performance statistics threshold based on measured performance	Adapts performance statistics threshold based on performance history	Preemptively changes TCP congestion control parameters to accommodate known changes in data service requirements (e.g., preplanned video connection)	Shares and uses TCP performance information with and from neighboring nodes
Application	Adaptive rate control	Measures application performance statistics	Adapts application data rate based on measured performance	Adapts performance statistics threshold based on measured performance	Adapts performance statistics threshold based on performance history	Preemptively plans for rate adjustment based on a priori knowledge of application performance needs	Shares and uses TCP performance information with and from neighboring nodes

[a] Planning can be the result of a priori knowledge of changes to environment or application performance needs.

networking community continues researching advanced problems such as wireless internetworking, mobile ad hoc networking (MANET), and sensor networking. The layer-independence model of the OSI reference model is not well equipped to support these network scenarios.

Recently, cross-layer design has become an important research focus. In this paradigm, layer independence does not hold. Rather, technologies at a particular layer are designed jointly with technologies at other layers in order to jointly optimize the design. In other cases, information from other layers is shared between the layers to facilitate optimal configuration. In Reference [274], a cross-layer design is proposed for the 802.11 MAC, where the layer 2 strategy is based upon the content of the packet to be transmitted (information from layers 4–7 are employed to configure layer 2). The conclusion of the study presented in Reference [275] claimed that cross-layer protocol design in which optimization across multiple layers of the protocol stack is necessary in ad hoc wireless networks where energy is a primary constraint. The study presented in Reference [276] showed that sharing physical and MAC layer information about the wireless medium is required in order to provide sufficiently adaptable networking methods.

Cross-layer design principles have been applied to enable efficient use of TCP across wireless channels. The TCP protocol was designed for well-behaved, low error rate wired networks. As a result, TCP assumes that all packet losses are due to network congestion. Packet losses resulting from the variable nature of the wireless medium cause TCP to react as if network congestion has occurred. Consequently, TCP begins lowering its transmission rate to assist the network in combating the presumed congestion effects. This leads to a significant degradation in throughput performance. This behavior is more pronounced when the bandwidth-delay product (BDP) of the channel is large [251]. A study was presented in Reference [274] that investigated methods to alleviate the "TCP-over-wireless" problem. In that study, it was shown that with the introduction of explicit congestion notification (ECN) (i.e., informing TCP explicitly of when the network is congested), there is no significant degradation of TCP throughput performance even in the presence of a dynamic wireless channel. This result was also analytically derived in Reference [277]. This is an example of how information sharing across layers of the protocol stack enables a more efficient solution. There are many more examples of how cross-layer design and information sharing across the layers of the protocol stack has enabled more efficient communications solutions. A more comprehensive survey of cross-layer technologies is available in Reference [275] and [276].

However, the design of a cross-layer method requires a significant amount of effort in developing the high-level architecture [278]. Cross-layer design intertwines the layers of the protocol stack, causing multifarious codependent relationships. In Reference [278], it is pointed out that if the architecture itself is not carefully designed when positing a cross-layer design, the result may produce "spaghetti" code that is unmaintainable and unupgradeable. It is also pointed out in Reference [278] that without a well-planned architecture, unintended interactions between layers can significantly degrade performance. It is this point that makes us believe that the disciplines of cross-layer design and cognitive radio will ultimately be

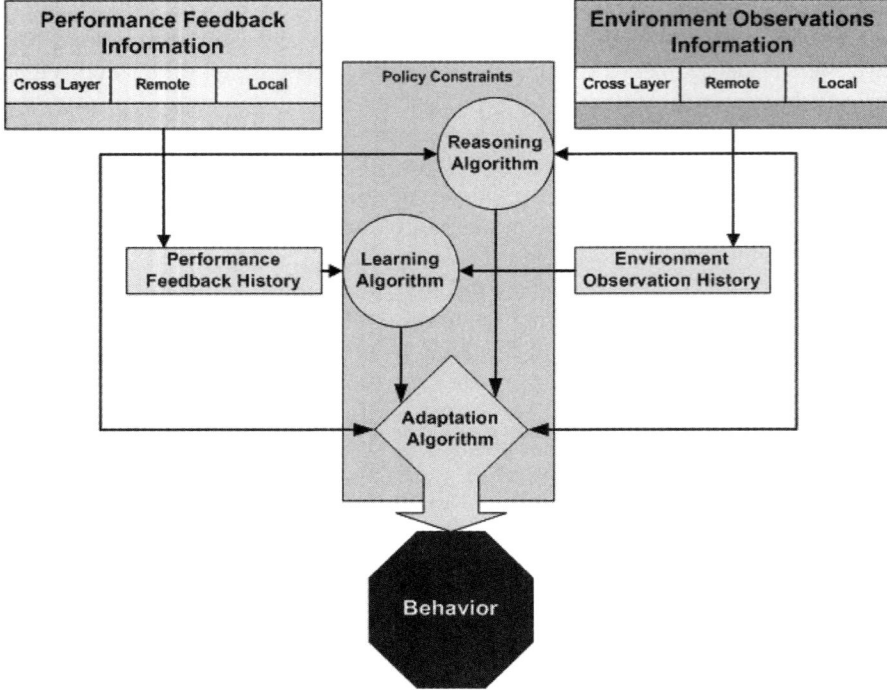

Figure 10.22   Envisioned framework for unification of cross-layer design and cognitive radio. Reprinted from Reference [279].

intertwined together into a single cognitive framework for adaptation and feedback across every layer of the protocol stack, as envisioned in Figure 10.22 [279].

10.5 NETWORK CODING

The concept of network coding was introduced at the turn of the millennium as an alternative to routing for the purpose of improving network throughput and overall system performance. The idea of allowing nodes to not only forward but also process the incoming independent data packets, as rudimentary as it was, created a paradigm shift in thinking about network communications [280].

Traditional computer networks employ a method known as "store and forward," where data are transmitted from the source node to each destination node via a chain of intermediate nodes. In this method, received data packets are stored at an intermediate node and a copy is forwarded to the next node [281].

According to Reference [281], the notion of network coding was first introduced for satellite communication networks in Reference [282] and later fully developed in Reference [283]. Considered a seminal paper within the realm of network coding,

Reference [283] actually coined the term "network coding," while successfully demonstrating the advantage of network coding over the store-and-forward method.

Due to its overall fundamentality and the limitlessness of conceivable applications, network coding has generated a lot of interest within numerous scientific areas, including wireless communications, information theory, networking, and switching. The digital communications industry is also becoming increasingly reliant on network coding with emergent applications in routing, relaying, network performance optimization, PHY, security, flooding, and error correction, to name a few.

Since the focus of this book is on wireless internetworking, this subsection only touches on a few existing and emerging network coding schemes and solutions as they apply to wireless communications. As such, network coding solutions and applications for wired networks are not discussed. Furthermore, if the reader is interested in learning the fundamentals of network coding theory, we highly recommend reading References [280, 281, 284].

So now a few questions arise. How can network coding be applied to wireless communications? Aside from the obviousness of system throughput, what other aspects unique to wireless networks with their fading channels, mobility, and security considerations can be optimized through network coding? Well, let us investigate.

As evidenced by a myriad of published scientific articles in the area of network coding for wireless networks (e.g., References [285–297]), there is a significant progress on the theoretical front. However, practical applications are only beginning to emerge. The network coding methods are quite attractive to the implementers because of their promise to substantially improve QoS, scalability, security, and robustness of a wireless network [284].

In the following subsections, we briefly overview several existing network coding approaches and applications for wireless systems. The overview is by no means exhaustive and only aims to give the reader an idea about the vastness of network coding possibilities within the wireless communications realm.

10.5.1   Network Coding Approaches

When it comes to digital communications, one could argue that there are two main approaches to network coding: intrasession coding and intersession coding [284].

10.5.1.1   Intrasession Network Coding

When this approach is implemented, the routers are allowed to mix packets addressed to the same destination. Here, the routers are not required to decode the packets since the destination eventually decodes them once a sufficient number of coded packets is received [283, 284].

Intrasession network coding has been applied to many areas (e.g., content distribution, secret communications, energy); however, when it comes to wireless networks, two protocols are worth mentioning, as their benefits have already been empirically demonstrated. Specifically, we address two protocols: (1) MAC-independent opportunistic routing and encoding (MORE) [304] and (2) "MIXIT" [305]. The former represents the very first implementation of intrasession network coding in a wireless system. The latter delivers a cross-layer approach to intrasession network coding in which the network actually takes advantage of all correctly received bytes, including correct bytes in corrupted packets [284].

10.5.1.2 Intersession Network Coding

Intersession network coding allows routers to mix or code packets belonging to different communications links that intersect at some intermediate node within the network. This type of network coding was first implemented in the "COPE" protocol, which demonstrated significant and, most importantly, practical throughput improvements [284, 300].

The practical success of intersession network coding resulted in an explosion of theoretical work characterizing the throughput bounds of this approach [284, 289, 294, 301, 302].

10.5.2 Network Coding Applications in Wireless Systems

In the previous subsection, we discussed two main approaches to network coding. The following subsections will now focus on practical implementations of these approaches.

10.5.2.1 MORE

MAC-independent Opportunistic Routing and Encoding (MORE) was the very first implementation of the intrasession network coding protocol. MORE is an opportunistic routing protocol, which randomly mixes packets before forwarding them. The opportunistic nature of MORE provides a much needed reliability to a wireless network. Opportunistic routing, in general, allows any node that hears a transmission and is closer to the destination to participate in forwarding the packet, which offers a significant resilience against packet loss [303]. However, one of the known drawbacks of traditional opportunistic routing protocols is that routers that hear the same transmission may unnecessarily forward the same packets [284]. The randomness in MORE was introduced to prevent routers from doing exactly that since too much redundancy may not always be a good thing [298].

MORE requires a special scheduler to coordinate routers and can be implemented directly above 802.11 MAC (but below the IP layer) [284, 298]. The reported experimental results from a 20-node wireless testbed demonstrated a 22% throughput increase over the Extremely Opportunistic Routing (ExOR) protocol (not discussed here), and the gains rose to 45% over ExOR when there was a chance of spatial reuse. For multicast, MORE's gains increased with the number of destinations and were 35–200% greater compared with ExOR protocol [298].

10.5.2.2 MIXIT

MIXIT protocol builds on ideas developed in prior research. Specifically, it draws from the work in opportunistic routing (ExOR [309] and MORE [298]) and partial packet recovery [304]. MIXIT was designed to improve the throughput of wireless mesh networks [300].

MIXIT protocol allows routers to use physical layer hints (SoftPHY hints [304]) to make their best estimate about which bits in a corrupted packet are likely to be correct, and then forward them onto the intended destination. MIXIT protocol operates on symbols (small groups of bits) and allows the nodes to opportunistically route said symbols to their destination with low overhead. In addition, MIXIT incorporates an end-to-end error correction component that the destination uses to correct any "escapee" errors [299].

MIXIT was implemented on a software radio platform running the ZigBee radio protocol (802.15.4). The experiments were conducted on a 25-node indoor testbed and produced a throughput gain of 2.8 times over MORE and approximately 3.9 times over traditional routing using the expected transmission count (ETX) metric [305].

10.5.2.3 COPE

COPE's wireless architecture is built on network coding for the purpose of improving the wireless system's throughput. COPE is the first general protocol to code packets across multiple connections irrespective of scheduling, topology, and traffic demands. COPE also represents the first design that integrates network coding into the current network protocol stack. In addition, COPE delivered the first successful implementation and evaluation of network coding in a testbed [284].

COPE's testbed consisted of a 20-node wireless network. The reported results demonstrated that using COPE protocol at the forwarding layer, without modifying routing and higher layers, increases network throughput. The gains varied from a few percent to several folds depending on the traffic pattern, congestion level, and transport protocol [300].

10.5.2.4 Network Coding in LTE: User Cooperation

Chapter 5 of Reference [284] proposes a new local retransmission scheme based on the user cooperation concept for LTE networks. This network coding scheme is to be used alongside evolved multimedia broadcast/multicast service (eMBMS) raptor codes. Raptor codes' main purpose is to provide a packet-level protection at the application layer to complement the bit-level FEC at PHY. Raptor codes reside on both the evolved broadcast multicast service center (eBMSC) and every single MS [284].

The proposed scheme applies network coding to user cooperation to improve the cooperation efficiency of the short-range links. The reported simulated results showed that the local transmission could save about 80% overhead in the LTE link as long as two or more MSs cooperate. Larger gains can be achieved with the increase in cooperating nodes. In addition, local retransmission may be able to employ smaller block size on the LTE link using raptor codes. This could lead, as perceived by MS, to shorter delays and improved QoS. Furthermore, the simulations results showed that network coding can save more that 50% of the short-range link traffic, as long as there are four MSs in the cooperation cluster. The resulting traffic reduction on the short-range link can significantly reduce the overall power consumption [284].

10.5.2.5 Control Over Network Coding for Enhanced Radio Transport Optimization Network Coding for Wireless MANETs

The BAE Systems' Control Over Network Coding for Enhanced Radio Transport Optimization (CONCERTO) is a MANET solution based on network coding. MANETs, as you may already know, represent one of the innovative emerging networking technologies, with limitless potential applications in personal area networks, military battlefield applications, emergency and rescue operations, and so on.

CONCERTO's design was successfully implemented and demonstrated in the field across multiple scenarios. Through the network coding of new and repair packets, benefiting multiple receive nodes, the CONCERTO system was able to achieve significant throughput gains in packet delivery performance and bandwidth efficiency. For more information about CONCERTO experiments the reader is referred to chapter 6 of Reference [284].

10.5.2.6 Network Scaling with Network Coding

Another active and exciting area of research is wireless network scalability associated with network coding. Chapter 9 of Reference [284] presents a few interesting findings on the network coding performance in unreliable wireless networks under several scaling regimes, including the coding window size, the network size, the number of network flows, and the application delay constraints. The reader is encouraged to examine the abovementioned reference if the topic of network scaling is of particular interest.

Chapter 11

Building the Wireless Internet: Putting It All Together

At this point, we have spent a lot of effort presenting the wireless networking landscape. We have provided a lot of information about a lot of technologies, discussing the "whats," "whys," and "hows" of technologies such as Bluetooth, Wi-Fi, WiMAX, GSM, UMTS, CDMA2000, and LTE. As should be clear by now, many of these technologies share many similarities. However, it should be equally clear that these technologies have fundamental differences. That is because these technologies have been and are being designed to solve a very particular subset of the overall problem space. This was qualitatively represented in Figure 1.8, which is shown in again in Figure 11.1 for convenience.

While this figure provides an accurate anecdotal portrayal of the wireless networking landscape, it is simplistic in nature. Qualitative representations of performance such as this can be found throughout literature of all shapes and sizes, where all types of aspects of performance are compared and contrasted. This is often done to make one technology seem superior or preferable to another technology. Unfortunately, this makes these types of qualitative characterizations almost as useless as any other type of technology taxonomy. The truth is that there is no ultimately superior technology. All wireless technologies, past, present, and future, have strengths and weaknesses.

As we move further into the era of the wireless Internet, this is an important reality to accept. There is no "silver bullet," no all-capable technology that will solve all our problems. Rather, the wireless Internet will require a multitude of technologies to serve as part of its eventual design. Consider the analogy of building a house. Does one build a house with a single building material using only a single tool? Of course not.

Wireless Networking: Understanding Internetworking Challenges, First Edition.
Jack L. Burbank, Julia Andrusenko, Jared S. Everett, and William T.M. Kasch.
© 2013 the Institute of Electrical and Electronics Engineers, Inc. Published 2013 by John Wiley & Sons, Inc.

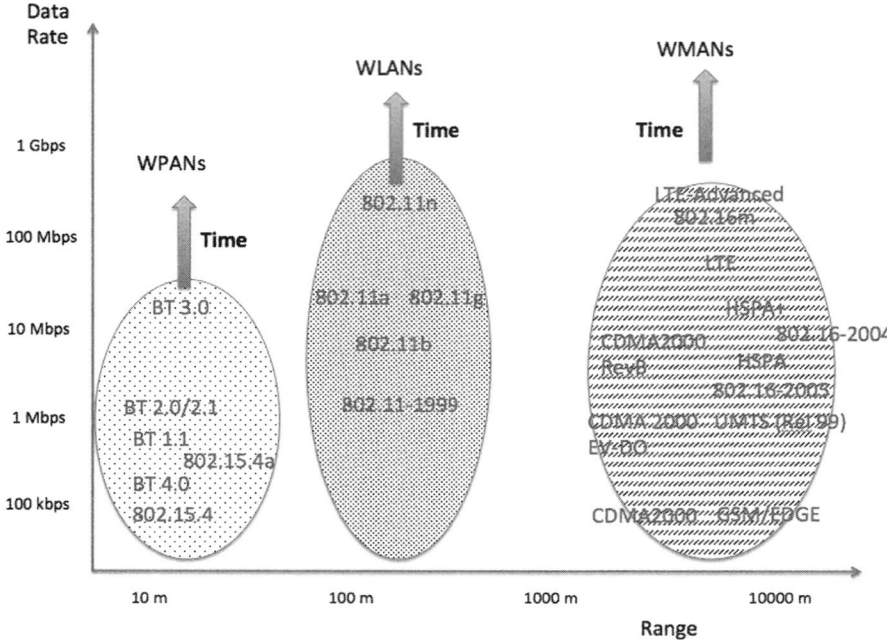

Figure 11.1   Qualitative summary of the many existing wireless networking technologies.

So let us quickly revisit and summarize what we have learned about the various technological tools that are at our disposal to build the wireless Internet. This will help us understand which technology may be right for a particular problem. This will subsequently help us understand how all these technologies can be put together to solve the overall problem of how we build the wireless Internet.

11.1 DIMENSIONS OF PERFORMANCE

There are many dimensions of performance that one must consider when understanding the true strengths and weaknesses of a particular wireless technology, consequently understanding the right problem against which to bring a particular technology to bear.

11.1.1 Data Rate

This dimension of performance is obvious, and seemingly the only one that most people care about. This is understandable as the need for increased capacity is undeniable. This is the dimension of performance that seems to be primarily driving most technology development.

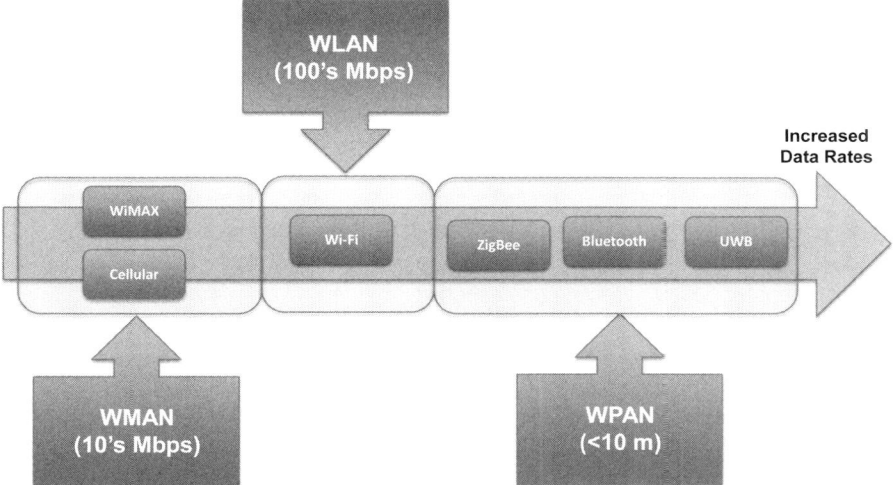

Figure 11.2   A simple taxonomy of wireless technologies based on data rate performance.

In this dimension, lower-range technologies tend to be at the state of the art. WPAN technologies tend to lead the way into higher data rates. WLAN technologies are typically the next to achieve equivalent data rates. WMAN and cellular technologies then follow. In other words, they all seem on the same trend, with WPAN ahead on the curve and with WMAN and cellular on the lower end of the curve. Therefore, while a somewhat simplified generalization, it can be said that at a specific point in time, WPAN technologies will provide the highest data rates, followed by WLAN technologies and then WMAN and cellular technologies. This trend is understandable, since lower ranges tend to make achieving higher data rates more plausible. Additionally, WMAN and cellular technologies typically have several other performance dimensions that are driving their overall design compared with lower-range technologies. But with all else equal, if you were solely looking for maximum capacity, WPAN technology is likely where you would find the leader, as qualitatively described in Figure 11.2 (we consider near-field communications [NFCs] as the extreme example of a high-rate WPAN technology).

11.1.2   Range

This dimension is perhaps the most obvious. Indeed, this dimension of performance is the basis for the most common taxonomy of technologies (i.e., WPAN, WLAN, WMAN), like the one illustrated in Chapter 1 and reshown in Figure 11.3.

Clearly a WMAN technology will generally achieve a greater range than a WPAN technology. However, this is not necessarily always the case across technologies and is sometimes overly simplistic. There are numerous examples of range extending

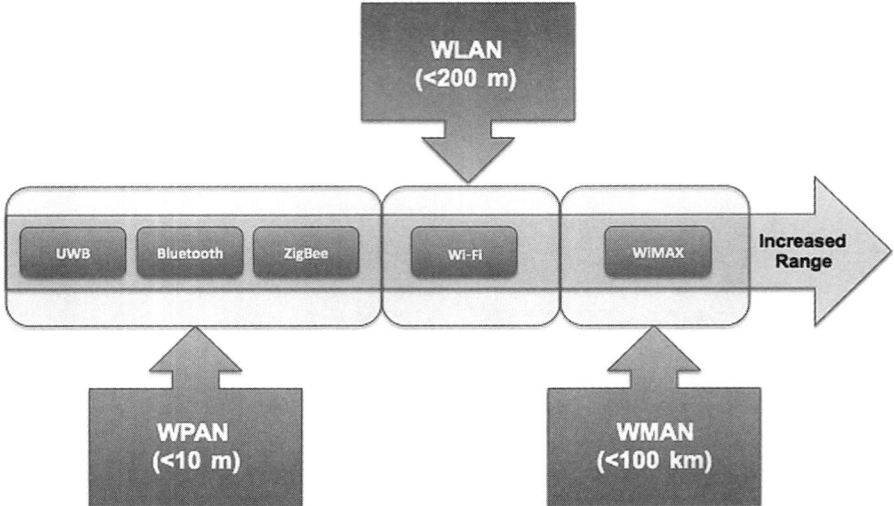

Figure 11.3   A simple taxonomy of wireless technologies based on range.

WLAN technology to distances far greater than typical WMAN technologies operate at. Just because it isn't designed for range, doesn't mean it cannot achieve it.

One must also understand that range performance, or even data rate performance, can often be misrepresented and subsequently misunderstood. You must understand that data rate and range are competing design goals. Both cannot always be simultaneously achieved. Next-generation technologies are achieving very high data rates. However, this accomplishment comes at the price of robustness. The highest data rates are typically a product of very high-order modulation methods and very high coding rates. This leads to poor link error rate performance that will effectively decrease the effective range of that communications system.

Furthermore, increased data rates and range performance often rely on higher-order MIMO approaches with many transmit and receive antennas. MIMO takes advantage of multipath propagation conditions to provide many benefits to a communications system. However, it is also dependent on multipath propagation conditions to provide many of those same benefits. In propagation environments where multipath fading is not a factor, the benefits of a MIMO system diminish. So it should be understood that the performance of any wireless technology is highly dependent on the conditions in which it will operate. Indeed, you can find (sometimes common) RF environments in which older wireless technologies might outperform much newer state-of-the-art technologies.

11.1.3   Power Efficiency

Every portion of the wireless communications community is concerned with power efficiency. No community wishes to develop a technology that will lead to consumer

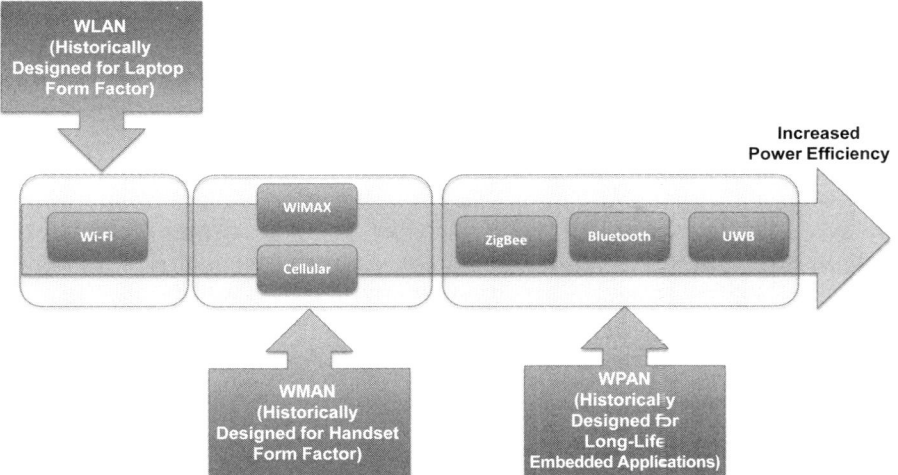

Figure 11.4 A simple taxonomy of wireless technologies based on power efficiency.

devices with poor battery life. However, the importance of power efficiency in a design does indeed vary across technologies. If you look at the WPAN community, much of their recent development efforts have been to improve power efficiency, even by sacrificing data rate performance. In these cases, they are often designing solutions for which highly disadvantaged users wish to use their radio system on a single battery for an extended period of time. Other communities, by contrast, have developed technologies that often times sacrifice power efficiency in the name of data rate performance, relying instead on advances in battery technology.

While this may be an oversimplification, WPAN technologies have typically proven the most power efficient. This makes sense intuitively since these are often designed for long lifetime applications. WMAN and cellular technologies have historically been somewhat better than WLAN technologies in terms of power efficiency, which again makes sense given their traditional focus on handsets. WLAN technologies have historically been developed for platforms that are more advantaged in terms of battery technology (e.g., laptop computer). This high-level relationship is depicted in Figure 11.4.

11.1.4 Spectral Efficiency

It is clear that spectral efficiency is important. Spectrum is a precious resource and to achieve higher data rates, we must find more efficient ways to use it. However, it should be realized that this is not always how higher data rates are being achieved. It is true that many next-generation technologies offer spectral efficiencies that are much improved over their predecessors. Nevertheless, many of the very high data rates we often hear claimed for various technologies are misleading. Many of the very high data rates of modern technologies rely on enormous amounts of bandwidth.

The basic problem of communications has remained unchanged since Claude Shannon so eloquently characterized it nearly 70 years ago. The achievable capacity through a channel is a function of the bandwidth and the quality of the channel. We have redefined "the channel" many times over the years to improve performance; MIMO is the latest redefinition. Advanced coding schemes have improved the quality of the channel. Moreover, increased channel bandwidths have improved the capacity of wireless technologies. We have moved from narrowband 200 kHz channels in GSM to 1.25 MHz channels in cdma2000 to 5 MHz channels in WCDMA to 10 MHz and larger channel bandwidths for LTE, LTE-Advanced, and IEEE 802.16m. IEEE 802.11 achieves higher data rates in part because of the larger channels. We have moved from 22 MHz with IEEE 802.11b to over 100 MHz bandwidth in the upcoming IEEE 802.11ac standard.

The problem with this model is that spectrum is a precious resource. How many entities hold the rights to 100+ MHz in bandwidth for their networks? It is important to understand that when data rate claims are made for a particular technology, those are typically theoretical maximum data rates. These data rates are calculated by typically making every best-case assumption possible, including maximum channel bandwidth. When understanding which technology is best for a particular application, it is important to try to understand the typical data rates of a technology as opposed to just its theoretical maximum.

11.1.5 Mobility Support

Wireless technologies offer a wide range of mobility support. WPAN technologies typically place a much lower value on mobility support because of their localized intended space of operation. On the other end of the spectrum, WMAN and cellular technologies place a high value on mobility support. WLAN technology initially provided little in the way of mobility support but has evolved over time to provide increasing amounts of mobility support (e.g., IEEE 802.11r). This relationship is depicted in Figure 11.5.

This makes sense intuitively given the portion of the overall problem space these technologies were designed to solve.

11.1.6 Security

Wireless networks have received a tarnished reputation over the past decade when it comes to network security, and much of this is well deserved. The past 10 years of technology development have seen many of the same security mistakes made repeatedly, with us collectively failing to learn from our past mistakes. How many times must we face a "rogue access point" or a "rogue base station" before we place a premium on two-way authentication? The answer is evidently far too many, as

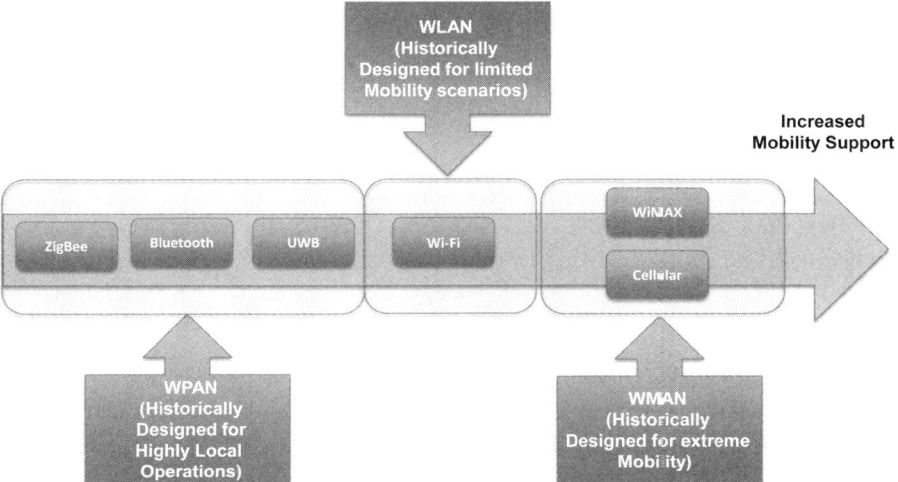

Figure 11.5   A simple taxonomy of wireless technologies based on mobility support.

technology after technology defined itself with only one-way authentication as a mandatory component. How many times will we define technologies that unforgivably reuse cryptographic credentials for far too many functions? How many times will we develop technologies that do not include automated key management functionality? Evidently, we still have not grown tired of this. Many emerging technologies still do not require automated key management functionality. This is a bit frustrating, since the strongest cryptographic system doesn't do much good if administrators never change keys. And that is exactly what happens when networks get large and there is no automated way to manage the security aspects of the network.

The sad truth is that security is the one subject that most designers do not want to face. Thus, far too often we tend to ignore it as long as we can. And when we cannot ignore the issue of security any longer, we hurriedly apply a solution, hoping to minimize the time we must spend on it. It is true that developing security architectures may not be as appealing as developing the cutting-edge modulation technique that will give a huge boost in system performance. And let us all admit that we usually don't like those security folks. Those are the folks who tell us what we can't do. Those are the folks who take capacity away from our system. Those are the folks who drive the complexity of our devices. Wireless network security has made monumental improvements over the past 10–15 years. We have indeed come a long way from the era of WEP security in IEEE 802.11 and the unauthenticated base stations of GSM. But the sad truth is that we have a long way to go. And unfortunately our jobs are only going to get worse. We are witnessing the rise of the "smart device." Smartphones and tablets are becoming the norm. Mobile devices increasingly resemble computers more than phones. And eventually that means that the security woes of computer platforms are soon to follow. What is required is a culture change, a fresh outlook and fresh perspective on the task of securing our wireless networks.

The problem is even worse in the case of ad hoc networks. The security aspects of ad hoc wireless networks have long been a research area. But far too often the answer is to apply the same solutions and approaches we would take with traditional wired networks. With the amount of time, resources, and money invested in network security for the traditional Internet model, it is only natural that people would wish to provide the same level of solution to the MANET environments. However, it is well understood that the existing security mechanisms do not operate well in ad hoc scenarios [306]. Traditional network security mechanisms have a dependency on dedicated infrastructure, which may or may not be present in the case of the ideal MANET.

11.1.7   Manageability

The realm of network management covers a vast spectrum. Issues such as spectrum allocation, security materials, IP configuration, and network monitoring fall under the management umbrella. While these components are not unique to wireless networks, they do become more difficult when nodal mobility, dynamic network membership, and unstable links are introduced to the network. There is ongoing work within both the research and standardization communities to address topics such as IP address assignment. However, solutions in this area remain insufficient.

Centralized network management architectures often fail in wireless networks, particularly in the extreme case of MANET. Not only is dynamic membership an issue, but more important, the MANET may be disconnected from the larger network for long periods of time. This renders the collection/monitoring of data in a centralized server infeasible. Dynamic networks call for management capabilities that are distributed in nature. Localized functionality reduces the dependency on infrastructure or constant connectivity to a centralized management station. However, these types of distributed management capabilities remain relatively immature. We as a community continue to try to manage wireless networks in the same way we manage wired networks. In large fixed-infrastructure wireless networks, that may be possible to some extent. But in the extreme case of the MANET, it is difficult to move your network operations center (NOC) with your network.

11.2   CONCLUDING REMARKS

It is clear that we have entered the wireless Internet era. This era is changing the way in which we acquire, share, and protect information. It is reshaping the global socioeconomic and cultural landscape. It is fundamentally changing the way in which we live our lives. The resulting technology landscape is complex and hard to understand, littered with a plethora of technologies of ever-increasing sophistication. In this book we have discussed many of the past, present, and future technologies that comprise this landscape. We have seen that there is no "silver bullet," no technology capable of solving all problems. Rather, all wireless technologies have strengths and weaknesses. We have also seen that even outside the scope of individual technologies there exist significant challenges in constructing larger networks using these technologies. There exist fundamental technical challenges at every layer of the protocol stack that will likely keep us busy for a long time.

As we move further into this new wireless Internet era, we believe that the technology landscape will become even more complex and harder to understand. We believe that an increasing number of technologies will emerge and that these technologies will continue to grow in complexity and sophistication. And as user demands continue to increase, supporting these needs will become only harder. Constructing larger networks will become an increasingly daunting task.

We are not trying to present this as an insurmountable problem. In fact, the exact opposite is true. While the challenges are indeed profound, we find this an extremely exciting time. We view these challenges as new opportunities for innovation and invention. We believe that the key to technical success in this complex landscape is knowledge and proper perspective. We hope that this book has helped provide a better understanding of this landscape.

References

[1] "Part 11: Wireless LAN Medium Access Control (MAC) and Physical Layer (PHY) Specifications," IEEE 802.11-1999, 1999.

[2] "Specification of the Bluetooth System," Specification volume 1, Core Version 1.0 A, July 26, 1999, Bluetooth SIG.

[3] "Specification of the Bluetooth System," Specification volume 0, Core Version 4.0, June 30, 2010, Bluetooth SIG.

[4] "IEEE Standard for Local and Metropolitan Area Networks—Part 16: Air Interface for Fixed Broadband Wireless Access Systems," IEEE 802.16-2001, December 6, 2001.

[5] T.K. SARKER, et al., *History of Wireless*, Wiley-IEEE Press, Piscataway, NJ, 2006.

[6] Recommendation ITU-R M.1645, "Framework and Overall Objectives of the Future Development of IMT-2000 and Systems beyond IMT-2000," June 2003.

[7] Resolution ITU-R 56.1, "Naming for International Mobile Telecommunications," 2012.

[8] "IEEE Standard for Telecommunications and Information Exchange Between Systems—LAN/MAN Specific Requirements—Part 15: Wireless Medium Access Control (MAC) and Physical Layer (PHY) Specifications for Wireless Personal Area Networks (WPANs)," IEEE 802.15.1-2002, 2002.

[9] "Specification of the Bluetooth System," Specification volume 0, Core Version 2.0+ EDR, November 4, 2004, Bluetooth SIG.

[10] "Specification of the Bluetooth System," Specification volume 0, Core Version 2.1+ EDR, July 26, 2007, Bluetooth SIG.

[11] "Specification of the Bluetooth System," Specification volume 0, Core Version 3.0+ HS, April 21, 2009, Bluetooth SIG.

[12] "IEEE Standard for Telecommunications and Information Exchange Between Systems—LAN/MAN Specific Requirements—Part 15: Wireless Medium Access Control (MAC) and Physical Layer (PHY) Specifications for Low Rate Wireless Personal Area Networks (WPAN)," IEEE 802.15.4-2003, October 1, 2003.

[13] "Part 15.4: Wireless Medium Access Control and Physical Layer Specifications for Low-Rate Wireless Personal Area Networks," IEEE 802.15.4-2006, September 8, 2006.

[14] "Part 15.4: Wireless MAC and PHY Specifications for Low-Rate Wireless Personal Area Networks Amendment 1: Add Alternate PHYs," IEEE 802.15.4a-2007, August 31, 2007.

[15] "Part 15.4: Amendment 2: Alternative Physical Layer Extension to support one or more of the Chinese 314–316 MHz, 430–434 MHz, and 779–787 MHz Bands," IEEE 802.15c-2009, April 17, 2009.

Wireless Networking: Understanding Internetworking Challenges, First Edition.
Jack L. Burbank, Julia Andrusenko, Jared S. Everett, and William T.M. Kasch.
© 2013 the Institute of Electrical and Electronics Engineers, Inc. Published 2013 by John Wiley & Sons, Inc.

[16] "IEEE Standard for Information Technology—Local and Metropolitan Area Networks—Specific Requirements—Part 15.4: Amendment 3: Alternative Physical Layer Extension to Support the Japanese 950 MHz Bands," IEEE 802.15.4d-2009, April 17, 2009.

[17] "Part 11: Wireless LAN Medium Access Control (MAC) and Physical Layer (PHY) Specifications: High-Speed Physical Layer in the 5 GHz Band," IEEE 802.11a-1999, 1999.

[18] "Part 11: Wireless LAN Medium Access Control (MAC) and Physical Layer (PHY) Specifications: Higher-Speed Physical Layer Extension in the 2.4 GHz Band," IEEE 802.11b-1999, 1999.

[19] "IEEE Standard for Local and Metropolitan Area Networks Media Access Control (MAC) Bridges," IEEE 802.1D-2004, June 9, 2004.

[20] "Part 11: Wireless Medium Access Control (MAC) and Physical Layer (PHY) Specifications Amendment 3: Specification for Operation in additional Regulatory Domains," IEEE 802.11d-2001, January 1, 2001.

[21] "Part 11: Wireless LAN Medium Access Control (MAC) and Physical Layer (PHY) Specifications—Amendment: Medium Access Method (MAC) Quality of Service Enhancements," IEEE 802.11e-2005, November 11, 2005.

[22] "Part 11: Wireless LAN Medium Access Control (MAC) and Physical Layer (PHY) Specifications: High-Speed Physical Layer in the 2.4 GHz Band," IEEE 802.11g-2003, 2003.

[23] "Part 11: Wireless LAN Medium Access Control (MAC) and Physical Layer (PHY) Specifications—Spectrum and Transmit Power Management Extensions in the 5 GHz Band in Europe," IEEE 802.11h-2003, October 14, 2003.

[24] "Part 11: Wireless LAN Medium Access Control (MAC) and Physical Layer (PHY) Specifications: Medium Access Control (MAC) Security Enhancements," IEEE 802.11i-2004, 2004.

[25] "Part 11: Wireless LAN Medium Access Control (MAC) and Physical Layer (PHY) Specifications—Amendment 7: 4.9 GHz–5 GHz Operation in Japan," IEEE-802.11j-2004, October 29, 2004.

[26] "Part 11: Wireless LAN Medium Access Control (MAC)and Physical Layer (PHY) Specifications Amendment 1: Radio Resource Measurement of Wireless LANs," IEEE 802.11k-2008, June 12, 2008.

[27] "Part 11: Wireless LAN Medium Access Control (MAC)and Physical Layer (PHY) Specifications Amendment 5: Enhancements for Higher Throughput," IEEE 802.11n-2009, October 29, 2009.

[28] "Part 11: Wireless LAN Medium Access Control (MAC) and Physical Layer (PHY) Specifications Amendment 6: Wireless Access in Vehicular Environments," IEEE 802.11p-2010, July 15, 2010.

[29] "Part 11: Wireless LAN Medium Access Control (MAC) and Physical Layer (PHY) Specifications Amendment 2: Fast Basic Service Set (BSS) Transition," IEEE 802.11r-2008, July 15, 2008.

[30] "Part 11: Wireless LAN Medium Access Control (MAC) and Physical Layer (PHY) Specifications Amendment 10: Mesh Networking," IEEE 802.11s-2011, September 10, 2011.

[31] "Part II: Wireless LAN Medium Access Control (MAC) and Physical Layer (PHY) Specifications: Amendment 9: Interworking with External Networks," IEEE 802.11u-2011.

[32] "Part 11: Wireless LAN Medium Access Control (MAC) and Physical Layer (PHY) Specifications Amendment 8: IEEE 802.11 Wireless Network Management," IEEE 802.11v-2011, February 9, 2011.

[33] "Part 11: Wireless LAN Medium Access Control (MAC) and Physical Layer (PHY) Specifications Amendment 4: Protected Management Frames," IEEE 802.11w-2009, September 30, 2009.

[34] "Part 11: Wireless LAN Medium Access Control (MAC) and Physical Layer (PHY) Specifications Amendment 3: 3650–3700 MHz Operation in USA," IEEE 802.11y-2008, November 3, 2008.

[35] "Part 11: Wireless LAN Medium Access Control (MAC) and Physical Layer (PHY) Specifications Amendment 7: Extensions to Direct-Link Setup (DLS)," IEEE 802.11z-2010, October 14, 2010.

[36] "IEEE Standard for Information Technology—Telecommunications and Information Exchange Between Systems—Local and Metropolitan Area Networks—Specific Requirements—Part 11: Wireless LAN Medium Access Control (MAC) and Physical Layer (PHY) Specifications," IEEE 802.11-2007, June 2007.

[37] "IEEE Standard for Information Technology—Telecommunications and information exchange between systems Local and metropolitan area networks—Specific Requirements Part 11: Wireless LAN Medium Access Control (MAC) and Physical Layer (PHY) Specifications," IEEE 802.11-2012, March 2012.

[38] B. Kraemer, "802.11 Wireless Local Area Networks," Presentation to the IEEE 802 March 2011 Workshop, March 2011.

[39] "IEEE Standard for Local and Metropolitan Area Networks—Part 16: Air Interface for Fixed Broadband Wireless Access Systems—Amendment 2: Medium Access Control Modifications and Additional Physical Layer Specifications for 2–11 GHz," IEEE 802.16a-2001, April 1, 2003.

[40] "IEEE Standard for Local and Metropolitan Area Networks—Amendment 1: Detailed System Profiles for 10–66 GHz," IEEE 802.16c-2001, January 15, 2003.

[41] "IEEE Standard for Local and Metropolitan Area Networks—Air Interface for Fixed Broadband Wireless Access Systems," IEEE 802.16-2004, October 1, 2004.

[42] "IEEE Standard for Local and Metropolitan Area Networks—Air Interface for Fixed and Mobile Broadband Wireless Access Systems," IEEE 802.16e-2005, February 28, 2006.

[43] "IEEE Standard for Local and Metropolitan Area Networks—Part 16: Air Interface for Fixed Broadband Wireless Access Systems—Amendment 1: Management Information Base," IEEE 802.16f, December 1, 2005.

[44] "IEEE Standard for Local and Metropolitan Area Networks—Part 16: Air Interface for Fixed and Mobile Broadband Wireless Access Systems—Amendment 3: Management Plane Procedures and Services," IEEE 802.16g-2007, December 31, 2007.

[45] "IEEE Standard for Local and Metropolitan Area Networks—IEEE Standard for Local and Metropolitan Area Networks Part 16: Air Interface for Broadband Wireless Access Systems," IEEE 802.16-2009, May 29, 2009.

[46] "IEEE Standard for Local and Metropolitan Area Networks—Part 16: Air Interface for Broadband Wireless Access Systems Amendment 2: Improved Coexistence Mechanisms for License-Exempt Operation," IEEE 802.16h-2010, July 30, 2010.

[47] "IEEE Standard for Local and Metropolitan Area Networks—Part 16: Air Interface for Broadband Wireless Access Systems Amendment 1: Multihop Relay Specification," IEEE 802.16j-2009, June 12, 2009.

[48] "IEEE Standard for Local and Metropolitan Area Networks—IEEE Standard for Local and Metropolitan Area Networks: Media Access Control (MAC) Bridges—Amendment 2: Bridging of IEEE 802.16," IEEE 802.16k-2007, August 14, 2007.

[49] "IEEE Standard for Local and Metropolitan Area Networks—Part 16: Air Interface for Broadband Wireless Access Systems Amendment 3: Advanced Air Interface," IEEE 802.16m-2011, May 6, 2011.

[50] "IEEE Standard for Local and Metropolitan Area Networks—Part 20: Air Interface for Mobile Broadband Wireless Access Systems Supporting Vehicular Mobility—Physical and Media Access Control Layer Specification," IEEE 802.20-2008, August 29, 2008.

[51] F. HILLEBRAND, *GSM & UMTS: The Creation of Global Mobile Communications*, John Wiley & Sons, Somerset, NJ, 2001.

[52] B. TURNER, "Wireless Tutorial," NMS Communications White Paper, October 2008.

[53] GSMA History Page, "About Us." Available online: http://www.gsmworld.com/about-us/history.htm. Accessed February 10, 2013.

[54] Global Mobile Suppliers Association Web site. Available online: http://www.gsacom.com. Accessed March 23, 2013.

[55] GSM World Web Site. http://www.gsmworld.com. Accessed March 23, 2013.

[56] "Overview and Guide to the IEEE 802 LMSC," IEEE 802, March 2008.

[57] P. HOFFMAN and S. HARRIS, "The Tao of IETF: A Novice's Guide to the Internet Engineering Task Force," RFC 4677, September 2006.

[58] R. HOVEY and S. BRADNER, "The Organizations Involved in the IETF Standards Process," RFC 2028, October 1996.

[59] Booklet, "ITU-R Study Groups," International Telecommunications Union, 2010.

[60] "WiFi Certified™ Makes it Wi-Fi: An Overview of the Wi-Fi Alliance Approach to Certification," Wi-Fi Alliance®, September 2006.

[61] "2010 Annual Report: Bluetooth v4.0 and Expanding New Markets," Bluetooth SIG, 2010.

[62] "Service Discovery Protocol (SDP) Universally Unique Identifier (UUID) short forms." Available online: http://www.bluetooth.org. Accessed February 10, 2013.

[63] R. RASHID and R. YUSOFF, "Bluetooth Performance Analysis in Personal Area Network (PAN)," 2006 International RF and Microwave Conference Proceedings, September 12–14, 2006.

[64] "ZigBee Specification," ZigBee Document 053474r17, Sponsored by ZigBee Alliance, January 17, 2008.

[65] J.-S. LEE, "Performance Evaluation of IEEE 802.15.4 for Low-Rate Wireless Personal Area Networks," IEEE Transactions on Consumer Electronics, 52(3), 742–749, 2006.

[66] FCC 02-48, "Revision of Part 15 of the Commission's Rules Regarding Ultra-Wideband Transmission Systems," Adopted February 14, 2002.

[67] Jd. Pavon et al., "The MBOA-WiMedia Specification for Ultra Wideband Distributed Networks," IEEE Communications Magazine, 44(6), pp. 128–134, June 2006.

[68] "Radio Equipment and Systems (RES); High Performance Radio Local Area Network (HIPERLAN) Type 1; Functional Specification," ETS 300652, June 1996.

[69] "Broadband Radio Access Networks (BRAN); HIPERLAN Type 2; Physical (PHY) Layer," ETSI TS 101 475, V1.2.2, 2001–2002.

[70] M.S. Gast, *802.11 Wireless Networks: The Definitive Guide*, 1st Edition, O'Reilly & Associated, Sebastopol, CA, 2002.

[71] G. Anastasi and L. Lenzini, "QoS Provided by the IEEE 802.11 Wireless LAN to Advanced Data Applications: A Simulation Analysis," Wireless Nets, 6(2), 99, 2000.

[72] G. Bianchi, "Performance Analysis of the IEEE 802.11 Distributed Coordination Function," IEEE Journal on Selected Areas in Communication, 18(3), 535, 2000.

[73] M.J.E. Golay, "Static Multi-slit Spectrometry and Its Applications to the Panoramic Display of Infrared Spectra," Journal of the Optical Society of America, 41(7), 468–472, July 1951.

[74] Federal Communications Commission (FCC), "Unlicensed NII Devices in the 5 GHz Frequency Range," Federal Register, 62(21), 4649–4657, 1997.

[75] Federal Communications Commission, "Unlicensed NII Devices in the 5 GHz Frequency Range," Federal Register, 63(147), 40831–40837, 1998.

[76] B. Brown, "802.11: The Security Difference between b and I," IEEE Potentials, 22(4), 23–27, October/November 2003.

[77] ETSI TS 101 761-1 V1.1.1, "Broadband Radio Access Networks (BRAN); HIPERLAN Type 2; Data Link Control (DLC) Layer Part 1: Basic Data Transport Functions," 2000–2004.

[78] A. Doufexi, et al., "A Comparison of the HIPERLAN/2 and IEEE 802.11a Wireless LAN Standards," IEEE Communications Magazine, 40, 172–180, May 2002.

[79] H. Li, et al., "Performance Comparison of the Radio Link Protocols of IEEE 802.11a and HIPERLAN/2," IEEE Fall Vehicular Technology Conference (VTC), 5(24–28), 2185, 2000.

[80] S. Simoens, et al., "The Evolution of 5 GHz WLAN Toward Higher Throughputs," IEEE Wireless Communications, 10, 6–13, December 2003.

[81] C. Heegard, et al., "High-Performance Wireless Ethernet," IEEE Communications Magazine, 39, 64–73, November 2001.

[82] M. Stahlberg, "Radio Jamming Attacks against Two Popular Mobile Networks," Helsinki University of Technology Seminar on Network Security, 2000.

[83] T. Karhima, et al., "IEEE 802.11B/G WLAN Tolerance to Jamming," Proceedings of the 2004 IEEE Military Communications (MILCOM) Conference, November 2004.

[84] M.-J. Ho, et al., "IEEE 802.11g OFDM WLAN Throughput Performance," Proceedings of the 2003 IEEE Vehicular Technology Conference (Fall), Vol. 4, pp. 2252–2256, October 2003.

[85] S. Khurana, et al., "Performance Evaluation of Distributed Co-Ordination Function for IEEE 802.11 Wireless LAN Protocol in Presence of Mobile and Hidden Terminals," Proceedings of the 1999 International Symposium on Modeling, Analysis, and Simulation of Computer and Telecommunication Systems, pp. 40–47, October 1999.

[86] S. Sibecas, et al., "On the Suitability of 802.11a/RA for High-Mobility DSRC," Proceedings of the 2002 IEEE Vehicular Technology Conference (Spring), pp. 229–234, May 2002.

[87] J. PAL SINGH, et al., "Wireless LAN Performance Under Varied Stress Conditions in Vehicular Traffic Scenarios," Proceedings of the 2002 IEEE Vehicular Technology Conference (Fall), Vol. 2, pp. 743–747, September 2002.

[88] T. COOKLEV, *Wireless Communications Standards: A Study of IEEE 802.11, 802.15, and 802.16*, IEEE Press, New York, 2004.

[89] M.S. GAST, *802.11 Wireless Networks: The Definite Guide*, 2nd Edition, O'Reilly, Sebastopol, CA, 2005.

[90] M.S. GAST, *802.11n: A Survival Guide*, O'Reilly, Sebastopol, CA, 2012.

[91] J.G. ANDREWS, A. GHOSH, and R. MUHAMED, *Fundamentals of WiMAX: Understanding Broadband Wireless Networking*, Prentice Hall, New York, 2007.

[92] "WiMAX Forum Working Groups," WiMAX Forum, 2013. Available online: http://www.wimaxforum.org/about/working-groups. Accessed March 23, 2013.

[93] "WiMAX Forum: Industry Research Report May 2011," WiMAX Forum, 2011. Available online: http://www.wimaxforum.org. Accessed February 10, 2013.

[94] "Federal Communications Commission: Auctions for Wireless Communications Service (WCS)," August 2011. Available online: http://wireless.fcc.gov/auctions/. Accessed February 10, 2013.

[95] C. EKLUND, R.B. MARKS, S. PONNUSWAMY, and N.J.M. van WAES, *WirelessMAN: Inside the IEEE 802.16 Standard for Wireless Metropolitan Area Networks*, IEEE, New York, 2006.

[96] R. PRASAD and F.J. VELEZ, *WiMAX Networks: Techno-Economic Vision and Challenges*, Springer Science+Business Media B. V., Dordrecht, 2010.

[97] H. KRAWCZYK, et al., "HMAC: Keyed-Hashing for Message Authentication," IETF RTC 2104, February 1997.

[98] "Secure Hash Standard," Federal Information Processing Standards Publication 180-1, April 1995.

[99] T. HAUG, "Overview of GSM: Philosophy and Results," International Journal of Wireless Information Networks, 1(1), 7–16, January 1994.

[100] 3GPP Document, "Overview of 3GPP Release 6," Version 0.1.1, February 2010. Available online: http://www.3gpp.org/ftp/Information/WORK_PLAN/Description_Releases/. Accessed February 10, 2013.

[101] 3GPP Document, "Overview of 3GPP Release 7," Version 0.9.15, September 2011. Available online: http://www.3gpp.org/ftp/Information/WORK_PLAN/Description_Releases/. Accessed February 10, 2013.

[102] 3GPP Document, "Overview of 3GPP Release 8," Version 0.2.4, September 2011. Available online: http://www.3gpp.org/ftp/Information/WORK_PLAN/Description_Releases/. Accessed February 10, 2013.

[103] 3GPP Document, "Overview of 3GPP Release 9," Version 0.2.3, September 2011. Available online: http://www.3gpp.org/ftp/Information/WORK_PLAN/Description_Releases/. Accessed February 10, 2013.

[104] 3GPP Document, "Overview of 3GPP Release 10," Version 0.1.2, September 2011. Available online: http://www.3gpp.org/ftp/Information/WORK_PLAN/Description_Releases/. Accessed February 10, 2013.

[105] 3GPP Document, "Overview of 3GPP Release 11," Version 0.0.8, September 2011. Available online: http://www.3gpp.org/ftp/Information/WORK_PLAN/Description_Releases/. Accessed February 10, 2013.

[106] 3GPP TS 23.002, "Network Architecture (Release 10)," Version 10.3.0, September 2011.

[107] 3GPP FAQ Web site, "What Is the Difference between a SIM and a USIM? What Is a UICC?". Available online: http://www.3gpp.org/FAQ#outil_sommaire_58. Accessed February 10, 2013.

[108] 3GPP TS 23.003, "Technical Specification Group Core Network and Terminals; Numbering, Addressing and Identification (Release 10)," Version 10.2.0, June 2011.

[109] 3GPP TS 42.017, "Technical Specification Group Terminals; Subscriber Identity Modules (SIM); Functional Characteristics (Release 4)," Version 4.0.0, March 2001.

[110] 3GPP TS 22.060, "Technical Specification Group Services and System Aspects; General Packet Radio Service (GPRS); Service Description; Stage 1 (Release 10)," Version 10.0.0, March 2011.

[111] 3GPP TS 43.055, "Technical Specification Group GERAN; Dual Transfer Mode (DTM); Stage 2 (Release 10)," Version 10.0.0, March 2011.

[112] 3GPP TS 45.002, "Multiplexing and Multiple Access on the Radio Path (Release 10)," Version 10.1.0, May 2011.

[113] G. HEINE, *GSM Networks: Protocols, Terminology, and Implementation*, Artech House, Boston, 1998.

[114] J. EBERSPACHER, et al., *GSM Architecture, Protocols, and Services*, 3rd Edition, John Wiley & Sons, Chichester, UK, 2009.

[115] 3GPP TS 23.060, "General Packet Radio System (GPRS); Service Description; Stage 2 (Release 1999)," Version 3.17.0, December 2006.

[116] T. HALONEN, et al., *GSM, GPRS, and EDGE Performance*, 2nd Edition, John Wiley & Sons, Chichester, UK, 2003.

[117] M. SAUTER, *Communications Systems for the Mobile Information Society*, John Wiley & Sons, Chichester, UK, 2006.

[118] 3GPP Standard, "Technical Specification Group Services and System Aspects; Security-Related Network Functions (3GPP TS 43.020 Version 10.1.0 Release 10)," June 2011.

[119] 3GPP TS 22.016, "Technical Specification Group Services and System Aspects; International Mobile station Equipment Identities (IMEI) (Release 10)," Version 10.0.0, March 2011.

[120] 3GPP TS 03.03, "Technical Specification Group Core Network; Numbering, Addressing and Identification (Release 1998)," Version 7.8.0, September 2003.

[121] ITU-T Recommendation E.212, "The International Identification Plan for Public Networks and Subscriptions," 3rd Edition, November 1998. (Updated from the original edition, published Oct. 1984.).

[122] ITU-T Recommendation E.213, "Telephone and ISDN Numbering Plan for Land Mobile Stations in Public Land Mobile Networks (PLMN)," 2nd Edition, November 1988. (Updated from the original edition, published Oct. 1984.).

[123] GSM TS 02.60, "General Packet Radio Service (GPRS); Service Description; Stage 1 (Release 1997)," Version 6.3.1, November 2000.

[124] GSM TS 03.60, "General Packet Radio System (GPRS); Service Description; Stage 2 (Release 1997)," Version 6.11.0, September 2002.

[125] GSM Association Official Document DG.06, "IMEI Allocation and Approval Guidelines," Version 5.0, September 2010.

[126] 3GPP TS 44.003, "Mobile Station—Base Station System (MS-BSS) Interface Channel Structures and Access Capabilities (Release 10)," Version 10.0.0, April 2011.

[127] 3GPP TS 45.001, "Physical Layer on the Radio Path; General Description (Release 10)," Version 10.0.0, December 2011.

[128] GSM TS 43.064, "Overall Description of the GPRS Radio Interface; Stage 2 (Release 10)," Version 10.0.0, September 2010.

[129] 3GPP TS 45.010, "Radio Subsystem Synchronization (Release 10)," Version 10.1.0, April 2011.

[130] 3GPP TS 45.005, "Radio Transmission and Reception (Release 10), Version 10.1.0, June 2011.

[131] ISO 4335, "Information Processing Systems—Data Communication—High-Level Logical Link Control Procedures—Consolidation of Elements of Procedures," 1987.

[132] ITU-T Q.921, "ISDN User-Network Interface—Data Link Layer Specification," 1988.

[133] 3GPP TS 04.04, "Layer 1; General Requirements (Release 1999)," v8.1.2, May 2002.

[134] 3GPP TS 03.60, "GPRS Service Description; Stage 2 (Release 1997)," Version 6.11.0, September 2002.

[135] G. Sanders, et al., *GPRS Networks*, John Wiley & Sons, Chichester, UK, 2003.

[136] 3GPP TS 23.107, "Quality of Service (QoS) Concept and Architecture (Release 1999)," Version 3.9.0, September 2002.

[137] 3GPP TS 03.64, "Overall Description of the GPRS Radio Interface; Stage 2 (Release 1999)," Version 8.12.0, April 2004.

[138] 3GPP TS 04.64, "General Packet Radio Service (GPRS); Mobile Station—Serving GPRS Support Node (MS-SGSN) Logical Link Control (LLC) Layer Specification (Release 99)," Version 8.7.0, January 2002.

[139] 3GPP TS 04.60, "GPRS MS-BSS Interface; RLC/MAC Protocol (Release 1999)," Version 8.27.0, September 2005.

[140] TIA IS-130, "800 MHz Cellular System—TDMA Radio Interface—Radio Link Protocol 1" Arlington: Telecommunications Industry Association. 1995.

[141] 3GPP TS 04.65, "General Packet Radio Service (GPRS); Mobile Station (MS)—Serving GPRS Support Node (SGSN); Subnetwork Dependent Convergence Protocol (SNDCP) (Release 1999)," Version 8.2.0, October 2001.

[142] 3GPP TS 03.20, "(GSM) Security Related Network Functions (Release 1999)," Version 8.6.0, December 2007.

[143] A. Furuskar, et al., "EDGE: Enhanced Data Rates for GSM and TDMA/136 Evolution," IEEE Personal Communications, June 1999.

[144] 3GPP TS 43.051, "GSM/EDGE Radio Access Network; Overall Description—Stage 2 (Release 5)," Version 5.10.0, August 2008.

[145] GSA White Paper, "EDGE Evolution—Benefits, Market Developments," August 23, 2010. Available online: http://www.gsacom.com. Accessed February 10, 2013.

[146] 3GPP TS 45.002, "Multiplexing and Multiple Access on the Radio Path (Release 7)," Version 7.8.0, April 2011.

[147] M. Säily, G. Sébire, and E. Riddington, *GSM/EDGE: Evolution and Performance*, John Wiley & Sons, Chichester, UK, 2011.

[148] 3GPP TR 45.914, "Circuit Switched Voice Capacity Evolution for GSM/EDGE Radio Access Network (GERAN) (Release 9)," Version 10.1.0, December 2011.

[149] 3GPP TS 45.004, "Modulation (Release 9)," Version 9.1.0, June 2010.

[150] GSA White Paper, "EDGE Fact Sheet," April 8, 2011. Available online: http://www. gsacom.com. Accessed February 10, 2013.

[151] "Fast Facts," December 9, 2011. Available online: http://www.gsacom.com/news/ gsa_fastfacts.php4. Accessed February 10, 2013.

[152] P. Possi, "UMTS/3G History and Future Milestones," 2009. Available online: http:// www.umtsworld.com/umts/history.htm. Accessed February 10, 2013.

[153] 3GPP TS 25.301, "Radio Interface Protocol Architecture (Release 10)," Version 10.0.0, March 2011.

[154] S. Su, *The UMTS Air-Interface in RF Engineering*, McGraw-Hill, New York, 2007.

[155] S. Kasera and N. Narang, *3G Networks: Architecture, Protocols and Procedures*, Tata McGraw-Hill Publishing Company Limited, New Delhi, India, 2004.

[156] H. Holma and A. Toskala, *WCDMA for UMTS: HSPA Evolution and LTE*, 4th Edition, John Wiley & Sons, Ltd., New York, 2007.

[157] 3GPP TS 25.309, "FDD Enhanced Uplink; Overall Description; Stage 2 (Release 6)," Version 6.6.0, April 2006.

[158] "HSPA," 2009. Available online: http://www.3gpp.org/HSPA. Accessed February 10, 2013.

[159] "Mobile WiMAX—Part I: A Technical Overview and Performance Evaluation," WiMAX Forum, August 2006.

[160] "WiMAX Forum Network Architecture: Architecture Tenets, Reference Model and Reference Points—Base Specification," WiMAX Forum, WMF-T32-001-R016v01, November 30, 2010.

[161] WMF-T23-001-R010v09, "WiMAX Forum® Mobile System Profile, Release 1," September 7, 2010.

[162] S. Ahmadi, *Mobile WiMAX: A Systems Approach to Understanding IEEE 802.16m Radio Access Technology*, Elsevier Inc., Burlington, MA, 2011.

[163] WMF-T23-005-R015v05, "WiMAX Forum® Air Interface Specifications/WiMAX Forum® Mobile Radio Specification," April 4, 2011.

[164] CDMA Development Group Web site, 2012. Available online: http://www.cdg.org/. Accessed February 10, 2013.

[165] S. Carew, "Qualcomm Halts UMB Project, Sees No Major Job Cuts," November 13, 2008. Available online: http://www.reuters.com/article/2008/11/13/qualcomm-umb-idUSN1335969420081113?rpc=401&. Accessed February 10, 2013.

[166] V. Vanghi, A. Damnjanovic, and B. Vojcic, *The CDMA2000 System for Mobile Communications*, Prentice Hall PTR, Upper Saddle River, NJ, 2004.

[167] "ZTE Completes Industry's First CDMA2000 1X Advanced IOT Testing," April 21, 2011. Available online: http://wwwen.zte.com.cn/en/press_center/news/201104/ t20110421_351096.html. Accessed March 23, 2013.

[168] "What's Next for CDMA?" October 2011. Available online: http://www.qualcomm. com/technology. Accessed February 10, 2013.

[169] "Interoperability Specification (IOS) for cdma2000® Access Network Interfaces Release C," Telecommunications Industry Association (TIA), TIA-2001-C, July 2003.

[170] "TIA/EIA-IS-2001 Inter-operability Specification (IOS) for CDMA2000 Access Network Interfaces," Telecommunications Industry Association, 2002.

[171] "TIA/EIA-IS-2000 Introduction to CDMA2000 Spread Spectrum System Release C," Telecommunications Industry Association, 2002.

[172] "TIA/EIA-IS-2000 Medium Access Control (MAC) Standard for CDMA2000 Spread Spectrum Systems Release C," Telecommunications Industry Association, 2002.

[173] "TIA/EIA-IS-2000 Physical Layer Standard for CDMA2000 Spread Spectrum Systems Release C," Telecommunications Industry Association, 2002.

[174] ANSI/TIA/EIA-41-D, "Cellular Radiotelecommunications Intersystem Operations," November 1997.

[175] "TIA/EIA-IS-93-B Wireless Telecommunications: Ai-Di Interfaces Standard," Telecommunication Industry Association, 2001.

[176] 3GPP2 X.S0011-001-C, "cdma2000 Wireless IP Network Standard: Introduction," Version 1.0.0, August 2003.

[177] M. GALLAGHER and R. SNYDER, *Mobile Telecommunications Networking*, McGraw-Hill, New York, 1997.

[178] Signals Research Group/CDMA Development Group, "The Impact of Mobile Computers and Smartphones on CDMA2000 Networks," January 2011. Available online: http://www.signalsresearch.com. Accessed February 19, 2013.

[179] K. ETEMAD, *CDMA2000 Evolution: System Concepts and Design Principles*, Wiley-Interscience, Hoboken, NJ, 2004.

[180] 3GPP2 C.S0024-0 v4.0, "CDMA2000 High Rate Packet Data Air Interface Specification," 2002.

[181] "2Q 2011 CDMA Subscribers," CDMA Development Group Quarterly Report, June 2011.

[182] 3GPP Document, "3GPP TSG RAN Future Evolution Work Shop 2–3 November 2004, Toronto, Canada; Compendium of Abstracts." 2004. Available online: http://www.3gpp.org. Accessed February 10, 2013.

[183] 3GPP TD RP-040461, "Proposed Study Item on Evolved UTRA and UTRAN." December 2004. Available online: http://www.3gpp.org. Accessed February 10, 2013.

[184] 3GPP TR 25.913, "Requirements for Evolved UTRA (E-UTRA) and Evolved UTRAN (E-UTRAN)," Release 7, Version 7.3.0, March 2006.

[185] 3GPP TR 25.912, "Feasibility Study for Evolved Universal Terrestrial Radio Access (UTRA) and Universal Terrestrial Radio Access Network (UTRAN) (Release 10)," Version 10.0.0, April 2012.

[186] 3GPP TR 25.978, "All-IP Network (AIPN) Feasibility Study (Release 10)," Version 10.0.0, April 2011.

[187] 3GPP TR 23.882, "3GPP System Architecture Evolution (SAE): Report on Technical Options and Conclusions (Release 8)," Version 8.0.0, September 2008.

[188] H. HOLMA and A. TOSKALA, *LTE for UMTS: OFDMA and SC-FDMA Based Radio Access*, John Wiley & Sons, Chichester, UK, 2009.

[189] 3GPP TS 31.102, "Characteristics of the Universal Subscriber Identity Module (USIM) Application (Release 8)," Version 8.14.0, January 2012.

[190] 3GPP TS 36.306, "Evolved Universal Terrestrial Radio Access (E-UTRA); User Equipment (UE) Radio Access Capabilities (Release 8)," Version 8.8.0, June 2011.

[191] 3GPP TS 23.002, "Network Architecture (Release 8)," Version 8.7.0, December 2010.

[192] 3GPP TS 36.300, "Evolved Universal Terrestrial Radio Access (E-UTRA) and Evolved Universal Terrestrial Radio Access Network (E-UTRAN); Overall Description; Stage 2 (Release 8)," Version 8.12.0, April 2010.

[193] 3GPP TS 23.203, "Policy and Charging Control Architecture (Release 8)," Version 8.14.0, June 2012.

[194] 3GPP TS 23.401, "General Packet Radio Service (GPRS) Enhancements for Evolved Universal Terrestrial Radio Access Network (E-UTRAN) Access (Release 8)," Version 8.16.0, March 2012.

[195] 3GPP TS 23.402, "Architecture Enhancements for Non-3GPP Access (Release 8)," Version 8.10.0, March 2012.

[196] 3GPP TS 23.234, "3GPP System to Wireless Local Area Network (WLAN) Interworking: System Description (Release 10)," Version 10.0.0, March 2011.

[197] 3GPP TS 23.003, "Numbering, Addressing, and Identification (Release 11)," Version 11.2.0, June 2012.

[198] S. SESIA, I. TOUFIK, and M. BAKER, *LTE The UMTS Long Term Evolution: From Theory to Practice*, 2nd Edition, John Wiley & Sons, New York, 2011.

[199] 3GPP TS 36.211, "Evolved Universal Terrestrial Radio Access (E-UTRA): Physical Channels and Modulation (Release 10)," Version 10.5.0, June 2012.

[200] 3GPP TS 36.211, "Evolved Universal Terrestrial Radio Access (E-UTRA): Physical Channels and Modulation (Release 8)," Version 8.9.0, December 2009.

[201] C. REINERS and H. ROHLING, "Multicarrier Transmission Technique in Cellular Mobile Communications Systems," in Proceedings IEEE Vehicular Technology Conference '94, Stockholm, Sweden, pp. 1660–1664, June 8–10, 1994.

[202] A. CZYLWIK, "Comparison between Adaptive OFDM and Single Carrier Modulation with Frequency Domain Equalization," IEEE 47th Vehicular Technology Conference, May 4–7, 1997.

[203] 3GPP TS 36.101, "Evolved Universal Terrestrial Radio Access (E-UTRA); User Equipment (UE) Radio Transmission and Reception (Release 8)," Version 8.18.0, July 2012.

[204] 3GPP TS 36.101, "Evolved Universal Terrestrial Radio Access (E-UTRA); User Equipment (UE) Radio Transmission and Reception (Release 9)," Version 9.12.0, July 2012.

[205] 3GPP TS 36.101, "Evolved Universal Terrestrial Radio Access (E-UTRA); User Equipment (UE) Radio Transmission and Reception (Release 10)," Version 10.7.0, July 2012.

[206] 3GPP TS 36.101, "Evolved Universal Terrestrial Radio Access (E-UTRA); User Equipment (UE) Radio Transmission and Reception (Release 11)," Version 11.1.0, July 2012.

[207] 3GPP TS 36.307, "Evolved Universal Terrestrial Radio Access (E-UTRA); Requirements on User Equipments (UEs) Supporting a Release-Independent Frequency Band (Release 8)," Version 8.6.0, July 2012.

[208] 3GPP TS 36.423, "Evolved Universal Terrestrial Radio Access Network (E-UTRAN); X2 Application Protocol (X2AP) (Release 8)," Version 8.9.0, April 2010.

[209] M. RAHMAN and H. YANIKOMEROGLU, "Enhancing Cell-Edge Performance: A Downlink Dynamic Interference Avoidance Scheme with Inter-Cel. Coordination," IEEE Transactions on Wireless Communications, 9(4), 1414–1425, April 2010.

[210] D. ASTÉLY, E. DAHLMAN, A. FURUSKÄR, Y. JADING, M. LINDSRÖM, and S. PARKVALL, "LTE: The Evolution of Mobile Broadband," IEEE Communications Magazine, 47(4),44–51, April 2009.

[211] 3GPP TS 36.213, "Evolved Universal Terrestrial Radio Access (E-UTRA): Services Provided by the Physical Layer (Release 9)," Version 9.3.1, December 2011.

[212] G. BAUCH and T. ABE, "On the Parameter Choice for Cyclic Delay Diversity Based Precoding with Spatial Multiplexing," IEEE Global Telecommunications Conference, November–December 2009.

[213] 3GPP TS 36.302, "Evolved Universal Terrestrial Radio Access (E-UTRA): Services Provided by the Physical Layer (Release 8)," Version 8.2.1, December 2011.

[214] 3GPP TS 36.201, "Evolved Universal Terrestrial Radio Access (E-UTRA): LTE Physical Layer General Description (Release 8)," Version 8.3 0, March 2009.

[215] C. COX, *An Introduction to LTE: LTE, LTE-Advanced, SAE and 4G Mobile Communications*, John Wiley & Sons, Chichester, UK, 2012.

[216] 3GPP TS 36.321, "Evolved Universal Terrestrial Radio Access (E-UTRA): Medium Access Control (MAC) Protocol Specification (Release 8)," Version 8.12.0, March 2012.

[217] 3GPP TS 36.322, "Evolved Universal Terrestrial Radio Access (E-UTRA): Radio Link Control (RLC) Protocol Specification (Release 8)," Version 8.8.0, June 2010.

[218] 3GPP TS 36.323, "Evolved Universal Terrestrial Radio Access (E-UTRA): Packet Data Convergence Protocol (PDCP) Specification (Release 8)," Version 8.6.0, June 2009.

[219] 4G Americas Web site. Available online: http://4gamericas.org. Accessed February 10, 2013.

[220] Global Mobile Suppliers Association White Paper Repor., "Status of the LTE Ecosystem," January 20, 2012. Available online: http://www.gsacom.com. Accessed February 10, 2013.

[221] Global Mobile Suppliers Association White Paper Repor., "Status of the LTE Ecosystem," July 3, 2012. Available online: http://www.gsacom.com. Accessed February 10, 2013.

[222] Global Mobile Suppliers Association White Paper Report, "GSA Evolution to LTE Report," July 11, 2012. Available online: http://www.gsacom.com. Accessed February 10, 2013.

[223] 3GPP TS 36.300, "Evolved Universal Terrestrial Radio Access (E-UTRA) and Evolved Universal Terrestrial Radio Access Network (E-UTRAN); Overall description; Stage 2 (Release 10)," Version 10.9.0, January 2013.

[224] 3GPP Document, "Carrier Aggregation for LTE V0.0.1 (2012-06)," June 2012. Available online: http://www.3gpp.org. Accessed February 10. 2013.

[225] S. PARKVALL et al., "LTE Evolution towards IMT-Advanced ard Commercial Network Performance," *Proceedings of the 2010 IEEE International Conference on Communication Systems (ICCS)*, pp. 151–155, November 2010.

[226] 3GPP TS 23.401, "Evolved Universal Terrestrial Radio Access Network (E-UTRAN): Architecture Description (Release 10)," Version 10.4.0, July 2012.

[227] A. DAMNJANOVIC, et al., "A Survey on 3GPP Heterogeneous Networks," IEEE Wireless Communications, 18(3), 10–21, June 2011.

[228] H. LI, J. HAJIPOUR, A. ATTAR, and V.C.M. LEUNG, "Efficient HetNet Implementation Using Broadband Wireless Access with Fiber-Connected Massively Distributed Antennas Architecture," IEEE Wireless Communications, 18(3), 72–78, June 2011.

[229] T. CAMP, J. BOLENG, and V. DAVIES, "A Survey of Mobility Models for Ad Hoc Network Research," Wireless Communication and Mobile Computing (WCMC): Special Issue on Mobile Ad Hoc Networking: Research, Trends and Applications, 2(5), 483–502, 2002.

[230] C. PERKINS, "IP Mobility Support for IPv4," RFC 3344, August 2002.

[231] D. JOHNSON, C. PERKINS, and J. ARKKO, "Mobility Support in IPv6," RFC 3775, June 2004.

[232] V. DEVARAPALLI, et al., "Network Mobility (NEMO) Basic Support Protocol," RFC 3963, January 2005.

[233] J.L. BURBANK, P.F. CHIMENTO, B.K. HABERMAN, and W.T. KASCH, "Key Challenges of Military Tactical Networking and the Elusive Promise of MANET Technology," IEEE Communications Magazine, 44(11), 39–45, November 2006.

[234] E. FLEISCHMAN and W. FURMANSKI, "Mobile Exterior Gateway Protocol: Extending IP Scalability," Military Communications Conference, October 2005.

[235] C. SIVA RAM MURTHY and B.S. MANOJ, *Ad-Hoc Wireless Networks: Architectures and Protocols*, Prentice Hall, Upper Saddle River, NJ, 2004.

[236] D. JOHNSON, Y. HU, and D. MALTZ, "The Dynamic Source Routing Protocol (DSR) for Mobile Ad Hoc Networks for IPv4," RFC 4728, February 2007.

[237] C. PERKINS, E. BELDING-ROYER, and S. DAS, "Ad-Hoc On-Demand Distance Vector (AODV) Routing," RFC 3561, July 2003.

[238] T. CLAUSEN and P. JACQUET, "Optimized Link State Routing Protocol (OLSR)," RFC 3626, October 2003.

[239] R. OGIER, F. TEMPLIN, and M. LEWIS, "Topology Dissemination Based on Reverse-Path Forwarding (TBRPF)," RFC 3684, February 2004.

[240] S. AGARWAL, A. AHUJA, J.P. SINGH, and R. SHOREY, "Route-Lifetime Assessment-Based Routing (RABR) Protocol for Mobile Ad Hoc Networks," Proceedings of IEEE ICC 2000, Vol. 3, pp. 1697–1701, June 2000.

[241] B.S. MANOJ, R. ANANTHAPADMANABHA, and C. SIVA RAM MURTHY, "Link Life-Based Routing Protocol for Ad Hoc Wireless Networks," Proceedings of IEEE IC-CCN 2001, pp. 573–576, October 2001.

[242] S. SINGH, M. WOO, and C.S. RAGHAVENDRA, "Power-Aware Routing in Mobile Ad Hoc Networks," Proceedings of ACM MOBICOM 1998, pp. 181–190, October 1998.

[243] Y. KO and N.H. VAIDYA, "Location-Aided Routing (LAR) in Mobile Ad Hoc Networks," Proceedings of ACM MOBICOM 1998, pp. 66–75, October 1998.

[244] A. VAHDAT and D. BECKER, "Epidemic Routing for Partially-Connected Ad Hoc Networks," Duke University Technical Report CS2000-06, 2000.

[245] A. LINDGREN, A. DORIA, and O. SCHELÉN, "Probabilistic Routing in Intermittently Connected Networks," Proceedings of the First International Workshop on Service Assurance with Partial and Intermittent Resources (SAPIR 2004), August 2004.

[246] Z.J. HAAS, J.Y. HALPERN, and L. LI, "Gossip-Based Ad Hoc Routing," IEEE 2002.

[247] J. Postel, "Transmission Control Protocol, DARPA Internet Program Protocol Specification," RFC 793, September 1981.

[248] M. Allman, et al., "TCP Congestion Control," RFC 5681, September 2009.

[249] M. Mathis, et al., "TCP Selective Acknowledgement Options," RFC 2018, October 1996.

[250] C. Caini, et al., "TCP Hybla: A TCP Enhancement for Heterogeneous Networks," International Journal of Satellite Communications and Networking, 22, 547–566, 2004.

[251] H. Balakrishnan, et al., "A Comparison of Mechanisms for Improving TCP Performance over Wireless Links," IEEE/ACM Transactions on Networking, 5(6), 756–769, December 1997.

[252] S. Ha, et al., "CUBIC: A New TCP-Friendly High-Speed TCP Variant," ACM SIGOPS Operating Systems Review, 42(5), 64–74, 2008.

[253] T.M. Duman, *Coding for MIMO Communication Systems*, 1st Edition, John Wiley & Sons, New York, 2008.

[254] D. Zarbouti, G. Tsoulos, and D. Kaklamani, "Theory and Practice of MIMO Wireless Communication Systems." In: *MIMO System Technology for Wireless Communications*, CRC Press, Taylor & Francis Group, Boca Raton, FL, 2006.

[255] A. Paulraj, R. Nabar, and D. Gore, *Introduction to Space-Time Wireless Communications*, Cambridge University Press, Cambridge, UK, 2003.

[256] J. Bryzek, K. Peterson, and W. McCulley, "Micro-Machines on the March," IEEE Spectrum, 31(5), 20–31, May 1994.

[257] E.R. Brown, "RF-MEMS Switches for Reconfigurable Integrated Circuits," IEEE Transactions on Microwave Theory and Techniques, 46(11), 1868–1880, Nov 1998.

[258] W.H. Weedon, W.J. Payne, and G.M. Rebeiz, "MEMS-Switched Reconfigurable Antennas," IEEE Antennas and Propagation Society International Symposium, 2001, Vol. 3, pp. 654–657, 2001.

[259] J.D. Boerman and J.T. Bernhard, "Performance Study of Pattern Reconfigurable Antennas in MIMO Communication Systems," IEEE Transactions on Antennas and Propagation, 56(1), 231–236, Jan 2008.

[260] Y. Jang, J. Choi, and S. Lim, "Frequency Reconfigurable Zero-Order Resonant Antenna Using RF MEMS Switch," Microwave and Optical Technology Letters, 54, 1266–1269, 2012.

[261] I. Develi and E.N. Yazlik, "Optimum Antenna Configuration in MIMO Systems: A Differential Evolution Based Approach," Wireless Communications and Mobile Computing, 12, 473–480, 2012.

[262] J. Zheng, X. Gao, Z. Zhang, and Z. Feng, "A Compact Eighteen-Port Antenna Cube for MIMO Systems," IEEE Transactions on Antennas and Propagation, 60(2), 445–455, February 2012.

[263] B.A. Cetiner, H. Jafarkhani, J.-Y. Qian, H.J. Yoo, A. Grau, and F. De Flaviis, "Multifunctional Reconfigurable MEMS Integrated Antennas for Adaptive MIMO Systems," IEEE Communications Magazine, 42(12), 62–70, December 2004.

[264] D. Piazza, N.J. Kirsch, A. Forenza, R.W. Heath, and K.R. Dandekar, "Design and Evaluation of a Reconfigurable Antenna Array for MIMO Systems," IEEE Transactions on Antennas and Propagation, 56(3), 869–881, March 2008.

[265] P.-Y. QIN, Y.J. GUO, A.R. WEILY, and C.-H. LIANG, "A Pattern Reconfigurable U-Slot Antenna and Its Applications in MIMO Systems," IEEE Transactions on Antennas and Propagation, 60(2), 516–528, February 2012.

[266] A. HONDA, I. IDA, Y. OISHI, Q. TUAN TRAN, S. HARA, and J.-I. TAKADA, "Experimental Evaluation of MIMO Antenna Selection System using RF-MEMS Switches on a Mobile Terminal," IEEE 18th International Symposium on Personal, Indoor and Mobile Radio Communications, 2007. PIMRC 2007., pp. 1–5, September 3–7, 2007.

[267] A. GRAU, H. JAFARKHANI, and F. DE FLAVIIS, "A Reconfigurable Multiple-Input Multiple-Output Communication System," IEEE Transactions on Wireless Communications, 7(5), 1719–1733, May 2008.

[268] D.J. CHUNG, D. ANAGNOSTOU, G. PONCHAK, M.M. TENTZERIS, and J. PAPAPOLYMEROU, "Light Weight MIMO Phased Arrays with Beam Steering Capabilities using RF MEMS," IEEE 18th International Symposium on Personal, Indoor and Mobile Radio Communications, 2007. PIMRC 2007. pp. 1–3, September 3–7, 2007.

[269] P.B. RUFFIN, J.C. HOLT, J.H. MULLINS, T. HUDSON, and J. ROCK, "MEMS-Based Phased Arrays for Army Applications," Proceedings of the SPIE, 6528, 652802, 2007.

[270] O.N. ALRABADI, J. PERRUISSEAU-CARRIER, and A. KALIS, "MIMO Transmission Using a Single RF Source: Theory and Antenna Design," IEEE Transactions on Antennas and Propagation, 60(2), 654–664, February 2012.

[271] F. KHAN, *LTE for 4G Mobile Broadband: Air Interface Technologies and Performance*, Cambridge University Press, Cambridge, UK, 2009.

[272] "Unlicensed Operations in the TV Broadcast Bands and Additional Spectrum for Unlicensed Devices Below 900 MHz and in the 3 GHz Band," Federal Communications Commission, FCC 08-260, November 14, 2008.

[273] "Unlicensed Operations in the TV Broadcast Bands and Additional Spectrum for Unlicensed Devices Below 900 MHz and in the 3 GHz Band," Second Memorandum Opinion and Order, Federal Communications Commission, FCC 10-174, September 14, 2010.

[274] G. PAU, et al., "A Cross-Layer Framework for Wireless LAN QoS Support," Proceedings of the International Information Technology: Research and Education (ITRE), pp. 331–334, August 2003. Proceedings. ITRE2003.

[275] A.J. GOLDSMITH and S.B. WICKER, "Design Challenges for Energy-Constrained Ad Hoc Wireless Networks," IEEE Wireless Communications Magazine, 9(4), 8–27, August 2002.

[276] S. SHAKKOTTAI and T.S. RAPPAPORT, "Cross-Layer Design for Wireless Networks," IEEE Communications Magazine, 41(10), 74–80, October 2003.

[277] S. KUNNIYUR and R. SRIKANT, "End-to-End Congestion Control: Utility Functions, Random Losses and ECN Marks," Proceedings of IEEE Infocom, Vol. 3, pp. 1323–1332, March 2000.

[278] V. KAWADIA and P.R. KUMAR, "A Cautionary Perspective on Cross-Layer Design." IEEE Wireless Communications Magazine, 12(1), 3–11, February 2005.

[279] J.L. BURBANK and W.T.M. KASCH, "The Application of Human and Social Behavior Inspired Security Models for Self-Aware Collaborative Cognitive Radio Networks," 2008 ACM Trusted Collaboration Workshop, 2008.

[280] C. Fragouli and E. Soljanin, "Network Coding Applications," Foundations and Trends in Networking, 2(2), 135, 2007.

[281] R.W. Yeung, S.-Y.R. Li, N. Cai, and Z. Zhang, "Network Coding Theory," Foundation and Trends in Communications and Information Theory, 2(4 and 5), 241–381, 2005.

[282] R.W. Yeung and Z. Zhang, "Distributed Source Coding for Satellite Communications," IEEE Transactions on Information Theory, 45, 1111–1120, 1999.

[283] R. Ahlswede, N. Cai, S.-Y.R. Li, and R.W. Yeung, "Network Information Flow," IEEE Transactions on Information Theory, 46, 1204–1216, 2000.

[284] M. Medard and A. Sprintson, eds., *Network Coding: Fundamentals and Applications*, Academic Press, Waltham, MA, 2012.

[285] A. Zhan and C. He, "Joint Design of Channel Coding and Physical Network Coding for Wireless Networks," 2008 International Conference on Neural Networks and Signal Processing, pp. 512–516, June 7–11, 2008.

[286] T. Xiaobin, Q. Guihong, and C. Wenfei, "Loss-Aware Linear Network Coding for Wireless Networks," Control Conference (CCC), 2011 30th Chinese, pp. 4382–4387, July 22–24, 2011.

[287] W. Chen, K.B. Letaief, and Z. Cao, "Opportunistic Network Coding for Wireless Networks," Communications, 2007. ICC '07. IEEE International Conference on, pp. 4634–4639, June 24–28, 2007.

[288] T. Cui, L. Chen, and T. Ho, "Energy Efficient Opportunistic Network Coding for Wireless Networks," INFOCOM 2008. The 27th Conference on Computer Communications. IEEE, pp. 361–365, April 13–18, 2008.

[289] S. Sengupta, S. Rayanchu, and S. Banerjee, "An Analysis of Wireless Network Coding for Unicast Sessions: The Case for Coding-Aware Routing," INFOCOM 2007, 26th IEEE International Conference on Computer Communications, pp. 1028–1036, May 6–12, 2007.

[290] M. Ghaderi, D. Towsley, and J. Kurose, "Reliability Gain of Network Coding in Lossy Wireless Networks," INFOCOM 2008. The 27th Conference on Computer Communications. IEEE, pp. 2171–2179, April 13–18, 2008.

[291] W. Pu, C. Luo, S. Li, and C. Wen Chen, "Continuous Network Coding in Wireless Relay Networks," INFOCOM 2008. The 27th Conference on Computer Communications. IEEE, pp. 1526–1534, April 13–18, 2008.

[292] Y.E. Sagduyu and A. Ephremides, "Crosslayer Design for Distributed MAC and Network Coding in Wireless Ad Hoc Networks," Information Theory, 2005. ISIT 2005. Proceedings. International Symposium on, pp. 1863–1867, September 4–9, 2005.

[293] S. Sengupta, S. Rayanchu, and S. Banerjee, "Network Coding-Aware Routing in Wireless Networks," IEEE/ACM Transactions on Networking, 18(4), 1158, August 2010.

[294] J. Liu, D. Goeckel, and D. Towsley, "Bounds on the Gain of Network Coding and Broadcasting in Wireless Networks," INFOCOM 2007, 26th IEEE International Conference on Computer Communications, pp. 724–732, May 6–12, 2007.

[295] H. Li, Q. Feng, G. Han, and W. Dou, "Performance Analysis in Wireless Network Coding: An Approach based on Service Curve Model," Communication Software and Networks (ICCSN), 2011 IEEE 3rd International Conference on, pp. 233–237, May 27–29, 2011.

[296] H. Li, X. Liu, W. He, J. Li, and W. Dou, "End-to-End Delay Analysis in Wireless Network Coding: A Network Calculus-Based Approach," Distributed Computing Systems (ICDCS), 2011 31st International Conference on, pp. 47–56, June 20–24, 2011.

[297] A. Khreishah, I.M. Khalil, P. Ostovari, and J. Wu, "Flow-Based XOR Network Coding for Lossy Wireless Networks," IEEE Transactions on Wireless Communications, 11(6), 2321, 2012.

[298] S. Chachulski, M. Jennings, S. Katti, and D. Katabi, "Trading Structure for Randomness in Wireless Opportunistic Routing," ACM SIGCOMM, 2007.

[299] S. Katti, D. Katabi, H. Balakrishnan, and M. Medard, "Symbol-Level Network Coding for Wireless Mesh Networks," ACM SIGCOMM, 2008.

[300] S. Katti, H. Rahul, D. Katabi, M. Medard, and J. Crowcroft, "XORs in the Air: Practical Wireless Network Coding," IEEE/ACM Transactions on Networking, 16(3), 497–510, June 2008.

[301] J. Liu, D. Goeckel, and D. Towsley, "Bounds on the Throughput Gain of Network Coding in Unicast and Multicast Wireless Networks," IEEE Journal on Selected Areas in Communications, (5), 582–592, June 2009.

[302] M. Yang and Y. Yang, "A Linear Inter-Session Network Coding Scheme for Multicast," Network Computing and Applications, 2008, NCA '08, Seventh IEEE International Symposium on, pp. 177–184, July 10–12, 2008.

[303] S. Biswas and R. Morris, "Opportunistic Routing in Multi-hop Wireless Networks," ACM SIGCOMM, 2005.

[304] K. Jamieson and H. Balakrishnan, "Partial Packet Recovery for Wireless Networks," ACM SIGCOMM, 2007.

[305] D. De Couto, D. Aguayo, J. Bicket, and R. Morris, "A High-Throughput Path Metric for Multi-Hop Wireless Routing," MobiCom '03 Proceedings of the 9th Annual International Conference on Mobile Computing and Networking, pp. 134–146, 2003.

[306] L. Zhou and Z. Haas, "Securing Ad-Hoc Networks," IEEE Network, 13(6), 24–30, 1999.

[307] ITU-T Recommendation X.200, "Information Technology—Open Systems Interconnection—Basic Reference Model: The Basic Model," July 1994, International Telecommunications Union.

[308] C.E. Perkins and P. Bhagwat, "Highly Dynamic Destination-Sequenced Distance-Vector Routing (DSDV) for Mobile Computers," Proceedings of ACM SIG-COMM 1994, pp. 234–244, August 1994.

[309] S. Murthy and J.J. Garcia-Luna-Aceves, "An Efficient Routing Protocol for Wireless Networks," ACM Mobile Networks and Applications Journal, Special Issue on Routing in Mobile Communication Networks, 1(2), 183, 1997.

[310] C.C. Chiang, H.K. Wu, W. Liu, and M. Gerla, "Routing in Clustered Multi-Hop Mobile Wireless Networks with Fading Channel," Proceedings of IEEE SICON 1997, pp. 197–211, April 1997.

[311] J.J. Garcia-Luna-Aceves and M. Spohn, "Source-Tree Routing in Wireless Networks," Proceedings of IEEE ICNP 1999, pp. 273–282, October 1999.

[312] A. Iwata, et al., "Scalable Routing Strategies for Ad-Hoc Wireless Networks," IEEE Journal on Selected Areas in Communications, 17(8), 1369–1379, August 1999.

[313] T.W. Chen and M. Gerla, "Global State Routing: A New Routing Scheme for Ad-Hoc Wireless Networks," Proceedings of IEEE ICC 1998, pp. 171–175, June 1998.

[314] C.K. Toh, "Associativity-Based Routing for Ad-Hoc Mobile Networks," Wireless Personal Communications, 4(2), 1–36, March 1997

[315] R. Dube, et al., "Signal Stability-Based Adaptive Routing for Ad-Hoc Mobile Networks," IEEE Personal Communications Magazine, 36–45, February 1997.

[316] W. Su and M. Gerla, "IPv6 Flow Handoff in Ad Hoc Wireless Networks Using Mobility Prediction," Proceedings of IEEE Globecom 1999, pp. 271–275, December 1999.

[317] R.S. Sisodia, B.S. Manoj, and C. Siva Ram Murthy, "A Preferred Link-Based Routing Protocol for Ad-Hoc Wireless Networks," Journal of Communications and Networks, 4(1), 14–21, March 2002.

[318] 3GPP TS 36.212, "Evolved Universal Terrestrial Radio Access (E-UTRA): Multiplexing and Channel Coding (Release 8)," Version 8.8.0, December 2009.

[319] 3GPP TS 36.306, "Evolved Universal Terrestrial Radio Access (E-UTRA) and Evolved Universal Terrestrial Radio Access Network (E-UTRAN): Overall Description Stage 2 (Release 10)," Version 10.8.0, July 2012.

[320] H. Holma and A. Toskala, *LTE Advanced: 3GPP Solution for IMT-Advanced*, John Wiley & Sons, Chichester, UK, 2012.

[321] 3GPP TS 36.331, "Evolved Universal Terrestrial Radio Access (E-UTRA): Radio Resource Control (RRC) Protocol Specification (Release 8)," Version 8.17.0, July 2012.

[322] 3GPP TS 25.212, "UMTS Multiplexing and Channel Coding (FDD)," Version 6.6.0, Copyright 2005, ETSI.

[323] 3GPP TS 25.306, "UMTS UE Radio Access Capabilities," Version 6.13.0, Copyright 2009, ETSI.

[324] 3GPP2 C.S0002-E, "Physical Layer Standard for cdma2000 Spread Spectrum Systems: Revision E," Version 1.0, September 2009.

[325] M. Ergen, *Mobile Broadband Including WiMAX and LTE* Springer Science and Business Media, New York, 2009.

[326] R. W. Chang, "Synthesis of Band-Limited Orthogonal Signals for Multi-Channel Data Transmission," Bell System Technical Journal 46, 1775–1796 1966.

[327] M. Wang, M. Georgiades, and R. Tafazolli, "Signalling Cost Evaluation of Mobility Management Schemes for Different Core Network Architectural Arrangements in 3GPP LTE/SAE," IEEE Vehicular Technology Conference Proceedings, pp. 2253–2258, May 11–14, 2008.

Index

Wireless Networking: Understanding Internetworking Challenges, First Edition.
Jack L. Burbank, Julia Andrusenko, Jared S. Everett, and William T.M. Kasch.
© 2013 the Institute of Electrical and Electronics Engineers, Inc. Published 2013 by John Wiley & Sons, Inc.